H-Transforms

Theory and Applications

ANALYTICAL METHODS AND SPECIAL FUNCTIONS

An International Series of Monographs in Mathematics

H-Transforms

Theory and Applications

Anatoly A. Kilbas

Belarusian State University, Belarus

Megumi Saigo

Fukuoka University, Japan

CRC Press
Taylor & Francis Group
Boca Raton London New York

CRC Press is an imprint of the
Taylor & Francis Group, an **informa** business
A CHAPMAN & HALL BOOK

First published 2004 by Chapman & Hall

Published 2019 by CRC Press
Taylor & Francis Group
6000 Broken Sound Parkway NW, Suite 300
Boca Raton, FL 33487-2742

First issued in paperback 2019

ISBN 13: 978-0-367-45439-5 (pbk)
ISBN 13: 978-0-415-29916-9 (hbk)

Visit the Taylor & Francis Web site at
http://www.taylorandfrancis.com

and the CRC Press Web site at
http://www.crcpress.com

Library of Congress Card Number 2003070007

Library of Congress Cataloging-in-Publication Data

Saigo, Megumi
H-transforms: theory and applications / Megumi Saigo, Anatoly A. Kilbas.
p. cm. -- (Analytical methods and special functions ; 9)
Includes bibliographical references and index.
ISBN 0-415-29916-0 (alk. paper)
1. H-functions. 2. Integral transforms. I Kilbas, A.A. (Anatolii Aleksandrovich) II. Title. III. Series.

QA353.H9S35 2004
515'.723—dc22 2003070007

CONTENTS

PREFACE

This book deals with integral transforms involving the so-called H-functions as kernels, or $\boldsymbol{H}$-transforms, and their applications. The H-function is defined by the Mellin–Barnes type integral with the integrand containing products and quotients of the Euler gamma functions. Such a function generalizes most of the known special functions, which means that almost all integral transforms can be put into the form of $\boldsymbol{H}$-transforms. Another generalization, though a special case of $\boldsymbol{H}$-transforms, was proposed as the $\boldsymbol{G}$-transform of the integral transform with the Meijer G-function as a kernel; this generalization includes the classical Laplace and Hankel transforms, the Riemann–Liouville fractional integral transforms, the even and odd Hilbert transforms, the integral transforms with the Gauss hypergeometric function, and others. However, there are transforms that cannot be reduced to a $\boldsymbol{G}$-transform but can be put into the form of $\boldsymbol{H}$-transforms: the modified Laplace and Hankel transforms, the Erdélyi–Kober type fractional integration operators, the modified transforms with the Gauss hypergeometric function as kernel, the Bessel type integral transforms, the Mittag-Leffler type integral transforms and others.

The classical Fourier, Laplace, Mellin and Hankel transforms have been widely used in various problems of mathematical physics and applied mathematics. Such applications in various fields have been covered in the books by I.N. Sneddon [1]–[5] and in his survey paper [6]. These transforms are usually involved in solutions of boundary value problems for models of ordinary and partial differential equations that characterize certain processes. There are many other problems whose solutions cannot be represented by the above classical integral transforms but may be characterized by integral transforms with various special functions as kernels.

Among the integral transforms with special function kernels, the integral transforms involving functions of hypergeometric and Bessel types, as well as fractional integral transforms, have often arisen in applied problems. For example, integral transforms with Bessel type kernels are important in many axially symmetric problems. When we want to get solutions that are regular on the axis of symmetry, such a solution involves the Hankel transform with the Bessel function of the first kind in the kernel (I.N. Sneddon [4]). When the solution has a singularity near the axis of symmetry, the Struve transform and its reciprocal containing Neumann and Struve functions are involved in this solution (P.G. Rooney [4]).

The interest in integral transforms with special function kernels is also motivated by the desire to study the corresponding integral equations of the first kind and of the so-called dual and triple equations that are also encountered in applications. The results in these fields were presented in a series of books including the above books by I.N. Sneddon. In particular, the results in these fields obtained by methods of fractional calculus were characterized in Chapter 7 of the book by S.G. Samko, A.A. Kilbas and O.I. Marichev [1]. We also note the book by H.M. Srivastava and R.G. Buschman [4], which contains many examples of the convolution integral equations with special function kernels.

Integral transforms with various special function kernels have been investigated by many authors. They have basically been studied in L^1 and L^2-spaces or in certain spaces of tested and generalized functions. The results in the former were presented as examples in the books by E.C. Titchmarsh [3], I.N. Sneddon [1], and V.A. Ditkin and A.P. Prudnikov [1] together with the theory of classical transforms. These publications involved the one-dimensional integral transform, while the book by Yu.A. Brychkov, H.-J. Glaeske, A.P. Prudnikov and Vu Kim Tuan [1] was devoted to the multidimensional case. The results for integral transforms in spaces of tested and generalized functions appeared in the books by Yu.A. Brychkov and A.P. Prudnikov [1] and R.S. Pathak [11].

It should be noted that many new results, obtained recently in the theory of integral transforms with special function kernels and with more general H-function kernels, were not reflected in the monographs in the above literature. As for $\boldsymbol{H}$-transforms, only some results in the theory of these transforms in L^1- and L^2-spaces were presented in Chapter 3 of the book by H.M. Srivastava, K.C. Gupta and S.P. Goyal [1].

The interest of the present authors in $\boldsymbol{H}$-transforms was motivated by generalizations of the integral transforms with hypergeometric and Bessel type kernels, which are involved in solutions of integral and differential equations of fractional order. Investigation of more general integral transforms with H-function kernels allows us to look more clearly at the problems with these equations, which are basically related to the mapping properties and asymptotic behavior of their solutions. These problems were solved by extending the range of parameters of the $\boldsymbol{H}$-transforms. Such a result was obtained, in turn, by the extension of the parameters of the H-function under its asymptotic behavior near zero and infinity.

In this book we present the results from the theory of integral transforms with H-function kernels in weighted spaces of Lebesgue r-summable functions $\mathfrak{L}_{\nu,r}$ (for real ν and $1 \leqq r \leqq \infty$, see Section 3.1) on the half-axis $\mathbb{R}_+ = (0, \infty)$. Properties such as the boundedness, including the one-to-one property of the map, the representation, and the ranges of $\boldsymbol{H}$-transforms in $\mathfrak{L}_{\nu,r}$-spaces are studied together with their inversion formulas. Applications are given to integral transforms with kernels involving the Meijer G-function and special functions of the hypergeometric and Bessel types.

Our investigations are based on the method of Mellin multipliers developed by Rooney [2] and on the asymptotic expansions of the H-function at zero and infinity. The latter were previously proved by the authors by developing the approach suggested by B.L.J. Braaksma [1]. These asymptotic estimates allow us not only to characterize the $\mathfrak{L}_{\nu,r}$-theory of the $\boldsymbol{H}$-transforms and integral transforms with special functions as kernels but also to extend and more precisely characterize the known statements in the theory of the H-function.

When $\nu = 1/r$, the spaces $\mathfrak{L}_{\nu,r}$ coincide with the spaces $L^r(\mathbb{R}_+)$ of r-summable functions on $\mathbb{R}_+$, and one may deduce the L_r-theory of the integral transforms with H- and G-functions and special functions of hypergeometric and Bessel types as kernels from our statements proved for the $\mathfrak{L}_{\nu,r}$-spaces. These results generalize the known assertions on the boundedness properties of integral transforms with special function kernels in L^2- and L^r-spaces $(1 < r < \infty)$ in the main.

The book consists of eight chapters. The results presented in the main body of the chapters are, as a rule, given with complete proofs. Distinctive historical survey sections complete every chapter, and are given in Sections 1.11, 2.10, 3.7, 4.11, 5.7, 6.8, 7.12 and 8.14. These sections provide historical comments on the content of the previous chapter and contain discussions and formulations of the results closely related to the subject matter of the chapter but not included in the main text.

The first two chapters contain various results from the theory of the H-function. Chapter 1 deals with the existence and representations of the H-function, its explicit power and power-logarithmic expansions, and the asymptotic expansions near zero and infinity. Chapter 2 is devoted to other properties of the H-function such as elementary formulas, differentiation and recurrence relations, expansion and multiplication formulas and the transformation formulas of Mellin, Laplace and Hankel transforms, and fractional integro-differentiation. General integral formulas involving the H-function are considered together with the special cases.

Chapters 3–5 present the $\mathfrak{L}_{\nu,r}$-theory of integral transforms with H-function kernels. The mapping properties such as the boundedness, the one-to-one properties of the map, and the representation of the $\boldsymbol{H}$-transforms in $\mathfrak{L}_{\nu,2}$-space are proved in Chapter 3. These results are extended to $\mathfrak{L}_{\nu,r}$-space with arbitrary r $(1 \leqq r \leqq \infty)$ in Chapter 4, in which the range $\mathbf{H}(\mathfrak{L}_{\nu,r})$ is also characterized and inversion relations are given. Chapter 5 deals with $\mathfrak{L}_{\nu,r}$-theory of the modified integral transform with H-function kernels that are obtained from $\boldsymbol{H}$-transforms by elementary transforms of translation, dilatation, reflection, and multiplication by a power function.

The last three chapters contain the results from the $\mathfrak{L}_{\nu,r}$-theory of integral transforms with special function kernels. The integral transforms with the Meijer G-function as kernel are considered in Chapter 6 together with the modified $\boldsymbol{G}$-transforms. Chapter 7 is devoted to the hypergeometric type integral transforms such as the Laplace and Stieltjes type transforms, the transforms that contain Whittaker and parabolic cylinder functions, hypergeometric functions and Wright functions as kernels. Bessel type integral transforms are studied in Chapter 8, in which Hankel type transforms including the Bessel function of the first kind and its generalizations in the kernels are treated, as well as cosine and sine transforms, even and odd Hilbert transforms, and transforms with Neumann, Struve and Macdonald functions, modified Macdonald functions and Lommel–Maitland functions as kernels.

We note that the asymptotic estimates in Chapter 1 may be extended considerably beyond the known range of parameters for the H-function. Furthermore, the statements in Chapter 2 clarify the known assertions, and the results from $\mathfrak{L}_{\nu,r}$-theory of the $\boldsymbol{H}$-transforms in Chapters 3 and 4 allow us to generalize the known investigations for the modified $\boldsymbol{H}$- and $\boldsymbol{G}$-transforms in Chapters 5 and 6 as well as for the integral transforms with kernel functions of hypergeometric and Bessel types in Chapters 7 and 8. We also note that most of the results in Chapters 5–7 were not published previous to our articles.

This book reflects the above investigations made by ourselves and in cooperation with other mathematicians. More detailed information can be found in the historical survey sections. We must note that the $\mathfrak{L}_{\nu,2}$-theory results of the $\boldsymbol{H}$-transforms and their inversions presented in Chapter 3 and Sections 4.9–4.10 were obtained together with S.A. Shlapakov, most of the results on the $\mathfrak{L}_{\nu,r}$-theory of the $\boldsymbol{H}$-transforms in Sections 4.1–4.8 were obtained together with H.-J. Glaeske and S.A. Shlapakov, and some of the results in Chapter 8 were proved together with H.-J. Glaeske, J.J. Trujillo, J. Rodriguez, M. Rivero, B. Bonilla, L. Rodriguez and

A.N. Borovco. We also note that the boundedness, the representation, and the ranges of ***H***-transforms in $\mathfrak{L}_{\nu,r}$-spaces were investigated by J.J. Betancor and C. Jerez Diaz, independently.

The bibliography consists of 578 entries, published up to 2002. We hope that we have listed all publications concerning one-dimensional integral transforms with H-function kernels. However, it cannot be considered a complete bibliography of the investigations in the fields of H-function and one-dimensional integral transforms with special function kernels. Interested readers may find many additional references in the books by A.M. Mathai and R.K. Saxena [2] and H.M. Srivastava, K.C. Gupta and S.P. Goyal [1] for the former, and in the books by H.M. Srivastava and R.G. Buschman [4], S.G. Samko, A.A. Kilbas and O.I. Marichev [1, Chapter 7], Yu.A. Brychkov and A.P. Prudnikov [1], R.S. Pathak [11], A.I. Zayed [1] and L. Debnath [1] for the latter. One can also find references for works on multidimensional integral transforms in the book by Yu.A. Brychkov, H.-J. Glaeske, A.P. Prudnikov and Vu Kim Tuan [1].

We would like to express our deep gratitude to the late Professor Anatolii Platonovich Prudnikov for his interest in our investigations and valuable discussions.

The preparation of this book was supported in part by the Belarusian Fundamental Research Fund, and by the Science Research Promotion Fund from the Japan Private School Promotion Foundation.

Chapter 1

DEFINITION, REPRESENTATIONS AND EXPANSIONS OF THE H-FUNCTION

1.1. Definition of the H-Function

For integers m, n, p, q such that $0 \leqq m \leqq q$, $0 \leqq n \leqq p$, for $a_i, b_j \in \mathbb{C}$ with $\mathbb{C}$, the set of complex numbers, and for $\alpha_i, \beta_j \in \mathbb{R}_+ = (0, \infty)$ $(i = 1, 2, \cdots, p; j = 1, 2, \cdots, q)$, the H-function $H_{p,q}^{m,n}(z)$ is defined via a Mellin–Barnes type integral in the form

$$H_{p,q}^{m,n}(z) \equiv H_{p,q}^{m,n}\left[z \,\middle|\, \begin{matrix} (a_i, \alpha_i)_{1,p} \\ (b_j, \beta_j)_{1,q} \end{matrix}\right] \equiv H_{p,q}^{m,n}\left[z \,\middle|\, \begin{matrix} (a_p, \alpha_p) \\ (b_q, \beta_q) \end{matrix}\right]$$

$$\equiv H_{p,q}^{m,n}\left[z \,\middle|\, \begin{matrix} (a_1, \alpha_1), \cdots, (a_p, \alpha_p) \\ (b_1, \beta_1), \cdots, (b_q, \beta_q) \end{matrix}\right] = \frac{1}{2\pi i}\int_{\mathfrak{L}} \mathcal{H}_{p,q}^{m,n}(s) z^{-s} ds \tag{1.1.1}$$

with

$$\mathcal{H}_{p,q}^{m,n}(s) \equiv \mathcal{H}_{p,q}^{m,n}\left[\begin{matrix} (a_i, \alpha_i)_{1,p} \\ (b_j, \beta_j)_{1,q} \end{matrix} \,\middle|\, s\right] \equiv \mathcal{H}_{p,q}^{m,n}\left[\begin{matrix} (a_p, \alpha_p) \\ (b_q, \beta_q) \end{matrix} \,\middle|\, s\right]$$

$$= \frac{\prod_{j=1}^{m} \Gamma(b_j + \beta_j s) \prod_{i=1}^{n} \Gamma(1 - a_i - \alpha_i s)}{\prod_{i=n+1}^{p} \Gamma(a_i + \alpha_i s) \prod_{j=m+1}^{q} \Gamma(1 - b_j - \beta_j s)}. \tag{1.1.2}$$

Here

$$z^{-s} = \exp[-s\{\log|z| + i \arg z\}], \quad z \neq 0, \quad i = \sqrt{-1}, \tag{1.1.3}$$

where $\log|z|$ represents the natural logarithm of $|z|$ and $\arg z$ is not necessarily the principal value. An empty product in (1.1.2), if it occurs, is taken to be one, and the poles

$$b_{jl} = \frac{-b_j - l}{\beta_j} \quad (j = 1, \cdots, m;\ l = 0, 1, 2, \cdots) \tag{1.1.4}$$

of the gamma functions $\Gamma(b_j + \beta_j s)$ and the poles

$$a_{ik} = \frac{1 - a_i + k}{\alpha_i} \quad (i = 1, \cdots, n;\ k = 0, 1, 2, \cdots) \tag{1.1.5}$$

of the gamma functions $\Gamma(1-a_i-\alpha_i s)$ do not coincide:

$$\alpha_i(b_j+l)\neq\beta_j(a_i-k-1)\quad(i=1,\cdots,n;\ j=1,\cdots,m;\ k,l=0,1,2,\cdots). \tag{1.1.6}$$

$\mathfrak{L}$ in (1.1.1) is the infinite contour which separates all the poles b_{jl} in (1.1.4) to the left and all the poles a_{ik} in (1.1.5) to the right of $\mathfrak{L}$, and has one of the following forms:

(i) $\mathfrak{L}=\mathfrak{L}_{-\infty}$ is a left loop situated in a horizontal strip starting at the point $-\infty+i\varphi_1$ and terminating at the point $-\infty+i\varphi_2$ with $-\infty<\varphi_1<\varphi_2<+\infty$;

(ii) $\mathfrak{L}=\mathfrak{L}_{+\infty}$ is a right loop situated in a horizontal strip starting at the point $+\infty+i\varphi_1$ and terminating at the point $+\infty+i\varphi_2$ with $-\infty<\varphi_1<\varphi_2<+\infty$.

(iii) $\mathfrak{L}=\mathfrak{L}_{i\gamma\infty}$ is a contour starting at the point $\gamma-i\infty$ and terminating at the point $\gamma+i\infty$, where $\gamma\in\mathbb{R}=(-\infty,+\infty)$.

The properties of the H-function $H^{m,n}_{p,q}(z)$ depend on the numbers a^*, Δ, δ, μ, a_1^* and a_2^* which are expressed via m, n,p, q, a_i, α_i $(i=1,2,\cdots,p)$ and b_j, β_j $(j=1,2,\cdots,q)$ by the following relations:

$$a^*=\sum_{i=1}^{n}\alpha_i-\sum_{i=n+1}^{p}\alpha_i+\sum_{j=1}^{m}\beta_j-\sum_{j=m+1}^{q}\beta_j; \tag{1.1.7}$$

$$\Delta=\sum_{j=1}^{q}\beta_j-\sum_{i=1}^{p}\alpha_i; \tag{1.1.8}$$

$$\delta=\prod_{i=1}^{p}\alpha_i^{-\alpha_i}\prod_{j=1}^{q}\beta_j^{\beta_j}; \tag{1.1.9}$$

$$\mu=\sum_{j=1}^{q}b_j-\sum_{i=1}^{p}a_i+\frac{p-q}{2}; \tag{1.1.10}$$

$$a_1^*=\sum_{j=1}^{m}\beta_j-\sum_{i=n+1}^{p}\alpha_i; \tag{1.1.11}$$

$$a_2^*=\sum_{i=1}^{n}\alpha_i-\sum_{j=m+1}^{q}\beta_j; \tag{1.1.12}$$

$$a_1^*+a_2^*=a^*,\ a_1^*-a_2^*=\Delta; \tag{1.1.13}$$

$$\xi=\sum_{j=1}^{m}b_j-\sum_{j=m+1}^{q}b_j+\sum_{i=1}^{n}a_i-\sum_{i=n+1}^{p}a_i; \tag{1.1.14}$$

$$c^*=m+n-\frac{p+q}{2}. \tag{1.1.15}$$

An empty sum in (1.1.7), (1.1.8), (1.1.10)–(1.1.12), (1.1.14) and an empty product in (1.1.9), if they occur, are taken to be zero and one, respectively.

1.2. Existence and Representations

The existence of the H-function $H_{p,q}^{m,n}(z)$ may be recognized by the convergence of the integral in (1.1.1) which depends on the asymptotic behavior of the function $\mathcal{H}_{p,q}^{m,n}(s)$ in (1.1.2) at infinity. Such an asymptotic is based on the following relations for the gamma function $\Gamma(z)$ of $z = x + iy$ $(x, y \in \mathbb{R})$:

$$|\Gamma(x+iy)| \sim \sqrt{2\pi}|x|^{x-1/2}e^{-x-\pi[1-\operatorname{sign}(x)]y/2} \quad (|x| \to \infty) \tag{1.2.1}$$

and

$$|\Gamma(x+iy)| \sim \sqrt{2\pi}|y|^{x-1/2}e^{-x-\pi|y|/2} \quad (|y| \to \infty), \tag{1.2.2}$$

which are proved by using the Stirling relation (see Erdélyi, Magnus, Oberhettinger and Tricomi [1, 1.18(2)])

$$\Gamma(z) \sim \sqrt{2\pi}e^{(z-1/2)\log z}e^{-z} \quad (z \to \infty). \tag{1.2.3}$$

Remark 1.1. The relation 1.18(6) in Erdélyi, Magnus, Oberhettinger and Tricomi [1] needs correction with the addition of the multiplier e^x in the left-hand side and must be replaced by

$$\lim_{|y|\to\infty} |\Gamma(x+iy)|\, e^{x+\pi|y|/2}|y|^{1/2-x} = \sqrt{2\pi}. \tag{1.2.4}$$

The following assertions follow from (1.1.2), (1.2.1) and (1.2.2).

Lemma 1.1. *For t, $\sigma \in \mathbb{R}$, there hold the estimates*

$$\left|\mathcal{H}_{p,q}^{m,n}(t+i\sigma)\right| \sim A\left(\frac{e}{t}\right)^{-\Delta t}\delta^t t^{\operatorname{Re}(\mu)} \quad (t \to +\infty) \tag{1.2.5}$$

with

$$A = (2\pi)^{c^*}e^{q-m-n}\,\frac{\prod_{j=1}^{q}\left[\beta_j^{\operatorname{Re}(b_j)-1/2}e^{-\operatorname{Re}(b_j)}\right]\prod_{i=1}^{n}e^{\pi[\sigma\alpha_i+\operatorname{Im}(a_i)]}}{\prod_{i=1}^{p}\left[\alpha_i^{\operatorname{Re}(a_i)-1/2}e^{-\operatorname{Re}(a_i)}\right]\prod_{j=m+1}^{q}e^{\pi[\sigma\beta_j+\operatorname{Im}(b_j)]}}, \tag{1.2.6}$$

and

$$\left|\mathcal{H}_{p,q}^{m,n}(t+i\sigma)\right| \sim B\left(\frac{e}{|t|}\right)^{\Delta|t|}\delta^{-|t|}|t|^{\operatorname{Re}(\mu)} \quad (t \to -\infty) \tag{1.2.7}$$

with

$$B = (2\pi)^{c^*}e^{q-m-n}\,\frac{\prod_{j=1}^{q}\left[\beta_j^{\operatorname{Re}(b_j)-1/2}e^{-\operatorname{Re}(b_j)}\right]\prod_{i=n+1}^{p}e^{\pi[\sigma\alpha_i+\operatorname{Im}(a_i)]}}{\prod_{i=1}^{p}\left[\alpha_i^{\operatorname{Re}(a_i)-1/2}e^{-\operatorname{Re}(a_i)}\right]\prod_{j=1}^{m}e^{\pi[\sigma\beta_j+\operatorname{Im}(b_j)]}}. \tag{1.2.8}$$

Lemma 1.2. *For $\sigma,\ t \in \mathbb{R}$, there holds the estimate*

$$\left|\mathcal{H}_{p,q}^{m,n}(\sigma+it)\right| \sim C|t|^{\Delta\sigma+\mathrm{Re}(\mu)}e^{-\pi[|t|a^*+\mathrm{Im}(\xi)\mathrm{sign}(t)]/2} \quad (|t|\to\infty) \tag{1.2.9}$$

uniformly in σ on any bounded interval in $\mathbb{R}$, where

$$C = (2\pi)^{c^*}e^{-c^*-\Delta\sigma-\mathrm{Re}(\mu)}\delta^{\sigma}\prod_{i=1}^{p}\alpha_i^{1/2-\mathrm{Re}(a_i)}\prod_{j=1}^{q}\beta_j^{\mathrm{Re}(b_j)-1/2} \tag{1.2.10}$$

and ξ is defined in (1.1.14).

Let $\varphi_1,\ \varphi_2,\ \gamma \in \mathbb{R}$ and $l_1,\ l_2$ and l_γ be the lines

$$l_1 = \{t+i\varphi_1:\ t\in\mathbb{R}\},\quad l_2=\{t+i\varphi_2:\ t\in\mathbb{R}\}\quad (\varphi_1<\varphi_2),\quad l_\gamma=\{\gamma+it: t\in\mathbb{R}\}.$$

From Lemmas 1.1 and 1.2 we obtain the following asymptotic relations at infinity for the integrand in (1.1.1) on the lines l_1, l_2 and l_γ:

$$\left|\mathcal{H}_{p,q}^{m,n}(s)z^{-s}\right| \sim B_i e^{\varphi_i \arg z}\left(\frac{e}{|t|}\right)^{\Delta|t|}\left(\frac{|z|}{\delta}\right)^{|t|}|t|^{\mathrm{Re}(\mu)} \tag{1.2.11}$$

$$(s = t+i\varphi_i \in l_i;\ i=1,2)$$

as $t\to-\infty$;

$$\left|\mathcal{H}_{p,q}^{m,n}(s)z^{-s}\right| \sim A_i e^{\varphi_i \arg z}\left(\frac{e}{t}\right)^{-\Delta t}\left(\frac{\delta}{|z|}\right)^{t}t^{\mathrm{Re}(\mu)} \tag{1.2.12}$$

$$(s = t+i\varphi_i \in l_i;\ i=1,2)$$

as $t\to+\infty$; and

$$\left|\mathcal{H}_{p,q}^{m,n}(s)z^{-s}\right| \sim C_1 e^{-[\gamma\log|z|+\pi\mathrm{Im}(\xi)\mathrm{sign}(t)/2]}|t|^{\Delta\gamma+\mathrm{Re}(\mu)}e^{-\pi|t|a^*/2+t\arg z} \tag{1.2.13}$$

$$(s=\gamma+it\in l_\gamma)$$

as $|t|\to\infty$. Here A_1 and A_2, B_1 and B_2 are given in (1.2.6) and (1.2.8) with σ being replaced by φ_1 and φ_2, respectively, and C_1 by (1.2.10) with σ being replaced by γ.

The conditions for the existence of the H-function follow by virtue of these relations.

Theorem 1.1. *Let a^*, Δ, δ and μ be given in* (1.1.7)–(1.1.10). *Then the H-function $H_{p,q}^{m,n}(z)$ defined by* (1.1.1) *makes sense in the following cases:*

$$\mathfrak{L}=\mathfrak{L}_{-\infty},\quad \Delta>0,\quad z\neq 0; \tag{1.2.14}$$

$$\mathfrak{L}=\mathfrak{L}_{-\infty},\quad \Delta=0,\quad 0<|z|<\delta; \tag{1.2.15}$$

$$\mathfrak{L}=\mathfrak{L}_{-\infty},\quad \Delta=0,\quad |z|=\delta,\quad \mathrm{Re}(\mu)<-1; \tag{1.2.16}$$

$$\mathfrak{L}=\mathfrak{L}_{+\infty},\quad \Delta<0,\quad z\neq 0; \tag{1.2.17}$$

$$\mathfrak{L}=\mathfrak{L}_{+\infty},\quad \Delta=0,\quad |z|>\delta; \tag{1.2.18}$$

$$\mathfrak{L} = \mathfrak{L}_{+\infty}, \quad \Delta = 0, \quad |z| = \delta, \quad \mathrm{Re}(\mu) < -1; \tag{1.2.19}$$

$$\mathfrak{L} = \mathfrak{L}_{i\gamma\infty}, \quad a^* > 0, \quad |\arg z| < \frac{a^*\pi}{2}, \quad z \neq 0; \tag{1.2.20}$$

$$\mathfrak{L} = \mathfrak{L}_{i\gamma\infty}, \quad a^* = 0, \quad \Delta\gamma + \mathrm{Re}(\mu) < -1, \quad \arg z = 0, \quad z \neq 0. \tag{1.2.21}$$

Remark 1.2. The results of Theorem 1.1 in the cases (1.2.16), (1.2.19) and (1.2.21) are more precise than those in Prudnikov, Brychkov and Marichev [3, 8.3.1].

The estimate (1.2.11) holds for $t \to -\infty$ uniformly on the set which has a positive distance to the points b_{jl} in (1.1.4) and which does not contain any point to the right of $\mathfrak{L}_{-\infty}$. The same is true for the estimate (1.2.12) being valid for $t \to +\infty$ uniformly on the set with a positive distance to the points a_{ik} in (1.1.5) and containing no point to the left of $\mathfrak{L}_{+\infty}$. Therefore using Theorem 1.1 and the theory of residues, we obtain the following result:

Theorem 1.2. **(i)** *If the conditions* (1.1.6) *and* (1.2.14) *or* (1.2.15) *are satisfied, then the H-function* (1.1.1) *is an analytic function of z and*

$$H_{p,q}^{m,n}(z) = \sum_{j=1}^{m}\sum_{l=0}^{\infty} \operatorname*{Res}_{s=b_{jl}} \left[\mathcal{H}_{p,q}^{m,n}(s)z^{-s}\right], \tag{1.2.22}$$

where b_{jl} are given in (1.1.4).

(ii) *If the conditions* (1.1.6) *and* (1.2.17) *or* (1.2.18) *are satisfied, then the H-function* (1.1.1) *is an analytic function of z and*

$$H_{p,q}^{m,n}(z) = -\sum_{i=1}^{n}\sum_{k=0}^{\infty} \operatorname*{Res}_{s=a_{ik}} \left[\mathcal{H}_{p,q}^{m,n}(s)z^{-s}\right], \tag{1.2.23}$$

where a_{ik} are given in (1.1.5).

(iii) *If the conditions in* (1.1.6) *and* (1.2.20) *are satisfied, then the H-function* (1.1.1) *is an analytic function of z in the sector $|\arg z| < a^*\pi/2$.*

1.3. Explicit Power Series Expansions

In view of the cases (i) and (ii) of Theorem 1.2 we may establish the series representations for $H_{p,q}^{m,n}(z)$, in the respective cases of the poles b_{jl} in (1.1.4) of the gamma functions $\Gamma(b_j + \beta_j s)$ $(j = 1, \cdots, m)$ being simple:

$$\beta_j(b_i + k) \neq \beta_i(b_j + l) \quad (i \neq j;\ i, j = 1, \cdots, m;\ k, l = 0, 1, 2, \cdots) \tag{1.3.1}$$

and the poles a_{ik} in (1.1.5) of $\Gamma(1 - a_i + \alpha_i s)$ $(i = 1, \cdots, n)$ also being simple:

$$\alpha_j(1 - a_i + k) \neq \alpha_i(1 - a_j + l) \quad (i \neq j;\ i, j = 1, \cdots, n;\ k, l = 0, 1, 2, \cdots). \tag{1.3.2}$$

First we consider the former case. To apply Theorem 1.2(i) we evaluate the residues of $\mathcal{H}_{p,q}^{m,n}(s)z^{-s}$ at the points $s = b_{jl}$. For this, we use the property in Marichev [1, (3.39)], that is, in a neighborhood of the poles $z = -k$ $(k = 0, -1, -2, \cdots)$ the gamma function $\Gamma(z)$ can be extended in powers of $z + k = \epsilon$:

$$\Gamma(z) = \Gamma(-k+\epsilon) = \frac{(-1)^k}{k!\epsilon}\left[1 + \epsilon\psi(1+k) + O\left(\epsilon^2\right)\right] \quad (\epsilon \to 0), \tag{1.3.3}$$

where

$$\psi(z) = \frac{\Gamma'(z)}{\Gamma(z)} \tag{1.3.4}$$

is the psi function. Then, since the poles b_{jl} are simple, we have

$$\operatorname*{Res}_{s=b_{jl}}\left[\mathcal{H}_{p,q}^{m,n}(s)z^{-s}\right] = h_{jl}^* z^{-b_{jl}} \quad (j = 1, \cdots, m;\ l = 0, 1, 2, \cdots), \tag{1.3.5}$$

where

$$h_{jl}^* = \lim_{s \to b_{jl}}\left[(s - b_{jl})\mathcal{H}_{p,q}^{m,n}(s)\right]$$

$$= \frac{(-1)^l}{l!\beta_j}\frac{\prod_{\substack{i=1\\ i\neq j}}^{m}\Gamma\left(b_i - [b_j + l]\frac{\beta_i}{\beta_j}\right)\prod_{i=1}^{n}\Gamma\left(1 - a_i + [b_j + l]\frac{\alpha_i}{\beta_j}\right)}{\prod_{i=n+1}^{p}\Gamma\left(a_i - [b_j + l]\frac{\alpha_i}{\beta_j}\right)\prod_{i=m+1}^{q}\Gamma\left(1 - b_i + [b_j + l]\frac{\beta_i}{\beta_j}\right)}. \tag{1.3.6}$$

Thus we obtain the result:

Theorem 1.3. *Let the conditions in* (1.1.6) *and* (1.3.1) *be satisfied and let either* $\Delta > 0$, $z \neq 0$ *or* $\Delta = 0$, $0 < |z| < \delta$. *Then the H-function* (1.1.1) *has the power series expansion*

$$H_{p,q}^{m,n}(z) = \sum_{j=1}^{m}\sum_{l=0}^{\infty} h_{jl}^* z^{(b_j+l)/\beta_j}, \tag{1.3.7}$$

where the constants h_{jl}^* *are given in* (1.3.6).

Similarly we consider the case (1.3.2) and from Theorem 1.2(ii) come to the statement:

Theorem 1.4. *Let the conditions in* (1.1.6) *and* (1.3.2) *be satisfied and let either* $\Delta < 0$, $z \neq 0$ *or* $\Delta = 0$, $|z| > \delta$. *Then the H-function* (1.1.1) *has the power series expansion*

$$H_{p,q}^{m,n}(z) = \sum_{i=1}^{n}\sum_{k=0}^{\infty} h_{ik} z^{(a_i-1-k)/\alpha_i}, \tag{1.3.8}$$

where the constants h_{ik} *have the forms*

$$h_{ik} = \lim_{s \to a_{ik}}\left[-(s - a_{ik})\mathcal{H}_{p,q}^{m,n}(s)\right]$$

$$= \frac{(-1)^k}{k!\alpha_i}\frac{\prod_{j=1}^{m}\Gamma\left(b_j + [1 - a_i + k]\frac{\beta_j}{\alpha_i}\right)\prod_{\substack{j=1\\ j\neq i}}^{n}\Gamma\left(1 - a_j - [1 - a_i + k]\frac{\alpha_j}{\alpha_i}\right)}{\prod_{j=n+1}^{p}\Gamma\left(a_j + [1 - a_i + k]\frac{\alpha_j}{\alpha_i}\right)\prod_{j=m+1}^{q}\Gamma\left(1 - b_j - [1 - a_i + k]\frac{\beta_j}{\alpha_i}\right)} \tag{1.3.9}$$

in view of the relation

$$\operatorname*{Res}_{s=a_{ik}}[\mathcal{H}^{m,n}_{p,q}(s)z^{-s}] = h_{ik}z^{-a_{ik}} = h_{ik}z^{(a_i-1-k)/\alpha_i}. \tag{1.3.10}$$

1.4. Explicit Power-Logarithmic Series Expansions

When the conditions in (1.1.6) are satisfied, but some poles b_{jl} in (1.1.4) or a_{ik} in (1.1.5) coincide, we can also prove series representations for $H^{m,n}_{p,q}(z)$ by using Theorem 1.2 in cases (i) and (ii). First we consider the case (i). Let $b \equiv b_{jl}$ be one of the points (1.1.4) for which some other poles of the gamma functions $\Gamma(b_j+\beta_j s)$ $(j=1,\cdots,m)$ coincide and let $N^* \equiv N^*_{jl}$ be its overlapping order. This means that there exist $j_1,\cdots,j_{N^*}\in\{1,\cdots,m\}$ and $l_{j_1},\cdots,l_{j_{N^*}}\in\{0,1,2,\cdots\}$ such that

$$b \equiv b_{jl} = -\frac{b_{j_1}+l_{j_1}}{\beta_{j_1}} = \cdots = -\frac{b_{j_{N^*}}+l_{j_{N^*}}}{\beta_{j_{N^*}}}. \tag{1.4.1}$$

Thus $\mathcal{H}^{m,n}_{p,q}(s)z^{-s}$ has a pole at b of order N^* and hence

$$\operatorname*{Res}_{s=b}\left[\mathcal{H}^{m,n}_{p,q}(s)z^{-s}\right] = \frac{1}{(N^*-1)!}\lim_{s\to b}\left[(s-b)^{N^*}\mathcal{H}^{m,n}_{p,q}(s)z^{-s}\right]^{(N^*-1)}. \tag{1.4.2}$$

Denoting

$$\mathcal{H}^*_1(s) = (s-b)^{N^*}\prod_{j=j_1}^{j_{N^*}}\Gamma(b_j+\beta_j s), \qquad \mathcal{H}^*_2(s) = \frac{\mathcal{H}^{m,n}_{p,q}(s)}{\prod\limits_{j=j_1}^{j_{N^*}}\Gamma(b_j+\beta_j s)} \tag{1.4.3}$$

and using the Leibniz rule, we have

$$\begin{aligned}
&\left[(s-b)^{N^*}\mathcal{H}^{m,n}_{p,q}(s)z^{-s}\right]^{(N^*-1)} \\
&= \sum_{n=0}^{N^*-1}\binom{N^*-1}{n}[\mathcal{H}^*_1(s)]^{(N^*-1-n)}\,[\mathcal{H}^*_2(s)z^{-s}]^{(n)} \\
&= \sum_{n=0}^{N^*-1}\binom{N^*-1}{n}[\mathcal{H}^*_1(s)]^{(N^*-1-n)}\sum_{i=0}^{n}\binom{n}{i}(-1)^i[\mathcal{H}^*_2(s)]^{(n-i)}\,z^{-s}[\log z]^i \\
&= z^{-s}\sum_{i=0}^{N^*-1}\left\{\sum_{n=i}^{N^*-1}(-1)^i\binom{N^*-1}{n}\binom{n}{i}[\mathcal{H}^*_1(s)]^{(N^*-1-n)}[\mathcal{H}^*_2(s)]^{(n-i)}\right\}[\log z]^i.
\end{aligned}$$

Substituting this into (1.4.2), we obtain

$$\operatorname*{Res}_{s=b_{jl}}\left[\mathcal{H}^{m,n}_{p,q}(s)z^{-s}\right] = z^{(b_j+l)/\beta_j}\sum_{i=0}^{N^*_{jl}-1}H^*_{jli}[\log z]^i, \tag{1.4.4}$$

where

$$\begin{aligned}
H^*_{jli} &\equiv H^*_{jli}(N^*_{jl};b_{jl}) \\
&= \frac{1}{(N^*_{jl}-1)!}\sum_{n=i}^{N^*_{jl}-1}(-1)^i\binom{N^*_{jl}-1}{n}\binom{n}{i}[\mathcal{H}^*_1(b_{jl})]^{(N^*_{jl}-1-n)}[\mathcal{H}^*_2(b_{jl})]^{(n-i)}.
\end{aligned} \tag{1.4.5}$$

In particular, if we pick the case $l = 0$ and $i = N_{j0}^* - 1$ which will be treated later (cf. Theorem 1.12), we have from (1.3.3) and (1.4.3) by setting $N_j^* \equiv N_{j0}^*$

$$H_j^* \equiv H_{j,0,N_j^*-1}^*(N_j^*; b_{j0}) = \frac{(-1)^{N_j^*-1}}{(N_j^*-1)!} \mathcal{H}_1^*(b_{j0})\mathcal{H}_2^*(b_{j0})$$

$$= \frac{(-1)^{N_j^*-1}}{(N_j^*-1)!} \left\{ \prod_{k=1}^{N_j^*} \frac{(-1)^{j_k}}{j_k!\beta_{j_k}} \right\} \frac{\prod\limits_{\substack{i=1 \\ i \neq j_1, \cdots, j_{N_j^*}}}^{m} \Gamma\left(b_i - b_j \frac{\beta_i}{\beta_j}\right) \prod\limits_{i=1}^{n} \Gamma\left(1 - a_i + b_j \frac{\alpha_i}{\beta_j}\right)}{\prod\limits_{i=n+1}^{p} \Gamma\left(a_i - b_j \frac{\alpha_i}{\beta_j}\right) \prod\limits_{i=m+1}^{q} \Gamma\left(1 - b_i + b_j \frac{\beta_i}{\beta_j}\right)}. \tag{1.4.6}$$

Thus in view of Theorem 1.2(i), we obtain:

Theorem 1.5. *Let the conditions in (1.1.6) be satisfied and let either $\Delta > 0, z \neq 0$ or $\Delta = 0, 0 < |z| < \delta$. Then the H-function (1.1.1) has the power-logarithmic series expansion*

$$H_{p,q}^{m,n}(z) = \sum_{j,l}{}' h_{jl}^* z^{(b_j+l)/\beta_j} + \sum_{j,l}{}'' \sum_{i=0}^{N_{jl}^*-1} H_{jli}^* z^{(b_j+l)/\beta_j} [\log z]^i. \tag{1.4.7}$$

Here $\sum'$ and $\sum''$ are summations taken over j, l $(j = 1, \cdots, m;\ l = 0, 1, \cdots)$ such that the gamma functions $\Gamma(b_j + \beta_j s)$ have simple poles and poles of order N_{jl}^* at the points b_{jl}, respectively, and the constants h_{jl}^* are given in (1.3.6) while the constants H_{jli}^* are given by (1.4.5).

Similarly we consider the case (ii) in Theorem 1.2. Let $a = a_{ik}$ be one of the points (1.1.5) for which some other poles of $\Gamma(1 - a_i - \alpha_i s)$ coincide and let $N = N_{ik}$ be its overlapping order. This means that there exist $i_1, \cdots, i_N \in \{1, \cdots, n\}$ and $k_{i_1}, \cdots, k_{i_N} \in \mathbb{N}_0 \equiv \{0, 1, 2, \cdots\}$ such that

$$a \equiv a_{ik} = \frac{1 - a_{i_1} + k_{i_1}}{\alpha_{i_1}} = \cdots = \frac{1 - a_{i_N} + k_{i_N}}{\alpha_{i_N}}. \tag{1.4.8}$$

Then the integrand $\mathcal{H}_{p,q}^{m,n}(s)z^{-s}$ in (1.1.1) has a pole at a of order N. Denoting

$$\mathcal{H}_1(s) = (s-a)^N \prod_{i=i_1}^{i_N} \Gamma(1 - a_i - \alpha_i s), \qquad \mathcal{H}_2(s) = \frac{\mathcal{H}_{p,q}^{m,n}(s)}{\prod\limits_{i=i_1}^{i_N} \Gamma(1 - a_i - \alpha_i s)}, \tag{1.4.9}$$

similarly to the previous case we obtain

$$\operatorname*{Res}_{s=a_{ik}} \left[\mathcal{H}_{p,q}^{m,n}(s)z^{-s}\right] = z^{(a_i-1-k)/\alpha_i} \sum_{j=0}^{N_{ik}-1} H_{ikj} [\log z]^j, \tag{1.4.10}$$

where

$$H_{ikj} \equiv H_{ikj}(N_{ik}; a_{ik})$$

$$= \frac{1}{(N_{ik}-1)!} \left\{ \sum_{n=j}^{N_{ik}-1} (-1)^j \binom{N_{ik}-1}{n} \binom{n}{j} [\mathcal{H}_1(a_{ik})]^{(N_{ik}-1-n)} [\mathcal{H}_2(a_{ik})]^{(n-j)} \right\}. \tag{1.4.11}$$

For the special choice of $k=0$ and $j=N_{i0}-1$, (1.4.9) and (1.3.3) yield, by setting $N_i \equiv N_{i0}$,

$$H_i \equiv H_{i,0,N_i-1}(N_i;a_{i0}) = \frac{(-1)^{N_i-1}}{(N_i-1)!}\mathcal{H}_1(a_{i0})\mathcal{H}_2(a_{i0})$$

$$= \frac{(-1)^{N_i-1}}{(N_i-1)!}\left(\prod_{k=1}^{N_i}\frac{(-1)^{i_k-1}}{i_k!\alpha_{i_k}}\right)$$

$$\cdot \frac{\prod_{j=1}^{m}\Gamma\left(b_j+[1-a_i]\frac{\beta_j}{\alpha_i}\right)\prod_{\substack{j=1\\ j\neq i_1,\cdots,i_{N_i}}}^{n}\Gamma\left(1-a_j-[1-a_i]\frac{\alpha_j}{\alpha_i}\right)}{\prod_{j=n+1}^{p}\Gamma\left(a_j+[1-a_i]\frac{\alpha_j}{\alpha_i}\right)\prod_{j=m+1}^{q}\Gamma\left(1-b_j-[1-a_i]\frac{\beta_j}{\alpha_i}\right)}, \tag{1.4.12}$$

which will appear in Theorem 1.8, below.

Therefore Theorem 1.2(ii) implies:

Theorem 1.6. *Let the conditions in (1.1.6) be satisfied and let either $\Delta<0, z\neq 0$ or $\Delta=0, |z|>\delta$. Then the H-function (1.1.1) has the power-logarithmic series expansion*

$$H_{p,q}^{m,n}(z) = \sum_{i,k}{}' h_{ik}z^{(a_i-1-k)/\alpha_i} + \sum_{i,k}{}'' \sum_{j=0}^{N_{ik}-1} H_{ikj}z^{(a_i-1-k)/\alpha_i}[\log z]^j. \tag{1.4.13}$$

Here $\sum'$ and $\sum''$ are summations taken over i,k $(i=1,\cdots,n;\ k=0,1,\cdots)$ such that the gamma functions $\Gamma(1-a_i-\alpha_i s)$ have simple poles and poles of order N_{ik} at the points a_{ik}, respectively, and the constants h_{ik} are given in (1.3.9) while the constants H_{ikj} are given by (1.4.11).

1.5. Algebraic Asymptotic Expansions at Infinity

When $\Delta \leqq 0$, the results in Theorems 1.4 and 1.6 give the power and power-logarithmic asymptotic expansions of the H-function near infinity, namely

$$H_{p,q}^{m,n}(z) \sim \sum_{i=1}^{n}\sum_{k=0}^{\infty} h_{ik}z^{(a_i-1-k)/\alpha_i} \quad (z\to\infty), \tag{1.5.1}$$

when the poles a_{ik} in (1.4.8) of the gamma functions $\Gamma(1-a_i-\alpha_i s)$ $(1\leqq i\leqq n)$ do not coincide, and

$$H_{p,q}^{m,n}(z) \sim \sum_{i,k}{}' h_{ik}z^{(a_i-1-k)/\alpha_i} + \sum_{i,k}{}'' \sum_{j=0}^{N_{ik}-1} H_{ikj}z^{(a_i-1-k)/\alpha_i}[\log z]^j \quad (z\to\infty), \tag{1.5.2}$$

when some poles a_{ik} coincide, where the summations in $\sum'$ and $\sum''$ are taken over i,k $(i=1,\cdots,n;\ k=0,1,2,\cdots)$ as in Theorem 1.6 and N_{ik} is the orders of the poles.

If $\Delta>0$, $a^*>0$ and the conditions in (1.1.6) hold, then $\mathfrak{L}=\mathfrak{L}_{-\infty}$ and the asymptotic expansion of the H-function at infinity has the form

$$H_{p,q}^{m,n}(z) \sim \sum_{i=1}^{n}\sum_{k=0}^{\infty}\operatorname*{Res}_{s=-a_{ik}}[\mathcal{H}_{p,q}^{m,n}(-s)z^{s}] = \sum_{i=1}^{n}\sum_{k=0}^{\infty}\operatorname*{Res}_{s=a_{ik}}[\mathcal{H}_{p,q}^{m,n}(s)z^{-s}] \tag{1.5.3}$$

$$\left(z\to\infty,\ |\arg z|<\frac{a^*\pi}{2}\right).$$

This result was proved by Braaksma [1, (2.16)]. He considered the H-function in the form

$$H_{p,q}^{m,n}(z) = \frac{1}{2\pi i}\int_{\mathfrak{L}_{+\infty}} \mathcal{H}_{p,q}^{m,n}(-s)z^s ds \tag{1.5.4}$$

and developed the method to find the asymptotic expansion of this integral in the cases $\Delta > 0$ and $\Delta = 0,\ 0 < |z| < \delta$. This method is based on the Cauchy theorem and the residue theory, according to which the contour $\mathfrak{L} = \mathfrak{L}_{+\infty}$ is replaced by two other paths $\mathfrak{L}_1$ and $\mathfrak{L}_2$, being contours surrounding $\mathfrak{L}$, and the integral (1.5.4) can be represented in the form

$$H_{p,q}^{m,n}(z) = Q_w(z) + \frac{1}{2\pi i}\int_{\mathfrak{L}_1} \mathcal{H}_{p,q}^{m,n}(-s)z^s ds - \frac{1}{2\pi i}\int_{\mathfrak{L}_2} \mathcal{H}_{p,q}^{m,n}(-s)z^s ds, \tag{1.5.5}$$

where $Q_w(z)$ is the sum of residues of the function $\mathcal{H}_{p,q}^{m,n}(-s)z^s$ at some points a_{ik} given in (1.1.5).

When $a^* > 0$, two paths of integration in (1.5.5) may be replaced by the line parallel to the imaginary axis:

$$H_{p,q}^{m,n}(z) = Q_w(z) + \frac{1}{2\pi i}\int_{w+i\infty}^{w-i\infty} \mathcal{H}_{p,q}^{m,n}(-s)z^s ds, \tag{1.5.6}$$

where w is a certain real constant. It follows from here that, if $\Delta = 0$, $H_{p,q}^{m,n}(z)$ can be analytically continued into the sector

$$|\arg(z)| < \frac{a^*\pi}{2}, \tag{1.5.7}$$

and if $\Delta > 0$, the estimate

$$H_{p,q}^{m,n}(z) = Q_w(z) + O(z^w) \tag{1.5.8}$$

and hence (1.5.3) holds for $|z| \to \infty$ uniformly on every closed sector of (1.5.7).

It follows from (1.5.3) that the H-function (with $\Delta > 0,\ a^* > 0$) also has the asymptotic expansions (1.5.1) and (1.5.2) in the cases when the poles a_{ik} of the gamma functions $\Gamma(1 - a_i - \alpha_i s)$ $(1 \leqq i \leqq n;\ k = 0, 1, 2, \cdots)$ do not coincide and coincide, respectively. By the Cauchy theorem the results above are also valid for the H-function with $\mathfrak{L} = \mathfrak{L}_{i\gamma\infty}$ and $a^* > 0$ in the sector $|\arg z| < a^*\pi/2$.

Theorem 1.7. *Let the conditions in (1.1.6) and (1.3.2) be satisfied and let either $\Delta \leqq 0$ or $\Delta > 0,\ a^* > 0$. Then the asymptotic expansion of $H_{p,q}^{m,n}(z)$ near infinity is given in (1.5.1) and the principal terms of this asymptotic have the form*

$$H_{p,q}^{m,n}(z) = \sum_{i=1}^{n}\left[h_i z^{(a_i-1)/\alpha_i} + o\left(z^{(a_i-1)/\alpha_i}\right)\right] \quad (z \to \infty), \tag{1.5.9}$$

where $|\arg z| < a^*\pi/2$ for $\Delta > 0,\ a^* > 0$ and

$$h_i \equiv h_{i0} = \frac{1}{\alpha_i}\frac{\prod_{j=1}^{m}\Gamma\left(b_j - (a_i - 1)\frac{\beta_j}{\alpha_i}\right)\prod_{j=1, j\neq i}^{n}\Gamma\left(1 - a_j + (a_i - 1)\frac{\alpha_j}{\alpha_i}\right)}{\prod_{j=n+1}^{p}\Gamma\left(a_j - (a_i - 1)\frac{\alpha_j}{\alpha_i}\right)\prod_{j=m+1}^{q}\Gamma\left(1 - b_j + (a_i - 1)\frac{\beta_j}{\alpha_i}\right)}. \tag{1.5.10}$$

Corollary 1.7.1. *Let the assumptions of Theorem 1.7 be satisfied and let i_0 $(1 \leqq i_0 \leqq n)$ be an integer such that*

$$\frac{\mathrm{Re}(a_{i_0})-1}{\alpha_{i_0}} = \max_{1\leqq i\leqq n}\left[\frac{\mathrm{Re}(a_i)-1}{\alpha_i}\right]. \tag{1.5.11}$$

Then there holds the asymptotic estimate

$$H_{p,q}^{m,n}(z) = h_{i_0} z^{(a_{i_0}-1)/\alpha_{i_0}} + o\left(z^{(a_{i_0}-1)/\alpha_{i_0}}\right) \quad (z \to \infty), \tag{1.5.12}$$

where $|\arg z| < a^*\pi/2$ for $\Delta > 0,\ a^* > 0$, and h_{i_0} is given in (1.5.10) with $i = i_0$. In particular,

$$H_{p,q}^{m,n}(z) = O\left(z^{\rho}\right) \quad (z \to \infty), \tag{1.5.13}$$

where $\rho = \max\limits_{1\leqq i\leqq n}\left[\{\mathrm{Re}(a_i)-1\}/\alpha_i\right]$ and $|\arg z| < a^*\pi/2$ when $\Delta > 0,\ a^* > 0$.

Theorem 1.8. *Let the conditions in (1.1.6) be satisfied and let either $\Delta \leqq 0$ or $\Delta > 0$, $a^* > 0$. Let some poles of the gamma functions $\Gamma(1 - a_i - \alpha_i s)$ $(i = 1, \cdots, n)$ coincide and N_{ik} be the orders of these poles. Then the asymptotic expansion of $H_{p,q}^{m,n}(z)$ near infinity is given in (1.5.2) and the principal terms of this asymptotic have the form*

$$H_{p,q}^{m,n}(z) = \sum_i{}' \left[h_i z^{(a_i-1)/\alpha_i} + o\left(z^{(a_i-1)/\alpha_i}\right)\right]$$
$$+\sum_i{}'' \left[H_i z^{(a_i-1)/\alpha_i}[\log z]^{N_i-1} + o\left(z^{(a_i-1)/\alpha_i}[\log z]^{N_i-1}\right)\right] \quad (z \to \infty), \tag{1.5.14}$$

where $|\arg z| < a^*\pi/2$ for $\Delta > 0,\ a^* > 0$. Here $\sum'$ and $\sum''$ are summations taken over i $(i = 1, \cdots, n)$ such that the gamma functions $\Gamma(1 - a_i - \alpha_i s)$ have simple poles and poles of order $N_i \equiv N_{i0}$ at the points a_{i0}, respectively, and h_i are given in (1.5.10) while H_i are given by (1.4.12).

Corollary 1.8.1. *Let the assumptions of Theorem 1.8 be satisfied and let i_{01} and i_{02} $(1 \leqq i_{01}, i_{02} \leqq n)$ be numbers such that*

$$\rho_1 \equiv \frac{\mathrm{Re}(a_{i_{01}})-1}{\alpha_{i_{01}}} = \max_{1\leqq i\leqq n}\left[\frac{\mathrm{Re}(a_i)-1}{\alpha_i}\right], \tag{1.5.15}$$

when the poles a_{ik} $(i = 1, \cdots, n;\ k = 0, 1, \cdots)$ are simple, and

$$\rho_2 \equiv \frac{\mathrm{Re}(a_{i_{02}})-1}{\alpha_{i_{02}}} = \max_{1\leqq i\leqq n}\left[\frac{\mathrm{Re}(a_i)-1}{\alpha_i}\right], \tag{1.5.16}$$

when the poles a_{ik} $(i = 1, \cdots, n;\ k = 0, 1, \cdots)$ coincide.

a) *If $\rho_1 > \rho_2$, then the first term in the asymptotic expansion of the H-function has the form*

$$H_{p,q}^{m,n}(z) = h_{i_{01}} z^{(a_{i_{01}}-1)/\alpha_{i_{01}}} + o\left(z^{(a_{i_{01}}-1)/\alpha_{i_{01}}}\right) \quad (z \to \infty), \tag{1.5.17}$$

where $|\arg z| < a^*\pi/2$ for $\Delta > 0,\ a^* > 0$, and $h_{i_{01}}$ is given in (1.5.10) with $i = i_{01}$. In particular, the relation (1.5.13) holds.

b) *If $\rho_1 \leqq \rho_2$ and and the gamma function $\Gamma(1-a_i-\alpha_i s)$ has the pole $a_{i_{02}}$ of order $N_{i_{02}}$, then the principal term in the asymptotic expansion of the H-function has the form*

$$H_{p,q}^{m,n}(z) = H_{i_{02}} z^{(a_{i_{02}}-1)/\alpha_{i_{02}}}[\log z]^{N_{i_{02}}-1} + o\left(z^{(a_{i_{02}}-1)/\alpha_{i_{02}}}[\log z]^{N_{i_{02}}-1}\right) \tag{1.5.18}$$
$$(z \to \infty),$$

where $|\arg z| < a^*\pi/2$ for $\Delta > 0,\ a^* > 0$ and $H_{i_{02}}$ is given in (1.4.12) with $i = i_{02}$. In particular, if N is the largest order of general poles of the gamma functions $\Gamma(1-a_i-\alpha_i s)$ $(i = 1, \cdots, n)$, then

$$H_{p,q}^{m,n}(z) = O\left(z^{\rho}[\log z]^{N-1}\right) \quad (z \to \infty), \tag{1.5.19}$$

where $\rho = \max_{1 \leqq i \leqq n}[\{\mathrm{Re}(a_i)-1\}/\alpha_i]$ and $|\arg z| < a^*\pi/2$ when $\Delta > 0,\ a^* > 0$.

When $\Delta > 0$ and $a^* \leqq 0$, the asymptotic behavior of the H-function with $\mathfrak{L} = \mathfrak{L}_{-\infty}$ at infinity is more complicated. Such asymptotic expansions have exponential form and can be obtained from the results by Braaksma [1]. The case $\Delta > 0,\ a^* = 0$ will be considered in the next section. It should be noted that these asymptotic expansions have different forms in sectors of changing $\arg z$ and contain the special function $E(z)$ defined in (1.6.3). Asymptotics of the same form are also obtained in the special case $n = 0$ presented in Section 1.7.

1.6. Exponential Asymptotic Expansions at Infinity in the Case $\Delta > 0,\ a^* = 0$

In the previous section we proved the asymptotic behavior of the H-function at infinity when $\Delta \leqq 0$ or $\Delta > 0,\ a^* > 0$. Now we consider the remaining case $\Delta > 0,\ a^* = 0$. Let

$$c_0 = (2\pi i)^{m+n-p} \exp\left[\left(\sum_{i=n+1}^{p} a_i - \sum_{j=1}^{m} b_j\right)\pi i\right]; \tag{1.6.1}$$

$$d_0 = (-2\pi i)^{m+n-p} \exp\left[-\left(\sum_{i=n+1}^{p} a_i - \sum_{j=1}^{m} b_j\right)\pi i\right]; \tag{1.6.2}$$

$$E(z) = \frac{1}{2\pi i \Delta}\sum_{j=0}^{\infty} A_j \left(\frac{\Delta^\Delta}{\delta} z\right)^{(\mu-j+1/2)/\Delta} \exp\left[\left(\frac{\Delta^\Delta}{\delta} z\right)^{1/\Delta}\right]. \tag{1.6.3}$$

Here the constants A_j $(j \in \mathbb{N}_0)$ depending on $p,\ q,\ a_i,\ \alpha_i$ $(i = 1, \cdots, p)$ and $b_j,\ \beta_j$ $(j = 1, \cdots, q)$ are defined by the relations

$$\frac{\prod_{i=1}^{p}\Gamma(1-a_i-\alpha_i s)}{\prod_{j=1}^{q}\Gamma(1-b_j-\beta_j s)}\left(\frac{\Delta^\Delta}{\delta}\right)^s$$

$$= \sum_{j=0}^{N-1}\frac{A_j}{\Gamma(-\Delta s + j - \mu + 1/2)} + \frac{r_N(s)}{\Gamma(-\Delta s + N - \mu + 1/2)}, \tag{1.6.4}$$

where the function $r_N(s)$ is analytic in its domain of definition and $r_N(s) = O(1)$ as $|s| \to \infty$ uniformly on $|\arg s| \leqq \pi - \epsilon_1$ $(0 < \epsilon_1 < \pi)$. In particular,

$$A_0 = (2\pi)^{(p-q+1)/2}\Delta^{-\mu}\prod_{i=1}^{p}\alpha_i^{-a_i+1/2}\prod_{j=1}^{q}\beta_j^{b_j-1/2}. \tag{1.6.5}$$

Theorem 1.9. *Let a^*, Δ, δ and μ be given in (1.1.7)–(1.1.10) and let the conditions in (1.1.6) and (1.3.2) be satisfied. Let $\Delta > 0,\ a^* = 0$ and ϵ be a constant such that*

$$0 < \epsilon < \frac{\pi}{2}\min_{n+1\leqq i\leqq p, 1\leqq j\leqq m}[\alpha_i, \beta_j]. \tag{1.6.6}$$

Then there hold the following assertions.

(i) *The H-function (1.1.1) has the asymptotic expansion*

$$H_{p,q}^{m,n}(z) \sim \sum_{i=1}^{n}\sum_{k=0}^{\infty} h_{ik}z^{(a_i-1-k)/\alpha_i} + c_0E\left(ze^{i\Delta\pi/2}\right) - d_0E\left(ze^{-i\Delta\pi/2}\right) \tag{1.6.7}$$

$$(z \to \infty)$$

uniformly on $|\arg z| \leqq \epsilon$. Here the constants h_{ik} $(1 \leqq i \leqq n;\ k = 0, 1, 2, \cdots)$, c_0 and d_0 are given in (1.3.9), (1.6.1) and (1.6.2) and the series $E(z)$ has the form (1.6.3) with coefficients A_j being defined by (1.6.4).

(ii) *The main terms of (1.6.8) are expressed in the form*

$$\begin{aligned} H_{p,q}^{m,n}(z) &= \sum_{i=1}^{n}\left[h_i z^{(a_i-1)/\alpha_i} + o\left(z^{(a_i-1)/\alpha_i}\right)\right] \\ &+ Az^{(\mu+1/2)/\Delta}\left(c_0\exp\left[(B + Cz^{1/\Delta})i\right] - d_0\exp\left[-(B + Cz^{1/\Delta})i\right]\right) \\ &+ O\left(z^{(\mu-1/2)/\Delta}\right) \quad (z \to \infty, \quad |\arg z| \leqq \epsilon), \end{aligned} \tag{1.6.8}$$

where h_i $(i = 1, \cdots, n)$ are given in (1.5.10) and

$$A = \frac{A_0}{2\pi i\Delta}\left(\frac{\Delta^\Delta}{\delta}\right)^{(\mu+1/2)/\Delta}, \quad B = \frac{(2\mu+1)\pi}{4}, \quad C = \left(\frac{\Delta^\Delta}{\delta}\right)^{1/\Delta} \tag{1.6.9}$$

with A_0 being given in (1.6.5).

Proof. To prove (1.6.8) we apply the results by Braaksma [1]. First we note that if $a^* = 0$, then by (1.1.13)

$$a_1^* = -a_2^* = \frac{\Delta}{2}. \tag{1.6.10}$$

Owing to (1.1.7), (1.1.8) and (1.1.11) the formulas (1.8), (1.10) and (2.12) in Braaksma [1] take the forms

$$\mu = \Delta > 0, \quad \beta = \frac{1}{\delta}, \quad \delta_0 = a_1^*\pi, \tag{1.6.11}$$

and hence by (1.6.10)

$$\delta_0 = \frac{\Delta\pi}{2}. \tag{1.6.12}$$

The relations (4.1), (4.2) and (4.4) in Braaksma [1] for the function $h_1(s)$ defined by

$$h_1(s) = \pi^{m+n-p} \frac{\prod_{i=n+1}^{p} \sin[\pi(a_i - \alpha_i s)]}{\prod_{j=1}^{m} \sin[\pi(b_j - \beta_j s)]} \tag{1.6.13}$$

have the forms

$$h_1(s) = c_0 e^{i\gamma_0 s} \prod_{j=n+1}^{p} \left[1 - e^{2\pi i(\alpha_j s - a_j)}\right] \prod_{j=1}^{m} \sum_{k=0}^{\infty} e^{2k\pi i(\beta_j s - b_j)} = \sum_{j=0}^{\infty} c_j e^{i\gamma_j s} \tag{1.6.14}$$

and

$$h_1(s) = d_0 e^{-i\gamma_0 s} \prod_{j=n+1}^{p} \left[1 - e^{-2\pi i(\alpha_j s - a_j)}\right] \prod_{j=1}^{m} \sum_{k=0}^{\infty} e^{-2k\pi i(\beta_j s - b_j)}$$

$$= \sum_{j=0}^{\infty} d_j e^{-i\gamma_j s}, \tag{1.6.15}$$

where c_0 and d_0 are given in (1.6.1) and (1.6.2) and

$$\gamma_0 = \delta_0 = \frac{\Delta\pi}{2}. \tag{1.6.16}$$

Following Definition I of Braaksma [1] we denote by $\{\delta_m\}$ $(m = 0, \pm 1, \pm 2, \cdots)$ the monotonic sequence which arises if we write down the set of numbers γ_i and $-\gamma_j$ $(i, j = 0, 1, 2, \cdots)$, in (1.6.14) and (1.6.15), respectively. It follows from (1.6.14) and (1.6.15) that

$$\delta_{-1} = -\delta_0 = -\frac{\Delta\pi}{2}, \tag{1.6.17}$$

$$\delta_1 = \delta_0 + 2\pi \min_{n+1 \leqq i \leqq p, 1 \leqq j \leqq m} [\alpha_i, \beta_j], \quad \delta_{-2} = -\delta_1, \tag{1.6.18}$$

and thus

$$\delta_0 - \delta_{-1} = \Delta\pi \tag{1.6.19}$$

and

$$\delta_1 - \delta_0 = \delta_{-1} - \delta_{-2} = 2\pi \min_{n+1 \leqq i \leqq p, 1 \leqq j \leqq m} [\alpha_i, \beta_j]. \tag{1.6.20}$$

Following Braaksma [1, p.265], for an arbitrary integer r we distinguish three different cases for δ_r:

a) there exists a non-negative integer $i = i_0$ such that $\delta_r = \gamma_{i_0}$, while $\delta_r \neq -\gamma_j$ for $j = 0, 1, 2, \cdots$;

b) there exists a non-negative integer $j = j_0$ such that $\delta_r = -\gamma_{j_0}$, while $\delta_r \neq \gamma_i$ for $i = 0, 1, 2, \cdots$;

c) there exist two non-negative integers $i = i_0$ and $j = j_0$ such that $\delta_r = \gamma_{i_0} = -\gamma_{j_0}$.

We define numbers C_r and D_r by $C_r = c_{i_0}$, $D_r = 0$ in case a), $C_r = 0$, $D_r = -d_{j_0}$ in case b), and $C_r = c_{i_0}$, $D_r = -d_{j_0}$ in case c). We also define the integer κ by the relation

$$\delta_\kappa = -\gamma_0 = -\delta_0 \tag{1.6.21}$$

and note that $\kappa \geqq -1$, $\kappa = -1$ for $\delta_0 > 0$, $C_r = 0$ if $r < 0$, and $D_r = 0$ if $r > \kappa$ (see Braaksma [1, Lemma 4a]). Using these facts, the numbers C_i $(i = 0, 1, 2, \cdots)$ and D_j $(j = \kappa, \kappa - 1, \cdots)$ are defined by

$$h_1(s) = \sum_{i=0}^{\infty} C_i e^{i\delta_i s}, \qquad h_1(s) = -\sum_{j=-\infty}^{\kappa} D_j e^{i\delta_j s}. \tag{1.6.22}$$

By Braaksma [1, Theorem 7], if the conditions in (1.1.6) and (1.3.2) are satisfied, if $\Delta > 0$ and $\delta_r - \delta_{r-1} = \Delta\pi$ for some integer r, and if

$$0 < \epsilon < \frac{1}{4}\min[\delta_{r-1} - \delta_{r-2}, \delta_{r+1} - \delta_r], \tag{1.6.23}$$

then

$$\begin{aligned} H_{p,q}^{m,n}(z) \sim &-\sum_{i=1}^{n}\sum_{l=0}^{\infty} \operatorname*{Res}_{s=a_{il}} \left[\mathcal{H}_{p,q}^{m,n}(s) z^{-s}\right] + \sum_{j=r}^{\kappa} D_j P\left(z e^{i\delta_j}\right) \\ &- \sum_{j=0}^{r-1} C_j P\left(z e^{i\delta_j}\right) + (C_r + D_r) E\left(z e^{i\delta_r}\right) + (C_{r-1} + D_{r-1}) E\left(z e^{i\delta_{r-1}}\right) \end{aligned} \tag{1.6.24}$$

as $|z| \to \infty$ uniformly on

$$\frac{1}{2}\Delta\pi - \epsilon \leqq \arg z + \delta_r \leqq \frac{1}{2}\Delta\pi + \epsilon, \tag{1.6.25}$$

where an empty sum is understood to be zero. Here $P(z)$ is defined via a sum of residues

$$P(z) = \sum_{i=1}^{p}\sum_{k=0}^{\infty} \operatorname*{Res}_{s=a_{ik}} [h_0(s) z^{-s}], \qquad h_0(s) = \frac{\prod_{i=1}^{p} \Gamma(1 - a_i - \alpha_i s)}{\prod_{j=1}^{q} \Gamma(1 - b_j - \beta_j s)} \tag{1.6.26}$$

at points a_{ik} in (1.1.5) and $E(z)$ is given in (1.6.3).

Now we apply Braaksma's theorem above to prove (1.6.8). If $a^* = 0$ and $\Delta > 0$, then in accordance with (1.6.19) and (1.6.11) we can take $r = 0$, by (1.6.20) the condition in (1.6.23) coincides with that in (1.6.6). Therefore the relation (1.6.24) holds for $|z| \to \infty$ uniformly on the domain in (1.6.25) which gives $|\arg z| \leqq \epsilon$ if we take into account that

$$\delta_0 = \frac{\Delta\pi}{2} > 0 \tag{1.6.27}$$

according to (1.6.16) and (1.6.11). The evaluation of the residues in the double sum in (1.6.24), similarly to those in Theorem 1.4, gives the double sum in (1.6.8). By virtue of (1.6.27) (and (1.6.21)), $\kappa = -1$ and so the second and the third sums in (1.6.24) vanish. So (1.6.24) is reduced to the form

$$H_{p,q}^{m,n}(z)$$
$$\sim \sum_{i=1}^{n}\sum_{k=0}^{\infty} h_{ik} z^{(a_i-1-k)/\alpha_i} + (C_0 + D_0)E\left(ze^{i\delta_0}\right) + (C_{-1} + D_{-1})E\left(ze^{i\delta_{-1}}\right), \tag{1.6.28}$$

where h_{ik} are given in (1.3.9). To evaluate C_0, D_0, C_{-1} and D_{-1}, we note that by (1.6.16) there exists $i_0 = 0$ such that $\delta_0 = \gamma_0$, and it follows from (1.6.14) and (1.6.15) that $\delta_0 \neq -\gamma_j$ for $j = 0, 1, 2, \cdots$, if we take into account (1.6.27). So, the case a) above is realized and $C_0 = C_r = c_{i_0} = c_0$ while $D_0 = D_r = 0$, where c_0 is given in (1.6.1). Further, since $C_r = 0$ if $r < 0$ (for $\delta_0 > 0$) $C_{-1} = 0$, and in view of (1.6.15) and the second relation in (1.6.22), we have $D_{-1} = -d_0$, where d_0 is given in (1.6.2). Substituting these values of C_0, D_0, C_{-1} and D_{-1} into (1.6.28), we obtain (1.6.8). This completes the proof of the first assertion (i) of Theorem 1.9.

The second one (ii) follows from (1.6.8), if we take into account (1.6.1)–(1.6.3) and (1.6.5).

Corollary 1.9.1. *Let the assumptions of Theorem* 1.9 *hold and* i_0 $(1 \leqq i_0 \leqq n)$ *be such number that* (1.5.11) *is satisfied. Then the H-function* (1.1.1) *has the form*

(i) *If* $[\mathrm{Re}(a_{i_0}) - 1]/\alpha_{i_0} > \mathrm{Re}(\mu + 1/2)/\Delta$,

$$H_{p,q}^{m,n}(z) = h_{i_0} z^{(a_{i_0}-1)/\alpha_{i_0}} + o\left(z^{(a_{i_0}-1)/\alpha_{i_0}}\right) \quad (z \to \infty), \tag{1.6.29}$$

where h_{i_0} *is given in* (1.5.10) *with* $i = i_0$.

(ii) *If* $[\mathrm{Re}(a_{i_0}) - 1]/\alpha_{i_0} < \mathrm{Re}(\mu + 1/2)/\Delta$,

$$H_{p,q}^{m,n}(z) = Az^{(\mu+1/2)/\Delta}\left(c_0 \exp\left[(B + Cz^{1/\Delta})i\right] - d_0 \exp\left[-(B + Cz^{1/\Delta})i\right]\right)$$
$$+ o\left(z^{(\mu+1/2)/\Delta}\right) \quad (z \to \infty,\ |\arg z| \leqq \epsilon), \tag{1.6.30}$$

where A, B *and* C *are given in* (1.6.9).

(iii) *If* $[\mathrm{Re}(a_{i_0}) - 1]/\alpha_{i_0} = \mathrm{Re}(\mu + 1/2)/\Delta$,

$$H_{p,q}^{m,n}(z) = h_{i_0} z^{(\mu+1/2)/\Delta}$$
$$+ Az^{(\mu+1/2)/\Delta}\left(c_0 \exp\left[(B + Cz^{1/\Delta})i\right] - d_0 \exp\left[-(B + Cz^{1/\Delta})i\right]\right)$$
$$+ o\left(z^{(\mu+1/2)/\Delta}\right) \quad (z \to \infty,\ |\arg z| \leqq \epsilon). \tag{1.6.31}$$

In particular,

$$H_{p,q}^{m,n}(z) = O\left(z^{\rho}\right) \quad (z \to \infty,\ |\arg z| \leqq \epsilon), \tag{1.6.32}$$

where

$$\rho = \max_{1 \leqq j \leqq m}\left[\frac{\mathrm{Re}(a_i) - 1}{\alpha_i}, \frac{\mathrm{Re}(\mu) + 1/2}{\Delta}\right]. \tag{1.6.33}$$

Remark 1.3. If the poles a_{ik} in (1.1.5) of the gamma functions $\Gamma(1 - a_i - \alpha_i s)$ $(i = 1, \cdots, n)$ are not simple (i.e. the conditions in (1.3.2) are not satisfied), then similarly to the results in Section 1.5 it can be proved that the logarithmic multipliers of the form $[\log z]^{N-1}$ may be added in the asymptotic expansions (1.6.8), (1.6.8), (1.6.29)–(1.6.32).

1.7. Exponential Asymptotic Expansions at Infinity in the Case $n = 0$

In this section we give the asymptotic behavior of the H-function in the special case $n = 0$ provided that $\Delta > 0$, $a^* > 0$. Let c_0, d_0 and A_0 be given in (1.6.1)–(1.6.2) and (1.6.5). We note that, in particular, when $n = 0$ and $q = m$, c_0 and d_0 take the forms

$$c_0 = (2\pi)^{q-p} e^{-\mu\pi i}, \quad d_0 = (2\pi)^{q-p} e^{\mu\pi i}, \tag{1.7.1}$$

where μ is given in (1.1.10). Let

$$C_1 = \frac{c_0 A_0}{2\pi i} \Delta^{\mu-1/2} \left(\frac{e^{ia_1^*\pi}}{\delta}\right)^{(\mu+1/2)/\Delta}, \quad D_1 = \Delta \left(\frac{e^{ia_1^*\pi}}{\delta}\right)^{1/\Delta}, \tag{1.7.2}$$

$$C_2 = \frac{d_0 A_0}{2\pi i} \Delta^{\mu-1/2} \left(\frac{e^{-ia_1^*\pi}}{\delta}\right)^{(\mu+1/2)/\Delta}, \quad D_2 = \Delta \left(\frac{e^{-ia_1^*\pi}}{\delta}\right)^{1/\Delta}. \tag{1.7.3}$$

The following statement holds.

Theorem 1.10. *Let $n = 0$, a^*, Δ, δ, μ and a_1^* be given in (1.1.7)–(1.1.10) and (1.1.11) and let the conditions in (1.1.6) and (1.3.2) be satisfied. Let $\Delta > 0$, $a^* \geqq 0$ and let ϵ be a constant such that*

$$0 < \epsilon < \frac{\Delta\pi}{2}. \tag{1.7.4}$$

Then there hold the following assertions.

(i) *If $m = q$ and $a^* > 0$, then the H-function $H^{q,0}_{p,q}(z)$ has the asymptotic expansions at infinity*

$$H^{q,0}_{p,q}(z) \sim c_0 E\left(z e^{ia_1^*\pi}\right) \quad (z \to \infty) \tag{1.7.5}$$

and

$$H^{q,0}_{p,q}(z) \sim -d_0 E\left(z^{-ia_1^*\pi}\right) \quad (z \to \infty) \tag{1.7.6}$$

uniformly on $|\arg z| \leqq \Delta\pi/2 - \epsilon$, where c_0 and d_0 are given in (1.7.1) and $E(z)$ in (1.6.3). The principal terms in these asymptotics have the forms

$$H^{q,0}_{p,q}(z) = C_1 e^{D_1 z^{1/\Delta}} z^{(\mu+1/2)/\Delta} \left[1 + O\left(\frac{1}{z}\right)^{1/\Delta}\right] \quad (z \to \infty) \tag{1.7.7}$$

and

$$H^{q,0}_{p,q}(z) = -C_2 e^{D_2 z^{1/\Delta}} z^{(\mu+1/2)/\Delta} \left[1 + O\left(\frac{1}{z}\right)^{1/\Delta}\right] \quad (z \to \infty), \tag{1.7.8}$$

respectively.

(ii) *If $m < q$ and $a^* \geqq 0$, then the H-function $H_{p,q}^{m,0}(z)$ has the asymptotic expansion*

$$H_{p,q}^{m,0}(z) \sim c_0 E\left(ze^{ia_1^*\pi}\right) - d_0 E\left(ze^{-ia_1^*\pi}\right) \quad (z \to \infty) \tag{1.7.9}$$

uniformly on $|\arg z| \leqq \epsilon$ with $0 < \epsilon < \pi a^*/4$ for $a^* > 0$ and

$$0 < \epsilon < \frac{\pi}{2} \min_{1 \leqq i \leqq p, 1 \leqq j \leqq m} [\alpha_i, \beta_j]$$

for $a^* = 0$, where c_0 and d_0 are given in (1.6.1) and (1.6.2) with $n = 0$. The main term in this representation has the form

$$H_{p,q}^{m,0}(z) = \left[C_1 e^{D_1 z^{1/\Delta}} - C_2 e^{D_2 z^{1/\Delta}}\right] z^{(\mu+1/2)/\Delta} \left[1 + O\left(\frac{1}{z}\right)^{1/\Delta}\right] \quad (z \to \infty). \tag{1.7.10}$$

Proof. When $m = q$, $a^* > 0$ and $m < q$, $a^* > 0$, the results in (1.7.5), (1.7.6) and (1.7.9) follow from those proved by Braaksma [1; Theorem 4, the relations (7.11), (7.12) and (7.15)] (see Section 1.11), if we take into account the formulas in (1.6.11) and the equality

$$\delta_0 - \frac{1}{2}\mu\pi = \frac{1}{2}a^*\pi, \tag{1.7.11}$$

which is valid due to (1.6.11) and (1.1.13). When $m < q$ and $a^* = 0$, by (1.6.10) the asymptoic estimate (1.7.9) follows from the relation (1.6.8) in Theorem 1.9 in the case $n = 0$. According to (1.7.1)–(1.7.3) the relations (1.7.7) and (1.7.8) are deduced from (1.7.5) and (1.7.6) while (1.7.10) is obtained from (1.7.9).

Corollary 1.10.1. *If the assumptions of Theorem 1.10 are satisfied, then for the H-function $H_{p,q}^{m,0}(z)$, the following asymptotic estimate holds at infinity:*

$$H_{p,q}^{m,0}(z) = O\Bigg(|z|^{[\mathrm{Re}(\mu)+1/2]/\Delta} \cdot \exp\left\{\Delta\left(\frac{|z|}{\delta}\right)^{1/\Delta} \max\left[\cos\frac{(a_1^*\pi + \arg z)}{\Delta}, \cos\frac{(a_1^*\pi - \arg z)}{\Delta}\right]\right\}\Bigg) \quad (z \to \infty) \tag{1.7.12}$$

uniformly on $|\arg z| \leqq (\Delta\pi/2) - \epsilon$ and on $|\arg z| \leqq \epsilon$ when $m = q$ and $m < q$, respectively.

Corollary 1.10.2. *If the assumptions of Theorem 1.10 are satisfied, then for the H-function $H_{p,q}^{m,0}(x)$ with real x the following asymptotic estimates hold at infinity:*

$$H_{p,q}^{m,0}(x) = O\left(x^{[\mathrm{Re}(\mu)+1/2]/\Delta} \exp\left[\cos\left(\frac{a_1^*\pi}{\Delta}\right)\Delta\delta^{-1/\Delta}x^{1/\Delta}\right]\right) \quad (x \to +\infty). \tag{1.7.13}$$

In particular,

$$H_{p,q}^{q,0}(x) = O\left(x^{[\mathrm{Re}(\mu)+1/2]/\Delta} \exp\left[-\Delta\delta^{-1/\Delta}x^{1/\Delta}\right]\right) \quad (x \to +\infty). \tag{1.7.14}$$

Remark 1.4. The asymptotic estimate (1.7.13) for the H-function $H_{p,q}^{m,0}(z)$ is more precise than those given by Mathai and R.K. Saxena [2, (1.6.3)] and Srivastava, Gupta and Goyal [1, (2.2.14)].

1.8. Algebraic Asymptotic Expansions at Zero

When $\Delta \geqq 0$ the results in Theorems 1.3 and 1.5 lead us to the power and power-logarithmic asymptotic expansions of the H-function near zero, namely,

$$H_{p,q}^{m,n}(z) \sim \sum_{j=1}^{m} \sum_{l=0}^{\infty} h_{jl}^{*} z^{(b_j+l)/\beta_j} \quad (z \to 0), \tag{1.8.1}$$

when any pole b_{jl} of the gamma functions $\Gamma(b_j + \beta_j s)$ $(1 \leqq j \leqq m)$ do not coincide, and

$$H_{p,q}^{m,n}(z) \sim \sum_{j,l}{}' h_{jl}^{*} z^{(b_j+l)/\beta_j} + \sum_{j,l}{}'' \sum_{i=0}^{N_{jl}^{*}-1} H_{jli}^{*} z^{(b_j+l)/\beta_j} [\log z]^i \quad (z \to 0), \tag{1.8.2}$$

when some of the poles b_{jl} of the gamma functions $\Gamma(b_j + \beta_j s)$ $(1 \leqq j \leqq m)$ coincide. Here the summation signs $\sum'$ and $\sum''$ are taken as in Theorem 1.5.

On the other hand if $\Delta < 0$, $a^* > 0$ and the conditions in (1.1.6) hold, then for $\mathfrak{L} = \mathfrak{L}_{+\infty}$ similar arguments to those in Section 1.5 lead us to the following asymptotic expansion of the H-function at zero:

$$H_{p,q}^{m,n}(z) \sim -\sum_{j=1}^{m} \sum_{l=0}^{\infty} \operatorname*{Res}_{s=-b_{jl}} [\mathcal{H}_{p,q}^{m,n}(-s) z^s] \quad \left(z \to 0, \ |\arg z| < \frac{a^*\pi}{2}\right) \tag{1.8.3}$$

with a^* being defined in (1.1.7). It follows from here that the H-function (1.1.1) also has the asymptotic expansions (1.8.1) and (1.8.2) in the respective cases when the poles b_{jl} of the gamma functions $\Gamma(b_j+\beta_j s)$ $(1 \leqq j \leqq m)$ do not coincide and coincide. By Cauchy's theorem the results above are also valid for the H-function (1.1.1) with $\mathfrak{L} = \mathfrak{L}_{i\gamma\infty}$ and $a^* > 0$. Thus we have:

Theorem 1.11. *Let the conditions in* (1.1.6) *and* (1.3.1) *be satisfied and let either* $\Delta \geqq 0$ *or* $\Delta < 0$, $a^* > 0$. *Then the asymptotic expansion of* $H_{p,q}^{m,n}(z)$ *near zero is given in* (1.8.1) *with the additional condition* $|\arg z| < a^*\pi/2$ *when* $\Delta < 0$, $a^* > 0$. *The principal terms of this asymptotic have the form*

$$H_{p,q}^{m,n}(z) = \sum_{j=1}^{m} \left[h_j^* z^{b_j/\beta_j} + o\left(z^{b_j/\beta_j}\right)\right] \quad (z \to 0), \tag{1.8.4}$$

where

$$h_j^* \equiv h_{j0}^* = \frac{1}{\beta_j} \frac{\prod\limits_{\substack{i=1\\ i\neq j}}^{m} \Gamma\left(b_i - b_j \frac{\beta_i}{\beta_j}\right) \prod\limits_{i=1}^{n} \Gamma\left(1 - a_i + b_j \frac{\alpha_i}{\beta_j}\right)}{\prod\limits_{i=n+1}^{p} \Gamma\left(a_i - b_j \frac{\alpha_i}{\beta_j}\right) \prod\limits_{i=m+1}^{q} \Gamma\left(1 - b_i + b_j \frac{\beta_i}{\beta_j}\right)} \quad (j = 1, \cdots, m). \tag{1.8.5}$$

Corollary 1.11.1. *Let the assumptions of Theorem 1.11 be satisfied and let j_0 $(1 \leqq j_0 \leqq m)$ be an integer such that*

$$\frac{\operatorname{Re}(b_{j_0})}{\beta_{j_0}} = \min_{1\leqq j\leqq m}\left[\frac{\operatorname{Re}(b_j)}{\beta_j}\right]. \tag{1.8.6}$$

Then there holds the asymptotic estimate

$$H^{m,n}_{p,q}(z) = h^*_{j_0} z^{b_{j_0}/\beta_{j_0}} + o\left(z^{b_{j_0}/\beta_{j_0}}\right) \quad (z \to 0), \tag{1.8.7}$$

where $|\arg z| < a^*\pi/2$ *when* $\Delta < 0, a^* > 0$, *and* $h^*_{j_0}$ *is given in (1.8.5) with* $j = j_0$. *In particular,*

$$H^{m,n}_{p,q}(z) = O\left(z^{\rho^*}\right) \quad (z \to 0), \tag{1.8.8}$$

where $\rho^* = \min_{1\leqq j\leqq m}[\operatorname{Re}(b_j)/\beta_j]$ *and* $|\arg z| < a^*\pi/2$ *when* $\Delta < 0, a^* > 0$.

Theorem 1.12. *Let the conditions in (1.1.6) be satisfied and let either* $\Delta \geqq 0$ *or* $\Delta < 0,\ a^* > 0$. *Let some poles of the gamma functions* $\Gamma(b_j + \beta_j s)$ $(j = 1, \cdots, m)$ *coincide and let* N^*_{jl} *be the orders of these poles. Then the asymptotic expansion of* $H^{m,n}_{p,q}(z)$ *near zero is given in (1.8.2) with the additional condition* $|\arg z| < a^*\pi/2$ *when* $\Delta < 0,\ a^* > 0$. *The principal terms of these asymptotics have the form*

$$H^{m,n}_{p,q}(z) = \sum_j{}' \left[h^*_j z^{b_j/\beta_j} + o\left(z^{b_j/\beta_j}\right)\right]$$
$$+ \sum_j{}'' \left[H^*_j z^{b_j/\beta_j}[\log z]^{N^*_j-1} + o\left(z^{b_j/\beta_j}[\log z]^{N^*_j-1}\right)\right] \quad (z \to 0), \tag{1.8.9}$$

where $|\arg z| < a^*\pi/2$ *when* $\Delta < 0,\ a^* > 0$. *Here* $\sum'$ *and* $\sum''$ *are summations taken over* j $(j = 1, \cdots, m)$ *such that the gamma functions* $\Gamma(b_j + \beta_j s)$ *have simple poles and poles of order* $N^*_j \equiv N^*_{j0}$ *at the points* b_{j0}, *respectively, and* h^*_j *are given in (1.8.5) while* H^*_j *are given by (1.4.6).*

Corollary 1.12.1. *Let the assumptions of Theorem 1.12 be satisfied and let* j_{01} *and* j_{02} $(1 \leqq j_{01}, j_{02} \leqq m)$ *be numbers such that*

$$\rho^*_1 \equiv \frac{\operatorname{Re}(b_{j_{01}})}{\beta_{j_{01}}} = \min_{1\leqq j\leqq m}\left[\frac{\operatorname{Re}(b_j)}{\beta_j}\right], \tag{1.8.10}$$

when the poles b_{jl} $(j = 1, \cdots, m;\ l = 0, 1, \cdots)$ *are simple, and*

$$\rho^*_2 \equiv \frac{\operatorname{Re}(b_{j_{02}})}{\beta_{j_{02}}} = \min_{1\leqq j\leqq m}\left[\frac{\operatorname{Re}(b_j)}{\beta_j}\right], \tag{1.8.11}$$

when the poles b_{jl} $(j = 1, \cdots, m;\ l = 0, 1, \cdots)$ *coincide.*

a) *If* $\rho^*_1 < \rho^*_2$, *then the principal term in the asymptotic expansion of the H-function has the form*

$$H^{m,n}_{p,q}(z) = h^*_{j_{01}} z^{b_{j_{01}}/\beta_{j_{01}}} + o\left(z^{b_{j_{01}}/\beta_{j_{01}}}\right) \quad (z \to 0), \tag{1.8.12}$$

where $|\arg z| < a^*\pi/2$ *for* $\Delta < 0$, $a^* > 0$ *and* $h^*_{j_{01}}$ *is given in* (1.8.5) *with* $j = j_{01}$. *In particular, the relation* (1.8.8) *holds with* ρ_1^* *instead of* ρ^* *there.*

b) *If* $\rho_1^* \geqq \rho_2^*$ *and the gamma function* $\Gamma(b_j + \beta_j s)$ *has the pole* $b_{j_{02}}$ *of order* $N^*_{j_{02}}$, *then the principal term in the asymptotic expansion of the* H*-function has the form*

$$H^{m,n}_{p,q}(z) = H^*_{j_{02}} z^{b_{j_{02}}/\beta_{j_{02}}} [\log z]^{N^*_{j_{02}}-1} + o\left(z^{b_{j_{02}}/\beta_{j_{02}}} [\log z]^{N^*_{j_{02}}-1}\right) \quad (z \to 0), \tag{1.8.13}$$

where $|\arg z| < a^*\pi/2$ *for* $\Delta < 0$, $a^* > 0$, *and* $H^*_{j_{02}}$ *is given in* (1.4.6) *with* $j = j_{02}$. *In particular, if* N^* *is the largest order of general poles of gamma functions* $\Gamma(b_j + \beta_j s)$ $(j = 1, \cdots, m)$, *then*

$$H^{m,n}_{p,q}(z) = O\left(z^{\rho^*} [\log z]^{N^*-1}\right) \quad (z \to 0), \tag{1.8.14}$$

where $\rho^* = \min_{1 \leqq j \leqq m} [\mathrm{Re}(b_j)/\beta_j]$ *and* $|\arg z| < a^*\pi/2$ *when* $\Delta < 0$, $a^* > 0$.

When $\Delta < 0$ and $a^* \leqq 0$, the asymptotic behavior of the H-function (1.1.1) near zero is more complicated. For $\mathfrak{L} = \mathfrak{L}_{+\infty}$ such asymptotic estimates have exponential forms and can be found by the method similar to finding asymptotics of the H-function with $\mathfrak{L} = \mathfrak{L}_{-\infty}$ for $\Delta > 0$ at infinity which was suggested by Braaksma in [1]. The former asymptotic expansions can also be deduced from the latter ones on the basis of the translation formula

$$H^{m,n}_{p,q}\left[z \,\middle|\, \begin{matrix} (a_i, \alpha_i)_{1,p} \\ (b_j, \beta_j)_{1,q} \end{matrix}\right] = H^{n,m}_{q,p}\left[\frac{1}{z} \,\middle|\, \begin{matrix} (1 - b_j, \beta_j)_{1,q} \\ (1 - a_i, \alpha_i)_{1,p} \end{matrix}\right], \tag{1.8.15}$$

if we take into account the relations

$$a_0^* = -a^*, \ \Delta_0 = -\Delta, \ \delta_0 = \frac{1}{\delta}, \ \mu_0 = \mu, \ a_{10}^* = a_2^*, \ a_{20}^* = a_1^*. \tag{1.8.16}$$

Here we use a_0^*, Δ_0, δ_0, μ_0, a_{10}^* and a_{20}^* for $H^{n,m}_{q,p}\left[\frac{1}{z}\right]$ instead of a^*, Δ, δ, μ, a_1^* and a_2^* for $H^{m,n}_{p,q}[z]$ defined by (1.1.7)–(1.1.12).

The case $\Delta < 0$, $a^* = 0$ will be considered in the next section. It should be noted that the asymptotic expansions for the H-function (1.1.1) have different forms in sectors of changing $\arg z$ and contain the special function $E^*(z)$ defined in (1.9.3) below. Asymptotics of the same form are also found in the special case $m = 0$ presented in Section 1.10.

1.9. Exponential Asymptotic Expansions at Zero in the Case $\Delta < 0$, $a^* = 0$

In the previous section we proved the asymptotic behavior of the H-function (1.1.1) when $\Delta \geqq 0$ or $\Delta < 0$, $a^* > 0$. Now we consider the remaining case $\Delta < 0$, $a^* = 0$. Let

$$c_0^* = (-2\pi i)^{n+m-q} \exp\left[-\left(\sum_{j=m+1}^{q} b_j - \sum_{i=1}^{n} a_i\right)\pi i\right]; \tag{1.9.1}$$

$$d_0^* = (2\pi i)^{n+m-q} \exp\left[\left(\sum_{j=m+1}^{q} b_j - \sum_{i=1}^{n} a_i\right)\pi i\right]; \tag{1.9.2}$$

$$E^*(z) = \frac{1}{2\pi i |\Delta|} \sum_{j=0}^{\infty} A_j^* \left(\frac{|\Delta|^{|\Delta|}\delta}{z}\right)^{(\mu-j+1/2)/|\Delta|} \exp\left[\left(\frac{|\Delta|^{|\Delta|}\delta}{z}\right)^{1/|\Delta|}\right]. \tag{1.9.3}$$

Here the constants A_j^*, depending on p, q, a_i, α_i, b_j and β_j $(i = 1, \cdots, p;\ j = 1, \cdots, q)$, are defined by the relations

$$\frac{\prod_{j=1}^{q} \Gamma(b_j + \beta_j s)}{\prod_{i=1}^{p} \Gamma(a_i + \alpha_i s)} \left(|\Delta|^{|\Delta|} \delta\right)^{-s}$$

$$= \sum_{j=0}^{N-1} \frac{A_j^*}{\Gamma(\Delta s + j - \mu + 1/2)} + \frac{r_N^*(s)}{\Gamma(\Delta s + N - \mu + 1/2)}, \tag{1.9.4}$$

where the function $r_N^*(s)$ is analytic in its domain and $r_N^*(s) = O(1)$ as $|s| \to \infty$ uniformly on $|\arg s| \leqq \pi - \epsilon_1^*$ $(0 < \epsilon_1^* < \pi)$. In particular,

$$A_0^* = (2\pi)^{(q-p+1)/2} |\Delta|^{-\mu} \prod_{i=1}^{p} \alpha_i^{a_i - 1/2} \prod_{j=1}^{q} \beta_j^{-b_j + 1/2}. \tag{1.9.5}$$

Theorem 1.13. *Let a^*, Δ, δ and μ be given in (1.1.7)–(1.1.10) and let the conditions in (1.1.6) and (1.3.1) be satisfied. Let $\Delta < 0$, $a^* = 0$ and ϵ^* be a constant such as*

$$0 < \epsilon^* < \frac{\pi}{2} \min_{1 \leqq i \leqq n, m+1 \leqq j \leqq q} [\alpha_i, \beta_j]. \tag{1.9.6}$$

Then there hold the following assertions:

(i) *The H-function (1.1.1) has the asymptotic expansion*

$$H_{p,q}^{m,n}(z) \sim \sum_{j=1}^{m} \sum_{l=0}^{\infty} h_{jl}^* z^{(b_j + l)/\beta_j} + c_0^* E^* \left(z e^{-i\Delta\pi/2}\right) - d_0^* E^* \left(z e^{i\Delta\pi/2}\right) \tag{1.9.7}$$

$$(z \to 0)$$

uniformly on $|\arg z| \leqq \epsilon^*$. Here the constants h_{jl}^* $(1 \leqq j \leqq m;\ l = 0, 1, 2, \cdots)$, c_0^* and d_0^* are given in (1.3.6), (1.9.1) and (1.9.2), and the series $E^*(z)$ has the form (1.9.3) with coefficients A_j^* being defined by (1.9.4).

(ii) *The main part of (1.9.8) is expressed in the form*

$$H_{p,q}^{m,n}(z) = \sum_{j=1}^{m} \left[h_j^* z^{b_j/\beta_j} + o\left(z^{b_j/\beta_j}\right)\right]$$

$$+ A^* z^{-(\mu+1/2)/|\Delta|} \left(c_0^* \exp\left[-(B^* + C^* z^{-1/|\Delta|}) i\right] - d_0^* \exp\left[(B^* + C^* z^{-1/|\Delta|}) i\right]\right)$$

$$+ O\left(z^{(-\mu+1/2)/|\Delta|}\right) \quad (z \to 0,\ |\arg z| \leqq \epsilon^*), \tag{1.9.8}$$

where h_j^* $(j = 1, \cdots, m)$ are given in (1.8.5) and

$$A^* = \frac{A_0^*}{2\pi i |\Delta|} \left(|\Delta|^{|\Delta|} \delta\right)^{(\mu+1/2)/|\Delta|}, \quad B^* = \frac{(2\mu + 1)\pi}{4}, \quad C^* = \left(|\Delta|^{|\Delta|} \delta\right)^{1/|\Delta|} \tag{1.9.9}$$

with A_0^* being given in (1.9.5).

Proof. The assertion (i) is deduced from Theorem 1.9(i) if we apply the translation formula (1.8.15) and the relations in (1.8.16). According to the latter, the conditions of Theorem 1.13 coincide with those in Theorem 1.9 with rearrangement of m and n and p and q, and with replacement of a^*, Δ and δ by $-a^*$, $-\Delta$ and $1/\delta$. The assertion (ii) follows from (i) according to (1.9.1)–(1.9.3) and (1.9.5).

Corollary 1.13.1. *Let the assumptions of Theorem* 1.13 *hold and let* j_0 $(1 \leqq j_0 \leqq m)$ *be a number such that* (1.8.6) *is satisfied. Then the H-function has the form*

(i) *If* $\mathrm{Re}(b_{j_0})/\beta_{j_0} + \mathrm{Re}(\mu + 1/2)/\,|\Delta| < 0$,

$$H^{m,n}_{p,q}(z) = h^*_{j_0} z^{b_{j_0}/\beta_{j_0}} + o\left(z^{b_{j_0}/\beta_{j_0}}\right) \quad (z \to 0), \tag{1.9.10}$$

where $h^*_{j_0}$ *is given in* (1.8.5) *with* $j = j_0$.

(ii) *If* $\mathrm{Re}(b_{j_0})/\beta_{j_0} + \mathrm{Re}(\mu + 1/2)/\,|\Delta| > 0$,

$$\begin{aligned} H^{m,n}_{p,q}(z) = {} & A^* z^{-(\mu+1/2)/|\Delta|} \left(c_0^* \exp\left[-(B^* + C^* z^{-1/|\Delta|})i\right]\right. \\ & \left. - d_0^* \exp\left[(B^* + C^* z^{-1/|\Delta|})i\right]\right) \\ & + o\left(z^{-(\mu+1/2)/|\Delta|}\right) \quad (z \to 0,\ |\arg z| \leqq \epsilon^*), \end{aligned} \tag{1.9.11}$$

where A^*, B^* *and* C^* *are given in* (1.9.9).

(iii) *If* $\mathrm{Re}(b_{j_0})/\beta_{j_0} + \mathrm{Re}(\mu + 1/2)/\,|\Delta| = 0$,

$$\begin{aligned} H^{m,n}_{p,q}(z) = {} & h^*_{j_0} z^{-(\mu+1/2)/|\Delta|} \\ & + A^* z^{-(\mu+1/2)/|\Delta|} \left(c_0^* \exp\left[-(B^* + C^* z^{-1/|\Delta|})i\right]\right. \\ & \left. - d_0^* \exp\left[(B^* + C^* z^{-1/|\Delta|})i\right]\right) \\ & + o\left(z^{-(\mu+1/2)/|\Delta|}\right) \quad (z \to 0,\ |\arg z| \leqq \epsilon^*). \end{aligned} \tag{1.9.12}$$

In particular,

$$H^{m,n}_{p,q}(z) = O\left(z^{\rho^*}\right) \quad (z \to 0,\ |\arg z| \leqq \epsilon^*), \tag{1.9.13}$$

where

$$\rho^* = \min_{1 \leqq j \leqq m} \left[\frac{\mathrm{Re}(b_j)}{\beta_j}, -\frac{\mathrm{Re}(\mu) + 1/2}{|\Delta|}\right]. \tag{1.9.14}$$

Remark 1.5. If the poles b_{jl} in (1.1.4) of the gamma functions $\Gamma(b_j + \beta_j s)$ $(j = 1, \cdots, m)$ are not simple (i.e. the conditions in (1.3.1) are not satisfied), then similarly to the results in Section 1.8 it can be proved that the logarithmic multipliers of the form $[\log z]^{N^*-1}$ may be added in the asymptotic expansions (1.9.7), (1.9.8), (1.9.10)–(1.9.13).

1.10. Exponential Asymptotic Expansions at Zero in the Case $m = 0$

In this section we give the asymptotic behavior of the H-function in the special case $m = 0$ provided that $\Delta < 0$, $a^* > 0$. Let c_0^*, d_0^* and A_0^* be given in (1.9.1), (1.9.2) and (1.9.5). We note that, in particular, when $m = 0$ and $p = n$, c_0^* and d_0^* take the forms

$$c_0^* = (2\pi)^{p-q}e^{-\mu\pi i}, \quad d_0^* = (2\pi)^{p-q}e^{\mu\pi i}, \tag{1.10.1}$$

where μ is given in (1.1.10). Let

$$C_1^* = \frac{c_0^* A_0^*}{2\pi i}|\Delta|^{\mu-1/2}\left(\delta e^{-ia_2^*\pi}\right)^{(\mu+1/2)/|\Delta|}, \quad D_1^* = |\Delta|\left(\delta e^{-ia_2^*\pi}\right)^{1/|\Delta|}, \tag{1.10.2}$$

$$C_2^* = \frac{d_0^* A_0^*}{2\pi i}|\Delta|^{\mu-1/2}\left(\delta e^{ia_2^*\pi}\right)^{(\mu+1/2)/|\Delta|}, \quad D_2^* = |\Delta|\left(\delta e^{ia_2^*\pi}\right)^{1/|\Delta|}. \tag{1.10.3}$$

The following statement holds.

Theorem 1.14. *Let $m = 0$, and let a^*, Δ, δ, μ and a_2^* be given in (1.1.7)–(1.1.10) and (1.1.12). Let us assume the conditions in (1.1.6) and (1.3.1). If $\Delta < 0$, $a^* \geqq 0$ and ϵ^* is a constant such that*

$$0 < \epsilon^* < \frac{|\Delta|\pi}{2}, \tag{1.10.4}$$

then there hold the following assertions:

(i) *If $n = p$ and $a^* > 0$, then the H-function $H_{p,q}^{0,p}(z)$ has the asymptotic expansions at zero*

$$H_{p,q}^{0,p}(z) \sim c_0^* E^*\left(ze^{ia_2^*\pi}\right) \quad (|z| \to 0) \tag{1.10.5}$$

and

$$H_{p,q}^{0,p}(z) \sim -d_0^* E^*\left(ze^{-ia_2^*\pi}\right) \quad (z \to 0) \tag{1.10.6}$$

uniformly on $|\arg z| \leqq |\Delta|\pi/2 - \epsilon^*$, where c_0^* and d_0^* are given in (1.10.1) and $E^*(z)$ in (1.9.3). The principal terms in these asymptotics have the forms

$$H_{p,q}^{0,p}(z) = C_1^* e^{D_1^* z^{-1/|\Delta|}} z^{-(\mu+1/2)/|\Delta|}\left[1 + O\left(z^{1/|\Delta|}\right)\right] \quad (z \to 0) \tag{1.10.7}$$

and

$$H_{p,q}^{0,p}(z) = -C_2^* e^{D_2^* z^{-1/|\Delta|}} z^{-(\mu+1/2)/|\Delta|}\left[1 + O\left(z^{1/|\Delta|}\right)\right] \quad (z \to 0), \tag{1.10.8}$$

respectively.

(ii) *If $n < p$ and $a^* \geqq 0$, then the H-function $H_{p,q}^{0,n}(z)$ has the asymptotic expansion*

$$H_{p,q}^{0,n}(z) \sim c_0^* E^*\left(ze^{ia_2^*\pi}\right) - d_0^* E^*\left(ze^{-ia_2^*\pi}\right) \quad (z \to 0) \tag{1.10.9}$$

uniformly on $|\arg z| \leqq \epsilon^*$ with $0 < \epsilon < \pi a^*/4$ for $a^* > 0$ and

$$0 < \epsilon < \frac{\pi}{2} \min_{1 \leqq i \leqq n, 1 \leqq j \leqq q} [\alpha_i, \beta_j]$$

for $a^* = 0$, where c_0^* and d_0^* are given in (1.9.1) *and* (1.9.2) *with $m = 0$. The main term in this representation has the form*

$$H_{p,q}^{0,n}(z) = \left[C_1^* e^{D_1^* z^{-1/|\Delta|}} - C_2^* e^{D_2^* z^{-1/|\Delta|}}\right] z^{-(\mu+1/2)/|\Delta|} \left[1 + O\left(z^{1/|\Delta|}\right)\right] \qquad (1.10.10)$$

$$(z \to 0).$$

Proof. The assertions (i) and (ii) are deduced from Theorem 1.10 on the basis of the translation formula (1.8.15) and of the relations in (1.8.16).

Corollary 1.14.1. *If the assumptions of Theorem* 1.14 *are satisfied, then the H-function $H_{p,q}^{0,n}(z)$ has the following asymptotic estimate at zero:*

$$H_{p,q}^{0,n}(z) = O\Bigg(|z|^{-[\mathrm{Re}(\mu)+1/2]/|\Delta|} \cdot \exp\left\{ |\Delta| \left(\frac{\delta}{|z|}\right)^{1/|\Delta|} \max\left[\cos\frac{(a_2^*\pi + \arg z)}{|\Delta|}, \cos\frac{(a_2^*\pi - \arg z)}{|\Delta|}\right]\right\}\Bigg) \qquad (1.10.11)$$

$$(z \to 0)$$

uniformly on $|\arg z| \leqq |\Delta|\pi/2 - \epsilon^*$ and on $|\arg z| \leqq \epsilon^*$ when $n = p$ and $n < p$, respectively.

Corollary 1.14.2. *If the assumptions of Theorem* 1.14 *are satisfied, then the H-function $H_{p,q}^{0,n}(x)$ for real x has the following asymptotic estimates at zero:*

$$H_{p,q}^{0,n}(x) = O\left(x^{-[\mathrm{Re}(\mu)+1/2]/|\Delta|} \exp\left[\cos\left(\frac{a_2^*\pi}{|\Delta|}\right) |\Delta|\, \delta^{1/|\Delta|} x^{-1/|\Delta|}\right]\right) \qquad (1.10.12)$$

$$(x \to +0).$$

In particular,

$$H_{p,q}^{0,p}(x) = O\left(x^{-[\mathrm{Re}(\mu)+1/2]/|\Delta|} \exp\left[-\,|\Delta|\, \delta^{1/|\Delta|} x^{-1/|\Delta|}\right]\right) \quad (x \to +0). \qquad (1.10.13)$$

1.11. Bibliographical Remarks and Additional Information on Chapter 1

For Sections 1.1 and 1.2. The so-called Mellin–Barnes integral of the gamma functions were first introduced by S. Pincherle [1] (1888), and the theory has been developed by Mellin [1] (1910), where references to earlier works were indicated in the book by Erdélyi, Magnus, Oberhettinger and Tricomi [1, Section 1.19]. Dixon and Ferrar [1] (1936) first investigated the H-function represented by the Mellin–Barnes integral (1.1.1) with $\mathfrak{L} = \mathfrak{L}_{i\gamma\infty}$. In the case $m = n = 1$ and $p = q = 2$ they proved the statements of Theorem 1.1 for $\mathfrak{L} = \mathfrak{L}_{i\gamma\infty}$ and of Theorem 1.2(iii), as well as the absolute convergence of the integral in (1.1.1) for any real $z = x > 0$ in the case $a^* = \Delta = 0$ and $\mathrm{Re}(\mu) < -1$, and of the convergence of $H_{2,2}^{1,1}(z)$ for $z = x > 0$ ($x \neq \delta$) in the case $a^* = \Delta = 0$ and $-1 \leqq \mathrm{Re}(\mu) < 0$. They also studied analytic continuation of $H_{2,2}^{1,1}(z)$ and indicated that these results can be extended to the general integrals $H_{p,q}^{m,n}(z)$ of the form (1.1.1). One may refer these results to Section 1.19 of the book by Erdélyi, Magnus, Oberhettinger and Tricomi [1].

Interest in the H-function (1.1.1) returned in 1961 when Fox [2] investigated a particular case of such a function with $\mathfrak{L} = \mathfrak{L}_{i\gamma\infty}$ in the form

$$H^{q,p}_{2p,2q}(x) \equiv H^{q,p}_{2p,2q}\left[x \,\middle|\, \begin{matrix} (1-a_i,\alpha_i)_{1,p}, (a_i-\alpha_i,\alpha_i)_{1,p} \\ (b_j,\beta_j)_{1,q}, (1-b_j-\beta_j,\beta_j)_{1,q} \end{matrix}\right] \quad (x>0) \tag{1.11.1}$$

as a symmetrical Fourier kernel in $L_2(\mathbb{R}_+)$, which means that for

$$H_1(x) = \int_0^x H^{q,p}_{2p,2q}(t)dt \tag{1.11.2}$$

the transforms

$$g(x) = \frac{d}{dx}\int_0^\infty H_1(xt)f(t)\frac{dt}{t}, \quad f(x) = \frac{d}{dx}\int_0^\infty H_1(xt)g(t)\frac{dt}{t}$$

hold for $f, g \in L_2(\mathbb{R}_+)$ provided that

$$\begin{gathered} \Delta = 2\sum_{j=1}^q \beta_j - 2\sum_{i=1}^p \alpha_i > 0, \\ \mathrm{Re}(a_i) > \frac{\alpha_i}{2}, \quad \mathrm{Re}(b_j) > -\frac{\beta_j}{2} \quad (1 \leqq i \leqq p; 1 \leqq j \leqq q). \end{gathered} \tag{1.11.3}$$

We also note that for such a function the constants in (1.1.7) and (1.1.9) take the forms

$$a^* = 0, \quad \delta = \prod_{j=1}^q \beta_j^{2\beta_j} \prod_{i=1}^p \alpha_i^{-2\alpha_i}. \tag{1.11.4}$$

Therefore $H^{m,n}_{p,q}(z)$ in (1.1.1) is sometimes called Fox's H-function.

Braaksma [1, Theorem 1] (1964) proved the existence of the H-function (1.1.1) with $\mathfrak{L} = \mathfrak{L}_{-\infty}$, its analyticity and representation (1.2.22), for the cases (1.2.14) and (1.2.15) (see also the books by Mathai and R.K. Saxena [2, Section 1.1] and Srivastava, Gupta and Goyal [1, Section 2.2]). We note that Braaksma probably first considered $H^{m,n}_{p,q}(z)$ as an analytic function that is multiple-valued in general, but one-valued on the Riemann surface of $\log(z)$. He suggested some extensions of the definition of the H-function by its analytic continuation and such extensions were also discussed by Skibinski [1] (1970).

The proofs of Theorems 1.1 and 1.2, which were based on Lemmas 1.1 and 1.2 presenting different behaviors (1.2.11) and (1.2.12) of the integrand (1.1.2) at infinity, were given by the authors, see Kilbas and Saigo [6, Theorems 1 and 2]. We indicate that the representation (1.2.23) can also be deduced from (1.2.22) by using the translation formula (1.8.15), which was probably first found by K.C. Gupta and U.C. Jain [1] (1966).

It should be noted that the H-function makes sense under different conditions when either $\mathfrak{L} = \mathfrak{L}_{i\gamma\infty}$, $\mathfrak{L} = \mathfrak{L}_{-\infty}$ or $\mathfrak{L} = \mathfrak{L}_{+\infty}$, as shown in Theorem 1.1. This fact was first noted by Prudnikov, Brychkov and Marichev [3, 8.3.1] (Russian edition in 1985). In this connection we notice that the remark in Mathai and R.K. Saxena [2, Section 1.1] is not correct, where it is mentioned that the existence of the H-function does not depend on the contour $\mathfrak{L}$. Prudnikov, Brychkov and Marichev [3, 8.3.1] also indicated that the parameter μ in the cases (1.2.16) and (1.2.19) of Theorem 1.1 can be extended from $\mathrm{Re}(\mu) < 0$ to $\mathrm{Re}(\mu) < -1$.

We also note that the notation for the H-function given in (1.1.1) was suggested by Gupta [2] (1965), while the notation of numbers in (1.1.7)–(1.1.10) and (1.1.15) is by Prudnikov, Brychkov and Marichev [3, 8.3.1].

For Sections 1.3 and 1.4. The power series representation (1.3.7) was first given by Braaksma [1, (6.5)]. The representations (1.3.7) as well as (1.3.8) were presented in the books by Mathai and R.K. Saxena [2, (3.7.1) and (3.7.2)], Srivastava, Gupta and Goyal [1, (2.2.4) and (2.2.7)] and Prudnikov, Brychkov and Marichev [3, 8.3.2(3) and 8.3.2(4)]. Their proofs were also given by the authors in Kilbas and Saigo [7, Theorems 3 and 4].

The power-logarithmic series representations (1.4.7) and (1.4.13) were obtained by the authors in the above paper [6, Theorems 5 and 6]. We also note that, in particular cases of the H-function of the forms $H^{p,0}_{0,p}(z)$ and $H^{p,0}_{p,p}(z)$, the power-logarithmic series representations, which are more complicated than in (1.4.7), were given by Mathai and R.K. Saxena [2, Section 3.7] (see also Mathai and R.K. Saxena [1, Section 5.8] and Mathai [1]).

For Sections 1.5–1.7. Bochner [1, p.351] (1951) probably first obtained the asymptotic estimate at a large positive z of the H-function with $\mathfrak{L} = \mathfrak{L}_{-\infty}$ in the case $p = 0$ and $m = q$. Braaksma [1] developed the method to find the asymptotic expansion of the general H-function with $\mathfrak{L} = \mathfrak{L}_{-\infty}$.

Fox [2] investigated the asymptotic behavior at infinity of the H-function $H^{q,p}_{2p,2q}(x)$ in (1.11.2) with $\mathfrak{L} = \mathfrak{L}_{i\gamma\infty}$ $(0 < \gamma < 1/2)$, provided that $\Delta > 0, q > p$ and a_i $(1 \leqq i \leqq p)$ satisfy the conditions in (1.11.3) while $\mathrm{Re}(b_j) > 0$ $(1 \leqq j \leqq q)$. His method was based on the representation of (1.11.1):

$$H^{q,p}_{2p,2q}(x) = \frac{1}{2\pi i}\int_{\mathfrak{L}_{i\gamma\infty}} \mathcal{H}^{q,p}_{2p,2q}(s)x^{-s}ds = \frac{1}{2\pi i}\int_{\mathfrak{L}_{i\gamma\infty}} Q(s)\left(\frac{x}{\beta}\right)^{-s} ds \tag{1.11.5}$$

for $\beta = \delta\Delta^{-\Delta}$ and the function $\mathcal{H}^{q,p}_{2p,2q}(s)$ being given in (1.1.2) and on the asymptotic estimate of $Q(s)$ for large s, $s = \gamma + it$ with fixed γ by employing residue theory. He proved [2, Theorem 9] the following asymptotic estimates for a large positive x:

$$H^{q,p}_{2p,2q}(x) = \left(\frac{x}{\beta}\right)^{(1-\Delta)/2\Delta} \sum_{n=0}^{r} c_n \left(\frac{x}{\beta}\right)^{-n/\Delta} \sin\left[\frac{\pi}{2}\left(K - n + \frac{1-\Delta}{2}\right) + \left(\frac{x}{\beta}\right)^{1/\Delta}\right]$$
$$+ \sum_{i=1}^{p}\sum_{j=0}^{u_i} A_{ij}x^{-(a_i+j)/\alpha_i} + O\left(x^{-\gamma_0}\right) \quad (x \to +\infty), \tag{1.11.6}$$

when $q - p$ is an odd positive integer, and

$$H^{q,p}_{2p,2q}(x) = \left(\frac{x}{\beta}\right)^{(1-\Delta)/2\Delta} \sum_{n=0}^{r} c_n \left(\frac{x}{\beta}\right)^{-n/\Delta} \cos\left[\frac{\pi}{2}\left(K - n + \frac{1-\Delta}{2}\right) + \left(\frac{x}{\beta}\right)^{1/\Delta}\right]$$
$$+ \sum_{i=1}^{p}\sum_{j=0}^{u_i} C_{ij}x^{-(a_i+j)/\alpha_i} + O\left(x^{-\gamma_0}\right) \quad (x \to +\infty), \tag{1.11.7}$$

when $q - p$ is an even positive integer. Here

$$K = 2\left[\sum_{j=1}^{q}(b_j + \beta_j) - \sum_{i=1}^{p} a_i\right], \tag{1.11.8}$$

Δ, δ and β are given in (1.11.3), (1.11.4) and (1.11.5), $\gamma_0 > 1/2$ is a given constant, r denotes the greatest integer of $[\Delta(2\gamma_0 - 1) + 3]/2$ and u_i denotes the greatest positive integer of $\alpha_i\gamma_0 - \mathrm{Re}(a_i)$ $(1 \leqq i \leqq p)$, while c_n $(0 \leqq n \leqq r)$ and C_{ij} $(1 \leqq i \leqq p;\ 0 \leqq j \leqq u_i)$ are constants which depend on a_i, α_i $(1 \leqq i \leqq p)$ and b_j, β_j $(1 \leqq j \leqq q)$ but are independent of x. The constants $c_0, c_1, \cdots, c_r$ are defined for large t, $s = \gamma + it$ with fixed γ, by

$$Q(s) - \sum_{n=0}^{r} c_n\Gamma\left(\Delta s - n + \frac{1-\Delta}{2}\right)\sin\left[\frac{(K - \Delta s)\pi}{2}\right]$$
$$= O\left(|s|^{(-2r-1+2\Delta\gamma-\Delta)/r}\right), \tag{1.11.9}$$

when $q - p$ is odd, and

$$Q(s) - \sum_{n=0}^{r} c_n\Gamma\left(\Delta s - n + \frac{1-\Delta}{2}\right)\cos\left[\frac{(K - \Delta s)\pi}{2}\right]$$
$$= O\left(|s|^{(-2r-1+2\Delta\gamma-\Delta)/r}\right), \tag{1.11.10}$$

when $q-p$ is even. The constants C_{ij} $(1 \leqq i \leqq p;\ 0 \leqq j \leqq u_i)$ are obtained by using residue theory to compute the integrals of the form (1.11.5) in which $Q(s)$ is replaced by the left-hand side of (1.11.9) and (1.11.10) for an odd and even $q-p$, respectively.

As was indicated in Section 1.5, Braaksma [1] (1964) developed the method, based on the Cauchy theorem and residue theory, to rewrite (1.5.4) in the form (1.5.5) and to find the asymptotic expansion of the general H-function $H_{p,q}^{m,n}(z)$ with $\mathfrak{L} = \mathfrak{L}_{+\infty}$ in the cases $\Delta > 0$ and $\Delta = 0,\ 0 < |z| < \delta$. When $\Delta > 0$ and $a^* > 0$, he proved the asymptotic expansion (1.5.3) (see [1, (2.16)]), which is called an algebraic asymptotic expansion.

In the general case, when $\Delta > 0$ but a^* is not necessary positive, Braaksma [1, Theorem 3] obtained the algebraic asymptotic expansion for $H_{p,q}^{m,n}(z)$ in the form

$$H_{p,q}^{m,n}(z) \sim \sum_{i=1}^{n}\sum_{k=0}^{\infty} \operatorname*{Res}_{s=a_{ik}} [\mathcal{H}_{p,q}^{m,n}(s)z^{-s}] + \sum_{j=r}^{\kappa} D_j P\left(ze^{i\delta_j}\right) - \sum_{j=0}^{r-1} C_j P\left(ze^{i\delta_j}\right), \tag{1.11.11}$$

which holds for $|z| \to \infty$ uniformly on

$$\frac{\Delta\pi}{2} - \delta_r + \epsilon \leqq \arg(z) \leqq -\frac{\Delta\pi}{2} - \delta_{r-1} - \epsilon \tag{1.11.12}$$

$$\left(0 < \epsilon < \frac{1}{2}(\delta_r - \delta_{r-1} - \Delta\pi)\right);$$

in particular, when $r = 0$, on

$$|\arg(z)| \leqq \frac{1}{2}a^*\pi - \epsilon \quad \left(0 < \epsilon < \frac{1}{2}a^*\pi\right). \tag{1.11.13}$$

Here the constants κ, δ_j $(j = 0, \cdots, \kappa)$, r, D_j and C_j are chosen as in the proof of Theorem 1.9. The proof of (1.11.11)–(1.11.13) was based on the representation of (1.5.5) in the form

$$H_{p,q}^{m,n}(z) = Q_w(z) + \sum_{j=r}^{\kappa} D_j P_w\left(ze^{i\delta_j}\right) - \sum_{j=0}^{r-1} C_j P_w\left(ze^{i\delta_j}\right)$$

$$+\frac{1}{2\pi i}\left(\int_L - \int_{L_1}\right) h_0(-s)\left[h_1(-s) + \sum_{j=r}^{\kappa} D_j e^{-i\delta_j s} - \sum_{j=0}^{r-1} C_j e^{-i\delta_j s}\right] z^{-s} ds, \tag{1.11.14}$$

where $h_1(s)$ and $h_0(s)$ are given in (1.6.13) and (1.6.26), $P_w(z)$ is the sum of the residues of the function $h_0(-s)z^s$ at the points a_{ik} $(i = 1, \cdots, n;\ k = 0, 1, 2, \cdots)$ given in (1.1.5), the constants δ_j, r, k, C_j and D_j are indicated in the proof of Theorem 1.9. Applying some preliminary lemmas, characterizing the terms in (1.11.14), Braaksma deduced the asymptotic expansion (1.11.11) from (1.11.14).

Braaksma noted, however, that in some cases, all coefficients of the above algebraic asymptotic expansion can be taken to be equal to zero, and showed that in these cases $H_{p,q}^{m,n}(z)$ may have exponential asymptotic expansions. When $n = 0$, Braaksma [1, Theorem 4] proved such asymptotic expansions (called exponentially small expansions) presented in (1.7.5), (1.7.6) for $H_{p,q}^{q,0}(z)$ and in (1.7.9) for $H_{p,q}^{m,0}(z)$ in Theorem 1.10. He also showed that these asymptotic expansions hold uniformly on the sectors in Theorem 1.10 and some other regions for $\arg(z)$. His proof was based on the representation of $H_{p,q}^{m,0}(z)$ in the form (1.5.6):

$$H_{p,q}^{m,0}(z) = \frac{1}{2\pi i}\int_{w-i\infty}^{w+i\infty} \mathcal{H}_{p,q}^{m,0}(-s)z^s ds, \tag{1.11.15}$$

which in the case $m = q$ is reduced to the integral

$$H_0(z) = \frac{1}{2\pi i}\int_{w-i\infty}^{w+i\infty} h_2(s)z^s ds, \qquad h_2(s) = \frac{\prod_{j=1}^{q}\Gamma(b_j - \beta_j s)}{\prod_{i=1}^{p}\Gamma(a_i - \alpha_i s)} \tag{1.11.16}$$

(closely connected with $E(z)$ in (1.6.3)), and in the case $0 < m < q$ is expressed in terms of $H_0(z)$ by

$$H^{m,0}_{p,q}(z) = \sum_{k=0}^{M} d_k H_0\left(ze^{\omega_k}\right). \tag{1.11.17}$$

Here the positive integer M, real ω_k and complex d_k are defined by the relations:

$$\frac{\mathcal{H}^{m,0}_{p,q}(s)}{h_2(s)} = \pi^{m-q} \prod_{j=m+1}^{q} \sin[(b_j - \beta_j s)\pi] = \sum_{k=0}^{M} d_k e^{i\omega_k s}. \tag{1.11.18}$$

In the general case of the H-function $H^{m,n}_{p,q}(z)$ with $\Delta > 0$, Braaksma proved that (1.5.5) is reduced to the relation (see [1, (2.50)])

$$H^{m,n}_{p,q}(z) = Q_w(z) + \sum_{j=\lambda}^{\kappa} D_j P_w\left(ze^{i\delta_j}\right) + \sum_{j=\lambda}^{\nu}(C_j + D_j)F\left(ze^{i\delta_j}\right) + O\left(z^{\omega}\right), \tag{1.11.19}$$

$$\text{with} \quad F(z) = -P_w(z) + \frac{1}{2\pi i}\int_{L_1} h_0(-s)z^s ds, \tag{1.11.20}$$

which holds for $|z| \to \infty$ uniformly on

$$\epsilon_0 - \frac{1}{2}(\delta_r + \delta_{r+1}) \leqq \arg(z) \leqq \epsilon_0 - \frac{1}{2}(\delta_{r-1} + \delta_r). \tag{1.11.21}$$

Here κ, r, δ_j, C_j and D_j are as in the proof of Theorem 1.9, ϵ_0 is a positive number independent of z and the integers λ, ν are defined by

$$\begin{gathered} \delta_{\lambda-1} \leqq \frac{1}{2}(\delta_r + \delta_{r-1} - \Delta\pi) - 2\epsilon_0, \quad \frac{1}{2}(\delta_r + \delta_{r+1} + \Delta\pi) \leqq \delta_{\nu+1}, \\ \lambda \leqq 0 \leqq \nu, \quad \lambda \leqq \kappa \leqq \nu, \quad \lambda < r \leqq \nu. \end{gathered} \tag{1.11.22}$$

Using estimates for some auxiliary functions, Braaksma [1, Theorem 7] obtained the asymptotic expansion (1.6.24) which holds uniformly on (1.6.25) and the following asymptotic estimates for the H-function are valid:

$$H^{m,n}_{p,q}(z) \sim (C_r + D_r)E\left(ze^{i\delta_r}\right) \tag{1.11.23}$$

for $|z| \to \infty$ uniformly on

$$\begin{gathered} \epsilon - \frac{1}{2}\min[\Delta\pi, \delta_{r+1} - \delta_r] \leqq \arg(z) + \delta_r \leqq \frac{1}{2}\min[\Delta\pi, \delta_r - \delta_{r-1}] - \epsilon \\ \left(0 < \epsilon < \frac{1}{4}\min[\Delta\pi, \delta_r - \delta_{r-1}, \delta_r - \delta_{r-1}]\right) \end{gathered} \tag{1.11.24}$$

(see Braaksma [1, Theorem 5]),

$$H^{m,n}_{p,q}(z) \sim (C_r + D_r)E\left(ze^{i\delta_r}\right) + (C_{r-1} + D_{r-1})E\left(ze^{i\delta_{r-1}}\right) \tag{1.11.25}$$

for $|z| \to \infty$ uniformly on

$$\begin{gathered} \frac{1}{2}(\delta_r - \delta_{r-1}) - \epsilon \leqq \arg(z) + \delta_r \leqq \frac{1}{2}(\delta_r - \delta_{r-1}) + \epsilon \\ \left(0 < \epsilon < \frac{1}{4}\min[\delta_{r-1} - \delta_{r-2}, \delta_r - \delta_{r-1}, \delta_{r+1} - \delta_r, \Delta\pi - \delta_r + \delta_{r-1}]\right), \end{gathered} \tag{1.11.26}$$

provided that $\delta_r - \delta_{r-1} < \Delta\pi$ (see Braaksma [1, Theorem 6]),

$$H_{p,q}^{m,n}(z) \sim \sum_{i=1}^{n}\sum_{k=0}^{\infty} \operatorname*{Res}_{s=a_{ik}} \left[\mathcal{H}_{p,q}^{m,n}(s)z^{-s}\right]$$

$$+\sum_{j=r}^{\kappa} D_j P\left(ze^{i\delta_j}\right) - \sum_{j=0}^{r-1} C_j P\left(ze^{i\delta_j}\right) + (C_r + D_r)E\left(ze^{i\delta_r}\right) \tag{1.11.27}$$

for $|z| \to \infty$ uniformly on (1.6.25) with

$$\frac{1}{2}\Delta\pi - \epsilon \leqq \arg(z) + \delta_r \leqq \frac{1}{2}\Delta\pi + \epsilon \tag{1.11.28}$$

$$\left(0 < \epsilon < \frac{1}{4}\min[\delta_{r+1} - \delta_r, \delta_r - \delta_{r-1} - \Delta\pi]\right),$$

provided that $\delta_r - \delta_{r-1} > \Delta\pi$ (see Braaksma [1, Theorem 8]), and

$$H_{p,q}^{m,n}(z) \sim \sum_{i=1}^{n}\sum_{k=0}^{\infty} \operatorname*{Res}_{s=a_{ik}} \left[\mathcal{H}_{p,q}^{m,n}(s)z^{-s}\right]$$

$$+\sum_{j=r}^{\kappa} D_j P\left(ze^{i\delta_j}\right) - \sum_{j=0}^{r-1} C_j P\left(ze^{i\delta_j}\right) + (C_{r-1} + D_{r-1})E\left(ze^{i\delta_{r-1}}\right) \tag{1.11.29}$$

for $|z| \to \infty$ uniformly on

$$-\frac{1}{2}\Delta\pi - \epsilon \leqq \arg(z) + \delta_{r-1} \leqq -\frac{1}{2}\Delta\pi + \epsilon \tag{1.11.30}$$

$$\left(0 < \epsilon < \frac{1}{4}\min[\delta_{r-1} - \delta_{r-2}, \delta_r - \delta_{r-1} - \Delta\pi]\right),$$

provided that $\delta_r - \delta_{r-1} > \Delta\pi$ (see Braaksma [1, Theorem 9]).

The results in Sections 1.5 and 1.6 were presented by the authors in Kilbas and Saigo [1]. The results in Sections 1.7 have not been published before. We also note that the asymptotic estimate (1.5.13) was earlier indicated in books by Mathai and R.K. Saxena [2, (1.6.2)] and by Srivastava, Gupta and Goyal [1, (2.2.12)].

In conclusion we indicate that Theorems 1.7 and 1.8 show the explicit power asymptotic expansion (1.5.9) and the power-logarithmic one (1.5.14) of the H-function $H_{p,q}^{m,n}(z)$ at infinity in the cases either $\Delta \leqq 0$ or $\Delta > 0$, $a^* > 0$, Theorem 1.9 gives the explicit asymptotic behavior of $H_{p,q}^{m,n}(z)$ in the exceptional case $\Delta > 0$ and $a^* = 0$, and Theorem 1.10 contains the explicit asymptotic estimate of $H_{p,q}^{m,0}(z)$ for $\Delta > 0$, and either $a^* \geqq 0$ with $m < q$ or $a^* > 0$ with $m = q$. The problem of finding an explicit asymptotic expansion of $H_{p,q}^{m,n}(z)$ in the case $\Delta > 0$ and $a^* < 0$ is open, as well as that for $H_{p,q}^{m,0}(z)$ in the case $\Delta > 0$, when either $a^* < 0$ with $m < q$ or $a^* \leqq 0$ with $m = q$. We hope these problems can be solved by using the estimates (1.11.23)–(1.11.26) which were established in Theorems 5, 6, 8 and 9 of Braaksma [1], in the same way as Theorem 1.9 was proved on the basis of Braaksma [1, Theorem 7].

For Sections 1.8–1.10. The power and power-logarithmic asymptotic expansions of the H-function at zero in the cases $\Delta \leqq 0$ and $\Delta < 0$, $a^* > 0$ in Sections 1.8 and 1.9 were given by the authors in Kilbas and Saigo [1]. The asymptotic estimate (1.8.8) was earlier shown by Mathai and R.K. Saxena [2, (1.6.1)] and by Srivastava, Gupta and Goyal [1, (2.2.9)].

We note that, in the paper [1] of the authors, the asymptotic estimate (1.6.8) was proved by using the modification of the method suggested by Braaksma in [1] for the investigation of the asymptotic behavior of the H-function at infinity, but the formulas (1.6.8) and (1.6.8) in [1] should be corrected in accordance with Section 1.9.

We also note that the asymptotic estimates for the H-function at zero presented in Sections 1.8 and 1.9 can be deduced from the corresponding estimates at infinity given in Sections 1.5 and 1.6 by using the translation formula (1.8.15) and the relations (1.8.16).

The results in Section 1.10 have not been published before.

Chapter 2
PROPERTIES OF THE *H*-FUNCTION

2.1. Elementary Properties

In Sections 2.1–2.4 we suppose that the H-functions considered make sense in accordance with Theorem 1.1. The results in this section follow directly from the definition of the H-function in Section 1.1. First we give the simplest formulas.

Property 2.1. The H-function (1.1.1) is symmetric in the set of pairs $(a_1, \alpha_1), \cdots, (a_n, \alpha_n)$; in $(a_{n+1}, \alpha_{n+1}), \cdots, (a_p, \alpha_p)$; in $(b_1, \beta_1), \cdots, (b_m, \beta_m)$ and in $(b_{m+1}, \beta_{m+1}), \cdots, (b_q, \beta_q)$.

Property 2.2. If one of (a_i, α_i) $(i = 1, \cdots, n)$ is equal to one of (b_j, β_j) $(j = m+1, \cdots, q)$ (or one of (a_i, α_i) $(i = n+1, \cdots, p)$ is equal to one of (b_j, β_j) $(j = 1, \cdots, m)$), then the H-function reduces to a lower order one, that is, p, q and n (or m) decrease by unity. We give two examples of such reduction formulas:

$$H^{m,n}_{p,q}\left[z \,\middle|\, \begin{matrix} (a_i, \alpha_i)_{1,p} \\ (b_j, \beta_j)_{1,q-1}, (a_1, \alpha_1) \end{matrix}\right] = H^{m,n-1}_{p-1,q-1}\left[z \,\middle|\, \begin{matrix} (a_i, \alpha_i)_{2,p} \\ (b_j, \beta_j)_{1,q-1} \end{matrix}\right], \tag{2.1.1}$$

provided $n \geqq 1$ and $q > m$; and

$$H^{m,n}_{p,q}\left[z \,\middle|\, \begin{matrix} (a_i, \alpha_i)_{1,p-1}, (b_1, \beta_1) \\ (b_j, \beta_j)_{1,q} \end{matrix}\right] = H^{m-1,n}_{p-1,q-1}\left[z \,\middle|\, \begin{matrix} (a_i, \alpha_i)_{1,p-1} \\ (b_j, \beta_j)_{2,q} \end{matrix}\right], \tag{2.1.2}$$

provided $m \geqq 1$ and $p > n$.

Property 2.3. There holds the relation

$$H^{m,n}_{p,q}\left[\frac{1}{z} \,\middle|\, \begin{matrix} (a_i, \alpha_i)_{1,p} \\ (b_j, \beta_j)_{1,q} \end{matrix}\right] = H^{n,m}_{q,p}\left[z \,\middle|\, \begin{matrix} (1 - b_j, \beta_j)_{1,q} \\ (1 - a_i, \alpha_i)_{1,p} \end{matrix}\right]. \tag{2.1.3}$$

Property 2.4. For $k > 0$, there holds the relation

$$H^{m,n}_{p,q}\left[z \,\middle|\, \begin{matrix} (a_i, \alpha_i)_{1,p} \\ (b_j, \beta_j)_{1,q} \end{matrix}\right] = kH^{m,n}_{p,q}\left[z^k \,\middle|\, \begin{matrix} (a_i, k\alpha_i)_{1,p} \\ (b_j, k\beta_j)_{1,q} \end{matrix}\right]. \tag{2.1.4}$$

Property 2.5. For $\sigma \in \mathbb{C}$, there holds the relation

$$z^{\sigma} H_{p,q}^{m,n}\left[z \,\middle|\, \begin{matrix} (a_i, \alpha_i)_{1,p} \\ (b_j, \beta_j)_{1,q} \end{matrix}\right] = H_{p,q}^{m,n}\left[z \,\middle|\, \begin{matrix} (a_i + \sigma\alpha_i, \alpha_i)_{1,p} \\ (b_j + \sigma\beta_j, \beta_j)_{1,q} \end{matrix}\right]. \tag{2.1.5}$$

The next six formulas follow from the definition of the H-function in (1.1.1) and the reflection formula for the gamma function (see Erdélyi, Magnus, Oberhettinger and Tricomi [1, 1.2(6)])

$$\Gamma(z)\Gamma(1-z) = \frac{\pi}{\sin(z\pi)}. \tag{2.1.6}$$

Property 2.6. For $c \in \mathbb{C}$, $\alpha > 0$ and $k = 0, \pm 1, \pm 2, \cdots$, there hold the relations

$$H_{p+1,q+1}^{m,n+1}\left[z \,\middle|\, \begin{matrix} (c, \alpha), (a_i, \alpha_i)_{1,p} \\ (b_j, \beta_j)_{1,q}, (c+k, \alpha) \end{matrix}\right] = (-1)^k H_{p+1,q+1}^{m+1,n}\left[z \,\middle|\, \begin{matrix} (a_i, \alpha_i)_{1,p}, (c, \alpha) \\ (c+k, \alpha), (b_j, \beta_j)_{1,q} \end{matrix}\right]; \tag{2.1.7}$$

$$H_{p+1,q+1}^{m+1,n}\left[z \,\middle|\, \begin{matrix} (a_i, \alpha_i)_{1,p}, (c, \alpha) \\ (c+k, \alpha), (b_j, \beta_j)_{1,q} \end{matrix}\right] = (-1)^k H_{p+1,q+1}^{m,n+1}\left[z \,\middle|\, \begin{matrix} (c, \alpha), (a_i, \alpha_i)_{1,p} \\ (b_j, \beta_j)_{1,q}, (c+k, \alpha) \end{matrix}\right]. \tag{2.1.8}$$

Property 2.7. For $a, b \in \mathbb{C}$, there hold the relations

$$H_{p,q}^{m,n}\left[z \,\middle|\, \begin{matrix} (a, 0), (a_i, \alpha_i)_{2,p} \\ (b_j, \beta_j)_{1,q} \end{matrix}\right] = \Gamma(1-a) H_{p-1,q}^{m,n-1}\left[z \,\middle|\, \begin{matrix} (a_i, \alpha_i)_{2,p} \\ (b_j, \beta_j)_{1,q} \end{matrix}\right], \tag{2.1.9}$$

when $\mathrm{Re}(1-a) > 0$ and $n \geqq 1$;

$$H_{p,q}^{m,n}\left[z \,\middle|\, \begin{matrix} (a_i, \alpha_i)_{1,p} \\ (b, 0), (b_j, \beta_j)_{2,q} \end{matrix}\right] = \Gamma(b) H_{p,q-1}^{m-1,n}\left[z \,\middle|\, \begin{matrix} (a_i, \alpha_i)_{1,p} \\ (b_j, \beta_j)_{2,q} \end{matrix}\right], \tag{2.1.10}$$

when $\mathrm{Re}(b) > 0$ and $m \geqq 1$;

$$H_{p,q}^{m,n}\left[z \,\middle|\, \begin{matrix} (a_i, \alpha_i)_{1,p-1}, (a, 0) \\ (b_j, \beta_j)_{1,q} \end{matrix}\right] = \frac{1}{\Gamma(a)} H_{p-1,q}^{m,n}\left[z \,\middle|\, \begin{matrix} (a_i, \alpha_i)_{1,p-1} \\ (b_j, \beta_j)_{1,q} \end{matrix}\right], \tag{2.1.11}$$

when $\mathrm{Re}(a) > 0$ and $p > n$; and

$$H_{p,q}^{m,n}\left[z \,\middle|\, \begin{matrix} (a_i, \alpha_i)_{1,p} \\ (b_j, \beta_j)_{1,q-1}, (b, 0) \end{matrix}\right] = \frac{1}{\Gamma(1-b)} H_{p,q-1}^{m,n}\left[z \,\middle|\, \begin{matrix} (a_i, \alpha_i)_{1,p} \\ (b_j, \beta_j)_{1,q-1} \end{matrix}\right], \tag{2.1.12}$$

when $\mathrm{Re}(1-b) > 0$ and $q > m$.

2.2. Differentiation Formulas

It is proved here that differentiation of the H-function also gives the same function of greater order. From the definition of the H-function we easily obtain the following relations:

Property 2.8.

$$\left(\frac{d}{dz}\right)^k \left\{ z^\omega H_{p,q}^{m,n} \left[cz^\sigma \middle| \begin{matrix} (a_i, \alpha_i)_{1,p} \\ (b_j, \beta_j)_{1,q} \end{matrix} \right] \right\}$$

$$= z^{\omega-k} H_{p+1,q+1}^{m,n+1} \left[cz^\sigma \middle| \begin{matrix} (-\omega, \sigma), (a_i, \alpha_i)_{1,p} \\ (b_j, \beta_j)_{1,q}, (k-\omega, \sigma) \end{matrix} \right] \tag{2.2.1}$$

$$= (-1)^k z^{\omega-k} H_{p+1,q+1}^{m+1,n} \left[cz^\sigma \middle| \begin{matrix} (a_i, \alpha_i)_{1,p}, (-\omega, \sigma) \\ (k-\omega, \sigma), (b_j, \beta_j)_{1,q} \end{matrix} \right] \tag{2.2.2}$$

for $\omega,\ c \in \mathbb{C},\ \sigma > 0$;

$$\prod_{j=1}^{k} \left(z\frac{d}{dz} - c_j \right) \left\{ z^\omega H_{p,q}^{m,n} \left[az^\sigma \middle| \begin{matrix} (a_i, \alpha_i)_{1,p} \\ (b_j, \beta_j)_{1,q} \end{matrix} \right] \right\}$$

$$= z^\omega H_{p+k,q+k}^{m,n+k} \left[az^\sigma \middle| \begin{matrix} (c_j - \omega, \sigma)_{1,k}, (a_i, \alpha_i)_{1,p} \\ (b_j, \beta_j)_{1,q}, (c_j + 1 - \omega, \sigma)_{1,k} \end{matrix} \right] \tag{2.2.3}$$

$$= (-1)^k z^\omega H_{p+k,q+k}^{m+k,n} \left[az^\sigma \middle| \begin{matrix} (a_i, \alpha_i)_{1,p}, (c_j - \omega, \sigma)_{1,k} \\ (c_j + 1 - \omega, \sigma)_{1,k}, (b_j, \beta_j)_{1,q} \end{matrix} \right] \tag{2.2.4}$$

for $\omega,\ a,\ c_j \in \mathbb{C}\ (j = 1, \cdots, k),\ \sigma > 0$; and

$$\left(\frac{d}{dz}\right)^k H_{p,q}^{m,n} \left[(cz+d)^\sigma \middle| \begin{matrix} (a_i, \alpha_i)_{1,p} \\ (b_j, \beta_j)_{1,q} \end{matrix} \right]$$

$$= \frac{c^k}{(cz+d)^k} H_{p+1,q+1}^{m,n+1} \left[(cz+d)^\sigma \middle| \begin{matrix} (0, \sigma), (a_i, \alpha_i)_{1,p} \\ (b_j, \beta_j)_{1,q}, (k, \sigma) \end{matrix} \right], \tag{2.2.5}$$

$$\left(\frac{d}{dz}\right)^k H_{p,q}^{m,n} \left[\frac{1}{(cz+d)^\sigma} \middle| \begin{matrix} (a_i, \alpha_i)_{1,p} \\ (b_j, \beta_j)_{1,q} \end{matrix} \right]$$

$$= \frac{c^k}{(cz+d)^k} H_{p+1,q+1}^{m+1,n} \left[\frac{1}{(cz+d)^\sigma} \middle| \begin{matrix} (a_i, \alpha_i)_{1,p}, (1-k, \sigma) \\ (1, \sigma), (b_j, \beta_j)_{1,q} \end{matrix} \right] \tag{2.2.6}$$

for $c, d \in \mathbb{C},\ \sigma > 0$.

Further differentiation formulas connect H-functions the same orders, but of different parameters:

Property 2.9.

$$\left(\frac{d}{dz}\right)^k \left(z^{-\sigma b_1/\beta_1} H_{p,q}^{m,n}\left[z^\sigma \,\middle|\, \begin{matrix} (a_i,\alpha_i)_{1,p} \\ (b_j,\beta_j)_{1,q} \end{matrix}\right]\right)$$

$$= \left(-\frac{\sigma}{\beta_1}\right)^k z^{-k-\sigma b_1/\beta_1} H_{p,q}^{m,n}\left[z^\sigma \,\middle|\, \begin{matrix} (a_i,\alpha_i)_{1,p} \\ (b_1+k,\beta_1),(b_j,\beta_j)_{2,q} \end{matrix}\right] \tag{2.2.7}$$

with $m \geqq 1$, while $\sigma = \beta_1$ for $k > 1$;

$$\left(\frac{d}{dz}\right)^k \left(z^{-\sigma b_q/\beta_q} H_{p,q}^{m,n}\left[z^\sigma \,\middle|\, \begin{matrix} (a_i,\alpha_i)_{1,p} \\ (b_j,\beta_j)_{1,q} \end{matrix}\right]\right)$$

$$= \left(\frac{\sigma}{\beta_q}\right)^k z^{-k-\sigma b_q/\beta_q} H_{p,q}^{m,n}\left[z^\sigma \,\middle|\, \begin{matrix} (a_i,\alpha_i)_{1,p} \\ (b_j,\beta_j)_{1,q-1},(b_q+k,\beta_q) \end{matrix}\right] \tag{2.2.8}$$

with $m < q$, while $\sigma = \beta_q$ for $k > 1$;

$$\left(\frac{d}{dz}\right)^k \left(z^{-\sigma(1-a_1)/\alpha_1} H_{p,q}^{m,n}\left[z^{-\sigma} \,\middle|\, \begin{matrix} (a_i,\alpha_i)_{1,p} \\ (b_j,\beta_j)_{1,q} \end{matrix}\right]\right)$$

$$= \left(-\frac{\sigma}{\alpha_1}\right)^k z^{-k-\sigma(1-a_1)/\alpha_1} H_{p,q}^{m,n}\left[z^{-\sigma} \,\middle|\, \begin{matrix} (a_1-k,\alpha_1),(a_i,\alpha_i)_{2,p} \\ (b_j,\beta_j)_{1,q} \end{matrix}\right] \tag{2.2.9}$$

with $n \geqq 1$, while $\sigma = \alpha_1$ for $k > 1$;

$$\left(\frac{d}{dz}\right)^k \left(z^{-\sigma(1-a_p)/\alpha_p} H_{p,q}^{m,n}\left[z^{-\sigma} \,\middle|\, \begin{matrix} (a_i,\alpha_i)_{1,p} \\ (b_j,\beta_j)_{1,q} \end{matrix}\right]\right)$$

$$= \left(\frac{\sigma}{\alpha_p}\right)^k z^{-k-\sigma(1-a_p)/\alpha_p} H_{p,q}^{m,n}\left[z^{-\sigma} \,\middle|\, \begin{matrix} (a_i,\alpha_i)_{1,p-1},(a_p-k,\alpha_p) \\ (b_j,\beta_j)_{1,q} \end{matrix}\right] \tag{2.2.10}$$

with $p > n$, while $\sigma = \alpha_p$ for $k > 1$.

The relations (2.2.7)–(2.2.10) follow from (2.2.1) and (2.2.2), if we take Property 2.2 into account.

Property 2.10.

$$z\frac{d}{dz}\left\{H_{p,q}^{m,n}\left[z^\sigma \,\middle|\, \begin{matrix} (a_i,\alpha_i)_{1,p} \\ (b_j,\beta_j)_{1,q} \end{matrix}\right]\right\}$$

$$= \frac{\sigma(a_1-1)}{\alpha_1} H_{p,q}^{m,n}\left[z^\sigma \,\middle|\, \begin{matrix} (a_i,\alpha_i)_{1,p} \\ (b_j,\beta_j)_{1,q} \end{matrix}\right]$$

$$+\frac{\sigma}{\alpha_1} H_{p,q}^{m,n}\left[z^\sigma \,\middle|\, \begin{matrix} (a_1-1,\alpha_1),(a_i,\alpha_i)_{2,p} \\ (b_j,\beta_j)_{1,q} \end{matrix}\right] \tag{2.2.11}$$

for $n \geqq 1$;

$$z\frac{d}{dz}\left\{H_{p,q}^{m,n}\left[z^{\sigma}\,\middle|\,\begin{matrix}(a_i,\alpha_i)_{1,p}\\(b_j,\beta_j)_{1,q}\end{matrix}\right]\right\}$$

$$=\frac{\sigma(a_p-1)}{\alpha_p}H_{p,q}^{m,n}\left[z^{\sigma}\,\middle|\,\begin{matrix}(a_i,\alpha_i)_{1,p}\\(b_j,\beta_j)_{1,q}\end{matrix}\right]$$

$$-\frac{\sigma}{\alpha_p}H_{p,q}^{m,n}\left[z^{\sigma}\,\middle|\,\begin{matrix}(a_i,\alpha_i)_{1,p-1},(a_p-1,\alpha_p)\\(b_j,\beta_j)_{1,q}\end{matrix}\right] \tag{2.2.12}$$

for $n \leqq p-1$;

$$z\frac{d}{dz}\left\{H_{p,q}^{m,n}\left[z^{\sigma}\,\middle|\,\begin{matrix}(a_i,\alpha_i)_{1,p}\\(b_j,\beta_j)_{1,q}\end{matrix}\right]\right\}$$

$$=\frac{\sigma b_1}{\beta_1}H_{p,q}^{m,n}\left[z^{\sigma}\,\middle|\,\begin{matrix}(a_i,\alpha_i)_{1,p}\\(b_j,\beta_j)_{1,q}\end{matrix}\right]$$

$$-\frac{\sigma}{\beta_1}H_{p,q}^{m,n}\left[z^{\sigma}\,\middle|\,\begin{matrix}(a_i,\alpha_i)_{1,p}\\(b_1+1,\beta_1),(b_j,\beta_j)_{2,q}\end{matrix}\right] \tag{2.2.13}$$

for $m \geqq 1$;

$$z\frac{d}{dz}\left\{H_{p,q}^{m,n}\left[z^{\sigma}\,\middle|\,\begin{matrix}(a_i,\alpha_i)_{1,p}\\(b_j,\beta_j)_{1,q}\end{matrix}\right]\right\}$$

$$=\frac{\sigma b_q}{\beta_q}H_{p,q}^{m,n}\left[z^{\sigma}\,\middle|\,\begin{matrix}(a_i,\alpha_i)_{1,p}\\(b_j,\beta_j)_{1,q}\end{matrix}\right]$$

$$+\frac{\sigma}{\beta_q}H_{p,q}^{m,n}\left[z^{\sigma}\,\middle|\,\begin{matrix}(a_i,\alpha_i)_{1,p}\\(b_j,\beta_j)_{1,q-1},(b_q+1,\beta_q)\end{matrix}\right] \tag{2.2.14}$$

for $m \leqq q-1$.

The formulas (2.2.11)–(2.2.14) are established by virtue of the following relations:

$$-\alpha_1 s\Gamma(1-a_1-\alpha_1 s)=(a_1-1)\Gamma(1-a_1-\alpha_1 s)+\Gamma(2-a_1-\alpha_1 s);$$

$$-\frac{\alpha_p s}{\Gamma(a_p+\alpha_p s)}=\frac{a_p-1}{\Gamma(a_p+\alpha_p s)}-\frac{1}{\Gamma(a_p-1+\alpha_p s)};$$

$$-\beta_1 s\Gamma(b_1+\beta_1 s)=b_1\Gamma(b_1+\beta_1 s)-\Gamma(b_1+1+\beta_1 s);$$

$$-\frac{\beta_q s}{\Gamma(1-b_q-\beta_q s)}=\frac{b_q}{\Gamma(1-b_q-\beta_q s)}+\frac{1}{\Gamma(-b_q-\beta_q s)},$$

respectively, which follow from the relation (see Erdélyi, Magnus, Oberhettinger and Tricomi [1, 1.2(1)])

$$z\Gamma(z)=\Gamma(z+1). \tag{2.2.15}$$

2.3. Recurrence Relations and Expansion Formulas

We give two three-term recurrence formulas which present linear combinations of the H-function with the same m, n, p and q, in which some a_i and b_j are replaced by $a_i \pm 1$ and $b_j \pm 1$ $(i = 1, \cdots, p;\ j = 1, \cdots, q)$. Such relations are called contiguous relations by Srivastava, Gupta and Goyal [1, Section 2.9]. These relations can be deduced directly by using the definition of the H-function given in (1.1.1) and (1.1.2), if we take (2.2.15) into account.

Property 2.11.

$$(b_1\alpha_p - a_p\beta_1 + \beta_1)H_{p,q}^{m,n}\left[z^\sigma \,\middle|\, \begin{matrix}(a_i,\alpha_i)_{1,p}\\ (b_j,\beta_j)_{1,q}\end{matrix}\right]$$

$$= \alpha_p H_{p,q}^{m,n}\left[z^\sigma \,\middle|\, \begin{matrix}(a_i,\alpha_i)_{1,p}\\ (b_1+1,\beta_1),(b_j,\beta_j)_{2,q}\end{matrix}\right]$$

$$-\beta_1 H_{p,q}^{m,n}\left[z^\sigma \,\middle|\, \begin{matrix}(a_i,\alpha_i)_{1,p-1},(a_p-1,\alpha_p)\\ (b_j,\beta_j)_{1,q}\end{matrix}\right] \tag{2.3.1}$$

for $m \geqq 1$ and $1 \leqq n \leqq p-1$;

$$(b_q\alpha_1 - a_1\beta_q + \beta_q)H_{p,q}^{m,n}\left[z^\sigma \,\middle|\, \begin{matrix}(a_i,\alpha_i)_{1,p}\\ (b_j,\beta_j)_{1,q}\end{matrix}\right]$$

$$= \beta_q H_{p,q}^{m,n}\left[z^\sigma \,\middle|\, \begin{matrix}(a_1-1,\alpha_1),(a_i,\alpha_i)_{2,p}\\ (b_j,\beta_j)_{1,q}\end{matrix}\right]$$

$$-\alpha_1 H_{p,q}^{m,n}\left[z^\sigma \,\middle|\, \begin{matrix}(a_i,\alpha_i)_{1,p}\\ (b_j,\beta_j)_{1,q-1},(b_q+1,\beta_q)\end{matrix}\right] \tag{2.3.2}$$

for $n \geqq 1$ and $1 \leqq m \leqq q-1$;

Remark 2.1. A complete list of contiguous relations of the H-function may be found in Buschman [2].

From (2.3.1) we come to the following finite series relations:

Property 2.12. For any $r \in \mathbb{N}$, $m \geqq 1$ and $1 \leqq n \leqq p-1$,

$$\frac{\beta_1}{\alpha_p}\sum_{k=1}^{r}\frac{1}{\Gamma\left(b_1+k-\dfrac{(a_p-1)\beta_1}{\alpha_p}\right)}H_{p,q}^{m,n}\left[z \,\middle|\, \begin{matrix}(a_i,\alpha_i)_{1,p-1},(a_p-1,\alpha_p)\\ (b_1+k-1,\beta_1),(b_j,\beta_j)_{2,q}\end{matrix}\right]$$

$$= \frac{1}{\Gamma\left(b_1+r-\dfrac{(a_p-1)\beta_1}{\alpha_p}\right)}H_{p,q}^{m,n}\left[z \,\middle|\, \begin{matrix}(a_i,\alpha_i)_{1,p}\\ (b_1+r,\beta_1),(b_j,\beta_j)_{2,q}\end{matrix}\right]$$

$$-\frac{1}{\Gamma\left(b_1-\dfrac{(a_p-1)\beta_1}{\alpha_p}\right)}H_{p,q}^{m,n}\left[z\,\middle|\,\begin{matrix}(a_i,\alpha_i)_{1,p}\\(b_j,\beta_j)_{1,q}\end{matrix}\right]; \tag{2.3.3}$$

$$\frac{\alpha_p}{\beta_1}\sum_{k=1}^{r}\frac{(-1)^k}{\Gamma\left(k-[a_p-1]+\dfrac{b_1\alpha_p}{\beta_1}\right)}H_{p,q}^{m,n}\left[z\,\middle|\,\begin{matrix}(a_i,\alpha_i)_{1,p-1},(a_p-k+1,\alpha_p)\\(b_1+1,\beta_1),(b_j,\beta_j)_{2,q}\end{matrix}\right]$$

$$=\frac{(-1)^r}{\Gamma\left(r-[a_p-1]+\dfrac{b_1\alpha_p}{\beta_1}\right)}H_{p,q}^{m,n}\left[z\,\middle|\,\begin{matrix}(a_i,\alpha_i)_{1,p-1},(a_p-r,\alpha_p)\\(b_j,\beta_j)_{1,q}\end{matrix}\right]$$

$$-\frac{1}{\Gamma\left(-[a_p-1]+\dfrac{b_1\alpha_p}{\beta_1}\right)}H_{p,q}^{m,n}\left[z\,\middle|\,\begin{matrix}(a_i,\alpha_i)_{1,p}\\(b_j,\beta_j)_{1,q}\end{matrix}\right]; \tag{2.3.4}$$

$$\sum_{k=1}^{r}\alpha_p^{k-1}\beta_1^{r-k}[(b_1+k)\alpha_p-(a_p+k-1)\beta_1]H_{p,q}^{m,n}\left[z\,\middle|\,\begin{matrix}(a_i,\alpha_i)_{1,p-1},(a_p+k,\alpha_p)\\(b_1+k,\beta_1),(b_j,\beta_j)_{2,q}\end{matrix}\right]$$

$$=\alpha_p^r H_{p,q}^{m,n}\left[z\,\middle|\,\begin{matrix}(a_i,\alpha_i)_{1,p-1},(a_p+r,\alpha_p)\\(b_1+r+1,\beta_1),(b_j,\beta_j)_{2,q}\end{matrix}\right]$$

$$-\beta_1^r H_{p,q}^{m,n}\left[z\,\middle|\,\begin{matrix}(a_i,\alpha_i)_{1,p}\\(b_1+1,\beta_1),(b_j,\beta_j)_{2,q}\end{matrix}\right]. \tag{2.3.5}$$

Proof. The formula (2.3.3) is proved on the basis of (2.3.1). For simplicity let us denote the determinant

$$d(b_1,a_p-1)=\begin{vmatrix}b_1 & a_p-1\\ \beta_1 & \alpha_p\end{vmatrix}=b_1\alpha_p-(a_p-1)\beta_1. \tag{2.3.6}$$

We write the H-function as H and use the notation $H[b_1+1]$ and $H[a_p-1]$ replacing b_1 and a_p by b_1+1 and a_p-1, respectively, without any change of all other parameters. Then the relations (2.3.1) and (2.3.3) are simplified as

$$d(b_1,a_p-1)H=\alpha_p H[b_1+1]-\beta_1 H[a_p-1] \tag{2.3.7}$$

and

$$\frac{\beta_1}{\alpha_p}\sum_{k=1}^{r}\frac{H[b_1+k-1,a_p-1]}{\Gamma\left(b_1+k-[a_p-1]\dfrac{\beta_1}{\alpha_p}\right)}$$

$$=\frac{H[b_1+r,a_p]}{\Gamma\left(b_1+r-[a_p-1]\dfrac{\beta_1}{\alpha_p}\right)}-\frac{H}{\Gamma\left(b_1-[a_p-1]\dfrac{\beta_1}{\alpha_p}\right)}. \tag{2.3.8}$$

Now applying (2.3.7) with b_1 being replaced by b_1+k-1 and using the relation (2.2.15), we have

$$\frac{\beta_1}{\alpha_p}\sum_{k=1}^{r}\frac{H[b_1+k-1,a_p-1]}{\Gamma\left(b_1+k-[a_p-1]\dfrac{\beta_1}{\alpha_p}\right)}$$

$$=\frac{1}{\alpha_p}\sum_{k=1}^{r}\frac{\alpha_pH[b_1+k]-d(b_1+k-1,a_p-1)H[b_1+k-1]}{\Gamma\left(b_1+k-[a_p-1]\dfrac{\beta_1}{\alpha_p}\right)}$$

$$=\sum_{k=1}^{r}\frac{H[b_1+k]}{\Gamma\left(b_1+k-[a_p-1]\dfrac{\beta_1}{\alpha_p}\right)}-\sum_{k=0}^{r-1}\frac{H[b_1+k]}{\Gamma\left(b_1+k-[a_p-1]\dfrac{\beta_1}{\alpha_p}\right)},$$

which implies (2.3.8) and hence (2.3.3).

The relation (2.3.4) is similarly proved by another use of (2.3.7).

As for (2.3.5), according to the notation (2.3.6) the left-hand side takes the form

$$\sum_{k=1}^{r}\alpha_p^{k-1}\beta_1^{r-k}d(b_1+k,a_p+k-1)H[b_1+k,a_p+k].$$

Applying (2.3.7) and again replacing the order of summation, we find

$$\sum_{k=1}^{r}\alpha_p^{k-1}\beta_1^{r-k}d(b_1+k,a_p+k-1)H[b_1+k,a_p+k]$$

$$=\sum_{k=1}^{r}\alpha_p^{k-1}\beta_1^{r-k}\left([\alpha_pH[b_1+k+1,a_p+k]-\beta_1H[b_1+k,a_p+k-1]\right)$$

$$=\sum_{k=1}^{r}\alpha_p^{k}\beta_1^{r-k}H[b_1+k+1,a_p+k]-\sum_{k=0}^{r-1}\alpha_p^{k}\beta_1^{r-k}H[b_1+k+1,a_p+k]$$

$$=\alpha_p^{r}H[b_1+r+1,a_p+r]-\beta_1^{r}H[b_1+1,a_p],$$

and hence (2.3.5) is proved.

Further, from (2.3.2) we similarly obtain the following formulas:

Property 2.13. For any $r\in\mathbb{N}$, $n\geqq 1$ and $1\leqq m\leqq q-1$,

$$\frac{\alpha_1}{\beta_q}\sum_{k=1}^{r}\frac{1}{\Gamma\left(1+k-a_1+b_q\dfrac{\alpha_1}{\beta_q}\right)}H_{p,q}^{m,n}\left[z\,\middle|\,\begin{matrix}(a_1-k+1,\alpha_1),(a_i,\alpha_i)_{2,p}\\(b_j,\beta_j)_{1,q-1},(b_q+1,\beta_q)\end{matrix}\right]$$

$$=\frac{1}{\Gamma\left(1+r-a_1+b_q\dfrac{\alpha_1}{\beta_q}\right)}H_{p,q}^{m,n}\left[z\,\middle|\,\begin{matrix}(a_1-r,\alpha_1),(a_i,\alpha_i)_{2,p}\\(b_j,\beta_j)_{1,q}\end{matrix}\right]$$

$$-\frac{1}{\Gamma\left(1-a_1+b_q\dfrac{\alpha_1}{\beta_q}\right)}H_{p,q}^{m,n}\left[z\left|\begin{array}{l}(a_i,\alpha_i)_{1,p}\\(b_j,\beta_j)_{1,q}\end{array}\right.\right]; \tag{2.3.9}$$

$$\frac{\beta_q}{\alpha_1}\sum_{k=1}^{r}\frac{(-1)^k}{\Gamma\left(b_q+k-[a_1-1]\dfrac{\beta_q}{\alpha_1}\right)}H_{p,q}^{m,n}\left[z\left|\begin{array}{l}(a_1-1,\alpha_1),(a_i,\alpha_i)_{2,p}\\(b_j,\beta_j)_{1,q-1},(b_q+k-1,\beta_q)\end{array}\right.\right]$$

$$=\frac{(-1)^r}{\Gamma\left(b_q+r-[a_1-1]\dfrac{\beta_q}{\alpha_1}\right)}H_{p,q}^{m,n}\left[z\left|\begin{array}{l}(a_i,\alpha_i)_{1,p}\\(b_j,\beta_j)_{1,q-1},(b_q+r,\beta_q)\end{array}\right.\right]$$

$$-\frac{1}{\Gamma\left(b_q-[a_1-1]\dfrac{\beta_q}{\alpha_1}\right)}H_{p,q}^{m,n}\left[z\left|\begin{array}{l}(a_i,\alpha_i)_{1,p}\\(b_j,\beta_j)_{1,q}\end{array}\right.\right]; \tag{2.3.10}$$

$$\sum_{k=1}^{r}\alpha_1^{k-1}\beta_q^{r-k}[(b_q+k-1)\alpha_1-(a_1+k-2)\beta_q]H_{p,q}^{m,n}\left[z\left|\begin{array}{l}(a_1+k-1,\alpha_1),(a_i,\alpha_i)_{2,p}\\(b_j,\beta_j)_{1,q-1},(b_q+k-1,\beta_q)\end{array}\right.\right]$$

$$=\beta_q^r\ H_{p,q}^{m,n}\left[z\left|\begin{array}{l}(a_1-1,\alpha_1),(a_i,\alpha_i)_{2,p}\\(b_j,\beta_j)_{1,q}\end{array}\right.\right]$$

$$-\alpha_1^r\ H_{p,q}^{m,n}\left[z\left|\begin{array}{l}(a_1+r-1,\alpha_1),(a_i,\alpha_i)_{2,p}\\(b_j,\beta_j)_{1,q-1},(b_q+r,\beta_q)\end{array}\right.\right]. \tag{2.3.11}$$

We also give the expansions known as multiplication theorems for the H-function (see Srivastava, Gupta and Goyal [1, 2.9.]).

Theorem 2.1. *Let $\lambda \in \mathbb{C}$ and let the conditions in* (1.1.6) *be satisfied. Then the following relations hold*:

$$H_{p,q}^{m,n}\left[\lambda z\left|\begin{array}{l}(a_i,\alpha_i)_{1,p}\\(b_j,\beta_j)_{1,q}\end{array}\right.\right]$$

$$=\lambda^{b_1/\beta_1}\sum_{k=0}^{\infty}\frac{\left(1-\lambda^{1/\beta_1}\right)^k}{k!}H_{p,q}^{m,n}\left[z\left|\begin{array}{l}(a_i,\alpha_i)_{1,p}\\(b_1+k,\beta_1),(b_j,\beta_j)_{2,q}\end{array}\right.\right], \tag{2.3.12}$$

where $m>0$, while $\left|\lambda^{1/\beta_1}-1\right|<1$ for $m>1$;

$$H_{p,q}^{m,n}\left[\lambda z\left|\begin{array}{l}(a_i,\alpha_i)_{1,p}\\(b_j,\beta_j)_{1,q}\end{array}\right.\right]$$

$$=\lambda^{b_q/\beta_q}\sum_{k=0}^{\infty}\frac{\left(\lambda^{1/\beta_q}-1\right)^k}{k!}H_{p,q}^{m,n}\left[z\left|\begin{array}{l}(a_i,\alpha_i)_{1,p}\\(b_j,\beta_j)_{1,q-1},(b_q+k,\beta_q)\end{array}\right.\right], \tag{2.3.13}$$

where $q > m$ and $\left|\lambda^{1/\beta_q} - 1\right| < 1$;

$$H_{p,q}^{m,n}\left[\lambda z \,\middle|\, \begin{matrix}(a_i,\alpha_i)_{1,p}\\(b_j,\beta_j)_{1,q}\end{matrix}\right]$$

$$= \lambda^{(a_1-1)/\alpha_1}\sum_{k=0}^{\infty}\frac{\left(1-\lambda^{-1/\alpha_1}\right)^k}{k!}H_{p,q}^{m,n}\left[z \,\middle|\, \begin{matrix}(a_1-k,\alpha_1),(a_i,\alpha_i)_{2,p}\\(b_j,\beta_j)_{1,q}\end{matrix}\right], \quad (2.3.14)$$

where $n > 0$ and $\mathrm{Re}\left(\lambda^{1/\alpha_1}\right) > 1/2$;

$$H_{p,q}^{m,n}\left[\lambda z \,\middle|\, \begin{matrix}(a_i,\alpha_i)_{1,p}\\(b_j,\beta_j)_{1,q}\end{matrix}\right]$$

$$= \lambda^{(a_p-1)/\alpha_p}\sum_{k=0}^{\infty}\frac{\left(\lambda^{-1/\alpha_p}-1\right)^k}{k!}H_{p,q}^{m,n}\left[z \,\middle|\, \begin{matrix}(a_i,\alpha_i)_{1,p-1},(a_p-k,\alpha_p)\\(b_j,\beta_j)_{1,q}\end{matrix}\right], \quad (2.3.15)$$

where $p > n$ and $\mathrm{Re}\left(\lambda^{1/\alpha_p}\right) > 1/2$.

Proof. By Theorem 2.2 the function $z^{-b_1}H_{p,q}^{m,n}\left[z^{\beta_1} \,\middle|\, \begin{matrix}(a_i,\alpha_i)_{1,p}\\(b_j,\beta_j)_{1,q}\end{matrix}\right]$ is analytic for $z \in \mathbb{C}$ $(z \neq 0)$. Therefore for $|\eta| < |z|$ there holds the Taylor formula

$$(z+\eta)^{-b_1}H_{p,q}^{m,n}\left[(z+\eta)^{\beta_1} \,\middle|\, \begin{matrix}(a_i,\alpha_i)_{1,p}\\(b_j,\beta_j)_{1,q}\end{matrix}\right]$$

$$= \sum_{k=0}^{\infty}\frac{\eta^k}{k!}\left(\frac{d}{dz}\right)^k\left\{(z+\eta)^{-b_1}H_{p,q}^{m,n}\left[(z+\eta)^{\beta_1} \,\middle|\, \begin{matrix}(a_i,\alpha_i)_{1,p}\\(b_j,\beta_j)_{1,q}\end{matrix}\right]\right\}\Bigg|_{\eta=0}.$$

Applying (2.2.7) with $\sigma = \beta_1$, we have

$$(z+\eta)^{-b_1}H_{p,q}^{m,n}\left[(z+\eta)^{\beta_1} \,\middle|\, \begin{matrix}(a_i,\alpha_i)_{1,p}\\(b_j,\beta_j)_{1,q}\end{matrix}\right]$$

$$= \sum_{k=0}^{\infty}\frac{(-\eta)^k}{k!}\left\{z^{-b_1-k}H_{p,q}^{m,n}\left[z^{\beta_1} \,\middle|\, \begin{matrix}(a_i,\alpha_i)_{1,p}\\(b_1+k,\beta_1),(b_j,\beta_j)_{2,q}\end{matrix}\right]\right\}.$$

Setting $\eta = z\left(1-\lambda^{1/\beta_1}\right)$ for $|1-\lambda^{1/\beta_1}| < 1$, we obtain

$$H_{p,q}^{m,n}\left[\lambda z^{\beta_1} \,\middle|\, \begin{matrix}(a_i,\alpha_i)_{1,p}\\(b_j,\beta_j)_{1,q}\end{matrix}\right] = \lambda^{b_1/\beta_1}\sum_{k=0}^{\infty}\frac{(1-\lambda^{1/\beta_1})^k}{k!}H_{p,q}^{m,n}\left[z^{\beta_1} \,\middle|\, \begin{matrix}(a_i,\alpha_i)_{1,p}\\(b_1+k,\beta_1),(b_j,\beta_j)_{2,q}\end{matrix}\right].$$

By replacing z^{β_1} by z we arrive at (2.3.12).

The relations (2.3.13)–(2.3.15) are proved similarly.

2.4. Multiplication and Transformation Formulas

In this section we present the multiplication formulas for the H-function. The first relation is an analog of the general Gauss–Legendre multiplication formula for the gamma function (see Erdélyi, Magnus, Oberhettinger and Tricomi [1, (1.2.11)])

$$\prod_{k=0}^{m-1}\Gamma\left(z+\frac{k}{m}\right)=(2\pi)^{(m-1)/2}m^{1/2-mz}\Gamma(mz)\quad(m=2,3,4,\cdots). \tag{2.4.1}$$

Let

$$c^*=m+n-\frac{p+q}{2}. \tag{2.4.2}$$

For $a_i, b_j \in \mathbb{C}$ and $\alpha_i, \beta_j > 0$ $(i=1,\cdots,p;\ j=1,\cdots,q)$ we shall use the symbols $\left(\Delta(k,a_i),\alpha_i\right)_{1,p}$ and $\left(\Delta(k,b_j),\beta_j\right)_{1,q}$ to abbreviate the parameter sequences

$$\left(\frac{a_i}{k},\alpha_i\right)_{1,p},\ \left(\frac{a_i+1}{k},\alpha_i\right)_{1,p},\cdots,\left(\frac{a_i+k-1}{k},\alpha_i\right)_{1,p} \tag{2.4.3}$$

and

$$\left(\frac{b_j}{k},\beta_j\right)_{1,q},\ \left(\frac{b_j+1}{k},\beta_j\right)_{1,q},\cdots,\left(\frac{b_j+k-1}{k},\beta_j\right)_{1,q}, \tag{2.4.4}$$

respectively. The following multiplication relation for the H-function follows directly from the definition (1.1.1)–(1.1.2):

Property 2.14.

$$H^{m,n}_{p,q}\left[z\,\middle|\,\begin{matrix}(a_i,\alpha_i)_{1,p}\\(b_j,\beta_j)_{1,q}\end{matrix}\right]=(2\pi)^{(1-k)c^*}k^{\mu+1}H^{km,kn}_{kp,kq}\left[\left(zk^{-\Delta}\right)^k\,\middle|\,\begin{matrix}\left(\Delta(k,a_i),\alpha_i\right)_{1,p}\\\left(\Delta(k,b_j),\beta_j\right)_{1,q}\end{matrix}\right], \tag{2.4.5}$$

where $k\in\mathbb{N}$, and Δ, c^* and μ are given in (1.1.15), (1.1.8) and (1.1.10).

To give another multiplication formula we introduce some notation. Let $n,m,r,s\in\mathbb{N}$, $a_i,b_j,c_k,d_l\in\mathbb{C}$ and $\alpha_i,\beta_j,\gamma_k,\delta_l\in\mathbb{R}_+$ and $N_i,M_j,R_k,S_l\in\mathbb{N}$ $(i=1,\cdots,n;\ j=1,\cdots,m;\ k=1,\cdots,r;\ l=1,\cdots,s)$. Let

$$N=\sum_{i=1}^{n}N_i,\ M=\sum_{j=1}^{m}M_j,\ R=\sum_{k=1}^{r}R_k,\ S=\sum_{l=1}^{s}S_l, \tag{2.4.6}$$

$$\alpha=\prod_{i=1}^{n}(N_i)^{\alpha_i},\ \beta=\prod_{j=1}^{m}(M_j)^{\beta_j},\ \gamma=\prod_{k=1}^{r}(R_k)^{\gamma_k},\ \delta=\prod_{l=1}^{s}(S_l)^{\delta_l}. \tag{2.4.7}$$

The symbols $(\Delta(N_i,a_i),\alpha_i/N_i)_{1,n}$ and $(\Delta(M_j,b_j),\beta_j/M_j)_{1,m}$ abbreviate the parameter sequences

$$\left(\frac{a_i}{N_i},\frac{\alpha_i}{N_i}\right)_{1,n},\ \left(\frac{a_i+1}{N_i},\frac{\alpha_i}{N_i}\right)_{1,n},\cdots,\left(\frac{a_i+N_i-1}{N_i},\frac{\alpha_i}{N_i}\right)_{1,n} \tag{2.4.8}$$

and

$$\left(\frac{b_j}{M_j}, \frac{\beta_j}{M_j}\right)_{1,m}, \left(\frac{b_j+1}{M_j}, \frac{\beta_j}{M_j}\right)_{1,m}, \cdots, \left(\frac{b_j+M_j-1}{M_j}, \frac{\beta_j}{M_j}\right)_{1,m}, \tag{2.4.9}$$

respectively, and similarly for $(\Delta(R_k, c_k), \gamma_k/R_k)_{1,r}$ and $(\Delta(S_l, d_l), \delta_l/S_l)_{1,s}$. Then there holds another multiplication relation:

Property 2.15.

$$H^{m,n}_{n+r,m+s}\left[z \,\middle|\, \begin{matrix}(a_i,\alpha_i)_{1,n}, (c_k,\gamma_k)_{1,r}\\ (b_j,\beta_j)_{1,m}, (d_l,\delta_l)_{1,s}\end{matrix}\right]$$

$$= (2\pi)^{(m+n-r-s-M-N+R+S)/2}\prod_{i=1}^{n}(N_i)^{-a_i+1/2}\prod_{j=1}^{m}(M_j)^{b_j-1/2}$$

$$\cdot \prod_{k=1}^{r}(R_k)^{-c_k+1/2}\prod_{l=1}^{s}(S_l)^{d_l-1/2}$$

$$\cdot H^{M,N}_{N+R,M+S}\left[\frac{\alpha\delta}{\beta\gamma}z \,\middle|\, \begin{matrix}\left(\Delta(N_i,a_i), \frac{\alpha_i}{N_i}\right)_{1,n}, \left(\Delta(R_k,c_k), \frac{\gamma_k}{R_k}\right)_{1,r}\\ \left(\Delta(M_j,b_j), \frac{\beta_j}{M_j}\right)_{1,m}, \left(\Delta(S_l,d_l), \frac{\delta_l}{S_l}\right)_{1,s}\end{matrix}\right]. \tag{2.4.10}$$

The next relation is a certain transformation of infinite series involving the H-function.

Property 2.16.

$$\sum_{k=0}^{\infty}\frac{(a)_k}{k!}x^k H^{m+1,n}_{p+1,q+1}\left[z \,\middle|\, \begin{matrix}(a_i,\alpha_i)_{1,p}, (c+k,\gamma)\\ (b+k,\gamma), (b_j,\beta_j)_{1,q}\end{matrix}\right]$$

$$= \frac{\Gamma(c-a-b)}{\Gamma(c-b)}\sum_{k=0}^{\infty}\frac{(a)_k}{(a+b-c+1)_k k!}(1-x)^k$$

$$\cdot H^{m+1,n}_{p+1,q+1}\left[z \,\middle|\, \begin{matrix}(a_i,\alpha_i)_{1,p}, (c-a,\gamma)\\ (b+k,\gamma), (b_j,\beta_j)_{1,q}\end{matrix}\right]$$

$$+\frac{\Gamma(a+b-c)}{\Gamma(a)}\sum_{k=0}^{\infty}\frac{(c-b)_k}{(c-a-b+1)_k k!}(1-x)^{c-a-b+k}$$

$$\cdot H^{m+1,n}_{p+1,q+1}\left[z \,\middle|\, \begin{matrix}(a_i,\alpha_i)_{1,p}, (c-a,\gamma)\\ (c-a+k,\gamma), (b_j,\beta_j)_{1,q}\end{matrix}\right], \tag{2.4.11}$$

where $a, b, c \in \mathbb{C}$, $\gamma > 0$, $|\arg(1-x)| < \pi$, and $\mathrm{Re}(c-a-b) > 0$ if $x = 1$. Here $(a)_k$ ($a \in \mathbb{C}$; $k = 0, 1, 2, \cdots$) is the Pochhammer symbol (see Erdélyi, Magnus, Oberhettinger and Tricomi [1, Section 2.1.1]) defined by

$$(a)_0 = 1, \quad (a)_k = a(a+1)\cdots(a+k-1) = \frac{\Gamma(a+k)}{\Gamma(a)} \quad (k = 1, 2, \cdots). \tag{2.4.12}$$

The formula (2.4.11) may be derived by using the transformation formula:

$$ {}_2F_1(a,b;c;x) = \frac{\Gamma(c)\Gamma(c-a-b)}{\Gamma(c-a)\Gamma(c-b)}\, {}_2F_1(a,b;a+b-c+1;1-x) $$

$$ + \frac{\Gamma(c)\Gamma(a+b-c)}{\Gamma(a)\Gamma(b)}(1-x)^{c-a-b}\, {}_2F_1(c-a,c-b;c-a-b+1;1-x) \quad (2.4.13) $$

$$ (|\arg(1-x)| < \pi) $$

for the Gauss hypergeometric function (2.9.2).

2.5. Mellin and Laplace Transforms of the H-Function

The Mellin and Laplace transforms of a function $f(x)$ $(x > 0)$ are defined by

$$ \left(\mathfrak{M}f\right)(s) = \int_0^\infty f(x)x^{s-1}dx \quad (s \in \mathbb{C}) \tag{2.5.1} $$

and

$$ \left(\mathbb{L}f\right)(t) = \int_0^\infty f(x)e^{-tx}dx \quad (t \in \mathbb{C}), \tag{2.5.2} $$

respectively. The theory of these transforms may be found in the books by Doetsch [1], [2], Ditkin and Prudnikov [1], Sneddon [1], Titchmarsh [3] and Widder [1]. In particular, the Mellin inversion formula is given by

$$ f(x) = \frac{1}{2\pi i}\int_{\gamma-i\infty}^{\gamma+i\infty}\left(\mathfrak{M}f\right)(s)x^{-s}ds \equiv \left(\mathfrak{M}^{-1}f\right)(x), \tag{2.5.3} $$

while the Laplace inversion has the form

$$ f(x) = \frac{1}{2\pi i}\int_{\gamma-i\infty}^{\gamma+i\infty}\left(\mathbb{L}f\right)(t)e^{xt}dt \equiv \left(\mathbb{L}^{-1}f\right)(x), \tag{2.5.4} $$

where $\gamma \in \mathbb{R}$ is specially chosen. We shall consider the H-function provided that the conditions in (1.1.6) are satisfied. The first result for the Mellin transform follows from Theorem 1.1 and the Mellin inversion theorem (see, for example, Titchmarsh [3, Section 1.5]).

Theorem 2.2. *Let $a^* \geqq 0$ and $s \in \mathbb{C}$ be such that*

$$ -\min_{1\leqq j\leqq m}\left[\frac{\mathrm{Re}(b_j)}{\beta_j}\right] < \mathrm{Re}(s) < \min_{1\leqq i\leqq n}\left[\frac{1-\mathrm{Re}(a_i)}{\alpha_i}\right], \tag{2.5.5} $$

when $a^* > 0$, and, additionally,

$$ \Delta\mathrm{Re}(s) + \mathrm{Re}(\mu) < -1, \tag{2.5.6} $$

when $a^* = 0$. Then the Mellin transform of the H-function exists and the relation

$$ \left(\mathfrak{M}H_{p,q}^{m,n}\left[x \,\middle|\, \begin{matrix}(a_i,\alpha_i)_{1,p}\\(b_j,\beta_j)_{1,q}\end{matrix}\right]\right)(s) = \mathcal{H}_{p,q}^{m,n}\left[\begin{matrix}(a_i,\alpha_i)_{1,p}\\(b_j,\beta_j)_{1,q}\end{matrix} \,\middle|\, s\right] \tag{2.5.7} $$

holds, where $\mathcal{H}_{p,q}^{m,n}(s)$ *is given in* (1.1.2).

Proof. By (1.1.4) and (1.1.5) for the poles b_{jl} $(1 \leqq j \leqq m;\ l = 0, 1, 2, \cdots)$ and a_{ik} $(1 \leqq i \leqq n;\ k = 0, 1, 2, \cdots)$ of the gamma functions $\Gamma(b_j + \beta_j s)$ $(1 \leqq j \leqq m)$ and $\Gamma(1 - a_i - \alpha_i s)$ $(1 \leqq i \leqq n)$, there hold the following estimates:

$$\operatorname{Re}(b_{jl}) \leqq \operatorname{Re}(b_{j0}) \leqq -\min_{1 \leqq j \leqq m} \left[\frac{\operatorname{Re}(b_j)}{\beta_j}\right] \quad (1 \leqq j \leqq m;\ l = 0, 1, 2, \cdots) \tag{2.5.8}$$

and

$$\operatorname{Re}(a_{ik}) \geqq \operatorname{Re}(a_{i0}) \geqq \min_{1 \leqq i \leqq n} \left[\frac{1 - \operatorname{Re}(a_i)}{\alpha_i}\right] \quad (1 \leqq i \leqq n;\ k = 0, 1, 2, \cdots). \tag{2.5.9}$$

In view of Section 1.1 we can choose the contour $\mathfrak{L} = \mathfrak{L}_{i\gamma\infty}$ with $\gamma = \operatorname{Re}(s)$ for the H-function in the left side of (2.5.7), because it follows from (2.5.8) and (2.5.9) that all poles b_{jl} $(1 \leqq j \leqq m;\ l = 0, 1, 2, \cdots)$ lie in the left of the contour $\mathfrak{L}_{i\gamma\infty}$ while all poles a_{ik} $(1 \leqq i \leqq n;\ k = 0, 1, 2, \cdots)$ lie in the right of $\mathfrak{L}_{i\gamma\infty}$. By Theorem 1.1 such an H-function exists. Hence (1.1.1) and (1.1.2) give the result in (2.5.7) according to the Mellin inversion theorem.

Corollary 2.2.1. *Let* $a^* \geqq 0$, $a \in \mathbb{C}$ *and* $\sigma \in \mathbb{R}$ $(\sigma \neq 0)$. *Let us assume that* $s, w \in \mathbb{C}$ *satisfy*

$$-\sigma \min_{1 \leqq j \leqq m} \left[\frac{\operatorname{Re}(b_j)}{\beta_j}\right] < \operatorname{Re}(s + w) < \sigma \min_{1 \leqq i \leqq n} \left[\frac{1 - \operatorname{Re}(a_i)}{\alpha_i}\right], \tag{2.5.10}$$

when $a^* > 0$, $|\arg z| < a^*\pi/2$, $z \neq 0$, *and additionally,*

$$\frac{\Delta}{\sigma}\operatorname{Re}(s + w) + \operatorname{Re}(\mu) < -1, \tag{2.5.11}$$

when $a^* = 0$, $\arg z = 0$ *and* $z \neq 0$. *Then the following relations hold:*

$$\left(\mathfrak{M} x^w H_{p,q}^{m,n}\left[ax^\sigma \,\middle|\, \begin{matrix}(a_i, \alpha_i)_{1,p} \\ (b_j, \beta_j)_{1,q}\end{matrix}\right]\right)(s)$$

$$= \frac{a^{-(s+w)/\sigma}}{\sigma} \mathcal{H}_{p,q}^{m,n}\left[\begin{matrix}(a_i, \alpha_i)_{1,p} \\ (b_j, \beta_j)_{1,q}\end{matrix} \,\middle|\, \frac{s + w}{\sigma}\right] \quad (\sigma > 0); \tag{2.5.12}$$

$$\left(\mathfrak{M} x^w H_{p,q}^{m,n}\left[ax^\sigma \,\middle|\, \begin{matrix}(a_i, \alpha_i)_{1,p} \\ (b_j, \beta_j)_{1,q}\end{matrix}\right]\right)(s)$$

$$= \frac{a^{-(s+w)/|\sigma|}}{|\sigma|} \mathcal{H}_{q,p}^{n,m}\left[\begin{matrix}(1 - b_j, \beta_j)_{1,q} \\ (1 - a_i, \alpha_i)_{1,p}\end{matrix} \,\middle|\, \frac{s + w}{|\sigma|}\right] \quad (\sigma < 0). \tag{2.5.13}$$

Proof. When $\sigma > 0$, the relation (2.5.12) follows from (2.5.7). For $\sigma < 0$, (2.5.13) is deduced by using (1.8.15) and (2.5.12). The general case $a \in \mathbb{C}$ follows by analytic continuation by virtue of Theorem 1.1.

The next statement presents the Laplace transform of the H-function.

Theorem 2.3. *Let either $a^* > 0$ or $a^* = 0$, $\mathrm{Re}(\mu) < -1$, and assume*

$$\min_{1\leqq j\leqq m}\left[\frac{\mathrm{Re}(b_j)}{\beta_j}\right] > -1, \tag{2.5.14}$$

when $a^* > 0$, or $a^* = 0$, $\Delta \geqq 0$, and

$$\min_{1\leqq j\leqq m}\left[\frac{\mathrm{Re}(b_j)}{\beta_j}, \frac{\mathrm{Re}(\mu)+1/2}{\Delta}\right] > -1, \tag{2.5.15}$$

when $a^* = 0$, $\Delta < 0$. Then, the Laplace transform of the H-function exists and the relation

$$\left(\mathbb{L}H^{m,n}_{p,q}\left[x\,\middle|\,\begin{matrix}(a_i,\alpha_i)_{1,p}\\(b_j,\beta_j)_{1,q}\end{matrix}\right]\right)(t) = \frac{1}{t}H^{m,n+1}_{p+1,q}\left[\frac{1}{t}\,\middle|\,\begin{matrix}(0,1),(a_i,\alpha_i)_{1,p}\\(b_j,\beta_j)_{1,q}\end{matrix}\right] \tag{2.5.16}$$

holds for $t \in \mathbb{C}$ ($\mathrm{Re}(t) > 0$).

Proof. First we indicate that, if we denote by a_0^* the constant (1.1.7) for the H-function in the right side of (2.5.16), then $a_0^* = a^* + 1 \geqq 1$ if $a^* \geqq 0$. By (1.1.1) and (1.1.2) for the H-functions in the left and right sides of (2.5.16) we take the contour $\mathfrak{L} = \mathfrak{L}_{i\gamma\infty}$, where the choice of γ will depend on the relations between the constant A, B and M defined by

$$A = \min_{1\leqq i\leqq n}\left[\frac{1-\mathrm{Re}(a_i)}{\alpha_i}\right], \quad B = -\min_{1\leqq j\leqq m}\left[\frac{\mathrm{Re}(b_j)}{\beta_j}\right], \quad M = -\frac{\mathrm{Re}(\mu)+1}{\Delta}, \tag{2.5.17}$$

for which in accordance with (1.1.1), (1.1.2), (2.5.8), (2.5.9) and (2.5.15) there hold the relations

$$\mathrm{Re}(b_{jl}) \leqq B < 1, \quad \mathrm{Re}(a_{ik}) \geqq A, \quad c_h \geqq 1 \tag{2.5.18}$$

for the poles b_{jl} ($1 \leqq j \leqq m$; $l = 0,1,2,\cdots$) of the gamma functions $\Gamma(b_j+\beta_j s)$ ($1 \leqq j \leqq m$), for the poles a_{ik} ($1 \leqq i \leqq n$; $k = 0,1,2,\cdots$) of $\Gamma(1-a_i-\alpha_i s)$ ($1 \leqq i \leqq n$) and for the poles $c_h = h+1$ ($h = 0,1,2,\cdots$) of $\Gamma(1-s)$.

If $a^* > 0$, we choose γ by the relations

$$\begin{aligned} &B < \gamma < \min[A,1] && (B < A);\\ &B < \gamma < 1 && (B = A);\\ &A < \gamma < B && (B > A). \end{aligned} \tag{2.5.19}$$

When $B < A$, then (2.5.18) implies all poles b_{jl} ($1 \leqq j \leqq m$; $l = 0,1,2,\cdots$) lie in the left of the contour $\mathfrak{L}_{i\gamma\infty}$, while all poles a_{ik} ($1 \leqq i \leqq n$; $k = 0,1,2,\cdots$) and all poles c_h ($h = 0,1,2,\cdots$) lie in the right of $\mathfrak{L}_{i\gamma\infty}$. When $B = A$, only a finite number of points $s = a_{ik}$ may lie in the strip $B = A < \mathrm{Re}(s) < 1$ and therefore we can choose $\mathfrak{L}_{i\gamma\infty}$ with $B < \gamma < 1$ such that all b_{jl} lie in the left of $\mathfrak{L}_{i\gamma\infty}$, while all a_{ik} and c_h lie in the right of $\mathfrak{L}_{i\gamma\infty}$. For example, if the points a_{ik} ($1 \leqq i \leqq n$; $0 \leqq k \leqq N_i$) lie in the strip $B = A < \mathrm{Re}(s) < 1$, putting

$$c = \min_{1\leqq i\leqq n, 0\leqq k\leqq N_i}[\mathrm{Im}(a_{ik})], \quad d = \max_{1\leqq i\leqq n, 0\leqq k\leqq N_i}[\mathrm{Im}(a_{ik})],$$

we consider the rectangle

$$\Pi = \{(x,y):\ B < x < 1,\ c-1 < y < d+1\}$$

and, then, we may choose $\mathfrak{L}_{i\gamma\infty}$ as a sum of the half-lines $L_1 = \{\gamma + it,\ -\infty < t \leqq c\}$ and $L_2 = \{\gamma + it,\ d \leqq t < +\infty\}$ and the curve L_3, lying in Π, with the beginning at $\gamma + ic$ and the end at $\gamma + id$ such that all points a_{ik} $(1 \leqq i \leqq n;\ 0 \leqq k \leqq N_i)$ lie in the right of L_3. In the case $B > A$ we can similarly choose $\mathfrak{L}_{i\gamma\infty}$ with $A < \gamma < B$ because only a finite number of points $s = b_{jl}$ and $s = a_{ik}$ may lie in the strip $A < \mathrm{Re}(s) < B$. Then by Theorem 1.1 the H-functions in the left and the right sides of (2.5.16) exist.

We treat the case $a^* = 0$. When $\Delta = 0$, γ can be chosen as above in (2.5.19). We note that since $\mathrm{Re}(\mu) < -1$, $M > 0$ when $\Delta > 0$, and $M < 0$ when $\Delta < 0$. Then if $a^* = 0$ and $\Delta > 0$, we choose γ by

$$\begin{array}{ll} \gamma < M & (M \leqq B,\ M \leqq A); \\ A < \gamma < M & (M \leqq B,\ M > A); \\ B < \gamma < \min[M,1] & (M > B,\ M \leqq A); \\ B < \gamma < \min[A,1] & (M > B,\ M > A,\ B < A); \\ B < \gamma < \min[M,1] & (M > B,\ M > A,\ B = A); \\ A < \gamma < B & (M > B,\ M > A,\ B > A). \end{array} \tag{2.5.20}$$

If $a^* = 0$ and $\Delta < 0$, γ is chosen as

$$\begin{array}{ll} M < \gamma & (M \geqq B,\ M \geqq A); \\ M < \gamma < \min[A,1] & (M \geqq B,\ M < A); \\ M < \gamma < B & (M < B,\ M \geqq A) \end{array} \tag{2.5.21}$$

and as in (2.5.19) for $M < B$, $M < A$. Similar arguments yield that, for the case $a^* = 0$, only a finite number of points $s = b_{jl}$ and $s = a_{ik}$ may lie in the corresponding strips and we can choose $\mathfrak{L}_{i\gamma\infty}$ with γ in these strips. According to (2.5.18), (2.5.20), (2.5.21) we have

$$\gamma < M \quad (\Delta > 0), \qquad \gamma > M \quad (\Delta < 0), \tag{2.5.22}$$

which is equivalent to $\Delta\gamma + \mathrm{Re}(\mu) < -1$. So in all cases above $a^* = 0$ and $\Delta\gamma + \mathrm{Re}(\mu) < -1$ (with different γ in the strips) and hence by Theorem 1.1 the H-functions in the left and the right sides of (2.5.16) exist.

Due to Corollaries 1.11.1, 1.12.1 and 1.13.1 and Remark 1.5, the H-function $H_{p,q}^{m,n}(x)$ has the asymptotic behavior near zero of the form (1.8.8), (1.8.14) or (1.9.14), where

$$\rho^* = \min_{1 \leqq j \leqq m}\left[\mathrm{Re}\left(\frac{b_j}{\beta_j}\right)\right]$$

in the cases $\Delta \geqq 0$ or $\Delta < 0, a^* > 0$, and ρ^* is given by (1.9.14) for $\Delta < 0$, $a^* = 0$. Therefore the integral in the left side of (2.5.16) exists.

Now the relation (2.5.16) is proved directly by using (2.5.2), (1.1.1), (1.1.2) and changing the order of integration:

$$\left(\mathbb{L}H_{p,q}^{m,n}\left[x \,\middle|\, \begin{matrix} (a_i,\alpha_i)_{1,p} \\ (b_j,\beta_j)_{1,q} \end{matrix}\right]\right)(t)$$

$$= \frac{1}{2\pi i} \int_{\mathfrak{L}_{i\gamma\infty}} \mathcal{H}_{p,q}^{m,n} \left[\begin{array}{c|} (a_i, \alpha_i)_{1,p} \\ (b_j, \beta_j)_{1,q} \end{array} \, s \right] ds \int_0^\infty x^{-s} e^{-tx} dx$$

$$= \frac{1}{t} \frac{1}{2\pi i} \int_{\mathfrak{L}_{i\gamma\infty}} \mathcal{H}_{p,q}^{m,n} \left[\begin{array}{c|} (a_i, \alpha_i)_{1,p} \\ (b_j, \beta_j)_{1,q} \end{array} \, s \right] \Gamma(1-s) \left(\frac{1}{t}\right)^{-s} ds,$$

which gives (2.5.16). This completes the proof of Theorem 2.3.

Corollary 2.3.1. *Let either $a^* > 0$ or $a^* = 0$, and $\mathrm{Re}(\mu) < -1$. Let us assume $\omega \in \mathbb{C}$, $a > 0$ and $\sigma > 0$ are such that*

$$\sigma \min_{1 \leqq j \leqq m} \left[\frac{\mathrm{Re}(b_j)}{\beta_j} \right] + \mathrm{Re}(\omega) > -1, \tag{2.5.23}$$

when $a^* > 0$, or $a^* = 0$, $\Delta \geqq 0$, and

$$\sigma \min_{1 \leqq j \leqq m} \left[\frac{\mathrm{Re}(b_j)}{\beta_j}, \frac{\mathrm{Re}(\mu) + 1/2}{\Delta} \right] + \mathrm{Re}(\omega) > -1, \tag{2.5.24}$$

when $a^* = 0$, $\Delta < 0$. Then the relation

$$\left(\mathbb{L} x^\omega H_{p,q}^{m,n} \left[a x^\sigma \,\middle|\, \begin{array}{c} (a_i, \alpha_i)_{1,p} \\ (b_j, \beta_j)_{1,q} \end{array} \right] \right)(t) = \frac{1}{t^{\omega+1}} H_{p+1,q}^{m,n+1} \left[\frac{a}{t^\sigma} \,\middle|\, \begin{array}{c} (-\omega, \sigma), (a_i, \alpha_i)_{1,p} \\ (b_j, \beta_j)_{1,q} \end{array} \right] \tag{2.5.25}$$

holds for $t \in \mathbb{C}$ $(\mathrm{Re}(t) > 0)$.

Remark 2.2. The relations (2.5.7) and (2.5.16) were indicated by Srivastava, Gupta and Goyal [1, (2.4.1) and (2.4.2)] provided that $\Delta > 0$, $a^* > 0$, and the formulas of the form (2.5.12) and (2.5.25) were listed in Prudnikov, Brychkov and Marichev [3, (2.25.2.1) and (2.25.2.3)] provided $a^* > 0$. The asymptotic estimates of $H_{p,q}^{m,n}(z)$ at zero in Sections 1.5 and 1.6 allow us to extend (in Theorems 2.2 and 2.3) these formulas to the case $a^* = 0$.

In what follows the so-called generalized Laplace transform will be treated, which is defined by

$$\left(\mathbb{L}_{k,\alpha} f \right)(t) = \int_0^\infty (xt)^{-\alpha} e^{-|k|(xt)^{1/k}} f(x) dx \quad (t > 0) \tag{2.5.26}$$

for a function $f(x)$ $(x > 0)$ and for $k, \alpha \in \mathbb{R}$ $(k \neq 0)$. For such a transform a similar result to Theorem 2.3 may be derived.

Theorem 2.4. *Let either $a^* > 0$ or $a^* = 0$, and $\mathrm{Re}(\mu) < -1$. Let $k, \alpha \in \mathbb{R}$ $(k \neq 0)$ be such that*

$$k \min_{1 \leqq j \leqq m} \left[\frac{\mathrm{Re}(b_j)}{\beta_j} \right] > k(\alpha - 1), \tag{2.5.27}$$

when $a^* > 0$ or $a^* = 0$, $\Delta \geqq 0$, while

$$k \min_{1 \leqq j \leqq m} \left[\frac{\mathrm{Re}(b_j)}{\beta_j}, -\frac{\mathrm{Re}(\mu) + 1/2}{\Delta} \right] > k(\alpha - 1), \tag{2.5.28}$$

when $a^* = 0$, $\Delta < 0$. Then the generalized Laplace transform of the H-function exists and the following relations hold for $t > 0$:

$$\left(\mathbb{L}_{k,\alpha} H_{p,q}^{m,n}\left[x \,\middle|\, \begin{matrix} (a_i,\alpha_i)_{1,p} \\ (b_j,\beta_j)_{1,q} \end{matrix}\right]\right)(t)$$

$$= \frac{k^{k(\alpha-1)+1}}{t} H_{p+1,q}^{m,n+1}\left[\frac{k^{-k}}{t} \,\middle|\, \begin{matrix} (k(\alpha-1)+1,k),(a_i,\alpha_i)_{1,p} \\ (b_j,\beta_j)_{1,q} \end{matrix}\right], \tag{2.5.29}$$

if $k > 0$ and

$$\left(\mathbb{L}_{k,\alpha} H_{p,q}^{m,n}\left[x \,\middle|\, \begin{matrix} (a_i,\alpha_i)_{1,p} \\ (b_j,\beta_j)_{1,q} \end{matrix}\right]\right)(t)$$

$$= \frac{|k|^{k(\alpha-1)+1}}{t} H_{p,q+1}^{m+1,n}\left[\frac{|k|^{-k}}{t} \,\middle|\, \begin{matrix} (a_i,\alpha_i)_{1,p} \\ (|k|(\alpha-1),|k|),(b_j,\beta_j)_{1,q} \end{matrix}\right], \tag{2.5.30}$$

if $k < 0$.

Proof. Theorem 2.4 is proved similarly to Theorem 2.3 by using (1.1.1)–(1.1.2), Theorem 1.1 and the asymptotic estimates near zero given in Sections 1.8 and 1.9.

2.6. Hankel Transforms of the *H*-Function

The Hankel transform of a function $f(x)$ $(x > 0)$ is defined by

$$\left(\mathbb{H}_\eta f\right)(x) = \int_0^\infty (xt)^{1/2} J_\eta(xt) f(t)\,dt, \tag{2.6.1}$$

where $J_\eta(z)$ is the Bessel function of the first kind of order $\eta \in \mathbb{C}$ $(\mathrm{Re}(\eta) > -1)$ (see Erdélyi, Magnus, Oberhettinger and Tricomi [2, 7.2(2)]) defined by

$$J_\eta(z) = \sum_{k=0}^\infty \frac{(-1)^k}{\Gamma(\eta+k+1)k!}\left(\frac{z}{2}\right)^{2k+\eta}. \tag{2.6.2}$$

One may find the theory of this transform in the books by Ditkin and Prudnikov [1] and Sneddon [1]. In this section we shall consider the H-function, provided that the conditions in (1.1.6) are fulfilled.

Using the asymptotic estimates of the H-function at zero and infinity given in Theorems 1.12 and 1.7 as well as the asymptotic estimates of $J_\eta(z)$ (see Erdélyi, Magnus, Oberhettinger and Tricomi [2, 7.13(3)])

$$J_\eta(z) = O(z^\eta) \quad (|z| \to 0), \qquad J_\eta(z) = O\left(z^{-1/2}\right) \quad (|z| \to \infty,\ |\arg(z)| < \pi), \tag{2.6.3}$$

and the known formula in Prudnikov, Brychkov and Marichev [2, 2.12.2.2]

$$\int_0^\infty x^\lambda J_\eta(ax)\,dx = 2^\lambda a^{-\lambda-1} \frac{\Gamma\left(\dfrac{\lambda+1+\eta}{2}\right)}{\Gamma\left(1+\dfrac{\eta-\lambda-1}{2}\right)} \tag{2.6.4}$$

$$(a > 0,\ -\mathrm{Re}(\eta) - 1 < \mathrm{Re}(\lambda) < 1/2),$$

we obtain the following result:

Theorem 2.5. *Let us assume either* $a^* > 0$ *or* $a^* = \Delta = 0$, *and* $\mathrm{Re}(\mu) < -1$. *Let* $\eta \in \mathbb{C}$ *be such that*

$$\mathrm{Re}(\eta) > -\frac{1}{2}, \quad \mathrm{Re}(\eta) + \min_{1 \leqq j \leqq m}\left[\frac{\mathrm{Re}(b_j)}{\beta_j}\right] > -\frac{3}{2}, \quad \min_{1 \leqq i \leqq n}\left[\frac{1-\mathrm{Re}(a_i)}{\alpha_i}\right] > 1. \tag{2.6.5}$$

Then the Hankel transform of the H-function exists and the relation

$$\left(\mathbb{H}_\eta H^{m,n}_{p,q}\left[t \,\middle|\, \begin{matrix}(a_i,\alpha_i)_{1,p}\\(b_j,\beta_j)_{1,q}\end{matrix}\right]\right)(x)$$

$$= \frac{\sqrt{2}}{x} H^{m,n+1}_{p+2,q}\left[\frac{2}{x} \,\middle|\, \begin{matrix}\left(\frac{1}{4}-\frac{\eta}{2},\frac{1}{2}\right),(a_i,\alpha_i)_{1,p},\left(\frac{1}{4}+\frac{\eta}{2},\frac{1}{2}\right)\\(b_j,\beta_j)_{1,q}\end{matrix}\right] \tag{2.6.6}$$

holds for $x > 0$.

Proof. By (2.6.3), (1.8.8) and (1.5.13) there hold the following asymptotic estimates near zero and infinity:

$$(xt)^{1/2}J_\eta(xt)H^{m,n}_{p,q}(xt) = O\left(t^{\rho^*+\eta+1/2}\right) \quad (t \to +0), \quad \rho^* = \min_{1 \leqq j \leqq m}\left[\frac{\mathrm{Re}(b_j)}{\beta_j}\right]$$

and

$$(xt)^{1/2}J_\eta(xt)H^{m,n}_{p,q}(xt) = O\left(t^{\rho}\right) \quad (t \to +\infty), \quad \rho = -\min_{1 \leqq i \leqq n}\left[\frac{1-\mathrm{Re}(a_i)}{\alpha_i}\right]$$

with addition of the multipliers $[\log(x)]^{N^*}$ and $[\log(x)]^N$, respectively, in the cases when the conditions in (1.3.1) and (1.3.2) are not valid, on which one may consult Theorems 1.8 and 1.12. Therefore in accordance with (2.6.5) the integral in the left side of (2.6.6) is convergent.

For b_{jl} in (1.1.4) and a_{ik} in (1.1.5), the assumption (2.6.5) implies

$$\mathrm{Re}(b_{jl}) \leqq -\min_{1 \leqq j \leqq m}\left[\frac{\mathrm{Re}(b_j)}{\beta_j}\right] < \mathrm{Re}(\eta) + \frac{3}{2} \quad (1 \leqq j \leqq m;\ l = 0,1,2,\cdots)$$

and

$$\mathrm{Re}(a_{ik}) \geqq \min_{1 \leqq i \leqq n}\left[\frac{1-\mathrm{Re}(a_i)}{\alpha_i}\right] > 1 \quad (1 \leqq i \leqq n;\ k = 0,1,2,\cdots).$$

Since only a finite number of points b_{jl} and a_{ik} can lie in the strip $1 < \mathrm{Re}(s) < \mathrm{Re}(\eta) + 3/2$, we can choose the contour $\mathfrak{L}_{i\gamma\infty}$ with

$$1 < \gamma < \mathrm{Re}(\eta) + \frac{3}{2} \tag{2.6.7}$$

such that all poles b_{jl} of the gamma functions $\Gamma(b_j + \beta_j s)$ $(1 \leqq j \leqq m)$ lie in the left of the contour $\mathfrak{L}_{i\gamma\infty}$ while all poles a_{ik} of $\Gamma(1 - a_i - \alpha_i s)$ $(1 \leqq i \leqq n)$ lie in the right of $\mathfrak{L}_{i\gamma\infty}$. Since $\mathrm{Re}(\eta) + 3/2 > 0$, all poles $c_h = \eta + 2h + 3/2$ $(h = 0,1,2,\cdots)$ of $\Gamma(3/4 + \eta/2 - s/2)$ also lie in the right of $\mathfrak{L}_{i\gamma\infty}$. Hence for the H-functions in the left and right sides of (2.6.6) the contour

$\mathfrak{L} = \mathfrak{L}_{i\gamma\infty}$ of the integral (1.1.1) can be choosen as $0 < \mathrm{Re}(s) < \mathrm{Re}(\eta) + 3/2$. Further, if we denote by a_0^*, Δ_0 and μ_0 the constants (1.1.7), (1.1.8) and (1.1.10) for the H-function in the right side of (2.6.6), then it is directly verified that $a_0^* = a^*$, $\Delta_0 = \Delta - 1$ and $\mu_0 = \mu + 1/2$. Then in accordance with (1.2.20) and (1.2.21) the H-functions in the left and right side of (2.6.6) exist provided that $a^* > 0$, or $a^* = \Delta = 0$ and $\mathrm{Re}(\mu) < -1$.

Now the relation (2.6.6) is proved directly by using (2.6.1) and (1.1.1)–(1.1.2), and by applying the relation (2.6.4), which holds for $\mathrm{Re}(s) = \gamma$ under (2.6.7), such that

$$\left(\mathbb{H}_\eta H^{m,n}_{p,q}\left[t\,\middle|\,\begin{matrix}(a_i,\alpha_i)_{1,p}\\(b_j,\beta_j)_{1,q}\end{matrix}\right]\right)(x)$$

$$= \frac{1}{2\pi i}\int_{\mathfrak{L}_{i\gamma\infty}} \mathcal{H}^{m,n}_{p,q}\left[\begin{matrix}(a_i,\alpha_i)_{1,p}\\(b_j,\beta_j)_{1,q}\end{matrix}\,\middle|\, s\right] ds \int_0^\infty (xt)^{1/2} J_\eta(xt) t^{-s} dt$$

$$= \frac{1}{2\pi i}\int_{\mathfrak{L}_{i\gamma\infty}} \mathcal{H}^{m,n}_{p,q}\left[\begin{matrix}(a_i,\alpha_i)_{1,p}\\(b_j,\beta_j)_{1,q}\end{matrix}\,\middle|\, s\right] x^{s-1} 2^{-s+1/2} \frac{\Gamma\left(\frac{3}{4}+\frac{\eta}{2}-\frac{s}{2}\right)}{\Gamma\left(\frac{1}{4}+\frac{\eta}{2}+\frac{s}{2}\right)} ds,$$

which gives (2.6.6).

Corollary 2.5.1. *Let $a^* > 0$ or $a^* = \Delta = 0$ and $\mathrm{Re}(\mu) < -1$. Let $\eta,\ \omega \in \mathbb{C}$, $\tau > 0$ and $\sigma > 0$ be such that*

$$\sigma \mathrm{Re}(\eta) + \mathrm{Re}(\omega) + \tau \min_{1\leqq j\leqq m}\left[\frac{\mathrm{Re}(b_j)}{\beta_j}\right] > -1, \tag{2.6.8}$$

$$\tau \min_{1\leqq i\leqq n}\left[\frac{1-\mathrm{Re}(a_i)}{\alpha_i}\right] > \mathrm{Re}(\omega) - \frac{\sigma}{2} + 1 \tag{2.6.9}$$

and

$$\mathrm{Re}(\eta) > -\frac{1}{2}. \tag{2.6.10}$$

Then for $a > 0$, $b > 0$

$$\int_0^\infty (xt)^\omega J_\eta\left[a(xt)^\sigma\right] H^{m,n}_{p,q}\left[bt^\tau\,\middle|\,\begin{matrix}(a_i,\alpha_i)_{1,p}\\(b_j,\beta_j)_{1,q}\end{matrix}\right] dt$$

$$= \frac{1}{2\sigma x}\left(\frac{2}{a}\right)^{(\omega+1)/\sigma} H^{m,n+1}_{p+2,q}\left[b\left(\frac{2}{a}\right)^{\tau/\sigma}\frac{1}{x^\tau}\,\middle|\,\begin{matrix}\left(1-\frac{\omega+1}{2\sigma}-\frac{\eta}{2},\frac{\tau}{2\sigma}\right),(a_i,\alpha_i)_{1,p},\left(1-\frac{\omega+1}{2\sigma}+\frac{\eta}{2},\frac{\tau}{2\sigma}\right)\\(b_j,\beta_j)_{1,q}\end{matrix}\right] \quad (x>0). \tag{2.6.11}$$

Proof. Corollary 2.5.1 may be established similarly to Theorem 2.5 if we use the asymptotic estimates (1.8.8), (1.5.13) and (2.6.3) and the relations

$$\mathrm{Re}(b_{jl}) \leqq -\min_{1\leqq j\leqq m}\left[\frac{\mathrm{Re}(b_j)}{\beta_j}\right] < \frac{\sigma\mathrm{Re}(\eta)+\mathrm{Re}(\omega)+1}{\tau} \quad (1 \leqq j \leqq m;\ l = 0,1,2,\cdots),$$

$$\mathrm{Re}(a_{ik}) \geqq \min_{1\leqq i\leqq n}\left[\frac{1-\mathrm{Re}(a_i)}{\alpha_i}\right] > \frac{\mathrm{Re}(\omega)-\sigma/2+1}{\tau} \quad (1 \leqq i \leqq n;\ k=0,1,2,\cdots),$$

following from (1.1.4), (1.1.5), (2.6.8) and (2.6.9), and choose γ by

$$\frac{\mathrm{Re}(\omega)-\sigma/2+1}{\tau} < \gamma < \frac{\sigma\mathrm{Re}(\eta)+\mathrm{Re}(\omega)+1}{\tau}$$

in view of (2.6.10). Note that, for the H-function in the right side of (2.6.11), $a_0^* = a^*$, $\Delta_0 = \Delta - \tau/\sigma$ and $\mu_0 = \mu + (\omega+1)/\sigma - 1$.

In what follows, the so-called generalized Hankel transform will be used which is defined, for $k \in \mathbb{R}$ $(k \neq 0)$ and $\eta \in \mathbb{C}$ $(\mathrm{Re}(\eta) > -3/2)$, by

$$\left(\mathbb{H}_{k,\eta} f\right)(x) = \int_0^\infty (xt)^{1/k-1/2} J_\eta(|k|\,(xt)^{1/k}) f(t)dt. \tag{2.6.12}$$

For such a transform a similar result to Theorem 2.5 holds.

Theorem 2.6. *Let $a^* > 0$ or $a^* = \Delta = 0$ and $\mathrm{Re}(\mu) < -1$. Let $\eta \in \mathbb{C}$, $k \in \mathbb{R}$ $(k \neq 0)$ be such that*

$$\frac{\mathrm{Re}(\eta)}{k} + \min_{1\leqq j\leqq m}\left[\frac{\mathrm{Re}(b_j)}{\beta_j}\right] > -\frac{1}{k} - \frac{1}{2}, \quad \min_{1\leqq i\leqq n}\left[\frac{1-\mathrm{Re}(a_i)}{\alpha_i}\right] > \frac{1}{2k} + \frac{1}{2}, \tag{2.6.13}$$

$$\mathrm{Re}(\eta) > -\frac{1}{2}. \tag{2.6.14}$$

Then the generalized Hankel transform of the H-function exists and the following relation holds

$$\left(\mathbb{H}_{k,\eta} H^{m,n}_{p,q}\left[t \,\middle|\, \begin{matrix}(a_i,\alpha_i)_{1,p}\\ (b_j,\beta_j)_{1,q}\end{matrix}\right]\right)(x)$$

$$= \left(\frac{2}{|k|}\right)^{k/2} \frac{1}{x} H^{m,n+1}_{p+2,q}\left[\left(\frac{2}{|k|}\right)^k \frac{1}{x} \,\middle|\, \begin{matrix}\left(\frac{1}{2}-\frac{k}{4}-\frac{\eta}{2},\frac{k}{2}\right),(a_i,\alpha_i)_{1,p},\left(\frac{1}{2}-\frac{k}{4}+\frac{\eta}{2},\frac{k}{2}\right)\\ (b_j,\beta_j)_{1,q}\end{matrix}\right] \tag{2.6.15}$$

for $x > 0$.

Proof. When $k > 0$, Theorem 2.6 follows from Corollary 2.5.1 when $\omega = 1/k - 1/2$, $a = k$, $\sigma = 1/k$ and $b = \tau = 1$. If $k < 0$, the result is obtained similarly to the proof of Theorem 2.5.

2.7. Fractional Integration and Differentiation of the *H*-Function

For $\alpha \in \mathbb{C}$ $(\mathrm{Re}(\alpha) > 0)$, the Riemann–Liouville fractional integrals and derivatives are defined as follows (see Samko, Kilbas and Marichev [1, Sections 2.3, 2.4 and 5.1]):

$$\left(I^\alpha_{0+} f\right)(x) = \frac{1}{\Gamma(\alpha)}\int_0^x \frac{f(t)}{(x-t)^{1-\alpha}}\,dt \quad (x>0); \tag{2.7.1}$$

$$\left(I^\alpha_{-} f\right)(x) = \frac{1}{\Gamma(\alpha)}\int_x^\infty \frac{f(t)}{(t-x)^{1-\alpha}}\,dt \quad (x>0), \tag{2.7.2}$$

and

$$\left(D_{0+}^{\alpha}f\right)(x) = \left(\frac{d}{dx}\right)^{[\mathrm{Re}(\alpha)]+1} \frac{1}{\Gamma(1-\alpha+[\mathrm{Re}(\alpha)])} \int_0^x \frac{f(t)}{(x-t)^{\alpha-[\mathrm{Re}(\alpha)]}}\,dt$$

$$= \left(\frac{d}{dx}\right)^{[\mathrm{Re}(\alpha)]+1} \left(I_{0+}^{1-\alpha+[\mathrm{Re}(\alpha)]}f\right)(x) \quad (x > 0); \tag{2.7.3}$$

$$\left(D_{-}^{\alpha}f\right)(x) = \left(-\frac{d}{dx}\right)^{[\mathrm{Re}(\alpha)]+1} \frac{1}{\Gamma(1-\alpha+[\mathrm{Re}(\alpha)])} \int_x^{\infty} \frac{f(t)}{(t-x)^{\alpha-[\mathrm{Re}(\alpha)]}}\,dt$$

$$= \left(-\frac{d}{dx}\right)^{[\mathrm{Re}(\alpha)]+1} \left(I_{-}^{1-\alpha+[\mathrm{Re}(\alpha)]}f\right)(x) \quad (x > 0), \tag{2.7.4}$$

respectively, where $[\mathrm{Re}(\alpha)]$ is the integral part of $\mathrm{Re}(\alpha)$. In particular, for real $\alpha > 0$, (2.7.3) and (2.7.4) take the simpler forms

$$\left(D_{0+}^{\alpha}f\right)(x) = \left(\frac{d}{dx}\right)^{[\alpha]+1} \frac{1}{\Gamma(1-\{\alpha\})} \int_0^x \frac{f(t)}{(x-t)^{\{\alpha\}}}\,dt \quad (x > 0); \tag{2.7.5}$$

$$\left(D_{-}^{\alpha}f\right)(x) = \left(-\frac{d}{dx}\right)^{[\alpha]+1} \frac{1}{\Gamma(1-\{\alpha\})} \int_x^{\infty} \frac{f(t)}{(t-x)^{\{\alpha\}}}\,dt \quad (x > 0), \tag{2.7.6}$$

where $[\alpha]$ and $\{\alpha\}$ are the integral and fractional parts of α, respectively.

As in Sections 2.5 and 2.6 we consider the H-function under the conditions in (1.1.6). Using (1.1.1), (1.1.2), Theorem 1.1 and the asymptotic estimates for the H-function at infinity and zero given in Sections 1.5, 1.6 and 1.8, 1.9, we obtain the result for fractional integration.

Theorem 2.7. *Let* $\alpha \in \mathbb{C}$ $(\mathrm{Re}(\alpha) > 0)$, $\omega \in \mathbb{C}$ *and* $\sigma > 0$. *Let us assume either* $a^* > 0$ *or* $a^* = 0$, $\mathrm{Re}(\mu) < -1$. *Then the following statements are valid:*

(i) *If*

$$\sigma \min_{1\leqq j\leqq m}\left[\frac{\mathrm{Re}(b_j)}{\beta_j}\right] + \mathrm{Re}(\omega) > -1 \tag{2.7.7}$$

for $a^* > 0$ *or* $a^* = 0$, $\Delta \geqq 0$, *while*

$$\sigma \min_{1\leqq j\leqq m}\left[\frac{\mathrm{Re}(b_j)}{\beta_j}, \frac{\mathrm{Re}(\mu)+1/2}{\Delta}\right] + \mathrm{Re}(\omega) > -1 \tag{2.7.8}$$

for $a^* = 0$ *and* $\Delta < 0$, *then the fractional integration transform* I_{0+}^{α} *of the* H*-function exists and there holds the relation*

$$\left(I_{0+}^{\alpha} t^{\omega} H_{p,q}^{m,n}\left[t^{\sigma} \,\middle|\, \begin{matrix}(a_i,\alpha_i)_{1,p}\\ (b_j,\beta_j)_{1,q}\end{matrix}\right]\right)(x)$$

$$= x^{\omega+\alpha} H_{p+1,q+1}^{m,n+1}\left[x^{\sigma} \,\middle|\, \begin{matrix}(-\omega,\sigma),(a_i,\alpha_i)_{1,p}\\ (b_j,\beta_j)_{1,q},(-\omega-\alpha,\sigma)\end{matrix}\right]. \tag{2.7.9}$$

(ii) *If*

$$\sigma \max_{1\leqq i\leqq n}\left[\frac{\mathrm{Re}(a_i)-1}{\alpha_i}\right]+\mathrm{Re}(\omega)+\mathrm{Re}(\alpha)<0 \tag{2.7.10}$$

for $a^*>0$ *or* $a^*=0$, $\Delta\leqq 0$, *while*

$$\sigma \max_{1\leqq i\leqq n}\left[\frac{\mathrm{Re}(a_i)-1}{\alpha_i},\frac{\mathrm{Re}(\mu)+1/2}{\Delta}\right]+\mathrm{Re}(\omega)+\mathrm{Re}(\alpha)<0 \tag{2.7.11}$$

for $a^*=0$ *and* $\Delta>0$, *then the fractional integration transform* I_-^α *of the* H*-function exists and there holds the relation*

$$\left(I_-^\alpha t^\omega H_{p,q}^{m,n}\left[t^\sigma \,\middle|\, \begin{matrix}(a_i,\alpha_i)_{1,p}\\(b_j,\beta_j)_{1,q}\end{matrix}\right]\right)(x)$$

$$=x^{\omega+\alpha}H_{p+1,q+1}^{m+1,n}\left[x^\sigma \,\middle|\, \begin{matrix}(a_i,\alpha_i)_{1,p},(-\omega,\sigma)\\(-\omega-\alpha,\sigma),(b_j,\beta_j)_{1,q}\end{matrix}\right]. \tag{2.7.12}$$

Proof. First we note that if we denote by a_1^*, Δ_1, μ_1 and by a_2^*, Δ_2, μ_2 the constants (1.1.7), (1.1.8) and (1.1.10) for the H-functions in the right sides of (2.7.9) and (2.7.12), then it is directly verified that

$$a_1^*=a_2^*=a^*,\quad \Delta_1=\Delta_2=\Delta,\quad \mu_1=\mu_2=\mu-\alpha. \tag{2.7.13}$$

First we prove (i). For the H-functions in both sides of (2.7.9) we take the contour $\mathfrak{L}=\mathfrak{L}_{i\gamma\infty}$, where the choice of γ will depend on the relations between the constant A, B, M given in (2.5.14) and the constant C defined by

$$C=\frac{\mathrm{Re}(\omega)+1}{\sigma}, \tag{2.7.14}$$

for which by (1.1.1), (1.1.2), (2.5.6), (2.5.7) and (2.7.7), (2.7.8) there hold the relations

$$\mathrm{Re}(b_{jl})\leqq B<C,\quad \mathrm{Re}(a_{ik})\geqq A,\quad \mathrm{Re}(c_h)\geqq C \tag{2.7.15}$$

for the poles b_{jl} $(1\leqq j\leqq m;\ l=0,1,2,\cdots)$ of the gamma functions $\Gamma(b_j+\beta_j s)$ $(1\leqq j\leqq m)$, for the poles a_{ik} $(1\leqq i\leqq n;\ k=0,1,2,\cdots)$ of $\Gamma(1-a_i-\alpha_i s)$ $(1\leqq i\leqq n)$ and for the poles $c_h=(\omega+1+h)/\sigma$ $(h=0,1,2,\cdots)$ of $\Gamma(1+\omega-\sigma s)$.

Now we can choose the contour $\mathfrak{L}_{i\gamma\infty}$ if we put γ similarly as in the proof of Theorem 2.3: if $a^*>0$

$$\begin{aligned}&B<\gamma<\min[A,C] && (B<A),\\ &B<\gamma<C && (B=A),\\ &A<\gamma<B && (B>A);\end{aligned} \tag{2.7.16}$$

if $a^*=0$ and $\Delta>0$

$$\begin{aligned}&\gamma<M && (M\leqq B,\ M\leqq A),\\ &A<\gamma<M && (M\leqq B,\ M>A),\\ &B<\gamma<\min[M,C] && (M>B,\ M\leqq A),\\ &B<\gamma<\min[A,C] && (M>B,\ M>A);\end{aligned} \tag{2.7.17}$$

if $a^* = 0$ and $\Delta < 0$,

$$\begin{aligned} &M < \gamma && (M \geqq B,\ M \geqq A), \\ &M < \gamma < \min[A, C] && (M \geqq B,\ M < A), \\ &M < \gamma < B && (M < B,\ M \geqq A), \end{aligned} \tag{2.7.18}$$

and as in (2.7.16) for $M < B$, $M < A$.

So by Theorem 1.1 in all cases above the H-function in the left side of (2.7.9) exists. The right side also exists, because in view of (2.7.13) $a_0^* = a^* > 0$ and the condition $a^* = 0$, $\Delta\gamma + \mathrm{Re}(\mu) < -1$ means $a_0^* = 0$, $\Delta\gamma + \mathrm{Re}(\mu - \alpha) < -1$.

Due to Corollaries 1.11.1, 1.12.1 and 1.13.1 and Remark 1.5, $H_{p,q}^{m,n}(x)$ has asymptotic behavior near zero of the form (1.8.8), (1.8.14) or (1.9.13), where

$$\rho^* = \min_{1 \leqq j \leqq m} \left[\frac{\mathrm{Re}(b_j)}{\beta_j} \right]$$

in the cases $\Delta \geqq 0$ or $\Delta < 0, a^* > 0$, and ρ^* is given by (1.9.14) for $\Delta < 0$, $a^* = 0$. Therefore the integral in the left side of (2.7.9) exists and this relation is proved directly by using (2.7.1) and (1.1.1), (1.1.2), by changing the order of integration and by applying the formula (Samko, Kilbas and Marichev [1, (2.44)]):

$$\left(I_{0+}^{\alpha} t^{\beta-1} \right)(x) = \frac{\Gamma(\beta)}{\Gamma(\alpha+\beta)} x^{\alpha+\beta-1} \quad (\mathrm{Re}(\beta) > 0). \tag{2.7.19}$$

Namely, we have

$$\begin{aligned} &\left(I_{0+}^{\alpha} t^{\omega} H_{p,q}^{m,n} \left[t^{\sigma} \,\middle|\, \begin{matrix} (a_i, \alpha_i)_{1,p} \\ (b_j, \beta_j)_{1,q} \end{matrix} \right] \right)(x) \\ &= \frac{1}{2\pi i} \int_{\mathfrak{L}_{i\gamma\infty}} \mathcal{H}_{p,q}^{m,n} \left[\begin{matrix} (a_i, \alpha_i)_{1,p} \\ (b_j, \beta_j)_{1,q} \end{matrix} \,\middle|\, s \right] \left(I_{0+}^{\alpha} t^{\omega - \sigma s} \right)(x) ds \\ &= x^{\alpha+\omega} \frac{1}{2\pi i} \int_{\mathfrak{L}_{i\gamma\infty}} \mathcal{H}_{p,q}^{m,n} \left[\begin{matrix} (a_i, \alpha_i)_{1,p} \\ (b_j, \beta_j)_{1,q} \end{matrix} \,\middle|\, s \right] \frac{\Gamma(\omega+1-\sigma s)}{\Gamma(\alpha+\omega+1-\sigma s)} x^{-\sigma s} ds, \end{aligned} \tag{2.7.20}$$

which gives (2.7.9).

The assertion (ii) is proved similarly. In fact, the existence of the H-functions in the left and right sides of (2.7.12) is proved by choosing $\mathfrak{L} = \mathfrak{L}_{i\gamma\infty}$ with γ in certain intervals and by using Theorem 1.1. For the left side integral of (2.7.12), the existence can be proved on the basis of the asymptotic behavior near infinity of the form (1.5.13), (1.5.19) or (1.6.33). The relation (2.7.12) is shown directly as (2.7.20) by using the relation (Samko, Kilbas and Marichev [1, Table 9.3(1)]):

$$\left(I_{-}^{\alpha} t^{\beta-1} \right)(x) = \frac{\Gamma(1-\alpha-\beta)}{\Gamma(1-\beta)} x^{\alpha+\beta-1} \quad (\mathrm{Re}(\alpha) > 0,\ \mathrm{Re}(\alpha+\beta) < 1). \tag{2.7.21}$$

A similar statement for fractional differentiation follows from Theorem 2.7 by using the relations (2.7.3), (2.7.4), if we take the formulas (2.2.1), (2.2.2), (2.1.1) and (2.1.2) into account.

Theorem 2.8. *Let $\alpha \in \mathbb{C}$ $(\mathrm{Re}(\alpha) > 0)$, $\omega \in \mathbb{C}$ and $\sigma > 0$. Let us assume either $a^* > 0$ or $a^* = 0$, $\mathrm{Re}(\mu) < -1$. Then the following statements are valid:*

(i) *If the condition in (2.7.7) is satisfied for $a^* > 0$ and $a^* = 0$, $\Delta \geqq 0$ and the condition in (2.7.8) holds for $a^* = 0$ and $\Delta < 0$, then the fractional differentiation transform D_{0+}^{α} of the H-function exists and there holds the relation*

$$\left(D_{0+}^{\alpha} t^{\omega} H_{p,q}^{m,n}\left[t^{\sigma} \,\middle|\, \begin{matrix} (a_i, \alpha_i)_{1,p} \\ (b_j, \beta_j)_{1,q} \end{matrix}\right]\right)(x)$$

$$= x^{\omega-\alpha} H_{p+1,q+1}^{m,n+1}\left[x^{\sigma} \,\middle|\, \begin{matrix} (-\omega, \sigma), (a_i, \alpha_i)_{1,p} \\ (b_j, \beta_j)_{1,q}, (-\omega+\alpha, \sigma) \end{matrix}\right]. \tag{2.7.22}$$

(ii) *If*

$$\sigma \max_{1 \leqq i \leqq n}\left[\frac{\mathrm{Re}(a_i) - 1}{\alpha_i}\right] + \mathrm{Re}(\omega) + 1 - \{\mathrm{Re}(\alpha)\} < 0 \tag{2.7.23}$$

for $a^* > 0$ and $a^* = 0$, $\Delta \leqq 0$, while

$$\sigma \max_{1 \leqq i \leqq n}\left[\frac{\mathrm{Re}(a_i) - 1}{\alpha_i}, \frac{\mathrm{Re}(\mu) + 1/2}{\Delta}\right] + \mathrm{Re}(\omega) + 1 - \{\mathrm{Re}(\alpha)\} < 0 \tag{2.7.24}$$

for $a^* = 0$ and $\Delta > 0$, where $\{\mathrm{Re}(\alpha)\}$ stands for the fractional part of $\mathrm{Re}(\alpha)$, then the fractional differentiation transform D_{-}^{α} of the H-function exists and there holds the relation

$$\left(D_{-}^{\alpha} t^{\omega} H_{p,q}^{m,n}\left[t^{\sigma} \,\middle|\, \begin{matrix} (a_i, \alpha_i)_{1,p} \\ (b_j, \beta_j)_{1,q} \end{matrix}\right]\right)(x)$$

$$= x^{\omega-\alpha} H_{p+1,q+1}^{m+1,n}\left[x^{\sigma} \,\middle|\, \begin{matrix} (a_i, \alpha_i)_{1,p}, (-\omega, \sigma) \\ (-\omega+\alpha, \sigma), (b_j, \beta_j)_{1,q} \end{matrix}\right]. \tag{2.7.25}$$

Proof. We first prove (2.7.25). Using (2.7.4), (2.7.12) with α being replaced by $\beta = 1 - \alpha + [\mathrm{Re}(\alpha)]$, (2.2.2) and (2.1.2) we have

$$\left(D_{-}^{\alpha} t^{\omega} H_{p,q}^{m,n}\left[t^{\sigma} \,\middle|\, \begin{matrix} (a_i, \alpha_i)_{1,p} \\ (b_j, \beta_j)_{1,q} \end{matrix}\right]\right)(x)$$

$$= \left(-\frac{d}{dx}\right)^{[\mathrm{Re}(\alpha)]+1}\left(I_{-}^{\beta} t^{\omega} H_{p,q}^{m,n}\left[t^{\sigma} \,\middle|\, \begin{matrix} (a_i, \alpha_i)_{1,p} \\ (b_j, \beta_j)_{1,q} \end{matrix}\right]\right)(x)$$

$$= \left(-\frac{d}{dx}\right)^{[\mathrm{Re}(\alpha)]+1}\left\{x^{\omega+\beta} H_{p+1,q+1}^{m+1,n}\left[x^{\sigma} \,\middle|\, \begin{matrix} (a_i, \alpha_i)_{1,p}, (-\omega, \sigma) \\ (-\omega-\beta, \sigma), (b_j, \beta_j)_{1,q} \end{matrix}\right]\right\}$$

$$= x^{\omega+\beta-[\mathrm{Re}(\alpha)]-1} H_{p+2,q+2}^{m+2,n}\left[x^{\sigma} \,\middle|\, \begin{matrix} (a_i, \alpha_i)_{1,p}, (-\omega, \sigma), (-\omega-\beta, \sigma) \\ ([\mathrm{Re}(\alpha)] + 1 - \omega - \beta, \sigma), (-\omega-\beta, \sigma), (b_j, \beta_j)_{1,q} \end{matrix}\right]$$

$$= x^{\omega-\alpha} H_{p+1,q+1}^{m+1,n}\left[x^{\sigma} \,\middle|\, \begin{matrix} (a_i, \alpha_i)_{1,p}, (-\omega, \sigma) \\ (-\omega+\alpha, \sigma), (b_j, \beta_j)_{1,q} \end{matrix}\right],$$

which is (2.7.25).

The relation (2.7.22) is proved similarly by using (2.7.3), (2.7.9), (2.2.1) and (2.1.1). This completes the proof of the theorem.

Remark 2.3. When $\alpha = k = 1, 2, \cdots$, the relations (2.7.22) and (2.7.25) coincide with (2.2.1) and (2.2.2).

Remark 2.4. In the case $a^* > 0$ a relation more general than (2.7.9) was indicated by Prudnikov, Brychkov and Marichev [3, (2.25.2.2)]. The formula (2.7.12) with real $\alpha > 0$ was given by Raina and Koul [3, (2.5)], but the conditions for its validity have to be corrected in accordance with (2.7.10) which for real α takes the form

$$\sigma \max_{1 \leqq i \leqq n} \left[\frac{\text{Re}(a_i) - 1}{\alpha_i} \right] + \text{Re}(\omega) + \alpha < 0. \tag{2.7.26}$$

The result in (2.7.22) for $a^* > 0$ was obtained by Srivastava, Gupta and Goyal [1, (2.7.13)], but the conditions should be corrected as (2.7.7), too. The relation of the form (2.7.25) being proved for $a^* > 0$ by Raina and Koul [2, (14a)] (see also Raina and Koul [3, (2.2)] and Srivastava, Gupta and Goyal [1, (2.7.9)]) contains mistakes, and it should be replaced by (2.7.25) under the condition

$$\sigma \max_{1 \leqq i \leqq n} \left[\frac{\text{Re}(a_i) - 1}{\alpha_i} \right] + \text{Re}(\omega) + 1 - \{\alpha\} < 0. \tag{2.7.27}$$

2.8. Integral Formulas Involving the *H*-Function

In Sections 2.5–2.7 we have proved various formulas which present the integrals of the H-function. Here we give two further integrals which generalize these relations. First we consider the integral involving the product of two H-functions. Let $H^{M,N}_{P,Q}\left[z \middle| \begin{matrix} (c_i, \gamma_i)_{1,P} \\ (d_j, \delta_j)_{1,Q} \end{matrix} \right]$ be the second H-function defined by (1.1.1), which also satisfies the conditions of the form (1.1.6). Let a_0^* be the constant a^* in (1.1.7) for the second H-function.

Theorem 2.9. *Let $a^* > 0$, $a_0^* > 0$ and let $\eta \in \mathbb{C}$, $\sigma > 0$, $z, w \in \mathbb{C}$ be such that* $|\arg(z)| < a^*\pi/2$,

$$-\min_{1 \leqq j \leqq m} \left[\frac{\text{Re}(b_j)}{\beta_j} \right] < \min_{1 \leqq i \leqq n} \left[\frac{1 - \text{Re}(a_i)}{\alpha_i} \right], \tag{2.8.1}$$

$$-\min_{1 \leqq i \leqq N} \left[\frac{1 - \text{Re}(c_i)}{\gamma_i} \right] < \min_{1 \leqq j \leqq M} \left[\frac{\text{Re}(d_j)}{\delta_j} \right] \tag{2.8.2}$$

and

$$-\sigma \min_{1 \leqq j \leqq m} \left[\frac{\text{Re}(b_j)}{\beta_j} \right] - \min_{1 \leqq j \leqq M} \left[\frac{\text{Re}(d_j)}{\delta_j} \right]$$
$$< \text{Re}(\eta) < \sigma \min_{1 \leqq i \leqq n} \left[\frac{1 - \text{Re}(a_i)}{\alpha_i} \right] + \min_{1 \leqq i \leqq N} \left[\frac{1 - \text{Re}(c_i)}{\gamma_i} \right]. \tag{2.8.3}$$

Then there holds the relation

$$\int_0^\infty t^{\eta-1} H_{p,q}^{m,n}\left[zt^\sigma \middle| \begin{matrix}(a_i,\alpha_i)_{1,p}\\(b_j,\beta_j)_{1,q}\end{matrix}\right] H_{P,Q}^{M,N}\left[wt \middle| \begin{matrix}(c_i,\gamma_i)_{1,P}\\(d_j,\delta_j)_{1,Q}\end{matrix}\right] dt$$

$$= w^{-\eta} H_{p+Q,q+P}^{m+N,n+M}\left[zw^{-\sigma} \middle| \begin{matrix}(a_i,\alpha_i)_{1,n},(1-d_j-\eta\delta_j,\sigma\delta_j)_{1,Q},(a_i,\alpha_i)_{n+1,p}\\(b_j,\beta_j)_{1,m},(1-c_i-\eta\gamma_i,\sigma\gamma_i)_{1,P},(b_j,\beta_j)_{m+1,q}\end{matrix}\right]. \tag{2.8.4}$$

Proof. First we note that, according to Corollaries 1.11.1, 1.12.1, 1.7.1 and 1.8.1, two H-functions $H_{p,q}^{m,n}(z)$ and $H_{P,Q}^{M,N}(z)$ have the asymptotic estimates at zero of the form (1.8.8) or (1.8.14) with

$$\varrho^* = \min_{1\leqq j\leqq m}\left[\frac{\mathrm{Re}(b_j)}{\beta_j}\right] \quad \text{and} \quad \varrho^* = \min_{1\leqq j\leqq M}\left[\frac{\mathrm{Re}(d_j)}{\delta_j}\right],$$

respectively, and asymptotics at infinity of the form (1.5.13) or (1.5.19) with

$$\varrho = -\min_{1\leqq i\leqq n}\left[\frac{1-\mathrm{Re}(a_i)}{\alpha_i}\right] \quad \text{and} \quad \varrho = -\min_{1\leqq i\leqq N}\left[\frac{1-\mathrm{Re}(c_i)}{\gamma_i}\right],$$

respectively. Then the integral in the left side of (2.8.4) exists provided that the condition (2.8.3) is satisfied.

The proof of the formula (2.8.4) is based on Theorem 2.2 and Corollary 2.2.1 and on the Mellin convolution relation (Titchmarsh [3, Theorem 44])

$$\left(\mathfrak{M}\{k*f\}\right)(s) = \left(\mathfrak{M}k\right)(s)\left(\mathfrak{M}f\right)(s) \quad \text{for} \quad (k*f)(x) = \int_0^\infty k\left(\frac{x}{t}\right) f(t)\frac{dt}{t}. \tag{2.8.5}$$

Applying the translation formula (1.8.15) to $H_{P,Q}^{M,N}\left[wt \middle| \begin{matrix}(c_i,\gamma_i)_{1,P}\\(d_j,\delta_j)_{1,Q}\end{matrix}\right]$ and making the change of variable $t=\tau^{-1/\sigma}$, we rewrite the left side of (2.8.4) in the form (2.8.5):

$$\int_0^\infty t^{\eta-1} H_{p,q}^{m,n}\left[zt^\sigma \middle| \begin{matrix}(a_i,\alpha_i)_{1,p}\\(b_j,\beta_j)_{1,q}\end{matrix}\right] H_{P,Q}^{M,N}\left[wt \middle| \begin{matrix}(c_i,\gamma_i)_{1,P}\\(d_j,\delta_j)_{1,Q}\end{matrix}\right] dt$$

$$= \int_0^\infty t^{\eta-1} H_{p,q}^{m,n}\left[zt^\sigma \middle| \begin{matrix}(a_i,\alpha_i)_{1,p}\\(b_j,\beta_j)_{1,q}\end{matrix}\right] H_{Q,P}^{N,M}\left[\frac{1}{wt} \middle| \begin{matrix}(1-d_j,\delta_j)_{1,Q}\\(1-c_i,\gamma_i)_{1,P}\end{matrix}\right] dt$$

$$= \frac{1}{\sigma}\int_0^\infty H_{p,q}^{m,n}\left[\frac{z}{\tau} \middle| \begin{matrix}(a_i,\alpha_i)_{1,p}\\(b_j,\beta_j)_{1,q}\end{matrix}\right] \tau^{-\eta/\sigma} H_{Q,P}^{N,M}\left[\frac{\tau^{1/\sigma}}{w} \middle| \begin{matrix}(1-d_j,\delta_j)_{1,Q}\\(1-c_i,\gamma_i)_{1,P}\end{matrix}\right] \frac{d\tau}{\tau}.$$

Then the Mellin transform of the left side of (2.8.4) leads to

$$\left(\mathfrak{M}\int_0^\infty t^{\eta-1} H_{p,q}^{m,n}\left[zt^\sigma \middle| \begin{matrix}(a_i,\alpha_i)_{1,p}\\(b_j,\beta_j)_{1,q}\end{matrix}\right] H_{P,Q}^{M,N}\left[wt \middle| \begin{matrix}(c_i,\gamma_i)_{1,P}\\(d_j,\delta_j)_{1,Q}\end{matrix}\right] dt\right)(s)$$

$$= \frac{1}{\sigma}\left(\mathfrak{M}H_{p,q}^{m,n}\left[z \middle| \begin{matrix}(a_i,\alpha_i)_{1,p}\\(b_j,\beta_j)_{1,q}\end{matrix}\right]\right)(s)$$

$$\cdot \left(\mathfrak{M}\tau^{-\eta/\sigma} H_{Q,P}^{N,M}\left[\frac{\tau^{1/\sigma}}{w} \middle| \begin{matrix}(1-d_j,\delta_j)_{1,Q}\\(1-c_i,\gamma_i)_{1,P}\end{matrix}\right]\right)(s). \tag{2.8.6}$$

According to (2.8.1) and (2.8.2) we can choose $s \in \mathbb{C}$ such that

$$-\min_{1\leqq j\leqq m}\left[\frac{\mathrm{Re}(b_j)}{\beta_j}\right] < \mathrm{Re}(s) < \min_{1\leqq i\leqq n}\left[\frac{1-\mathrm{Re}(a_i)}{\alpha_i}\right] \tag{2.8.7}$$

and

$$-\min_{1\leqq i\leqq N}\left[\frac{1-\mathrm{Re}(c_i)}{\gamma_i}\right] < \sigma\mathrm{Re}(s) - \mathrm{Re}(\eta) < \min_{1\leqq j\leqq M}\left[\frac{\mathrm{Re}(d_j)}{\delta_j}\right]. \tag{2.8.8}$$

Now we can use Theorem 2.2 for the first term and Corollary 2.2.1 for the second, where the conditions (2.5.5) and (2.5.10) are certified by (2.8.7) and (2.8.8). Applying the relations (2.5.7) and (2.5.12), we obtain

$$\left(\mathfrak{M}\int_0^\infty t^{\eta-1}H_{p,q}^{m,n}\left[zt^\sigma \,\middle|\, \begin{matrix}(a_i,\alpha_i)_{1,p}\\ (b_j,\beta_j)_{1,q}\end{matrix}\right] H_{P,Q}^{M,N}\left[wt \,\middle|\, \begin{matrix}(c_i,\gamma_i)_{1,P}\\ (d_j,\delta_j)_{1,Q}\end{matrix}\right]dt\right)(s)$$

$$= w^{\sigma s-\eta}\mathcal{H}_{p,q}^{m,n}\left[\begin{matrix}(a_i,\alpha_i)_{1,p}\\ (b_j,\beta_j)_{1,q}\end{matrix}\,\middle|\, s\right]\mathcal{H}_{Q,P}^{N,M}\left[\begin{matrix}(1-d_j,\delta_j)_{1,Q}\\ (1-c_i,\gamma_i)_{1,P}\end{matrix}\,\middle|\,\sigma s-\eta\right]$$

$$= w^{\sigma s-\eta}\mathcal{H}_{p+Q,q+P}^{m+N,n+M}\left[\begin{matrix}(a_i,\alpha_i)_{1,n},(1-d_j-\eta\delta_j,\sigma\delta_j)_{1,Q},(a_i,\alpha_i)_{n+1,p}\\ (b_j,\beta_j)_{1,m}(1-c_i-\eta\gamma_i,\sigma\gamma_i)_{1,P},(b_j,\beta_j)_{m+1,q}\end{matrix}\,\middle|\, s\right]. \tag{2.8.9}$$

If we denote by a_r^* the constant a^* in (1.1.7) for the H-function in the right side of (2.8.4), then it is directly verified that

$$a_r^* = a^* + \sigma a_0^* \tag{2.8.10}$$

and hence $a_r^* > 0$. By the conditions (2.8.1) and (2.8.2) we can choose $s \in \mathbb{C}$ such that

$$-\sigma\min_{1\leqq j\leqq m}\left[\frac{\mathrm{Re}(b_j)}{\beta_j}\right] - \min_{1\leqq i\leqq N}\left[\frac{1-\mathrm{Re}(c_i)}{\gamma_i}\right]$$

$$< \mathrm{Re}(s) < \sigma\min_{1\leqq i\leqq n}\left[\frac{1-\mathrm{Re}(a_i)}{\alpha_i}\right] + \min_{1\leqq j\leqq M}\left[\frac{\mathrm{Re}(d_j)}{\delta_j}\right]$$

and we can apply Corollary 2.2.1 to the right side of (2.8.4) and a direct calculation, similar to the above by using (2.5.13), shows that the Mellin transform of the right side of (2.8.4) coincides with the right side of (2.8.9). Hence by the Mellin inversion theorem, we arrive at the relation (2.8.4), which completes the proof of the theorem.

Corollary 2.9.1. *If the conditions of Theorem 2.9 are satisfied, then there holds the relation*

$$\int_0^\infty t^{\eta-1}H_{p,q}^{m,n}\left[zt^{-\sigma} \,\middle|\, \begin{matrix}(a_i,\alpha_i)_{1,p}\\ (b_j,\beta_j)_{1,q}\end{matrix}\right] H_{P,Q}^{M,N}\left[wt \,\middle|\, \begin{matrix}(c_i,\gamma_i)_{1,P}\\ (d_j,\delta_j)_{1,Q}\end{matrix}\right]dt$$

$$= w^{\eta}H_{p+P,q+Q}^{m+M,n+N}\left[zw^\sigma \,\middle|\, \begin{matrix}(a_i,\alpha_i)_{1,n},(c_i+\eta\gamma_i,\sigma\gamma_i)_{1,P},(a_i,\alpha_i)_{n+1,p}\\ (b_j,\beta_j)_{1,m},(d_j+\eta\delta_j,\sigma\delta_j)_{1,Q},(b_j,\beta_j)_{m+1,q}\end{matrix}\right]. \tag{2.8.11}$$

Proof. By virtue of the translation formula (1.8.15) we have

$$I = \int_0^\infty t^{\eta-1} H_{p,q}^{m,n}\left[zt^{-\sigma} \left| \begin{matrix} (a_i,\alpha_i)_{1,p} \\ (b_j,\beta_j)_{1,q} \end{matrix} \right.\right] H_{P,Q}^{M,N}\left[wt \left| \begin{matrix} (c_i,\gamma_i)_{1,P} \\ (d_j,\delta_j)_{1,Q} \end{matrix} \right.\right] dt$$

$$= \int_0^\infty t^{\eta-1} H_{q,p}^{n,m}\left[z^{-1}t^{\sigma} \left| \begin{matrix} (1-b_j,\beta_j)_{1,q} \\ (1-a_i,\alpha_i)_{1,p} \end{matrix} \right.\right] H_{P,Q}^{M,N}\left[wt \left| \begin{matrix} (c_i,\gamma_i)_{1,P} \\ (d_j,\delta_j)_{1,Q} \end{matrix} \right.\right] dt.$$

It is easy to see that the conditions of the form (2.8.1) and (2.8.2) for the function $H_{q,p}^{n,m}(z^{-1}t^{\sigma})$ coincide with (2.8.1) and (2.8.2). Therefore applying Theorem 2.9 and using (2.8.4), we have

$$I = w^{\eta} H_{p+P,q+Q}^{n+N,m+M}\left[z^{-1}w^{-\sigma} \left| \begin{matrix} (1-b_j,\beta_j)_{1,m},(1-d_j-\eta\delta_j,\sigma\delta_j)_{1,Q},(1-b_j,\beta_j)_{m+1,q} \\ (1-a_i,\alpha_i)_{1,n},(1-c_i-\eta\gamma_i,\sigma\gamma_i)_{1,P},(1-a_i,\alpha_i)_{n+1,p} \end{matrix} \right.\right]$$

and, then, again by the translation formula (1.8.15), the relation (2.8.11) is obtained.

For $z = x \in \mathbb{R}$ and $w = y \in \mathbb{R}$, Theorem 2.9 remains true in the case when $a^* = 0$ or $a_0^* = 0$. We note that, if $a^* = 0$, from Sections 1.9 and 1.6 the asymptotic estimates of the H-function $H_{p,q}^{m,n}(z)$ at zero and infinity are known when $\Delta > 0$ and $\Delta < 0$, respectively. Since for the convergence of the integral in the left side of (2.8.4) we have to take into account the asymptotic estimates both at zero and infinity, only the case $\Delta = 0$ is excepted. In this way we obtain the following result, where Δ_0 and μ_0 denote the constants in (1.1.8) and (1.1.10) for the H-function $H_{P,Q}^{M,N}(z)$.

Theorem 2.10. *Let either*

(a) $a^* = \Delta = 0$, $\mathrm{Re}(\mu) < -1$ *and* $a_0^* > 0$;

(b) $a^* > 0$ *and* $a_0^* = \Delta_0 = 0$, $\mathrm{Re}(\mu_0) < -1$; *or*

(c) $a^* = \Delta = 0$, $\mathrm{Re}(\mu) < -1$ *and* $a_0^* = \Delta_0 = 0$, $\mathrm{Re}(\mu_0) < -1$.

Let $\eta \in \mathbb{C}$, $\sigma > 0$ *and* $x \in \mathbb{R}$ *be such that the conditions in* (2.8.1)–(2.8.3) *are satisfied. Then there holds the relation* (2.8.4) *for real* x *and* y:

$$\int_0^\infty t^{\eta-1} H_{p,q}^{m,n}\left[xt^{\sigma} \left| \begin{matrix} (a_i,\alpha_i)_{1,p} \\ (b_j,\beta_j)_{1,q} \end{matrix} \right.\right] H_{P,Q}^{M,N}\left[yt \left| \begin{matrix} (c_i,\gamma_i)_{1,P} \\ (d_j,\delta_j)_{1,Q} \end{matrix} \right.\right] dt$$

$$= y^{-\eta} H_{p+Q,q+P}^{m+N,n+M}\left[xy^{-\sigma} \left| \begin{matrix} (a_i,\alpha_i)_{1,n},(1-d_j-\eta\delta_j,\sigma\delta_j)_{1,Q},(a_i,\alpha_i)_{n+1,p} \\ (b_j,\beta_j)_{1,m},(1-c_i-\eta\gamma_i,\sigma\gamma_i)_{1,P},(b_j,\beta_j)_{m+1,q} \end{matrix} \right.\right]. \tag{2.8.12}$$

Proof. When $a^* = \Delta = 0$ or $a_0^* = \Delta_0 = 0$, then by Corollaries 1.11.1, 1.12.1, 1.7.1 and 1.8.1 we have asymptotic estimates for the H-funtions $H_{p,q}^{m,n}(z)$ and $H_{P,Q}^{M,N}(z)$ at zero and infinity, as indicated in the proof of Theorem 2.9 and hence the integral in the left side of (2.8.12) converges. The next part of the proof is similar to that in Theorem 2.9, if we take into account the relations

$$\Delta_r = \Delta - \sigma\Delta_0, \quad \mu_r = \mu + \mu_0 + \eta\Delta_0, \tag{2.8.13}$$

where Δ_r and μ_r denote the constants in (1.1.8) and (1.1.10) for the H-function in the right side of (2.8.12). Applying the Mellin convolution theorem (2.8.5), Theorem 2.2 and Corollary 2.2.1 for the functions in the left side of (2.8.12) in the cases $a^* = 0$ and $a_0^* = 0$, respectively, we come to the relation (2.8.9) with $w = y$. The Mellin transform of the H-function in the right side of (2.8.12) gives the same result by using Corollary 2.2.1 if we take into account (2.8.10) and (2.8.13), $a_r^* > 0$ in the cases (a) and (b), while in the case (c) $a_r^* = \Delta_r = 0$ and $\mathrm{Re}(\mu_r) = \mathrm{Re}(\mu + \mu_0) < -2$.

Corollary 2.10.1. *If the conditions of Theorem 2.10 are satisfied, then there holds the relation for real x and y:*

$$\int_0^\infty t^{\eta-1} H_{p,q}^{m,n}\left[xt^{-\sigma} \,\middle|\, \begin{matrix} (a_i,\alpha_i)_{1,p} \\ (b_j,\beta_j)_{1,q} \end{matrix}\right] H_{P,Q}^{M,N}\left[yt \,\middle|\, \begin{matrix} (c_i,\gamma_i)_{1,P} \\ (d_j,\delta_j)_{1,Q} \end{matrix}\right] dt$$

$$= y^\eta H_{p+P,q+Q}^{m+M,n+N}\left[xy^\sigma \,\middle|\, \begin{matrix} (a_i,\alpha_i)_{1,n}, (c_i+\eta\gamma_i,\sigma\gamma_i)_{1,P}, (a_i,\alpha_i)_{n+1,p} \\ (b_j,\beta_j)_{1,m}, (d_j+\eta\delta_j,\sigma\delta_j)_{1,Q}, (b_j,\beta_j)_{m+1,q} \end{matrix}\right]. \tag{2.8.14}$$

Proof. The proof is the same as in Corollary 2.9.1 by using (1.8.15) and Theorem 2.10.

The results in Theorems 2.9 and 2.10 generalize those in Sections 2.5 and 2.6 (see Remark 2.5 below). Now we present a generalization of the result in Section 2.7.

Theorem 2.11. *Let either $a^* > 0$ or $a^* = 0$, $\mathrm{Re}(\mu) < -1$. Let $\alpha, w, a \in \mathbb{C}$, $\sigma > 0$ and $\gamma > 0$ be such that*

$$\sigma \min_{1\leqq j\leqq m}\left[\frac{\mathrm{Re}(b_j)}{\beta_j}\right] + \mathrm{Re}(w) > -1, \quad \gamma \min_{1\leqq j\leqq m}\left[\frac{\mathrm{Re}(b_j)}{\beta_j}\right] + \mathrm{Re}(\alpha) > 0 \tag{2.8.15}$$

for $a^* > 0$ and $a^* = 0$, $\Delta \geqq 0$, while

$$\begin{gathered} \sigma \min_{1\leqq j\leqq m}\left[\frac{\mathrm{Re}(b_j)}{\beta_j}, \frac{\mathrm{Re}(\mu)+1/2}{\Delta}\right] + \mathrm{Re}(w) > -1, \\ \gamma \min_{1\leqq j\leqq m}\left[\frac{\mathrm{Re}(b_j)}{\beta_j}, \frac{\mathrm{Re}(\mu)+1/2}{\Delta}\right] + \mathrm{Re}(\alpha) > 0 \end{gathered} \tag{2.8.16}$$

for $a^* = 0$ and $\Delta < 0$. Then the relation

$$\int_0^x t^w (x-t)^{\alpha-1} H_{p,q}^{m,n}\left[at^\sigma (x-t)^\gamma \,\middle|\, \begin{matrix} (a_i,\alpha_i)_{1,p} \\ (b_j,\beta_j)_{1,q} \end{matrix}\right] dt$$

$$= x^{w+\alpha} H_{p+2,q+1}^{m,n+2}\left[ax^{\sigma+\gamma} \,\middle|\, \begin{matrix} (-w,\sigma), (1-\alpha,\gamma), (a_i,\alpha_i)_{1,p} \\ (b_j,\beta_j)_{1,q}, (-w-\alpha,\sigma+\gamma) \end{matrix}\right] \tag{2.8.17}$$

holds for $x > 0$.

Proof. First we note that if we denote by a_0^*, Δ_0, μ_0 the constants (1.1.7), (1.1.8) and (1.1.10) for the H-function in the right side of (2.8.17), then

$$a_0^* = a^*, \quad \Delta_0 = \Delta, \quad \mu_0 = \mu - \frac{1}{2}. \tag{2.8.18}$$

By (1.1.1) and (1.1.2) for the H-functions in the left and right sides of (2.8.17) we take the contour $\mathfrak{L} = \mathfrak{L}_{i\gamma\infty}$, where the choice of γ will depend on the relations between the constants A, B and M given in (2.5.17), C in (2.7.14) and D by

$$D = \frac{\mathrm{Re}(\alpha)}{\gamma}, \tag{2.8.19}$$

for which (2.7.15) and (2.8.15) imply the relations

$$\mathrm{Re}(b_{jl}) \leqq B < C, \quad \mathrm{Re}(a_{ik}) \geqq A, \quad \mathrm{Re}(c_m) \geqq C, \quad \mathrm{Re}(d_n) \geqq D > B \tag{2.8.20}$$

for the poles b_{jl} $(1 \leqq j \leqq m;\ l = 0, 1, 2, \cdots)$ of the gamma functions $\Gamma(b_j + \beta_j s)$ $(1 \leqq j \leqq m)$, for a_{ik} $(1 \leqq i \leqq n;\ k = 0, 1, 2, \cdots)$ of $\Gamma(1 - a_i - \alpha_i s)$ $(1 \leqq i \leqq n)$, for $c_m = (w + 1 + m)/\sigma$ $(m = 0, 1, 2, \cdots)$ of $\Gamma(1 + w - \sigma s)$ and for $d_n = (\alpha + n)/\gamma$ $(n = 0, 1, 2, \cdots)$ of $\Gamma(\alpha - \gamma s)$. Since $B < \min[C, D]$, we can choose $\mathfrak{L} = \mathfrak{L}_{i\gamma\infty}$ in the same way as was done in the proof of Theorem 2.7(i), In fact,

$$\begin{aligned} &B < \gamma < \min[A, C, D] && (B < A);\\ &B < \gamma < \min[C, D] && (B = A);\\ &A < \gamma < B && (B > A). \end{aligned} \tag{2.8.21}$$

From Corollaries 1.11.1 and 1.12.1 and Remark 1.5 the H-function $H_{p,q}^{m,n}(x)$ has the asymptotic behavior at zero of the form (1.8.8) or (1.8.14), where $\varrho^* = \min_{1 \leqq j \leqq m}[\mathrm{Re}(b_j/\beta_j]$ in the cases $\Delta \geqq 0$ or $\Delta > 0$, $a^* = 0$, and ϱ^* is given by (1.9.14) for $\Delta < 0$ and $a^* = 0$. Therefore the integral in the left side of (2.8.17) exists and this relation is proved directly by using (1.1.1), (1.1.2), changing the order of integration and applying (2.7.19):

$$\int_0^x t^w (x-t)^{\alpha-1} H_{p,q}^{m,n}\left[a t^\sigma (x-t)^\gamma \,\middle|\, \begin{matrix} (a_i, \alpha_i)_{1,p} \\ (b_j, \beta_j)_{1,q} \end{matrix} \right] dt$$

$$= \frac{1}{2\pi i} \int_{\mathfrak{L}_{i\gamma\infty}} \mathcal{H}_{p,q}^{m,n}\left[\begin{matrix} (a_i, \alpha_i)_{1,p} \\ (b_j, \beta_j)_{1,q} \end{matrix} \,\middle|\, s \right] \Gamma(\alpha - \gamma s) \left(I_{0+}^{\alpha-\gamma s} t^{w-\sigma s} \right)(x) a^{-s} ds$$

$$= x^{\alpha+w} \frac{1}{2\pi i} \int_{\mathfrak{L}_{i\gamma\infty}} \mathcal{H}_{p,q}^{m,n}\left[\begin{matrix} (a_i, \alpha_i)_{1,p} \\ (b_j, \beta_j)_{1,q} \end{matrix} \,\middle|\, s \right] \frac{\Gamma(\alpha - \gamma s)\Gamma(w + 1 - \sigma s)}{\Gamma(\alpha + w + 1 - \sigma s - \gamma s)} \left(a x^{\sigma+\gamma} \right)^{-s} ds.$$

Remark 2.5. The relations (2.5.7), (2.5.12), (2.5.16), (2.5.25), (2.5.29), (2.5.30), (2.6.6), (2.6.11) and (2.6.15) can be obtained from (2.8.4) if we take the function $H_{P,Q}^{M,N}[wt]$ with special parameters to coincide with one of the functions x^{w+s-1}, $x^w e^{-px}$ and $x^w J_\eta(x)$ (see the relations (2.9.4), (2.9.6) and (2.9.18) below). It should also be noted that many known integrals presented in the books by Prudnikov, Brychkov and Marichev [1]–[3], can be obtained from (2.8.4) by taking H-functions as certain elementary or special functions.

Remark 2.6. The relation (2.8.4) was proved by K.C. Gupta and U.C. Jain [1] provided that $a^* > 0$, $a_0^* > 0$, $\Delta > 0$ and $\Delta_0 > 0$, while the formula (2.8.17) was obtained by G.K. Goyal [1] for $a^* > 0$, $\Delta > 0$ (see also Srivastava, Gupta and Goyal [1, (5.1.1) and (5.2.3)]). The relations (2.8.4) and (2.8.17) were indicated in Prudnikov, Brychkov and Marichev [3, 2.25.1.1 and 2.25.2.2] provided that $a^* > 0$, $a_0^* > 0$ and $a^* > 0$, respectively. But the conditions in the above papers and books have to be corrected in accordance with the conditions of Theorems 2.9 and 2.11.

2.9. Special Cases of the *H*-Function

The H-function, being in generalized form, contains a vast number of special functions as its particular cases. First of all, when $\alpha_i = \beta_j = 1$ $(i = 1, \cdots, p; j = 1, \cdots, q)$, it reduces to the Meijer G-function. Namely,

$$H_{p,q}^{m,n}\left[z \,\middle|\, \begin{matrix} (a_i, 1)_{1,p} \\ (b_j, 1)_{1,q} \end{matrix}\right] = G_{p,q}^{m,n}\left[z \,\middle|\, \begin{matrix} (a_i)_{1,p} \\ (b_j)_{1,q} \end{matrix}\right] = G_{p,q}^{m,n}\left[z \,\middle|\, \begin{matrix} a_1, \cdots, a_p \\ b_1, \cdots, b_q \end{matrix}\right]$$

$$\equiv \frac{1}{2\pi i}\int_{\mathfrak{L}} \frac{\prod_{j=1}^{m}\Gamma(b_j+s)\prod_{i=1}^{n}\Gamma(1-a_i-s)}{\prod_{i=n+1}^{p}\Gamma(a_i+s)\prod_{j=m+1}^{q}\Gamma(1-b_j-s)} z^{-s}ds, \tag{2.9.1}$$

where $\mathfrak{L}$ is the same contour taken for the H-function in Section 1.2.

The theory of the G-function (see Erdélyi, Magnus, Oberhettinger and Tricomi [1, Sections 5.3–5.6]) can be found in Luke [1, Chapter V], Mathai and R.K. Saxena [2], Prudnikov, Brychkov and Marichev [3, Sections 8.2 and 8.4]. We only note that some properties of such a function can be deduced from the corresponding properties of the H-function given in Sections 1.2–1.8 and in 2.1–2.8. The G-function is a generalization of a number of known special functions, in particular of the Gauss hypergeometric function ${}_2F_1(a, b; c; z)$ or the more general hypergeometric function ${}_pF_q(a_1, \cdots, a_p; b_1, \cdots, b_q; z)$ defined by the series of z

$${}_2F_1(a, b; c; z) = \sum_{k=0}^{\infty} \frac{(a)_k(b)_k}{(c)_k}\frac{z^k}{k!} \quad (a, b, c \in \mathbb{C},\ c \neq 0, -1, -2, \cdots); \tag{2.9.2}$$

$${}_pF_q(a_1, \cdots, a_p; b_1, \cdots, b_q; z) = \sum_{k=0}^{\infty} \frac{(a_1)_k \cdots (a_p)_k}{(b_1)_k \cdots (b_q)_k}\frac{z^k}{k!} \tag{2.9.3}$$

$$(a_i,\ b_j \in \mathbb{C},\ a_i,\ b_j \neq 0, -1, -2, \cdots \quad (i = 1, \cdots, p;\ j = 1, \cdots, q)),$$

where the Pochhammer symbol $(a)_k$ $(k \in \mathbb{N}_0)$ is given in (2.4.12). The theory of these functions is given in Erdélyi, Magnus, Oberhettinger and Tricomi [1, Chapters II and IV]. In particular, the series (2.9.2) converges for $|z| < 1$ and for $|z| = 1$, $\text{Re}(c - a - b) > 0$ and the series (2.9.3) converges for all finite z if $p \leqq q$ and for $|z| < 1$ if $p = q + 1$.

A detailed account of the special cases of the G-function is available in Erdélyi, Magnus, Oberhettinger and Tricomi [1, Section 5.6], Luke [1, Section 6.4] and Mathai and R.K. Saxena [2, Chapter II]. Since all elementary and special functions as special cases of the G-function are also considered as special cases of the H-function due to the relation (2.9.1), we can write all special functions in terms of the H-function. We list here a few interesting cases of the

H-function which may be useful for the workers on integral transforms and special functions. First we begin with the elementary functions:

$$H_{0,1}^{1,0}\left[z\,\middle|\,\begin{matrix}\text{---}\\(b,\beta)\end{matrix}\right]=\frac{1}{\beta}z^{b/\beta}\exp(-z^{1/\beta});\tag{2.9.4}$$

$$H_{1,1}^{1,1}\left[z\,\middle|\,\begin{matrix}(1-a,1)\\(0,1)\end{matrix}\right]=\Gamma(a)(1+z)^{-a}=\Gamma(a)\ {}_1F_0(a;-z);\tag{2.9.5}$$

$$H_{1,1}^{1,0}\left[z\,\middle|\,\begin{matrix}(\alpha+\beta+1,1)\\(\alpha,1)\end{matrix}\right]=z^{\alpha}(1-z)^{\beta};\tag{2.9.6}$$

$$H_{0,2}^{1,0}\left[\frac{z^2}{4}\,\middle|\,\begin{matrix}\text{---}\\\left(\frac{1}{2},1\right),(0,1)\end{matrix}\right]=\frac{1}{\sqrt{\pi}}\sin z;\tag{2.9.7}$$

$$H_{0,2}^{1,0}\left[\frac{z^2}{4}\,\middle|\,\begin{matrix}\text{---}\\(0,1),\left(\frac{1}{2},1\right)\end{matrix}\right]=\frac{1}{\sqrt{\pi}}\cos z;\tag{2.9.8}$$

$$H_{0,2}^{1,0}\left[-\frac{z^2}{4}\,\middle|\,\begin{matrix}\text{---}\\\left(\frac{1}{2},1\right),(0,1)\end{matrix}\right]=\frac{i}{\sqrt{\pi}}\sinh z;\tag{2.9.9}$$

$$H_{0,2}^{1,0}\left[-\frac{z^2}{4}\,\middle|\,\begin{matrix}\text{---}\\(0,1),\left(\frac{1}{2},1\right)\end{matrix}\right]=\frac{1}{\sqrt{\pi}}\cosh z;\tag{2.9.10}$$

$$\pm H_{2,2}^{1,0}\left[z\,\middle|\,\begin{matrix}(1,1),(1,1)\\(1,1),(0,1)\end{matrix}\right]=\log(1\pm z);\tag{2.9.11}$$

$$H_{2,2}^{1,2}\left[-z^2\,\middle|\,\begin{matrix}\left(\frac{1}{2},1\right),\left(\frac{1}{2},1\right)\\(0,1),\left(-\frac{1}{2},1\right)\end{matrix}\right]=2\arcsin z;\tag{2.9.12}$$

$$H_{2,2}^{1,2}\left[z^2\,\middle|\,\begin{matrix}\left(\frac{1}{2},1\right),(1,1)\\\left(\frac{1}{2},1\right),(0,1)\end{matrix}\right]=2\arctan z.\tag{2.9.13}$$

Next we present the H-functions reducible to the hypergeometric functions (2.9.2) and (2.9.3):

$$H_{1,2}^{1,1}\left[z\,\middle|\,\begin{matrix}(1-a,1)\\(0,1),(1-c,1)\end{matrix}\right]=\frac{\Gamma(a)}{\Gamma(c)}\ {}_1F_1(a;c;-z),\tag{2.9.14}$$

which is the so-called confluent hypergeometric function of Kummer (see Erdélyi, Magnus, Oberhettinger and Tricomi [1, 6.1]),

$$H_{2,2}^{1,2}\left[z\,\middle|\,\begin{matrix}(1-a,1),(1-b,1)\\(0,1),(1-c,1)\end{matrix}\right]=\frac{\Gamma(a)\Gamma(b)}{\Gamma(c)}\ {}_2F_1(a,b;c;-z);\tag{2.9.15}$$

$$H^{1,p}_{p,q+1}\left[z\,\middle|\,\begin{matrix}(1-a_i,1)_{1,p}\\(0,1),(1-b_j,1)_{1,q}\end{matrix}\right]=\frac{\prod\limits_{i=1}^{p}\Gamma(a_i)}{\prod\limits_{j=1}^{q}\Gamma(b_j)}\,{}_pF_q(a_1,\cdots,a_p;b_1,\cdots,b_q;-z). \tag{2.9.16}$$

The following H-function is also reduced to a function of generalized hypergeometric type:

$$H^{p,1}_{q+1,p}\left[z\,\middle|\,\begin{matrix}(1,1),(b_j,1)_{1,q}\\(a_i,1)_{1,p}\end{matrix}\right]=E(a_1,\cdots,a_p:b_1,\cdots,b_q:z), \tag{2.9.17}$$

where $E(a_1,\cdots,a_p:b_1,\cdots,b_q:z)$ is the MacRobert E-function (see Erdélyi, Magnus, Oberhettinger and Tricomi [1, Section 5.2]).

Now we give the H-functions which are reduced to Bessel type functions:

$$H^{1,0}_{0,2}\left[\frac{z^2}{4}\,\middle|\,\begin{matrix}\overline{\qquad\qquad}\\\left(\frac{a+\eta}{2},1\right),\left(\frac{a-\eta}{2},1\right)\end{matrix}\right]=\left(\frac{z}{2}\right)^a J_\eta(z), \tag{2.9.18}$$

where $J_\eta(z)$ is the Bessel function of the first kind (2.6.2);

$$H^{2,0}_{0,2}\left[\frac{z^2}{4}\,\middle|\,\begin{matrix}\overline{\qquad\qquad}\\\left(\frac{a-\eta}{2},1\right),\left(\frac{a+\eta}{2},1\right)\end{matrix}\right]=2\left(\frac{z}{2}\right)^a K_\eta(z), \tag{2.9.19}$$

where $K_\eta(z)$ is the modified Bessel function of the third kind or Macdonald function (see Erdélyi, Magnus, Oberhettinger and Tricomi [2, Section 7.2.2] and (8.9.2));

$$H^{2,0}_{1,3}\left[\frac{z^2}{4}\,\middle|\,\begin{matrix}\left(\frac{a-\eta-1}{2},1\right)\\\left(\frac{a-\eta}{2},1\right),\left(\frac{a+\eta}{2},1\right),\left(\frac{a-\eta-1}{2},1\right)\end{matrix}\right]=\left(\frac{z}{2}\right)^a Y_\eta(z), \tag{2.9.20}$$

where $Y_\eta(z)$ is the Bessel function of the second kind or the Neumann function (see Erdélyi, Magnus, Oberhettinger and Tricomi [2, Section 7.2.1] and (8.7.2));

$$H^{2,0}_{1,2}\left[z\,\middle|\,\begin{matrix}(a-\lambda+1,1)\\\left(a+\mu+\frac{1}{2},1\right),\left(a-\mu+\frac{1}{2},1\right)\end{matrix}\right]=z^a e^{-z/2}W_{\lambda,\mu}(z), \tag{2.9.21}$$

where $W_{\lambda,\mu}(z)$ is the Whittaker function (see Erdélyi, Magnus, Oberhettinger and Tricomi [1, Section 6.9] and (7.2.2));

$$H^{3,1}_{1,3}\left[\frac{z^2}{4}\,\middle|\,\begin{matrix}\left(\frac{1+\mu}{2},1\right)\\\left(\frac{1+\mu}{2},1\right),\left(\frac{\eta}{2},1\right),\left(-\frac{\eta}{2},1\right)\end{matrix}\right]$$

$$=2^{1-\mu}\Gamma\left(\frac{1}{2}-\frac{\mu}{2}-\frac{\eta}{2}\right)\Gamma\left(\frac{1}{2}-\frac{\mu}{2}+\frac{\eta}{2}\right)s_{\mu,\eta}(z), \tag{2.9.22}$$

where $s_{\mu,\eta}(z)$ is the Lommel function (see Erdélyi, Magnus, Oberhettinger and Tricomi [2, Section 7.5.5]).

Finally we present the special cases of the H-function which cannot be obtained from the G-function:

$$H_{0,2}^{1,0}\left[z \,\middle|\, \begin{matrix} \overline{\qquad\qquad} \\ (0,1),(-\nu,\mu) \end{matrix}\right] = J^{\mu,\nu}(z), \tag{2.9.23}$$

where

$$J^{\mu,\nu}(z) = \sum_{k=0}^{\infty} \frac{(-z)^k}{\Gamma(k\mu+\nu+1)\ k!} \tag{2.9.24}$$

is the Bessel–Maitland function (see Marichev [1, (11.63)]);

$$H_{1,3}^{1,1}\left[\frac{z^2}{4} \,\middle|\, \begin{matrix} \left(\lambda+\frac{\nu}{2},1\right) \\ \left(\lambda+\frac{\nu}{2},1\right),\left(\frac{\nu}{2},1\right),\left(\mu\left[\lambda+\frac{\nu}{2}\right]-\lambda-\nu,\mu\right) \end{matrix}\right] = J_{\nu,\lambda}^{\mu}(z), \tag{2.9.25}$$

where

$$J_{\nu,\lambda}^{\mu}(z) = \sum_{k=0}^{\infty} \frac{(-1)^k}{\Gamma(k\mu+\nu+\lambda+1)\Gamma(k+\lambda+1)} \left(\frac{z}{2}\right)^{\nu+2\lambda+2k} \tag{2.9.26}$$

is the generalized Bessel–Maitland function (see Marichev [1, (8.2)]);

$$H_{1,2}^{1,1}\left[-z \,\middle|\, \begin{matrix} (0,1) \\ (0,1),(1-\mu,\rho) \end{matrix}\right] = E_{\rho,\mu}(z), \tag{2.9.27}$$

where

$$E_{\rho,\mu}(z) = \sum_{k=0}^{\infty} \frac{z^k}{\Gamma(\rho k+\mu)} \tag{2.9.28}$$

is the Mittag–Leffler function (see Erdélyi, Magnus, Oberhettinger and Tricomi [3, Section 18.2]);

$$H_{p,q+1}^{1,p}\left[z \,\middle|\, \begin{matrix} (1-a_i,\alpha_i)_{1,p} \\ (0,1),(1-b_j,\beta_j)_{1,q} \end{matrix}\right] = {}_p\Psi_q\left[\begin{matrix} (a_i,\alpha_i)_{1,p}; \\ (b_j,\beta_j)_{1,q}; \end{matrix}\ z\right], \tag{2.9.29}$$

where

$${}_p\Psi_q\left[\begin{matrix} (a_i,\alpha_i)_{1,p}; \\ (b_j,\beta_j)_{1,q}; \end{matrix}\ z\right] = \sum_{k=0}^{\infty} \frac{\prod_{i=1}^{p}\Gamma(a_i+k\alpha_i)}{\prod_{j=1}^{q}\Gamma(b_j+k\beta_j)} \frac{z^k}{k!} \tag{2.9.30}$$

is Wright's generalized hypergeometric function (see Erdélyi, Magnus, Oberhettinger and Tricomi [1, Section 4.1]);

$$H_{0,2}^{2,0}\left[z \,\middle|\, \begin{matrix} \overline{\qquad\qquad} \\ (0,1),\left(\frac{\gamma}{\beta},\frac{1}{\beta}\right) \end{matrix}\right] = \beta Z_{\beta}^{\gamma}(z) \tag{2.9.31}$$

with

$$Z_\beta^\gamma(z) = \int_0^\infty t^{\gamma-1} \exp\left(-t^\beta - \frac{z}{t}\right) dt \quad (\gamma \in \mathbb{C},\ \beta > 0); \tag{2.9.32}$$

$$H_{1,2}^{2,0}\left[z \,\middle|\, \begin{matrix} \left(\gamma + 1 - \frac{1}{n}, \frac{1}{n}\right) \\ (\gamma n, 1), \left(0, \frac{1}{n}\right) \end{matrix}\right] = (2\pi)^{(1-n)/2} n^{\gamma n + 1/2} \lambda_\gamma^{(n)}(z) \tag{2.9.33}$$

with

$$\lambda_\gamma^{(n)}(z) = \frac{(2\pi)^{(n-1)/2}\sqrt{n}}{\Gamma\left(\gamma + 1 - \frac{1}{n}\right)} \left(\frac{z}{n}\right)^{\gamma n} \int_1^\infty (t^n - 1)^{\gamma - 1/n} e^{-zt} dt \tag{2.9.34}$$

$$\left(n \in \mathbb{N};\ \mathrm{Re}(\gamma) > \frac{1}{n} - 1,\ \mathrm{Re}(z) > 0\right);$$

$$H_{1,2}^{2,0}\left[z \,\middle|\, \begin{matrix} \left(1 - \frac{\sigma+1}{\beta}, \frac{1}{\beta}\right) \\ (0, 1), \left(-\gamma - \frac{\sigma}{\beta}, \frac{1}{\beta}\right) \end{matrix}\right] = \lambda_{\gamma,\sigma}^{(\beta)}(z) \tag{2.9.35}$$

with

$$\lambda_{\gamma,\sigma}^{(\beta)}(z) = \frac{\beta}{\Gamma\left(\gamma + 1 - \frac{1}{\beta}\right)} \int_1^\infty (t^\beta - 1)^{\gamma - 1/\beta} t^\sigma e^{-zt} dt \tag{2.9.36}$$

$$\left(\beta > 0;\ \mathrm{Re}(\gamma) > \frac{1}{\beta} - 1; \sigma \in \mathbb{C},\ \mathrm{Re}(z) > 0\right).$$

According to the following integral representations in Erdélyi, Magnus, Oberhettinger and Tricomi [2, 7.12(23) and 7.12(19)] for the modified Bessel function of the third kind or the Macdonald function $K_\nu(z)$:

$$K_\nu(z) = \frac{1}{2} \int_0^\infty e^{-z(t+1/t)/2} t^{-\nu-1} dt \tag{2.9.37}$$

$$= \frac{\sqrt{\pi}}{\Gamma\left(\frac{1}{2} - \nu\right)} \left(\frac{2}{z}\right)^\nu \int_1^\infty e^{-zt} (t^2 - 1)^{-\nu - 1/2} dt \quad (\mathrm{Re}(z) > 0), \tag{2.9.38}$$

the functions (2.9.34), (2.9.32) and (2.9.36) are connected with the function $K_{-\gamma}(z)$ by the relations

$$Z_1^\gamma\left(\frac{z^2}{4}\right) = 2\left(\frac{z}{2}\right)^\gamma K_{-\gamma}(z); \tag{2.9.39}$$

$$\lambda_\gamma^{(2)}(z) = 2\left(\frac{z}{2}\right)^\gamma K_{-\gamma}(z); \tag{2.9.40}$$

$$\lambda_{\gamma,0}^{(2)}(z) = \frac{2}{\sqrt{\pi}} \left(\frac{2}{z}\right)^\gamma K_{-\gamma}(z). \tag{2.9.41}$$

We also note that the function $\lambda_\gamma^{(n)}(z)$ in (2.9.34) is expressed via $\lambda_{\gamma,\sigma}^{(\beta)}(z)$ in (2.9.36) when $\sigma = 0$ and $\beta = n \in \mathbb{N}$:

$$\lambda_\gamma^{(n)}(z) = (2\pi)^{(n-1)/2} n^{-(\gamma n+1/2)} z^{\gamma n} \lambda_{\gamma,0}^{(n)}(z). \tag{2.9.42}$$

2.10. Bibliographical Remarks and Additional Information on Chapter 2

For Section 2.1. The elememtry properties of the H-function given in Section 2.1 were shown by many authors (see the books by Mathai and R.K. Saxena [2, Section 1.2] and Srivastava, Gupta and Goyal [1, Section 2.3]). Other various elementary properties of the H-function were given by Braaksma [1], Gupta [2], K.C. Gupta and U.C. Jain [1], [3], Lawrynowicz [1], Anandani [2], [3], Bajpai [1], Skibinski [1] and Chaurasia [1].

For Section 2.2. The differentiation formulas (2.2.1)–(2.2.5) were proved by K.C. Gupta and U.C. Jain [2], Nair [1], [2] and Oliver and Kalla [1], respectively. The relations (2.2.4), (2.2.5) and (2.2.6) were given in the book by Prudnikov, Brychkov and Marichev [3, (8.3.2.20), (8.3.2.18) and (8.3.2.19)]. The formulas (2.2.7)–(2.2.10) were obtained by Lawrynowicz [1] (see also (2.2.7) and (2.2.9) in Prudnikov, Brychkov and Marichev [3, (8.3.2.17) and (8.3.2.16)]). Other differentiation relations were also presented in the papers above and in the papers by A.N. Goyal and G.K. Goyal [1], Anandani [6] (see Mathai and R.K. Saxena [2, Section 1.3] and Srivastava, Gupta and Goyal [1, Section 2.7]). We also note that partial derivatives of the H-function with respect to parameters were investigated by Buschman [3], while Pathak [4] established the differential equations for the H-function (1.1) with positive rational α_i and β_j.

For Section 2.3. A number of recurrence relations for the H-function were obtained by Gupta [2], U.C. Jain [1], Agrawal [1], Anandani [2], [3], Mathur [1], Bora and Kalla [1], A. Srivastava and Gupta [1], Raina [1] and other authors (see the bibliography in the books of Mathai and R.K. Saxena [2] and of Srivastava, Gupta and Goyal [1]). They used methods based on the evaluation of an integral of a product of the H-function and some other special functions and have applied the known identities for the latter. Buschman [2] has used another simpler method, obtained 30 so-called contiguous relations and gave applications to find finite series expansions. We present two such formulas in (2.3.1) and (2.3.2) and six finite series expansions (2.3.3)–(2.3.5) and (2.3.9)–(2.3.11), which hold for any $r \in \mathbb{N}$. We also indicate that the particular case $r = n$ of the relations (2.3.3)–(2.3.5) was given by Srivastava, Gupta and Goyal [1, (2.9.6)–(2.9.8)].

In the proof of Theorem 2.1 we follow the method analogous to that adopted by Meijer [5] for the G-function (2.9.1). The relations (2.3.12)–(2.3.15) were first proved by Lawrynowicz [1] under some additional conditions to those given in Theorem 2.1. Shah [1]–[3] and A. Srivastava and Gupta [1] have obtained some other elementary expansion formulas (see Mathai and R.K. Saxena [2, Section 1.5] and Srivastava, Gupta and Goyal [1, Section 2.9]).

Many authors have investigated the expansions of the H-functions in series of orthogonal functions such as Fourier series, Fourier–Bessel series, Fourier–Jacobi series, Neumann series, etc. Bajpai [1]–[3] has obtained expansions of the H-functions in terms of Jacobi polynomials, sine- and cosine-functions and Bessel functions. Shah [1] and Anandani [8] have proved the expansions of the H-functions in terms of the associated Legendre functions, while S.L. Soni [1] and Shah [2] derived such results in terms of Gegenbauer polynomials (see, for the formulas, other results and bibliography, Mathai and R.K. Saxena [2, Chapter 3] and Srivastava, Gupta and Goyal [1, Chapter 5]). We only note that Shah [4] considered some extensions of Theorem 2.1 and Anandani [1] has established the general formulas which give the expansion of the H-function in series of products of the generalized hypergeometric function ${}_pF_q(a_1, \cdots, a_p; b_1, \cdots, b_q; z)$ and the H-functions of lower order (see Srivastava, Gupta and Goyal [1, (5.6.2)–(5.6.3)]).

For Section 2.4. The multiplication formulas (2.4.5) and (2.4.10) were proved by K.C. Gupta and U.C. Jain [1], [3]. In [3] from (2.4.10) they deduced a relation between the H-function and the Meijer G-function (see Srivastava, Gupta and Goyal [1, (2.5.4)]). Such a result for rational α_i β_j $(i = 1, \cdots, p;\ j = 1, \cdots, q)$ was proved earlier by Boersma [1], and on the basis of this result Pathak [4] established the differential equation for the H-function when α_i and β_j $(i = 1, \cdots, p;\ j = 1, \cdots, q)$ are positive rational numbers.

Bajpai [4] and S.P. Goyal [1] have proved certain transformations of infinite series involving the H-function. In (2.4.11) we present one such result. Various finite and infinite series for the H-functions were given by R.N. Jain [1], M.M. Srivastava [1], Anandani [3]–[5], [7], Olkha [1], S.P. Goyal [1], R.K. Saxena and Mathur [1], Skibinski [1], Gupta and A. Srivastava [1], Taxak [1], and others (see Srivastava, Gupta and Goyal [1, Section 2.11] in this connection).

It should be noted that many authors have investigated finite and infinite series for the H-function (see the results and bibliography in the books by Mathai and R.K. Saxena [2, Chapter 3] and Srivastava, Gupta and Goyal [1, Chapter 5]).

For Section 2.5. The relation (2.5.7) for the Mellin transform is well known and can be formally deduced from the Mellin inversion theorem (see, for example, the books by Titchmarsh [3], Sneddon [1], Ditkin and Prudnikov [1]). The formulas (2.5.12) and (2.5.25) for the Mellin and Laplace transforms of the H-function were presented in the book by Srivastava, Gupta and Goyal [1, (2.4.1) and (2.4.2)] under the conditions (2.5.10) and (2.5.23), respectively in the case when $a^* > 0$ and $\Delta > 0$. For $a^* > 0$ the relation (2.5.12) with $w = 0$, $\sigma = 1$ and (2.5.25) were also given by Prudnikov, Brychkov and Marichev [3, (2.25.2.1) and (2.25.2.3)].

The asymptotic results for the H-function at infinity and zero, given in Sections 1.5, 1.6 and 1.8, 1.9, allow us to extend (2.5.7), (2.5.12), (2.5.16) as well as (2.5.25) to the case when $a^* = 0$.

For Section 2.6. The formula of the form (2.6.11) with $a = \sigma = 1$ was given in Prudnikov, Brychkov and Marichev [3, (2.25.3.2)] without any condition of the form (2.6.10).

The asymptotic estimates for the H-function at zero and infinity given in Sections 1.5 and 1.8 allow us to extend (2.6.11) as well as (2.6.6) to the case when $a^* = \Delta = 0$.

We also indicate that O.P. Sharma [1] obtained the Hankel transform of the H-function (4.11.4) considered by Fox [2], and showed that the function

$$f(x) = x^{1/2-\eta-2\sigma} H^{n+1,n}_{p,p+1}\left[\frac{x^2}{2}\,\middle|\,\begin{matrix}(a_i,\alpha_i)_{1,p-1},(a_p,\alpha_1)\\(\eta+\sigma,1),(b_j,\alpha_j)_{1,p-1},(b_p,\alpha_1)\end{matrix}\right] \tag{2.10.1}$$

is a self-reciprocal function (see (8.14.30)) with respect to the Hankel transform:

$$\left(\mathbb{H}_\eta f\right)(x) = f(x) \quad (x > 0), \tag{2.10.2}$$

where $b_j = 1 - a_j + (2\sigma + \eta - 1)\alpha_j$ $(1 \leqq j \leqq n - 1)$, $b_p = 1 - a_p + (2\sigma + \eta - 1)\alpha_1$, provided that $p > n \geqq 0$, $\eta \in \mathbb{C}$ $(\mathrm{Re}(\eta) > -1)$, $\sigma \in \mathbb{C}$ and

$$\begin{gathered}2\left(\sum_{i=2}^{n}\alpha_i - \sum_{i=n+1}^{p-1}\alpha_i\right) > -1, \quad \mathrm{Re}\{2 - a_p + (2\sigma+\eta-1)\alpha_1\} > 0;\\ \mathrm{Re}\left(\eta + \sigma + \frac{1-a_i}{\alpha_i}\right) > 0 \quad (i = 1, 2, \cdots, n).\end{gathered} \tag{2.10.3}$$

For Section 2.7. In the case $a^* > 0$, fractional integration and differentiation of the H-function were considered by Raina and Koul [1]–[3], Srivastava, Gupta and Goyal [1] and Prudnikov, Brychkov and Marichev [3]. Raina and Koul [2] probably first considered fractional differentiation D^α_- in (2.7.6) with real $\alpha > 0$ of the H-function. However their result in [2, (14a)] (presented also in Raina and Koul [3, (2.2)] and Srivastava, Gupta and Goyal [1, (2.7.9)]) contains a mistake and should be replaced by (2.7.25) with the conditions in (2.7.23). For real $\alpha > 0$ the formula (2.7.12) was given by Raina and Koul [3, (2.5)], but the conditions for its validity should be corrected in line with (2.7.26). The relation (2.7.22) was obtained by Srivastava, Gupta and Goyal [1, (2.7.13)], but the conditions should also be corrected in accordance with (2.7.7). A formula more general than (2.7.9) was indicated by Prudnikov, Brychkov and Marichev [3, (2.25.2.2)].

Due to the asymptotic behavior of the H-function at infinity and zero given in Sections 1.5, 1.6 and 1.8, 1.9, the formulas (2.7.9), (2.7.11), (2.7.22) and (2.7.25) can be extended to the case when $a^* = 0$. The results of this section are based on those obtained in our papers Kilbas and Saigo [4], [5], where, in the case $a^* = 0$, instead of the condition $\mathrm{Re}(\mu) < -1$ we have used the condition $\Delta\gamma + \mathrm{Re}(\mu) < -1$ while considering the H-function (1.1.1) with $\mathfrak{L} = \mathfrak{L}_{i\gamma\infty}$. In our paper Saigo and Kilbas [4], these results were extended to generalized fractional integrals and derivatives with the Gauss hypergeometric function (2.9.2) in the kernel, introduced by Saigo [1] (see (7.12.45), (7.12.46) and Samko, Kilbas and Marichev [1, Section 23.2, note 18.6]). Such a generalized fractional integration and differentiation of the H-function $H^{q,p}_{2p,2q}(x)$ given by (1.11.1) was considered by Saigo, R.K. Saxena and Ram [1]. R.K. Saxena and Saigo [2] extended the results in Saigo and Kilbas [4] to generalized fractional integration and differentiation operators with the Appell function $F_3(a, a', b, b'; c; x, y)$ in the kernel (see Erdélyi, Magnus, Oberhettinger and Tricomi [1, 5.7.1(8)]):

$$F_3(a, a', b, b'; c; x, y) = \sum_{k,l=0}^{\infty} \frac{(a)_k (a')_l (b)_k (b')_l}{(c)_{k+l}\, k!\, l!} x^k y^l \quad (0 < x, y < 1), \tag{2.10.4}$$

where $(a)_k$ is the Pochhammer symbol (2.4.12).

For Section 2.8. The integral formula (2.8.4) was proved by K.C. Gupta and U.C. Jain [1] provided that $a^* > 0$, $a_0^* > 0$, $\Delta > 0$ and $\Delta_0 > 0$, and the relation (2.8.17) was obtained by G.K. Goyal [1] for $a^* > 0$ and $\Delta > 0$ (see also Srivastava, Gupta and Goyal [1, (5.2.1) and (5.2.3)]). Prudnikov, Brychkov and Marichev [3, (2.25.1.1) and (2.25.2.2)] indicated that the formulas (2.8.4) and (2.8.17) hold for $a^* > 0$, $a_0^* > 0$ and $a^* > 0$, respectively.

By the asymptotic estimates of the H-function at infinity and zero given in Sections 1.5, 1.6 and 1.8, 1.9, the relations (2.8.4) and (2.8.17) are extended to the case $a^* = 0$, though the former is given in Theorem 2.10 only for positive $z = x > 0$.

Many known integrals presented in the three-volume work by Prudnikov, Brychkov and Marichev [1]–[3] can be obtained from the integrals in (2.8.4) and (2.8.17) by taking the H-function as certain elementary or special function. It should be noted that there is a wide list of papers devoted to various integral formulas (finite and infinite) involving the H-function. See the books by Mathai and R.K. Saxena [2, Chapter 2], Srivastava, Gupta and Goyal [1, Chapter 5] and Prudnikov, Brychkov and Marichev [3].

We also note that R.K. Saxena and Saigo [1] evaluated the integral of the form

$$\int_a^x (t-a)^{\omega}(x-t)^{\alpha-1}(ct+d)^{\gamma} H^{m,n}_{p,q}\left[a(t-a)^{\sigma}(x-t)^{\gamma}(ct+d)^{-\nu} \,\middle|\, \begin{matrix} (a_i, \alpha_i)_{1,p} \\ (b_j, \alpha_j)_{1,q} \end{matrix} \right] \tag{2.10.5}$$

more general than that in (2.8.17). Laddha [1] also calculated the Erdélyi–Kober type integrals generalizing the operators in (3.3.1) and (3.3.2) and involving the generalized polynomial sets of the H-function.

For Section 2.9. The particular cases of the H-function as the Meijer G-function were first given by Erdélyi, Magnus, Oberhettinger and Tricomi [1, Section 5.6] and then presented in the books by Mathai and R.K. Saxena [1, Chapter II], Mathai and R.K. Saxena [2, Section 1.7], Srivastava, Gupta and Goyal [1, Section 2.6] and Prudnikov, Brychkov and Marichev [3, Section 8.4.52]. Some of these cases were presented in the formulas (2.9.4)–(2.9.22). The relations (2.9.23) and (2.9.29), when the H-function is not reduced to the G-function, were also considered in Mathai and R.K. Saxena [2, (1.7.8)–(1.7.9)], Srivastava, Gupta and Goyal [1, (2.6.10)–(2.6.11)] and Prudnikov, Brychkov and Marichev [3, (8.4.51.4)], while the relations (2.9.25) and (2.9.27) are found in Prudnikov, Brychkov and Marichev [3, (8.4.51.7) and (8.4.51.6)].

The formulas (2.9.31), (2.9.33) and (2.9.35) are never mentioned in the monograph literature. The modified Bessel type functions (2.9.32) and (2.9.34) were introduced by Krätzel in [5] and [1], respectively, who also investigated the properties of the former function in [5] and of the latter in [1]–[4]. The function in (2.9.36) was introduced by the authors and Glaeske in Kilbas, Saigo and Glaeske [1] and Glaeske, Kilbas and Saigo [1], where the relations (2.9.40)–(2.9.42) were proved. Such a function was used by Bonilla, Kilbas, Rivero, Rodriguez and Trujillo [2] to solve some homogeneous differential equations of fractional order and Volterra integral equations.

Chapter 3
H-TRANSFORM ON THE SPACE $\mathfrak{L}_{\nu,2}$

3.1. The H-Transform and the Space $\mathfrak{L}_{\nu,r}$

This and the next chapters deal with the integral transform of the form

$$\left(\boldsymbol{H}f\right)(x) = \int_0^\infty H_{p,q}^{m,n}\left[xt \,\middle|\, \begin{matrix}(a_i,\alpha_i)_{1,p}\\(b_j,\beta_j)_{1,q}\end{matrix}\right] f(t)dt, \tag{3.1.1}$$

where $H_{p,q}^{m,n}\left[z \,\middle|\, \begin{matrix}(a_p,\alpha_p)\\(b_q,\beta_q)\end{matrix}\right]$ is the H-function defined in (1.1.1). Such a transform is called an $\boldsymbol{H}$-transform.

Most of the known integral transforms can be put into the form (3.1.1). In fact, for $\alpha_i = \beta_j = 1$ $(1 \leqq i \leqq p;\ 1 \leqq j \leqq q)$ the function $H_{p,q}^{m,n}\left[z \,\middle|\, \begin{matrix}(a_i,\alpha_i)_{1,p}\\(b_j,\beta_j)_{1,q}\end{matrix}\right]$ is interpreted as the Meijer G-function $G_{p,q}^{m,n}\left[z \,\middle|\, \begin{matrix}(a_i)_{1,p}\\(b_j)_{1,q}\end{matrix}\right]$ given in (2.9.1). Then (3.1.1) is reduced to the so-called integral transform with G-function kernel or $\boldsymbol{G}$-transform

$$\left(\boldsymbol{G}f\right)(x) = \int_0^\infty G_{p,q}^{m,n}\left[xt \,\middle|\, \begin{matrix}(a_i)_{1,p}\\(b_j)_{1,q}\end{matrix}\right] f(t)dt. \tag{3.1.2}$$

Such a transform includes the classical Laplace and Hankel transforms given in (2.5.2) and (2.6.1). The Riemann–Liouville fractional integrals (2.7.1) and (2.7.2) and other integral transforms can be reduced to this $\boldsymbol{G}$-transform. For the theory and historical notes the reader is referred to Samko, Kilbas and Marichev [1, §§36, 39]. There are other transforms which cannot be reduced to $\boldsymbol{G}$-transforms but can be put into the $\boldsymbol{H}$-transforms given in (3.1.1). They are the generalized Laplace and Hankel transforms, the Erdélyi–Kober type fractional integration operators, the transforms with the Gauss hypergeometric function as kernel, the Bessel-type integral transforms, etc. (see Sections 3.3, 3.7 and Chapters 7 and 8 below). In Chapter 5 we consider such former and latter particular cases of the $\boldsymbol{H}$-transform (3.1.1) more precisely.

In the present chapter, we study the $\boldsymbol{H}$-transform (3.1.1) on the summable space $\mathfrak{L}_{\nu,2}$ $(\nu \in \mathbb{R})$, while the next chapter is devoted to that on $\mathfrak{L}_{\nu,r}$ $(\nu \in \mathbb{R},\ 1 \leqq r < \infty)$. Here, the space $\mathfrak{L}_{\nu,r}$ consist of those Lebesgue measurable complex valued functions f for which

$$\|f\|_{\nu,r} = \left\{\int_0^\infty |t^\nu f(t)|^r \frac{dt}{t}\right\}^{1/r} < \infty \quad (1 \leqq r < \infty,\ \nu \in \mathbb{R}). \tag{3.1.3}$$

In particular, when $\nu = 1/r$ the space $\mathfrak{L}_{1/r,r}$ coincides with the space $L_r(\mathbb{R}_+)$ of r-summable functions on $\mathbb{R}_+$ with the norm

$$\|f\|_r = \left\{\int_0^\infty |f(t)|^r \, dt\right\}^{1/r} < \infty \quad (1 \leqq r < \infty). \tag{3.1.4}$$

Further we shall investigate the particular aspects of the $\boldsymbol{H}$-transforms (3.1.1) on $\mathfrak{L}_{\nu,r}$ together with the main problems such as the existence, boundedness, representation, range and invertibility.

We have said that the transform $\boldsymbol{H}$ under discussion is "formally" defined by (3.1.1), and we must make this more precise. If we formally take the Mellin transform $\mathfrak{M}$ of (3.1.1), we obtain

$$\left(\mathfrak{M}\boldsymbol{H}f\right)(s) = \mathcal{H}_{p,q}^{m,n}\left[\begin{array}{c|c} (a_i, \alpha_i)_{1,p} & \\ & s \\ (b_j, \beta_j)_{1,q} & \end{array}\right]\left(\mathfrak{M}f\right)(1-s), \tag{3.1.5}$$

where $\mathcal{H}_{p,q}^{m,n}\left[\begin{array}{c|c} (a_i, \alpha_i)_{1,p} & \\ (b_j, \beta_j)_{1,q} & s \end{array}\right]$ is given in (1.1.2). When there is no confusion, we denote such a function simply by $\mathcal{H}(s)$. It transpires that, for certain ranges of the parameters in $\mathcal{H}(s)$, (3.1.5) can be used to define the transform $\boldsymbol{H}$ on $\mathfrak{L}_{\nu,r}$. Namely, for certain $h \in \mathbb{R} \setminus \{0\}$ and $\lambda \in \mathbb{C}$, we have

$$\left(\boldsymbol{H}f\right)(x)$$
$$= hx^{1-(\lambda+1)/h}\frac{d}{dx}x^{(\lambda+1)/h}\int_0^\infty H_{p+1,q+1}^{m,n+1}\left[xt\,\middle|\begin{array}{c} (-\lambda, h), (a_i, \alpha_i)_{1,p} \\ (b_j, \beta_j)_{1,q}, (-\lambda-1, h) \end{array}\right]f(t)dt \tag{3.1.6}$$

or

$$\left(\boldsymbol{H}f\right)(x)$$
$$= -hx^{1-(\lambda+1)/h}\frac{d}{dx}x^{(\lambda+1)/h}\int_0^\infty H_{p+1,q+1}^{m+1,n}\left[xt\,\middle|\begin{array}{c} (a_i, \alpha_i)_{1,p}, (-\lambda, h) \\ (-\lambda-1, h), (b_j, \beta_j)_{1,q} \end{array}\right]f(t)dt. \tag{3.1.7}$$

Due to (2.2.1), (2.2.2), (2.1.1) and (2.1.2), formal differentiation with respect to x under the integral sign in (3.1.6) and (3.1.7) yields (3.1.1). Later we give conditions for the representability of (3.1.6), (3.1.7) as well as (3.1.1) being valid.

Our main tool for studying the $\boldsymbol{H}$-transform on $\mathfrak{L}_{\nu,r}$-space is based on the Mellin transform to which more precise investigation should be made for such spaces.

3.2. The Mellin Transform on $\mathfrak{L}_{\nu,r}$

The Mellin transform $\mathfrak{M}$ is usually defined by (2.5.1), namely

$$\left(\mathfrak{M}f\right)(s) = \int_0^\infty f(x)x^{s-1}dx \tag{3.2.1}$$

for complex $s = \nu + it$ $(\nu, t \in \mathbb{R})$. However, unless $f \in \mathfrak{L}_{\nu,1}$, the integral in (3.2.1) does not exist. But if we denote by $\mathfrak{F}$ the Fourier transform of a function F on $\mathbb{R}$ defined by

$$\left(\mathfrak{F}F\right)(x) = \int_{-\infty}^{+\infty} F(t)e^{itx}dt \tag{3.2.2}$$

and take $f \in \mathfrak{L}_{\nu,1}$, then by an elementary change of variable we obtain

$$\left(\mathfrak{M}f\right)(\nu + it) = \left(\mathfrak{F}C_\nu f\right)(t), \tag{3.2.3}$$

where

$$\left(C_\nu f\right)(x) = e^{\nu x}f(e^x). \tag{3.2.4}$$

It is shown in Rooney [2, Lemma 2.1] that C_ν is an isometric isomorphism of $\mathfrak{L}_{\nu,r}$ onto $L_r(\mathbb{R})$. Since the Fourier transform is defined on $L_r(\mathbb{R})$ for $1 \leqq r \leqq 2$, the right-hand side of (3.2.3) makes sense for $f \in \mathfrak{L}_{\nu,r}$ $(\nu \in \mathbb{R}, 1 \leqq r \leqq 2)$. From here we come to the following definition.

Definition 3.1. For $f \in \mathfrak{L}_{\nu,r}$ $(\nu \in \mathbb{R}, 1 \leqq r \leqq 2)$, we define the Mellin transform $\mathfrak{M}$ of f by the relation

$$\left(\mathfrak{M}f\right)(\nu + it) = \int_{-\infty}^{\infty} e^{(\nu+it)\tau}f(e^\tau)d\tau. \tag{3.2.5}$$

The relation (3.2.5) is also written as $\left(\mathfrak{M}f\right)(s)$ with $\mathrm{Re}(s) = \nu$. In particular, if $f \in \mathfrak{L}_{\nu,2} \cap \mathfrak{L}_{\nu,1}$, the Mellin transform $\left(\mathfrak{M}f\right)(s)$ is given by the usual expression (2.5.1) for $\mathrm{Re}(s) = \nu$.

Property 3.1. The Mellin transform (3.2.5) has the properties

(a) $\mathfrak{M}$ is a unitary mapping of $\mathfrak{L}_{\nu,2}$ onto $L_2(\mathbb{R})$.

(b) For $f \in \mathfrak{L}_{\nu,2}$, there holds

$$f(x) = \frac{1}{2\pi i} \lim_{R\to\infty} \int_{\nu-iR}^{\nu+iR} \left(\mathfrak{M}f\right)(s)x^{-s}ds, \tag{3.2.6}$$

where the limit is taken in the topology of $\mathfrak{L}_{\nu,2}$ and where, if $F(\nu + it) \in L_1(-R, R)$,

$$\int_{\nu-iR}^{\nu+iR} F(s)ds = i\int_{-R}^{R} F(\nu + it)dt. \tag{3.2.7}$$

(c) For $f \in \mathfrak{L}_{\nu,2}$ and $g \in \mathfrak{L}_{1-\nu,2}$, there holds

$$\int_0^\infty f(x)g(x)dx = \frac{1}{2\pi i}\int_{\nu-i\infty}^{\nu+i\infty} \left(\mathfrak{M}f\right)(s)\left(\mathfrak{M}g\right)(1-s)ds. \tag{3.2.8}$$

In the theory of $\boldsymbol{H}$-transforms on $\mathfrak{L}_{\nu,r}$-spaces we require a multiplier theorem for the Mellin transform. First we give a definition.

Definition 3.2. We say that a function m belongs to the class $\mathcal{A}$, if there are extended real number $\alpha(m)$ and $\beta(m)$ with $\alpha(m) < \beta(m)$ such that

(a) $m(s)$ is analytic in the strip $\alpha(m) < \mathrm{Re}(s) < \beta(m)$.

(b) $m(s)$ is bounded in every closed substrip $\sigma_1 \leqq \mathrm{Re}(s) \leqq \sigma_2$, where $\alpha(m) \leqq \sigma_1 \leqq \sigma_2 \leqq \beta(m)$.

(c) $|m'(\sigma + it)| = O(|t|^{-1})$ as $|t| \to \infty$ for $\alpha(m) < \sigma < \beta(m)$.

For two Banach spaces X and Y we use the notation $[X, Y]$ to denote the collection of bounded linear operators from X to Y, and $[X, X]$ is abbreviated to $[X]$. There holds the following multiplier theorem in Rooney [2, Theorem 1].

Theorem 3.1. *Suppose $m \in \mathcal{A}$. Then there is a transform $T_m \in [\mathfrak{L}_{\nu,r}]$ with $\alpha(m) < \nu < \beta(m)$ and $1 < r < \infty$ so that, if $f \in \mathfrak{L}_{\nu,r}$ with $\alpha(m) < \nu < \beta(m)$ and $1 < r \leqq 2$, there holds the relation*

$$\left(\mathfrak{M}T_m f\right)(s) = m(s)\left(\mathfrak{M}f\right)(s) \quad (\mathrm{Re}(s) = \nu). \tag{3.2.9}$$

For $\alpha(m) < \nu < \beta(m)$ and $1 < r \leqq 2$, the transform T_m is one-to-one on $\mathfrak{L}_{\nu,r}$, except when $m = 0$. If $1/m \in \mathcal{A}$, then for $\max[\alpha(m), \alpha(1/m)] < \nu < \min[\beta(m), \beta(1/m)]$ and for $1 < r < \infty$, T_m maps $\mathfrak{L}_{\nu,r}$ one-to-one onto itself, and for the inverse operator T_m^{-1} there holds the formula

$$T_m^{-1} = T_{1/m}. \tag{3.2.10}$$

3.3. Some Auxiliary Operators

In the discussions of the next chapter we shall use special integral operators. The first of them are the Erdélyi–Kober type fractional integrals (see Samko, Kilbas and Marichev [1, §18.1]) defined by

$$\left(I^{\alpha}_{0+;\sigma,\eta}f\right)(x) = \frac{\sigma x^{-\sigma(\alpha+\eta)}}{\Gamma(\alpha)}\int_0^x (x^{\sigma} - t^{\sigma})^{\alpha-1} t^{\sigma\eta+\sigma-1} f(t)dt, \tag{3.3.1}$$

$$\left(I^{\alpha}_{-;\sigma,\eta}f\right)(x) = \frac{\sigma x^{\sigma\eta}}{\Gamma(\alpha)}\int_x^{\infty} (t^{\sigma} - x^{\sigma})^{\alpha-1} t^{\sigma(1-\alpha-\eta)-1} f(t)dt \tag{3.3.2}$$

for $\mathrm{Re}(\alpha) > 0$, $\sigma > 0$, $\eta \in \mathbb{C}$ and $x \in \mathbb{R}_+$. We shall also work with the generalized Laplace transform defined in (2.5.23), that is

$$\left(\mathbb{L}_{k,\alpha}f\right)(x) = \int_0^{\infty} (xt)^{-\alpha} e^{-|k|(xt)^{1/k}} f(t)dt \tag{3.3.3}$$

for $k \in \mathbb{R}\backslash\{0\}$, $\alpha \in \mathbb{C}$, $x \in \mathbb{R}_+$, and with the generalized Hankel transform defined in (2.6.11)

$$\left(\mathbb{H}_{k,\eta}f\right)(x) = \int_0^{\infty} (xt)^{1/k-1/2} J_{\eta}\left(|k|(xt)^{1/k}\right) f(t)dt \tag{3.3.4}$$

for $k \in \mathbb{R} \setminus \{0\}$, $\mathrm{Re}(\eta) > -3/2$, $x \in \mathbb{R}_+$. The transforms (3.3.1)–(3.3.4) are defined for continuous functions f with compact support on $\mathbb{R}_+$ under the parameters indicated.

Definition 3.3. For $\nu \in \mathbb{R}_+$ we denote by $\mathfrak{L}_{\nu,\infty}$ the collection of functions f, measurable on $\mathbb{R}_+$, such that

$$||f||_{\nu,\infty} = \operatorname*{ess\,sup}_{x\in\mathbb{R}} |x^\nu f(x)| < \infty. \tag{3.3.5}$$

The boundedness properties of the transforms (3.3.1)–(3.3.4) and their Mellin transforms are given by the following results in Rooney [6, Theorem 5.1].

Theorem 3.2. *The following statements are valid:*

(a) *If $1 \leqq r \leqq \infty$, $\mathrm{Re}(\alpha) > 0$ and $\nu < \sigma(1+\mathrm{Re}(\eta))$, then for all $s \geqq r$ such that $1/s > 1/r - \mathrm{Re}(\alpha)$, the operator $I^{\alpha}_{0+;\sigma,\eta}$ belongs to $[\mathfrak{L}_{\nu,r}, \mathfrak{L}_{\nu,s}]$ and is one-to-one. For $1 \leqq r \leqq 2$ and $f \in \mathfrak{L}_{\nu,r}$, there holds the relation*

$$\left(\mathfrak{M}I^{\alpha}_{0+;\sigma,\eta}f\right)(s) = \frac{\Gamma\left(1+\eta-\dfrac{s}{\sigma}\right)}{\Gamma\left(1+\eta+\alpha-\dfrac{s}{\sigma}\right)}\left(\mathfrak{M}f\right)(s) \quad (\mathrm{Re}(s)=\nu). \tag{3.3.6}$$

(b) *If $1 \leqq r \leqq \infty$, $\mathrm{Re}(\alpha) > 0$ and $\nu > -\sigma\mathrm{Re}(\eta)$, then for all $s \geqq r$ such that $1/s > 1/r - \mathrm{Re}(\alpha)$, the operator $I^{\alpha}_{-;\sigma,\eta}$ belongs to $[\mathfrak{L}_{\nu,r}, \mathfrak{L}_{\nu,s}]$ and is one-to-one. For $1 \leqq r \leqq 2$ and $f \in \mathfrak{L}_{\nu,r}$, there holds the relation*

$$\left(\mathfrak{M}I^{\alpha}_{-;\sigma,\eta}f\right)(s) = \frac{\Gamma\left(\eta+\dfrac{s}{\sigma}\right)}{\Gamma\left(\eta+\alpha+\dfrac{s}{\sigma}\right)}\left(\mathfrak{M}f\right)(s) \quad (\mathrm{Re}(s)=\nu). \tag{3.3.7}$$

(c) *If $1 \leqq r \leqq s \leqq \infty$, and if $\nu < 1-\mathrm{Re}(\alpha)$ for $k > 0$ and $\nu > 1-\mathrm{Re}(\alpha)$ for $k < 0$, then the operator $\mathbb{L}_{k,\alpha}$ belongs to $[\mathfrak{L}_{\nu,r}, \mathfrak{L}_{1-\nu,s}]$ and is one-to-one. If $1 \leqq r \leqq 2$ and $f \in \mathfrak{L}_{\nu,r}$, then there holds the relation*

$$\left(\mathfrak{M}\mathbb{L}_{k,\alpha}f\right)(s) = \Gamma[k(s-\alpha)]|k|^{1-k(s-\alpha)}\left(\mathfrak{M}f\right)(1-s) \quad (\mathrm{Re}(s)=1-\nu). \tag{3.3.8}$$

(d) *If $1 < r < \infty$ and $\gamma(r) \leqq k(\nu-1/2)+1/2 < \mathrm{Re}(\eta)+3/2$, where*

$$\gamma(r) = \max\left[\frac{1}{r},\frac{1}{r'}\right], \quad \frac{1}{r}+\frac{1}{r'}=1, \tag{3.3.9}$$

then for all $s \geqq r$ such that $s' \geqq (k(\nu-1/2)+1/2)^{-1}$ and $1/s+1/s'=1$, the operator $\mathbb{H}_{k,\eta}$ belongs to $[\mathfrak{L}_{\nu,r}, \mathfrak{L}_{1-\nu,s}]$ and is one-to-one. If $1 < r \leqq 2$ and $f \in \mathfrak{L}_{\nu,r}$, then there holds the relation

$$\left(\mathfrak{M}\mathbb{H}_{k,\eta}f\right)(s) = \left(\frac{2}{|k|}\right)^{k(s-1/2)}\frac{\Gamma\left(\dfrac{1}{2}\left[\eta+k\left(s-\dfrac{1}{2}\right)+1\right]\right)}{\Gamma\left(\dfrac{1}{2}\left[\eta-k\left(s-\dfrac{1}{2}\right)+1\right]\right)}\left(\mathfrak{M}f\right)(1-s) \tag{3.3.10}$$

$$(\mathrm{Re}(s)=1-\nu).$$

For further investigation we also need certain elementary operators. For a function f being defined almost everywhere on $\mathbb{R}_+$ we denote the operators M_ζ, W_δ and R as follows:

$$\left(M_\zeta f\right)(x) = x^\zeta f(x) \quad (\zeta \in \mathbb{C}); \tag{3.3.11}$$

$$\left(W_\delta f\right)(x) = f\left(\frac{x}{\delta}\right) \quad (\delta \in \mathbb{R}_+); \tag{3.3.12}$$

$$\left(Rf\right)(x) = \frac{1}{x} f\left(\frac{1}{x}\right). \tag{3.3.13}$$

It is easy to check that these operators have the following properties:

Lemma 3.1. *Let $\nu \in \mathbb{R}$ and $1 \leqq r < \infty$. Then we have the following statements:*

(i) *M_ζ is an isometric isomorphism of $\mathfrak{L}_{\nu,r}$ onto $\mathfrak{L}_{\nu-\mathrm{Re}(\zeta),r}$, and if $f \in \mathfrak{L}_{\nu,r}$ $(1 \leqq r \leqq 2)$, then*

$$\left(\mathfrak{M} M_\zeta f\right)(s) = \left(\mathfrak{M} f\right)(s+\zeta) \qquad (\mathrm{Re}(s) = \nu - \mathrm{Re}(\zeta)). \tag{3.3.14}$$

(ii) *W_δ is a bounded isomorphism of $\mathfrak{L}_{\nu,r}$ onto itself, and if $f \in \mathfrak{L}_{\nu,r}$ $(1 \leqq r \leqq 2)$, then*

$$\left(\mathfrak{M} W_\delta f\right)(s) = \delta^s \left(\mathfrak{M} f\right)(s) \qquad (\mathrm{Re}(s) = \nu). \tag{3.3.15}$$

(iii) *R is an isometric isomorphism of $\mathfrak{L}_{\nu,r}$ onto $\mathfrak{L}_{1-\nu,r}$, and if $f \in \mathfrak{L}_{\nu,r}$ $(1 \leqq r \leqq 2)$, then*

$$\left(\mathfrak{M} R f\right)(s) = \left(\mathfrak{M} f\right)(1-s) \qquad (\mathrm{Re}(s) = 1 - \nu). \tag{3.3.16}$$

The compositions between the operators $R, W_\delta, M_\zeta, I^\alpha_{0+;\sigma,\eta}, I^\alpha_{-;\sigma,\eta}, \mathbb{L}_{k,\alpha}$ and $\mathbb{H}_{k,\eta}$ will be employed in the next chapters. Note that the operator $\mathbb{H}_{k,\eta}$ cannot be commutated with M_ζ.

Lemma 3.2. *Under the existence conditions of the operators $I^\alpha_{0+;\sigma,\eta}$, $I^\alpha_{-;\sigma,\eta}$, $\mathbb{L}_{k,\alpha}$ and $\mathbb{H}_{k,\eta}$, there hold the following formulas:*

$$RI^\alpha_{0+;\sigma,\eta} = I^\alpha_{-;\sigma,\eta+1-1/\sigma}R; \qquad RI^\alpha_{-;\sigma,\eta} = I^\alpha_{0+;\sigma,\eta-1+1/\sigma}R; \tag{3.3.17}$$

$$R\mathbb{L}_{k,\alpha} = \mathbb{L}_{-k,1-\alpha}R; \qquad R\mathbb{H}_{k,\eta} = \mathbb{H}_{-k,\eta}R; \tag{3.3.18}$$

$$W_\delta I^\alpha_{0+;\sigma,\eta} = I^\alpha_{0+;\sigma,\eta}W_\delta; \qquad W_\delta I^\alpha_{-;\sigma,\eta} = I^\alpha_{-;\sigma,\eta}W_\delta; \tag{3.3.19}$$

$$W_\delta \mathbb{L}_{k,\alpha} = \delta \mathbb{L}_{k,\alpha} W_{1/\delta}; \qquad W_\delta \mathbb{H}_{k,\eta} = \delta \mathbb{H}_{k,\eta} W_{1/\delta}; \tag{3.3.20}$$

$$RW_\delta = \delta W_{1/\delta} R; \qquad RM_\zeta = M_{-\zeta} R; \tag{3.3.21}$$

$$W_\delta M_\zeta = \delta^{-\zeta} M_\zeta W_\delta; \qquad M_\zeta \mathbb{L}_{k,\alpha} = \mathbb{L}_{k,\alpha-\zeta} M_{-\zeta}; \tag{3.3.22}$$

$$M_\zeta I^\alpha_{0+;\sigma,\eta} = I^\alpha_{0+;\sigma,\eta-\zeta/\sigma} M_\zeta; \qquad M_\zeta I^\alpha_{-;\sigma,\eta} = I^\alpha_{-;\sigma,\eta+\zeta/\sigma} M_\zeta. \tag{3.3.23}$$

Since $I_{0+}^{\alpha} = M^{\alpha} I_{0+;1,0}^{\alpha}$, $I_{-}^{\alpha} = I_{-;1,0}^{\alpha} M^{\alpha}$ and $\mathbb{L} = \mathbb{L}_{1,0}$, then from Theorem 3.2(a),(b) and Lemma 3.1(c) we have:

Corollary 3.2.1. *The following statements are valid:*

(a) *If* $1 \leqq r \leqq \infty$, $\mathrm{Re}(\alpha) > 0$ *and* $\nu < 1$, *then for all* $s \geqq r$ *such that* $1/s > 1/r - \mathrm{Re}(\alpha)$, *the operator* I_{0+}^{α} *belongs to* $[\mathfrak{L}_{\nu,r}, \mathfrak{L}_{\nu-\mathrm{Re}(\alpha),s}]$ *and is one-to-one. For* $1 \leqq r \leqq 2$ *and* $f \in \mathfrak{L}_{\nu,r}$, *there holds the relation*

$$\left(\mathfrak{M} I_{0+}^{\alpha} f\right)(s) = \frac{\Gamma(1-\alpha-s)}{\Gamma(1-s)} \left(\mathfrak{M} f\right)(s+\alpha) \quad (\mathrm{Re}(s+\alpha) = \nu). \tag{3.3.24}$$

(b) *If* $1 \leqq r \leqq \infty$, $\mathrm{Re}(\alpha) > 0$ *and* $\nu > \mathrm{Re}(\alpha)$, *then for all* $s \geqq r$ *such that* $1/s > 1/r - \mathrm{Re}(\alpha)$, *the operator* I_{-}^{α} *belongs to* $[\mathfrak{L}_{\nu,r}, \mathfrak{L}_{\nu-\mathrm{Re}(\alpha),s}]$ *and is one-to-one. For* $1 \leqq r \leqq 2$ *and* $f \in \mathfrak{L}_{\nu,r}$, *there holds the relation*

$$\left(\mathfrak{M} I_{-}^{\alpha} f\right)(s) = \frac{\Gamma(s)}{\Gamma(\alpha+s)} \left(\mathfrak{M} f\right)(s+\alpha) \quad (\mathrm{Re}(s+\alpha) = \nu). \tag{3.3.25}$$

(c) *If* $1 \leqq r \leqq s \leqq \infty$, *and if* $\nu < 1$, *then the operator* $\mathbb{L}$ *belongs to* $[\mathfrak{L}_{\nu,r}, \mathfrak{L}_{1-\nu,s}]$ *and is one-to-one. If* $1 \leqq r \leqq 2$ *and* $f \in \mathfrak{L}_{\nu,r}$, *then there holds the relation*

$$\left(\mathfrak{M}\mathbb{L} f\right)(s) = \Gamma(s)\left(\mathfrak{M} f\right)(1-s) \quad (\mathrm{Re}(s) = 1-\nu). \tag{3.3.26}$$

3.4. Integral Representations for the *H*-Function

In this section we give the integral representation for the H-function. When there is no confusion, we denote (1.1.1) and (1.1.2) simply by $H_{p,q}^{m,n}(x)$, $\mathcal{H}_{p,q}^{m,n}(s)$ or $H(x)$, $\mathcal{H}(s)$. Let a^*, Δ, δ, μ be given in (1.1.7)–(1.1.10) and let α and β be defined by

$$\alpha = \begin{cases} -\min\limits_{1 \leqq i \leqq m} \left[\dfrac{\mathrm{Re}(b_i)}{\beta_i}\right] & \text{if} \quad m > 0; \\ -\infty & \text{if} \quad m = 0, \end{cases} \tag{3.4.1}$$

and

$$\beta = \begin{cases} \min\limits_{1 \leqq j \leqq n} \left[\dfrac{1-\mathrm{Re}(a_j)}{\alpha_j}\right] & \text{if} \quad n > 0; \\ \infty & \text{if} \quad n = 0. \end{cases} \tag{3.4.2}$$

Theorem 3.3. *Let* $\alpha < \gamma < \beta$. *If any of the following conditions holds:*

(i) $a^* > 0$,

(ii) $a^* = 0, \Delta \neq 0$ *and* $\mathrm{Re}(\mu) + \Delta\gamma \leqq 0$,

(iii) *$a^* = 0, \Delta = 0$ and $\mathrm{Re}(\mu) < 0$,*

then for all $x > 0$ there holds the relation

$$H^{m,n}_{p,q}\left[x \left| \begin{matrix} (a_i, \alpha_i)_{1,p} \\ (b_j, \beta_j)_{1,q} \end{matrix} \right.\right] = \frac{1}{2\pi i} \lim_{R\to\infty} \int_{\gamma-iR}^{\gamma+iR} \mathcal{H}^{m,n}_{p,q}\left[\begin{matrix} (a_i, \alpha_i)_{1,p} \\ (b_j, \beta_j)_{1,q} \end{matrix} \left| s \right.\right] x^{-s} ds, \tag{3.4.3}$$

except for $x = \delta$ in the case (iii) *(when $H(x)$ is not defined).*

Proof. When $a^* > 0$, then (3.4.3) follows from the definition of the H-function (1.1.1) with (1.1.2) and (1.2.20). We prove (3.4.3) for $a^* = 0$ and either $\Delta < 0$, or $\Delta = 0$ and $x > \delta$. The proof for $a^* = 0$ and either $\Delta > 0$, or $\Delta = 0$ and $0 < x < \delta$ is exactly similar. In the case under consideration, it follows from (1.1.1)–(1.1.2) that

$$H(x) = \frac{1}{2\pi i} \int_{\mathfrak{L}} x^{-s} \mathcal{H}(s) ds,$$

where $\mathfrak{L}$ is a loop starting and ending at ∞ and encircling all of the poles of $\Gamma(1 - a_i - \alpha_i s)$ $(i = 1, 2, \cdots, n)$ once in the negative direction, but encircling none of the poles of $\Gamma(b_j + \beta_j s)$ $(j = 1, 2, \cdots, m)$.

Let

$$\tau > \max\left[\left|\mathrm{Im}\left(\frac{a_1}{\alpha_1}\right)\right|, \cdots, \left|\mathrm{Im}\left(\frac{a_n}{\alpha_n}\right)\right|\right]$$

and choose $k, \gamma < k < \beta$. We choose $\mathfrak{L}$ to be the loop consisting of the half-line $\mathrm{Im}(s) = -\tau$ from $\infty - i\tau$ to $k - i\tau$, the segment $\mathrm{Re}(s) = k$ from $k - i\tau$ to $k + i\tau$ and the half-line $\mathrm{Im}(s) = \tau$ from $k + i\tau$ to $\infty + i\tau$. For $R > k - \gamma$, let $\mathfrak{L}_R$ denote the portion of $\mathfrak{L}$ on which $|s - \gamma| \leqq R$. It is clear that

$$H(x) = \frac{1}{2\pi i} \lim_{R\to\infty} \int_{\mathfrak{L}_R} x^{-s} \mathcal{H}(s) ds. \tag{3.4.4}$$

For such an R we denote by Λ the closed curve composed of the segment from $\gamma - iR$ to $\gamma + iR$ on the line $\mathrm{Re}(s) = \gamma$, the portion Λ_1 of the circle $|s - \gamma| = R$ clockwise from $\gamma + iR$ to the intersecting point with $\mathfrak{L}_R$, and the portion Λ_2 of the circle $|s - \gamma| = R$ clockwise from the intersecting point with $\mathfrak{L}_R$ to $\gamma - iR$. Since $\alpha < \gamma < \beta$, Cauchy's theorem implies that

$$\begin{aligned} 0 &= \int_{\Lambda} x^{-s} \mathcal{H}(s) ds \\ &= \int_{\gamma-iR}^{\gamma+iR} x^{-s} \mathcal{H}(s) ds - \int_{\mathfrak{L}_R} x^{-s} \mathcal{H}(s) ds + \int_{\Lambda_1} x^{-s} \mathcal{H}(s) ds + \int_{\Lambda_2} x^{-s} \mathcal{H}(s) ds \end{aligned}$$

for $x > 0$. In view of (3.4.4) it is sufficient to show that

$$\lim_{R\to\infty} \int_{\Lambda_i} x^{-s} \mathcal{H}(s) ds = 0 \qquad (i = 1, 2). \tag{3.4.5}$$

Applying the relation (2.1.6), we represent $\mathcal{H}(s)$ as

$$\mathcal{H}(s) = \mathcal{H}_1(s) \mathcal{H}_2(s), \tag{3.4.6}$$

where

$$\mathcal{H}_1(s)=\frac{\prod_{j=1}^{q}\Gamma(b_j+\beta_j s)}{\prod_{i=1}^{p}\Gamma(a_i+\alpha_i s)},\qquad \mathcal{H}_2(s)=\pi^{n+m-q}\frac{\prod_{j=m+1}^{q}\sin[(b_j+\beta_j s)\pi]}{\prod_{i=1}^{n}\sin[(a_i+\alpha_i s)\pi]}. \tag{3.4.7}$$

Now let us prove (3.4.5) for $i=1$. We first estimate $\mathcal{H}_1(s)$. From Stirling's formula (1.2.3) we have that, if $\zeta=Re^{i\theta}$ $(-\pi/2\leqq\theta\leqq\pi/2)$ and $c\in\mathbb{C}$, then uniformly in θ, as $R\to\infty$,

$$\Gamma(c+\zeta)\sim\sqrt{2\pi}e^{-\theta\mathrm{Im}(c)}R^{R\cos\theta+\mathrm{Re}(c)-1/2}e^{-R[\cos\theta+\theta\sin\theta]-\mathrm{Re}(c)}.$$

Hence, on Λ_1, putting $s=\gamma+\zeta,\zeta=Re^{i\theta}$, we have uniformly in θ, as $R\to\infty$,

$$\mathcal{H}_1(s)=\mathcal{H}_1(\gamma+\zeta)\sim(2\pi)^{(q-p)/2}\delta^{\gamma}\frac{\prod_{j=1}^{q}\beta_j^{\mathrm{Re}(b_j)-1/2}}{\prod_{i=1}^{p}\alpha_i^{\mathrm{Re}(a_i)-1/2}}$$

$$\cdot\exp\left\{-\theta\ \mathrm{Im}\left(\sum_{j=1}^{q}b_j-\sum_{i=1}^{p}a_i\right)\right\}R^{\mathrm{Re}(\mu)+(\gamma+R\cos\theta)\Delta}$$

$$\cdot\ \delta^{R\cos\theta}\exp\left\{-R\Delta(\cos\theta+\theta\sin\theta)-\mathrm{Re}\left(\sum_{j=1}^{q}b_j-\sum_{i=1}^{p}a_i\right)-\gamma\Delta\right\},$$

where δ,Δ and μ are given in (1.1.9), (1.1.8) and (1.1.10). Since $|\theta|\leqq\pi/2$ on Λ_1, then

$$\exp\left\{-\theta\ \mathrm{Im}\left(\sum_{j=1}^{q}b_j-\sum_{i=1}^{p}a_i\right)\right\}\leqq\exp\left\{\frac{\pi}{2}\left|\mathrm{Im}\left(\sum_{j=1}^{q}b_j-\sum_{i=1}^{p}a_i\right)\right|\right\}.$$

Hence there is a constant A_1 such that

$$|\mathcal{H}_1(s)|\leqq A_1R^{\mathrm{Re}(\mu)+(\gamma+R\cos\theta)\Delta}\ \delta^{R\cos\theta}\ e^{-R\Delta(\cos\theta+\theta\sin\theta)}, \tag{3.4.8}$$

if $s\in\Lambda_1$ and R is sufficiently large.

Now we estimate $\mathcal{H}_2(s)$. If $s\in\Lambda_1$, $s=\gamma+Re^{i\theta}=\gamma+\xi+i\eta$, then by invoking the estimate

$$\sinh y\leqq|\sin(x+iy)|\leqq\cosh y \tag{3.4.9}$$

we have

$$|\sin[(b_j+\beta_j s)\pi]|\leqq\cosh[(\mathrm{Im}(b_j)+\beta_j\eta)\pi]\leqq B_je^{\pi\beta_jR\sin\theta}$$

$$\text{with}\quad B_j=e^{\pi|\mathrm{Im}(b_j)|}\quad(j=m+1,\cdots,q).$$

By virtue of the left inequality of (3.4.9) and $\eta\geqq\tau$, we have

$$|\sin[(a_i+\alpha_i s)\pi]|\geqq\sinh[(\mathrm{Im}(a_i)+\alpha_i\eta)\pi]>0\quad(i=1,2,\cdots,n).$$

Since

$$\frac{d}{d\eta}e^{-\pi\alpha_i\eta}\sinh[(\mathrm{Im}(a_i)+\alpha_i\eta)\pi] = \pi\alpha_i e^{-\pi(2\alpha_i\eta+\mathrm{Im}(a_i))} > 0,$$

then for $\eta \geqq \tau$ we have

$$e^{-\pi\alpha_i\eta}\sinh[(\mathrm{Im}(a_i)+\alpha_i\eta)\pi] \geqq C_i$$

with

$$C_i = e^{-\pi\tau\alpha_i}\sinh[(\mathrm{Im}(a_i)+\alpha_i\tau)\pi] \quad (i = 1, 2, \cdots, n),$$

and therefore

$$|\sin[(a_i+\alpha_i s)\pi]| \geqq C_i e^{\pi\alpha_i R\sin\theta} \quad (i = 1, 2, \cdots, n).$$

Substituting these estimates into $\mathcal{H}_2(s)$ and taking into account the relations

$$a^* = 0, \quad \sum_{j=m+1}^{q}\beta_j - \sum_{i=1}^{n}\alpha_i = \sum_{j=1}^{m}\beta_j - \sum_{i=n+1}^{p}\alpha_i = \frac{\Delta}{2},$$

we obtain

$$|\mathcal{H}_2(s)| \leqq A_2 e^{\pi R\Delta\sin\theta/2}, \quad A_2 = \pi^{m+n-q}\frac{\prod_{j=m+1}^{q} B_j}{\prod_{i=1}^{n} C_i}. \tag{3.4.10}$$

Since $\Delta \leqq 0$ and $0 < \theta \leqq \pi/2$, then $R\Delta(\theta-\pi/2)\sin\theta \geqq 0$. Therefore from (3.4.6), (3.4.8) and (3.4.10) we have

$$\begin{aligned} |\mathcal{H}(s)| &\leqq AR^{\mathrm{Re}(\mu)+(\gamma+R\cos\theta)\Delta}\,\delta^{R\cos\theta}\,e^{-R\Delta[\cos\theta+(\theta-\pi/2)\sin\theta]} \\ &\leqq AR^{\mathrm{Re}(\mu)+(\gamma+R\cos\theta)\Delta}\,\delta^{R\cos\theta}\,e^{-R\Delta\cos\theta} \end{aligned} \tag{3.4.11}$$

for $s \in \Lambda_1$ and for sufficiently large R, say $R > R_0$, with $A = A_1A_2$.

Let us consider the case $\Delta < 0$. Remembering that $\mathrm{Re}(\mu)+\gamma\Delta \leqq 0$ in the hypothesis (ii) of the theorem, if $x > 0$ and $R > \max[R_0, K]$ with $K = e(x/\delta)^{1/\Delta}$, we have

$$\begin{aligned} \left|\int_{\Lambda_1} x^{-s}\mathcal{H}(s)ds\right| &\leqq Ax^{-\gamma}\,R^{\mathrm{Re}(\mu)+\gamma\Delta+1}\int_0^{\pi/2} R^{R\Delta\cos\theta}\,e^{-R\Delta\cos\theta}\left(\frac{x}{\delta}\right)^{-R\cos\theta} d\theta \\ &= Ax^{-\gamma}\,R^{\mathrm{Re}(\mu)+\gamma\Delta+1}\int_0^{\pi/2} e^{R\Delta(\log R-\log K)\cos\theta}d\theta \\ &= Ax^{-\gamma}\,R^{\mathrm{Re}(\mu)+\gamma\Delta+1}\int_0^{\pi/2} e^{R\Delta(\log R-\log K)\sin\theta}d\theta \\ &\leqq Ax^{-\gamma}\,R^{\mathrm{Re}(\mu)+\gamma\Delta+1}\int_0^{\pi/2} e^{2R\Delta(\log R-\log K)\theta/\pi}d\theta \\ &= \frac{A\pi}{2\Delta(\log R-\log K)}x^{-\gamma}\,R^{\mathrm{Re}(\mu)+\gamma\Delta}(e^{R\Delta(\log R-\log K)}-1) \to 0 \end{aligned}$$

as $R \to \infty$, and thus (3.4.5) for $i = 1$ is proved when $\Delta < 0$.

When $\Delta = 0$ and $x > \delta$, we assumed $\mathrm{Re}(\mu) < 0$ in the hypothesis (iii) of the theorem. Therefore if $R > R_0$, then we have from (3.4.11) that

$$\begin{aligned}\left|\int_{\Lambda_1} x^{-s}\mathcal{H}(s)ds\right| &\leqq Ax^{-\gamma}R^{\mathrm{Re}(\mu)+1}\int_0^{\pi/2}\left(\frac{x}{\delta}\right)^{-R\cos\theta}d\theta \\ &= Ax^{-\gamma}R^{\mathrm{Re}(\mu)+1}\int_0^{\pi/2}e^{-R\cos\theta\log(x/\delta)}d\theta \\ &= Ax^{-\gamma}R^{\mathrm{Re}(\mu)+1}\int_0^{\pi/2}e^{-R\sin\theta\log(x/\delta)}d\theta \\ &\leqq Ax^{-\gamma}R^{\mathrm{Re}(\mu)+1}\int_0^{\pi/2}e^{-2R\theta\log(x/\delta)/\pi}d\theta \\ &= \frac{\pi A}{2\log\left(\frac{x}{\delta}\right)}\,x^{-\gamma}R^{\mathrm{Re}(\mu)}\left(1-e^{-R\log(x/\delta)}\right) \to 0\end{aligned}$$

as $R \to \infty$. The proof of (3.4.5) for $i = 1$ is completed. The proof for the case $i = 2$ is similar. Thus the theorem is proved.

Theorem 3.4. *Suppose that $\alpha < \gamma < \beta$ and that either of the conditions $a^* > 0$ or $a^* = 0$ and $\Delta\gamma + \mathrm{Re}(\mu) < -1$ holds. Then for $x > 0$, except for $x = \delta$ when $a^* = 0$ and $\Delta = 0$, the relation*

$$H^{m,n}_{p,q}\left[x\,\middle|\,\begin{matrix}(a_p,\alpha_p)\\(b_q,\beta_q)\end{matrix}\right] = \frac{1}{2\pi i}\int_{\gamma-i\infty}^{\gamma+i\infty}\mathcal{H}^{m,n}_{p,q}\left[\begin{matrix}(a_p,\alpha_p)\\(b_q,\beta_q)\end{matrix}\,\middle|\,t\right]x^{-t}dt \tag{3.4.12}$$

holds and the estimate

$$\left|H^{m,n}_{p,q}\left[x\,\middle|\,\begin{matrix}(a_p,\alpha_p)\\(b_q,\beta_q)\end{matrix}\right]\right| \leqq A_\gamma\, x^{-\gamma} \tag{3.4.13}$$

is valid, where A_γ is a positive constant depending only on γ.

Proof. By virtue of the assumptions, we find that the relation (1.2.9) implies $\mathcal{H}(\gamma+it) \in L_1(-\infty,\infty)$. So (3.4.12) follows from (3.4.3). The estimate (3.4.13) can be seen from the proof of Theorem 3.3.

To conclude this section, we give the asymptotic estimate at infinity for the derivative of the function $\mathcal{H}(s)$ defined in (1.1.2) which we need in the next chapter.

Lemma 3.3. *There holds the estimate as $|t| \to \infty$,*

$$\mathcal{H}'(\sigma+it) = \mathcal{H}(\sigma+it)\left[\log\delta + a_1^*\log(it) - a_2^*\log(-it) + \frac{\mu+\Delta\sigma}{it} + O\left(\frac{1}{t^2}\right)\right]. \tag{3.4.14}$$

Proof. It follows from (1.1.2) that for $s = \sigma + it$

$$\mathcal{H}'(s) = \mathcal{H}(s)\left[\sum_{j=1}^{m}\beta_j\psi(b_j+\beta_j s) - \sum_{i=1}^{n}\alpha_i\psi(1-a_i-\alpha_i s)\right.$$

$$+\sum_{j=m+1}^{q}\beta_j\psi(1-b_j-\beta_j s)-\sum_{i=n+1}^{p}\alpha_i\psi(a_i+\alpha_i s)\Bigg], \qquad (3.4.15)$$

where $\psi(z)=\Gamma'(z)/\Gamma(z)$ is the psi function defined in (1.3.4). There exists the following asymptotic expansion of $\psi(z)$ at infinity (see Erdélyi, Magnus, Oberhettinger and Tricomi [1, 1.18(7)])

$$\psi(z)=\log z-\frac{1}{2z}+O\left(\frac{1}{z^2}\right) \quad (|z|\to\infty). \qquad (3.4.16)$$

In accordance with (3.4.16) and (3.4.18) for $c\in\mathbb{C}$, we have as $|t|\to\infty$,

$$\psi(c+\sigma\pm it)=\log(\pm it)\pm\frac{c+\sigma-1/2}{it}+O\left(\frac{1}{t^2}\right). \qquad (3.4.17)$$

Substituting this into (3.4.15), we arrive at (3.4.14) and the lemma is proved.

3.5. $\mathfrak{L}_{\nu,2}$-Theory of the General Integral Transform

In this section we consider the general integral transform $\boldsymbol{K}$ of the form

$$\left(\boldsymbol{K}f\right)(x)=hx^{1-(\lambda+1)/h}\frac{d}{dx}x^{(\lambda+1)/h}\int_0^\infty k(xt)f(t)dt, \qquad (3.5.1)$$

where the kernel $k\in\mathfrak{L}_{1-\nu,2}$, $\lambda\in\mathbb{C}$ and $h\in\mathbb{R}\setminus\{0\}$.

Theorem 3.5. **(a)** *Let the transform $\boldsymbol{K}$ of the form* (3.5.1) *be in* $[\mathfrak{L}_{\nu,2},\mathfrak{L}_{1-\nu,2}]$, *then* $k\in\mathfrak{L}_{1-\nu,2}$. *If we set for* $\nu\neq 1-(\mathrm{Re}(\lambda)+1)/h$

$$\left(\mathfrak{M}k\right)(1-\nu+it)=\frac{\omega(t)}{\lambda+1-(1-\nu+it)h} \quad \text{a.e.} \qquad (3.5.2)$$

then $\omega\in L_\infty(\mathbb{R})$, *and for* $f\in\mathfrak{L}_{\nu,2}$ *there holds the formula*

$$\left(\mathfrak{M}\boldsymbol{K}f\right)(1-\nu+it)=\omega(t)\left(\mathfrak{M}f\right)(\nu-it) \quad \text{a.e.} \qquad (3.5.3)$$

(b) *Conversely, for given* $\omega\in L_\infty(\mathbb{R})$, $\nu\in\mathbb{R}$ *and* $h\in\mathbb{R}_+$, *there is a transform* $\boldsymbol{K}\in[\mathfrak{L}_{\nu,2},\mathfrak{L}_{1-\nu,2}]$ *so that* (3.5.3) *holds for* $f\in\mathfrak{L}_{\nu,2}$. *Moreover, if* $\nu\neq 1-(\mathrm{Re}(\lambda)+1)/h$, *then* $\boldsymbol{K}f$ *is representable in the form* (3.5.1) *with the kernel* k *given in* (3.5.2).

(c) *Under the hypotheses of* (a) *or* (b) *with* $\omega\neq 0$ a.e., $\boldsymbol{K}$ *is a one-to-one transform from* $\mathfrak{L}_{\nu,2}$ *into* $\mathfrak{L}_{1-\nu,2}$, *and if in addition* $1/\omega\in L_\infty(\mathbb{R})$, *then* $\boldsymbol{K}$ *maps* $\mathfrak{L}_{\nu,2}$ *onto* $\mathfrak{L}_{1-\nu,2}$. *Further, for* $f,g\in\mathfrak{L}_{\nu,2}$, *the relation*

$$\int_0^\infty f(x)\left(\boldsymbol{K}g\right)(x)dx=\int_0^\infty\left(\boldsymbol{K}f\right)(x)g(x)dx \qquad (3.5.4)$$

is valid.

Proof. First we treat (a). We suppose that $\boldsymbol{K}$, given in (3.5.1), is in $[\mathfrak{L}_{\nu,2},\mathfrak{L}_{1-\nu,2}]$, where $\nu\neq 1-(\mathrm{Re}(\lambda)+1)/h$. We consider the case $\nu>1-(\mathrm{Re}(\lambda)+1)/h$. For $a>0$, let

$$g_a(t)=\begin{cases} t^{(\lambda+1)/h-1} & \text{if } 0<t<a;\\ 0 & \text{if } t>a.\end{cases} \qquad (3.5.5)$$

Then

$$\|g_a\|_{\nu,2} = \left\{ \int_0^a t^{2\{[(\mathrm{Re})(\lambda)+1]/h+\nu-1\}-1} dt \right\}^{1/2} < \infty,$$

which means $g_a \in \mathfrak{L}_{\nu,2}$. Hence

$$\left(\boldsymbol{K} g_1\right)(x) = hx^{1-(\lambda+1)/h} \frac{d}{dx} x^{(\lambda+1)/h} \int_0^1 k(xt) t^{(\lambda+1)/h-1} dt$$
$$= hx^{1-(\lambda+1)/h} \frac{d}{dx} \int_0^x \tau^{(\lambda+1)/h-1} k(\tau) d\tau = hk(x)$$

almost everywhere, so that $\boldsymbol{K} g_1 = hk$. Therefore since $\boldsymbol{K} \in [\mathfrak{L}_{\nu,2}, \mathfrak{L}_{1-\nu,2}]$, we have that $k \in \mathfrak{L}_{1-\nu,2}$.

Since $f \in \mathfrak{L}_{\nu,2}$ and $k \in \mathfrak{L}_{1-\nu,2}$, by using the Schwartz inequality

$$\left| \int_a^b f(x) g(x) dx \right| \leqq \left(\int_a^b |f(x)|^2 dx \right)^{1/2} \left(\int_a^b |g(x)|^2 dx \right)^{1/2} \tag{3.5.6}$$

$$(-\infty \leqq a < b \leqq \infty),$$

we have

$$\left| x^{(\lambda+1)/h} \int_0^\infty k(xt) f(t) dt \right| = \left| x^{(\lambda+1)/h} \int_0^\infty \left\{ t^{1/2-\nu} k(xt) \right\} \left\{ t^{-1/2+\nu} f(t) \right\} dt \right|$$
$$\leqq x^{[\mathrm{Re}(\lambda)+1]/h} \left\{ \int_0^\infty \left| t^{1-\nu} k(xt) \right|^2 \frac{dt}{t} \right\}^{1/2} \|f\|_{\nu,2}$$
$$= x^{\nu-1+[\mathrm{Re}(\lambda)+1]/h} \|k\|_{1-\nu,2} \|f\|_{\nu,2} = o(1)$$

as $x \to +0$. Hence after integrating both sides of (3.5.1), we obtain for $x > 0$

$$\int_0^x t^{(\lambda+1)/h-1} \left(\boldsymbol{K} f\right)(t) dt = hx^{(\lambda+1)/h} \int_0^\infty k(xt) f(t) dt. \tag{3.5.7}$$

Now for $x > 0$ and $\mathrm{Re}(s) + [\mathrm{Re}(\lambda) + 1]/h > 1$, we have

$$\left(\mathfrak{M} g_x\right)(s) = \frac{hx^{(\lambda+1)/h+s-1}}{\lambda + 1 - h(1-s)}. \tag{3.5.8}$$

Since $f \in \mathfrak{L}_{\nu,2}$ and $g_x \in \mathfrak{L}_{\nu,2}$, from (3.2.8), we obtain

$$\int_0^x t^{(\lambda+1)/h-1} \left(\boldsymbol{K} f\right)(t) dt = \int_0^\infty g_x(t) \left(\boldsymbol{K} f\right)(t) dt$$
$$= \frac{1}{2\pi i} \int_{\nu-i\infty}^{\nu+i\infty} \left(\mathfrak{M} g_x\right)(s) \left(\mathfrak{M} \boldsymbol{K} f\right)(1-s) ds$$
$$= \frac{hx^{(\lambda+1)/h}}{2\pi i} \int_{1-\nu-i\infty}^{1-\nu+i\infty} x^{-s} \left(\mathfrak{M} \boldsymbol{K} f\right)(s) \frac{ds}{\lambda + 1 - hs}$$
$$= \frac{hx^{\nu-1+(\lambda+1)/h}}{2\pi} \int_{-\infty}^{\infty} x^{-it} \frac{\left(\mathfrak{M} \boldsymbol{K} f\right)(1-\nu+it)}{\lambda + 1 - (1-\nu+it)h} dt. \tag{3.5.9}$$

Similarly, from (3.2.8) and (3.3.15) we find

$$hx^{(\lambda+1)/h}\int_0^\infty k(xt)f(t)dt = hx^{(\lambda+1)/h}\int_0^\infty \left(W_{1/x}k\right)(t)f(t)dt$$
$$= \frac{hx^{(\lambda+1)/h}}{2\pi i}\int_{1-\nu-i\infty}^{1-\nu+i\infty} x^{-s}\left(\mathfrak{M}k\right)(s)\left(\mathfrak{M}f\right)(1-s)ds$$
$$= \frac{hx^{\nu-1+(\lambda+1)/h}}{2\pi}\int_{-\infty}^{\infty} x^{-it}\left(\mathfrak{M}k\right)(1-\nu+it)\left(\mathfrak{M}f\right)(\nu-it)dt. \tag{3.5.10}$$

Substituting (3.5.9) and (3.5.10) into (3.5.7), denoting

$$F(t) = \frac{\left(\mathfrak{M}\boldsymbol{K}f\right)(1-\nu+it)}{\lambda+1-(1-\nu+it)h} - \left(\mathfrak{M}k\right)(1-\nu+it)\left(\mathfrak{M}f\right)(\nu-it) \tag{3.5.11}$$

and writing $x = e^{-y}$, we obtain for all $y \in \mathbb{R}$ that

$$\int_{-\infty}^{+\infty} e^{iyt}F(t)dt = 0. \tag{3.5.12}$$

According to property (a) of the Mellin transform in Section 3.2, $\mathfrak{M} \in [\mathfrak{L}_{\sigma,2}, L_2(\mathbb{R})]$ for any $\sigma \in \mathbb{R}$. Therefore

$$\left(\mathfrak{M}\boldsymbol{K}f\right)(1-\nu+it), \quad \left(\mathfrak{M}g_1\right)(\nu-it) = \frac{h}{\lambda+1-(1-\nu+it)h},$$
$$\left(\mathfrak{M}k\right)(1-\nu+it), \quad \left(\mathfrak{M}f\right)(\nu-it)$$

belong to $L_2(\mathbb{R})$, and $F(t)$ in (3.5.11) is also in $L_2(\mathbb{R})$. Hence (3.5.12) means that $F(t) = 0$ a.e. Defining ω by (3.5.2), we obtain (3.5.3) in view of (3.5.11).

It remains to show that $\omega \in L_\infty(\mathbb{R})$. It follows from (3.5.3) that, if $f \in \mathfrak{L}_{\nu,2}$, then $\omega(t)\left(\mathfrak{M}f\right)(\nu-it) \in L_2(\mathbb{R})$. Due to the property (a) in Section 3.2, $\mathfrak{M}$ maps $\mathfrak{L}_{\nu,2}$ onto $L_2(\mathbb{R})$ and thus $\omega(t)\phi(t) \in L_2(\mathbb{R})$ for any $\phi \in L_2(\mathbb{R})$. Therefore from Halmos [1, Problem 51] $\omega \in L_\infty(\mathbb{R})$. This completes the proof of (a) for $\nu > 1 - (\mathrm{Re}(\lambda)+1)/h$.

If $\nu < 1 - (\mathrm{Re}(\lambda)+1)/h$, the proof is similar if we replace $g_a(t)$ in (3.5.5) by the function $h_a(t)$ defined for $a > 0$ by

$$h_a(t) = \begin{cases} 0 & \text{if } 0 < t < a; \\ t^{(\lambda+1)/h-1} & \text{if } t > a. \end{cases} \tag{3.5.13}$$

Now we prove (b). We suppose that $\omega \in L_\infty(\mathbb{R})$ and $f \in \mathfrak{L}_{\nu,2}$. By the fact in Section 3.2 that $\mathfrak{M}$ is a unitary mapping of $\mathfrak{L}_{1-\nu,2}$ onto $L_2(\mathbb{R})$, there is a unique function $g \in \mathfrak{L}_{1-\nu,2}$ such that $\left(\mathfrak{M}g\right)(1-\nu+it) = \omega(t)\left(\mathfrak{M}f\right)(\nu-it)$. We define $\boldsymbol{K}$ by $\boldsymbol{K}f = g$. Then (3.5.3) holds. $\boldsymbol{K}$ is also a linear operator, namely, if $f_i \in \mathfrak{L}_{\nu,2}$ and $c_i \in \mathbb{R}$ $(i = 1, 2)$, then

$$\left(\mathfrak{M}\boldsymbol{K}[c_1f_1+c_2f_2]\right)(1-\nu+it) = \omega(t)\left(\mathfrak{M}[c_1f_1+c_2f_2]\right)(\nu-it)$$
$$= c_1\omega(t)\left(\mathfrak{M}f_1\right)(\nu-it) + c_2\omega(t)\left(\mathfrak{M}f_2\right)(\nu-it)$$
$$= c_1\left(\mathfrak{M}f_1\right)(1-\nu+it) + c_2\left(\mathfrak{M}f_2\right)(1-\nu+it)$$
$$= \left(\mathfrak{M}[c_1\boldsymbol{K}f_1+c_2\boldsymbol{K}f_2]\right)(1-\nu+it),$$

which implies that $\boldsymbol{K}(c_1f_1 + c_2f_2) = c_1\boldsymbol{K}f_1 + c_2\boldsymbol{K}f_2$.

Further, it follows from the same property in Section 3.2 that, taking $\omega^*(t) = \omega(-t)$, we have

$$\|\boldsymbol{K}f\|_{1-\nu,2} = \|\mathfrak{M}\boldsymbol{K}f\|_2 = \|\omega^*\mathfrak{M}f\|_2 \leqq \|\omega^*\|_\infty \|\mathfrak{M}f\|_2 = \|\omega\|_\infty \|f\|_{\nu,2},$$

where $\|\omega\|_\infty$ is the $L_\infty(\mathbb{R})$-norm of ω. This means that $\boldsymbol{K} \in [\mathfrak{L}_{\nu,2}, \mathfrak{L}_{1-\nu,2}]$.

We suppose that $\nu \neq 1 - (\mathrm{Re}(\lambda) + 1)/h$ and let the function $k(t)$ be defined by (3.5.2). Then $k \in \mathfrak{L}_{1-\nu,2}$ by Property 3.1(c), because of $1/(p + it) \in L_2(\mathbb{R})$ for a constant $p \neq 0$. If $\nu < 1 - (\mathrm{Re}(\lambda) + 1)/h$ and $h_a(t)$ is given in (3.5.13), then

$$\left(\mathfrak{M}h_x\right)(s) = \frac{-hx^{(\lambda+1)/h+s-1}}{\lambda + 1 - h(1 - s)}. \tag{3.5.14}$$

From (3.5.13), (3.2.8), (3.5.14), (3.5.3), (3.5.2) and (3.3.15), if $x > 0$ then similarly to (3.5.9) we have

$$\begin{aligned}
&\int_x^\infty t^{(\lambda+1)/h-1}\left(\boldsymbol{K}f\right)(t)dt \\
&= \int_0^\infty h_x(t)\left(\boldsymbol{K}f\right)(t)dt \\
&= \frac{1}{2\pi i}\int_{\nu-i\infty}^{\nu+i\infty} \left(\mathfrak{M}h_x\right)(s)\left(\mathfrak{M}\boldsymbol{K}f\right)(1-s)ds \\
&= \frac{1}{2\pi}\int_{-\infty}^{\infty} \frac{-hx^{(\lambda+1)/h+\nu+it-1}}{\lambda+1-h(1-\nu-it)}\left(\mathfrak{M}\boldsymbol{K}f\right)(1-\nu-it)dt \\
&= \frac{-h}{2\pi}\, x^{(\lambda+1)/h+\nu-1}\int_{-\infty}^{\infty} \frac{x^{it}}{\lambda+1-h(1-\nu-it)}\,\omega^*(t)\left(\mathfrak{M}f\right)(\nu+it)dt \\
&= \frac{-h}{2\pi}\, x^{(\lambda+1)/h+\nu-1}\int_{-\infty}^{\infty} x^{it}\left(\mathfrak{M}k\right)(1-\nu-it)\left(\mathfrak{M}f\right)(\nu+it)dt \\
&= \frac{-h}{2\pi i}\, x^{(\lambda+1)/h+\nu-1}\int_{1-\nu-i\infty}^{1-\nu+i\infty} x^{1-\nu-s}\left(\mathfrak{M}k\right)(s)\left(\mathfrak{M}f\right)(1-s)ds \\
&= \frac{-h}{2\pi i}\, x^{(\lambda+1)/h}\int_{1-\nu-i\infty}^{1-\nu+i\infty} \left(\mathfrak{M}W_{1/x}k\right)(s)\left(\mathfrak{M}f\right)(1-s)ds \\
&= -\frac{h}{2\pi i}\, x^{(\lambda+1)/h}\int_{1-\nu-i\infty}^{1-\nu+i\infty} \left(\mathfrak{M}k(xt)\right)(s)\left(\mathfrak{M}f\right)(1-s)ds \\
&= -hx^{(\lambda+1)/h}\int_0^\infty k(xt)f(t)dt.
\end{aligned}$$

Differentiating this relation, we arrive at (3.5.1). Similarly for the case $\nu > 1 - (\mathrm{Re}(\lambda) + 1)/h$, the formula (3.5.8) for the function $g_a(t)$ in (3.5.5) is used, and the precise calculations are omitted.

Finally we prove (c). We suppose that $\omega \neq 0$ a.e. Then if $f \in \mathfrak{L}_{\nu,2}$ and $\boldsymbol{K}f = 0$, it follows from (3.5.3) that $\omega(t)\left(\mathfrak{M}f\right)(\nu - it) = 0$ a.e., and hence $\left(\mathfrak{M}f\right)(\nu - it) = 0$ a.e. This implies that $f = 0$ a.e. which means that $\boldsymbol{K}$ is one-to-one. We suppose that $1/\omega \in L_\infty(\mathbb{R})$.

In accordance with (b), there is a transform $T \in [\mathfrak{L}_{1-\nu,2}, \mathfrak{L}_{\nu,2}]$ such that if $g \in \mathfrak{L}_{1-\nu,2}$, then

$$\left(\mathfrak{M}Tg\right)(\nu+it) = \frac{1}{\omega(-t)}\left(\mathfrak{M}g\right)(1-\nu-it) \quad \text{a.e.}$$

Thus by (3.5.3), we have

$$\left(\mathfrak{M}\boldsymbol{K}Tg\right)(1-\nu+it) = \omega(t)\left(\mathfrak{M}Tg\right)(\nu-it) = \left(\mathfrak{M}g\right)(1-\nu+it).$$

So for all $g \in \mathfrak{L}_{1-\nu,2}$, we have the identity $\boldsymbol{K}Tg = g$ which means that $\boldsymbol{K}$ maps $\mathfrak{L}_{\nu,2}$ onto $\mathfrak{L}_{1-\nu,2}$.

Finally, if $f, g \in \mathfrak{L}_{\nu,2}$, from (3.2.8) and (3.5.3) we obtain

$$\begin{aligned}
\int_0^\infty f(x)\left(\boldsymbol{K}g\right)(x)dx &= \frac{1}{2\pi i}\int_{\nu-i\infty}^{\nu+i\infty}\left(\mathfrak{M}f\right)(s)\left(\mathfrak{M}\boldsymbol{K}g\right)(1-s)ds \\
&= \frac{1}{2\pi}\int_{-\infty}^{+\infty}\left(\mathfrak{M}f\right)(\nu+it)\left(\mathfrak{M}\boldsymbol{K}g\right)(1-\nu-it)dt \\
&= \frac{1}{2\pi}\int_{-\infty}^{+\infty}\left(\mathfrak{M}f\right)(\nu+it)\omega(-t)\left(\mathfrak{M}g\right)(\nu+it)dt \\
&= \frac{1}{2\pi}\int_{-\infty}^{+\infty}\omega(t)\left(\mathfrak{M}f\right)(\nu-it)\left(\mathfrak{M}g\right)(\nu-it)dt \\
&= \frac{1}{2\pi}\int_{-\infty}^{+\infty}\left(\mathfrak{M}\boldsymbol{K}f\right)(1-\nu+it)\left(\mathfrak{M}g\right)(1-[1-\nu+it])dt \\
&= \frac{1}{2\pi i}\int_{1-\nu-i\infty}^{1-\nu+i\infty}\left(\mathfrak{M}\boldsymbol{K}f\right)(s)\left(\mathfrak{M}g\right)(1-s)ds \\
&= \int_0^\infty\left(\boldsymbol{K}f\right)(x)g(x)dx.
\end{aligned}$$

This completes the proof of Theorem 3.5.

3.6. $\mathfrak{L}_{\nu,2}$-Theory of the $\boldsymbol{H}$-Transform

The results of Theorem 3.5 may be applied to obtain the $\mathfrak{L}_{\nu,2}$-theory of the $\boldsymbol{H}$-transform (3.1.1). For this we need a definition.

Definition 3.4. Let the function $\mathcal{H}(s) = \mathcal{H}_{p,q}^{m,n}(s)$ be given in (1.1.2) and let the real numbers α and β be defined by (3.4.1) and (3.4.2), respectively. We call the exceptional set $\mathcal{E}_{\mathcal{H}}$ of $\mathcal{H}$ the set of real numbers ν such that $\alpha < 1-\nu < \beta$ and $\mathcal{H}(s)$ has a zero on the line $\mathrm{Re}(s) = 1-\nu$.

Theorem 3.6. *We suppose that*

(a) $\alpha < 1-\nu < \beta$;

and that either of the conditions

(b) $a^* > 0$; *or*

(c) $a^* = 0,\ \Delta(1-\nu) + \mathrm{Re}(\mu) \leqq 0$

holds. Then we have the following results:

(i) *There is a one-to-one transform* $\boldsymbol{H} \in [\mathfrak{L}_{\nu,2},\ \mathfrak{L}_{1-\nu,2}]$ *so that* (3.1.5) *holds for* $\mathrm{Re}(s) = 1-\nu$ *and* $f \in \mathfrak{L}_{\nu,2}$. *If* $a^* = 0$, $\Delta(1-\nu) + \mathrm{Re}(\mu) = 0$ *and* $\nu \notin \mathcal{E}_{\mathcal{H}}$, *then the operator* $\boldsymbol{H}$ *maps* $\mathfrak{L}_{\nu,2}$ *onto* $\mathfrak{L}_{1-\nu,2}$.

(ii) *If* $f, g \in \mathfrak{L}_{\nu,2}$, *then the relation* (3.5.4) *holds for* $\boldsymbol{H}$:

$$\int_0^\infty f(x)\left(\boldsymbol{H}g\right)(x)dx = \int_0^\infty \left(\boldsymbol{H}f\right)(x)g(x)dx. \tag{3.6.1}$$

(iii) *Let* $f \in \mathfrak{L}_{\nu,2}$, $\lambda \in \mathbb{C}$ *and* $h > 0$. *If* $\mathrm{Re}(\lambda) > (1-\nu)h - 1$, *then* $\boldsymbol{H}f$ *is given in* (3.1.6), *namely,*

$$\left(\boldsymbol{H}f\right)(x)$$
$$= hx^{1-(\lambda+1)/h}\frac{d}{dx}x^{(\lambda+1)/h}\int_0^\infty H^{m,n+1}_{p+1,q+1}\left[xt \,\middle|\, \begin{matrix}(-\lambda,h),(a_i,\alpha_i)_{1,p}\\ (b_j,\beta_j)_{1,q},(-\lambda-1,h)\end{matrix}\right]f(t)dt. \tag{3.6.2}$$

When $\mathrm{Re}(\lambda) < (1-\nu)h - 1$, $\boldsymbol{H}f$ *is given in* (3.1.7) :

$$\left(\boldsymbol{H}f\right)(x)$$
$$= -hx^{1-(\lambda+1)/h}\frac{d}{dx}x^{(\lambda+1)/h}\int_0^\infty H^{m+1,n}_{p+1,q+1}\left[xt \,\middle|\, \begin{matrix}(a_i,\alpha_i)_{1,p},(-\lambda,h)\\ (-\lambda-1,h),(b_j,\beta_j)_{1,q}\end{matrix}\right]f(t)dt. \tag{3.6.3}$$

(iv) *The transform* $\boldsymbol{H}$ *is independent of* ν *in the sense that, for* ν *and* $\widetilde{\nu}$ *satisfying the assumptions* (a), *and either* (b) *or* (c), *and for the respective transforms* $\boldsymbol{H}$ *on* $\mathfrak{L}_{\nu,2}$ *and* $\widetilde{\boldsymbol{H}}$ *on* $\mathfrak{L}_{\widetilde{\nu},2}$ *given in* (3.1.5), *then* $\boldsymbol{H}f = \widetilde{\boldsymbol{H}}f$ *for* $f \in \mathfrak{L}_{\nu,2} \cap \mathfrak{L}_{\widetilde{\nu},2}$.

Proof. Let $\omega(t) = \mathcal{H}(1-\nu+it)$. By virtue of (1.1.2), (3.4.1), (3.4.2) and the condition (a) the function $\mathcal{H}(s)$ is analytic in the strip $\alpha < \mathrm{Re}(s) < \beta$. In accordance with (1.2.9) and the condition (b) or (c), $\omega(t) = O(1)$ as $|t| \to \infty$. Therefore $\omega \in L_\infty(\mathbb{R})$, and hence we obtain from Theorem 3.5(b) that there exists a transform $\boldsymbol{H} \in [\mathfrak{L}_{\nu,2},\ \mathfrak{L}_{1-\nu,2}]$ such that

$$\left(\mathfrak{M}\boldsymbol{H}f\right)(1-\nu+it) = \mathcal{H}(1-\nu+it)\left(\mathfrak{M}f\right)(\nu-it)$$

for $f \in \mathfrak{L}_{\nu,2}$. This means that the equality (3.1.5) holds for $\mathrm{Re}(s) = 1-\nu$. Since $\mathcal{H}(s)$ is analytic in the strip $\alpha < \mathrm{Re}(s) < \beta$ and has isolated zeros, then $\omega(t) \neq 0$ almost everywhere. Thus we obtain from Theorem 3.5(c) that $\boldsymbol{H} \in [\mathfrak{L}_{\nu,2},\ \mathfrak{L}_{1-\nu,2}]$ is a one-to-one transform. If $a^* = 0$, $\Delta(1-\nu) + \mathrm{Re}(\mu) = 0$ and ν is not in the exceptional set of $\mathcal{H}$, then $1/\omega \in L_\infty(\mathbb{R})$, and again from Theorem 3.5(c) we obtain that $\boldsymbol{H}$ transforms $\mathfrak{L}_{\nu,2}$ onto $\mathfrak{L}_{1-\nu,2}$. This completes the proof of the first assertion (i) of the theorem.

Further, if $f \in \mathfrak{L}_{\nu,2}$ and $g \in \mathfrak{L}_{\nu,2}$, then the relation (3.6.1) is valid according to Theorem 3.5(c), which is the assertion (ii).

Let us prove (3.6.2). Let $f \in \mathfrak{L}_{\nu,2}$ and $\mathrm{Re}(\lambda) > (1-\nu)h - 1$. To show the relation (3.6.2), it is sufficient to calculate the kernel k in the transform (3.5.1) for such λ. From (3.5.2) we have the equality

$$\left(\mathfrak{M}k\right)(1-\nu+it) = \mathcal{H}(1-\nu+it)\ \frac{1}{\lambda+1-(1-\nu+it)h}$$

or, for $\mathrm{Re}(s) = 1 - \nu$

$$\left(\mathfrak{M}k\right)(s) = \mathcal{H}(s)\,\frac{1}{\lambda + 1 - hs}.$$

Then from (3.2.6) we obtain the expression for the kernel k, namely

$$k(x) = \frac{1}{2\pi i}\lim_{R\to\infty}\int_{1-\nu-iR}^{1-\nu+iR} x^{-s}\mathcal{H}(s)\,\frac{1}{\lambda+1-hs}ds, \tag{3.6.4}$$

where the limit is taken in the topology of $\mathfrak{L}_{\nu,2}$.

According to (1.1.2) we have

$$\mathcal{H}(s)\,\frac{1}{\lambda+1-hs} = \mathcal{H}(s)\frac{\Gamma(1-(-\lambda)-hs)}{\Gamma(1-(-\lambda-1)-hs)}$$

$$= \mathcal{H}^{m,n+1}_{p+1,q+1}\left[\begin{array}{c|c} (-\lambda,h),(a_i,\alpha_i)_{1,p} & \\ & s \\ (b_j,\beta_j)_{1,q},(-\lambda-1,h) & \end{array}\right]. \tag{3.6.5}$$

We denote by α_1, β_1, a_1^*, Δ_1, μ_1 the constants in (3.4.1), (3.4.2), (1.1.7), (1.1.8), (1.1.10) for this $\mathcal{H}^{m,n+1}_{p+1,q+1}$. Then $\alpha_1 = \alpha$, $\beta_1 = \min[\beta,(1+\mathrm{Re}(\lambda))/h]$, $a_1^* = a^*$, $\Delta_1 = \Delta$, $\mu_1 = \mu - 1$. Thus, it follows that

(a)′ $\alpha_1 < 1 - \nu < \beta_1$;

from $\mathrm{Re}(\lambda) > (1-\nu)h - 1$, and either of the conditions:

(b)′ $a_1^* > 0$; or

(c)′ $a_1^* = 0$, $\Delta_1(1-\nu) + \mathrm{Re}(\mu_1) = \Delta(1-\nu) + \mathrm{Re}(\mu) - 1 \leqq -1$

holds. Applying Theorem 3.4 for $x > 0$, except possibly for $x = \delta$, we obtain that the equality

$$H^{m,n+1}_{p+1,q+1}\left[x\,\middle|\,\begin{array}{c}(-\lambda,h),(a_i,\alpha_i)_{1,p}\\ (b_j,\beta_j)_{1,q},(-\lambda-1,h)\end{array}\right]$$

$$= \frac{1}{2\pi i}\lim_{R\to\infty}\int_{1-\nu-iR}^{1-\nu+iR} x^{-s}(\mathcal{H}f)(s)\,\frac{1}{\lambda+1-hs}\,ds \tag{3.6.6}$$

holds almost everywhere. Then, (3.6.4) and (3.6.6) yield that the kernel k is given by

$$k(x) = H^{m,n+1}_{p+1,q+1}\left[x\,\middle|\,\begin{array}{c}(-\lambda,h),(a_i,\alpha_i)_{1,p}\\ (b_j,\beta_j)_{1,q},(-\lambda-1,h)\end{array}\right],$$

and (3.6.2) is proved.

The relation (3.6.3) is proved similarly to (3.6.2), if we use the equality

$$\frac{\mathcal{H}(s)}{\lambda+1-hs} = -\mathcal{H}(s)\,\frac{\Gamma(hs-\lambda-1)}{\Gamma(hs-\lambda)}$$

$$= -\mathcal{H}^{m+1,n}_{p+1,q+1}\left[\begin{array}{c|c} (a_i,\alpha_i)_{1,p},(-\lambda,h) & \\ & s \\ (-\lambda-1,h),(b_j,\beta_j)_{1,q} & \end{array}\right] \tag{3.6.7}$$

instead of (3.6.5), and hence (iii) is proved.

Lastly, let us prove (iv). If $f \in \mathfrak{L}_{\nu,2} \cap \mathfrak{L}_{\widetilde{\nu},2}$ and $\operatorname{Re}(\lambda) > \max[(1-\nu)h-1, (1-\widetilde{\nu})h-1]$ or $\operatorname{Re}(\lambda) < \min[(1-\nu)h-1, (1-\widetilde{\nu})h-1]$, then both transforms $\boldsymbol{H}f$ and $\widetilde{\boldsymbol{H}}f$ are given in (3.6.2) or (3.6.3), respectively, which shows that they are independent of ν.

Corollary 3.6.1. *Let $\alpha < \beta$ and let one of the following conditions hold:*

(b) $a^* > 0$;

(e) $a^* = 0,\ \Delta > 0$ *and* $\alpha < -\dfrac{\operatorname{Re}(\mu)}{\Delta}$;

(f) $a^* = 0,\ \Delta < 0$ *and* $\beta > -\dfrac{\operatorname{Re}(\mu)}{\Delta}$;

(g) $a^* = 0,\ \Delta = 0$ *and* $\operatorname{Re}(\mu) \leqq 0$.

Then the $\boldsymbol{H}$-transform can be defined on $\mathfrak{L}_{\nu,2}$ with $\alpha < \nu < \beta$.

Proof. When $1-\beta < \nu < 1-\alpha$, by Theorem 3.6, if either $a^* > 0$ or $a^* = 0,\ \Delta(1-\nu) + \operatorname{Re}(\mu) \leqq 0$ is satisfied, then the $\boldsymbol{H}$-transform can be defined on $\mathfrak{L}_{\nu,2}$, which is also valid when $\alpha < \nu < \beta$. Hence the corollary is clear in cases (b) and (g). When $\Delta > 0$ and $\alpha < -\operatorname{Re}(\mu)/\Delta$, the assumption $\alpha < \beta$ yields that there exists a number ν such that $\alpha < 1-\nu \leqq -\operatorname{Re}(\mu)/\Delta$ and $1-\nu < \beta$, which are required. For the case (f) the situation is similar, that is, there exists ν of the forms $\beta > 1-\nu \geqq -\operatorname{Re}(\mu)/\Delta$ and $\alpha < 1-\nu$. Thus the proof is completed.

Theorem 3.7. *Let $\alpha < 1-\nu < \beta$ and suppose either of the the following conditions holds:*

(b) $a^* > 0$;

(d) $a^* = 0,\ \Delta(1-\nu) + \operatorname{Re}(\mu) < 0$.

Then for $x > 0$, $\left(\boldsymbol{H}f\right)(x)$ is given in (3.1.1) for $f \in \mathfrak{L}_{\nu,2}$, namely,

$$\left(\boldsymbol{H}f\right)(x) = \int_0^\infty H_{p,q}^{m,n}\left[xt \,\middle|\, \begin{matrix} (a_i,\alpha_i)_{1,p} \\ (b_j,\beta_j)_{1,q} \end{matrix}\right] f(t)dt. \tag{3.6.8}$$

Proof. We denote by $\widetilde{\mu}$ the function $H_{p+1,q+1}^{m,n+1}$ in (3.6.2) instead of μ for the H-function (1.1.1). By (1.1.10) $\widetilde{\mu} = \mu - 1$ and since $\Delta(1-\nu) + \operatorname{Re}(\mu) < 0$, then $\Delta(1-\nu) + \operatorname{Re}(\widetilde{\mu}) < -1$. It follows from Theorem 1.2 in Section 1.2 that if $\operatorname{Re}(\lambda) > (1-\nu)h - 1$, then $H_{p+1,q+1}^{m,n+1}$ in (3.6.2) is continuously differentiable on $\mathbb{R}_+$. Therefore we can differentiate under the integral sign in (3.6.2). Applying the relations (2.2.7) and (2.1.1), we arrive at (3.6.8) provided that the integral in (3.6.8) exists.

The existence of this integral is proved on the basis of (3.4.13). Indeed, we choose γ_1 and γ_2 so that $\alpha < \gamma_1 < 1-\nu < \gamma_2 < \beta$. According to (3.4.13) there are constants A_1 and A_2 such that for almost all $t > 0$, the inequalities

$$|H_{p,q}^{m,n}(t)| \leqq A_i t^{-\gamma_i} \quad (i = 1, 2)$$

hold. Therefore, using the Schwartz inequality (3.5.6), we have

$$\int_0^\infty \left|H_{p,q}^{m,n}(xt)f(t)\right| dt \leqq \left(\int_0^\infty \left|H(xt)t^{1/2-\nu}\right|^2 dt\right)^{1/2} \|f\|_{\nu,2}$$

$$\leqq \left[A_1 x^{-\gamma_1}\left(\int_0^{1/x} t^{2(1-\nu-\gamma_1)-1}dt\right)^{1/2}\right.$$

$$\left.+A_2 x^{-\gamma_2}\left(\int_{1/x}^\infty t^{2(1-\nu-\gamma_2)-1}dt\right)^{1/2}\right] \|f\|_{\nu,2}$$

$$= Cx^{\nu-1} < \infty,$$

where

$$C = \left\{A_1[2(1-\nu-\gamma_1)]^{-1/2} + A_2[2(\gamma_2+\nu-1)]^{-1/2}\right\} \|f\|_{\nu,2},$$

and the theorem is proved.

In conclusion of this section we indicate the conditions for the **H**-transform (3.6.8) to be defined on some $\mathfrak{L}_{\nu,2}$-space.

Corollary 3.7.1. *Let $\alpha < \beta$ and let one of the following conditions hold:*

(b) $a^* > 0$;

(h) $a^* = 0,\ \Delta > 0$ and $\alpha < -\dfrac{\mathrm{Re}(\mu)+1}{\Delta}$;

(i) $a^* = 0,\ \Delta < 0$ and $\beta > -\dfrac{\mathrm{Re}(\mu)+1}{\Delta}$;

(j) $a^* = 0,\ \Delta = 0$ and $\mathrm{Re}(\mu) < 0$.

*Then the **H**-transform can be defined* (3.6.8) *on $\mathfrak{L}_{\nu,2}$ with $\alpha < \nu < \beta$.*

Proof. The proof of this statement is similar to those of Corollary 3.6.1.

3.7. Bibliographical Remarks and Additional Information on Chapter 3

For Section 3.1. The general integral transforms with the H-functions as kernels were introduced by Gupta and P.K. Mittal [1] and R. Singh [1] in 1970 in the forms

$$\left(\boldsymbol{H}f\right)(x) = x\int_0^\infty H_{p,q}^{m,n}\left[xt\left|\begin{matrix}(a_i,\alpha_i)_{1,p}\\(b_j,\beta_j)_{1,q}\end{matrix}\right.\right]f(t)dt \quad (x>0) \tag{3.7.1}$$

and

$$\left(\boldsymbol{H}f\right)(x) = \lambda x\int_0^\infty H_{p,q}^{m,n}\left[cxt\left|\begin{matrix}(a_i,\alpha_i)_{1,p}\\(b_j,\beta_j)_{1,q}\end{matrix}\right.\right]f(t)dt \quad (x>0), \tag{3.7.2}$$

respectively. It should be noted that the form (3.1.1) for the **H**-transform was first suggested by C.B.L. Verma [1] (1966), and that some **H**-transforms were implicitly studied by Fox [2] (1961) and

Kesarwani [11] (1965) while investigating the H-functions as symmetrical and unsymmetrical Fourier kernels, respectively, and by R.K. Saxena [8] while considering the inversion of the Mellin–Barnes integrals of the form (1.1.2).

As indicated in Section 3.1, most of the known integral transforms can be put into the form (3.1.1). These are the $\boldsymbol{G}$-transform (3.1.2) and its particular cases such as the classical Laplace and Hankel transforms (2.5.2) and (2.6.1), the Riemann–Liouville fractional integrals (2.7.1) and (2.7.2), the even and odd Hilbert transforms (8.3.10) and (8.3.11), the Varma-type integral transforms (7.2.1), (7.2.15), (7.3.1) and (7.3.13) with the Whittaker functions as kernels, the integral transform (7.4.1) with parabolic cylinder function in the kernel, the integral transforms (7.5.1), (7.5.13), (7.6.1) and (7.6.12) with hypergeometric functions ${}_1F_1$ and ${}_1F_2$ in the kernels, the integral transforms (7.7.1), (7.7.2), (7.7.21), (7.7.22), (7.8.1), (7.8.2), (7.8.23) and (7.8.24) with the Gauss hypergeometric function in the kernels, the generalized Stieltjes transform (7.9.1) and the ${}_pF_q$-transforms (7.10.1).

There also exist transforms which cannot be reduced to $\boldsymbol{G}$-transforms but can put into the $\boldsymbol{H}$-transforms given in (3.1.1). Such examples are the cosine- and sine-transforms (8.1.2) and (8.1.3), the Erdélyi–Kober type fractional integration operators (3.3.1) and (3.3.2), the transforms with the Gauss hypergeometric function as kernel proved by R.K. Saxena [9], R.K. Saxena and Kumbhat [1], [3], McBride [1], Kumbhat [1] (see (7.12.37), (7.12.38), (7.12.43) and (4.11.25), (4.11.26)), the Wright transform ${}_p\boldsymbol{\Psi}_q$ in (7.11.1), the Laplace type transform $\mathbb{L}^*_{\gamma,k}$ in (7.1.1), the generalized Laplace and Hankel transforms $\mathbb{L}_{k,\alpha}$, $\mathbb{H}_{k,\eta}$ in (3.3.3) and (3.3.4), the extended Hankel transform $\mathbb{H}_{\eta,l}$ in (8.4.1), Hankel type transforms $\mathbb{H}_{\eta;\sigma,\varpi;k,\lambda}$, $\mathbf{h}_{\eta;1}$, $\mathbf{h}_{\eta;2}$, $\mathbf{b}_{\eta;1}$, $\mathbf{b}_{\eta;2}$ in (8.5.1) and (8.6.1)–(8.6.4), $\mathbb{Y}_\eta$- and $\mathfrak{H}_\eta$-transforms in (8.7.1) and (8.8.1), the Meijer transform $\mathfrak{K}_\eta$ in (8.9.1), the Bessel type transforms $\mathbf{K}^\rho_\eta$, $\mathbf{L}^{(m)}_\eta$, $\boldsymbol{L}^{(\beta)}_{\eta,\sigma}$ in (8.10.1), (8.10.2) and (8.11.1), the modified Hardy– and Hardy–Titchmarsh transforms $\mathbb{J}_{\eta,\sigma}$, $\mathbb{J}_{a,b,c;\omega}$ in (8.12.6) and (8.12.1) and the Lommel–Maitland transform $\mathbb{J}^\gamma_{\eta,\sigma}$ in (8.13.1).

It should be noted that, though many authors studied $\boldsymbol{H}$-transforms of the forms (3.1.1), (3.7.1), (3.7.2) and modified $\boldsymbol{H}$-transforms, most of the investigations were devoted to studying the inversion of $\boldsymbol{H}$-transforms in spaces $L_r(\mathbb{R}_+)$ $(1 \leqq r < \infty)$, their Mellin transform and the relation of fractional integration by parts of the form (3.6.1). The boundedness of such $\boldsymbol{H}$-transforms in the space $L_r(\mathbb{R}_+)$ were studied in several papers (see Sections 4.11 and 5.7 in this connection).

The mapping properties such as the boundedness and the respresentation of the $\boldsymbol{H}$-transform (3.1.1) in the space $\mathfrak{L}_{\nu,2}$ were presented in Section 3.6. Such properties together with the range and invertibilty of such an $\boldsymbol{H}$-transform in the spaces $\mathfrak{L}_{\nu,r}$ $(1 \leqq r \leqq \infty)$ will be given in Chapter 4 by extending results in this chapter. As for further results we indicate that Vu Kim Tuan [1] studied the factorization properties of the $\boldsymbol{H}$-transform (3.1.1) in the special functional space L_2^Φ, and Nguyen Thanh Hai and Yakubovich [1] investigated general integral convolution for these integral transforms.

A series of papers was devoted to studying integral transforms with H-function kernels in spaces of generalized functions and Abelian theorems for them. Classical Abelian theorems give the asymptotic behavior of the integral transform $(\boldsymbol{H}f)(x)$ near zero or infinity on the basis of the known asymptotic behavior of $f(t)$ as $t \to \infty$ or $t \to 0$, respectively. R.K. Saxena [10] obtained classical Abelian theorems for more general integral transforms than (3.1.1) with H-function kernel of the form

$$\left(\boldsymbol{H}_\omega f\right)(x) = \int_0^\infty e^{-\omega x t} H^{m,n}_{p,q}\left[xt \,\middle|\, \begin{matrix} (a_i,\alpha_i)_{1,p} \\ (b_j,\beta_j)_{1,q} \end{matrix}\right] f(t)dt \quad (x > 0) \tag{3.7.3}$$

with positive $\omega > 0$ and extended the results to generalized functions. R.K. Saxena [10] proved that if δ $(\delta > -1)$ and α are given real numbers such that $t^{-\delta}f(t)$ is absolutely continuous on $[0,\infty)$ and

$$\lim_{t\to+\infty} f(t) = 0, \quad \lim_{t\to+0} t^{-\delta}f(t) = \alpha, \tag{3.7.4}$$

then

$$\lim_{x\to+\infty} \frac{(\omega x)^{\delta+1}}{H(1/\omega)} \left(\boldsymbol{H}_\omega f\right)(x) = \alpha, \tag{3.7.5}$$

where the function in the denominator is

$$H(z) = H^{m,n+1}_{p+1,q}\left[z \,\middle|\, \begin{matrix} (-\delta,1), (a_i,\alpha_i)_{1,p} \\ (b_j,\beta_j)_{1,q} \end{matrix}\right]. \tag{3.7.6}$$

He also extended this result to generalized functions and gave other sufficient conditions for (3.7.5) to be valid. Joshi and Raj.K. Saxena [1]–[3] investigated the $\boldsymbol{H}$-transform of the form (3.7.3), where the variable xt of the H-function in the integrand is replaced by $\rho(xt)^{\mu}$ with $\rho > 0$ and $\mu > 0$. They obtained in [1] the same assertions as those given above by R.K. Saxena [10]. They further proved in [2] two structure theorems for the generalized functions (represented by such generalized $\boldsymbol{H}$-transforms) in terms of differential operators acting on functions or on measures, and used such structures in [3] to establish the complex inversion formula for this generalized $\boldsymbol{H}$-transform and to discuss its uniqueness. See also Brychkov and Prudnikov [1, Section 8.2] in this connection.

Carmichael and Pathak [2] extended the above results of R.K. Saxena [10] to complex x by proving Abelian theorems in a wedge domain in the right half-plane, defined quasi-asymptotic behavior of such $\boldsymbol{H}$-transformable generalized functions, proved a structure theorem for generalized functions having such quasi-asymptotic behavior and gave applications to obtain the asymptotic behavior of the $\boldsymbol{H}$-transform of these generalized functions. Similar results for the $\boldsymbol{H}$-transform (3.1.1) were obtained by Carmichael and Pathak [1].

Malgonde and Raj.K. Saxena extended the $\boldsymbol{H}$-transform of the form

$$\left(\boldsymbol{H}f\right)(x) = \int_0^{\infty} H^{m+1,0}_{m,m+1}\left[xt \,\middle|\, \begin{matrix} (a_i + b_i, \alpha_i)_{1,m} \\ (b_j, \beta_j)_{1,m}, (\rho, \beta_{m+1}) \end{matrix}\right] f(t)dt \quad (x > 0) \tag{3.7.7}$$

to generalized functions and proved the representation for such a transform in [1], [3] and an inversion and a uniqueness theorem in [2], [5]. They established some Abelian theorems for such a distributional transform and for more general distributional transforms with the kernel $H^{m,0}_{p,q}$ in [6] and [4], respectively. See also Raj.K. Saxena, Koranne and Malgonde [1] in this connection.

We also indicate that, though applications of the H-function (1.1.1) in statistics and other disciplines are well known (see Mathai and R.K. Saxena [2]), only one paper by Raina [2] was devoted to application of the $\boldsymbol{H}$-transform concerning the problem of absolute moments of arbitrary order of a probability distribution function.

For Sections 3.2 and 3.3. The results presented in these sections belong to Rooney [6, Sections 2 and 5].

We only note that the Erdélyi–Kober type fractional integrals (3.3.1) and (3.3.2) were introduced by Kober [2] for $\sigma = 1$ and by Erdélyi [3] in the general case, and they gave the conditions for the boundedness of these operators in the space $\mathfrak{L}_{1/r,r} = L^r(\mathbb{R}_+)$ for $1 < r < \infty$. When $\sigma = 2$, the operators (3.3.1) and (3.3.2) are known as Erdélyi–Kober operators, and Sneddon [2] was the first to name them in this way.

For Sections 3.4–3.6. The results of these sections were proved by the authors together with Shlapakov in Kilbas, Saigo and Shlapakov [1], and some of these assertions were indicated in Kilbas and Shlapakov [2] (see also the papers of Shlapakov [1] and Kilbas, Saigo and Shlapakov [2], [3]). Theorems 3.3 and 3.4 in Section 3.4 establish the integral representations (3.4.3) and (3.4.12) for the H-function $H^{m,n}_{p,q}$ in (1.1.1) on the spaces $\mathfrak{L}_{\nu,2}$. They, together with Theorem 3.5 in Section 3.5, play the main role in constructing $\mathfrak{L}_{\nu,2}$-theory for the $\boldsymbol{H}$-transforms in Theorems 3.6 and 3.7 in Section 3.6. The obtained results generalize the corresponding results by Rooney [6, Lemmas 3.2, 3.3 and 4.1 and Theorem 4.1], being proved for the $\boldsymbol{G}$-transform (3.1.2) (see also Section 5.1 in this connection).

We also note that the statements presented in Sections 3.4–3.6 were independently obtained by Betancor and Jerez Diaz [1].

Chapter 4
$\boldsymbol{H}$-TRANSFORM ON THE SPACE $\mathfrak{L}_{\nu,r}$

In Chapter 3 we constructed so-called $\mathfrak{L}_{\nu,2}$-theory of the $\boldsymbol{H}$-transform (3.1.1), where we characterize the existence, the boundedness and representation properties of the transforms $\boldsymbol{H}$ on the space $\mathfrak{L}_{\nu,2}$ given in (3.1.3). The present chapter is devoted to extending the above results from $r = 2$ to any $r \geqq 1$. Moreover, we shall deal with the study of properties such as the range and the invertibility of the $\boldsymbol{H}$-transform on the space $\mathfrak{L}_{\nu,r}$ with any $1 \leqq r < \infty$. The results will be different in nine cases:

1) $a^* = \Delta = \text{Re}(\mu) = 0$; **2)** $a^* = \Delta = 0$, $\text{Re}(\mu) < 0$; **3)** $a^* = 0$, $\Delta > 0$;
4) $a^* = 0$, $\Delta < 0$; **5)** $a_1^* > 0$, $a_2^* > 0$; **6)** $a_1^* > 0$, $a_2^* = 0$;
7) $a_1^* = 0$, $a_2^* > 0$; **8)** $a^* > 0$, $a_1^* > 0$, $a_2^* < 0$; **9)** $a^* > 0$, $a_1^* < 0$, $a_2^* > 0$.

Here a^*, Δ, μ, a_1^* and a_2^* are given in (1.1.7), (1.1.8), (1.1.10), (1.1.11) and (1.1.12), respectively. We shall also use the constants α and β defined by (3.4.1) and (3.4.2), respectively.

4.1. $\mathfrak{L}_{\nu,r}$-Theory of the $\boldsymbol{H}$-Transform When $a^* = \Delta = 0$ and $\text{Re}(\mu) = 0$

In this and the next sections, based on the existence of the transform $\boldsymbol{H}$ on the space $\mathfrak{L}_{\nu,2}$ which is guaranteed in Theorem 3.6 for some $\nu \in \mathbb{R}$ and $a^* = \Delta = 0$, $\text{Re}(\mu) \leqq 0$, we prove that such a transform can be extended to $\mathfrak{L}_{\nu,r}$ for $1 < r < \infty$ such that $\boldsymbol{H} \in [\mathfrak{L}_{\nu,r}, \mathfrak{L}_{1-\nu,s}]$ for a certain range of the value s. We also characterize the range of $\boldsymbol{H}$ on $\mathfrak{L}_{\nu,r}$ in terms of the Erdélyi–Kober type fractional integral operators $I_{0+;\sigma,\eta}^{\alpha}$ and $I_{-;\sigma,\eta}^{\alpha}$ given in (3.3.1) and (3.3.2) except for its isolated values $\nu \in \mathcal{E}_{\mathcal{H}}$. The results will be different in the cases $\text{Re}(\mu) = 0$ and $\text{Re}(\mu) \neq 0$. In this section we consider the former case.

Theorem 4.1. *Let $a^* = \Delta = 0, \text{Re}(\mu) = 0$ and $\alpha < 1 - \nu < \beta$. Let $1 < r < \infty$.*

(a) *The transform $\boldsymbol{H}$ defined on $\mathfrak{L}_{\nu,2}$ can be extended to $\mathfrak{L}_{\nu,r}$ as an element of $[\mathfrak{L}_{\nu,r}, \mathfrak{L}_{1-\nu,r}]$.*

(b) *If $1 < r \leqq 2$, the transform $\boldsymbol{H}$ is one-to-one on $\mathfrak{L}_{\nu,r}$ and there holds the equality* (3.1.5), *namely,*

$$\left(\mathfrak{M}\boldsymbol{H}f\right)(s) = \mathcal{H}(s)\left(\mathfrak{M}f\right)(1-s) \quad (\text{Re}(s) = 1 - \nu). \tag{4.1.1}$$

(c) *If $\nu \notin \mathcal{E}_{\mathcal{H}}$, then $\boldsymbol{H}$ is a one-to-one transform on $\mathfrak{L}_{\nu,r}$ onto $\mathfrak{L}_{1-\nu,r}$, i.e.*

$$\boldsymbol{H}(\mathfrak{L}_{\nu,r}) = \mathfrak{L}_{1-\nu,r}. \tag{4.1.2}$$

(d) *If $f \in \mathfrak{L}_{\nu,r}$ and $g \in \mathfrak{L}_{\nu,r'}$ and $r' = r/(r-1)$, then the relation* (3.6.1) *holds:*

$$\int_0^\infty f(x)\left(\boldsymbol{H}g\right)(x)dx = \int_0^\infty \left(\boldsymbol{H}f\right)(x)g(x)dx. \tag{4.1.3}$$

(e) *If $f \in \mathfrak{L}_{\nu,r}$, $\lambda \in \mathbb{C}$ and $h > 0$, then $\boldsymbol{H}f$ is given by*

$$\left(\boldsymbol{H}f\right)(x)$$

$$= hx^{1-(\lambda+1)/h}\frac{d}{dx}x^{(\lambda+1)/h}\int_0^\infty H^{m,n+1}_{p+1,q+1}\left[xt \,\middle|\, \begin{matrix}(-\lambda,h),(a_i,\alpha_i)_{1,p}\\(b_j,\beta_j)_{1,q},(-\lambda-1,h)\end{matrix}\right]f(t)dt \qquad (4.1.4)$$

for $\mathrm{Re}(\lambda) > (1-\nu)h - 1$, *while*

$$\left(\boldsymbol{H}f\right)(x)$$

$$= -hx^{1-(\lambda+1)/h}\frac{d}{dx}x^{(\lambda+1)/h}\int_0^\infty H^{m+1,n}_{p+1,q+1}\left[xt \,\middle|\, \begin{matrix}(a_i,\alpha_i)_{1,p},(-\lambda,h)\\(-\lambda-1,h),(b_j,\beta_j)_{1,q}\end{matrix}\right]f(t)dt \qquad (4.1.5)$$

for $\mathrm{Re}(\lambda) < (1-\nu)h - 1$.

Proof. Since $\alpha < 1-\nu < \beta$ and $\Delta(1-\nu) + \mathrm{Re}(\mu) \leqq 0$, then according to Theorem 3.6 the transform $\boldsymbol{H}$ is defined on $\mathfrak{L}_{\nu,2}$. We denote by $\mathcal{H}_0(s)$ the function

$$\mathcal{H}_0(s) = \delta^{-s}\mathcal{H}(s), \qquad (4.1.6)$$

where δ is defined in (1.1.9). It follows from (1.2.9) that

$$\mathcal{H}_0(\sigma+it) \sim \prod_{j=1}^{q}\beta_j^{\mathrm{Re}(b_j)-1/2}\prod_{i=1}^{p}\alpha_i^{1/2-\mathrm{Re}(a_i)}(2\pi)^{c^*}e^{-c^*}e^{-\pi\mathrm{Im}(\xi)\mathrm{sign}(t)/2} \qquad (4.1.7)$$

$$(|t| \to \infty)$$

is uniformly in σ in any bounded interval of $\mathbb{R}$. Therefore $\mathcal{H}_0(s)$ is analytic in the strip $\alpha < \mathrm{Re}(s) < \beta$, and if $\alpha < \sigma_1 \leqq \sigma_2 < \beta$, then $\mathcal{H}_0(s)$ is bounded in the strip $\sigma_1 \leqq \mathrm{Re}(s) \leqq \sigma_2$. Since $a^* = \Delta = 0$, then in accordance with (1.1.13) $a_1^* = -a_2^* = \Delta/2 = 0$. Then from (4.1.6) and (3.4.14) we have

$$\mathcal{H}_0'(\sigma+it) = \mathcal{H}_0(\sigma+it)\left[-\log\delta + \frac{\mathcal{H}'(\sigma+it)}{\mathcal{H}(\sigma+it)}\right]$$

$$= \mathcal{H}_0(\sigma+it)\left[\frac{\mathrm{Im}(\mu)}{t} + O\left(\frac{1}{t^2}\right)\right] = O\left(\frac{1}{t}\right) \quad (|t| \to \infty) \qquad (4.1.8)$$

for $\alpha < \sigma < \beta$. Thus $\mathcal{H}_0(s)$ belongs to the class $\mathcal{A}$ (see Definition 3.2) with $\alpha(\mathcal{H}_0) = \alpha$ and $\beta(\mathcal{H}_0) = \beta$. Therefore by virtue of Theorem 3.1, there is a transform $T \in [\mathfrak{L}_{1-\nu,r}]$ for $1 < r < \infty$ and $\alpha < 1-\nu < \beta$. When $1 < r \leqq 2$, then T is a one-to-one transform on $\mathfrak{L}_{1-\nu,r}$ and the relation

$$\left(\mathfrak{M}Tf\right)(s) = \mathcal{H}_0(s)\left(\mathfrak{M}f\right)(s) \quad (\mathrm{Re}(s) = 1-\nu) \qquad (4.1.9)$$

holds for $f \in \mathfrak{L}_{1-\nu,r}$. Let

$$\boldsymbol{H}_0 = W_\delta TR, \qquad (4.1.10)$$

where W_δ and R are given in (3.3.12) and (3.3.13). According to Lemma 3.1(ii) and Lemma 3.1(iii), $R \in [\mathfrak{L}_{\nu,r}, \mathfrak{L}_{1-\nu,r}]$, $W_\delta \in [\mathfrak{L}_{1-\nu,r}]$ and hence $\boldsymbol{H}_0 \in [\mathfrak{L}_{\nu,r}, \mathfrak{L}_{1-\nu,r}]$ for $\alpha < 1-\nu < \beta$ and $1 < r < \infty$, too. When $\alpha < 1-\nu < \beta$, $1 < r \leqq 2$ and $f \in \mathfrak{L}_{\nu,r}$, it follows from (4.1.10), (3.3.15), (4.1.9), (3.3.16) and (4.1.6) that

$$\begin{aligned}\left(\mathfrak{M}\boldsymbol{H}_0 f\right)(s) &= \left(\mathfrak{M}W_\delta TRf\right)(s) = \delta^s\left(\mathfrak{M}TRf\right)(s)\\ &= \delta^s\mathcal{H}_0(s)\left(\mathfrak{M}Rf\right)(s) = \delta^s\mathcal{H}_0(s)\left(\mathfrak{M}f\right)(1-s)\\ &= \mathcal{H}(s)\left(\mathfrak{M}f\right)(1-s) \qquad (4.1.11)\end{aligned}$$

for $\mathrm{Re}(s) = 1-\nu$. In particular, for $f \in \mathfrak{L}_{\nu,2}$ Theorem 3.6(i), (3.1.5) and (4.1.11) imply the equality

$$\left(\mathfrak{M}\boldsymbol{H}_0 f\right)(s) = \left(\mathfrak{M}\boldsymbol{H}f\right)(s) \quad (\mathrm{Re}(s) = 1-\nu). \qquad (4.1.12)$$

Thus $\boldsymbol{H}_0 f = \boldsymbol{H}f$ for $f \in \mathfrak{L}_{\nu,2}$ and therefore, if $\alpha < 1-\nu < \beta$, $\boldsymbol{H} = \boldsymbol{H}_0$ on $\mathfrak{L}_{\nu,2}$ by Theorem 3.6(iv). Since $\mathfrak{L}_{\nu,2} \bigcap \mathfrak{L}_{\nu,r}$ is dense in $\mathfrak{L}_{\nu,r}$ (see Rooney [2, Lemma 2.2]), $\boldsymbol{H}$ can be extended to $\mathfrak{L}_{\nu,r}$ and, if we denote it there by $\boldsymbol{H}$ again, $\boldsymbol{H} \in [\mathfrak{L}_{\nu,r}, \mathfrak{L}_{1-\nu,r}]$. This completes the proof of assertion (a) of the theorem.

The property (b) is clear from the fact that the operator T above and the operators W_δ and R are one-to-one and (4.1.1) follows from (4.1.11).

Let us prove (c). Since $R(\mathfrak{L}_{\nu,r}) = \mathfrak{L}_{1-\nu,r}$ and $W_\delta(\mathfrak{L}_{1-\nu,r}) = \mathfrak{L}_{1-\nu,r}$, then the onto map property $\boldsymbol{H}(\mathfrak{L}_{\nu,r}) = \mathfrak{L}_{1-\nu,r}$ holds if and only if $T(\mathfrak{L}_{1-\nu,r}) = \mathfrak{L}_{1-\nu,r}$. To prove this, it should be noted that the abscissas of the zeros of $\mathcal{H}(s)$ divide the interval (α,β) into disjoint open intervals, where and thereafter $\mathcal{H}_0(s)$ in (4.1.6) is renamed $\mathcal{H}(s)$. Let (α_1,β_1) be one such interval. Then the function $1/\mathcal{H}(s)$ is analytic in $\alpha_1 < \mathrm{Re}(s) < \beta_1$. In view of (4.1.7) we have

$$\frac{1}{\mathcal{H}(\sigma+it)} \sim \prod_{j=1}^{q} \beta_j^{1/2-\mathrm{Re}(b_j)} \prod_{i=1}^{p} \alpha_i^{\mathrm{Re}(a_i)-1/2}(2\pi)^{-c^*}e^{c^*}e^{\pi\mathrm{Im}(\xi)\mathrm{sign}(t)/2} \quad (|t| \to \infty).$$

So if we take $\alpha_1 < \sigma_1 \leqq \sigma_2 < \beta_1$, then $1/\mathcal{H}(s)$ is bounded in the strip $\sigma_1 \leqq \mathrm{Re}(s) \leqq \sigma_2$. The equality

$$\left\{\frac{1}{\mathcal{H}}\right\}'(\sigma+it) = -\frac{\mathcal{H}'(\sigma+it)}{\mathcal{H}^2(\sigma+it)}$$

implies by (4.1.7) and (4.1.8) that

$$\left\{\frac{1}{\mathcal{H}}\right\}'(\sigma+it) = O\left(\frac{1}{t}\right) \quad (|t| \to \infty)$$

for $\alpha_1 < \sigma < \alpha_2$. Thus we have that $1/\mathcal{H} \in \mathcal{A}$ with $\alpha(1/\mathcal{H}) = \alpha_1$ and $\beta(1/\mathcal{H}) = \beta_1$. Then for $\alpha_1 < \nu < \beta_1$ and $1 < r < \infty$ it follows from Theorem 3.1 that the transform T is one-to-one on $\mathfrak{L}_{\nu,r}$ and $T(\mathfrak{L}_{\nu,r}) = \mathfrak{L}_{\nu,r}$. But if $\nu \notin \mathcal{E}_{\mathcal{H}}$, then the value $1-\nu$ does not coincide with the abscissa of any zero of $\mathcal{H}(s)$, and hence $1-\nu$ lies in such (α_1,β_1). Therefore $\boldsymbol{H}$ is a one-to-one transform on $\mathfrak{L}_{\nu,r}$ and $\boldsymbol{H}(\mathfrak{L}_{\nu,r}) = \mathfrak{L}_{1-\nu,r}$. The assertion (c) of the theorem is thus proved.

Now we prove (4.1.3). If $\alpha < 1-\nu < \beta$, then by using the Hölder inequality

$$\left|\int_a^b f(x)g(x)dx\right| \leqq \left(\int_a^b |f(x)|^p dx\right)^{1/p}\left(\int_a^b |g(x)|^{p'} dx\right)^{1/p'} \tag{4.1.13}$$

$$\left(\frac{1}{p}+\frac{1}{p'}=1,\ -\infty \leqq a < b \leqq \infty\right)$$

we have

$$\left|\int_0^\infty f(x)\left(\boldsymbol{H}g\right)(x)dx\right| = \left|\int_0^\infty [x^{\nu-1/r}f(x)][x^{1/r-\nu}\left(\boldsymbol{H}g\right)(x)]dx\right|$$

$$\leqq \|f\|_{\nu,r}\|\boldsymbol{H}g\|_{1-\nu,r'} \leqq K\|f\|_{\nu,r}\|g\|_{\nu,r'} \quad \left(\frac{1}{r}+\frac{1}{r'}=1\right),$$

where K is a bound for $\boldsymbol{H} \in [\mathfrak{L}_{\nu,r'}, \mathfrak{L}_{1-\nu,r'}]$. Hence the left-hand side of (4.1.3) represents a bounded bilinear functional on $\mathfrak{L}_{\nu,r} \times \mathfrak{L}_{\nu,r'}$. Similarly it is proved that the right-hand side of (4.1.3) represents such a functional on $\mathfrak{L}_{\nu,r} \times \mathfrak{L}_{\nu,r'}$. By virtue of Theorem 3.6(ii), if $f \in \mathfrak{L}_{\nu,2}$ and $g \in \mathfrak{L}_{\nu,2}$, (3.6.1) is also true. By Rooney [2, Lemma 2.2] $\mathfrak{L}_{\nu,r} \bigcap \mathfrak{L}_{\nu,2}$ is dense in $\mathfrak{L}_{\nu,r}$ and hence (3.6.1) is true for $f \in \mathfrak{L}_{\nu,r}$ and $g \in \mathfrak{L}_{\nu,r'}$ with $1 < r < \infty$ and $\alpha < 1-\nu < \beta$. This completes the proof of the assertion (d) of the theorem.

Finally we prove (e). If $\mathrm{Re}(\lambda) > (1-\nu)h - 1$, then the function

$$g_x(t) = \begin{cases} t^{(\lambda+1)/h-1} & \text{if } 0 < t < x; \\ 0 & \text{if } t > x \end{cases} \tag{4.1.14}$$

belongs to $\mathfrak{L}_{\nu,s}$ for $1 \leqq s < \infty$. When $s = 2$, we may apply Theorem 3.6(iii) for $g_x \in \mathfrak{L}_{\nu,2}$ and we have

$$\left(\boldsymbol{H}g_x\right)(y) = hy^{1-(\lambda+1)/h}\frac{d}{dy}y^{(\lambda+1)/h}\int_0^x H^{m,n+1}_{p+1,q+1}\left[yt \,\middle|\, \begin{matrix}(-\lambda,h),(a_i,\alpha_i)_{1,p}\\ (b_j,\beta_j)_{1,q},(-\lambda-1,h)\end{matrix}\right]$$

$$\cdot\, t^{(\lambda+1)/h-1}dt$$

$$= hy^{1-(\lambda+1)/h}\frac{d}{dy}\int_0^{xy} H^{m,n+1}_{p+1,q+1}\left[t \,\middle|\, \begin{matrix}(-\lambda,h),(a_i,\alpha_i)_{1,p}\\ (b_j,\beta_j)_{1,q},(-\lambda-1,h)\end{matrix}\right] t^{(\lambda+1)/h-1}dt$$

$$= hx^{(\lambda+1)/h}H^{m,n+1}_{p+1,q+1}\left[xy \,\middle|\, \begin{matrix}(-\lambda,h),(a_i,\alpha_i)_{1,p}\\ (b_j,\beta_j)_{1,q},(-\lambda-1,h)\end{matrix}\right]$$

almost everywhere. For $f \in \mathfrak{L}_{\nu,r}$ with $\alpha < 1-\nu < \beta$ and $1 < r < \infty$ and for the above $g_x \in \mathfrak{L}_{\nu,r'}$, we have from the previous result (d) that

$$\int_0^x t^{(\lambda+1)/h-1}\left(\boldsymbol{H}f\right)(t)dt = \int_0^\infty \left(\boldsymbol{H}f\right)(t)g_x(t)dt = \int_0^\infty f(t)\left(\boldsymbol{H}g_x\right)(t)dt$$

$$= hx^{(\lambda+1)/h}\int_0^\infty H^{m,n+1}_{p+1,q+1}\left[xt \,\middle|\, \begin{matrix}(-\lambda,h),(a_i,\alpha_i)_{1,p}\\ (b_j,\beta_j)_{1,q},(-\lambda-1,h)\end{matrix}\right] f(t)dt.$$

From here, after differentiation with respect to x, we arrive at (4.1.4). In the case $\mathrm{Re}(\lambda) < (1-\nu)h - 1$ the relation (4.1.5) is proved similarly if we use the function

$$h_x(t) = \begin{cases} 0 & \text{if } 0 < t < x; \\ t^{(\lambda+1)/h-1} & \text{if } t > x \end{cases}$$

instead of the function $g_x(t)$. Thus the theorem is proved.

4.2. $\mathfrak{L}_{\nu,r}$-Theory of the $\boldsymbol{H}$-Transform When $a^* = \Delta = 0$ and $\mathrm{Re}(\mu) < 0$

Now we consider the $\boldsymbol{H}$-transform in the case $a^* = \Delta = 0$ and $\mathrm{Re}(\mu) < 0$.

Theorem 4.2. *Let $a^* = \Delta = 0, \mathrm{Re}(\mu) < 0$ and $\alpha < 1 - \nu < \beta$, and let either $m > 0$ or $n > 0$. Let $1 < r < \infty$.*

(a) *The transform $\boldsymbol{H}$ defined on $\mathfrak{L}_{\nu,2}$ can be extended to $\mathfrak{L}_{\nu,r}$ as an element of $[\mathfrak{L}_{\nu,r}, \mathfrak{L}_{1-\nu,s}]$ for all $s \geqq r$ such that $1/s > 1/r + \mathrm{Re}(\mu)$.*

(b) *If $1 < r \leqq 2$, then the transform $\boldsymbol{H}$ is one-to-one on $\mathfrak{L}_{\nu,r}$ and there holds the equality (4.1.1).*

(c) *Let $k > 0$. If $\nu \notin \mathcal{E}_{\mathcal{H}}$, then the transform $\boldsymbol{H}$ is one-to-one on $\mathfrak{L}_{\nu,r}$ and there hold*

$$\boldsymbol{H}\left(\mathfrak{L}_{\nu,r}\right) = I^{-\mu}_{-;k,-\alpha/k}\left(\mathfrak{L}_{1-\nu,r}\right) \tag{4.2.1}$$

for $m > 0$, and

$$\boldsymbol{H}\left(\mathfrak{L}_{\nu,r}\right) = I^{-\mu}_{0+;k,\beta/k-1}\left(\mathfrak{L}_{1-\nu,r}\right) \tag{4.2.2}$$

for $n > 0$. If $\nu \in \mathcal{E}_{\mathcal{H}}$, $\boldsymbol{H}\left(\mathfrak{L}_{\nu,r}\right)$ is a subset of the right-hand sides of (4.2.1) and (4.2.2) in the respective cases.

(d) *If $f \in \mathfrak{L}_{\nu,r}$ and $g \in \mathfrak{L}_{\nu,s}$ with $1 < r < \infty, 1 < s < \infty$ and $1 \leqq 1/r + 1/s < 1 - \mathrm{Re}(\mu)$, then the relation (4.1.3) holds.*

(e) *If $f \in \mathfrak{L}_{\nu,r}, \lambda \in \mathbb{C}$ and $h > 0$, then $\boldsymbol{H}f$ is given in (4.1.4) for $\mathrm{Re}(\lambda) > (1-\nu)h - 1$, while in (4.1.5) for $\mathrm{Re}(\lambda) < (1-\nu)h - 1$. Furthermore $\boldsymbol{H}f$ is given in (3.1.1), namely,*

$$\left(\boldsymbol{H}f\right)(x) = \int_0^\infty H^{m,n}_{p,q}\left[xt \,\middle|\, \begin{matrix} (a_i, \alpha_i)_{1,p} \\ (b_j, \beta_j)_{1,q} \end{matrix}\right] f(t)dt. \tag{4.2.3}$$

Proof. Since $\alpha < 1 - \nu < \beta$, $a^* = 0$ and $\Delta(1-\nu) + \mathrm{Re}(\mu) = \mathrm{Re}(\mu) < 0$, then from Theorem 3.6 the transform $\boldsymbol{H}$ is defined on $\mathfrak{L}_{\nu,2}$.

If $m > 0$ or $n > 0$, then α or β is finite in view of (3.4.1) and (3.4.2). We set

$$\mathcal{H}_1(s) = \frac{\Gamma\left(\dfrac{s-\alpha}{k} - \mu\right)}{\Gamma\left(\dfrac{s-\alpha}{k}\right)}\mathcal{H}(s) = \mathcal{H}^{m+1,n}_{p+1,q+1}\left[\begin{matrix} (a_i, \alpha_i)_{1,p}, \left(-\dfrac{\alpha}{k}, \dfrac{1}{k}\right) \\ \left(-\mu - \dfrac{\alpha}{k}, \dfrac{1}{k}\right), (b_j, \beta_j)_{1,q} \end{matrix} \,\middle|\, s\right] \tag{4.2.4}$$

for $m > 0$, and

$$\mathcal{H}_2(s) = \frac{\Gamma\left(\dfrac{\beta-s}{k} - \mu\right)}{\Gamma\left(\dfrac{\beta-s}{k}\right)}\mathcal{H}(s) = \mathcal{H}^{m,n+1}_{p+1,q+1}\left[\begin{matrix} \left(1 + \mu - \dfrac{\beta}{k}, \dfrac{1}{k}\right), (a_i, \alpha_i)_{1,p} \\ (b_j, \beta_j)_{1,q}, \left(1 - \dfrac{\beta}{k}, \dfrac{1}{k}\right) \end{matrix} \,\middle|\, s\right] \tag{4.2.5}$$

for $n > 0$. We denote α_1, β_1, $\tilde{a}_1^*$, Δ_1, δ_1 and μ_1 for $\mathcal{H}_1$, and α_2, β_2, $\tilde{a}_2^*$, Δ_2, δ_2 and μ_2 for $\mathcal{H}_2$ instead of that for $\mathcal{H}$. Then we find that

$$\begin{aligned} &\alpha_1 = \max[\alpha, \alpha + k\mathrm{Re}(\mu)] = \alpha, \quad \beta_1 = \beta, \\ &\tilde{a}_1^* = a^* = 0, \quad \Delta_1 = \Delta = 0, \quad \delta_1 = \delta, \quad \mu_1 = 0, \\ &\alpha_2 = \alpha, \quad \beta_2 = \min[\beta, \beta - k\mathrm{Re}(\mu)] = \beta, \\ &\tilde{a}_2^* = a^* = 0, \quad \Delta_2 = \Delta = 0, \quad \delta_2 = \delta, \quad \mu_2 = 0, \end{aligned} \tag{4.2.6}$$

and the exceptional sets $\mathcal{E}_{\mathcal{H}_1}$ and $\mathcal{E}_{\mathcal{H}_2}$ of $\mathcal{H}_1$ and $\mathcal{H}_2$ in Definition 3.4 coincide with $\mathcal{E}_{\mathcal{H}}$ for $\mathcal{H}$. Since $\mu_1 = \mu_2 = 0$, Theorem 4.1 guarantees the existence of two transforms $\widetilde{\boldsymbol{H}}_1$ and $\widetilde{\boldsymbol{H}}_2$ in $[\mathfrak{L}_{\nu,r}, \mathfrak{L}_{1-\nu,r}]$ for $m > 0$ and for $n > 0$, respectively. Further, if $f \in \mathfrak{L}_{\nu,r}$ with $1 < r \leqq 2$, then by (4.1.1)

$$\left(\mathfrak{M}\widetilde{\boldsymbol{H}}_i f\right)(s) = \mathcal{H}_i(s)\left(\mathfrak{M}f\right)(1-s) \quad (\mathrm{Re}(s) = 1-\nu;\ i = 1, 2). \tag{4.2.7}$$

We set

$$\boldsymbol{H}_1 = I^{-\mu}_{-;k,-\alpha/k}\widetilde{\boldsymbol{H}}_1 \tag{4.2.8}$$

for $m > 0$, and

$$\boldsymbol{H}_2 = I^{-\mu}_{0+;k,\beta/k-1}\widetilde{\boldsymbol{H}}_2 \tag{4.2.9}$$

for $n > 0$.

Let (i) $m > 0$ or (ii) $n > 0$. For $1 \leqq r \leqq \infty$, $s \geqq r$, $1/s > 1/r + \mathrm{Re}(\mu)$ Theorem 3.2(b) and Theorem 3.2(a) imply that $I^{-\mu}_{-;k,-\alpha/k} \in [\mathfrak{L}_{1-\nu,r}, \mathfrak{L}_{1-\nu,s}]$ for $\alpha < 1-\nu$ and $I^{-\mu}_{0+;k,\beta/k-1} \in [\mathfrak{L}_{1-\nu,r}, \mathfrak{L}_{1-\nu,s}]$ for $\beta > 1-\nu$, respectively. Therefore, if $\alpha < 1-\nu < \beta$ and $1 < r < \infty$, we have $\widetilde{\boldsymbol{H}}_i \in [\mathfrak{L}_{\nu,r}, \mathfrak{L}_{1-\nu,s}]$ $(i = 1, 2)$ for all $s \geqq r$ with $1/s > 1/r + \mathrm{Re}(\mu)$. Since from Theorem 3.2(a),(b) $I^{-\mu}_{-;k,-\alpha/k}$ and $I^{-\mu}_{0+;k,\beta/k-1}$ are one-to-one on $\mathfrak{L}_{1-\nu,r}$ and then by Theorem 4.1(b),(c), if $\nu \notin \mathcal{E}_{\mathcal{H}}$, or if $1 < r \leqq 2$, $\boldsymbol{H}_i$ $(i = 1, 2)$ are one-to-one on $\mathfrak{L}_{\nu,r}$. For $f \in \mathfrak{L}_{\nu,r}$ with $\alpha < 1-\nu < \beta$ and $1 < r \leqq 2$, (4.2.8), (3.3.7), (4.2.7), (4.2.4) and (4.2.9), (3.3.6), (4.2.7), (4.2.5) imply

$$\begin{aligned} \left(\mathfrak{M}\boldsymbol{H}_1 f\right)(s) &= \left(\mathfrak{M}I^{-\mu}_{-;k,-\alpha/k}\widetilde{\boldsymbol{H}}_1 f\right)(s) = \frac{\Gamma\left(\dfrac{s-\alpha}{k}\right)}{\Gamma\left(\dfrac{s-\alpha}{k} - \mu\right)}\left(\mathfrak{M}\widetilde{\boldsymbol{H}}_1 f\right)(s) \\ &= \frac{\Gamma\left(\dfrac{s-\alpha}{k}\right)}{\Gamma\left(\dfrac{s-\alpha}{k} - \mu\right)}\mathcal{H}_1(s)\left(\mathfrak{M}f\right)(1-s) \\ &= \mathcal{H}(s)\left(\mathfrak{M}f\right)(1-s) \quad (\mathrm{Re}(s) = 1-\nu) \end{aligned} \tag{4.2.10}$$

for the case (i) and

$$\left(\mathfrak{M}\boldsymbol{H}_2 f\right)(s) = \left(\mathfrak{M}I^{-\mu}_{0+;k,\beta/k-1}\widetilde{\boldsymbol{H}}_2 f\right)(s) = \frac{\Gamma\left(\dfrac{\beta-s}{k}\right)}{\Gamma\left(\dfrac{\beta-s}{k} - \mu\right)}\left(\mathfrak{M}\widetilde{\boldsymbol{H}}_2 f\right)(s)$$

$$= \frac{\Gamma\left(\dfrac{\beta - s}{k}\right)}{\Gamma\left(\dfrac{\beta - s}{k} - \mu\right)} \mathcal{H}_2(s)\left(\mathfrak{M}f\right)(1 - s)$$

$$= \mathcal{H}(s)\left(\mathfrak{M}f\right)(1 - s) \quad (\mathrm{Re}(s) = 1 - \nu) \tag{4.2.11}$$

for the case (ii). In particular, for $f \in \mathfrak{L}_{\nu,2}$ with $\alpha < 1 - \nu < \beta$ there hold

$$\left(\mathfrak{M}\boldsymbol{H}_i f\right)(s) = \left(\mathfrak{M}\boldsymbol{H} f\right)(s) \quad (\mathrm{Re}(s) = 1 - \nu;\ i = 1, 2)$$

for the cases (i) and (ii), respectively. Hence, $\boldsymbol{H} = \boldsymbol{H}_i$ on $\mathfrak{L}_{\nu,2}$ $(i = 1, 2)$. Thus $\boldsymbol{H}$ can be extended to $\mathfrak{L}_{\nu,r}$ by defining it as $\boldsymbol{H}_1$ for (i) and by $\boldsymbol{H}_2$ for (ii), and $\boldsymbol{H} \in [\mathfrak{L}_{\nu,r}, \mathfrak{L}_{1-\nu,s}]$ for all $s \geqq r$ such that $1/s > 1/r + \mathrm{Re}(\mu)$. This completes the proof of (a).

The statement (b) follows from the fact that $\widetilde{\boldsymbol{H}}_i$ $(i = 1, 2)$ are one-to-one by Theorem 4.1(b) and that $I^{-\mu}_{-;k,-\alpha/k}$ and $I^{-\mu}_{0+;k,\beta/k-1}$ are one-to-one, too.

Now we proceed to prove (c). The one-to-one property can be obtained similar to (b). If the conditions of our theorem hold, then due to Theorem 4.1 $\widetilde{\boldsymbol{H}}_i(\mathfrak{L}_{\nu,r}) \subseteq \mathfrak{L}_{1-\nu,r}$, and if $\nu \notin \mathcal{E}_{\mathcal{H}}$, then $\widetilde{\boldsymbol{H}}_i(\mathfrak{L}_{\nu,r}) = \mathfrak{L}_{1-\nu,r}$ $(i = 1, 2)$. Therefore, it follows from (4.2.6), (4.2.10) and (4.2.9), (4.2.11) the following:

For (i), $\boldsymbol{H}(\mathfrak{L}_{\nu,r}) \subseteq I^{-\mu}_{-;k,-\alpha/k}(\mathfrak{L}_{1-\nu,r})$ and, if further $\nu \notin \mathcal{E}_{\mathcal{H}}$, $\boldsymbol{H}(\mathfrak{L}_{\nu,r}) = I^{-\mu}_{-;k,-\alpha/k}$ $(\mathfrak{L}_{1-\nu,r})$, i.e. (4.2.1) holds.

For (ii), $\boldsymbol{H}(\mathfrak{L}_{\nu,r}) \subseteq I^{-\mu}_{0+;k,\beta/k-1}(\mathfrak{L}_{1-\nu,r})$ and, if further $\nu \notin \mathcal{E}_{\mathcal{H}}$, $\boldsymbol{H}(\mathfrak{L}_{\nu,r}) = I^{-\mu}_{0+;k,\beta/k-1}$ $(\mathfrak{L}_{1-\nu,r})$, i.e. (4.2.2) holds.

These prove (c).

To establish (4.1.3) for $f \in \mathfrak{L}_{\nu,r}$ and $g \in \mathfrak{L}_{\nu,s}$ with $1 < r < \infty, 1 < s < \infty$ and $1 \leqq 1/r + 1/s < 1 - \mathrm{Re}(\mu)$, we show as in the proof of Theorem 4.1(d) that both sides of (4.1.3) are bounded bilinear functionals on $\mathfrak{L}_{\nu,r} \times \mathfrak{L}_{\nu,s}$. From the assumption we have $r' \geqq s$ $(1/r + 1/r' = 1)$ and $1/r' > 1/s + \mathrm{Re}(\mu)$, and hence $\boldsymbol{H} \in [\mathfrak{L}_{\nu,s}, \mathfrak{L}_{1-\nu,r'}]$ by assertion (a). Applying Hölder's inequality (4.1.13), we have

$$\left|\int_0^\infty f(x)\left(\boldsymbol{H}g\right)(x)dx\right| = \left|\int_0^\infty x^{\nu-1/r} f(x) x^{1/r-\nu}\left(\boldsymbol{H}g\right)(x)dx\right|$$
$$\leqq \|f\|_{\nu,r}\,\|\boldsymbol{H}g\|_{1-\nu,r'} \leqq K\|f\|_{\nu,r}\|g\|_{\nu,s},$$

where K is a bound for $\boldsymbol{H} \in [\mathfrak{L}_{\nu,s}, \mathfrak{L}_{1-\nu,r'}]$. Hence the left-hand side of (4.1.3) represents a bounded bilinear functional on $\mathfrak{L}_{\nu,r} \times \mathfrak{L}_{\nu,s}$. Similarly it can be proved that the right-hand side of (4.1.3) represents such a functional on $\mathfrak{L}_{\nu,r} \times \mathfrak{L}_{\nu,s}$. Thus the assertion (d) is obtained.

Lastly we prove (e). Let $f \in \mathfrak{L}_{\nu,r}$ with $1 < r < \infty$. Then the representations (4.1.4) and (4.1.5), when $\mathrm{Re}(\lambda) > (1 - \nu)h - 1$ and $\mathrm{Re}(\lambda) < (1 - \nu)h - 1$, respectively, are deduced from Theorem 3.6(iii) in the same way as was done in the proof of Theorem 4.1. Now denote $\widetilde{\mu}$ for the function $H^{m,n+1}_{p+1,q+1}(t)$ in (4.1.4) instead of μ for the H-function (1.1.1). By (1.1.10) $\widetilde{\mu} = \mu - 1$, then the assumption $\mathrm{Re}(\mu) < 0$ implies $\mathrm{Re}(\widetilde{\mu}) < -1$. Since $\alpha < 1 - \nu < \beta$, it follows from Theorem 1.2 that if $\mathrm{Re}(\lambda) > (1 - \nu)h - 1$, then $H^{m,n+1}_{p+1,q+1}(t)$ in (4.1.4) is continuously differentiable with respect to t on $\mathbb{R}_+$. Therefore we can differentiate under the integral sign in (4.1.4). Applying the relations (2.2.7) and (2.1.1), we have

$$\left(\boldsymbol{H}f\right)(x) = hx^{1-(\lambda+1)/h}\frac{d}{dx}x^{(\lambda+1)/h}\int_0^\infty H^{m,n+1}_{p+1,q+1}\left[xt \,\middle|\, \begin{matrix}(-\lambda, h), (a_i, \alpha_i)_{1,p} \\ (b_j, \beta_j)_{1,q}, (-\lambda - 1, h)\end{matrix}\right] f(t)dt$$

$$= hx^{1-(\lambda+1)/h}\int_0^\infty t^{1-(\lambda+1)/h}$$

$$\cdot \frac{d}{d(xt)}\left\{(xt)^{(\lambda+1)/h}H^{m,n+1}_{p+1,q+1}\left[xt\,\middle|\,\begin{matrix}(-\lambda,h),(a_i,\alpha_i)_{1,p}\\(b_j,\beta_j)_{1,q},(-\lambda-1,h)\end{matrix}\right]\right\}f(t)dt$$

$$= \int_0^\infty H^{m,n+1}_{p+1,q+1}\left[xt\,\middle|\,\begin{matrix}(-\lambda-1,h),(a_i,\alpha_i)_{1,p}\\(b_j,\beta_j)_{1,q},(-\lambda-1,h)\end{matrix}\right]f(t)dt$$

$$= \int_0^\infty H^{m,n}_{p,q}\left[xt\,\middle|\,\begin{matrix}(a_i,\alpha_i)_{1,p}\\(b_j,\beta_j)_{1,q}\end{matrix}\right]f(t)dt \tag{4.2.12}$$

and we obtain (4.2.3), provided that the last integral in (4.2.12) exists. The existence of this integral is proved on the basis of (3.4.13) similarly to those in Theorem 3.7. Indeed, we choose γ_1 and γ_2 so that $\alpha < \gamma_1 < 1-\nu < \gamma_2 < \beta$. According to (3.4.13) there are constants A_1 and A_2 such that for almost all $t > 0$, the inequalities

$$\left|H^{m,n}_{p,q}(t)\right| \leqq A_i t^{-\gamma_i} \quad (i = 1, 2)$$

hold. Therefore, using the Hölder inequality (4.1.13), we have

$$\int_0^\infty \left|H^{m,n+1}_{p+1,q+1}(xt)f(t)\right|dt$$

$$\leqq \left(\int_0^\infty \left|H^{m,n+1}_{p+1,q+1}(xt)t^{1/r-\nu}\right|^{r'}dt\right)^{1/r}\|f\|_{\nu,r}$$

$$\leqq \left[A_1x^{-\gamma_1}\left(\int_0^{1/x} t^{r'(1-\nu-\gamma_1)-1}dt\right)^{1/r'} + A_2x^{-\gamma_2}\left(\int_{1/x}^\infty t^{r'(1-\nu-\gamma_2)-1}dt\right)^{1/r'}\right]\|f\|_{\nu,r}$$

$$\leqq Cx^{\nu-1} < \infty,$$

where

$$C = (A_1[r'(1-\nu-\gamma_1)]^{-1/r'} + A_2[r'(\gamma_2+\nu-1)]^{-1/r'})\|f\|_{\nu,r}.$$

The proof for the case $\mathrm{Re}(\lambda) < (1-\nu)h - 1$ is similar. Thus the theorem is proved.

4.3. $\mathfrak{L}_{\nu,r}$-Theory of the $\boldsymbol{H}$-Transform When $a^* = 0, \Delta > 0$

In this and the next sections we prove that, if $a^* = 0$ and $\Delta \neq 0$, the transform $\boldsymbol{H}$ defined on $\mathfrak{L}_{\nu,2}$ can be extended to $\mathfrak{L}_{\nu,r}$ such as $\boldsymbol{H} \in [\mathfrak{L}_{\nu,r}, \mathfrak{L}_{1-\nu,s}]$ for some range of values s. Then we characterize the range $\boldsymbol{H}$ on $\mathfrak{L}_{\nu,r}$ except for its isolated values $\nu \in \mathcal{E}_{\mathcal{H}}$ in terms of the modified Hankel transform $\mathbb{H}_{k,\eta}$ and the elementary transform M_ζ given in (3.3.4) and (3.3.11). The result will be different in the cases $\Delta > 0$ and $\Delta < 0$. In this section we consider the case $\Delta > 0$.

Theorem 4.3. *Let $a^* = 0, m > 0, \Delta > 0, \alpha < 1-\nu < \beta, 1 < r < \infty$ and $\Delta(1-\nu) + \mathrm{Re}(\mu) \leqq 1/2 - \gamma(r)$, where $\gamma(r)$ is defined in (3.3.9).*

(a) *The transform* $\boldsymbol{H}$ *defined on* $\mathfrak{L}_{\nu,2}$ *can be extended to* $\mathfrak{L}_{\nu,r}$ *as an element of* $[\mathfrak{L}_{\nu,r}, \mathfrak{L}_{1-\nu,s}]$ *for all* s *with* $r \leqq s < \infty$ *such that* $s' \geqq [1/2 - \Delta(1-\nu) - \mathrm{Re}(\mu)]^{-1}$ *with* $1/s + 1/s' = 1$.

(b) *If* $1 < r \leqq 2$, *then the transform* $\boldsymbol{H}$ *is one-to-one on* $\mathfrak{L}_{\nu,r}$ *and there holds the equality* (4.1.1).

(c) *If* $\nu \notin \mathcal{E}_{\mathcal{H}}$, *then the transform* $\boldsymbol{H}$ *is one-to-one on* $\mathfrak{L}_{\nu,r}$. *If we set* $\eta = -\Delta\alpha - \mu - 1$, *then* $\mathrm{Re}(\eta) > -1$ *and there holds*

$$\boldsymbol{H}(\mathfrak{L}_{\nu,r}) = \left(M_{\mu/\Delta+1/2}\mathbb{H}_{\Delta,\eta}\right)\left(\mathfrak{L}_{\nu-1/2-\mathrm{Re}(\mu)/\Delta,r}\right). \tag{4.3.1}$$

When $\nu \in \mathcal{E}_{\mathcal{H}}$, $\boldsymbol{H}(\mathfrak{L}_{\nu,r})$ *is a subset of the right-hand side of* (4.3.1).

(d) *If* $f \in \mathfrak{L}_{\nu,r}$ *and* $g \in \mathfrak{L}_{\nu,s}$ $1 < s < \infty, 1/r + 1/s \geqq 1$ *and* $\Delta(1-\nu) + \mathrm{Re}(\mu) \leqq 1/2 - \max[\gamma(r), \gamma(s)]$, *then the relation* (4.1.3) *holds.*

(e) *If* $f \in \mathfrak{L}_{\nu,r}, \lambda \in \mathbb{C}, h > 0$ *and* $\Delta(1-\nu) + \mathrm{Re}(\mu) \leqq 1/2 - \gamma(r)$, *then* $\boldsymbol{H}f$ *is given in* (4.1.4) *for* $\mathrm{Re}(\lambda) > (1-\nu)h - 1$, *while in* (4.1.5) *for* $\mathrm{Re}(\lambda) < (1-\nu)h - 1$. *If* $\Delta(1-\nu) + \mathrm{Re}(\mu) < 0$, $\boldsymbol{H}f$ *is given in* (4.2.3).

Proof. Since $\gamma(r) \geqq 1/2$, we have $\Delta(1-\nu) + \mathrm{Re}(\mu) \leqq 0$ by assumption, and hence from Theorem 3.6 the transform $\boldsymbol{H}$ is defined on $\mathfrak{L}_{\nu,2}$. The conditions $\Delta > 0$, $\alpha < 1 - \nu$ and the relation $\Delta(1-\nu) + \mathrm{Re}(\mu) \leqq 0$ imply $\nu \geqq 1 + \mathrm{Re}(\mu)/\Delta$ and $\alpha < -\mathrm{Re}(\mu)/\Delta$.

Since $a^* = 0$, then by (1.1.11)–(1.1.13)

$$a_1^* = -a_2^* = \frac{\Delta}{2} > 0. \tag{4.3.2}$$

We denote by $\mathcal{H}_3(s)$ the function

$$\mathcal{H}_3(s) = \delta^{s-1}(a_1^*)^{(1-s)\Delta+\mu}\,\frac{\Gamma[-\mu + a_1^*(s-1-\alpha)]}{\Gamma[a_1^*(1-\alpha-s)]}\,\mathcal{H}(1-s). \tag{4.3.3}$$

As already known, the function $\mathcal{H}(1-s)$ is analytic in the strip $1 - \beta < \mathrm{Re}(s) < 1 - \alpha$ and the function $\Gamma[-\mu + a_1^*(s-1-\alpha)]$ is analytic in the half-plane

$$\mathrm{Re}(s) > \alpha + 1 + \frac{\mathrm{Re}(\mu)}{a_1^*} = \alpha + 1 + \frac{2\mathrm{Re}(\mu)}{\Delta}.$$

Since $\alpha < -\mathrm{Re}(\mu)/\Delta$, $1 - \alpha > \alpha + 1 + 2\mathrm{Re}(\mu)/\Delta$, and, if we take

$$\alpha_1 = \max\left[1 - \beta, \alpha + 1 + \frac{2\mathrm{Re}(\mu)}{\Delta}\right], \quad \beta_1 = 1 - \alpha,$$

then $\alpha_1 < \beta_1$ and $\mathcal{H}_3(s)$ is analytic in the strip $\alpha_1 < \mathrm{Re}(s) < \beta_1$.

Setting $s = \sigma + it$ and a complex constant $k = c + id$, in accordance with (1.2.2) we have the asymptotic behavior

$$\Gamma(s+k) = \Gamma[c + \sigma + i(d+t)] \sim \sqrt{2\pi}\; e^{-c-\sigma}|t|^{c+\sigma-1/2}e^{-\pi|t|/2-\pi d\,\mathrm{sign}(t)/2} \tag{4.3.4}$$
$$(|t| \to \infty).$$

Then by taking $a^* = 0$, $\Delta = 2a_1^*$ and (1.2.9) into account from (4.3.4) we have

$$\mathcal{H}_3(\sigma + it) \sim (2\pi)^{c^*}e^{-c^*}\prod_{j=1}^{q}\beta_j^{\mathrm{Re}(b_j)-1/2}\prod_{i=1}^{p}\alpha_i^{1/2-\mathrm{Re}(a_i)}e^{\pi[\mathrm{Im}(\mu+\xi)]\mathrm{sign}(t)/2}$$
$$= \kappa \neq 0 \tag{4.3.5}$$

as $|t| \to \infty$. Therefore, for $\alpha_1 < \sigma_1 \leqq \sigma_2 < \beta_1$, $\mathcal{H}_3(s)$ is bounded in $\sigma_1 \leqq \mathrm{Re}(s) \leqq \sigma_2$.

If $\alpha_1 < \sigma < \beta_1$, then

$$\mathcal{H}_3'(\sigma+it) = \mathcal{H}_3(\sigma+it)\Bigg\{\log\delta - \Delta\log a_1^* + a_1^*\psi[a_1^*(s-\alpha-1)-\mu] \\ + a_1^*\psi[a_1^*(1-\alpha-s)] - \frac{\mathcal{H}'(1-s)}{\mathcal{H}(1-s)}\Bigg\}. \tag{4.3.6}$$

Using (3.4.14) with $a^* = 0$, (4.3.2) and (4.3.5), and applying the estimate (3.4.17) for the psi function $\psi(z)$, we have from (4.3.6), as $|t| \to \infty$,

$$\mathcal{H}_3'(\sigma+it) \\ = \mathcal{H}_3(\sigma+it)\Bigg\{\log\delta - 2a_1^*\log a_1^* + a_1^*\left[\log(ia_1^*t) + \frac{a_1^*(\sigma-\alpha-1)-\mu-\frac{1}{2}}{ia_1^*t}\right] \\ + a_1^*\left[\log(-ia_1^*t) - \frac{a_1^*(1-\alpha-\sigma)-\frac{1}{2}}{ia_1^*t}\right] \\ - \left[\log\delta + a_1^*\log(-it) + a_1^*\log(it) - \frac{\mu+\Delta(1-\sigma)}{it}\right] + O\left(\frac{1}{t^2}\right)\Bigg\} \\ = O\left(\frac{1}{t^2}\right). \tag{4.3.7}$$

So $\mathcal{H}_3 \in \mathcal{A}$ with $\alpha(\mathcal{H}_3) = \alpha_1$ and $\beta(\mathcal{H}_3) = \beta_1$. Hence due to Theorem 3.1 there is a transform T_3 corresponding to $\mathcal{H}_3$ with $1 < r < \infty$ and $\alpha_1 < \nu < \beta_1$, and if $1 < r \leqq 2$, then T_3 is one-to-one on $\mathfrak{L}_{\nu,r}$ into itself and

$$(\mathfrak{M}T_3f)(s) = \mathcal{H}_3(s)(\mathfrak{M}f)(s) \quad (\mathrm{Re}(s) = \nu). \tag{4.3.8}$$

In particular, if ν and r satisfy the hypothesis of this theorem, it is directly verified that $\alpha_1 < \nu < \beta_1$ and hence the relation (4.3.8) is true. Indeed, from $\alpha < -\mathrm{Re}(\mu)/\Delta$ we find

$$\alpha + 1 + \frac{2\mathrm{Re}(\mu)}{\Delta} < -\frac{\mathrm{Re}(\mu)}{\Delta} + 1 + \frac{2\mathrm{Re}(\mu)}{\Delta} \leqq \nu$$

by virtue of $\Delta(1-\nu) + \mathrm{Re}(\mu) \leqq 0$. Together with this and $\alpha < 1-\nu < \beta$, the relation $\alpha_1 < \nu < \beta_1$ follows.

Let $\eta = -\Delta\alpha - \mu - 1$ and let $\boldsymbol{H}_3$ be the operator

$$\boldsymbol{H}_3 = W_\delta M_{\mu/\Delta+1/2}\mathbb{H}_{\Delta,\eta}M_{\mu/\Delta+1/2}T_3 \tag{4.3.9}$$

composed by the operators W_δ in (3.3.12) and M_ζ in (3.3.11), the modified Hankel transform (3.3.4) and the transform T_3 above, where $\alpha < -\mathrm{Re}(\mu)/\Delta$ so that $\mathrm{Re}(\eta) > -1$. For $1 < r < \infty$ and $\alpha_1 < \nu < \beta_1$, Lemma 3.1(i) implies that $M_{\mu/\Delta+1/2}T_3$ maps $\mathfrak{L}_{\nu,r}$ boundedly into $\mathfrak{L}_{\nu-\mathrm{Re}(\mu)/\Delta-1/2,r}$. Since $\nu < 1-\alpha$ and $\Delta(1-\nu) + \mathrm{Re}(\mu) \leqq 1/2 - \gamma(r)$, we have

$$\gamma(r) \leqq \Delta\left(\nu - \frac{\mathrm{Re}(\mu)}{\Delta} - 1\right) + \frac{1}{2} < -\Delta\alpha - \mathrm{Re}(\mu) + \frac{1}{2} = \mathrm{Re}(\eta) + \frac{3}{2}.$$

Hence Theorem 3.2(d), and Lemma 3.1(i),(ii) again yield $\boldsymbol{H}_3 \in [\mathfrak{L}_{\nu,r}, \mathfrak{L}_{1-\nu,s}]$ for all $s \geqq r$ such that

$$s' \geqq \left[\Delta\left(\nu - \frac{\operatorname{Re}(\mu)}{\Delta} - 1\right) + \frac{1}{2}\right]^{-1} = \left[\frac{1}{2} - \Delta(1-\nu) - \operatorname{Re}(\mu)\right]^{-1}. \tag{4.3.10}$$

In particular, $\boldsymbol{H}_3 \in [\mathfrak{L}_{\nu,r}, \mathfrak{L}_{1-\nu,r}]$, since r can be taken instead of s due to the fact that $\gamma(r) \geqq 1/r'$.

If $f \in \mathfrak{L}_{\nu,r}$ with $1 < r \leqq 2$, then applying (3.3.15), (3.3.14), (3.3.10), (3.3.14), (4.3.8) and (4.3.3) and using the relation $\eta = -\Delta\alpha - \mu - 1$, we have for $\operatorname{Re}(s) = 1 - \nu$

$$\begin{aligned}
\left(\mathfrak{M}\boldsymbol{H}_3 f\right)(s) &= \left(\mathfrak{M}W_\delta M_{\mu/\Delta+1/2}\mathbb{H}_{\Delta,\eta}M_{\mu/\Delta+1/2}T_3 f\right)(s)\\
&= \delta^s\left(\mathfrak{M}M_{\mu/\Delta+1/2}\mathbb{H}_{\Delta,\eta}M_{\mu/\Delta+1/2}T_3 f\right)(s)\\
&= \delta^s\left(\mathfrak{M}\mathbb{H}_{\Delta,\eta}M_{\mu/\Delta+1/2}T_3 f\right)\left(s + \frac{\mu}{\Delta} + \frac{1}{2}\right)\\
&= \delta^s\left(\frac{2}{\Delta}\right)^{\Delta s+\mu}\frac{\Gamma\left(\dfrac{\eta+\Delta s+\mu+1}{2}\right)}{\Gamma\left(\dfrac{\eta-\Delta s-\mu+1}{2}\right)}\left(\mathfrak{M}M_{\mu/\Delta+1/2}T_3 f\right)\left(1 - s - \frac{\mu}{\Delta} - \frac{1}{2}\right)\\
&= \delta^s\left(\frac{2}{\Delta}\right)^{\Delta s+\mu}\frac{\Gamma\left(\dfrac{\Delta(s-\alpha)}{2}\right)}{\Gamma\left(-\mu - \dfrac{\Delta(s+\alpha)}{2}\right)}\left(\mathfrak{M}T_3 f\right)(1-s)\\
&= \delta^s\left(a_1^*\right)^{-\mu-\Delta s}\frac{\Gamma[a_1^*(s-\alpha)]}{\Gamma[-\mu - a_1^*(s+\alpha)]}\mathcal{H}_3(1-s)\left(\mathfrak{M}f\right)(1-s)\\
&= \mathcal{H}(s)\left(\mathfrak{M}f\right)(1-s).
\end{aligned} \tag{4.3.11}$$

In particular, if we take $r = 2$ and $f \in \mathfrak{L}_{\nu,2}$, (4.3.11) and Theorem 3.6(i) imply $(\mathfrak{M}\boldsymbol{H}_3 f)(s) = (\mathfrak{M}\boldsymbol{H}f)(s)$ for $\operatorname{Re}(s) = 1-\nu$. Therefore $\boldsymbol{H}_3 f = \boldsymbol{H}f$ and hence $\boldsymbol{H}_3 = \boldsymbol{H}$ on $\mathfrak{L}_{\nu,2}$. Thus, for all ν and r satisfying the hypotheses of this theorem, $\boldsymbol{H}$ can be extended from $\mathfrak{L}_{\nu,2}$ to $\mathfrak{L}_{\nu,r}$ if we define it by $\boldsymbol{H}_3$ given in (4.3.9) as an operator on $[\mathfrak{L}_{\nu,r}, \mathfrak{L}_{1-\nu,s}]$. This completes the proof of the statement (a) of the theorem.

The assertion (b) is already obtained.

Now we prove (c). By Theorem 3.2(d) and the properties (i) and (ii) in Lemma 3.1, we find that $H_{\Delta,\eta}$, M_ζ and W_δ are one-to-one in the corresponding spaces. Therefore, $\boldsymbol{H}_3$ is a one-to-one transform if and only if T_3 is one-to-one. Further, we have that

$$\boldsymbol{H}(\mathfrak{L}_{\nu,r}) \subseteq \left(W_\delta M_{\mu/\Delta+1/2}\mathbb{H}_{\Delta,\eta}M_{\mu/\Delta+1/2}\right)(\mathfrak{L}_{\nu,r}), \tag{4.3.12}$$

and that

$$\boldsymbol{H}(\mathfrak{L}_{\nu,r}) = \left(W_\delta M_{\mu/\Delta+1/2}\mathbb{H}_{\Delta,\eta}M_{\mu/\Delta+1/2}\right)(\mathfrak{L}_{\nu,r}), \tag{4.3.13}$$

if and only if $T_3(\mathfrak{L}_{\nu,r}) = \mathfrak{L}_{\nu,r}$. The relation (4.3.13) can be written as

$$\boldsymbol{H}(\mathfrak{L}_{\nu,r}) = \left(M_{\mu/\Delta+1/2}\mathbb{H}_{\Delta,\eta}\right)\left(\mathfrak{L}_{\nu-1/2-\operatorname{Re}(\mu)/\Delta,r}\right) \tag{4.3.14}$$

by virtue of Theorem 3.2, Lemmas 3.1, 3.2, and the fact such a range does not depend on a non-zero constant multiplier.

The abscissas of zero of $\mathcal{H}_3$ divide the interval (α_1, β_1) into disjoint open subintervals. If we take (α_0, β_0) as such a subinterval, then $1/\mathcal{H}_3 \in \mathcal{A}$ with $\alpha(1/\mathcal{H}_3) = \alpha_0, \beta(1/\mathcal{H}_3) = \beta_0$. For $1 < r < \infty$ and $\alpha_0 < \nu < \beta_0$ by Theorem 3.1 we find that $\boldsymbol{H}_3$ is one-to-one transform on $\mathfrak{L}_{\nu,r}$ and $\boldsymbol{H}_3(\mathfrak{L}_{\nu,r}) = \mathfrak{L}_{\nu,r}$. If r and ν satisfy the hypotheses of this theorem and if $\nu \notin \mathcal{E}_{\mathcal{H}}$, then ν does not coincide with any abscissa of zeros of $\mathcal{H}_3(s)$. It follows from Definition 3.4 that $1 - \nu$ is not on the abscissa of any zero of $\mathcal{H}(s)$ and therefore ν is not on the abscissa of any zero of $\mathcal{H}_3(1 - s)$. We also have that ν is not a zero of $1/\Gamma[a_1^*(1 - s - \alpha)]$, because $1 - \nu > \alpha$. So if $\nu \notin \mathcal{E}_{\mathcal{H}}$, then $\boldsymbol{H}_3$ is a one-to-one transform on $\mathfrak{L}_{\nu,r}$, and $\boldsymbol{H}_3(\mathfrak{L}_{\nu,r}) = \mathfrak{L}_{\nu,r}$. Therefore the statement (c) already follows from (4.3.12) and (4.3.13).

The proofs of assertions (d) and (e) of this theorem are carried out in the same way as they were done in the proofs of the statements (d) and (e) in Theorem 4.2. This completes the proof of the theorem.

4.4. $\mathfrak{L}_{\nu,r}$-Theory of the $\boldsymbol{H}$-Transform When $a^* = 0, \Delta < 0$

Now we consider the $\boldsymbol{H}$-transform with $a^* = 0$ and $\Delta < 0$. Our arguments are based on reducing the case above to the one investigated in Theorem 4.3 for $\Delta > 0$.

Theorem 4.4. *Let* $a^* = 0, \Delta < 0, n > 0, \alpha < 1 - \nu < \beta, 1 < r < \infty$ *and* $\Delta(1 - \nu) + \mathrm{Re}(\mu) \leqq 1/2 - \gamma(r)$.

(a) *The transform* $\boldsymbol{H}$ *defined on* $\mathfrak{L}_{\nu,2}$ *can be extended to* $\mathfrak{L}_{\nu,r}$ *as an element of* $[\mathfrak{L}_{\nu,r}, \mathfrak{L}_{1-\nu,s}]$ *for all* s *with* $r \leqq s < \infty$ *such that* $s' \geqq [1/2 - \Delta(1 - \nu) - \mathrm{Re}(\mu)]^{-1}$ *with* $1/s + 1/s' = 1$.

(b) *If* $1 < r \leqq 2$, *then the transform* $\boldsymbol{H}$ *is one-to-one on* $\mathfrak{L}_{\nu,r}$ *and there holds the equality* (4.1.1).

(c) *If* $\nu \notin \mathcal{E}_{\mathcal{H}}$, *then the transform* $\boldsymbol{H}$ *is one-to-one on* $\mathfrak{L}_{\nu,r}$. *If we set* $\eta = -\Delta\beta - \mu - 1$, *then* $\mathrm{Re}(\eta) > -1$ *and the relation*

$$\boldsymbol{H}(\mathfrak{L}_{\nu,r}) = \left(M_{\mu/\Delta+1/2}\mathbb{H}_{\Delta,\eta}\right)\left(\mathfrak{L}_{\nu-1/2-\mathrm{Re}(\mu)/\Delta,r}\right) \tag{4.4.1}$$

holds. When $\nu \in \mathcal{E}_{\mathcal{H}}$, $\boldsymbol{H}(\mathfrak{L}_{\nu,r})$ *is a subset of the set on the right-hand side of* (4.4.1).

(d) *If* $f \in \mathfrak{L}_{\nu,r}$ *and* $g \in \mathfrak{L}_{\nu,s}$ *with* $1 < s < \infty$, $1/r + 1/s \geqq 1$ *and* $\Delta(1-\nu) + \mathrm{Re}(\mu) \leqq 1/2 - \max[\gamma(r), \gamma(s)]$, *then the relation* (4.1.3) *holds.*

(e) *If* $f \in \mathfrak{L}_{\nu,r}, \lambda \in \mathbb{C}, h > 0$ *and* $\Delta(1-\nu) + \mathrm{Re}(\mu) \leqq 1/2 - \gamma(r)$, *then* $\boldsymbol{H}f$ *is given in* (4.1.4) *for* $\mathrm{Re}(\lambda) > (1 - \nu)h - 1$, *while in* (4.1.5) *for* $\mathrm{Re}(\lambda) < (1 - \nu)h - 1$. *If* $\Delta(1 - \nu) + \mathrm{Re}(\mu) < 0$, *then* $\boldsymbol{H}f$ *is given in* (4.2.3).

Proof. By similar arguments to that in the proof of Theorem 4.3 we find that the transform $\boldsymbol{H}$ is defined on $\mathfrak{L}_{\nu,2}$.

Let $\boldsymbol{H}_0 = R\boldsymbol{H}R$, where R is given in (3.3.13). Then according to Lemma 3.1(iii), $\boldsymbol{H}_0 \in [\mathfrak{L}_{1-\nu,2}, \mathfrak{L}_{\nu,2}]$. If $f \in \mathfrak{L}_{1-\nu,2}$, then using (3.3.16) and (3.1.5), we have

$$\left(\mathfrak{M}\boldsymbol{H}_0 f\right)(s) = \left(\mathfrak{M}R\boldsymbol{H}Rf\right)(s) = \left(\mathfrak{M}\boldsymbol{H}Rf\right)(1 - s) = \mathcal{H}(1 - s)\left(\mathfrak{M}Rf\right)(s)$$

$$= \mathcal{H}(1 - s)\left(\mathfrak{M}f\right)(1 - s) = \mathcal{H}_0(s)\left(\mathfrak{M}f\right)(1 - s) \quad (\mathrm{Re}(s) = \nu), \tag{4.4.2}$$

where

$$\mathcal{H}_0(s) = \mathcal{H}(1-s) = \mathcal{H}_{q,p}^{n,m}\left[\begin{array}{c|c} (1-b_j-\beta_j,\beta_j)_{1,q} & \\ & s \\ (1-a_i-\alpha_i,\alpha_i)_{1,p} & \end{array}\right]. \tag{4.4.3}$$

We denote α_0, β_0, a_0^*, δ_0, Δ_0 and μ_0 for $\mathcal{H}_0$ instead of that for $\mathcal{H}$. Then taking (3.4.1), (3.4.2) and (1.1.7)–(1.1.10) into account, we have

$$\alpha_0 = 1-\beta,\ \beta_0 = 1-\alpha,\ a_0^* = a^* = 0,\ \delta_0 = 1/\delta,\ \Delta_0 = -\Delta,\ \mu_0 = \mu + \Delta. \tag{4.4.4}$$

If we denote $1-\nu$ by ν_0, then from (4.4.4) and the hypotheses of this theorem we have $\alpha_0 < 1-\nu_0 < \beta_0$ and

$$\Delta_0(1-\nu_0) + \mathrm{Re}(\mu_0) = \Delta(1-\nu) + \mathrm{Re}(\mu) \leqq \frac{1}{2} - \gamma(r).$$

Since $a_0^* = 0$ and $\Delta_0 > 0$, we can apply Theorem 4.3(a) to obtain that $\boldsymbol{H}_0$ can be extended from $\mathfrak{L}_{\nu_0,2} = \mathfrak{L}_{1-\nu,2}$ to $\mathfrak{L}_{\nu_0,r} = \mathfrak{L}_{1-\nu,r}$ as an element of $[\mathfrak{L}_{\nu_0,r}, \mathfrak{L}_{1-\nu_0,s}] = [\mathfrak{L}_{1-\nu,r}, \mathfrak{L}_{\nu,s}]$ for all s with $r \leqq s < \infty$ such that

$$s' \geqq \left[\frac{1}{2} - \Delta_0(1-\nu_0) - \mathrm{Re}(\mu_0)\right]^{-1} = \left[\frac{1}{2} - \Delta(1-\nu) - \mathrm{Re}(\mu)\right]^{-1}.$$

Let us notice that

$$\boldsymbol{H}f = R\boldsymbol{H}_0Rf \quad (f \in \mathfrak{L}_{\nu,2}). \tag{4.4.5}$$

In accordance with Lemma 3.1(iii),

$$R(\mathfrak{L}_{\nu,r}) = \mathfrak{L}_{1-\nu,r} = \mathfrak{L}_{\nu_0,r}. \tag{4.4.6}$$

So we can extend $\boldsymbol{H}$ from $\mathfrak{L}_{\nu,2}$ to $\mathfrak{L}_{\nu,r}$ if it is defined by (4.4.5), and we obtain the boundedness of the operator $\boldsymbol{H}$, which establishes the statement (a).

The one-to-one property for $1 < r \leqq 2$ is already found, and the equality (4.1.1) is seen by

$$\begin{aligned}\left(\mathfrak{M}\boldsymbol{H}f\right)(s) &= \left(\mathfrak{M}R\boldsymbol{H}_0Rf\right)(s) = \left(\mathfrak{M}\boldsymbol{H}_0Rf\right)(1-s) = \mathcal{H}_0(1-s)\left(\mathfrak{M}Rf\right)(s) \\ &= \mathcal{H}_0(1-s)\left(\mathfrak{M}f\right)(1-s) = \mathcal{H}(s)\left(\mathfrak{M}f\right)(1-s) \quad (\mathrm{Re}(s) = 1-\nu),\end{aligned}$$

which proves (b).

That $\nu \notin \mathcal{E}_{\mathcal{H}}$ is equivalent to $\nu_0 \notin \mathcal{E}_{\mathcal{H}_0}$ implies the one-to-one property by Theorem 4.3(c).

To prove (4.4.1) we note that from (4.4.5),

$$\boldsymbol{H}(\mathfrak{L}_{\nu,r}) = \left(R\boldsymbol{H}_0R\right)(\mathfrak{L}_{\nu,r}) = \mathfrak{L}_{\nu_0,s}.$$

Then using (4.4.6) and (4.3.1) with μ being replaced by $\mu_0 = \mu + \Delta$, δ by δ_0 and Δ by $\Delta_0 = -\Delta$, while $\eta_0 = \eta = -\Delta_0\alpha_0 - \mu_0 - 1 = -\Delta\beta - \mu - 1$, and applying the relations in

(3.3.21) and (3.3.18), and again (4.4.6) we have

$$\begin{aligned}\boldsymbol{H}(\mathfrak{L}_{\nu,r}) &= \left(R\boldsymbol{H}_0R\right)(\mathfrak{L}_{\nu,r}) = \left(R\boldsymbol{H}_0\right)(\mathfrak{L}_{1-\nu,r})\\ &= \left(RM_{\mu_0/\Delta_0+1/2}\mathbb{H}_{\Delta_0,\eta_0}\right)(\mathfrak{L}_{1/2-\nu-\mathrm{Re}(\mu_0)/\Delta_0,r})\\ &= \left(RM_{-\mu/\Delta-1/2}\mathbb{H}_{-\Delta,\eta}\right)(\mathfrak{L}_{3/2-\nu+\mathrm{Re}(\mu)/\Delta,r})\\ &= \left(M_{\mu/\Delta+1/2}R\mathbb{H}_{-\Delta,\eta}\right)(\mathfrak{L}_{3/2-\nu+\mathrm{Re}(\mu)/\Delta,r})\\ &= \left(M_{\mu/\Delta+1/2}\mathbb{H}_{\Delta,\eta}R\right)(\mathfrak{L}_{3/2-\nu+\mathrm{Re}(\mu)/\Delta,r})\\ &= \left(M_{\mu/\Delta+1/2}\mathbb{H}_{\Delta,\eta}R\right)(\mathfrak{L}_{\nu-1/2-\mathrm{Re}(\mu)/\Delta,r}).\end{aligned}$$

Thus (4.4.1) is proved.

The assertions (d) and (e) can be established in the same way as in the statements (d) and (e) of Theorem 4.2, which completes the proof of the theorem.

To conclude this section we give useful statements followed from Theorems 4.1–4.4.

Corollary 4.4.1. *Let* $1 < r < \infty$, $\alpha < \beta$, $a^* = 0$ *and let one of the following conditions hold:*

(a) $\Delta > 0$, $\alpha < \dfrac{1}{\Delta}\left(\dfrac{1}{2} - \mathrm{Re}(\mu) - \gamma(r)\right)$,

(b) $\Delta < 0$, $\beta > \dfrac{1}{\Delta}\left(\dfrac{1}{2} - \mathrm{Re}(\mu) - \gamma(r)\right)$,

(c) $\Delta = 0$, $\mathrm{Re}(\mu) \leqq 0$,

where $\gamma(r)$ *is given in* (3.3.9). *Then the transform* $\boldsymbol{H}$ *can be defined on* $\mathfrak{L}_{\nu,r}$ *with* $\alpha < 1-\nu < \beta$.

Proof. Let $a^* = 0$, $\Delta \neq 0$. Then by Theorems 4.3 and 4.4 the condition

$$\Delta(1-\nu) + \mathrm{Re}(\mu) \leqq \frac{1}{2} - \gamma(r) \tag{4.4.7}$$

must be satisfied together with $\alpha < 1-\nu < \beta$ in order that the transform $\boldsymbol{H}$ can be defined on $\mathfrak{L}_{\nu,r}$. If $\Delta > 0$, the assumption of (a) and $\alpha < \beta$ yield the existence of a number ν such that

$$\alpha < 1-\nu \leqq \frac{1}{\Delta}\left(\frac{1}{2} - \mathrm{Re}(\mu) - \gamma(r)\right), \qquad 1-\nu < \beta,$$

which show the requirement. Similarly, for the case (b) we can deduce the existence of ν of the form

$$\frac{1}{\Delta}\left(\frac{1}{2} - \mathrm{Re}(\mu) - \gamma(r)\right) \leqq 1-\nu < \beta, \qquad \alpha < 1-\nu.$$

When $a^* = \Delta = 0$ and $\mathrm{Re}(\mu) \leqq 0$, by Theorems 4.1 and 4.2 the transform $\boldsymbol{H}$ can be defined on $\mathfrak{L}_{\nu,r}$. This completes the proof.

Remark 4.1. For $r = 2$, $\gamma(r) = 1/2$ and hence the cases (e), (f) and (g) of Corollary 3.6.1 follow from Corollary 4.4.1.

4.5. $\mathfrak{L}_{\nu,r}$-Theory of the $\boldsymbol{H}$-Transform When $a^* > 0$

When $a^* > 0$ and $\alpha < 1 - \nu < \beta$, Theorems 3.6 and 3.7 guarantee the existence of the $\boldsymbol{H}$-transform on $\mathfrak{L}_{\nu,2}$ which is given in (3.1.1). Now let us prove that it can be extended to $\mathfrak{L}_{\nu,r}$ for any $1 \leqq r \leqq \infty$. To do so we first state a preliminary result due to Rooney [6, Lemma 5.1].

Lemma 4.1. **(a)** *Suppose $f \in \mathfrak{L}_{\nu,r}$ and $k \in \mathfrak{L}_{\nu,q}$, where $1 \leqq r \leqq \infty$, $1 \leqq q \leqq \infty$ and $1/r + 1/q \geqq 1$. Then the transform*

$$\left(\boldsymbol{K}f\right)(x) = \int_0^\infty k\left(\frac{x}{t}\right) f(t)\frac{dt}{t} \tag{4.5.1}$$

exists for almost all $x \in \mathbb{R}_+$ and $\boldsymbol{K} \in [\mathfrak{L}_{\nu,r}, \mathfrak{L}_{\nu,s}]$, where $1/s = 1/r + 1/q - 1$. If, in addition, $r \leqq 2$, $q \leqq 2$ and $s \leqq 2$, then the Mellin convolution relation

$$\left(\mathfrak{M}\boldsymbol{K}f\right)(\nu + it) = \left(\mathfrak{M}k\right)(\nu + it)\left(\mathfrak{M}f\right)(\nu + it) \tag{4.5.2}$$

holds.

(b) *Suppose $f \in \mathfrak{L}_{\nu,r}$ and $k \in \mathfrak{L}_{1-\nu,q}$, where $1 \leqq r \leqq \infty$, $1 \leqq q \leqq \infty$ and $1/r + 1/q \geqq 1$. Then the transform*

$$\left(\boldsymbol{T}f\right)(x) = \int_0^\infty k(xt)f(t)dt \tag{4.5.3}$$

exists for almost all $x \in \mathbb{R}_+$, and $\boldsymbol{T} \in [\mathfrak{L}_{\nu,r}, \mathfrak{L}_{1-\nu,s}]$, where $1/s = 1/r + 1/q - 1$. If, in addition, $r \leqq 2$, $q \leqq 2$ and $s \leqq 2$, then the modified Mellin convolution relation

$$\left(\mathfrak{M}\boldsymbol{T}f\right)(1 - \nu + it) = \left(\mathfrak{M}k\right)(1 - \nu + it)\left(\mathfrak{M}f\right)(\nu - it) \tag{4.5.4}$$

holds.

The next theorem presents the $\mathfrak{L}_{\nu,r}$-theory of the $\boldsymbol{H}$-transform with $a^* > 0$ in $\mathfrak{L}_{\nu,r}$-spaces for any $\nu \in \mathbb{C}$ and $1 \leqq r \leqq \infty$.

Theorem 4.5. *Let $a^* > 0, \alpha < 1 - \nu < \beta$ and $1 \leqq r \leqq s \leqq \infty$.*

(a) *The transform $\boldsymbol{H}$ defined on $\mathfrak{L}_{\nu,2}$ can be extended to $\mathfrak{L}_{\nu,r}$ as an element of $[\mathfrak{L}_{\nu,r}, \mathfrak{L}_{1-\nu,s}]$. When $1 \leqq r \leqq 2$, $\boldsymbol{H}$ is a one-to-one transform from $\mathfrak{L}_{\nu,r}$ onto $\mathfrak{L}_{1-\nu,s}$.*

(b) *If $f \in \mathfrak{L}_{\nu,r}$ and $g \in \mathfrak{L}_{\nu,s'}$ with $1/s + 1/s' = 1$, then the relation* (4.1.3) *holds.*

Proof. It follows from Theorem 3.6 that the $\boldsymbol{H}$-transform is defined on $\mathfrak{L}_{\nu,2}$, and by Theorem 3.7 it is given in (3.1.1).

Define a number Q by the equality $1/Q = 1/s - 1/r + 1$. Then $1 \leqq Q \leqq \infty$, and $1/r + 1/Q \geqq 1$. We choose γ_1 and γ_2 so that $\alpha < \gamma_1 < 1 - \nu < \gamma_2 < \beta$. By the relation (3.4.13) in Theorem 3.4 there are constants A_1 and A_2 such that

$$\left|H_{p,q}^{m,n}(t)\right| \leqq A_i t^{-\gamma_i} \quad (i = 1, 2)$$

hold for $t \in \mathbb{R}_+$. Using this, we have

$$\begin{aligned}
\int_0^\infty \left|x^{1-\nu} H_{p,q}^{m,n}(x)\right|^Q \frac{dx}{x} &= \int_0^1 \left|x^{1-\nu} H_{p,q}^{m,n}(x)\right|^Q \frac{dx}{x} + \int_1^\infty \left|x^{1-\nu} H_{p,q}^{m,n}(x)\right|^Q \frac{dx}{x} \\
&\leqq \int_0^1 x^{(1-\nu-\gamma_1)Q-1} dx + \int_1^\infty x^{(1-\nu-\gamma_2)Q-1} dx \\
&= \frac{1}{(1-\nu-\gamma_1)Q} + \frac{1}{(\gamma_2+\nu-1)Q} < \infty.
\end{aligned}$$

Therefore $H_{p,q}^{m,n} \in \mathfrak{L}_{1-\nu,Q}$. Now we set

$$\left(\boldsymbol{H}_1 f\right)(x) = \int_0^\infty H_{p,q}^{m,n}(xt) f(t) dt. \tag{4.5.5}$$

By Lemma 4.1(b), $\boldsymbol{H}_1 \in [\mathfrak{L}_{\nu,r}, \mathfrak{L}_{1-\nu,s}]$. But $\boldsymbol{H}_1 = \boldsymbol{H}$ on $\mathfrak{L}_{\nu,2}$. So we can extend $\boldsymbol{H}$ from $\mathfrak{L}_{\nu,2}$ to $\mathfrak{L}_{\nu,r}$ if we define it by (4.5.5), and $\boldsymbol{H} \in [\mathfrak{L}_{\nu,r}, \mathfrak{L}_{1-\nu,s}]$.

In particular, if $s = r$, $\boldsymbol{H} \in [\mathfrak{L}_{\nu,r}, \mathfrak{L}_{1-\nu,r}]$. Now let $1 \leqq r \leqq 2$. From (4.5.4), (3.4.12) and Titchmarsh [3, Theorem 29], we obtain the relation for $f \in \mathfrak{L}_{\nu,r}$

$$\left(\mathfrak{M}\boldsymbol{H} f\right)(s) = \mathcal{H}_{p,q}^{m,n}(s)\left(\mathfrak{M} f\right)(1-s) \quad (\mathrm{Re}(s) = 1-\nu), \tag{4.5.6}$$

where $\mathcal{H}_{p,q}^{m,n}(s)$ is given in (1.1.2). Since the poles of the gamma function are isolated (see Erdélyi, Magunus, Oberhettiger and Tricomi [1, Section 1.1]), the zeros of the function $\mathcal{H}_{p,q}^{m,n}(s)$ are isolated, too. Therefore, if $\boldsymbol{H} f = 0$ and $\mathrm{Im}(s) = t$, then $(\mathfrak{M} f)(\nu - it) = 0$ almost everywhere. Hence for $1 \leqq r \leqq 2$, $\boldsymbol{H}$ is a one-to-one transform from $\mathfrak{L}_{\nu,r}$ onto $\mathfrak{L}_{1-\nu,s}$. The assertion (a) of the theorem is proved.

To prove (4.1.3) for $f \in \mathfrak{L}_{\nu,r}$ and $g \in \mathfrak{L}_{\nu,s'}$ with $1/s + 1/s' = 1$ we show that both sides of (4.1.3) are bounded bilinear functionals on $\mathfrak{L}_{\nu,r} \times \mathfrak{L}_{\nu,s'}$. By the assumption $r \leqq s$ we have $s' \leqq r'$ and $\boldsymbol{H} \in [\mathfrak{L}_{\nu,s'}, \mathfrak{L}_{1-\nu,r'}]$. Applying the Hölder inequality (4.1.13), we have

$$\begin{aligned}
\left|\int_0^\infty f(x)\left(\boldsymbol{H} g\right)(x) dx\right| &= \left|\int_0^\infty x^{\nu-1/r} f(x) x^{1-\nu-1/r'}\left(\boldsymbol{H} g\right)(x) dx\right| \\
&\leqq \|f\|_{\nu,r} \|\boldsymbol{H} g\|_{1-\nu,r'} \leqq K\|f\|_{\nu,r}\|g\|_{\nu,s'},
\end{aligned}$$

where K is a bound for $\boldsymbol{H} \in [\mathfrak{L}_{\nu,s'}, \mathfrak{L}_{1-\nu,r'}]$. Hence the left-hand side of (4.1.3) is a bounded bilinear functional on $\mathfrak{L}_{\nu,r} \times \mathfrak{L}_{\nu,s'}$. Similarly it is proved that the right-hand side of (4.1.3) is a bounded bilinear functional on $\mathfrak{L}_{\nu,r} \times \mathfrak{L}_{\nu,s'}$. This completes the proof of Theorem 4.5.

4.6. Boundedness and Range of the $\boldsymbol{H}$-Transform When $a_1^* > 0$ and $a_2^* > 0$

In this and the next sections we consider the boundedness and range of the $\boldsymbol{H}$-transform in the cases when $a^* > 0$ and $a_1^* \geqq 0$ or $a_2^* \geqq 0$. We give conditions for the transform $\boldsymbol{H}$ to be one-to-one on $\mathfrak{L}_{\nu,r}$ and to characterize its range on $\mathfrak{L}_{\nu,r}$ except for its isolated values $\nu \in \mathcal{E}_{\mathcal{H}}$, in terms of the Erdélyi–Kober type fractional integration operators $I_{0+;\sigma,\eta}^\alpha$, $I_{-;\sigma,\eta}^\alpha$, the modified Laplace transform $\mathbb{L}_{k,\alpha}$ and the operator W_δ given in (3.3.1)–(3.3.3) and (3.3.12). The results will be different depending on combinations of the signs of a_1^* and a_2^*. In this section we consider the case when $a_1^* > 0$ and $a_2^* > 0$ and hence $a^* > 0$ by (1.1.13).

Theorem 4.6. *Let $a_1^* > 0, a_2^* > 0, m > 0, n > 0, \alpha < 1-\nu < \beta$ and $\omega = \mu + a_1^*\alpha - a_2^*\beta + 1$ and let $1 < r < \infty$.*

(a) *If $\nu \notin \mathcal{E}_{\mathcal{H}}$, or if $1 \leqq r \leqq 2$, then the transform $\boldsymbol{H}$ is one-to-one on $\mathfrak{L}_{\nu,r}$.*

(b) *If $\mathrm{Re}(\omega) \geqq 0$ and $\nu \notin \mathcal{E}_{\mathcal{H}}$, then*

$$\boldsymbol{H}(\mathfrak{L}_{\nu,r}) = \left(\mathbb{L}_{a_1^*,\alpha}\mathbb{L}_{a_2^*,1-\beta-\omega/a_2^*}\right)(\mathfrak{L}_{1-\nu,r}). \tag{4.6.1}$$

When $\nu \in \mathcal{E}_{\mathcal{H}}$, $\boldsymbol{H}(\mathfrak{L}_{\nu,r})$ is a subset of the right-hand side of (4.6.1).

(c) *If $\mathrm{Re}(\omega) < 0$ and $\nu \notin \mathcal{E}_{\mathcal{H}}$, then*

$$\boldsymbol{H}(\mathfrak{L}_{\nu,r}) = \left(I^{-\omega}_{-;1/a_1^*,-a_1^*\alpha}\mathbb{L}_{a_1^*,\alpha}\mathbb{L}_{a_2^*,1-\beta}\right)(\mathfrak{L}_{1-\nu,r}). \tag{4.6.2}$$

When $\nu \in \mathcal{E}_{\mathcal{H}}$, $\boldsymbol{H}(\mathfrak{L}_{\nu,r})$ is a subset of the right-hand side of (4.6.2).

Proof. We first consider the case $\mathrm{Re}(\omega) \geqq 0$. We define $\mathcal{H}_4(s)$ by

$$\mathcal{H}_4(s) = \frac{(a_1^*)^{a_1^*(s-\alpha)-1}(a_2^*)^{a_2^*(\beta-s)+\omega-1}}{\Gamma[a_1^*(s-\alpha)]\Gamma[a_2^*(\beta-s)+\omega]}\,\delta^{-s}\mathcal{H}(s). \tag{4.6.3}$$

Since $\alpha < 1-\nu < \beta$, the function $\mathcal{H}_4(s)$ is analytic in the strip $\alpha < \mathrm{Re}(s) < \beta$. According to (1.2.9) and (4.3.4) we have the estimate, as $|t| \to \infty$,

$$\begin{aligned}\mathcal{H}_4(\sigma+it) &\sim (a_1^*)^{a_1^*(\sigma-\alpha)-1}(a_2^*)^{a_2^*(\beta-\sigma)+\mathrm{Re}(\omega)-1}\frac{1}{2\pi}(a_1^*|t|)^{a_1^*(\alpha-\sigma)+1/2}\\ &\quad\cdot(a_2^*|t|)^{a_2^*(\sigma-\beta)-\mathrm{Re}(\omega)+1/2}e^{\Delta\sigma+\mathrm{Re}(\mu)+1}e^{[a^*|t|+\mathrm{Im}(\omega)\mathrm{sign}(t)]\pi/2}\prod_{j=1}^{q}\beta_j^{\mathrm{Re}(b_j)-1/2}\\ &\quad\cdot\prod_{i=1}^{p}\alpha_i^{1/2-\mathrm{Re}(a_i)}(2\pi)^{c^*}e^{-\Delta\sigma-\mathrm{Re}(\mu)-c^*}|t|^{\Delta\sigma+\mathrm{Re}(\mu)}e^{-\pi[a^*|t|+\mathrm{Im}(\xi)\mathrm{sign}(t)]/2}\\ &\sim\prod_{j=1}^{q}\beta_j^{\mathrm{Re}(b_j)-1/2}\prod_{i=1}^{p}\alpha_i^{1/2-\mathrm{Re}(a_i)}(2\pi)^{c^*-1}(a_1^*a_2^*)^{-1/2}e^{-c^*+1}e^{\pi\mathrm{Im}(\omega-\xi)\mathrm{sign}(t)/2}\end{aligned} \tag{4.6.4}$$

uniformly in σ on any bounded interval of $\mathbb{R}$, where ξ and c^* are given in (1.1.14) and (1.1.15). Further, in accordance with (3.4.17) and (3.4.14)

$$\begin{aligned}\mathcal{H}_4'(\sigma+it) &= \mathcal{H}_4(\sigma+it)\Bigg\{a_1^*\log a_1^* - a_2^*\log a_2^* - a_1^*\psi[a_1^*(\sigma+it-\alpha)]\\ &\quad+a_2^*\psi[a_2^*(\beta-\sigma-it)+\omega] - \log\delta + \frac{\mathcal{H}'(\sigma+it)}{\mathcal{H}(\sigma+it)}\Bigg\}\\ &= \mathcal{H}_4(\sigma+it)\Bigg\{a_1^*\log a_1^* - a_2^*\log a_2^* - a_1^*\left[\log(ia_1^*t) + \frac{a_1^*(\sigma-\alpha)-1/2}{ia_1^*t}\right]\\ &\quad+a_2^*\left[\log(-ia_2^*t) - \frac{a_2^*(\beta-\sigma)+\omega-1/2}{ia_2^*t}\right] - \log\delta\\ &\quad+\left[\log\delta + a_1^*\log(it) - a_2^*\log(-it) + \frac{\mathrm{Re}(\mu)+\Delta\sigma}{it}\right] + O\left(\frac{1}{t^2}\right)\Bigg\}\end{aligned}$$

$$= O\left(\frac{1}{t^2}\right) \quad (|t| \to \infty). \tag{4.6.5}$$

So $\mathcal{H}_4 \in \mathcal{A}$ with $\alpha(\mathcal{H}_4) = \alpha$ and $\beta(\mathcal{H}_4) = \beta$, and Theorem 3.1 implies that there is a transform $T_4 \in [\mathfrak{L}_{\nu,r}]$ for $1 < r < \infty$ and $\alpha < \nu < \beta$ so that, if $1 < r \leqq 2$, the relation

$$(\mathfrak{M}T_4 f)(s) = \mathcal{H}_4(s)(\mathfrak{M}f)(s) \quad (\mathrm{Re}(s) = \nu) \tag{4.6.6}$$

holds. Let

$$\boldsymbol{H}_4 = W_\delta \mathbb{L}_{a_1^*,\alpha} \mathbb{L}_{a_2^*,1-\beta-\omega/a_2^*} T_4 R, \tag{4.6.7}$$

where W_δ and R are defined in (3.3.12) and (3.3.13). Then it follows from Lemma 3.1(ii),(iii) and Theorem 3.2(c) that if $1 < r \leqq s < \infty$ and $\alpha < 1-\nu < \beta$, then $\boldsymbol{H}_4 \in [\mathfrak{L}_{\nu,r}, \mathfrak{L}_{1-\nu,s}]$.

For $f \in \mathfrak{L}_{\nu,2}$, applying (4.6.7), (3.3.15), (3.3.8), (4.6.6) and (3.3.16), and noting that $\mathrm{Re}(\omega) > 0$, we have for $\mathrm{Re}(s) = 1-\nu$

$$\begin{aligned}
\left(\mathfrak{M}\boldsymbol{H}_4 f\right)(s) &= \left(\mathfrak{M}W_\delta \mathbb{L}_{a_1^*,\alpha} \mathbb{L}_{a_2^*,1-\beta-\omega/a_2^*} T_4 R f\right)(s) \\
&= \delta^s \frac{\Gamma[a_1^*(s-\alpha)]}{(a_1^*)^{a_1^*(s-\alpha)-1}} \left(\mathfrak{M}\mathbb{L}_{a_2^*,1-\beta-\omega/a_2^*} T_4 R f\right)(1-s) \\
&= \delta^s \frac{\Gamma[a_1^*(s-\alpha)]}{(a_1^*)^{a_1^*(s-\alpha)-1}} \frac{\Gamma\{a_2^*[1-s-(1-\beta-\omega/a_2^*)]\}}{(a_2^*)^{a_2^*[1-s-(1-\beta-\omega/a_2^*)]-1}} \left(\mathfrak{M}T_4 R f\right)(s) \\
&= \delta^s \frac{\Gamma[a_1^*(s-\alpha)]}{(a_1^*)^{a_1^*(s-\alpha)-1}} \frac{\Gamma[a_2^*(\beta-s)+\omega]}{(a_2^*)^{a_2^*(\beta-s)+\omega-1}} \mathcal{H}_4(s)\left(\mathfrak{M}R f\right)(s) \\
&= \mathcal{H}(s)\left(\mathfrak{M}f\right)(1-s).
\end{aligned} \tag{4.6.8}$$

So we obtain that, if $f \in \mathfrak{L}_{\nu,2}$, $\left(\mathfrak{M}\boldsymbol{H}_4 f\right)(s) = \left(\mathfrak{M}\boldsymbol{H} f\right)(s)$ with $\mathrm{Re}(s) = 1-\nu$. Hence $\boldsymbol{H}_4 = \boldsymbol{H}$ on $\mathfrak{L}_{\nu,2}$ and $\boldsymbol{H}$ can be extended from $\mathfrak{L}_{\nu,2}$ to $\mathfrak{L}_{\nu,r}$ if we define it by (4.6.7).

The operator R is a one-to-one transform of $\mathfrak{L}_{\nu,r}$ onto $\mathfrak{L}_{1-\nu,r}$ by Lemma 3.1(iii), and $\mathbb{L}_{a_1^*,\alpha}$ and $\mathbb{L}_{a_2^*,1-\beta-\omega/a_2^*}$ are also one-to-one by Theorem 3.2(c). Therefore $\boldsymbol{H}_4 \in [\mathfrak{L}_{\nu,r}, \mathfrak{L}_{1-\nu,s}]$ is one-to-one if and only if T_4 is one-to-one. Since $T_4 \in [\mathfrak{L}_{1-\nu,r}]$, we have the imbedding

$$\boldsymbol{H}(\mathfrak{L}_{\nu,r}) \subseteq \left(W_\delta \mathbb{L}_{a_1^*,\alpha} \mathbb{L}_{a_2^*,1-\beta-\omega/a_2^*}\right)(\mathfrak{L}_{1-\nu,r}) = \left(\mathbb{L}_{a_1^*,\alpha} \mathbb{L}_{a_2^*,1-\beta-\omega/a_2^*}\right)(\mathfrak{L}_{1-\nu,r})$$

for $\alpha < 1-\nu < \beta$ and $1 < r < \infty$ (here we take into account (3.3.20) and the independence of the range on a non-zero constant milptiplier). Moreover, the equality (4.6.1) holds if and only if $T_4(\mathfrak{L}_{1-\nu,r}) = \mathfrak{L}_{1-\nu,r}$. According to Theorem 3.1, T_4 is one-to-one on $\mathfrak{L}_{1-\nu,r}$ and $T_4(\mathfrak{L}_{1-\nu,r}) = \mathfrak{L}_{1-\nu,r}$ if and only if $1/\mathcal{H}_4 \in \mathcal{A}$ and $\alpha(1/\mathcal{H}_4) < 1-\nu < \beta(1/\mathcal{H}_4)$. It is proved similarly, as was done in the proof of Theorem 4.3 that, if $\nu \notin \mathcal{E}_{\mathcal{H}}$, these conditions are satisfied. This completes the proof of the theorem for $\mathrm{Re}(\omega) \geqq 0$.

Now we assume $\mathrm{Re}(\omega) < 0$ and denote by $\mathcal{H}_5(s)$ the function

$$\mathcal{H}_5(s) = \frac{(a_1^*)^{a_1^*(s-\alpha)-1}(a_2^*)^{a_2^*(\beta-s)-1}\Gamma[a_1^*(s-\alpha)-\omega]}{\Gamma^2[a_1^*(s-\alpha)]\Gamma[a_2^*(\beta-s)]} \delta^{-s}\mathcal{H}(s), \tag{4.6.9}$$

which is analytic in the strip $\alpha < \mathrm{Re}(s) < \beta$. Similar arguments to (4.6.4) and (4.6.5) show that the estimates

$$\mathcal{H}_5(\sigma+it) \sim \prod_{j=1}^{q} \beta_j^{\mathrm{Re}(b_j)-1/2} \prod_{i=1}^{p} \alpha_i^{1/2-\mathrm{Re}(a_i)} \cdot (2\pi)^{c^*-1}(a_1^*)^{-\mathrm{Re}(\omega)-1/2}(a_2^*)^{-1/2}e^{1-c^*}e^{\pi\mathrm{Im}(\omega-\xi)\mathrm{sign}(t)/2} \tag{4.6.10}$$

and

$$\mathcal{H}_5'(\sigma+it) = \mathcal{H}_5(\sigma+it)\Bigg\{a_1^*\log a_1^* - a_2^*\log a_2^* + a_1^*\psi[a_1^*(\sigma+it-\alpha)-\omega] -2a_1^*\psi[a_1^*(\sigma+it-\alpha)] + a_2^*\psi[a_2^*(\beta-\sigma-it)] - \log\delta + \frac{\mathcal{H}'(\sigma+it)}{\mathcal{H}(\sigma+it)}\Bigg\} = O\left(\frac{1}{t^2}\right) \quad (|t|\to\infty) \tag{4.6.11}$$

hold uniformly in σ on any bounded interval of $\mathbb{R}$. So $\mathcal{H}_5 \in \mathcal{A}$ with $\alpha(\mathcal{H}_5) = \alpha$ and $\beta(\mathcal{H}_5) = \beta$. By Theorem 3.1 there is a transform $T_5 \in [\mathfrak{L}_{\nu,r}]$ for $1 < r < \infty$ and $\alpha < \nu < \beta$ such that, if $1 < r \leqq 2$,

$$\left(\mathfrak{M}T_5 f\right)(s) = \mathcal{H}_5(s)\left(\mathfrak{M}f\right)(s) \quad (\mathrm{Re}(s) = \nu). \tag{4.6.12}$$

Let

$$\boldsymbol{H}_5 = W_\delta I^{-\omega}_{-;1/a_1^*,-a_1^*\alpha}\mathbb{L}_{a_1^*,\alpha}\mathbb{L}_{a_2^*,1-\beta}T_5 R. \tag{4.6.13}$$

Using again Lemma 3.1(ii),(iii) and Theorem 3.2(b),(c), we have that if $\alpha < 1-\nu < \beta$ and $1 < r \leqq s < \infty$, then $\boldsymbol{H}_5 \in [\mathfrak{L}_{\nu,r}, \mathfrak{L}_{1-\nu,s}]$. For $f \in \mathfrak{L}_{\nu,2}$ and $\mathrm{Re}(s) = 1-\nu$ applying (4.6.13), (3.3.15), (3.3.7), (3.3.8), (4.6.12), (3.3.16) and (4.6.9) we obtain similarly to (4.6.8) that

$$\begin{aligned}
\left(\mathfrak{M}\boldsymbol{H}_5 f\right)(s) &= \delta^s\left(\mathfrak{M}I^{-\omega}_{-;1/a_1^*,-a_1^*\alpha}\mathbb{L}_{a_1^*,\alpha}\mathbb{L}_{a_2^*,1-\beta}T_5Rf\right)(s)\\
&= \delta^s\frac{\Gamma[a_1^*(s-\alpha)]}{\Gamma[a_1^*(s-\alpha)-\omega]}\left(\mathfrak{M}\mathbb{L}_{a_1^*,\alpha}\mathbb{L}_{a_2^*,1-\beta}T_5Rf\right)(s)\\
&= \delta^s\frac{\Gamma[a_1^*(s-\alpha)]}{\Gamma[a_1^*(s-\alpha)-\omega]}\frac{\Gamma[a_1^*(s-\alpha)]}{(a_1^*)^{a_1^*(s-\alpha)-1}}\left(\mathfrak{M}\mathbb{L}_{a_2^*,1-\beta}T_5Rf\right)(1-s)\\
&= \delta^s\frac{\Gamma[a_1^*(s-\alpha)]}{\Gamma[a_1^*(s-\alpha)-\omega]}\frac{\Gamma[a_1^*(s-\alpha)]}{(a_1^*)^{a_1^*(s-\alpha)-1}}\frac{\Gamma[a_2^*(1-s)-a_2^*(1-\beta)]}{(a_2^*)^{a_2^*[(1-s)-(1-\beta)]}}\left(\mathfrak{M}T_5Rf\right)(s)\\
&= \delta^s\frac{\Gamma^2[a_1^*(s-\alpha)]\Gamma[a_2^*(\beta-s)]}{\Gamma[a_1^*(s-\alpha)-\omega](a_1^*)^{a_1^*(s-\alpha)-1}(a_2^*)^{a_2^*(\beta-s)}}\mathcal{H}_5(s)\left(\mathfrak{M}Rf\right)(s)\\
&= \mathcal{H}(s)\left(\mathfrak{M}f\right)(1-s).
\end{aligned} \tag{4.6.14}$$

Applying this equality and using similar arguments to those in the case $\mathrm{Re}(\omega) \geqq 0$, we complete the proof for $\mathrm{Re}(\omega) < 0$. Hence the theorem is proved.

4.7. Boundedness and Range of the $\boldsymbol{H}$-Transform When $a^* > 0$ and $a_1^* = 0$ or $a_2^* = 0$

In this section we consider the cases when $a_1^* > 0,\ a_2^* = 0$ and $a_1^* = 0,\ a_2^* > 0$. First we treat the former case.

Theorem 4.7. *Let $a_1^* > 0, a_2^* = 0, m > 0, \alpha < 1-\nu < \beta$ and $\omega = \mu + a_1^*\alpha + 1/2$ and let $1 < r < \infty$.*

(a) *If $\nu \notin \mathcal{E}_{\mathcal{H}}$, or if $1 < r \leqq 2$, then the transform $\boldsymbol{H}$ is one-to-one on $\mathfrak{L}_{\nu,r}$.*

(b) *If $\mathrm{Re}(\omega) \geqq 0$ and $\nu \notin \mathcal{E}_{\mathcal{H}}$, then*

$$\boldsymbol{H}(\mathfrak{L}_{\nu,r}) = \mathbb{L}_{a_1^*,\alpha-\omega/a_1^*}(\mathfrak{L}_{\nu,r}). \tag{4.7.1}$$

When $\nu \in \mathcal{E}_{\mathcal{H}}$, $\boldsymbol{H}(\mathfrak{L}_{\nu,r})$ is a subset of the right-hand side of (4.7.1).

(c) *If $\mathrm{Re}(\omega) < 0$ and $\nu \notin \mathcal{E}_{\mathcal{H}}$, then*

$$\boldsymbol{H}(\mathfrak{L}_{\nu,r}) = \left(I^{-\omega}_{-;1/a_1^*,-a_1^*\alpha}\mathbb{L}_{a_1^*,\alpha}\right)(\mathfrak{L}_{\nu,r}). \tag{4.7.2}$$

When $\nu \in \mathcal{E}_{\mathcal{H}}$, $\boldsymbol{H}(\mathfrak{L}_{\nu,r})$ is a subset of the right-hand side of (4.7.2).

Proof. We first consider the case $\mathrm{Re}(\omega) \geqq 0$. We define $\mathcal{H}_6(s)$ by

$$\mathcal{H}_6(s) = \frac{(a_1^*)^{a_1^*(s-\alpha)+\omega-1}}{\Gamma[a_1^*(s-\alpha)+\omega]}\,\delta^{-s}\mathcal{H}(s). \tag{4.7.3}$$

Since $\mathrm{Re}(\omega) \geqq 0$, $\mathcal{H}_6(s)$ is analytic in the strip $\alpha < \mathrm{Re}(s) < \beta$. Arguments similar to those in (4.6.4) and (4.6.5) lead to the estimates

$$\mathcal{H}_6(\sigma+it) \sim \prod_{j=1}^{q}\beta_j^{\mathrm{Re}(b_j)-1/2}\prod_{i=1}^{p}\alpha_i^{1/2-\mathrm{Re}(a_i)}$$
$$\cdot\,(2\pi)^{c^*-1/2}(a_1^*)^{-1/2}e^{-c^*+1/2}e^{\pi \mathrm{Im}(\omega-\xi)\mathrm{sign}(t)/2} \tag{4.7.4}$$

and

$$\mathcal{H}_6'(\sigma+it) = \mathcal{H}_6(\sigma+it)\left\{a_1^*\log a_1^* - a_1^*\psi[a_1^*(s-\alpha)+\omega] - \log\delta + \frac{\mathcal{H}'(\sigma+it)}{\mathcal{H}(\sigma+it)}\right\}$$
$$= \mathcal{H}_6(\sigma+it)\left\{a_1^*\log a_1^* - a_1^*\left[\log(ia_1^*t) + \frac{a_1^*(\sigma-\alpha)+\mathrm{Re}(\omega)-\frac{1}{2}}{ia_1^*t}\right]\right.$$
$$\left. -\log\delta + \left[\log\delta + a_1^*\log(it) + \frac{\mathrm{Re}(\mu)+\Delta\sigma}{it}\right] + O\left(\frac{1}{t^2}\right)\right\}$$
$$= O\left(\frac{1}{t^2}\right) \quad (|t|\to\infty) \tag{4.7.5}$$

uniformly in σ in any bounded interval of $\mathbb{R}$. So $\mathcal{H}_6 \in \mathcal{A}$ with $\alpha(\mathcal{H}_6) = \alpha$ and $\beta(\mathcal{H}_6) = \beta$ and Theorem 3.1 implies that there is a transform $T_6 \in [\mathfrak{L}_{\nu,r}]$ for $1 < r < \infty$ and $\alpha < \nu < \beta$, and if $1 < r \leqq 2$, then

$$\left(\mathfrak{M}T_6 f\right)(s) = \mathcal{H}_6(s)\left(\mathfrak{M}f\right)(s) \quad (\mathrm{Re}(s) = \nu). \tag{4.7.6}$$

We set

$$\boldsymbol{H}_6 = W_\delta \mathbb{L}_{a_1^*,\alpha-\omega/a_1^*} R T_6 R. \tag{4.7.7}$$

Then it follows from Lemma 3.1(ii),(iii) and Theorem 3.2(c) that, if $\alpha < \nu < \beta$ and $1 < r \leqq s < \infty$, $\boldsymbol{H}_6 \in [\mathfrak{L}_{\nu,r}, \mathfrak{L}_{1-\nu,s}]$.

For $f \in \mathfrak{L}_{\nu,2}$ applying (4.7.7), (3.3.15), (3.3.8), (4.7.6), (3.3.16) and (4.7.3), we obtain for $\mathrm{Re}(s) = 1 - \nu$

$$\begin{aligned}
\left(\mathfrak{M}\boldsymbol{H}_6 f\right)(s) &= \left(\mathfrak{M}W_\delta \mathbb{L}_{a_1^*,\alpha-\omega/a_1^*} R T_6 R f\right)(s) \\
&= \delta^s \left(\mathfrak{M}\mathbb{L}_{a_1^*,\alpha-\omega/a_1^*} R T_6 R f\right)(s) \\
&= \delta^s \frac{\Gamma[a_1^*(s-\alpha+\omega/a_1^*)]}{(a_1^*)^{a_1^*(s-\alpha+\omega/a_1^*)-1}} \left(\mathfrak{M}RT_6Rf\right)(1-s) \\
&= \delta^s \frac{\Gamma[a_1^*(s-\alpha)+\omega]}{(a_1^*)^{a_1^*(s-\alpha)+\omega-1}} \left(\mathfrak{M}T_6Rf\right)(s) \\
&= \delta^s \frac{\Gamma[a_1^*(s-\alpha)+\omega]}{(a_1^*)^{a_1^*(s-\alpha)+\omega-1}} \mathcal{H}_6(s) \left(\mathfrak{M}Rf\right)(s) \\
&= \delta^s \frac{\Gamma[a_1^*(s-\alpha)+\omega]}{(a_1^*)^{a_1^*(s-\alpha)+\omega-1}} \mathcal{H}_6(s) \left(\mathfrak{M}f\right)(1-s) \\
&= \mathcal{H}(s) \left(\mathfrak{M}f\right)(1-s).
\end{aligned} \tag{4.7.8}$$

Applying this relation and using arguments similar to those in the case $\mathrm{Re}(\omega) \geqq 0$ of Theorem 4.6 we complete the proof of the theorem for $\mathrm{Re}(\omega) \geqq 0$.

Now we consider the case $\mathrm{Re}(\omega) < 0$. Let us define $\mathcal{H}_7(s)$ by

$$\mathcal{H}_7(s) = \frac{(a_1^*)^{a_1^*(s-\alpha)-1}\Gamma[a_1^*(s-\alpha)-\omega]}{\Gamma^2[a_1^*(s-\alpha)]} \delta^{-s}\mathcal{H}(s). \tag{4.7.9}$$

$\mathcal{H}_7(s)$ is analytic in the strip $\alpha < \mathrm{Re}(s) < \beta$, and in accordance with (1.2.9), (3.4.14), (4.3.4) and (3.4.19), we have

$$\begin{aligned}
\mathcal{H}_7(\sigma+it) \sim \prod_{j=1}^{q} \beta_j^{\mathrm{Re}(b_j)-1/2} \prod_{i=1}^{p} \alpha_i^{1/2-\mathrm{Re}(a_i)} (2\pi)^{c^*-1/2} e^{-c^*+1/2} \\
\cdot (a_1^*)^{-\mathrm{Re}(\mu)-a_1^*\alpha-1} e^{\pi \mathrm{Im}(\omega-\xi)\mathrm{sign}(t)/2},
\end{aligned} \tag{4.7.10}$$

$$\begin{aligned}
\mathcal{H}_7'(\sigma+it) = \mathcal{H}_7(\sigma+it)\Bigg[a_1^* \log a_1^* + a_1^*\psi[a_1^*(s-\alpha)-\omega] \\
-2a_1^*\psi[a_1^*(s-\alpha)] - \log\delta + \frac{\mathcal{H}'(\sigma+it)}{\mathcal{H}(\sigma+it)}\Bigg] = O\left(\frac{1}{t^2}\right)
\end{aligned} \tag{4.7.11}$$

as $|t| \to \infty$ uniformly in σ on any bounded interval of $\mathbb{R}$. So $\mathcal{H}_7 \in \mathcal{A}$ with $\alpha(\mathcal{H}_7) = \alpha$ and $\beta(\mathcal{H}_7) = \beta$. By Theorem 3.1, there is $T_7 \in [\mathfrak{L}_{\nu,r}]$ for $1 < r < \infty$ and $\alpha < \nu < \beta$, so that, if $1 < r \leqq 2$,

$$\left(\mathfrak{M}T_7 f\right)(s) = \mathcal{H}_7(s)\left(\mathfrak{M}f\right)(s) \quad (\mathrm{Re}(s) = \nu). \tag{4.7.12}$$

Setting

$$\boldsymbol{H}_7 = W_\delta I^{-\omega}_{-;1/a_1^*,-a_1^*\alpha}\mathbb{L}_{a_1^*,\alpha}RT_7R, \tag{4.7.13}$$

we see $\boldsymbol{H}_7 \in [\mathfrak{L}_{\nu,r}, \mathfrak{L}_{1-\nu,s}]$, if $\alpha < 1-\nu < \beta$ and $1 < r \leqq s < \infty$, by virtue of Lemma 3.1(ii),(iii) and Theorem 3.2(b),(c). For $f \in \mathfrak{L}_{\nu,2}$ and $\mathrm{Re}(s) = 1-\nu$, applying (4.7.13), (3.3.15), (3.3.7), (3.3.8), (3.3.16) and (4.7.12), we obtain similarly to (4.6.8) that

$$\begin{aligned}
\left(\mathfrak{M}\boldsymbol{H}_7 f\right)(s) &= \left(\mathfrak{M}W_\delta I^{-\omega}_{-;1/a_1^*,-a_1^*\alpha}\mathbb{L}_{a_1^*,\alpha}RT_7Rf\right)(s)\\
&= \delta^s\left(\mathfrak{M}I^{-\omega}_{-;1/a_1^*,-a_1^*\alpha}\mathbb{L}_{a_1^*,\alpha}RT_7Rf\right)(s)\\
&= \delta^s\frac{\Gamma[a_1^*(s-\alpha)]}{\Gamma[a_1^*(s-\alpha)-\omega]}\left(\mathfrak{M}\mathbb{L}_{a_1^*,\alpha}RT_7Rf\right)(s)\\
&= \delta^s\frac{\Gamma^2[a_1^*(s-\alpha)]}{\Gamma[a_1^*(s-\alpha)-\omega](a_1^*)^{a_1^*(s-\alpha)-1}}\left(\mathfrak{M}RT_7Rf\right)(1-s)\\
&= \delta^s\frac{\Gamma^2[a_1^*(s-\alpha)]}{\Gamma[a_1^*(s-\alpha)-\omega](a_1^*)^{a_1^*(s-\alpha)-1}}\left(\mathfrak{M}T_7Rf\right)(s)\\
&= \delta^s\frac{\Gamma^2[a_1^*(s-\alpha)]}{\Gamma[a_1^*(s-\alpha)-\omega](a_1^*)^{a_1^*(s-\alpha)-1}}\mathcal{H}_7(s)\left(\mathfrak{M}Rf\right)(s)\\
&= \mathcal{H}(s)\left(\mathfrak{M}f\right)(1-s).
\end{aligned} \tag{4.7.14}$$

Using this relation and arguments similar to those in the case $\mathrm{Re}(\omega) < 0$ of Theorem 4.6, we obtain the result for $\mathrm{Re}(\omega) < 0$.

In the case $a_1^* = 0$ and $a_2^* > 0$ the following statement is valid.

Theorem 4.8. *Let $a_1^* = 0, a_2^* > 0, n > 0, \alpha < 1-\nu < \beta$ and $\omega = \mu - a_2^*\beta + 1/2$ and let $1 < r < \infty$.*

(a) *If $\nu \notin \mathcal{E}_{\mathcal{H}}$, or if $1 < r \leqq 2$, then the transform $\boldsymbol{H}$ is one-to-one on $\mathfrak{L}_{\nu,r}$.*

(b) *If $\mathrm{Re}(\omega) \geqq 0$ and $\nu \notin \mathcal{E}_{\mathcal{H}}$, then*

$$\boldsymbol{H}(\mathfrak{L}_{\nu,r}) = \mathbb{L}_{-a_2^*,\beta+\omega/a_2^*}(\mathfrak{L}_{\nu,r}). \tag{4.7.15}$$

When $\nu \in \mathcal{E}_{\mathcal{H}}$, $\boldsymbol{H}(\mathfrak{L}_{\nu,r})$ is a subset of the right-hand side of (4.7.15).

(c) *If $\mathrm{Re}(\omega) < 0$ and $\nu \notin \mathcal{E}_{\mathcal{H}}$, then*

$$\boldsymbol{H}(\mathfrak{L}_{\nu,r}) = \left(I^{-\omega}_{0+;1/a_2^*,a_2^*\beta-1}\mathbb{L}_{-a_2^*,\beta}\right)(\mathfrak{L}_{\nu,r}). \tag{4.7.16}$$

When $\nu \in \mathcal{E}_{\mathcal{H}}$, $\boldsymbol{H}(\mathfrak{L}_{\nu,r})$ is a subset of the right-hand side of (4.7.16).

Proof. The proof is derived from Theorem 4.7 by examing RTR and by invoking (3.3.17), (3.3.18) similarly to what was done in the proof of Theorem 4.4.

4.8. Boundedness and Range of the $\boldsymbol{H}$-Transform When $a^* > 0$ and $a_1^* < 0$ or $a_2^* < 0$

In this section we give conditions for the transform $\boldsymbol{H}$ to be one-to-one on $\mathfrak{L}_{\nu,r}$ and characterize its range on $\mathfrak{L}_{\nu,r}$ except for its isolated values $\nu \in \mathcal{E}_{\mathcal{H}}$ in terms of the modified Laplace transform $\mathbb{L}_{k,\alpha}$, modified Hankel transform $\mathbb{H}_{k,\eta}$ and elementary transform M_ζ given in (3.3.3), (3.3.4) and (3.3.11). The results will be different in the cases $a_1^* > 0, a_2^* < 0$ and $a_1^* < 0, a_2^* > 0$. We first consider the former case.

Theorem 4.9. *Let $a^* > 0, a_1^* > 0, a_2^* < 0, \alpha < 1 - \nu < \beta$ and let $1 < r < \infty$.*
(a) *If $\nu \notin \mathcal{E}_{\mathcal{H}}$, or if $1 < r \leqq 2$, then the transform $\boldsymbol{H}$ is one-to-one on $\mathfrak{L}_{\nu,r}$.*
(b) *Let $\omega, \eta, \zeta \in \mathbb{C}$ be chosen as*

$$\omega = a^*\eta - \mu - \frac{1}{2}; \tag{4.8.1}$$

$$a^*\mathrm{Re}(\eta) \geqq \gamma(r) + 2a_2^*(\nu - 1) + \mathrm{Re}(\mu); \tag{4.8.2}$$

$$\mathrm{Re}(\eta) > \nu - 1; \tag{4.8.3}$$

$$\mathrm{Re}(\zeta) < 1 - \nu, \tag{4.8.4}$$

where $\gamma(r)$ is given in (3.3.9). *If $\nu \notin \mathcal{E}_{\mathcal{H}}$, then*

$$\boldsymbol{H}(\mathfrak{L}_{\nu,r}) = \left(M_{1/2+\omega/(2a_2^*)}\mathbb{H}_{-2a_2^*,2a_2^*\zeta+\omega-1}\mathbb{L}_{-a^*,1/2+\eta-\omega/(2a_2^*)}\right)\left(\mathfrak{L}_{3/2-\nu+\mathrm{Re}(\omega)/(2a_2^*),r}\right). \tag{4.8.5}$$

When $\nu \in \mathcal{E}_{\mathcal{H}}$, $\boldsymbol{H}(\mathfrak{L}_{\nu,r})$ is a subset of the right-hand side of (4.8.5).

Proof. We denote by $\mathcal{H}_8(s)$ the function

$$\mathcal{H}_8(s) = \frac{(a^*)^{a^*(s+\eta)-1}|a_2^*|^{-2a_2^*s-\omega}\Gamma[a_2^*(s+\zeta)+\omega]}{\Gamma[a^*(s+\eta)]\Gamma[a_2^*(\zeta - s)]}\,\delta^{-s}\mathcal{H}(s). \tag{4.8.6}$$

For $\mathrm{Re}(s) = 1-\nu$, by virtue of the assumptions (4.8.1), (4.8.2), (4.8.4) and relations $\gamma(r) \geqq 1/2$, $a_2^* < 0$, we have

$$\begin{aligned}
\mathrm{Re}[a_2^*(s+\zeta)+\omega] &= a_2^*[1-\nu+\mathrm{Re}(\zeta)] + a^*\mathrm{Re}(\eta) - \mathrm{Re}(\mu) - \frac{1}{2} \\
&\geqq a_2^*[1-\nu+\mathrm{Re}(\zeta)] + [\gamma(r) + 2a_2^*(\nu-1) + \mathrm{Re}(\mu)] - \mathrm{Re}(\mu) - \frac{1}{2} \\
&= a_2^*[\nu - 1 + \mathrm{Re}(\zeta)] + \gamma(r) - \frac{1}{2} \geqq a_2^*[\nu - 1 + \mathrm{Re}(\zeta)] > 0,
\end{aligned}$$

and hence the function $\mathcal{H}_8(s)$ is analytic in the strip $\alpha < \mathrm{Re}(s) < \beta$. Applying (1.2.9), (3.4.14), (4.3.4) and (3.4.19) we obtain the estimates

$$\mathcal{H}_8(\sigma+it) \sim \prod_{j=1}^{q}\beta_j^{\mathrm{Re}(b_j)-1/2}\prod_{i=1}^{p}\alpha_i^{1/2-\mathrm{Re}(b_j)}$$

$$\cdot\,(2\pi)^{c^*-1/2}(a^*)^{-1/2}e^{-c^*+1/2}e^{-\pi\mathrm{Im}(\omega-a^*\eta+\xi)\mathrm{sign}(t)/2} \tag{4.8.7}$$

and

$$\mathcal{H}_8'(\sigma+it) = \mathcal{H}_8(\sigma+it)\Bigg\{a^*\log a^* - 2a_2^*\log|a_2^*| + a_2^*\psi[a_2^*(s+\zeta)+\omega]$$

$$-a^*\psi[a^*(s+\eta)] + a_2^*\psi[a_2^*(\zeta-s)] - \log\delta + \frac{\mathcal{H}'(\sigma+it)}{\mathcal{H}(\sigma+it)}\Bigg\}$$

$$= O\left(\frac{1}{t^2}\right) \tag{4.8.8}$$

as $|t| \to \infty$, uniformly in σ on any bounded interval of $\mathbb{R}$.

So $\mathcal{H}_8 \in \mathcal{A}$ with $\alpha(\mathcal{H}_8) = \alpha$ and $\beta(\mathcal{H}_8) = \beta$ and by Theorem 3.1, there is a transform $T_8 \in [\mathfrak{L}_{\nu,r}]$ for $1 < r < \infty$ and $\alpha < \nu < \beta$ so that, if $1 < r \leqq 2$,

$$\left(\mathfrak{M}T_8 f\right)(s) = \mathcal{H}_8(s)\left(\mathfrak{M}f\right)(s) \quad (\mathrm{Re}(s) = \nu). \tag{4.8.9}$$

Let

$$\boldsymbol{H}_8 = W_\delta M_{1/2+\omega/(2a_2^*)}\mathbb{H}_{-2a_2^*,2a_2^*\zeta+\omega-1}\mathbb{L}_{-a^*,1/2+\eta-\omega/(2a_2^*)}M_{-1/2-\omega/(2a_2^*)}T_8 R. \tag{4.8.10}$$

It is directly verified that under the conditions of the theorem, Lemma 3.1 and Theorem 3.2(c),(d) yield $\boldsymbol{H}_8 \in [\mathfrak{L}_{\nu,r}, \mathfrak{L}_{1-\nu,r}]$.

If $f \in \mathfrak{L}_{\nu,2}$, then applying (4.8.10), (3.3.15), (3.3.14), (3.3.10), (3.3.9), (3.3.14), (4.8.9), (3.3.16) and (4.8.6), we have for $\mathrm{Re}(s) = 1-\nu$

$$\left(\mathfrak{M}\boldsymbol{H}_8 f\right)(s)$$

$$= \left(\mathfrak{M}W_\delta M_{1/2+\omega/(2a_2^*)}\mathbb{H}_{-2a_2^*,2a_2^*\zeta+\omega-1}\mathbb{L}_{-a^*,1/2+\eta-\omega/(2a_2^*)}M_{-1/2-\omega/(2a_2^*)}T_8 Rf\right)(s)$$

$$= \delta^s\left(\mathfrak{M}M_{1/2+\omega/(2a_2^*)}\mathbb{H}_{-2a_2^*,2a_2^*\zeta+\omega-1}\mathbb{L}_{-a^*,1/2+\eta-\omega/(2a_2^*)}M_{-1/2-\omega/(2a_2^*)}T_8 Rf\right)(s)$$

$$= \delta^s\left(\mathfrak{M}\mathbb{H}_{-2a_2^*,2a_2^*\zeta+\omega-1}\mathbb{L}_{-a^*,1/2+\eta-\omega/(2a_2^*)}M_{-1/2-\omega/(2a_2^*)}T_8 Rf\right)\left(s+\frac{1}{2}+\frac{\omega}{2a_2^*}\right)$$

$$= \delta^s\,|a_2^*|^{2a_2^*s+\omega}\,\frac{\Gamma[a_2^*(\zeta-s)]}{\Gamma[a_2^*(s+\zeta)+\omega]}$$

$$\cdot\left(\mathfrak{M}\mathbb{L}_{-a^*,1/2+\eta-\omega/(2a_2^*)}M_{-1/2-\omega/(2a_2^*)}T_8 Rf\right)\left(\frac{1}{2}-s-\frac{\omega}{2a_2^*}\right)$$

$$= \delta^s\,|a_2^*|^{2a_2^*s+\omega}\,(a^*)^{1-a^*(s+\eta)}\frac{\Gamma[a_2^*(\zeta-s)]\Gamma[a^*(s+\eta)]}{\Gamma[a_2^*(s+\zeta)+\omega]}$$

$$\cdot\left(\mathfrak{M}M_{-1/2-\omega/(2a_2^*)}T_8 Rf\right)\left(\frac{1}{2}+s+\frac{\omega}{2a_2^*}\right)$$

$$= \delta^s\,|a_2^*|^{2a_2^*s+\omega}\,(a^*)^{1-a^*(s+\eta)}\frac{\Gamma[a_2^*(\zeta-s)]\Gamma[a^*(s+\eta)]}{\Gamma[a_2^*(s+\zeta)+\omega]}\left(\mathfrak{M}T_8 Rf\right)(s)$$

$$= \delta^s\,|a_2^*|^{2a_2^*s+\omega}\,(a^*)^{1-a^*(s+\eta)}\frac{\Gamma[a_2^*(\zeta-s)]\Gamma[a^*(s+\eta)]}{\Gamma[a_2^*(s+\zeta)+\omega]}\,\mathcal{H}_8(s)\left(\mathfrak{M}Rf\right)(s)$$

$$= \mathcal{H}(s)\left(\mathfrak{M}f\right)(1-s). \tag{4.8.11}$$

So we have that, for $f \in \mathfrak{L}_{\nu,2}$, $\left(\mathfrak{M}\boldsymbol{H}_8 f\right)(s) = \left(\mathfrak{M}\boldsymbol{H}f\right)(s)$ with $\mathrm{Re}(s) = 1-\nu$. Thus $\boldsymbol{H}_8 f = \boldsymbol{H}f$ and $\boldsymbol{H}_8 = \boldsymbol{H}$ on $\mathfrak{L}_{\nu,2}$. By the fact that $\mathfrak{L}_{\nu,2} \cap \mathfrak{L}_{\nu,r}$ is dense in $\mathfrak{L}_{\nu,r}$ (see Rooney [2, Lemma 2.2]), $\boldsymbol{H}$ can be extended from $\mathfrak{L}_{\nu,2}$ to $\mathfrak{L}_{\nu,r}$, if we define it by (4.8.10).

Moreover, (4.8.5) holds if and only if $T_8(\mathfrak{L}_{1-\nu,r}) = \mathfrak{L}_{1-\nu,r}$. Taking arguments similar to those in Theorem 4.7, we can show that the latter holds if $\nu \notin \mathcal{E}_{\mathcal{H}}$. Further, the transforms

$$M_{-1/2-\omega/(2a_2^*)}T_8R, \quad \mathbb{L}_{-a^*,1/2+\eta-\omega/(2a_2^*)}, \quad \mathbb{H}_{-2a_2^*,2a_2^*\zeta+\omega-1}$$

are one-to-one as respective operators in the corresponding spaces. Then, $\boldsymbol{H}$ is one-to-one if and only if T_8 is one-to-one. This is so, if $1 < r \leqq 2$, or if $\nu \notin \mathcal{E}_{\mathcal{H}}$, which can be proved as in Theorem 4.7. This completes the proof of the theorem.

Corollary 4.9.1. *Let $a^* > 0, a_1^* > 0, a_2^* < 0, \alpha < 1-\nu < \beta$ and let $1 < r < \infty$.*

(a) *If $\nu \notin \mathcal{E}_{\mathcal{H}}$, or if $1 < r \leqq 2$, then the transform $\boldsymbol{H}$ is one-to-one on $\mathfrak{L}_{\nu,r}$.*

(b) *Let $\omega = a^*\eta - \mu - 1/2$ and let η and ζ be chosen such that either of the following conditions holds:*

(i) $a^*\mathrm{Re}(\eta) \geqq \gamma(r) - 2a_2^*\beta + \mathrm{Re}(\mu)$, $\mathrm{Re}(\eta) \geqq -\alpha$, $\mathrm{Re}(\zeta) \leqq \alpha$, *if* $m > 0, n > 0$;

(ii) $a^*\mathrm{Re}(\eta) \geqq \gamma(r) - 2a_2^*\beta + \mathrm{Re}(\mu)$, $\mathrm{Re}(\eta) > \nu - 1$, $\mathrm{Re}(\zeta) < 1-\nu$, *if* $m = 0, n > 0$;

(iii) $a^*\mathrm{Re}(\eta) \geqq \gamma(r) + 2a_2^*(\nu-1) + \mathrm{Re}(\mu)$, $\mathrm{Re}(\eta) \geqq -\alpha$, $\mathrm{Re}(\zeta) \leqq \alpha$, *if* $m > 0, n = 0$.

Then, if $\nu \notin \mathcal{E}_{\mathcal{H}}$, $\boldsymbol{H}(\mathfrak{L}_{\nu,r})$ can be represented by the relation (4.8.5). When $\nu \in \mathcal{E}_{\mathcal{H}}$, $\boldsymbol{H}(\mathfrak{L}_{\nu,r})$ is a subset of the right-hand side of (4.8.5).

Proof. By assumptions $\alpha < 1-\nu < \beta$ and $a_2^* < 0$, if $\beta < \infty$ and the relation

$$a^*\mathrm{Re}(\eta) \geqq \gamma(r) - 2a_2^*\beta + \mathrm{Re}(\mu) \tag{4.8.12}$$

holds, then (4.8.2) is valid. If $\alpha > -\infty$ and the relations

$$\mathrm{Re}(\eta) \geqq -\alpha, \qquad \mathrm{Re}(\zeta) \leqq \alpha \tag{4.8.13}$$

hold, then (4.8.3) and (4.8.4) are true. Therefore the corollary follows from Theorem 4.9.

The next assertion follows from Corollary 4.9.1(i), if we set $\eta = -\alpha$.

Corollary 4.9.2. *Let $a^* > 0, a_1^* > 0, a_2^* < 0, m > 0, n > 0, \alpha < 1-\nu < \beta$ and let $1 < r < \infty$.*

(a) *If $\nu \notin \mathcal{E}_{\mathcal{H}}$, or if $1 < r \leqq 2$, then the transform $\boldsymbol{H}$ is one-to-one on $\mathfrak{L}_{\nu,r}$.*

(b) *Let $a^*\alpha - 2a_2^*\beta + \mathrm{Re}(\mu) + \gamma(r) \leqq 0$, $\omega = -a^*\alpha - \mu - 1/2$ and let ζ be chosen such that $\mathrm{Re}(\zeta) \leqq \alpha$. Then if $\nu \notin \mathcal{E}_{\mathcal{H}}$, $\boldsymbol{H}(\mathfrak{L}_{\nu,r})$ can be represented in the form (4.8.5). When $\nu \in \mathcal{E}_{\mathcal{H}}$, $\boldsymbol{H}(\mathfrak{L}_{\nu,r})$ is a subset of the right-hand side of (4.8.5).*

Finally we consider the case when $a^* > 0, a_1^* < 0$ and $a_2^* > 0$.

Theorem 4.10. *Let $a^* > 0, a_1^* < 0, a_2^* > 0, \alpha < 1-\nu < \beta$ and let $1 < r < \infty$.*

(a) *If $\nu \notin \mathcal{E}_{\mathcal{H}}$, or if $1 < r \leqq 2$, then the transform $\boldsymbol{H}$ is one-to-one on $\mathfrak{L}_{\nu,r}$.*

(b) *Let $\omega, \eta, \zeta \in \mathbb{C}$ be chosen as*

$$\omega = a^*\eta - \Delta - \mu - \frac{1}{2}; \tag{4.8.14}$$

$$a^*\mathrm{Re}(\eta) \geqq \gamma(r) - 2a_1^*\nu + \Delta + \mathrm{Re}(\mu); \tag{4.8.15}$$

$$\mathrm{Re}(\eta) > -\nu; \tag{4.8.16}$$

$$\mathrm{Re}(\zeta) < \nu. \tag{4.8.17}$$

If $\nu \notin \mathcal{E}_{\mathcal{H}}$, then

$$\boldsymbol{H}(\mathfrak{L}_{\nu,r}) = \left(M_{-1/2-\omega/(2a_1^*)}\mathbb{H}_{2a_1^*,2a_1^*\zeta+\omega-1}\mathbb{L}_{a^*,1/2-\eta+\omega/(2a_1^*)}\right)\left(\mathfrak{L}_{1/2-\nu-\mathrm{Re}(\omega)/(2a_1^*),r}\right). \tag{4.8.18}$$

When $\nu \in \mathcal{E}_{\mathcal{H}}$, $\boldsymbol{H}(\mathfrak{L}_{\nu,r})$ is a subset of the right-hand side of (4.8.18).

Proof. The proof is derived from Theorem 4.9 by examing RTR similarly to that which was done in Theorem 4.4.

Corollary 4.10.1. *Let $a^* > 0, a_1^* < 0, a_2^* > 0, \alpha < 1 - \nu < \beta$ and let $1 < r < \infty$.*

(a) *If $\nu \notin \mathcal{E}_{\mathcal{H}}$, or if $1 < r \leqq 2$, then the transform $\boldsymbol{H}$ is one-to-one on $\mathfrak{L}_{\nu,r}$.*

(b) *Let $\omega = a^*\eta - \Delta - \mu - 1/2$ and let η and ζ be chosen such that either of the following conditions holds:*

(i) $a^*\mathrm{Re}(\eta) \geqq \gamma(r) - 2a_1^*(1-\alpha) + \Delta + \mathrm{Re}(\mu)$, $\mathrm{Re}(\eta) \geqq \beta - 1$, $\mathrm{Re}(\zeta) \leqq 1 - \beta$, *if $m > 0, n > 0$;*

(ii) $a^*\mathrm{Re}(\eta) \geqq \gamma(r) - 2a_1^*\beta + \Delta + \mathrm{Re}(\mu)$, $\mathrm{Re}(\eta) \geqq \beta - 1$, $\mathrm{Re}(\zeta) \leqq 1 - \beta$, *if $m = 0, n > 0$;*

(iii) $a^*\mathrm{Re}(\eta) \geqq \gamma(r) - 2a_1^*(1-\alpha) + \Delta + \mathrm{Re}(\mu)$, $\mathrm{Re}(\eta) > -\nu$, $\mathrm{Re}(\zeta) < \nu$, *if $m > 0, n = 0$.*

Then, if $\nu \notin \mathcal{E}_{\mathcal{H}}$, $\boldsymbol{H}(\mathfrak{L}_{\nu,r})$ can be represented by the relation (4.8.18). *When $\nu \in \mathcal{E}_{\mathcal{H}}$, $\boldsymbol{H}(\mathfrak{L}_{\nu,r})$ is a subset of the set in the right-hand side of* (4.8.18).

Corollary 4.10.2. *Let $a^* > 0, a_1^* < 0, a_2^* > 0, m > 0, n > 0, \alpha < 1 - \nu < \beta$ and let $1 < r < \infty$.*

(a) *If $\nu \notin \mathcal{E}_{\mathcal{H}}$, or if $1 < r \leqq 2$, then the transform $\boldsymbol{H}$ is one-to-one on $\mathfrak{L}_{\nu,r}$.*

(b) *Let $2a_1^*\alpha - a^*\beta + \mathrm{Re}(\mu) + \gamma(r) \leqq 0$, $\omega = a^*\beta - 2a_1^* - \mu - 1/2$ and let ζ be chosen such that $\mathrm{Re}(\zeta) \leqq 1 - \beta$. Then, if $\nu \notin \mathcal{E}_{\mathcal{H}}$, $\boldsymbol{H}(\mathfrak{L}_{\nu,r})$ can be represented by the relation* (4.8.18). *When $\nu \in \mathcal{E}_{\mathcal{H}}$, $\boldsymbol{H}(\mathfrak{L}_{\nu,r})$ is a subset of the right-hand side of* (4.8.18).

4.9. Inversion of the $\boldsymbol{H}$-Transform When $\Delta = 0$

In Sections 4.1–4.4 and 4.6–4.8 we have proved that for certain ranges of parameters, the $\boldsymbol{H}$-transform (3.1.1) has the representation (4.1.4) or (4.1.5). In this and the next sections we show that the inversion of the $\boldsymbol{H}$-transform has the respective forms:

$$f(x) = hx^{1-(\lambda+1)/h}\frac{d}{dx}x^{(\lambda+1)/h}$$

$$\cdot \int_0^\infty H_{p+1,q+1}^{q-m,p-n+1}\left[xt \left| \begin{matrix} (-\lambda, h), (1-a_i-\alpha_i, \alpha_i)_{n+1,p}, (1-a_i-\alpha_i, \alpha_i)_{1,n} \\ (1-b_j-\beta_j, \beta_j)_{m+1,q}, (1-b_j-\beta_j, \beta_j)_{1,m}, (-\lambda-1, h) \end{matrix} \right.\right]$$
$$\cdot (\boldsymbol{H}f)(t)dt \tag{4.9.1}$$

or

$$f(x) = -hx^{1-(\lambda+1)/h}\frac{d}{dx}x^{(\lambda+1)/h}$$
$$\cdot \int_0^\infty H_{p+1,q+1}^{q-m+1,p-n}\left[xt \left| \begin{matrix} (1-a_i-\alpha_i, \alpha_i)_{n+1,p}, (1-a_i-\alpha_i, \alpha_i)_{1,n}, (-\lambda, h) \\ (-\lambda-1, h), (1-b_j-\beta_j, \beta_j)_{m+1,q}, (1-b_j-\beta_j, \beta_j)_{1,m} \end{matrix} \right.\right]$$
$$\cdot (\boldsymbol{H}f)(t)dt, \tag{4.9.2}$$

provided that $a^* = 0$. The conditions for the validity of the relations (4.9.1) and (4.9.2) will be different in the cases $\Delta = 0$ and $\Delta \neq 0$. Here we consider the first one.

If $f \in \mathfrak{L}_{\nu,2}$, and $\boldsymbol{H}$ is defined on $\mathfrak{L}_{\nu,r}$, then according to Theorem 4.1, the equality (4.1.1) holds under the assumption there. This fact implies the relation

$$\left(\mathfrak{M}f\right)(s) = \frac{\left(\mathfrak{M}\boldsymbol{H}f\right)(1-s)}{\mathcal{H}(1-s)} \tag{4.9.3}$$

for $\mathrm{Re}(s) = \nu$. By (1.1.2) we have

$$\frac{1}{\mathcal{H}(1-s)} = \mathcal{H}_{p,q}^{q-m,p-n}\left[\begin{matrix} (1-a_i-\alpha_i, \alpha_i)_{n+1,p}, (1-a_i-\alpha_i, \alpha_i)_{1,n} \\ (1-b_j-\beta_j, \beta_j)_{m+1,q}, (1-b_j-\beta_j, \beta_j)_{1,m} \end{matrix} \right| s \Bigg]$$
$$\equiv \mathcal{H}_0(s), \tag{4.9.4}$$

and hence (4.9.3) takes the form

$$(\mathfrak{M}f)(s) = (\mathfrak{M}\boldsymbol{H}f)(1-s)\mathcal{H}_0(s) \quad (\mathrm{Re}(s) = \nu). \tag{4.9.5}$$

We denote $\alpha_0, \beta_0, a_0^*, a_{01}^*, a_{02}^*, \delta_0, \Delta_0$ and μ_0 for $\mathcal{H}_0$ instead of those for $\mathcal{H}$. Then we find

$$\alpha_0 = \begin{cases} \max\left[\dfrac{\mathrm{Re}(b_{m+1})-1}{\beta_{m+1}}+1, \cdots, \dfrac{\mathrm{Re}(b_q)-1}{\beta_q}+1\right] & \text{if } q > m; \\ -\infty & \text{if } q = m, \end{cases} \tag{4.9.6}$$

$$\beta_0 = \begin{cases} \min\left[\dfrac{\mathrm{Re}\,(a_{n+1})}{\alpha_{n+1}}+1, \cdots, \dfrac{\mathrm{Re}\,(a_p)}{\alpha_p}+1\right] & \text{if } p > n; \\ \infty & \text{if } p = n, \end{cases} \tag{4.9.7}$$

$$a_0^* = -a^*; \quad a_{01}^* = -a_2^*; \quad a_{02}^* = -a_1^*; \quad \delta_0 = \delta; \quad \Delta_0 = \Delta; \quad \mu_0 = -\mu - \Delta. \tag{4.9.8}$$

We also note that if $\alpha_0 < \nu < \beta_0$, ν is not in the exceptional set of $\mathcal{H}_0$.

First we consider the case $r = 2$.

Theorem 4.11. *Let $\alpha < 1-\nu < \beta, \alpha_0 < \nu < \beta_0, a^* = 0$ and $\Delta(1-\nu) + \text{Re}(\mu) = 0$. Let $f \in \mathfrak{L}_{\nu,2}$. Then the relation* (4.9.1) *holds for* $\text{Re}(\lambda) > \nu h - 1$ *and* (4.9.2) *for* $\text{Re}(\lambda) < \nu h - 1$.

Proof. We apply Theorem 3.6 with $\mathcal{H}$ being replaced by $\mathcal{H}_0$ and ν by $1-\nu$. By the assumption and (4.9.8) we have

$$a_0^* = -a^* = 0, \tag{4.9.9}$$

$$\Delta_0[1-(1-\nu)] + \text{Re}(\mu_0) = \Delta\nu - \text{Re}(\mu) - \Delta = -[\Delta(1-\nu) + \text{Re}(\mu)] = 0 \tag{4.9.10}$$

and $\alpha_0 < 1-(1-\nu) < \beta_0$, and thus Theorem 3.6(i) applies. Then there is a one-to-one transform $\boldsymbol{H}_0 \in [\mathfrak{L}_{1-\nu,2}, \mathfrak{L}_{\nu,2}]$ so that the relation

$$\left(\mathfrak{M}\boldsymbol{H}_0 f\right)(s) = \mathcal{H}_0(s)\left(\mathfrak{M}f\right)(1-s) \tag{4.9.11}$$

holds for $f \in \mathfrak{L}_{1-\nu,2}$ and $\text{Re}(s) = \nu$. Further, if $f \in \mathfrak{L}_{\nu,2}$, $\boldsymbol{H}f \in \mathfrak{L}_{1-\nu,2}$ and it follows from (4.9.11), (4.1.1) and (4.9.4) that

$$\left(\mathfrak{M}\boldsymbol{H}_0\boldsymbol{H}f\right)(s) = \mathcal{H}_0(s)\left(\mathfrak{M}\boldsymbol{H}f\right)(1-s) = \mathcal{H}_0(s)\mathcal{H}(1-s)\left(\mathfrak{M}f\right)(s) = \left(\mathfrak{M}f\right)(s),$$

if $\text{Re}(s) = \nu$. Hence $\mathfrak{M}\boldsymbol{H}_0\boldsymbol{H}f = \mathfrak{M}f$ and

$$\boldsymbol{H}_0\boldsymbol{H}f = f \quad \text{for} \quad f \in \mathfrak{L}_{\nu,2}. \tag{4.9.12}$$

Applying Theorem 3.6(ii) again, we obtain for $f \in \mathfrak{L}_{1-\nu,2}$ that

$$\left(\boldsymbol{H}_0 f\right)(x) = hx^{1-(\lambda+1)/h}\frac{d}{dx}x^{(\lambda+1)/h}$$
$$\cdot \int_0^\infty H_{p+1,q+1}^{q-m,p-n+1}\left[xt \,\middle|\, \begin{matrix} (-\lambda,h), (1-a_i-\alpha_i,\alpha_i)_{n+1,p}, (1-a_i-\alpha_i,\alpha_i)_{1,n} \\ (1-b_j-\beta_j,\beta_j)_{m+1,q}, (1-b_j-\beta_j,\beta_j)_{1,m}, (-\lambda-1,h) \end{matrix}\right]$$
$$\cdot f(t)dt, \tag{4.9.13}$$

if $\text{Re}(\lambda) > [1-(1-\nu)]h - 1$ and

$$\left(\boldsymbol{H}_0 f\right)(x) = -hx^{1-(\lambda+1)/h}\frac{d}{dx}x^{(\lambda+1)/h}$$
$$\cdot \int_0^\infty H_{p+1,q+1}^{q-m+1,p-n}\left[xt \,\middle|\, \begin{matrix} (1-a_i-\alpha_i,\alpha_i)_{n+1,p}, (1-a_i-\alpha_i,\alpha_i)_{1,n}, (-\lambda,h) \\ (-\lambda-1,h), (1-b_j-\beta_j,\beta_j)_{m+1,q}, (1-b_j-\beta_j,\beta_j)_{1,m} \end{matrix}\right]$$
$$\cdot f(t)dt, \tag{4.9.14}$$

if $\text{Re}(\lambda) < [1-(1-\nu)]h - 1$. Replacing f by $\boldsymbol{H}f$ and using (4.9.12), we have the relations (4.9.1) and (4.9.2) for $f \in \mathfrak{L}_{\nu,2}$, if $\text{Re}(\lambda) > \nu h - 1$ and $\text{Re}(\lambda) < \nu h - 1$, respectively, which completes the proof of the theorem.

The next result is an extension of Theorem 4.11 to $\mathfrak{L}_{\nu,r}$-spaces for any $1 < r < \infty$, provided that $\Delta = 0$ and $\text{Re}(\mu) = 0$.

Theorem 4.12. *Let $\alpha < 1-\nu < \beta, \alpha_0 < \nu < \beta_0, a^* = 0, \Delta = 0$ and $\mathrm{Re}(\mu) = 0$. If $f \in \mathfrak{L}_{\nu,r}$ $(1 < r < \infty)$, the relation (4.9.1) holds for $\mathrm{Re}(\lambda) > \nu h - 1$ and (4.9.2) for $\mathrm{Re}(\lambda) < \nu h - 1$.*

Proof. We apply Theorem 4.1 with $\mathcal{H}$ replaced by $\mathcal{H}_0$ in (4.9.4) and ν by $1-\nu$. From the assumption and (4.9.8), we have $a_0^* = \Delta_0 = 0, \mathrm{Re}(\mu_0) = 0$ and $\alpha_0 < 1-(1-\nu) < \beta_0$, and thus Theorem 4.1(i) can be applied. Due to the theorem, $\boldsymbol{H}_0$ can be extended to $\mathfrak{L}_{1-\nu,r}$ as an element of $[\mathfrak{L}_{1-\nu,r}, \mathfrak{L}_{\nu,r}]$. By virtue of (4.9.12) $\boldsymbol{H}_0\boldsymbol{H}$ is an identical operator in $\mathfrak{L}_{\nu,2}$. By Rooney [2, Lemma 2.2] $\mathfrak{L}_{\nu,2}$ is dense in $\mathfrak{L}_{\nu,r}$ and since $\boldsymbol{H} \in [\mathfrak{L}_{\nu,r}, \mathfrak{L}_{1-\nu,r}]$ and $\boldsymbol{H}_0 \in [\mathfrak{L}_{1-\nu,r}, \mathfrak{L}_{\nu,r}]$, the operator $\boldsymbol{H}_0\boldsymbol{H}$ is identical on $\mathfrak{L}_{\nu,r}$ and hence

$$\boldsymbol{H}_0\boldsymbol{H}f = f \quad \text{for} \quad f \in \mathfrak{L}_{\nu,r}. \tag{4.9.15}$$

Applying Theorem 4.1(e) with $\mathcal{H}$ being replaced by $\mathcal{H}_0$ and ν by $1-\nu$, we obtain that the relations (4.9.13) and (4.9.14) hold for $f \in \mathfrak{L}_{1-\nu,r}$, when $\mathrm{Re}(\lambda) > [1-(1-\nu)]h - 1$ and $\mathrm{Re}(\lambda) < [1-(1-\nu)]h-1$, respectively. Replacing f by $\boldsymbol{H}f$ and using (4.9.15), we arrive at (4.9.1) and (4.9.2) for $f \in \mathfrak{L}_{1-\nu,r}$, if $\mathrm{Re}(\lambda) > \nu h - 1$ and $\mathrm{Re}(\lambda) < \nu h - 1$, respectively, which completes the proof of the theorem.

4.10. Inversion of the $\boldsymbol{H}$-Transform When $\Delta \neq 0$

We now investigate under what conditions the $\boldsymbol{H}$-transform with $\Delta \neq 0$ will have an inverse of the form (4.9.1) or (4.9.2). First, we consider the case $\Delta > 0$. To obtain the inversion of the $\boldsymbol{H}$-transform on $\mathfrak{L}_{\nu,r}$ we use the relation (4.1.3).

Theorem 4.13. *Let $m > 0, a^* = 0, \Delta > 0, \alpha < 1-\nu < \beta, \alpha_0 < \nu < \min\{\beta_0, [\mathrm{Re}(\mu + 1/2)/\Delta]+1\}, \Delta(1-\nu)+\mathrm{Re}(\mu) \leqq 1/2-\gamma(r)$ and let $1 < r < \infty$. If $f \in \mathfrak{L}_{\nu,r}$, then the relations (4.9.1) and (4.9.2) hold for $\mathrm{Re}(\lambda) > \nu h - 1$ and $\mathrm{Re}(\lambda) < \nu h - 1$, respectively.*

Proof. According to Theorem 4.3(a), the $\boldsymbol{H}$-transform is defined on $\mathfrak{L}_{\nu,r}$. First we consider the case $\mathrm{Re}(\lambda) > \nu h - 1$. Let $H_1(t)$ be the function

$$H_1(t) = H^{q-m,p-n+1}_{p+1,q+1}\left[t \left| \begin{matrix} (-\lambda, h), (1-a_i-\alpha_i, \alpha_i)_{n+1,p}, (1-a_i-\alpha_i, \alpha_i)_{1,n} \\ (1-b_j-\beta_j, \beta_j)_{m+1,q}, (1-b_j-\beta_j, \beta_j)_{1,m}, (-\lambda-1, h) \end{matrix} \right.\right]. \tag{4.10.1}$$

If we denote $\widetilde{a}^*, \widetilde{\delta}, \widetilde{\Delta}$ and $\widetilde{\mu}$ for H_1 instead of those for H, then

$$\widetilde{a}^* = -a^* = 0; \quad \widetilde{\delta} = \delta; \quad \widetilde{\Delta} = \Delta > 0; \quad \widetilde{\mu} = -\mu - \Delta - 1. \tag{4.10.2}$$

We prove that $H_1 \in \mathfrak{L}_{\nu,s}$ for any s $(1 \leqq s < \infty)$. For this, we first apply the results in Sections 1.5, 1.6 and 1.8, 1.9 to $H_1(t)$ to find its asymptotic behavior at zero and infinity. According to (4.9.6), (4.9.7) and the assumption, we find

$$\frac{\mathrm{Re}(b_j)-1}{\beta_j} + 1 \leqq \alpha_0 < \beta_0 \leqq \frac{\mathrm{Re}(a_i)}{\alpha_i} + 1 \quad (j = m+1, \cdots, q;\ i = n+1, \cdots, p);$$

$$\frac{\mathrm{Re}(b_j)-1}{\beta_j} + 1 \leqq \alpha_0 < \nu < \frac{\mathrm{Re}(\lambda)+1}{h} \quad (j = m+1, \cdots, q).$$

Then it follows that the poles

$$a_{ik} = \frac{a_i + k}{\alpha_i} + 1 \quad (i = n+1, \cdots, p;\ k = 0, 1, 2, \cdots),$$

$$\lambda_n = \frac{\lambda + 1 + n}{h} \quad (n = 0, 1, 2, \cdots)$$

of the gamma functions $\Gamma(a_i + \alpha_i - \alpha_i s)$ $(i = n+1, \cdots, p)$ and $\Gamma(1 + \lambda - hs)$, and the poles

$$b_{jl} = \frac{b_j - 1 - l}{\beta_j} + 1 \quad (j = m+1, \cdots, q;\ l = 0, 1, 2, \cdots)$$

of the gamma functions $\Gamma(1 - b_j - \beta_j + \beta_j s)$ $(j = m+1, \cdots, q)$ do not coincide. Hence by Theorems 1.11–1.13 and Remark 1.5, we have

$$H_1(t) = O(t^{-\alpha_0}) \quad (t \to 0) \quad \text{with} \quad \alpha_0 = \max_{m+1 \leqq j \leqq q} \left[\frac{\mathrm{Re}(b_j) - 1}{\beta_j}\right] + 1, \tag{4.10.3}$$

if the poles b_{jl} $(j = m+1, \cdots, q;\ l = 0, 1, 2, \cdots)$ are all simple, or

$$H_1(t) = O(t^{-\alpha_0}[\log t]^{N-1}) \quad (t \to 0),$$

if the gamma functions $\Gamma(1 - b_j - \beta_j + \beta_j s)$ $(j = m+1, \cdots, q)$ have general poles of order $N \geqq 2$ at some point.

Further by Theorems 1.7–1.9 and Remark 1.3,

$$H_1(t) = O\left(t^{\varrho_1}\right) \quad (t \to \infty) \quad \text{with} \quad \gamma_0 = \min\left[\beta_0, \frac{\mathrm{Re}(\mu) + \frac{1}{2}}{\Delta} + 1, \frac{\mathrm{Re}(\lambda) + 1}{h}\right], \tag{4.10.4}$$

if the poles a_{ik} $(i = n+1, \cdots, p;\ k = 0, 1, 2, \cdots)$ are all simple, or

$$H_1(t) = O\left(t^{-\gamma_0}[\log(t)]^{M-1}\right) \quad (t \to \infty),$$

if the gamma functions $\Gamma(1 + \lambda - hs), \Gamma(a_i + \alpha_i - \alpha_i s)$ $(i = n+1, \cdots, p)$ have general poles of order $M \geqq 2$ at some point.

Let the gamma functions $\Gamma(1 - b_j - \beta + \beta_j s)$ $(j = m+1, \cdots, q)$ and $\Gamma(1 + \lambda - hs), \Gamma(a_i + \alpha_i - \alpha_i s)$ $(i = n+1, \cdots, p)$ have simple poles. Then from (4.10.3) and (4.10.4) we see that, for $1 \leqq s < \infty$, $H_1(t) \in \mathfrak{L}_{\nu,s}$ if and only if, for some R_1 and R_2, $0 < R_1 < R_2 < \infty$, the integrals

$$\int_0^{R_1} t^{s(\nu - \alpha_0) - 1} dt, \qquad \int_{R_2}^{\infty} t^{s(\nu - \gamma_0) - 1} dt \tag{4.10.5}$$

are convergent. Since by the assumption $\nu > \alpha_0$, the first integral in (4.10.5) converges. In view of our assumptions

$$\nu < \beta_0, \qquad \nu < \frac{\mathrm{Re}(\mu) + \frac{1}{2}}{\Delta} + 1, \qquad \nu < \frac{\mathrm{Re}(\lambda) + 1}{h}$$

we find $\nu - \gamma_0 < 0$ and the second integral in (4.10.5) converges, too.

If the gamma functions $\Gamma(1-b_j-\beta_j+\beta_j s)$ $(j=m+1,\cdots,q)$ or $\Gamma(1+\lambda-hs), \Gamma(a_i+\alpha_i-\alpha_i s)$ $(i=n+1,\cdots,p)$ have general poles, then the logarithmic multipliers $[\log(t)]^{N-1}$ $(N=2,3,\cdots)$ may be added in the first integral in (4.10.5), but they do not influence its convergence. This is similar to the case when the gamma functions $\Gamma(1-\lambda-hs)$ and $\Gamma(a_i+\alpha_i-\alpha_i s)$ $(i=n+1,\cdots,p)$ have general poles of order $M \geqq 2$. Hence, we have

$$H_1(t) \in \mathfrak{L}_{\nu,s} \quad (1 \leqq s < \infty). \tag{4.10.6}$$

Let a be a positive number and Π_a denote the operator

$$\left(\Pi_a f\right)(x) = f(ax) \quad (x>0) \tag{4.10.7}$$

for a function f defined almost everywhere on $(0,\infty)$. By (3.3.12), $\left(\Pi_a f\right)(x) = W_{1/a}f(x)$ and hence in accordance with Lemma 3.1(ii), Π_a is a bounded isomorphism of $\mathfrak{L}_{\nu,r}$ onto itself, and if $f \in \mathfrak{L}_{\nu,r}$ $(1 \leqq r \leqq 2)$, there holds the relation for the Mellin transform $\mathfrak{M}$

$$\left(\mathfrak{M}\Pi_a f\right)(s) = a^{-s}\left(\mathfrak{M}f\right)(s) \quad (\mathrm{Re}(s)=\nu). \tag{4.10.8}$$

By virtue of Theorem 4.3(d) and (4.10.6), if $f \in \mathfrak{L}_{\nu,r}$ and $H_1 \in \mathfrak{L}_{\nu,r'}$ (and hence $\Pi_x H_1 \in \mathfrak{L}_{\nu,r'}$), then

$$\int_0^\infty H_1(xt)\left(\boldsymbol{H}f\right)(t)dt = \int_0^\infty \left(\Pi_x H_1\right)(t)\left(\boldsymbol{H}f\right)(t)dt = \int_0^\infty \left(\boldsymbol{H}\Pi_x H_1\right)(t)f(t)dt. \tag{4.10.9}$$

From the assumption $\Delta(1-\nu)+\mathrm{Re}(\mu) \leqq 1/2-\gamma(r) \leqq 0$, Theorem 4.3(b) and (4.10.8) imply that

$$\left(\mathfrak{M}\boldsymbol{H}\Pi_x H_1\right)(s) = \mathcal{H}(s)\left(\mathfrak{M}\Pi_x H_1\right)(1-s) = x^{-(1-s)}\mathcal{H}(s)\left(\mathfrak{M}H_1\right)(1-s) \tag{4.10.10}$$

for $\mathrm{Re}(s)=1-\nu$. Now from (4.10.6), $H_1(t) \in \mathfrak{L}_{\nu,1}$. Then by the definitions of the H-function (1.1.1) and the direct and inverse Mellin transforms (2.5.1) and (2.5.3), we have

$$\begin{aligned}\left(\mathfrak{M}H_1\right)(s) &= \mathcal{H}^{q-m,p-n+1}_{p+1,q+1}\left[\begin{array}{c}(-\lambda,h),(1-a_i-\alpha_i,\alpha_i)_{n+1,p},(1-a_i-\alpha_i,\alpha_i)_{1,n}\\ (1-b_j-\beta_j,\beta_j)_{m+1,q},(1-b_j-\beta_j,\beta_j)_{1,m},(-\lambda-1,h)\end{array}\middle| s\right]\\ &= \mathcal{H}^{q-m,p-n}_{p,q}\left[\begin{array}{c}(1-a_i-\alpha_i,\alpha_i)_{n+1,p},(1-a_i-\alpha_i,\alpha_i)_{1,n}\\ (1-b_j-\beta_j,\beta_j)_{m+1,q},(1-b_j-\beta_j,\beta_j)_{1,m}\end{array}\middle| s\right]\frac{\Gamma(1+\lambda-hs)}{\Gamma(2+\lambda-hs)}\\ &= \frac{\mathcal{H}_0(s)}{1+\lambda-hs}\end{aligned}$$

for $\mathrm{Re}(s)=\nu$, where $\mathcal{H}_0$ is given in (4.9.4). Thus for $\mathrm{Re}(s)=1-\nu$,

$$\left(\mathfrak{M}H_1\right)(1-s) = \frac{\mathcal{H}_0(1-s)}{1+\lambda-h(1-s)} = \frac{1}{\mathcal{H}(s)[1+\lambda-h(1-s)]}.$$

Substituting this into (4.10.10) we obtain

$$\left(\mathfrak{M}\boldsymbol{H}\Pi_x H_1\right)(s) = \frac{x^{-(1-s)}}{1+\lambda-h(1-s)} \quad (\mathrm{Re}(s)=1-\nu). \tag{4.10.11}$$

For $x > 0$ let us consider the function

$$g_x(t) = \begin{cases} \dfrac{1}{h}\, t^{(\lambda+1)/h-1} & \text{if } 0 < t < x; \\ 0 & \text{if } t > x. \end{cases} \tag{4.10.12}$$

Then we have for $\mathrm{Re}(\lambda) > h - h\mathrm{Re}(s) - 1$

$$\left(\mathfrak{M} g_x\right)(s) = \frac{x^{s+(\lambda+1)/h-1}}{1+\lambda-h(1-s)},$$

and (4.10.11) takes the form

$$\left(\mathfrak{M}\boldsymbol{H}\Pi_x H_1\right)(s) = \left(\mathfrak{M}[x^{-(\lambda+1)/h} g_x)]\right)(s),$$

which implies

$$\left(\boldsymbol{H}\Pi_x H_1\right)(t) = x^{-(\lambda+1)/h} g_x(t). \tag{4.10.13}$$

Substituting (4.10.13) into (4.10.9), we have

$$\int_0^\infty H_1(xt)\left(\boldsymbol{H}f\right)(t)dt = x^{-(\lambda+1)/h}\int_0^\infty g_x(t)f(t)dt$$

or, by virtue of (4.10.12),

$$\int_0^x t^{(\lambda+1)/h-1} f(t)dt = hx^{(\lambda+1)/h}\int_0^\infty H_1(xt)\left(\boldsymbol{H}f\right)(t)dt.$$

Differentiating this relation, we obtain

$$f(x) = hx^{1-(\lambda+1)/h}\frac{d}{dx}x^{(\lambda+1)/h}\int_0^\infty H_1(xt)\left(\boldsymbol{H}f\right)(t)dt,$$

which shows (4.9.1).

When $\mathrm{Re}(\lambda) < \nu h - 1$, the relation (4.9.2) is proved similarly to (4.9.1), by taking the function

$$H_2(t) = H^{q-m+1,p-n}_{p+1,q+1}\left[t \,\middle|\, \begin{matrix} (1-a_i-\alpha_i,\alpha_i)_{n+1,p}, (1-a_i-\alpha_i,\alpha_i)_{1,n}, (-\lambda,h) \\ (-\lambda-1,h), (1-b_j-\beta_j,\beta_j)_{m+1,q}, (1-b_j-\beta_j,\beta_j)_{1,m} \end{matrix}\right] \tag{4.10.14}$$

instead of the function $H_1(t)$ of (4.10.1). This completes the proof of the theorem.

In the case $\Delta < 0$ the following statement gives the inversion of the $\boldsymbol{H}$-transform on $\mathfrak{L}_{\nu,r}$.

Theorem 4.14. *Let $n > 0, a^* = 0, \Delta < 0, \alpha < 1-\nu < \beta, \max[\alpha_0, \{\mathrm{Re}(\mu+1/2)/\Delta\}+1] < \nu < \beta_0, \Delta(1-\nu) + \mathrm{Re}(\mu) \leqq 1/2 - \gamma(r)$ and let $1 < r < \infty$ If $f \in \mathfrak{L}_{\nu,r}$, then the relations (4.9.1) and (4.9.2) hold for $\mathrm{Re}(\lambda) > \nu h - 1$ and for $\mathrm{Re}(\lambda) < \nu h - 1$, respectively.*

This theorem can be proved similarly to Theorem 4.13, if we apply Theorem 4.4 instead of Theorem 4.3 and take into account the asymptotics of the H-function at infinity and zero given in Sections 1.5, 1.6 and 1.8, 1.9.

Remark 4.2. Formal differentiation of the right sides of (4.9.1) and (4.9.2) under the integral sign yields the relation

$$f(x) = \int_0^\infty H_{p,q}^{q-m,p-n}\left[xt \,\middle|\, \begin{matrix} (1-a_i-\alpha_i,\alpha_i)_{n+1,p},(1-a_i-\alpha_i,\alpha_i)_{1,n} \\ (1-b_j-\beta_j,\beta_j)_{m+1,q},(1-b_j-\beta_j,\beta_j)_{1,m} \end{matrix}\right] \left(\boldsymbol{H}f\right)(t)dt. \quad (4.10.15)$$

In fact, applying (2.2.1) to (4.9.1), we obtain

$$\begin{aligned} f(x) &= hx^{1-(\lambda+1)/h}\frac{d}{d(xt)}(xt)^{(\lambda+1)/h} \\ &\quad\cdot \int_0^\infty H_{p+1,q+1}^{q-m,p-n+1}\left[xt \,\middle|\, \begin{matrix} (-\lambda,h),(1-a_i-\alpha_i,\alpha_i)_{n+1,p},(1-a_i-\alpha_i,\alpha_i)_{1,n} \\ (1-b_j-\beta_j,\beta_j)_{m+1,q},(1-b_j-\beta_j,\beta_j)_{1,m},(-\lambda-1,h) \end{matrix}\right] \\ &\qquad \cdot t^{1-(\lambda+1)/h}\left(\boldsymbol{H}f\right)(t)dt \\ &= h\int_0^\infty H_{p+2,q+2}^{q-m,p-n+2}\left[xt \,\middle|\, \begin{matrix} (-(\lambda+1)/h,1),(-\lambda,h), \\ (1-b_j-\beta_j,\beta_j)_{m+1,q}, \end{matrix}\right. \\ &\qquad \left.\begin{matrix} (1-a_i-\alpha_i,\alpha_i)_{n+1,p},(1-a_i-\alpha_i,\alpha_i)_{1,n} \\ (1-b_j-\beta_j,\beta_j)_{1,m},(-\lambda-1,h),(1-(\lambda+1)/h,1) \end{matrix}\right] \left(\boldsymbol{H}f\right)(t)dt. \end{aligned} \quad (4.10.16)$$

According to (1.1.1)–(1.1.2) and (2.2.14) we have

$$\begin{aligned} &hH_{p+2,q+2}^{q-m,p-n+2}\left[xt \,\middle|\, \begin{matrix} \left(-\frac{\lambda+1}{h},1\right),(-\lambda,h),(1-a_i-\alpha_i,\alpha_i)_{n+1,p}, \\ (1-b_j-\beta_j,\beta_i)_{m+1,q},(1-b_j-\beta_j,\beta_j)_{1,m}, \end{matrix}\right. \\ &\qquad\qquad \left.\begin{matrix} (1-a_i-\alpha_i,\alpha_i)_{1,n} \\ (-\lambda-1,h),\left(1-\frac{\lambda+1}{h},1\right) \end{matrix}\right] \\ &= \frac{1}{2\pi i}\int_{\mathfrak{L}} h\frac{\Gamma\left(1+\frac{\lambda+1}{h}-s\right)\Gamma(1+\lambda-hs)}{\Gamma(2+\lambda-hs)\Gamma\left(\frac{\lambda+1}{h}-s\right)} \\ &\qquad \cdot \mathcal{H}_{p,q}^{q-m,p-n}\left[\begin{matrix} (1-a_i-\alpha_i,\alpha_i)_{n+1,p},(1-a_i-\alpha_i,\alpha_i)_{1,n} \\ (1-b_j-\beta_j,\beta_j)_{m+1,q},(1-b_j-\beta_j,\beta_j)_{1,m} \end{matrix} \,\middle|\, s\right](xt)^{-s}ds \\ &= \frac{1}{2\pi i}\int_{\mathfrak{L}} \mathcal{H}_{p,q}^{m,n}\left[\begin{matrix} (1-a_i-\alpha_i,\alpha_i)_{n+1,p},(1-a_i-\alpha_i,\alpha_i)_{1,n} \\ (1-b_j-\beta_j,\beta_j)_{m+1,q},(1-b_j-\beta_j,\beta_j)_{1,m} \end{matrix} \,\middle|\, s\right](xt)^{-s}ds \\ &= H_{p,q}^{q-m,p-n}\left[xt \,\middle|\, \begin{matrix} (1-a_i-\alpha_i,\alpha_i)_{n+1,p},(1-a_i-\alpha_i,\alpha_i)_{1,n} \\ (1-b_j-\beta_j,\beta_j)_{m+1,q},(1-b_j-\beta_j,\beta_j)_{1,m} \end{matrix}\right]. \end{aligned}$$

Substituting this relation into (4.10.16), we arrive at (4.10.15).

Similar calculations for (4.9.2) also yields the same result (4.10.15) by virtue of (2.2.2).

4.11. Bibliographical Remarks and Additional Information on Chapter 4

For Sections 4.1–4.8. As was indicated in Section 3.7, several papers were devoted to the mapping properties of the integral transforms (3.1.1) with H-function kernel in the space $L_r(\mathbb{R}_+)$ $(1 < r < \infty)$. Fox [2] first studied such properties in the space $L_2(\mathbb{R}_+)$. He [2, Theorem 5] proved that the special $\boldsymbol{H}$-transform

$$\left(\boldsymbol{H}f\right)(x) = \frac{d}{dx}\int_0^\infty H_1(xt)f(t)\frac{dt}{t} \quad \text{with} \quad H_1(x) = \int_0^x H^{q,p}_{2p,2q}(t)dt, \tag{4.11.1}$$

where the kernel $H^{q,p}_{2p,2q}(z)$ is given by (1.11.1), belong to $L_2(\mathbb{R}_+)$ provided that $f(x) \in L_2(\mathbb{R}_+)$ and the conditions in (1.11.3) hold. Kesarwani [11] obtained such a result for two integral transforms of the form (4.11.1)

$$\left(\boldsymbol{H}f\right)(x) = \frac{d}{dx}\int_0^\infty H_2(xt)f(t)\frac{dt}{t} \quad \text{with} \quad H_2(x) = \int_0^x H^{m,p}_{p+q,m+n}(t)dt \tag{4.11.2}$$

and

$$\left(\boldsymbol{H}f\right)(x) = \frac{d}{dx}\int_0^\infty H_3(xt)f(t)\frac{dt}{t} \quad \text{with} \quad H_3(z) = \int_0^x H^{n,q}_{q+p,n+m}(t)dt, \tag{4.11.3}$$

where the H-functions $H^{m,p}_{p+q,m+n}(x)$ and $H^{n,q}_{q+p,n+m}(x)$ are specially given by

$$H^{m,p}_{p+q,m+n}\left[x \left| \begin{matrix} \left(1-a_i-\frac{\alpha_i}{2},\alpha_i\right)_{1,p}, \left(b_i-\frac{\beta_i}{2},\beta_i\right)_{1,q} \\ \left(c_j-\frac{\gamma_j}{2},\gamma_j\right)_{1,m}, \left(1-d_j-\frac{\delta_j}{2},\delta_j\right)_{1,n} \end{matrix} \right.\right] \quad (x>0) \tag{4.11.4}$$

and

$$H^{n,q}_{q+p,n+m}\left[x \left| \begin{matrix} \left(1-b_i-\frac{\beta_i}{2},\beta_i\right)_{1,q}, \left(a_i-\frac{\alpha_i}{2},\alpha_i\right)_{1,p} \\ \left(d_j-\frac{\delta_j}{2},\delta_j\right)_{1,n}, \left(1-c_j-\frac{\gamma_j}{2},\gamma_j\right)_{1,m} \end{matrix} \right.\right] \quad (x>0), \tag{4.11.5}$$

respectively. Kesarwani [11, Theorem 1] proved that if $\alpha_i > 0$ $(1 \leqq i \leqq p)$, $\beta_i > 0$ $(1 \leqq i \leqq q)$, $\gamma_j > 0$ $(1 \leqq j \leqq m)$, $\delta_j > 0$ $(1 \leqq j \leqq n)$ and if the following relations are satisfied

$$\begin{gathered} m-q=n-p>0, \quad a^*=0, \quad \Delta>0, \quad \sum_{i=1}^q b_i - \sum_{i=1}^p a_i = \sum_{j=1}^m c_j - \sum_{j=1}^n d_j, \\ \mathrm{Re}(a_i)>0 \quad (1 \leqq i \leqq p), \quad \mathrm{Re}(b_i)>0 \quad (1 \leqq i \leqq q), \\ \mathrm{Re}(c_j)>0 \quad (1 \leqq j \leqq m), \quad \mathrm{Re}(d_j)>0 \quad (1 \leqq j \leqq n), \end{gathered} \tag{4.11.6}$$

then the $\boldsymbol{H}$-transforms (4.11.2) and (4.11.3) belong to $L_2(\mathbb{R}_+)$ provided that $f(x) \in L_2(\mathbb{R}_+)$. Formal differentiation show that (4.11.1), (4.11.2) and (4.11.3) may be represented in the form (3.1.1) with the H-functions (1.11.1), (4.11.4) and (4.11.5) as kernels.

R.K. Saxena [8, Lemma 1] showed that a particular case of the $\boldsymbol{H}$-transform (3.1.1) in the form

$$\left(\boldsymbol{H}f\right)(x) = \int_0^\infty H^{0,p+q}_{p+q,q}\left[xt \left| \begin{matrix} \left(\frac{c_i}{m_i},\frac{1}{m_i}\right)_{1,q} \\ \left(\frac{a_j}{p},\frac{1}{p}\right)_{1,p}, \left(\frac{b_j}{m_j},\frac{1}{m_j}\right)_{1,q} \end{matrix} \right.\right] f(t)dt \quad (x>0) \tag{4.11.7}$$

with $p > 0$, $q \geqq 0$, $m_i > 0$ $(1 \leqq i \leqq q)$ belongs to $L_2(\mathbb{R}_+)$ provided that $f(x) \in L_2(\mathbb{R}_+)$, and obtained the formula (4.1.1) of its Mellin transform. Kalla [8, Lemma 1] proved such a result for the more general $\boldsymbol{H}$-transform considered by R.K. Saxena [8, (32)–(34)].

Bhise and Dighe [2] (see also Dighe [1]) studied in the space $L_r(\mathbb{R}_+)$ the compositon of the Erdélyi–Kober operator $I_{0+;2k,\eta}^{a+1/2}$ with the generalized $\boldsymbol{H}$-transform of the form:

$$\left(\boldsymbol{H}_\sigma^\nu f\right)(x) = x^{\nu+\sigma} \int_0^\infty t^\nu H_{p,q}^{m,n} \left[c(xt)^{2k} \,\middle|\, \begin{matrix} (a_i, \alpha_i)_{1,p} \\ (b_j, \beta_j)_{1,q} \end{matrix} \right] f(t)dt \quad (x > 0) \tag{4.11.8}$$

with

$$\begin{gathered} \eta = \frac{1-\sigma-\nu-2k-ka}{2k}, \quad a^* > 0, \quad \Delta > 0, \quad \left|\arg c(xt)^{2k}\right| < \frac{a^*\pi}{2}, \\ \min_{1 \leqq j \leqq m} \frac{\operatorname{Re}(b_j)}{\beta_j} + \frac{1}{2k} > 0, \quad c > 0, \quad k > 0, \end{gathered} \tag{4.11.9}$$

and proved that, if $a > -1/2$ and $x^{\nu+ak} f(x) \in L^1(\mathbb{R}_+)$, the composition again becomes the $\boldsymbol{H}$-transform of the same form but of greater order:

$$\begin{aligned} &\left(I_{0+;2k,\eta}^{a+1/2} t^{\nu+ka} \boldsymbol{H}_k^\nu f\right)(x) \\ &= x^{\nu+\sigma} c^{-a/2} \int_0^\infty t^\nu H_{p+1,q+1}^{m,n+1} \left[c(xt)^{2k} \,\middle|\, \begin{matrix} \left(1+\frac{a}{2}-\frac{1}{2k}, 1\right), \left(a_i + \frac{a\alpha_i}{2}, \alpha_i\right)_{1,p} \\ \left(b_j + \frac{a\beta_j}{2}, \beta_j\right)_{1,q}, \left(\frac{1-a}{2} - \frac{1}{2k}, 1\right) \end{matrix} \right] f(t)dt. \end{aligned} \tag{4.11.10}$$

Using this relation, Bhise and Dighe [2] proved that the operator in (4.11.8) can be extended as a bounded operator from $L_r(\mathbb{R}_+)$ into $L_s(\mathbb{R}_+)$, provided that

$$\begin{gathered} r \geqq 1, \ s \geqq 1, \ \frac{1}{s} = 1 - \frac{1}{r} - \sigma > 0, \ 0 < \sigma + \frac{2}{r} - 1 \leqq \frac{1}{r}, \\ -2k \min_{1 \leqq j \leqq m} \frac{\operatorname{Re}(b_j)}{\beta_j} < 1 + \nu + ka - \frac{1}{p} < 2k \left(1 - \max_{1 \leqq i \leqq n} \left[\frac{\operatorname{Re}(a_i)}{\alpha_i}, \frac{2k-1}{2k}\right]\right). \end{gathered} \tag{4.11.11}$$

The results presented in Sections 4.1–4.8 concerning the existence, boundedness and representation properties of the $\boldsymbol{H}$-transforms in the weighted spaces $\mathfrak{L}_{\nu,r}$ for any $1 < r < \infty$ were proved by the authors together with S.A. Shlapakov and H.-J. Glaeske in the papers by Kilbas, Saigo and Shlapakov [2]–[3] and Glaeske, Kilbas, Saigo and Shlapakov [1]–[2], and by Betancor and Jerez Diaz [1] independently. We only indicate that Theorems 4.1–4.4 and Theorems 4.5–4.10 were first given by the authors and Shlapakov in [2] and [3], respectively in the particular case when $\delta = 1$.

It should be noted that our main tool was based on a technique of Mellin transforms developed by Rooney [2] and applied by him in [6] to construct the $\mathfrak{L}_{\nu,r}$-theory of $\boldsymbol{G}$-transforms (3.1.2). The latter is a particular case of the $\boldsymbol{H}$-transforms (3.1.1) when $\alpha_1 = \cdots \alpha_p = \beta_1 = \cdots = \beta_q = 1$. Therefore the results in Sections 4.1–4.8 are generalizations of the corresponding results by Rooney [6]. Namely Theorems 6.1–6.4 and 7.1–7.6 in [6] follow from Theorems 4.1–4.10 if we take into account the relation $\Delta = q - p$ (see Section 6.1 in this connection).

We also indicate the paper by Betancor and Jerez Diaz [2] in which the conditions were given on a set of positive Borel measure Ω on $\mathbb{R}_+$ and on non-negative functions $v(x)$ on Ω which are sufficient for the validity of the relation

$$\left[\int_0^\infty \left|\left(\boldsymbol{H}_\eta f\right)(x)\right|^s d\Omega(x)\right]^{1/s} \leqq K \left[\int_0^\infty |v(x) f(x)|^r dx\right]^{1/r} \tag{4.11.12}$$

for the $\boldsymbol{H}$-transform (3.3.1) and a complex-valued function $f(x)$ on $\mathbb{R}_+$ having a compact support, where $1 \leqq r \leqq \infty$, $1 \leqq s \leqq \infty$ and K is a suitable positive constant. The factorization of the

$\boldsymbol{H}$-transform (3.1.1), its mapping properties and inversion formulas in special spaces of functions were presented in the book by Yakubovich and Luchko [1].

For Sections 4.9 and 4.10. The results presented in these sections on the invertibility of the $\boldsymbol{H}$-transforms (3.1.1) in the space $\mathfrak{L}_{\nu,r}$ with any $1 < \nu < \infty$ for $a^* = 0$ and $\Delta = 0$ or $\Delta \neq 0$ were proved by the authors together with Shlapakov in Shlapakov, Saigo and Kilbas [1]. It should be noted that the main difficulty here is caused by the investigation of the invertibility of the $\boldsymbol{H}$-transform in the case $\Delta \neq 0$. This problem is closely connected with finding the explicit asymptotic expansions of the H-function near zero and infinity in the exceptional cases when $a^* = 0$ and $\Delta > 0$ or $\Delta < 0$. Such asymptotic expansions in Sections 1.6 and 1.9 were first obtained by the authors in Kilbas and Saigo [1] (see Section 1.11).

The results in Sections 4.9 and 4.10 generalize the corresponding statements by Rooney [6, Theorems 8.1–8.4] which follow from Theorems 4.11–4.14 (see Chapter 6 in this connection).

We also characterize other results on the invertability of the $\boldsymbol{H}$-transforms (3.1.1). Fox [2] first obtained the inversion formula for the integral transform (4.11.1) with the function $H^{q,p}_{2p,2q}$ (1.11.1) as a kernel in the space $L_2(\mathbb{R}_+)$. In [2, Theorem 5] he proved that if the conditions in (1.11.3) are satisfied and $f(x) \in L_2(\mathbb{R}_+)$, then the first relation in (1.11.2) defines almost everywhere a function $g(x) \in L_2(\mathbb{R}_+)$ and the second relation also holds almost everywhere, and the Parseval formula holds:

$$\int_0^\infty |f(x)|^2 dx = \int_0^\infty |g(x)|^2 dx. \tag{4.11.13}$$

Fox also considered the case of ordinary convergence and showed [2, Theorem 7], that, for $f(t)t^{(1-\Delta)2\Delta} \in L_1(\mathbb{R}_+)$ and $f(t)$ being of bounded variation near $t = x$ $(x > 0)$, if

$$\left(\boldsymbol{H}f\right)(x) = \int_0^\infty H^{q,p}_{2p,2q}(xt)f(t)dt, \tag{4.11.14}$$

then

$$\frac{f(x+0)+f(x-0)}{2} = \int_0^\infty H^{q,p}_{2p,2q}(xt)(\boldsymbol{H}f)(t)dt, \tag{4.11.15}$$

where $H^{q,p}_{2p,2q}(x)$ is given by (1.11.1),

$$\Delta > 0, \quad \mathrm{Re}(a_i) > \frac{\alpha_i(1+\Delta)}{2\Delta}, \quad \mathrm{Re}(b_j) > \frac{\beta_j(1-\Delta)}{2\Delta} \quad (1 \leqq i \leqq p, 1 \leqq j \leqq q). \tag{4.11.16}$$

Kesarwani [11] extended the above results of Fox [2] to the $\boldsymbol{H}$-transform with an unsymmetrical Fourier kernel $H^{m,p}_{p+q,m+n}(x)$ given by (4.11.4). In [11, Theorem 1] he derived the relations of the form (1.11.2)

$$g(x) = \frac{d}{dx}\int_0^\infty H_2(xt)f(t)\frac{dt}{t}, \quad f(x) = \frac{d}{dx}\int_0^\infty H_3(xt)g(t)\frac{dt}{t} \tag{4.11.17}$$

for $H_2(x)$ and $H_3(x)$ being given in (4.11.2) and (4.11.3), and for $f \in L_2(\mathbb{R}_+)$ provided that the conditions in (4.11.6) are satisfied, and he showed the analog of the Parseval relation (4.11.13):

$$\int_0^\infty |f(x)|^2 dx = \int_0^\infty g_2(x)g_3(x)dx, \quad g_i(x) = \frac{d}{dx}\int_0^\infty H_i(xt)f(t)dt \quad (i = 2, 3). \tag{4.11.18}$$

In [11, Theorem 2] Kesarwani established for the $\boldsymbol{H}$-transform

$$\left(\boldsymbol{H}f\right)(x) = \int_0^\infty H^{m,p}_{p+q,m+n}(xt)f(t)dt \tag{4.11.19}$$

the inversion formula of the form (4.11.15):

$$\frac{f(x+0)+f(x-0)}{2} = \int_0^\infty H^{n,q}_{q+p,n+m}(xt)\left(\boldsymbol{H}f\right)(t)dt, \tag{4.11.20}$$

where $H^{n,q}_{q+p,n+m}(x)$ is given by (4.11.5), under the conditions in (4.11.6) and additionally the assumptions

$$\begin{aligned} &\operatorname{Re}(a_i) > \frac{\alpha_i}{2\Delta} \quad (1 \leqq i \leqq p), \quad \operatorname{Re}(b_i) > \frac{\beta_i}{2\Delta} \ (1 \leqq i \leqq q), \\ &\operatorname{Re}(c_j) > \frac{\gamma_j}{2\Delta} \quad (1 \leqq j \leqq m), \quad \operatorname{Re}(d_j) > \frac{\delta_j}{2\Delta} \ (1 \leqq j \leqq n), \end{aligned} \tag{4.11.21}$$

$f(t)t^{(1-\Delta)2\Delta} \in L_1(\mathbb{R}_+)$ and $f(t)$ being of bounded variation near $t = x$ $(x > 0)$.

We note that R.U. Verma also used the same arguments as the above to present in [4] the results which were essentially proved by Fox [2], while in [3] R.U. Verma showed that the $\boldsymbol{H}$-transform

$$\left(\boldsymbol{H}f\right)(x) = \int_0^\infty H^{m,0}_{m,m}(xt)f(t)dt \tag{4.11.22}$$

with the symmetrical kernel

$$H^{m,0}_{m,m}(x) = H^{m,0}_{m,m}\left[x \,\middle|\, \begin{matrix} \left(1 - a_i + \dfrac{\alpha_i}{2}, \alpha_i\right)_{1,m} \\ \left(a_j - \dfrac{\alpha_j}{2}, \alpha_j\right)_{1,m} \end{matrix}\right]$$

has an inversion of the form (4.11.1)

$$f(x) = \frac{d}{dx}\int_0^\infty H_4(xt)f(t)\frac{dt}{t} \quad \text{with} \quad H_4(x) = \int_0^x H^{m,0}_{m,m}(t)dt \tag{4.11.23}$$

under certain conditions. We note that the transform (4.11.22) is reduced to the transform (4.11.19) (considered by Kesarwani [11]) if we put $p = n = 0$, $q = m$, $b_i = 1 - a_i$, $c_i = a_i$ $\gamma_i = \alpha_i$ $(1 \leqq i \leqq m)$ in (4.11.19) and take into acount (4.11.4).

On the basis of the Mellin transform of the composition of the Erdélyi–Kober type fractional integral $I^{\alpha}_{-;m,\beta/m} = \mathcal{R}(\alpha, \beta; m)$ with the $\boldsymbol{H}$-transform (4.11.7) R.K. Saxena [8, Theorem 1] obtained the inversion formula for such a transform in terms of the inverse Laplace transform $\mathbb{L}^{-1}$:

$$f(x) = p^{-1/2}(2\pi)^{(1-p)/2}$$

$$\cdot\, \mathfrak{L}^{-1}\left[\prod_{i=1}^{p}\mathcal{R}\left(\frac{a_i + i - 1}{p}, k-1; p\right)\prod_{j=1}^{q}\mathcal{R}\left(\frac{b_j - c_j}{m_j}, c_j; m_j\right)\left(\boldsymbol{H}f\right)\left(\frac{x}{p}\right)\right], \tag{4.11.24}$$

provided that $f(x) \in L_2(\mathbb{R}_+)$, $(\boldsymbol{H}f)(x) \in L_2(\mathbb{R}_+)$ and some other conditions are satisfied. R.K. Saxena [8, Theorem 2] also indicated the inversion of a more general $\boldsymbol{H}$-transform than (4.11.7) on the basis of the Mellin transform of its compositions with the Erdélyi–Kober type operators (3.3.1) and (3.3.2). These results were also indicated by Kalla [8]. Kumbhat [1] obtained a result similar to (4.11.24) on the basis of the Mellin transform of the generalized fractional integrals

$$\left(\mathcal{R}f\right)(x) = \frac{\mu x^{-\eta-1}}{\Gamma(1-\alpha)}\int_0^x {}_2F_1\left(\alpha, \beta + m; \gamma; a\left(\frac{t}{x}\right)^{\mu}\right)t^{\eta}f(t)dt; \tag{4.11.25}$$

$$\left(\mathcal{S}f\right)(x) = \frac{\mu x^{\delta}}{\Gamma(1-\alpha)}\int_0^x {}_2F_1\left(\alpha, \beta + m; \gamma; a\left(\frac{t}{x}\right)^{\mu}\right)t^{-\delta-1}f(t)dt \tag{4.11.26}$$

with $0 < \alpha < 1$ involving the Gauss hypergeometric function (2.9.2) in the kernel (considered by Kalla and R.K. Saxena [1], [3]) with a more general $\boldsymbol{H}$-transform than the one considered by R.K. Saxena [8].

R. Singh [1] and K.C. Gupta and P.K. Mittal [1] used the Mellin transform to obtain the inversion formulas for the integral transform with H-function (1.1.2) in the kernel with $\mathfrak{L} = \mathfrak{L}_{i\gamma\infty}$. R. Singh [1] considered the $\boldsymbol{H}$-transform

$$\left(\boldsymbol{H}_{\lambda,c}f\right)(x) = \lambda x\int_0^\infty H^{m,n}_{p,q}\left[cxt \,\middle|\, \begin{matrix}(a_i, \alpha_i)_{1,p} \\ (b_j, \beta_j)_{1,q}\end{matrix}\right]f(t)dt \quad (x > 0) \tag{4.11.27}$$

with constants $c \neq 0$ and $\lambda \neq 0$, and proved the inversion formula for $x > 0$ in the form

$$f(x) = \frac{c}{\lambda} \int_0^\infty H_{p,q}^{q-m,p-n} \left[cxt \,\middle|\, \begin{matrix} (1-a_i-\alpha_i, \alpha_i)_{n+1,p}, (1-a_i-\alpha_i, \alpha_i)_{1,n} \\ (1-b_j-\beta_j, \beta_j)_{m+1,q}, (1-b_j-\beta_j, \beta_j)_{1,m} \end{matrix} \right] \cdot \left(\boldsymbol{H}_{\lambda,c} f \right)(t) \frac{dt}{t}, \tag{4.11.28}$$

provided that $0 \leqq n \leqq p$, $0 \leqq q \leqq m$, $a^* > 0$, $\Delta > 0$, $f(x)$ is continuous on $\mathbb{R}_+$ with

$$f(x) = O\left(x^a\right) \quad (x \to 0;\ a + \varrho^* > -1),$$
$$f(x) = O\left(x^b\right) \quad (x \to \infty;\ b + \varrho > -1), \tag{4.11.29}$$
$$x^{\gamma-1} f(x) \in L_1(\mathbb{R}_+), \quad \varrho + 1 < \gamma < \varrho^* + 1$$

for

$$\varrho \equiv \max_{1 \leqq i \leqq n} \left[\frac{\mathrm{Re}(a_i) - 1}{\alpha_i} \right], \quad \varrho^* \equiv \min_{1 \leqq j \leqq m} \left[\frac{\mathrm{Re}(b_j)}{\beta_j} \right], \tag{4.11.30}$$

and the integral in (4.11.24) exists and $x^{-\gamma-1}(\boldsymbol{H}_{\lambda,c}f)(x) \in L_1(\mathbb{R}_+)$. We note that when $c = \lambda = 1$ and $(\boldsymbol{H}f)(x)$ is replaced by $x(\boldsymbol{H}f)(x)$, then (4.11.28) coincides with the relation (4.10.15) obtained by formal differentiation of (4.9.1) and (4.9.2) under the integral sign (see Remark 4.2 in Section 4.10).

K.C. Gupta and P.K. Mittal [1] obtained the inversion relation

$$\frac{1}{2}[f(x+0) + f(x-0)]$$
$$= \frac{1}{2\pi i} \int_{\mathfrak{L}_{i\gamma\infty}} \mathcal{H}_{p,q}^{m,n} \left[\begin{matrix} (1-a_i-\alpha_i, \alpha_i)_{n+1,p}, (1-a_i-\alpha_i, \alpha_i)_{1,n} \\ (1-b_j-\beta_j, \beta_j)_{m+1,q}, (1-b_j-\beta_j, \beta_j)_{1,m} \end{matrix} \,\middle|\, s \right] \cdot s^{-\gamma} \left(\mathfrak{M} \boldsymbol{H}_{1,1} f \right)(-s) ds \tag{4.11.31}$$

for the integral trasform $\boldsymbol{H}_{1,1}$ in (4.11.27) with $c = \lambda = 1$, provided that

$$a^* > 0, \quad |\arg(x)| < \frac{a^* \pi}{2}, \tag{4.11.32}$$

the conditions in (4.11.29) are satisfied, $f(x)$ is of bounded variation in the neighborhood of the point x and the $\boldsymbol{H}_{1,1}$-transform of $|f|$ exists. In [2] K.C. Gupta and P.K. Mittal proved the uniqueness result for the $\boldsymbol{H}$-transform (3.1.1): if

$$\int_0^\infty t^\sigma H_{p,q}^{m,n} \left[xt \,\middle|\, \begin{matrix} (a_i, \alpha_i)_{1,p} \\ (b_j, \beta_j)_{1,q} \end{matrix} \right] f(t) dt = 0 \quad (\sigma \in \mathbb{C}), \tag{4.11.33}$$

then $f(x) = 0$ provided that $f(x)$ is continuous on $\mathbb{R}_+$,

$$f(x) = O\left(x^a\right) \quad (x \to 0;\ \mathrm{Re}(a + \sigma) + \varrho^* > -1),$$
$$f(x) = O\left(x^b e^{-\lambda x}\right) \quad (x \to \infty;\ \mathrm{Re}(\lambda) > 0)),$$

the conditions in (4.11.32) are satisfied and

$$\mathrm{Re}\left[\frac{(a_i - 1)}{\alpha_i} - \frac{a_j}{\alpha_j} \right] < 0 \quad (i = 1, \cdots, n;\ j = n+1, \cdots, p),$$

$$\mathrm{Re}\left[\frac{b_i}{\beta_i} + \frac{(1 - b_j)}{\beta_j} \right] > 0 \quad (i = 1, \cdots, m;\ j = m+1, \cdots, q).$$

Using the relation (4.1.1) V.P. Saxena [1] and Buschman and Srivastava [1] discussed the inversion formulas for the $\boldsymbol{H}_{p,q}^{m,n}$-transform (3.1.1) on the basis of its representation as compositions of the $\boldsymbol{H}_{P,Q}^{M,N}$-transform of lower order ($M \leqq m$, $N \leqq n$, $P \leqq p$ and $Q \leqq q$) with the known integral transforms. V.P. Saxena [1] considered such compositions with the Laplace transform (2.5.2), the Meijer transform (8.9.1) and the modified Varma transform (7.2.15), while Buschman and Srivastava [1] considered the Eredélyi–Kober type fractional integral operators (3.3.1) and (3.3.2). We also mention N. Joshi and J.M.C. Joshi [1] who obtained the real inversion theorem for a certain $\boldsymbol{H}$-transform.

R.K. Saxena and Kushwaha [1] investigated the $\boldsymbol{H}$-transform

$$\left(\boldsymbol{H}f\right)(x) = \int_0^\infty H_{2p+m+n,2q+n+m}^{q+n,p+m}(xt)f(t)dt \quad (x > 0) \tag{4.11.34}$$

with the H-function kernel

$$H_{2p+m+n,2q+n+m}^{q+n,p+m}\left[x \,\middle|\, \begin{matrix} (1-a_i,\alpha_i)_{1,p}, (1-c_i,\gamma_i)_{1,m}, (a_i-\alpha_i,\alpha_i)_{1,p}, (f_i,\delta_i)_{1,n} \\ (b_j,\beta_j)_{1,q}, (g_j,\delta_j)_{1,n}, (1-b_j-\beta_j,\beta_j)_{1,q}, (1-d_j,\gamma_j)_{1,m} \end{matrix}\right] \tag{4.11.35}$$

showing that the composition of several Erdélyi–Kober type fractional integral operators (3.3.1) and (3.3.2) with the $\boldsymbol{H}$-transform (4.11.34) yields the $\boldsymbol{H}$-transform (3.1.1) with the function $H_{2p,2q}^{q,p}$ as a kernel and applied this result to obtain the inversion formula for the transform (4.11.34).

Using the technique of the Laplace transform $\mathbb{L}$ and its inverse $\mathbb{L}^{-1}$ developed by Fox [4], [5], R.U. Verma obtained in [5] and [9] the inversion formulas for the $\boldsymbol{H}$-transforms of the forms

$$\left(\boldsymbol{H}f\right)(x) = \int_0^\infty H_{2p,q}^{0,p}\left[xt \,\middle|\, \begin{matrix} (1-a_i,\alpha_i)_{1,p}, (a_i-\alpha_i,\alpha_i)_{1,p} \\ (1-b_j-\beta_j,\beta_j)_{1,q} \end{matrix}\right] f(t)dt \quad (x > 0) \tag{4.11.36}$$

and

$$\left(\boldsymbol{H}f\right)(x) = \int_0^\infty H_{p,2q}^{q,0}\left[xt \,\middle|\, \begin{matrix} (a_i-c,c)_{1,p} \\ (b_j,c)_{1,q}, (1-b_j-c,c)_{1,q} \end{matrix}\right] f(t)dt \quad (x > 0;\ c > 0), \tag{4.11.37}$$

respectively.

Nasim [2] treated the special $\boldsymbol{H}$-transform (3.1.1) of the form

$$\left(\boldsymbol{H}f\right)(x) = \int_0^\infty t^\sigma H_{0,2n}^{n,0}\left[xt \,\middle|\, \begin{matrix} \overline{\qquad\qquad} \\ (b_j,\beta_j)_{1,n}, (1-c_j,\beta_j)_{1,n} \end{matrix}\right] f(t)dt \quad (\sigma \in \mathbb{C}) \tag{4.11.38}$$

and suggested a method for its inversion on the basis of successive application of linear differential operators of infinite order of the type $\Gamma(\alpha+\gamma\theta)$, and operators of the type $1/\Gamma(\alpha+\gamma\theta)$, where $\theta = -xd/dx$.

Moharir and Raj.K. Saxena [1] established the Abelian theorems for the generalized $\boldsymbol{H}$-transform of the form

$$\left(\boldsymbol{H}f\right)(x) = \int_0^\infty e^{\mu xt} H_{p,q}^{m,0}\left[a(xt)^\lambda \,\middle|\, \begin{matrix} (a_i,\alpha_i)_{1,p} \\ (b_j,\beta_j)_{1,q} \end{matrix}\right] f(t)dt \quad (x > 0) \tag{4.11.39}$$

with real $\mu > 0$ and $\lambda > 0$ provided that $a^* > 0$.

Chapter 5

MODIFIED H-TRANSFORMS ON THE SPACE $\mathfrak{L}_{\nu,r}$

5.1. Modified H-Transforms

The present chapter is devoted to studying the following modifications of the $\boldsymbol{H}$-transform (3.1.1):

$$\left(\boldsymbol{H}^1 f\right)(x) = \int_0^\infty H_{p,q}^{m,n}\left[\frac{x}{t}\,\middle|\,\begin{matrix}(a_i,\alpha_i)_{1,p}\\(b_j,\beta_j)_{1,q}\end{matrix}\right] f(t)\frac{dt}{t}; \tag{5.1.1}$$

$$\left(\boldsymbol{H}^2 f\right)(x) = \int_0^\infty H_{p,q}^{m,n}\left[\frac{t}{x}\,\middle|\,\begin{matrix}(a_i,\alpha_i)_{1,p}\\(b_j,\beta_j)_{1,q}\end{matrix}\right] f(t)\frac{dt}{x}; \tag{5.1.2}$$

$$\left(\boldsymbol{H}_{\sigma,\kappa} f\right)(x) = x^\sigma \int_0^\infty H_{p,q}^{m,n}\left[xt\,\middle|\,\begin{matrix}(a_i,\alpha_i)_{1,p}\\(b_j,\beta_j)_{1,q}\end{matrix}\right] t^\kappa f(t)dt; \tag{5.1.3}$$

$$\left(\boldsymbol{H}^1_{\sigma,\kappa} f\right)(x) = x^\sigma \int_0^\infty H_{p,q}^{m,n}\left[\frac{x}{t}\,\middle|\,\begin{matrix}(a_i,\alpha_i)_{1,p}\\(b_j,\beta_j)_{1,q}\end{matrix}\right] t^\kappa f(t)\frac{dt}{t}; \tag{5.1.4}$$

$$\left(\boldsymbol{H}^2_{\sigma,\kappa} f\right)(x) = x^\sigma \int_0^\infty H_{p,q}^{m,n}\left[\frac{t}{x}\,\middle|\,\begin{matrix}(a_i,\alpha_i)_{1,p}\\(b_j,\beta_j)_{1,q}\end{matrix}\right] t^\kappa f(t)\frac{dt}{x}, \tag{5.1.5}$$

where $\sigma, \kappa \in \mathbb{C}$.

These transforms are connected with the $\boldsymbol{H}$-transform (3.1.1) by the relations

$$\left(\boldsymbol{H}^1 f\right)(x) = \left(\boldsymbol{H}\mathrm{R}f\right)(x); \tag{5.1.6}$$

$$\left(\boldsymbol{H}^2 f\right)(x) = \left(R\boldsymbol{H}f\right)(x); \tag{5.1.7}$$

$$\left(\boldsymbol{H}_{\sigma,\kappa} f\right)(x) = \left(M_\sigma \boldsymbol{H} M_\kappa f\right)(x); \tag{5.1.8}$$

$$\left(\boldsymbol{H}^1_{\sigma,\kappa} f\right)(x) = \left(M_\sigma \boldsymbol{H} R M_\kappa f\right)(x) = \left(M_\sigma \boldsymbol{H}^1 M_\kappa f\right)(x); \tag{5.1.9}$$

$$\left(\boldsymbol{H}^2_{\sigma,\kappa} f\right)(x) = \left(M_\sigma R \boldsymbol{H} M_\kappa f\right)(x) = \left(M_\sigma \boldsymbol{H}^2 M_\kappa f\right)(x), \tag{5.1.10}$$

where R and M_η are operators given in (3.3.13) and (3.3.11). By virtue of relations (5.1.6)–(5.1.10), (4.1.1), (3.3.14) and (3.3.16), the Mellin transforms of the modified $\boldsymbol{H}$-transforms (5.1.1)–(5.1.5) for "sufficiently good" functions f are given by the relations

$$\left(\mathfrak{M}\boldsymbol{H}^1 f\right)(s) = \mathcal{H}_{p,q}^{m,n}\left[\begin{matrix}(a_i,\alpha_i)_{1,p}\\(b_j,\beta_j)_{1,q}\end{matrix}\,\middle|\,s\right]\left(\mathfrak{M}f\right)(s); \tag{5.1.11}$$

$$\left(\mathfrak{M}\boldsymbol{H}^2 f\right)(s) = \mathcal{H}_{p,q}^{m,n}\left[\begin{array}{c|c} (a_i,\alpha_i)_{1,p} & 1-s \\ (b_j,\beta_j)_{1,q} & \end{array}\right]\left(\mathfrak{M}f\right)(s); \tag{5.1.12}$$

$$\left(\mathfrak{M}\boldsymbol{H}_{\sigma,\kappa} f\right)(s) = \mathcal{H}_{p,q}^{m,n}\left[\begin{array}{c|c} (a_i,\alpha_i)_{1,p} & s+\sigma \\ (b_j,\beta_j)_{1,q} & \end{array}\right]\left(\mathfrak{M}f\right)(1-s-\sigma+\kappa); \tag{5.1.13}$$

$$\left(\mathfrak{M}\boldsymbol{H}_{\sigma,\kappa}^1 f\right)(s) = \mathcal{H}_{p,q}^{m,n}\left[\begin{array}{c|c} (a_i,\alpha_i)_{1,p} & s+\sigma \\ (b_j,\beta_j)_{1,q} & \end{array}\right]\left(\mathfrak{M}f\right)(s+\sigma+\kappa); \tag{5.1.14}$$

$$\left(\mathfrak{M}\boldsymbol{H}_{\sigma,\kappa}^2 f\right)(s) = \mathcal{H}_{p,q}^{m,n}\left[\begin{array}{c|c} (a_i,\alpha_i)_{1,p} & 1-s-\sigma \\ (b_j,\beta_j)_{1,q} & \end{array}\right]\left(\mathfrak{M}f\right)(s+\sigma+\kappa), \tag{5.1.15}$$

where $\mathcal{H}_{p,q}^{m,n}\left[\begin{array}{c|c} (a_i,\alpha_i)_{1,p} & s \\ (b_j,\beta_j)_{1,q} & \end{array}\right]$ is the function defined in (1.1.2).

It is directly verified that for the "sufficiently good" functions f and g the following formulas hold

$$\int_0^\infty f(x)\left(\boldsymbol{H}^1 g\right)(x)dx = \int_0^\infty \left(\boldsymbol{H}^2 f\right)(x)g(x)dx; \tag{5.1.16}$$

$$\int_0^\infty f(x)\left(\boldsymbol{H}^2 g\right)(x)dx = \int_0^\infty \left(\boldsymbol{H}^1 f\right)(x)g(x)dx; \tag{5.1.17}$$

$$\int_0^\infty f(x)\left(\boldsymbol{H}_{\sigma,\kappa} g\right)(x)dx = \int_0^\infty \left(\boldsymbol{H}_{\kappa,\sigma} f\right)(x)g(x)dx; \tag{5.1.18}$$

$$\int_0^\infty f(x)\left(\boldsymbol{H}_{\sigma,\kappa}^1 g\right)(x)dx = \int_0^\infty \left(\boldsymbol{H}_{\kappa,\sigma}^2 f\right)(x)g(x)dx; \tag{5.1.19}$$

$$\int_0^\infty f(x)\left(\boldsymbol{H}_{\sigma,\kappa}^2 g\right)(x)dx = \int_0^\infty \left(\boldsymbol{H}_{\kappa,\sigma}^1 f\right)(x)g(x)dx. \tag{5.1.20}$$

Remark 5.1 The relations (5.1.16)–(5.1.20) are the analog of the formulas of fractional integration by parts for the fractional integration operators (2.7.1) , (2.7.2) and (3.3.1), (3.3.2) (see (2.20) and (18.18) in the book by Samko, Kilbas and Marichev [1]).

By Lemma 3.1(i),(iii) the operators M_ζ and R are isometric isomorphisms of $\mathfrak{L}_{\nu,r}$ onto $\mathfrak{L}_{\nu-\mathrm{Re}(\zeta),r}$ and $\mathfrak{L}_{1-\nu,r}$, respectively. Then (5.1.6)–(5.1.10) imply that the properties of the transforms (5.1.1)–(5.1.5) in $\mathfrak{L}_{\nu,r}$-spaces are similar to those for the $\boldsymbol{H}$-transforms (3.1.1) investigated in Chapters 3 and 4. We shall prove these properties in the following sections. We begin from the transform (5.1.1).

5.2. $\boldsymbol{H}^1$-Transform on the Space $\mathfrak{L}_{\nu,r}$

We consider the transform $\boldsymbol{H}^1$ defined in (5.1.1). Since the operator R is an isometric isomorphism of $\mathfrak{L}_{\nu,r}$ onto $\mathfrak{L}_{1-\nu,r}$: $R(\mathfrak{L}_{\nu,r}) = \mathfrak{L}_{1-\nu,r}$, (5.1.6) and (5.1.11) show that we can apply the results in Chapters 3 and 4 for the transform $\boldsymbol{H}$ to obtain the corresponding results for the transform $\boldsymbol{H}^1$.

We still use the notation a^*, Δ, δ, μ, a_1^*, a_2^*, α and β in (1.1.7), (1.1.8), (1.1.9), (1.1.10), (1.1.11), (1.1.12), (3.4.1) and (3.4.2), and let $\mathcal{E}_{\mathcal{H}}$ be the exceptional set of the function $\mathcal{H}_{p,q}^{m,n}(s)$ in (1.1.2) given in Definition 3.4. ¿From Theorems 3.6, 3.7 and 4.1–4.10 we obtain the $\mathfrak{L}_{\nu,2}$- and $\mathfrak{L}_{\nu,r}$-theory of the modified tranform $\boldsymbol{H}^1$ in (5.1.1). First we give the former.

Theorem 5.1. *Suppose that* **(a)** $\alpha < \nu < \beta$ *and that either of conditions* **(b)** $a^* > 0$, *or* **(c)** $a^* = 0$, $\Delta\nu + \mathrm{Re}(\mu) \leqq 0$ *holds. Then we have the following results:*

(i) *There is a one-to-one transform* $\boldsymbol{H}^1 \in [\mathfrak{L}_{\nu,2}, \mathfrak{L}_{\nu,2}]$ *such that* (5.1.11) *holds for* $\mathrm{Re}(s) = \nu$ *and* $f \in \mathfrak{L}_{\nu,2}$. *If* $a^* = 0$, $\Delta\nu + \mathrm{Re}(\mu) = 0$ *and* $1 - \nu \notin \mathcal{E}_{\mathcal{H}}$, *then the transform* $\boldsymbol{H}^1$ *maps* $\mathfrak{L}_{\nu,2}$ *onto* $\mathfrak{L}_{\nu,2}$.

(ii) *If* $f \in \mathfrak{L}_{\nu,2}$ *and* $g \in \mathfrak{L}_{1-\nu,2}$, *then the relation* (5.1.16) *holds for* $\boldsymbol{H}^1$.

(iii) *Let* $\lambda \in \mathbb{C}, h > 0$ *and* $f \in \mathfrak{L}_{\nu,2}$. *When* $\mathrm{Re}(\lambda) > \nu h - 1$, $\boldsymbol{H}^1 f$ *is given by*

$$\left(\boldsymbol{H}^1 f\right)(x)$$

$$= h x^{1-(\lambda+1)/h} \frac{d}{dx} x^{(\lambda+1)h} \int_0^\infty H_{p+1,q+1}^{m,n+1}\left[\frac{x}{t} \,\middle|\, \begin{matrix} (-\lambda, h), (a_i, \alpha_i)_{1,p} \\ (b_j, \beta_j)_{1,q}, (-\lambda - 1, h) \end{matrix}\right] \frac{f(t)}{t} dt. \tag{5.2.1}$$

When $\mathrm{Re}(\lambda) < \nu h - 1$,

$$\left(\boldsymbol{H}^1 f\right)(x)$$

$$= -h x^{1-(\lambda+1)/h} \frac{d}{dx} x^{(\lambda+1)/h} \int_0^\infty H_{p+1,q+1}^{m+1,n}\left[\frac{x}{t} \,\middle|\, \begin{matrix} (a_i, \alpha_i)_{1,p}, (-\lambda, h) \\ (-\lambda - 1, h), (b_j, \beta_j)_{1,q} \end{matrix}\right] \frac{f(t)}{t} dt. \tag{5.2.2}$$

(iv) *The transform* $\boldsymbol{H}^1$ *is independent of* ν *in the sense that, if* ν *and* $\widetilde{\nu}$ *satisfy* (a), *and either* (b) *or* (c), *and if the transforms* $\boldsymbol{H}^1$ *and* $\widetilde{\boldsymbol{H}^1}$ *are defined in* $\mathfrak{L}_{\nu,2}$ *and* $\mathfrak{L}_{\widetilde{\nu},2}$ *respectively by* (5.1.11), *then* $\boldsymbol{H}^1 f = \widetilde{\boldsymbol{H}^1} f$ *for* $f \in \mathfrak{L}_{\nu,2} \cap \mathfrak{L}_{\widetilde{\nu},2}$.

(v) *If* $a^* > 0$ *or if* $a^* = 0$, $\Delta\nu + \mathrm{Re}(\mu) < 0$, *then for* $f \in \mathfrak{L}_{\nu,2}$, $\boldsymbol{H}^1 f$ *is given in* (5.1.1).

Proof. Due to (5.1.6) and $R(\mathfrak{L}_{\nu,2}) = \mathfrak{L}_{1-\nu,2}$, the results in (i), (iv) and (v) follow by virtue of the corresponding statements for the $\boldsymbol{H}$-transform (3.1.1) given in Theorems 3.6 and 3.7. For "sufficiently good" functions f and g, (5.1.16) is verified directly. In fact, for $f \in \mathfrak{L}_{\nu,2}$ and $g \in \mathfrak{L}_{1-\nu,2}$ it is sufficient to show that both sides of (5.1.16) represent bounded linear functionals on $\mathfrak{L}_{\nu,2} \times \mathfrak{L}_{1-\nu,2}$. From (i) and the Schwarz inequality (3.5.6) we have

$$\left|\int_0^\infty f(x)\left(\boldsymbol{H}^1 g\right)(x) dx\right| = \left|\int_0^\infty [x^{\nu-1/2} f(x)][x^{1/2-\nu}\left(\boldsymbol{H}^1 g\right)(x)] dx\right|$$

$$\leqq \|f\|_{\nu,2} \left\|\boldsymbol{H}^1 g\right\|_{1-\nu,2} \leqq K \|f\|_{\nu,2} \|g\|_{1-\nu,2},$$

where K is a bound for $\boldsymbol{H} \in [\mathfrak{L}_{\nu,2}]$. Hence, the left-hand side in (5.1.16) represents a bounded bilinear functional on $\mathfrak{L}_{\nu,2} \times \mathfrak{L}_{1-\nu,2}$. It is similar for the right side of (5.1.16), which proves (ii). The formulas (5.2.1) and (5.2.2) are proved on the basis of the corresponding representations for the $\boldsymbol{H}$-transform (3.1.1) given in (3.6.2) and (3.6.3). For $f \in \mathfrak{L}_{\nu,2}$ and $\mathrm{Re}(\lambda) > \nu h - 1$, in view of $R(\mathfrak{L}_{\nu,2}) = \mathfrak{L}_{1-\nu,2}$, (5.1.6) and (3.6.2) imply

$$\left(\boldsymbol{H}^1 f\right)(x) = \left(\boldsymbol{H} R f\right)(x)$$

$$= hx^{1-(\lambda+1)/h}\frac{d}{dx}x^{(\lambda+1)/h}\int_0^\infty H^{m,n+1}_{p+1,q+1}\left[x\left|\begin{array}{c}(-\lambda,h),(a_i,\alpha_i)_{1,p}\\(b_j,\beta_j)_{1,q},(-1-\lambda,h)\end{array}\right.\right]\frac{1}{t}f\left(\frac{1}{t}\right)dt,$$

which yields (5.2.1) after replacing t by $1/t$ in the integrand. (5.2.2) is proved similarly by using (5.1.6) and (3.6.3). Thus (iii) is established, which completes the proof of the theorem.

Now on the basis of the results in Sections 4.1–4.4 and by using (5.1.6) and Lemma 3.1(iii), we present the $\mathfrak{L}_{\nu,r}$-theory of the transform $\boldsymbol{H}^1$ in (5.1.1) when $a^* = 0$. From Theorems 4.1–4.4 taking into account the isomorphic property $R(\mathfrak{L}_{\nu,r}) = \mathfrak{L}_{1-\nu,r}$, we obtain the mapping properties and the range of $\boldsymbol{H}^1$ on $\mathfrak{L}_{\nu,r}$ in three different cases when either $\Delta = \mathrm{Re}(\mu) = 0$ or $\Delta = 0$, $\mathrm{Re}(\mu) < 0$ or $\Delta \neq 0$.

Theorem 5.2. *Let $a^* = \Delta = 0, \mathrm{Re}(\mu) = 0, \alpha < \nu < \beta$ and let $1 < r < \infty$.*

(a) *The transform $\boldsymbol{H}^1$ defined on $\mathfrak{L}_{\nu,2}$ can be extended to $\mathfrak{L}_{\nu,r}$ as an element of $[\mathfrak{L}_{\nu,r}, \mathfrak{L}_{\nu,r}]$.*

(b) *If $1 < r \leqq 2$, then the transform $\boldsymbol{H}^1$ is one-to-one on $\mathfrak{L}_{\nu,r}$ and there holds the equality (5.1.11) for $f \in \mathfrak{L}_{\nu,r}$ and $\mathrm{Re}(s) = \nu$.*

(c) *If $f \in \mathfrak{L}_{\nu,r}$ and $g \in \mathfrak{L}_{1-\nu,r'}$ with $r' = r/(r-1)$, then the relation (5.1.16) holds.*

(d) *If $1-\nu \notin \mathcal{E}_{\mathcal{H}}$, then the transform $\boldsymbol{H}^1$ is one-to-one on $\mathfrak{L}_{\nu,r}$ and there holds*

$$\boldsymbol{H}^1(\mathfrak{L}_{\nu,r}) = \mathfrak{L}_{\nu,r}. \tag{5.2.3}$$

(e) *If $f \in \mathfrak{L}_{\nu,r}, \lambda \in \mathbb{C}$ and $h > 0$, then $\boldsymbol{H}^1 f$ is given in (5.2.1) for $\mathrm{Re}(\lambda) > \nu h - 1$, while in (5.2.2) for $\mathrm{Re}(\lambda) < \nu h - 1$.*

Theorem 5.3. *Let $a^* = \Delta = 0, \mathrm{Re}(\mu) < 0, \alpha < \nu < \beta$, and let either $m > 0$ or $n > 0$. Let $1 < r < \infty$.*

(a) *The transform $\boldsymbol{H}^1$ defined on $\mathfrak{L}_{\nu,2}$ can be extended to $\mathfrak{L}_{\nu,r}$ as an element of $[\mathfrak{L}_{\nu,r}, \mathfrak{L}_{\nu,s}]$ for all $s \geqq r$ such that $1/s > 1/r + \mathrm{Re}(\mu)$.*

(b) *If $1 < r \leqq 2$, then the transform $\boldsymbol{H}^1$ is one-to-one on $\mathfrak{L}_{\nu,r}$ and there holds the equality (5.1.11) for $f \in \mathfrak{L}_{\nu,r}$ and $\mathrm{Re}(s) = \nu$.*

(c) *If $f \in \mathfrak{L}_{\nu,r}$ and $g \in \mathfrak{L}_{1-\nu,s}$ with $1 < s < \infty$ and $1 \leqq 1/r + 1/s < 1 - \mathrm{Re}(\mu)$, then the relation (5.1.16) holds.*

(d) *Let $k > 0$. If $1-\nu \notin \mathcal{E}_{\mathcal{H}}$, then the transform $\boldsymbol{H}^1$ is one-to-one on $\mathfrak{L}_{\nu,r}$ and there hold*

$$\boldsymbol{H}^1(\mathfrak{L}_{\nu,r}) = I^{-\mu}_{-;k,-\alpha/k}(\mathfrak{L}_{\nu,r}) \tag{5.2.4}$$

for $m > 0$, and

$$\boldsymbol{H}^1(\mathfrak{L}_{\nu,r}) = I^{-\mu}_{0+;k,\beta/k-1}(\mathfrak{L}_{\nu,r}) \tag{5.2.5}$$

for $n > 0$. When $1-\nu \in \mathcal{E}_{\mathcal{H}}$, $\boldsymbol{H}^1(\mathfrak{L}_{\nu,r})$ is a subset of the right-hand sides of (5.2.4) and (5.2.5) in the respective cases.

(e) *If $f \in \mathfrak{L}_{\nu,r}, \lambda \in \mathbb{C}$ and $h > 0$, then $\boldsymbol{H}^1 f$ is given in (5.2.1) for $\mathrm{Re}(\lambda) > \nu h - 1$, while in (5.2.2) for $\mathrm{Re}(\lambda) < \nu h - 1$. Furthermore $\boldsymbol{H}^1 f$ is given in (5.1.1).*

Theorem 5.4. *Let $a^* = 0, \Delta \neq 0, \alpha < \nu < \beta, 1 < r < \infty$ and $\Delta\nu + \mathrm{Re}(\mu) \leqq 1/2 - \gamma(r)$. Assume that $m > 0$ if $\Delta > 0$ and $n > 0$ if $\Delta < 0$.*

(a) *The transform* $\boldsymbol{H}^1$ *defined on* $\mathfrak{L}_{\nu,2}$ *can be extended to* $\mathfrak{L}_{\nu,r}$ *as an element of* $[\mathfrak{L}_{\nu,r}, \mathfrak{L}_{\nu,s}]$ *for all* s *with* $r \leqq s < \infty$ *such that* $s' \geqq [1/2 - \Delta\nu - \mathrm{Re}(\mu)]^{-1}$ *with* $1/s + 1/s' = 1$.

(b) *If* $1 < r \leqq 2$, *then the transform* $\boldsymbol{H}^1$ *is one-to-one on* $\mathfrak{L}_{\nu,r}$ *and there holds the equality* (5.1.11) *for* $f \in \mathfrak{L}_{\nu,r}$ *and* $\mathrm{Re}(s) = \nu$.

(c) *If* $f \in \mathfrak{L}_{\nu,r}$ *and* $g \in \mathfrak{L}_{1-\nu,s}$ *with* $1 < s < \infty, 1/r + 1/s \geqq 1$ *and* $\Delta\nu + \mathrm{Re}(\mu) \leqq 1/2 - \max[\gamma(r), \gamma(s)]$, *then the relation* (5.1.16) *holds.*

(d) *If* $1 - \nu \notin \mathcal{E}_{\mathcal{H}}$, *then the transform* $\boldsymbol{H}^1$ *is one-to-one on* $\mathfrak{L}_{\nu,r}$. *If we set* $\eta = -\Delta\alpha - \mu - 1$ *for* $\Delta > 0$ *and* $\eta = -\Delta\beta - \mu - 1$ *for* $\Delta < 0$, *then* $\mathrm{Re}(\eta) > -1$ *and there holds*

$$\boldsymbol{H}^1(\mathfrak{L}_{\nu,r}) = \left(M_{\mu/\Delta+1/2}\mathbb{H}_{\Delta,\eta}\right)\left(\mathfrak{L}_{1/2-\nu-\mathrm{Re}(\mu)/\Delta,r}\right). \tag{5.2.6}$$

When $1 - \nu \in \mathcal{E}_{\mathcal{H}}$, $\boldsymbol{H}^1(\mathfrak{L}_{\nu,r})$ *is a subset of the right-hand side of* (5.2.6).

(e) *If* $f \in \mathfrak{L}_{\nu,r}$, $\lambda \in \mathbb{C}$, $h > 0$ *and* $\Delta\nu + \mathrm{Re}(\mu) \leqq 1/2 - \gamma(r)$, *then* $\boldsymbol{H}^1 f$ *is given in* (5.2.1) *for* $\mathrm{Re}(\lambda) > \nu h - 1$, *while in* (5.2.2) *for* $\mathrm{Re}(\lambda) < \nu h - 1$. *If furthermore* $\Delta\nu + \mathrm{Re}(\mu) < 0$, $\boldsymbol{H}^1 f$ *is given in* (5.1.1).

From Theorem 4.5 we obtain the $\mathfrak{L}_{\nu,r}$-theory of the transform $\boldsymbol{H}^1$ in (5.1.1) with $a^* > 0$ in $\mathfrak{L}_{\nu,r}$-spaces for any $\nu \in \mathbb{R}$ and $1 \leqq r \leqq \infty$.

Theorem 5.5. *Let* $a^* > 0$, $\alpha < \nu < \beta$ *and* $1 \leqq r \leqq s \leqq \infty$.

(a) *The transform* $\boldsymbol{H}^1$ *defined on* $\mathfrak{L}_{\nu,2}$ *can be extended to* $\mathfrak{L}_{\nu,r}$ *as an element of* $[\mathfrak{L}_{\nu,r}, \mathfrak{L}_{\nu,s}]$. *If* $1 \leqq r \leqq 2$, *then* $\boldsymbol{H}^1$ *is a one-to-one transform from* $\mathfrak{L}_{\nu,r}$ *onto* $\mathfrak{L}_{\nu,s}$.

(b) *If* $f \in \mathfrak{L}_{\nu,r}$ *and* $g \in \mathfrak{L}_{1-\nu,s'}$ *with* $1/s + 1/s' = 1$, *then the relation* (5.1.16) *holds.*

According to Theorems 4.6–4.10 and the isomorphic property $R(\mathfrak{L}_{\nu,r}) = \mathfrak{L}_{1-\nu,r}$ we characterize the boundedness and the range of $\boldsymbol{H}^1$ on $\mathfrak{L}_{\nu,r}$ which will be different in five cases: $a_1^* > 0$, $a_2^* > 0$; $a_1^* > 0$, $a_2^* = 0$; $a_1^* = 0$, $a_2^* > 0$; $a^* > 0$, $a_1^* > 0$, $a_2^* < 0$; and $a^* > 0$, $a_1^* < 0$, $a_2^* > 0$.

Theorem 5.6. *Let* $a_1^* > 0$, $a_2^* > 0$, $m > 0, n > 0$, $\alpha < \nu < \beta$ *and* $\omega = \mu + a_1^*\alpha - a_2^*\beta + 1$ *and let* $1 < r < \infty$.

(a) *If* $1 - \nu \notin \mathcal{E}_{\mathcal{H}}$, *or if* $1 \leqq r \leqq 2$, *then the transform* $\boldsymbol{H}^1$ *is one-to-one on* $\mathfrak{L}_{\nu,r}$.

(b) *If* $\mathrm{Re}(\omega) \geqq 0$ *and* $1 - \nu \notin \mathcal{E}_{\mathcal{H}}$, *then*

$$\boldsymbol{H}^1(\mathfrak{L}_{\nu,r}) = \left(\mathbb{L}_{a_1^*,\alpha}\mathbb{L}_{a_2^*,1-\beta-\omega/a_2^*}\right)(\mathfrak{L}_{\nu,r}). \tag{5.2.7}$$

When $1 - \nu \in \mathcal{E}_{\mathcal{H}}$, $\boldsymbol{H}^1(\mathfrak{L}_{\nu,r})$ *is a subset of the right-hand side of* (5.2.7).

(c) *If* $\mathrm{Re}(\omega) < 0$ *and* $1 - \nu \notin \mathcal{E}_{\mathcal{H}}$, *then*

$$\boldsymbol{H}^1(\mathfrak{L}_{\nu,r}) = \left(I^{-\omega}_{-;1/a_1^*,-a_1^*\alpha}\mathbb{L}_{a_1^*,\alpha}\mathbb{L}_{a_2^*,1-\beta}\right)(\mathfrak{L}_{\nu,r}). \tag{5.2.8}$$

When $1 - \nu \in \mathcal{E}_{\mathcal{H}}$, $\boldsymbol{H}^1(\mathfrak{L}_{\nu,r})$ *is a subset of the right-hand side of* (5.2.8).

Theorem 5.7. *Let* $a_1^* > 0, a_2^* = 0, m > 0$, $\alpha < \nu < \beta$ *and* $\omega = \mu + a_1^*\alpha + 1/2$ *and let* $1 < r < \infty$.

(a) *If* $1 - \nu \notin \mathcal{E}_{\mathcal{H}}$, *or if* $1 < r \leqq 2$, *then the transform* $\boldsymbol{H}^1$ *is one-to-one on* $\mathfrak{L}_{\nu,r}$.

(b) *If* $\mathrm{Re}(\omega) \geqq 0$ *and* $1-\nu \notin \mathcal{E}_{\mathcal{H}}$, *then*

$$\boldsymbol{H}^1(\mathfrak{L}_{\nu,r}) = \mathbb{L}_{a_1^*,\alpha-\omega/a_1^*}(\mathfrak{L}_{1-\nu,r}). \tag{5.2.9}$$

When $1-\nu \in \mathcal{E}_{\mathcal{H}}$, $\boldsymbol{H}^1(\mathfrak{L}_{\nu,r})$ *is a subset of the right-hand side of* (5.2.9).

(c) *If* $\mathrm{Re}(\omega) < 0$ *and* $1-\nu \notin \mathcal{E}_{\mathcal{H}}$, *then*

$$\boldsymbol{H}^1(\mathfrak{L}_{\nu,r}) = \left(I^{-\omega}_{-;1/a_1^*,-a_1^*\alpha}\mathbb{L}_{a_1^*,\alpha}\right)(\mathfrak{L}_{1-\nu,r}). \tag{5.2.10}$$

When $1-\nu \in \mathcal{E}_{\mathcal{H}}$, $\boldsymbol{H}^1(\mathfrak{L}_{\nu,r})$ *is a subset of the right-hand side of* (5.2.10).

Theorem 5.8. *Let* $a_1^* = 0, a_2^* > 0, n > 0$, $\alpha < \nu < \beta$ *and* $\omega = \mu - a_2^*\beta + 1/2$ *and let* $1 < r < \infty$.

(a) *If* $1-\nu \notin \mathcal{E}_{\mathcal{H}}$, *or if* $1 < r \leqq 2$, *then the transform* $\boldsymbol{H}^1$ *is one-to-one on* $\mathfrak{L}_{\nu,r}$.

(b) *If* $\mathrm{Re}(\omega) \geqq 0$ *and* $1-\nu \notin \mathcal{E}_{\mathcal{H}}$, *then*

$$\boldsymbol{H}^1(\mathfrak{L}_{\nu,r}) = \mathbb{L}_{-a_2^*,\beta+\omega/a_2^*}(\mathfrak{L}_{1-\nu,r}). \tag{5.2.11}$$

When $1-\nu \in \mathcal{E}_{\mathcal{H}}$, $\boldsymbol{H}^1(\mathfrak{L}_{\nu,r})$ *is a subset of the right-hand side of* (5.2.11).

(c) *If* $\mathrm{Re}(\omega) < 0$ *and* $1-\nu \notin \mathcal{E}_{\mathcal{H}}$, *then*

$$\boldsymbol{H}^1(\mathfrak{L}_{\nu,r}) = \left(I^{-\omega}_{0+;1/a_2^*,a_2^*\beta-1}\mathbb{L}_{-a_2^*,\beta}\right)(\mathfrak{L}_{1-\nu,r}). \tag{5.2.12}$$

When $1-\nu \in \mathcal{E}_{\mathcal{H}}$, $\boldsymbol{H}^1(\mathfrak{L}_{\nu,r})$ *is a subset of the right-hand side of* (5.2.12).

Theorem 5.9. *Let* $a^* > 0$, $a_1^* > 0$, $a_2^* < 0$, $\alpha < \nu < \beta$ *and let* $1 < r < \infty$.

(a) *If* $1-\nu \notin \mathcal{E}_{\mathcal{H}}$, *or if* $1 < r \leqq 2$, *then the transform* $\boldsymbol{H}^1$ *is one-to-one on* $\mathfrak{L}_{\nu,r}$.

(b) *Let* $\omega, \eta, \zeta \in \mathbb{C}$ *be chosen as*

$$\omega = a^*\eta - \mu - \frac{1}{2}; \tag{5.2.13}$$

$$a^*\mathrm{Re}(\eta) \geqq \gamma(r) - 2a_2^*\nu + \mathrm{Re}(\mu); \tag{5.2.14}$$

$$\mathrm{Re}(\eta) > -\nu; \tag{5.2.15}$$

$$\mathrm{Re}(\zeta) < \nu. \tag{5.2.16}$$

If $1-\nu \notin \mathcal{E}_{\mathcal{H}}$, *then*

$$\boldsymbol{H}^1(\mathfrak{L}_{\nu,r}) = \left(M_{1/2+\omega/(2a_2^*)}\mathbb{H}_{-2a_2^*,2a_2^*\zeta+\omega-1}\mathbb{L}_{-a^*,1/2+\eta-\omega/(2a_2^*)}\right)\left(\mathfrak{L}_{\nu+1/2+\mathrm{Re}(\omega)/(2a_2^*),r}\right). \tag{5.2.17}$$

When $1-\nu \in \mathcal{E}_{\mathcal{H}}$, $\boldsymbol{H}^1(\mathfrak{L}_{\nu,r})$ *is a subset of the right-hand side of* (5.2.17).

Theorem 5.10. *Let* $a^* > 0$, $a_1^* < 0$, $a_2^* > 0$, $\alpha < \nu < \beta$ *and let* $1 < r < \infty$.

(a) *If* $1-\nu \notin \mathcal{E}_{\mathcal{H}}$, *or if* $1 < r \leqq 2$, *then the transform* $\boldsymbol{H}^1$ *is one-to-one on* $\mathfrak{L}_{\nu,r}$.

(b) *Let* $\omega, \eta, \zeta \in \mathbb{C}$ *be chosen as*

$$\omega = a^*\eta - \Delta - \mu - \frac{1}{2}; \tag{5.2.18}$$

$$a^*\mathrm{Re}(\eta) \geqq \gamma(r) + 2a_1^*(\nu-1) + \Delta + \mathrm{Re}(\mu); \tag{5.2.19}$$

$$\mathrm{Re}(\eta) > \nu - 1; \tag{5.2.20}$$

$$\mathrm{Re}(\zeta) < 1 - \nu. \tag{5.2.21}$$

If $1-\nu \notin \mathcal{E}_{\mathcal{H}}$, *then*

$$\boldsymbol{H}^1(\mathfrak{L}_{\nu,r}) = \left(M_{-1/2-\omega/(2a_1^*)}\mathbb{H}_{2a_1^*,2a_1^*\zeta+\omega-1}\mathbb{L}_{a^*,1/2-\eta+\omega/(2a_1^*)}\right)\left(\mathfrak{L}_{\nu-1/2-\mathrm{Re}(\omega)/(2a_1^*),r}\right). \quad (5.2.22)$$

When $1-\nu \in \mathcal{E}_{\mathcal{H}}$, $\boldsymbol{H}^1(\mathfrak{L}_{\nu,r})$ *is a subset of the right-hand side of* (5.2.22).

To obtain the inversion formulas for the transform $\boldsymbol{H}^1 f$ of (5.1.1) with $f \in \mathfrak{L}_{\nu,r}$, we note that, in view of (5.1.6) and Lemma 3.1(iii), this transform is just the transform $\boldsymbol{H}g$ of $g = Rf \in \mathfrak{L}_{\nu,1-r}$:

$$\boldsymbol{H}g = \boldsymbol{H}^1 f. \quad (5.2.23)$$

Therefore, if $a^* = 0$, from (4.9.1) and (4.9.2) we come to the inversion formula for Rf of the form

$$\left(Rf\right)(x) = hx^{1-(\lambda+1)/h}\frac{d}{dx}x^{(\lambda+1)/h} \cdot \int_0^\infty H_{p+1,q+1}^{q-m,p-n+1}\left[xt \,\middle|\, \begin{matrix} (-\lambda,h),(1-a_i-\alpha_i,\alpha_i)_{n+1,p},(1-a_i-\alpha_i,\alpha_i)_{1,n} \\ (1-b_j-\beta_j,\beta_j)_{m+1,q},(1-b_j-\beta_j,\beta_j)_{1,m},(-\lambda-1,h) \end{matrix}\right] \cdot \left(\boldsymbol{H}^1 f\right)(t)dt \quad (5.2.24)$$

or

$$\left(Rf\right)(x) = -hx^{1-(\lambda+1)/h}\frac{d}{dx}x^{(\lambda+1)/h} \cdot \int_0^\infty H_{p+1,q+1}^{q-m+1,p-n}\left[xt \,\middle|\, \begin{matrix} (1-a_i-\alpha_i,\alpha_i)_{n+1,p},(1-a_i-\alpha_i,\alpha_i)_{1,n},(-\lambda,h) \\ (-\lambda-1,h),(1-b_j-\beta_j,\beta_j)_{m+1,q},(1-b_j-\beta_j,\beta_j)_{1,m} \end{matrix}\right] \cdot \left(\boldsymbol{H}^1 f\right)(t)dt. \quad (5.2.25)$$

By noting that the operator R^{-1}, the inverse to the operator R in (3.3.13), coincides with R, then the directly verified relation for $D = d/dx$ and for M_ζ given in (3.3.11)

$$RM_{-\zeta}DM_{\zeta+1} = -M_{\zeta+1}DM_{-\zeta}R \quad (5.2.26)$$

implies the inversion formulas for the transform $\boldsymbol{H}^1$ in (1.5.1) in the form

$$f(x) = -hx^{(\lambda+1)/h}\frac{d}{dx}x^{-(\lambda+1)/h} \cdot \int_0^\infty H_{p+1,q+1}^{q-m,p-n+1}\left[\frac{t}{x} \,\middle|\, \begin{matrix} (-\lambda,h),(1-a_i-\alpha_i,\alpha_i)_{n+1,p},(1-a_i-\alpha_i,\alpha_i)_{1,n} \\ (1-b_j-\beta_j,\beta_j)_{m+1,q},(1-b_j-\beta_j,\beta_j)_{1,m},(-\lambda-1,h) \end{matrix}\right] \cdot \left(\boldsymbol{H}^1 f\right)(t)dt \quad (5.2.27)$$

or

$$f(x) = hx^{(\lambda+1)/h}\frac{d}{dx}x^{-(\lambda+1)/h} \cdot \int_0^\infty H_{p+1,q+1}^{q-m,p-n+1}\left[\frac{t}{x} \,\middle|\, \begin{matrix} (1-a_i-\alpha_i,\alpha_i)_{n+1,p},(1-a_i-\alpha_i,\alpha_i)_{1,n},(-\lambda,h) \\ (-\lambda-1,h),(1-b_j-\beta_j,\beta_j)_{m+1,q},(1-b_j-\beta_j,\beta_j)_{1,m} \end{matrix}\right] \cdot \left(\boldsymbol{H}^1 f\right)(t)dt. \quad (5.2.28)$$

The conditions for the validity of (5.2.27) and (5.2.28) follow from Theorems 4.11–4.14 by taking the numbers α_0 and β_0 in (4.9.6) and (4.9.7). Thus we obtain the following inversion theorems for the transform $\boldsymbol{H}^1$ in (5.1.1) in the cases when $a^* = 0$ and either $\Delta = 0$ or $\Delta \neq 0$.

Theorem 5.11. *Let* $a^* = 0$, $\alpha < \nu < \beta$ *and* $\alpha_0 < 1 - \nu < \beta_0$, *and let* $\lambda \in \mathbb{C}, h > 0$.

(a) *If* $\Delta\nu + \mathrm{Re}(\mu) = 0$ *and* $f \in \mathfrak{L}_{\nu,2}$, *then the inversion formula* (5.2.27) *holds for* $\mathrm{Re}(\lambda) > (1 - \nu)h - 1$ *and* (5.2.28) *for* $\mathrm{Re}(\lambda) < (1 - \nu)h - 1$.

(b) *If* $\Delta = \mathrm{Re}(\mu) = 0$ *and* $f \in \mathfrak{L}_{\nu,r}$ $(1 < r < \infty)$, *then the inversion formula* (5.2.27) *holds for* $\mathrm{Re}(\lambda) > (1 - \nu)h - 1$ *and* (5.2.28) *for* $\mathrm{Re}(\lambda) < (1 - \nu)h - 1$.

Theorem 5.12. *Let* $a^* = 0$, $1 < r < \infty$ *and* $\Delta\nu + \mathrm{Re}(\mu) \leqq 1/2 - \gamma(r)$, *and let* $\lambda \in \mathbb{C}$, $h > 0$.

(a) *If* $\Delta > 0, m > 0, \alpha < \nu < \beta$, $\alpha_0 < 1 - \nu < \min[\beta_0, \{\mathrm{Re}(\mu + 1/2)/\Delta\} + 1]$ *and if* $f \in \mathfrak{L}_{\nu,r}$, *then the inversion formulas* (5.2.27) *and* (5.2.28) *hold for* $\mathrm{Re}(\lambda) > (1 - \nu)h - 1$ *and for* $\mathrm{Re}(\lambda) < (1 - \nu)h - 1$, *respectively.*

(b) *If* $\Delta < 0, n > 0, \alpha < \nu < \beta$, $\max[\alpha_0, \{\mathrm{Re}(\mu + 1/2)/\Delta\} + 1] < 1 - \nu < \beta_0$ *and if* $f \in \mathfrak{L}_{\nu,r}$, *then the inversion formulas* (5.2.27) *and* (5.2.28) *hold for* $\mathrm{Re}(\lambda) > (1 - \nu)h - 1$ *and for* $\mathrm{Re}(\lambda) < (1 - \nu)h - 1$, *respectively.*

5.3. $\boldsymbol{H}^2$-Transform on the Space $\mathfrak{L}_{\nu,r}$

We consider the transform $\boldsymbol{H}^2$ defined in (5.1.2). It was shown in Lemma 3.1(iii) that R in (3.3.11) is an isometric isomorphism of $\mathfrak{L}_{\nu,r}$ onto $\mathfrak{L}_{1-\nu,r}$: $R(\mathfrak{L}_{\nu,r}) = \mathfrak{L}_{1-\nu,r}$. Therefore, as in the previous section, on the basis of (5.1.7) and (5.1.12) we can apply the results for $\boldsymbol{H}$ in Chapters 3 and 4 to obtain the corresponding results for the transform $\boldsymbol{H}^2$. From Theorems 3.6, 3.7 and 4.1–4.10 we obtain the $\mathfrak{L}_{\nu,2}$- and $\mathfrak{L}_{\nu,r}$-theory of the modified $\boldsymbol{H}$-transform $\boldsymbol{H}^2$. First we present the former.

Theorem 5.13. *Suppose that* **(a)** $\alpha < 1 - \nu < \beta$ *and that either of conditions* **(b)** $a^* > 0$, *or* **(c)** $a^* = 0$, $\Delta(1 - \nu) + \mathrm{Re}(\mu) \leqq 0$ *holds. Then we have the following results:*

(i) *There is a one-to-one transform* $\boldsymbol{H}^2 \in [\mathfrak{L}_{\nu,2}, \mathfrak{L}_{\nu,2}]$ *such that* (5.1.12) *holds for* $\mathrm{Re}(s) = \nu$ *and* $f \in \mathfrak{L}_{\nu,2}$. *If* $a^* = 0$, $\Delta(1 - \nu) + \mathrm{Re}(\mu) = 0$ *and* $\nu \notin \mathcal{E}_{\mathcal{H}}$, *then the transform* $\boldsymbol{H}^2$ *maps* $\mathfrak{L}_{\nu,2}$ *onto* $\mathfrak{L}_{\nu,2}$.

(ii) *If* $f \in \mathfrak{L}_{\nu,2}$ *and* $g \in \mathfrak{L}_{1-\nu,2}$, *then the relation* (5.1.17) *holds for* $\boldsymbol{H}^2$.

(iii) *Let* $\lambda \in \mathbb{C}, h > 0$ *and* $f \in \mathfrak{L}_{\nu,2}$. *When* $\mathrm{Re}(\lambda) > (1 - \nu)h - 1$, $\boldsymbol{H}^2 f$ *is given by*

$$\left(\boldsymbol{H}^2 f\right)(x)$$

$$= -hx^{(\lambda+1)/h} \frac{d}{dx} x^{-(\lambda+1)/h} \int_0^\infty H_{q+1,p+1}^{n,m+1} \left[\frac{t}{x} \middle| \begin{matrix} (-\lambda, h), (a_i, \alpha_i)_{1,p} \\ (b_j, \beta_j)_{1,q}, (-\lambda - 1, h) \end{matrix} \right] f(t)dt. \quad (5.3.1)$$

When $\mathrm{Re}(\lambda) < (1 - \nu)h - 1$,

$$\left(\boldsymbol{H}^2 f\right)(x)$$

$$= hx^{(\lambda+1)/h}\frac{d}{dx}x^{-(\lambda+1)/h}\int_0^\infty H^{m+1,n}_{p+1,q+1}\left[\frac{t}{x}\,\middle|\,\begin{matrix}(a_i,\alpha_i)_{1,p},(-\lambda,h)\\(-\lambda-1,h),(b_j,\beta_j)_{1,q}\end{matrix}\right]f(t)dt. \tag{5.3.2}$$

(iv) *The transform $\boldsymbol{H}^2$ is independent of ν in the sense that, if ν and $\widetilde{\nu}$ satisfy* (a), *and either* (b) *or* (c), *and if the transforms $\boldsymbol{H}^2$ and $\widetilde{\boldsymbol{H}^2}$ are defined in $\mathfrak{L}_{\nu,2}$ and $\mathfrak{L}_{\widetilde{\nu},2}$ respectively by* (5.1.12), *then $\boldsymbol{H}^2 f = \widetilde{\boldsymbol{H}^2} f$ for $f \in \mathfrak{L}_{\nu,2} \cap \mathfrak{L}_{\widetilde{\nu},2}$.*

(v) *If $a^* > 0$ or if $a^* = 0$, $\Delta(1-\nu) + \mathrm{Re}(\mu) < 0$, then for $f \in \mathfrak{L}_{\nu,2}$, $\boldsymbol{H}^2 f$ is given in* (5.1.2).

Proof. Statements (i), (iii), (iv) and (v) follow from Theorem 3.6 and 3.7, and we only note that the representations (5.3.1) and (5.3.2) are proved similar to (5.2.27) and (5.2.28) on the basis of the representations (3.6.2) and (3.6.3) by using (5.2.26). (ii) is proved similarly to that in Theorem 5.1.

Applying the results in Sections 4.1–4.4 by using (5.1.7) and Lemma 3.1(iii) we present the $\mathfrak{L}_{\nu,r}$-theory of the transform $\boldsymbol{H}^2$ in (5.1.2) when $a^* = 0$. From Theorems 4.1–4.4 we obtain the mapping properties and the range of $\boldsymbol{H}^2$ on $\mathfrak{L}_{\nu,r}$ in three different cases when either $\Delta = \mathrm{Re}(\mu) = 0$ or $\Delta = 0$, $\mathrm{Re}(\mu) < 0$ or $\Delta \neq 0$. Note that (5.3.3), (5.3.4), (5.3.5) and (5.3.6) follow from (5.1.7) and (4.1.1), (4.2.1), (4.2.2) and (4.3.1), (4.4.1), if we take into account the isomorphic property $R(\mathfrak{L}_{\nu,r}) = \mathfrak{L}_{1-\nu,r}$ and the relations (3.3.17), (3.3.21) and (3.3.18) in Lemma 3.2.

Theorem 5.14. *Let $a^* = \Delta = 0, \mathrm{Re}(\mu) = 0, \alpha < 1-\nu < \beta$ and let $1 < r < \infty$.*

(a) *The transform $\boldsymbol{H}^2$ defined on $\mathfrak{L}_{\nu,2}$ can be extended to $\mathfrak{L}_{\nu,r}$ as an element of $[\mathfrak{L}_{\nu,r}, \mathfrak{L}_{\nu,r}]$.*

(b) *If $1 < r \leqq 2$, then the transform $\boldsymbol{H}^2$ is one-to-one on $\mathfrak{L}_{\nu,r}$ and there holds the equality* (5.1.12) *for $f \in \mathfrak{L}_{\nu,r}$ and $\mathrm{Re}(s) = \nu$.*

(c) *If $f \in \mathfrak{L}_{\nu,r}$ and $g \in \mathfrak{L}_{1-\nu,r'}$ with $r' = r/(r-1)$, then the relation* (5.1.17) *holds.*

(d) *If $\nu \notin \mathcal{E}_{\mathcal{H}}$, then the transform $\boldsymbol{H}^2$ is one-to-one on $\mathfrak{L}_{\nu,r}$ and there holds*

$$\boldsymbol{H}^2(\mathfrak{L}_{\nu,r}) = \mathfrak{L}_{\nu,r}. \tag{5.3.3}$$

(e) *If $f \in \mathfrak{L}_{\nu,r}, \lambda \in \mathbb{C}$ and $h > 0$ then $\boldsymbol{H}^2 f$ is given in* (5.3.1) *for $\mathrm{Re}(\lambda) > (1-\nu)h - 1$, while in* (5.3.2) *for $\mathrm{Re}(\lambda) < (1-\nu)h - 1$.*

Theorem 5.15. *Let $a^* = \Delta = 0, \mathrm{Re}(\mu) < 0$ and $\alpha < 1-\nu < \beta$, and let either $m > 0$ or $n > 0$. Let $1 < r < \infty$.*

(a) *The transform $\boldsymbol{H}^2$ defined on $\mathfrak{L}_{\nu,2}$ can be extended to $\mathfrak{L}_{\nu,r}$ as an element of $[\mathfrak{L}_{\nu,r}, \mathfrak{L}_{\nu,s}]$ for all $s \geqq r$ such that $1/s > 1/r + \mathrm{Re}(\mu)$.*

(b) *If $1 < r \leqq 2$, then the transform $\boldsymbol{H}^2$ is one-to-one on $\mathfrak{L}_{\nu,r}$ and there holds the equality* (5.1.12) *for $f \in \mathfrak{L}_{\nu,r}$ and $\mathrm{Re}(s) = \nu$.*

(c) *If $f \in \mathfrak{L}_{\nu,r}$ and $g \in \mathfrak{L}_{1-\nu,s}$ with $1 < s < \infty$ and $1 \leqq 1/r + 1/s < 1 - \mathrm{Re}(\mu)$, then the relation* (5.1.17) *holds.*

(d) *Let $k > 0$. If $\nu \notin \mathcal{E}_{\mathcal{H}}$, then the transform $\boldsymbol{H}^2$ is one-to-one on $\mathfrak{L}_{\nu,r}$ and there hold*

$$\boldsymbol{H}^2(\mathfrak{L}_{\nu,r}) = I^{-\mu}_{0+;k,(1-\alpha)/k-1}(\mathfrak{L}_{\nu,r}) \tag{5.3.4}$$

for $m > 0$, *and*

$$\boldsymbol{H}^2(\mathfrak{L}_{\nu,r}) = I^{-\mu}_{-;k,(\beta-1)/k}(\mathfrak{L}_{\nu,r}) \tag{5.3.5}$$

for $n > 0$. *When* $\nu \in \mathcal{E}_{\mathcal{H}}$, $\boldsymbol{H}^2(\mathfrak{L}_{\nu,r})$ *is a subset of the right-hand sides of* (5.3.4) *and* (5.3.5) *in the respective cases.*

(e) *If* $f \in \mathfrak{L}_{\nu,r}, \lambda \in \mathbb{C}$ *and* $h > 0$, *then* $\boldsymbol{H}^2 f$ *is given in* (5.3.1) *for* $\mathrm{Re}(\lambda) > (1-\nu)h - 1$, *while in* (5.3.2) *for* $\mathrm{Re}(\lambda) < (1-\nu)h - 1$. *Furthermore* $\boldsymbol{H}^2 f$ *is given in* (5.1.2).

Theorem 5.16. *Let* $a^* = 0, \Delta \neq 0, \alpha < 1-\nu < \beta, 1 < r < \infty$ *and* $\Delta(1-\nu) + \mathrm{Re}(\mu) \leqq 1/2 - \gamma(r)$. *Assume that* $m > 0$ *if* $\Delta > 0$ *and* $n > 0$ *if* $\Delta < 0$.

(a) *The transform* $\boldsymbol{H}^2$ *defined on* $\mathfrak{L}_{\nu,2}$ *can be extended to* $\mathfrak{L}_{\nu,r}$ *as an element of* $[\mathfrak{L}_{\nu,r}, \mathfrak{L}_{\nu,s}]$ *for all* s *with* $r \leqq s < \infty$ *such that* $s' \geqq [1/2 - \Delta(1-\nu) - \mathrm{Re}(\mu)]^{-1}$ *with* $1/s + 1/s' = 1$.

(b) *If* $1 < r \leqq 2$, *then the transform* $\boldsymbol{H}^2$ *is one-to-one on* $\mathfrak{L}_{\nu,r}$ *and there holds the equality* (5.1.12) *for* $f \in \mathfrak{L}_{\nu,r}$ *and* $\mathrm{Re}(s) = \nu$.

(c) *If* $f \in \mathfrak{L}_{\nu,r}$ *and* $g \in \mathfrak{L}_{1-\nu,s}$ *with* $1 < s < \infty, 1/r + 1/s \geqq 1$ *and* $\Delta(1-\nu) + \mathrm{Re}(\mu) \leqq 1/2 - \max[\gamma(r), \gamma(s)]$, *then the relation* (5.1.17) *holds.*

(d) *If* $\nu \notin \mathcal{E}_{\mathcal{H}}$, *then the transform* $\boldsymbol{H}^2$ *is one-to-one on* $\mathfrak{L}_{\nu,r}$. *If we set* $\eta = -\Delta\alpha - \mu - 1$ *for* $\Delta > 0$ *and* $\eta = -\Delta\beta - \mu - 1$ *for* $\Delta < 0$, *then* $\mathrm{Re}(\eta) > -1$ *and there holds*

$$\boldsymbol{H}^2(\mathfrak{L}_{\nu,r}) = \left(M_{-\mu/\Delta-1/2}\mathbb{H}_{-\Delta,\eta}\right)\left(\mathfrak{L}_{3/2-\nu+\mathrm{Re}(\mu)/\Delta,r}\right). \tag{5.3.6}$$

When $\nu \in \mathcal{E}_{\mathcal{H}}$, $\boldsymbol{H}^2(\mathfrak{L}_{\nu,r})$ *is a subset of the right-hand side of* (5.3.6).

(e) *If* $f \in \mathfrak{L}_{\nu,r}, \lambda \in \mathbb{C}, h > 0$ *and* $\Delta(1-\nu) + \mathrm{Re}(\mu) \leqq 1/2 - \gamma(r)$, *then* $\boldsymbol{H}^2 f$ *is given in* (5.3.1) *for* $\mathrm{Re}(\lambda) > (1-\nu)h - 1$, *while in* (5.3.2) *for* $\mathrm{Re}(\lambda) < (1-\nu)h - 1$. *If furthermore* $\Delta(1-\nu) + \mathrm{Re}(\mu) < 0$, $\boldsymbol{H}^2 f$ *is given in* (5.1.2).

From Theorem 4.5 we obtain the $\mathfrak{L}_{\nu,r}$-theory of the transform $\boldsymbol{H}^2$ in (5.1.2) with $a^* > 0$ in $\mathfrak{L}_{\nu,r}$-space for any $\nu \in \mathbb{R}$ and $1 \leqq r \leqq \infty$.

Theorem 5.17. *Let* $a^* > 0$, $\alpha < 1-\nu < \beta$ *and* $1 \leqq r \leqq s \leqq \infty$.

(a) *The transform* $\boldsymbol{H}^2$ *defined on* $\mathfrak{L}_{\nu,2}$ *can be extended to* $\mathfrak{L}_{\nu,r}$ *as an element of* $[\mathfrak{L}_{\nu,r}, \mathfrak{L}_{\nu,s}]$. *If* $1 \leqq r \leqq 2$, *then* $\boldsymbol{H}^2$ *is a one-to-one transform from* $\mathfrak{L}_{\nu,r}$ *onto* $\mathfrak{L}_{\nu,s}$.

(b) *If* $f \in \mathfrak{L}_{\nu,r}$ *and* $g \in \mathfrak{L}_{1-\nu,s'}$ *with* $1/s + 1/s' = 1$, *then the relation* (5.1.17) *holds.*

Due to Theorems 4.6–4.10, we characterize the boundedness and the range of $\boldsymbol{H}^2$ on $\mathfrak{L}_{\nu,r}$ which are different in five cases: $a_1^* > 0$, $a_2^* > 0$; $a_1^* > 0$, $a_2^* = 0$; $a_1^* = 0$, $a_2^* > 0$; $a^* > 0$, $a_1^* > 0$, $a_2^* < 0$; and $a^* > 0$, $a_1^* < 0$, $a_2^* > 0$. For these five cases, such results on the transform $\boldsymbol{H}^2$ can be obtained by virtue of (5.1.7) and by invoking Theorems 4.6–4.10 and Lemma 3.2. Note that (5.3.7)–(5.3.8), (5.3.9)–(5.3.10), (5.3.11)–(5.3.12) and (5.3.17), (5.3.22) follow from (5.1.7) and (4.6.1)–(4.6.2), (4.7.1)–(4.7.2), (4.7.15)–(4.7.16) and (4.8.5), (4.8.17) in accordance with the isomorphic property $R(\mathfrak{L}_{\nu,r}) = \mathfrak{L}_{1-\nu,r}$ and (3.3.18), (3.3.17) and (3.3.21).

Theorem 5.18. *Let* $a_1^* > 0$, $a_2^* > 0$, $m > 0$, $n > 0$, $\alpha < 1-\nu < \beta$ *and* $\omega = \mu + a_1^*\alpha - a_2^*\beta + 1$ *and let* $1 < r < \infty$.

(a) *If* $\nu \notin \mathcal{E}_{\mathcal{H}}$, *or if* $1 \leqq r \leqq 2$, *then the transform* $\boldsymbol{H}^2$ *is one-to-one on* $\mathfrak{L}_{\nu,r}$.
(b) *If* $\mathrm{Re}(\omega) \geqq 0$ *and* $\nu \notin \mathcal{E}_{\mathcal{H}}$, *then*

$$\boldsymbol{H}^2(\mathfrak{L}_{\nu,r}) = \left(\mathbb{L}_{-a_1^*,1-\alpha}\mathbb{L}_{-a_2^*,\beta+\omega/a_2^*}\right)(\mathfrak{L}_{\nu,r}). \tag{5.3.7}$$

When $\nu \in \mathcal{E}_{\mathcal{H}}$, $\boldsymbol{H}^2(\mathfrak{L}_{\nu,r})$ *is a subset of the right-hand side of* (5.3.7).
(c) *If* $\mathrm{Re}(\omega) < 0$ *and* $\nu \notin \mathcal{E}_{\mathcal{H}}$, *then*

$$\boldsymbol{H}^2(\mathfrak{L}_{\nu,r}) = \left(I^{-\omega}_{0+;1/a_1^*,(1-\alpha)a_1^*-1}\mathbb{L}_{-a_1^*,1-\alpha}\mathbb{L}_{-a_2^*,\beta}\right)(\mathfrak{L}_{\nu,r}). \tag{5.3.8}$$

When $\nu \in \mathcal{E}_{\mathcal{H}}$, $\boldsymbol{H}^2(\mathfrak{L}_{\nu,r})$ *is a subset of the right-hand side of* (5.3.8).

Theorem 5.19. *Let* $a_1^* > 0, a_2^* = 0, m > 0$, $\alpha < 1-\nu < \beta$ *and* $\omega = \mu + a_1^*\alpha + 1/2$ *and let* $1 < r < \infty$.
(a) *If* $\nu \notin \mathcal{E}_{\mathcal{H}}$, *or if* $1 < r \leqq 2$, *then the transform* $\boldsymbol{H}^2$ *is one-to-one on* $\mathfrak{L}_{\nu,r}$.
(b) *If* $\mathrm{Re}(\omega) \geqq 0$ *and* $\nu \notin \mathcal{E}_{\mathcal{H}}$, *then*

$$\boldsymbol{H}^2(\mathfrak{L}_{\nu,r}) = \mathbb{L}_{-a_1^*,1-\alpha+\omega/a_1^*}(\mathfrak{L}_{1-\nu,r}). \tag{5.3.9}$$

When $\nu \in \mathcal{E}_{\mathcal{H}}$, $\boldsymbol{H}^2(\mathfrak{L}_{\nu,r})$ *is a subset of the right-hand side of* (5.3.9).
(c) *If* $\mathrm{Re}(\omega) < 0$ *and* $\nu \notin \mathcal{E}_{\mathcal{H}}$, *then*

$$\boldsymbol{H}^2(\mathfrak{L}_{\nu,r}) = \left(I^{-\omega}_{0+;1/a_1^*,(1-\alpha)a_1^*-1}\mathbb{L}_{-a_1^*,1-\alpha}\right)(\mathfrak{L}_{1-\nu,r}). \tag{5.3.10}$$

When $\nu \in \mathcal{E}_{\mathcal{H}}$, $\boldsymbol{H}^2(\mathfrak{L}_{\nu,r})$ *is a subset of the right-hand side of* (5.3.10).

Theorem 5.20. *Let* $a_1^* = 0, a_2^* > 0, n > 0$, $\alpha < 1-\nu < \beta$ *and* $\omega = \mu - a_2^*\beta + 1/2$ *and let* $1 < r < \infty$.
(a) *If* $\nu \notin \mathcal{E}_{\mathcal{H}}$, *or if* $1 < r \leqq 2$, *then the transform* $\boldsymbol{H}^2$ *is one-to-one on* $\mathfrak{L}_{\nu,r}$.
(b) *If* $\mathrm{Re}(\omega) \geqq 0$ *and* $\nu \notin \mathcal{E}_{\mathcal{H}}$, *then*

$$\boldsymbol{H}^2(\mathfrak{L}_{\nu,r}) = \mathbb{L}_{a_2^*,1-\beta-\omega/a_2^*}(\mathfrak{L}_{1-\nu,r}). \tag{5.3.11}$$

When $\nu \in \mathcal{E}_{\mathcal{H}}$, $\boldsymbol{H}^2(\mathfrak{L}_{\nu,r})$ *is a subset of the right-hand side of* (5.3.11).
(c) *If* $\mathrm{Re}(\omega) < 0$ *and* $\nu \notin \mathcal{E}_{\mathcal{H}}$, *then*

$$\boldsymbol{H}^2(\mathfrak{L}_{\nu,r}) = \left(I^{-\omega}_{-;1/a_2^*,a_2^*(\beta-1)}\mathbb{L}_{a_2^*,1-\beta}\right)(\mathfrak{L}_{1-\nu,r}). \tag{5.3.12}$$

When $\nu \in \mathcal{E}_{\mathcal{H}}$, $\boldsymbol{H}^2(\mathfrak{L}_{\nu,r})$ *is a subset of the right-hand side of* (5.3.12).

Theorem 5.21. *Let* $a^* > 0$, $a_1^* > 0$, $a_2^* < 0$, $\alpha < 1-\nu < \beta$ *and let* $1 < r < \infty$.
(a) *If* $\nu \notin \mathcal{E}_{\mathcal{H}}$, *or if* $1 < r \leqq 2$, *then the transform* $\boldsymbol{H}^2$ *is one-to-one on* $\mathfrak{L}_{\nu,r}$.
(b) *Let* $\omega, \eta, \zeta \in \mathbb{C}$ *be chosen as*

$$\omega = a^*\eta - \mu - \frac{1}{2}; \tag{5.3.13}$$

$$a^*\mathrm{Re}(\eta) \geqq \gamma(r) + 2a_2^*(\nu-1) + \mathrm{Re}(\mu); \tag{5.3.14}$$

$$\mathrm{Re}(\eta) > \nu - 1; \tag{5.3.15}$$

$$\mathrm{Re}(\zeta) < 1 - \nu. \tag{5.3.16}$$

If $\nu \notin \mathcal{E}_{\mathcal{H}}$, *then*

$$\boldsymbol{H}^2(\mathfrak{L}_{\nu,r}) = \left(M_{-1/2-\omega/(2a_2^*)}\mathbb{H}_{2a_2^*,2a_2^*\zeta+\omega-1}\mathbb{L}_{a^*,1/2-\eta+\omega/(2a_2^*)}\right)\left(\mathfrak{L}_{\nu-1/2-\mathrm{Re}(\omega)/(2a_2^*),r}\right). \quad (5.3.17)$$

When $\nu \in \mathcal{E}_{\mathcal{H}}$, $\boldsymbol{H}^2(\mathfrak{L}_{\nu,r})$ *is a subset of the right-hand side of* (5.3.17).

Theorem 5.22. *Let* $a^* > 0$, $a_1^* < 0$, $a_2^* > 0$, $\alpha < 1-\nu < \beta$ *and let* $1 < r < \infty$.

(a) *If* $\nu \notin \mathcal{E}_{\mathcal{H}}$, *or if* $1 < r \leqq 2$, *then the transform* $\boldsymbol{H}^2$ *is one-to-one on* $\mathfrak{L}_{\nu,r}$.

(b) *Let* $\omega, \eta, \zeta \in \mathbb{C}$ *be chosen as*

$$\omega = a^*\eta - \Delta - \mu - \frac{1}{2}; \quad (5.3.18)$$

$$a^*\mathrm{Re}(\eta) \geqq \gamma(r) - 2a_1^*\nu + \Delta + \mathrm{Re}(\mu); \quad (5.3.19)$$

$$\mathrm{Re}(\eta) > -\nu; \quad (5.3.20)$$

$$\mathrm{Re}(\zeta) < \nu. \quad (5.3.21)$$

If $\nu \notin \mathcal{E}_{\mathcal{H}}$, *then*

$$\boldsymbol{H}^2(\mathfrak{L}_{\nu,r}) = \left(M_{1/2+\omega/(2a_1^*)}\mathbb{H}_{-2a_1^*,2a_1^*\zeta+\omega-1}\mathbb{L}_{-a^*,1/2+\eta-\omega/(2a_1^*)}\right)\left(\mathfrak{L}_{\nu+1/2+\mathrm{Re}(\omega)/(2a_1^*),r}\right). \quad (5.3.22)$$

When $\nu \in \mathcal{E}_{\mathcal{H}}$, $\boldsymbol{H}^2(\mathfrak{L}_{\nu,r})$ *is a subset of the right-hand side of* (5.3.22).

To obtain the inversion formulas for the transform $\boldsymbol{H}^2$ when $a^* = 0$, we note that (5.1.2) is equivalent to

$$\boldsymbol{H}f = R\boldsymbol{H}^2 f \qquad (f \in \mathfrak{L}_{\nu,r}) \quad (5.3.23)$$

by virtue of the isometric property $R(\mathfrak{L}_{\nu,r}) = \mathfrak{L}_{1-\nu,r}$ and the fact that the inverse of the operator R^{-1} of R coincides with R. Then, if $a^* = 0$, from the results in Sections 4.9 and 4.10 we obtain the following inversion by replacing $\boldsymbol{H}f$ by $R\boldsymbol{H}^2 f$ in (4.9.1) and (4.9.2), and we have

$$f(x) = hx^{1-(\lambda+1)/h}\frac{d}{dx}x^{(\lambda+1)/h} \cdot \int_0^\infty H_{p+1,q+1}^{q-m,p-n+1}\left[\frac{x}{t}\left|\begin{matrix}(-\lambda,h),(1-a_i-\alpha_i,\alpha_i)_{n+1,p},(1-a_i-\alpha_i,\alpha_i)_{1,n}\\(1-b_j-\beta_j,\beta_j)_{m+1,q},(1-b_j-\beta_j,\beta_j)_{1,m},(-\lambda-1,h)\end{matrix}\right.\right] \cdot \frac{1}{t}\left(\boldsymbol{H}^2 f\right)(t)dt \quad (5.3.24)$$

or

$$f(x) = -hx^{1-(\lambda+1)/h}\frac{d}{dx}x^{(\lambda+1)/h} \cdot \int_0^\infty H_{p+1,q+1}^{q-m+1,p-n}\left[\frac{x}{t}\left|\begin{matrix}(1-a_i-\alpha_i,\alpha_i)_{n+1,p},(1-a_i-\alpha_i,\alpha_i)_{1,n},(-\lambda,h)\\(-\lambda-1,h),(1-b_j-\beta_j,\beta_j)_{m+1,q},(1-b_j-\beta_j,\beta_j)_{1,m}\end{matrix}\right.\right] \cdot \frac{1}{t}\left(\boldsymbol{H}^2 f\right)(t)dt. \quad (5.3.25)$$

The conditions for the validity of (5.5.27) and (5.5.28) follow from Theorems 4.11–4.14. Thus we obtain the following inversion theorems for the transform $\boldsymbol{H}^2$ for $a^* = 0$ and either $\Delta = 0$ or $\Delta \neq 0$, where α_0 and β_0 are given in (4.9.6) and (4.9.7).

Theorem 5.23. *Let $a^* = 0$, $\alpha < 1 - \nu < \beta$ and $\alpha_0 < \nu < \beta_0$, and let $\lambda \in \mathbb{C}, h > 0$.*

(a) *If $\Delta(1-\nu) + \mathrm{Re}(\mu) = 0$ and $f \in \mathfrak{L}_{\nu,2}$, then the inversion formula (5.3.24) holds for $\mathrm{Re}(\lambda) > \nu h - 1$ and (5.3.25) for $\mathrm{Re}(\lambda) < \nu h - 1$.*

(b) *If $\Delta = \mathrm{Re}(\mu) = 0$ and $f \in \mathfrak{L}_{\nu,r}$ $(1 < r < \infty)$, then the inversion formula (5.3.24) holds for $\mathrm{Re}(\lambda) > \nu h - 1$ and (5.3.25) for $\mathrm{Re}(\lambda) < \nu h - 1$.*

Theorem 5.24. *Let $a^* = 0$, $1 < r < \infty$ and $\Delta(1-\nu) + \mathrm{Re}(\mu) \leqq 1/2 - \gamma(r)$, and let $\lambda \in \mathbb{C}, h > 0$.*

(a) *If $\Delta > 0, m > 0, \alpha < 1 - \nu < \beta$, $\alpha_0 < \nu < \min[\beta_0, \{\mathrm{Re}(\mu + 1/2)/\Delta\} + 1]$ and if $f \in \mathfrak{L}_{\nu,r}$, then the inversion formulas (5.3.24) and (5.3.25) hold for $\mathrm{Re}(\lambda) > \nu h - 1$ and for $\mathrm{Re}(\lambda) < \nu h - 1$, respectively.*

(b) *If $\Delta < 0, n > 0, \alpha < 1 - \nu < \beta$, $\max[\alpha_0, \{\mathrm{Re}(\mu + 1/2)/\Delta\} + 1] < \nu < \beta_0$ and if $f \in \mathfrak{L}_{\nu,r}$, then the inversion formulas (5.3.24) and (5.3.25) hold for $\mathrm{Re}(\lambda) > \nu h - 1$ and for $\mathrm{Re}(\lambda) < \nu h - 1$, respectively.*

5.4. $\boldsymbol{H}_{\sigma,\kappa}$-Transform on the Space $\mathfrak{L}_{\nu,r}$

We proceed to the transform $\boldsymbol{H}_{\sigma,\kappa}$ defined in (5.1.3). Due to Lemma 3.1(i), M_κ is an isometric isomorphism of $\mathfrak{L}_{\nu,r}$ onto $\mathfrak{L}_{\nu-\mathrm{Re}(\kappa),r}$:

$$M_\kappa(\mathfrak{L}_{\nu,r}) = \mathfrak{L}_{\nu-\mathrm{Re}(\kappa),r}. \tag{5.4.1}$$

Therefore (5.1.8) and (5.1.13) show that we can apply the results which were proved in Chapters 3 and 4 for the transform $\boldsymbol{H}$, by replacing ν by $\nu - \mathrm{Re}(\kappa)$ to obtain the corresponding results for the transform $\boldsymbol{H}_{\sigma,\kappa}$.

From Theorems 3.6, 3.7 and 4.1 - 4.10 we obtain the $\mathfrak{L}_{\nu,2}$- and $\mathfrak{L}_{\nu,r}$-theory of the modified tranform $\boldsymbol{H}_{\sigma,\kappa}$. First we give the former.

Theorem 5.25. *Suppose that* **(a)** $\alpha < 1 - \nu + \mathrm{Re}(\kappa) < \beta$ *and that either of conditions* **(b)** $a^* > 0$, *or* **(c)** $a^* = 0$, $\Delta[1 - \nu + \mathrm{Re}(\kappa)] + \mathrm{Re}(\mu) \leqq 0$ *holds. Then we have the following results:*

(i) *There is a one-to-one transform $\boldsymbol{H}_{\sigma,\kappa} \in [\mathfrak{L}_{\nu,2}, \mathfrak{L}_{1-\nu+\mathrm{Re}(\kappa-\sigma),2}]$ such that (5.1.13) holds for $\mathrm{Re}(s) = 1 - \nu + \mathrm{Re}(\kappa - \sigma)$ and $f \in \mathfrak{L}_{\nu,2}$. If $a^* = 0$, $\Delta[1 - \nu + \mathrm{Re}(\kappa)] + \mathrm{Re}(\mu) = 0$ and $\nu - \mathrm{Re}(\kappa) \notin \mathcal{E}_{\mathcal{H}}$, then the transform $\boldsymbol{H}_{\sigma,\kappa}$ maps $\mathfrak{L}_{\nu,2}$ onto $\mathfrak{L}_{1-\nu+\mathrm{Re}(\kappa-\sigma),2}$.*

(ii) *If $f \in \mathfrak{L}_{\nu,2}$ and $g \in \mathfrak{L}_{\nu+\mathrm{Re}(\kappa-\sigma),2}$, then the relation (5.1.18) holds for $\boldsymbol{H}_{\sigma,\kappa}$.*

(iii) *Let $\lambda \in \mathbb{C}, h > 0$ and $f \in \mathfrak{L}_{\nu,2}$. When $\mathrm{Re}(\lambda) > [1 - \nu + \mathrm{Re}(\kappa)]h - 1$, $\boldsymbol{H}_{\sigma,\kappa} f$ is given by*

$$\left(\boldsymbol{H}_{\sigma,\kappa} f\right)(x) = h x^{\sigma+1-(\lambda+1)/h} \frac{d}{dx} x^{(\lambda+1)h}$$

$$\cdot \int_0^\infty H_{p+1,q+1}^{m,n+1}\left[xt \left| \begin{matrix} (-\lambda,h),(a_i,\alpha_i)_{1,p} \\ (b_j,\beta_j)_{1,q},(-\lambda-1,h) \end{matrix} \right. \right] t^\kappa f(t)dt. \tag{5.4.2}$$

When $\mathrm{Re}(\lambda) < [1-\nu+\mathrm{Re}(\kappa)]h - 1$,

$$\left(\boldsymbol{H}_{\sigma,\kappa} f\right)(x) = -hx^{\sigma+1-(\lambda+1)/h}\frac{d}{dx}x^{(\lambda+1)/h}$$

$$\cdot \int_0^\infty H_{p+1,q+1}^{m+1,n}\left[xt \left| \begin{matrix} (a_i,\alpha_i)_{1,p},(-\lambda,h) \\ (-\lambda-1,h),(b_j,\beta_j)_{1,q} \end{matrix} \right. \right] t^\kappa f(t)dt. \tag{5.4.3}$$

(iv) *The transform $\boldsymbol{H}_{\sigma,\kappa}$ is independent of ν in the sense that, if ν and $\widetilde{\nu}$ satisfy* (a), *and* (b) *or* (c), *and if the transforms $\boldsymbol{H}_{\sigma,\kappa}$ and $\widetilde{\boldsymbol{H}}_{\sigma,\kappa}$ are defined in $\mathfrak{L}_{\nu,2}$ and $\mathfrak{L}_{\widetilde{\nu},2}$ respectively by* (5.1.13), *then $\boldsymbol{H}_{\sigma,\kappa} f = \widetilde{\boldsymbol{H}}_{\sigma,\kappa} f$ for $f \in \mathfrak{L}_{\nu,2} \cap \mathfrak{L}_{\widetilde{\nu},2}$.*

(v) *If $a^* > 0$ or if $a^* = 0$, $\Delta[1-\nu+\mathrm{Re}(\kappa)] + \mathrm{Re}(\mu) < 0$, then for $f \in \mathfrak{L}_{\nu,2}$, $\boldsymbol{H}_{\sigma,\kappa} f$ is given in* (5.1.3).

Proof. Statements (i), (iii), (iv) and (v) follow from Theorems 3.6 and 3.7 on the basis of the relations (5.1.8) and (5.1.13) and the isometric property (5.4.1). (ii) is proved similarly to that in Theorem 5.1 by using the Schwartz inequality (3.5.6).

Now the results in Sections 4.1–4.4, (1.5.7) and Lemma 3.1(i) yield the $\mathfrak{L}_{\nu,r}$-theory of the transform $\boldsymbol{H}_{\sigma,\kappa}$ in (5.1.3) when $a^* = 0$. From Theorems 4.1–4.4, we obtain the mapping properties and the range of $\boldsymbol{H}_{\sigma,\kappa}$ on $\mathfrak{L}_{\nu,r}$ in three different cases when either $\Delta = \mathrm{Re}(\mu) = 0$ or $\Delta = 0$, $\mathrm{Re}(\mu) < 0$ or $\Delta \neq 0$.

Note that the relations (5.4.5), (5.4.6) and (5.4.7) below follow from (4.2.1), (4.2.2) and (4.3.1), (4.4.1) taking into account (5.1.8), (5.4.1) and Lemma 3.2 for the Erdélyi–Kober type fractional integration operators $I_{0+;\sigma,\eta}^{\alpha}$ and $I_{-;\sigma,\eta}^{\alpha}$.

Theorem 5.26. *Let $a^* = \Delta = 0, \mathrm{Re}(\mu) = 0, \alpha < 1-\nu+\mathrm{Re}(\kappa) < \beta$ and let $1 < r < \infty$.*

(a) *The transform $\boldsymbol{H}_{\sigma,\kappa}$ defined on $\mathfrak{L}_{\nu,2}$ can be extended to $\mathfrak{L}_{\nu,r}$ as an element of $[\mathfrak{L}_{\nu,r}, \mathfrak{L}_{1-\nu+\mathrm{Re}(\kappa-\sigma),r}]$.*

(b) *If $1 < r \leqq 2$, then the transform $\boldsymbol{H}_{\sigma,\kappa}$ is one-to-one on $\mathfrak{L}_{\nu,r}$ and there holds the equality* (5.1.13) *for $f \in \mathfrak{L}_{\nu,r}$ and $\mathrm{Re}(s) = 1-\nu+\mathrm{Re}(\kappa-\sigma)$.*

(c) *If $f \in \mathfrak{L}_{\nu,r}$ and $g \in \mathfrak{L}_{\nu+\mathrm{Re}(\kappa-\sigma),r'}$ with $r' = r/(r-1)$, then the relation* (5.1.18) *holds.*

(d) *If $\nu - \mathrm{Re}(\kappa) \notin \mathcal{E}_{\mathcal{H}}$, then the transform $\boldsymbol{H}_{\sigma,\kappa}$ is one-to-one on $\mathfrak{L}_{\nu,r}$ and there holds*

$$\boldsymbol{H}_{\sigma,\kappa}(\mathfrak{L}_{\nu,r}) = \mathfrak{L}_{1-\nu+\mathrm{Re}(\kappa-\sigma),r}. \tag{5.4.4}$$

(e) *If $f \in \mathfrak{L}_{\nu,r}, \lambda \in \mathbb{C}$ and $h > 0$, then $\boldsymbol{H}_{\sigma,\kappa} f$ is given in* (5.4.2) *for $\mathrm{Re}(\lambda) > [1-\nu+\mathrm{Re}(\kappa)]h - 1$, while in* (5.4.3) *for $\mathrm{Re}(\lambda) < [1-\nu+\mathrm{Re}(\kappa)]h - 1$.*

Theorem 5.27. *Let $a^* = \Delta = 0, \mathrm{Re}(\mu) < 0$ and $\alpha < 1-\nu+\mathrm{Re}(\kappa) < \beta$, and let either $m > 0$ or $n > 0$. Let $1 < r < \infty$.*

(a) *The transform $\boldsymbol{H}_{\sigma,\kappa}$ defined on $\mathfrak{L}_{\nu,2}$ can be extended to $\mathfrak{L}_{\nu,r}$ as an element of $[\mathfrak{L}_{\nu,r}, \mathfrak{L}_{1-\nu+\mathrm{Re}(\kappa-\sigma),s}]$ for all $s \geqq r$ such that $1/s > 1/r + \mathrm{Re}(\mu)$.*

(b) *If $1 < r \leqq 2$, then the transform $\boldsymbol{H}_{\sigma,\kappa}$ is one-to-one on $\mathfrak{L}_{\nu,r}$ and there holds the equality* (5.1.13) *for $f \in \mathfrak{L}_{\nu,r}$ and $\mathrm{Re}(s) = 1-\nu+\mathrm{Re}(\kappa-\sigma)$.*

(c) *If $f \in \mathfrak{L}_{\nu,r}$ and $g \in \mathfrak{L}_{\nu+\mathrm{Re}(\kappa-\sigma),s}$ with $1 < s < \infty$ and $1 \leqq 1/r + 1/s < 1 - \mathrm{Re}(\mu)$, then the relation (5.1.18) holds.*

(d) *Let $k > 0$. If $\nu - \mathrm{Re}(\kappa) \notin \mathcal{E}_{\mathcal{H}}$, then the transform $\boldsymbol{H}_{\sigma,\kappa}$ is one-to-one on $\mathfrak{L}_{\nu,r}$ and there hold*

$$\boldsymbol{H}_{\sigma,\kappa}(\mathfrak{L}_{\nu,r}) = I^{-\mu}_{-;k,(\sigma-\alpha)/k}\left(\mathfrak{L}_{1-\nu+\mathrm{Re}(\kappa-\sigma),r}\right) \tag{5.4.5}$$

for $m > 0$, and

$$\boldsymbol{H}_{\sigma,\kappa}(\mathfrak{L}_{\nu,r}) = I^{-\mu}_{0+;k,(\beta-\sigma)/k-1}\left(\mathfrak{L}_{1-\nu+\mathrm{Re}(\kappa-\sigma),r}\right) \tag{5.4.6}$$

for $n > 0$. When $\nu - \mathrm{Re}(\kappa) \in \mathcal{E}_{\mathcal{H}}$, $\boldsymbol{H}_{\sigma,\kappa}(\mathfrak{L}_{\nu,r})$ is a subset of the right-hand sides of (5.4.5) and (5.4.6) in the respective cases.

(e) *If $f \in \mathfrak{L}_{\nu,r}, \lambda \in \mathbb{C}$ and $h > 0$, then $\boldsymbol{H}_{\sigma,\kappa}f$ is given in (5.4.2) for $\mathrm{Re}(\lambda) > [1 - \nu + \mathrm{Re}(\kappa)]h - 1$, while in (5.4.3) for $\mathrm{Re}(\lambda) < [1 - \nu + \mathrm{Re}(\kappa)]h - 1$. Furthermore $\boldsymbol{H}_{\sigma,\kappa}f$ is given in (5.1.3).*

Theorem 5.28. *Let $a^* = 0, \Delta \neq 0, \alpha < 1 - \nu + \mathrm{Re}(\kappa) < \beta$, $1 < r < \infty$ and $\Delta[1 - \nu + \mathrm{Re}(\kappa)] + \mathrm{Re}(\mu) \leqq 1/2 - \gamma(r)$. Assume that $m > 0$ if $\Delta > 0$ and $n > 0$ if $\Delta < 0$.*

(a) *The transform $\boldsymbol{H}_{\sigma,\kappa}$ defined on $\mathfrak{L}_{\nu,2}$ can be extended to $\mathfrak{L}_{\nu,r}$ as an element of $[\mathfrak{L}_{\nu,r}, \mathfrak{L}_{1-\nu+\mathrm{Re}(\kappa-\sigma),s}]$ for all s with $r \leqq s < \infty$ such that $s' \geqq [1/2 - \Delta\{1-\nu+\mathrm{Re}(\kappa)\} - \mathrm{Re}(\mu)]^{-1}$ with $1/s + 1/s' = 1$.*

(b) *If $1 < r \leqq 2$, then the transform $\boldsymbol{H}_{\sigma,\kappa}$ is one-to-one on $\mathfrak{L}_{\nu,r}$ and there holds the equality (5.1.13) for $f \in \mathfrak{L}_{\nu,r}$ and $\mathrm{Re}(s) = 1 - \nu + \mathrm{Re}(\kappa - \sigma)$.*

(c) *If $f \in \mathfrak{L}_{\nu,r}$ and $g \in \mathfrak{L}_{\nu+\mathrm{Re}(\kappa-\sigma),s}$ with $1 < s < \infty, 1/r + 1/s \geqq 1$ and $\Delta[1-\nu+\mathrm{Re}(\kappa)] + \mathrm{Re}(\mu) \leqq 1/2 - \max[\gamma(r), \gamma(s)]$, then the relation (5.1.18) holds.*

(d) *If $\nu - \mathrm{Re}(\kappa) \notin \mathcal{E}_{\mathcal{H}}$, then the transform $\boldsymbol{H}_{\sigma,\kappa}$ is one-to-one on $\mathfrak{L}_{\nu,r}$. If we set $\eta = -\Delta\alpha - \mu - 1$ for $\Delta > 0$ and $\eta = -\Delta\beta - \mu - 1$ for $\Delta < 0$, then $\mathrm{Re}(\eta) > -1$ and there holds*

$$\boldsymbol{H}_{\sigma,\kappa}(\mathfrak{L}_{\nu,r}) = \left(M_{\sigma+\mu/\Delta+1/2}\mathbb{H}_{\Delta,\eta}\right)\left(\mathfrak{L}_{\nu-1/2-\mathrm{Re}(\mu)/\Delta-\mathrm{Re}(\kappa),r}\right). \tag{5.4.7}$$

When $\nu - \mathrm{Re}(\kappa) \in \mathcal{E}_{\mathcal{H}}$, $\boldsymbol{H}_{\sigma,\kappa}(\mathfrak{L}_{\nu,r})$ is a subset of the right-hand side of (5.4.7).

(e) *If $f \in \mathfrak{L}_{\nu,r}$, $\lambda \in \mathbb{C}, h > 0$ and $\Delta[1 - \nu + \mathrm{Re}(\kappa)] + \mathrm{Re}(\mu) \leqq 1/2 - \gamma(r)$, then $\boldsymbol{H}_{\sigma,\kappa}f$ is given in (5.4.2) for $\mathrm{Re}(\lambda) > [1-\nu+\mathrm{Re}(\kappa)]h-1$, while in (5.4.3) for $\mathrm{Re}(\lambda) < [1-\nu+\mathrm{Re}(\kappa)]h-1$. If $\Delta[1 - \nu + \mathrm{Re}(\kappa)] + \mathrm{Re}(\mu) < 0$, $\boldsymbol{H}_{\sigma,\kappa}f$ is given in (5.1.3).*

According to (5.1.8) and (5.4.1), Theorem 4.5 deduces the $\mathfrak{L}_{\nu,r}$-theory of the transform $\boldsymbol{H}_{\sigma,\kappa}$ in (5.1.3) with $a^* > 0$ in $\mathfrak{L}_{\nu,r}$-spaces for any $\nu \in \mathbb{R}$ and $1 \leqq r \leqq \infty$.

Theorem 5.29. *Let $a^* > 0$, $\alpha < 1 - \nu + \mathrm{Re}(\kappa) < \beta$ and $1 \leqq r \leqq s \leqq \infty$.*

(a) *The transform $\boldsymbol{H}_{\sigma,\kappa}$ defined on $\mathfrak{L}_{\nu,2}$ can be extended to $\mathfrak{L}_{\nu,r}$ as an element of $[\mathfrak{L}_{\nu,r}, \mathfrak{L}_{1-\nu+\mathrm{Re}(\kappa-\sigma),s}]$. If $1 \leqq r \leqq 2$, then $\boldsymbol{H}_{\sigma,\kappa}$ is a one-to-one transform from $\mathfrak{L}_{\nu,r}$ onto $\mathfrak{L}_{1-\nu+\mathrm{Re}(\kappa-\sigma),s}$.*

(b) *If $f \in \mathfrak{L}_{\nu,r}$ and $g \in \mathfrak{L}_{\nu+\mathrm{Re}(\kappa-\sigma),s'}$ with $1/s + 1/s' = 1$, then the relation (5.1.18) holds.*

Due to (5.1.8) and (5.4.1), Theorems 4.6–4.10 and Lemma 3.2, we may characterize the boundedness and the range of $\boldsymbol{H}_{\sigma,\kappa}$ on $\mathfrak{L}_{\nu,r}$ which will be different in various combinations of signs of a^*, a_1^* and a_2^*.

Theorem 5.30. *Let $a_1^* > 0, a_2^* > 0, m > 0, n > 0, \alpha < 1 - \nu + \mathrm{Re}(\kappa) < \beta$ and $\omega = \mu + a_1^*\alpha - a_2^*\beta + 1$ and let $1 < r < \infty$.*

(a) *If $\nu - \mathrm{Re}(\kappa) \notin \mathcal{E}_{\mathcal{H}}$, or if $1 \leqq r \leqq 2$, then the transform $\boldsymbol{H}_{\sigma,\kappa}$ is one-to-one on $\mathfrak{L}_{\nu,r}$.*

(b) *If $\mathrm{Re}(\omega) \geqq 0$ and $\nu - \mathrm{Re}(\kappa) \notin \mathcal{E}_{\mathcal{H}}$, then*

$$\boldsymbol{H}_{\sigma,\kappa}(\mathfrak{L}_{\nu,r}) = \left(\mathbb{L}_{a_1^*,\alpha-\sigma}\mathbb{L}_{a_2^*,1-\beta+\sigma-\omega/a_2^*}\right)\left(\mathfrak{L}_{1-\nu+\mathrm{Re}(\kappa-\sigma),r}\right). \tag{5.4.8}$$

When $\nu - \mathrm{Re}(\kappa) \in \mathcal{E}_{\mathcal{H}}$, $\boldsymbol{H}_{\sigma,\kappa}(\mathfrak{L}_{\nu,r})$ is a subset of the right-hand side of (5.4.8).

(c) *If $\mathrm{Re}(\omega) < 0$ and $\nu - \mathrm{Re}(\kappa) \notin \mathcal{E}_{\mathcal{H}}$, then*

$$\boldsymbol{H}_{\sigma,\kappa}(\mathfrak{L}_{\nu,r}) = \left(I^{-\omega}_{-;1/a_1^*,a_1^*(\sigma-\alpha)}\mathbb{L}_{a_1^*,\alpha-\sigma}\mathbb{L}_{a_2^*,1-\beta+\sigma}\right)\left(\mathfrak{L}_{1-\nu+\mathrm{Re}(\kappa-\sigma),r}\right). \tag{5.4.9}$$

When $\nu - \mathrm{Re}(\kappa) \in \mathcal{E}_{\mathcal{H}}$, $\boldsymbol{H}_{\sigma,\kappa}(\mathfrak{L}_{\nu,r})$ is a subset of the right-hand side of (5.4.9).

Theorem 5.31. *Let $a_1^* > 0, a_2^* = 0, m > 0, \alpha < 1-\nu+\mathrm{Re}(\kappa) < \beta$ and $\omega = \mu + a_1^*\alpha + 1/2$ and let $1 < r < \infty$.*

(a) *If $\nu - \mathrm{Re}(\kappa) \notin \mathcal{E}_{\mathcal{H}}$, or if $1 < r \leqq 2$, then the transform $\boldsymbol{H}_{\sigma,\kappa}$ is one-to-one on $\mathfrak{L}_{\nu,r}$.*

(b) *If $\mathrm{Re}(\omega) \geqq 0$ and $\nu - \mathrm{Re}(\kappa) \notin \mathcal{E}_{\mathcal{H}}$, then*

$$\boldsymbol{H}_{\sigma,\kappa}(\mathfrak{L}_{\nu,r}) = \mathbb{L}_{a_1^*,\alpha-\sigma-\omega/a_1^*}\left(\mathfrak{L}_{\nu-\mathrm{Re}(\kappa-\sigma),r}\right). \tag{5.4.10}$$

When $\nu - \mathrm{Re}(\kappa) \in \mathcal{E}_{\mathcal{H}}$, $\boldsymbol{H}_{\sigma,\kappa}(\mathfrak{L}_{\nu,r})$ is a subset of the right-hand side of (5.4.10).

(c) *If $\mathrm{Re}(\omega) < 0$ and $\nu - \mathrm{Re}(\kappa) \notin \mathcal{E}_{\mathcal{H}}$, then*

$$\boldsymbol{H}_{\sigma,\kappa}(\mathfrak{L}_{\nu,r}) = \left(I^{-\omega}_{-;1/a_1^*,a_1^*(\sigma-\alpha)}\mathbb{L}_{a_1^*,\alpha-\sigma}\right)\left(\mathfrak{L}_{\nu-\mathrm{Re}(\kappa-\sigma),r}\right). \tag{5.4.11}$$

When $\nu - \mathrm{Re}(\kappa) \in \mathcal{E}_{\mathcal{H}}$, $\boldsymbol{H}_{\sigma,\kappa}(\mathfrak{L}_{\nu,r})$ is a subset of the right-hand side of (5.4.11).

Theorem 5.32. *Let $a_1^* = 0, a_2^* > 0, n > 0, \alpha < 1-\nu+\mathrm{Re}(\kappa) < \beta$ and $\omega = \mu - a_2^*\beta + 1/2$ and let $1 < r < \infty$.*

(a) *If $\nu - \mathrm{Re}(\kappa) \notin \mathcal{E}_{\mathcal{H}}$, or if $1 < r \leqq 2$, then the transform $\boldsymbol{H}_{\sigma,\kappa}$ is one-to-one on $\mathfrak{L}_{\nu,r}$.*

(b) *If $\mathrm{Re}(\omega) \geqq 0$ and $\nu - \mathrm{Re}(\kappa) \notin \mathcal{E}_{\mathcal{H}}$, then*

$$\boldsymbol{H}_{\sigma,\kappa}(\mathfrak{L}_{\nu,r}) = \mathbb{L}_{-a_2^*,\beta-\sigma+\omega/a_2^*}\left(\mathfrak{L}_{\nu-\mathrm{Re}(\kappa-\sigma),r}\right). \tag{5.4.12}$$

When $\nu - \mathrm{Re}(\kappa) \in \mathcal{E}_{\mathcal{H}}$, $\boldsymbol{H}_{\sigma,\kappa}(\mathfrak{L}_{\nu,r})$ is a subset of the right-hand side of (5.4.12).

(c) *If $\mathrm{Re}(\omega) < 0$ and $\nu - \mathrm{Re}(\kappa) \notin \mathcal{E}_{\mathcal{H}}$, then*

$$\boldsymbol{H}_{\sigma,\kappa}(\mathfrak{L}_{\nu,r}) = \left(I^{-\omega}_{0+;1/a_2^*,a_2^*(\beta-\sigma)-1}\mathbb{L}_{-a_2^*,\beta-\sigma}\right)\left(\mathfrak{L}_{\nu-\mathrm{Re}(\kappa-\sigma),r}\right). \tag{5.4.13}$$

When $\nu - \mathrm{Re}(\kappa) \in \mathcal{E}_{\mathcal{H}}$, $\boldsymbol{H}_{\sigma,\kappa}(\mathfrak{L}_{\nu,r})$ is a subset of the right-hand side of (5.4.13).

Theorem 5.33. *Let $a^* > 0$, $a_1^* > 0$, $a_2^* < 0$, $\alpha < 1-\nu+\mathrm{Re}(\kappa) < \beta$ and let $1 < r < \infty$.*

(a) *If $\nu - \mathrm{Re}(\kappa) \notin \mathcal{E}_{\mathcal{H}}$, or if $1 < r \leqq 2$, then the transform $\boldsymbol{H}_{\sigma,\kappa}$ is one-to-one on $\mathfrak{L}_{\nu,r}$.*

(b) *Let $\omega, \eta, \zeta \in \mathbb{C}$ be chosen as*

$$\omega = a^*\eta - \mu - \frac{1}{2}; \tag{5.4.14}$$

$$a^*\mathrm{Re}(\eta) \geqq \gamma(r) + 2a_2^*[\nu - \mathrm{Re}(\kappa) - 1] + \mathrm{Re}(\mu); \tag{5.4.15}$$

$$\mathrm{Re}(\eta) > \nu - \mathrm{Re}(\kappa) - 1; \tag{5.4.16}$$

$$\mathrm{Re}(\zeta) < 1 - \nu + \mathrm{Re}(\kappa). \tag{5.4.17}$$

If $\nu - \mathrm{Re}(\kappa) \notin \mathcal{E}_{\mathcal{H}}$, then

$$\boldsymbol{H}_{\sigma,\kappa}(\mathfrak{L}_{\nu,r}) = \left(M_{\sigma+1/2+\omega/(2a_2^*)}\mathbb{H}_{-2a_2^*,2a_2^*\zeta+\omega-1}\mathbb{L}_{-a^*,1/2+\eta-\omega/(2a_2^*)}\right) \left(\mathfrak{L}_{3/2-\nu+\mathrm{Re}(\omega)/(2a_2^*)-\mathrm{Re}(\kappa),r}\right). \tag{5.4.18}$$

When $\nu - \mathrm{Re}(\kappa) \in \mathcal{E}_{\mathcal{H}}$, $\boldsymbol{H}_{\sigma,\kappa}(\mathfrak{L}_{\nu,r})$ is a subset of the right-hand side of (5.4.18).

Theorem 5.34. *Let $a^* > 0$, $a_1^* < 0$, $a_2^* > 0$, $\alpha < 1 - \nu + \mathrm{Re}(\kappa) < \beta$ and let $1 < r < \infty$.*
(a) *If $\nu - \mathrm{Re}(\kappa) \notin \mathcal{E}_{\mathcal{H}}$, or if $1 < r \leqq 2$, then the transform $\boldsymbol{H}_{\sigma,\kappa}$ is one-to-one on $\mathfrak{L}_{\nu,r}$.*
(b) *Let $\omega, \eta, \zeta \in \mathbb{C}$ be chosen as*

$$\omega = a^*\eta - \Delta - \mu - \frac{1}{2}; \tag{5.4.19}$$

$$a^*\mathrm{Re}(\eta) \geqq \gamma(r) - 2a_1^*[\nu - \mathrm{Re}(\kappa)] + \Delta + \mathrm{Re}(\mu); \tag{5.4.20}$$

$$\mathrm{Re}(\eta) > -\nu + \mathrm{Re}(\kappa); \tag{5.4.21}$$

$$\mathrm{Re}(\zeta) < \nu - \mathrm{Re}(\kappa). \tag{5.4.22}$$

If $\nu - \mathrm{Re}(\kappa) \notin \mathcal{E}_{\mathcal{H}}$, then

$$\boldsymbol{H}_{\sigma,\kappa}(\mathfrak{L}_{\nu,r}) = \left(M_{\sigma-1/2-\omega/(2a_1^*)}\mathbb{H}_{2a_1^*,2a_1^*\zeta+\omega-1}\mathbb{L}_{a^*,1/2-\eta+\omega/(2a_1^*)}\right) \left(\mathfrak{L}_{1/2-\nu-\mathrm{Re}(\omega)/(2a_1^*)-\mathrm{Re}(\kappa),r}\right). \tag{5.4.23}$$

When $\nu - \mathrm{Re}(\kappa) \in \mathcal{E}_{\mathcal{H}}$, $\boldsymbol{H}_{\sigma,\kappa}(\mathfrak{L}_{\nu,r})$ is a subset of the right-hand side of (5.4.23).

The relation (5.1.8) and Lemma 3.1(i) imply that (5.1.3) is equaivalent to

$$\boldsymbol{H}M_\kappa f = M_{-\sigma}\boldsymbol{H}_{\sigma,\kappa}$$

by taking into account the isometric property (5.4.1) and the relation

$$M_\zeta^{-1} = M_{-\zeta}. \tag{5.4.24}$$

Thus, if $a^* = 0$, from the results of Sections 4.9 and 4.10 by applying (4.9.1), (4.9.2) and (5.4.24), we obtain the inversion formulas for the transform $\boldsymbol{H}_{\sigma,\kappa}$ in the form

$$f(x) = hx^{1-\kappa-(\lambda+1)/h}\frac{d}{dx}x^{(\lambda+1)/h} \cdot \int_0^\infty H_{p+1,q+1}^{q-m,p-n+1}\left[xt \,\middle|\, \begin{matrix} (-\lambda, h), (1 - a_i - \alpha_i, \alpha_i)_{n+1,p}, (1 - a_i - \alpha_i, \alpha_i)_{1,n} \\ (1 - b_j - \beta_j, \beta_j)_{m+1,q}, (1 - b_j - \beta_j, \beta_j)_{1,m}, (-\lambda - 1, h) \end{matrix}\right] \cdot t^{-\sigma}\left(\boldsymbol{H}_{\sigma,\kappa}f\right)(t)dt \tag{5.4.25}$$

or

$$f(x) = -hx^{1-\kappa-(\lambda+1)/h}\frac{d}{dx}x^{(\lambda+1)/h}$$

$$\cdot \int_0^\infty H^{q-m+1,p-n}_{p+1,q+1}\left[xt \,\middle|\, \begin{matrix}(1-a_i-\alpha_i,\alpha_i)_{n+1,p},(1-a_i-\alpha_i,\alpha_i)_{1,n},(-\lambda,h)\\(-\lambda-1,h),(1-b_j-\beta_j,\beta_j)_{m+1,q},(1-b_j-\beta_j,\beta_j)_{1,m}\end{matrix}\right]$$

$$\cdot\, t^{-\sigma}\left(\boldsymbol{H}_{\sigma,\kappa}f\right)(t)dt. \tag{5.4.26}$$

The conditions for the validity of (5.4.25) and (5.4.26) follow from Theorems 4.11–4.14 and we obtain the following inversion theorems for the transform $\boldsymbol{H}^1$ in the cases when $a^* = 0$ and either $\Delta = 0$ or $\Delta \neq 0$.

Theorem 5.35. *Let $a^* = 0$, $\alpha < 1-\nu+\mathrm{Re}(\kappa) < \beta$ and $\alpha_0 < \nu - \mathrm{Re}(\kappa) < \beta_0$, and let $\lambda \in \mathbb{C}, h > 0$.*

(a) *If $\Delta[1-\nu+\mathrm{Re}(\kappa)] + \mathrm{Re}(\mu) = 0$ and $f \in \mathfrak{L}_{\nu,2}$, then the inversion formula (5.4.25) holds for $\mathrm{Re}(\lambda) > [\nu - \mathrm{Re}(\kappa)]h - 1$ and (5.4.26) for $\mathrm{Re}(\lambda) < [\nu - \mathrm{Re}(\kappa)]h - 1$.*

(b) *If $\Delta = \mathrm{Re}(\mu) = 0$ and $f \in \mathfrak{L}_{\nu,r}$ $(1 < r < \infty)$, then the inversion formula (5.4.25) holds for $\mathrm{Re}(\lambda) > [\nu - \mathrm{Re}(\kappa)]h - 1$ and (5.4.26) for $\mathrm{Re}(\lambda) < [\nu - \mathrm{Re}(\kappa)]h - 1$.*

Theorem 5.36. *Let $a^* = 0$, $1 < r < \infty$ and $\Delta[1-\nu+\mathrm{Re}(\kappa)] + \mathrm{Re}(\mu) \leqq 1/2 - \gamma(r)$, and let $\lambda \in \mathbb{C}, h > 0$.*

(a) *If $\Delta > 0, m > 0, \alpha < 1-\nu+\mathrm{Re}(\kappa) < \beta$, $\alpha_0 < \nu - \mathrm{Re}(\kappa) < \min[\beta_0, \{\mathrm{Re}(\mu + 1/2)/\Delta\} + 1]$ and if $f \in \mathfrak{L}_{\nu,r}$, then the inversion formulas (5.4.25) and (5.4.26) hold for $\mathrm{Re}(\lambda) > [\nu - \mathrm{Re}(\kappa)]h - 1$ and for $\mathrm{Re}(\lambda) < [\nu - \mathrm{Re}(\kappa)]h - 1$, respectively.*

(b) *If $\Delta < 0, n > 0, \alpha < 1-\nu+\mathrm{Re}(\kappa) < \beta$, $\max[\alpha_0, \{\mathrm{Re}(\mu + 1/2)/\Delta\} + 1] < \nu - \mathrm{Re}(\kappa) < \beta_0$ and if $f \in \mathfrak{L}_{\nu,r}$, then the inversion formulas (5.4.25) and (5.4.26) hold for $\mathrm{Re}(\lambda) > [\nu - \mathrm{Re}(\kappa)]h - 1$ and for $\mathrm{Re}(\lambda) < [\nu - \mathrm{Re}(\kappa)]h - 1$, respectively.*

Remark 5.2. The results in this section generalize those in Sections 3.6 and 4.1–4.10. In fact, when $\sigma = \kappa = 0$, Theorems 3.6 and 3.7 follow from Theorem 5.25, Theorems 4.1 and 4.2 from Theorems 5.26 and 5.27, Theorems 4.3 and 4.4 from Theorem 5.28, Theorems 4.5–4.10 from Theorems 5.29–5.34, Theorems 4.11 and 4.12 from Theorem 5.35, and Theorems 4.13 and 4.14 from Theorem 5.36.

5.5. $\boldsymbol{H}^1_{\sigma,\kappa}$-Transform on the Space $\mathfrak{L}_{\nu,r}$

Now we proceed to the transform $\boldsymbol{H}^1_{\sigma,\kappa}$ defined in (5.1.4). In view of (5.1.9) this transform is connected with the transform $\boldsymbol{H}^1$ via the elementary operators M_σ and M_κ in the same way as the $\boldsymbol{H}_{\sigma,\kappa}$ in (5.1.3) is connected with the $\boldsymbol{H}$-transform in (3.1.1). Therefore by virtue of (5.1.9) and the isometric property (5.4.1), we apply the results in Section 5.2 for the transform $\boldsymbol{H}^1$ by replacing ν by $\nu - \mathrm{Re}(\kappa)$.

Theorem 5.37. *Suppose that* **(a)** *$\alpha < \nu - \mathrm{Re}(\kappa) < \beta$ and that either of conditions* **(b)** *$a^* > 0$, or* **(c)** *$a^* = 0$, $\Delta[\nu - \mathrm{Re}(\kappa)] + \mathrm{Re}(\mu) \leqq 0$ holds. Then we have the following results:*

(i) *There is a one-to-one transform* $\boldsymbol{H}^1_{\sigma,\kappa} \in [\mathfrak{L}_{\nu,2}, \mathfrak{L}_{\nu-\mathrm{Re}(\kappa+\sigma),2}]$ *such that* (5.1.14) *holds for* $\mathrm{Re}(s) = \nu - \mathrm{Re}(\kappa+\sigma)$ *and* $f \in \mathfrak{L}_{\nu,2}$. *If* $a^* = 0$, $\Delta[\nu - \mathrm{Re}(\kappa)] + \mathrm{Re}(\mu) = 0$ *and* $1-\nu+\mathrm{Re}(\kappa) \notin \mathcal{E}_{\mathcal{H}}$, *then the transform* $\boldsymbol{H}^1_{\sigma,\kappa}$ *maps* $\mathfrak{L}_{\nu,2}$ *onto* $\mathfrak{L}_{\nu-\mathrm{Re}(\kappa+\sigma),2}$.

(ii) *If* $f \in \mathfrak{L}_{\nu,2}$ *and* $g \in \mathfrak{L}_{1-\nu+\mathrm{Re}(\kappa+\sigma),2}$, *then the relation* (5.1.19) *holds for* $\boldsymbol{H}^1_{\sigma,\kappa}$.

(iii) *Let* $\lambda \in \mathbb{C}$, $h > 0$ *and* $f \in \mathfrak{L}_{\nu,2}$. *When* $\mathrm{Re}(\lambda) > [\nu - \mathrm{Re}(\kappa)]h - 1$, $\boldsymbol{H}^1_{\sigma,\kappa} f$ *is given by*

$$\left(\boldsymbol{H}^1_{\sigma,\kappa} f\right)(x) = h x^{\sigma+1-(\lambda+1)/h} \frac{d}{dx} x^{(\lambda+1)h} \cdot \int_0^\infty H^{m,n+1}_{p+1,q+1}\left[\frac{x}{t} \,\middle|\, \begin{matrix} (-\lambda, h), (a_i, \alpha_i)_{1,p} \\ (b_j, \beta_j)_{1,q}, (-\lambda-1, h) \end{matrix}\right] t^{\kappa-1} f(t)\,dt. \tag{5.5.1}$$

When $\mathrm{Re}(\lambda) < [\nu - \mathrm{Re}(\kappa)]h - 1$,

$$\left(\boldsymbol{H}^1_{\sigma,\kappa} f\right)(x) = -h x^{\sigma+1-(\lambda+1)/h} \frac{d}{dx} x^{(\lambda+1)/h} \cdot \int_0^\infty H^{m+1,n}_{p+1,q+1}\left[\frac{x}{t} \,\middle|\, \begin{matrix} (a_i, \alpha_i)_{1,p}, (-\lambda, h) \\ (-\lambda-1, h), (b_j, \beta_j)_{1,q} \end{matrix}\right] t^{\kappa-1} f(t)\,dt. \tag{5.5.2}$$

(iv) *The transform* $\boldsymbol{H}^1_{\sigma,\kappa}$ *is independent of* ν *in the sense that, if* ν *and* $\widetilde{\nu}$ *satisfy* (a), *and either* (b) *or* (c), *and if the transforms* $\boldsymbol{H}^1_{\sigma,\kappa}$ *and* $\widetilde{\boldsymbol{H}^1}_{\sigma,\kappa}$ *are defined in* $\mathfrak{L}_{\nu,2}$ *and* $\mathfrak{L}_{\widetilde{\nu},2}$ *respectively by* (5.1.14), *then* $\boldsymbol{H}^1_{\sigma,\kappa} f = \widetilde{\boldsymbol{H}^1}_{\sigma,\kappa} f$ *for* $f \in \mathfrak{L}_{\nu,2} \cap \mathfrak{L}_{\widetilde{\nu},2}$.

(v) *If* $a^* > 0$ *or if* $a^* = 0$, $\Delta[\nu - \mathrm{Re}(\kappa)] + \mathrm{Re}(\mu) < 0$, *then for* $f \in \mathfrak{L}_{\nu,2}$, $\boldsymbol{H}^1_{\sigma,\kappa} f$ *is given in* (5.1.4).

We now present the $\mathfrak{L}_{\nu,r}$-theory of the transform $\boldsymbol{H}^1_{\sigma,\kappa}$ when $a^* = 0$ by virtue of the results in Section 5.2, (5.1.9) and Lemmas 3.1(i) and 3.2. From Theorems 5.2–5.4 we deduce the mapping properties and the range of $\boldsymbol{H}^1_{\sigma,\kappa}$ on $\mathfrak{L}_{\nu,r}$ in three different cases when either $\Delta = \mathrm{Re}(\mu) = 0$ or $\Delta = 0$, $\mathrm{Re}(\mu) < 0$ or $\Delta \neq 0$.

Theorem 5.38. *Let* $a^* = \Delta = 0, \mathrm{Re}(\mu) = 0, \alpha < \nu - \mathrm{Re}(\kappa) < \beta$ *and let* $1 < r < \infty$.

(a) *The transform* $\boldsymbol{H}^1_{\sigma,\kappa}$ *defined on* $\mathfrak{L}_{\nu,2}$ *can be extended to* $\mathfrak{L}_{\nu,r}$ *as an element of* $[\mathfrak{L}_{\nu,r}, \mathfrak{L}_{\nu-\mathrm{Re}(\kappa+\sigma),r}]$.

(b) *If* $1 < r \leqq 2$, *then the transform* $\boldsymbol{H}^1_{\sigma,\kappa}$ *is one-to-one on* $\mathfrak{L}_{\nu,r}$ *and there holds the equality* (5.1.14) *for* $f \in \mathfrak{L}_{\nu,r}$ *and* $\mathrm{Re}(s) = \nu - \mathrm{Re}(\kappa+\sigma)$.

(c) *If* $f \in \mathfrak{L}_{\nu,r}$ *and* $g \in \mathfrak{L}_{1-\nu+\mathrm{Re}(\kappa+\sigma),r'}$ *with* $r' = r/(r-1)$, *then the relation* (5.1.19) *holds.*

(d) *If* $1 - \nu + \mathrm{Re}(\kappa) \notin \mathcal{E}_{\mathcal{H}}$, *then the transform* $\boldsymbol{H}^1_{\sigma,\kappa}$ *is one-to-one on* $\mathfrak{L}_{\nu,r}$ *and there holds*

$$\boldsymbol{H}^1_{\sigma,\kappa}(\mathfrak{L}_{\nu,r}) = \mathfrak{L}_{\nu-\mathrm{Re}(\kappa+\sigma),r}. \tag{5.5.3}$$

(e) *If* $f \in \mathfrak{L}_{\nu,r}$, $\lambda \in \mathbb{C}$ *and* $h > 0$, *then* $\boldsymbol{H}^1_{\sigma,\kappa} f$ *is given in* (5.5.1) *for* $\mathrm{Re}(\lambda) > [\nu - \mathrm{Re}(\kappa)]h - 1$, *while in* (5.5.2) *for* $\mathrm{Re}(\lambda) < [\nu - \mathrm{Re}(\kappa)]h - 1$.

Theorem 5.39. *Let* $a^* = \Delta = 0, \mathrm{Re}(\mu) < 0, \alpha < \nu - \mathrm{Re}(\kappa) < \beta$, *and let either* $m > 0$ *or* $n > 0$. *Let* $1 < r < \infty$.

(a) *The transform $\boldsymbol{H}^1_{\sigma,\kappa}$ defined on $\mathfrak{L}_{\nu,2}$ can be extended to $\mathfrak{L}_{\nu,r}$ as an element of $[\mathfrak{L}_{\nu,r}, \mathfrak{L}_{\nu-\mathrm{Re}(\kappa+\sigma),s}]$ for all $s \geqq r$ such that $1/s > 1/r + \mathrm{Re}(\mu)$.*

(b) *If $1 < r \leqq 2$, then the transform $\boldsymbol{H}^1_{\sigma,\kappa}$ is one-to-one on $\mathfrak{L}_{\nu,r}$ and there holds the equality (5.1.14) for $f \in \mathfrak{L}_{\nu,r}$ and $\mathrm{Re}(s) = \nu - \mathrm{Re}(\kappa+\sigma)$.*

(c) *If $f \in \mathfrak{L}_{\nu,r}$ and $g \in \mathfrak{L}_{1-\nu+\mathrm{Re}(\kappa+\sigma),s}$ with $1 < s < \infty$ and $1 \leqq 1/r + 1/s < 1 - \mathrm{Re}(\mu)$, then the relation (5.1.19) holds.*

(d) *Let $k > 0$. If $1 - \nu + \mathrm{Re}(\kappa) \notin \mathcal{E}_{\mathcal{H}}$, then the transform $\boldsymbol{H}^1_{\sigma,\kappa}$ is one-to-one on $\mathfrak{L}_{\nu,r}$ and there hold*

$$\boldsymbol{H}^1_{\sigma,\kappa}(\mathfrak{L}_{\nu,r}) = I^{-\mu}_{-;k,(\sigma-\alpha)/k}\left(\mathfrak{L}_{\nu-\mathrm{Re}(\kappa+\sigma),r}\right) \tag{5.5.4}$$

for $m > 0$, and

$$\boldsymbol{H}^1_{\sigma,\kappa}(\mathfrak{L}_{\nu,r}) = I^{-\mu}_{0+;k,(\beta-\sigma)/k-1}\left(\mathfrak{L}_{\nu-\mathrm{Re}(\kappa+\sigma),r}\right) \tag{5.5.5}$$

for $n > 0$. When $1 - \nu + \mathrm{Re}(\kappa) \in \mathcal{E}_{\mathcal{H}}$, $\boldsymbol{H}^1_{\sigma,\kappa}(\mathfrak{L}_{\nu,r})$ is a subset of the right-hand sides of (5.5.4) and (5.5.5) in the respective cases.

(e) *If $f \in \mathfrak{L}_{\nu,r}$, $\lambda \in \mathbb{C}$ and $h > 0$, then $\boldsymbol{H}^1_{\sigma,\kappa}f$ is given in (5.5.1) for $\mathrm{Re}(\lambda) > [\nu - \mathrm{Re}(\kappa)]h - 1$, while in (5.5.2) for $\mathrm{Re}(\lambda) < [\nu - \mathrm{Re}(\kappa)]h - 1$. Furthermore $\boldsymbol{H}^1_{\sigma,\kappa}f$ is given in (5.1.4).*

Theorem 5.40. *Let $a^* = 0, \Delta \neq 0, \alpha < \nu - \mathrm{Re}(\kappa) < \beta$, $1 < r < \infty$ and $\Delta[\nu - \mathrm{Re}(\kappa)] + \mathrm{Re}(\mu) \leqq 1/2 - \gamma(r)$. Assume that $m > 0$ if $\Delta > 0$ and $n > 0$ if $\Delta < 0$.*

(a) *The transform $\boldsymbol{H}^1_{\sigma,\kappa}$ defined on $\mathfrak{L}_{\nu,2}$ can be extended to $\mathfrak{L}_{\nu,r}$ as an element of $[\mathfrak{L}_{\nu,r}, \mathfrak{L}_{\nu-\mathrm{Re}(\kappa+\sigma),s}]$ for all s with $r \leqq s < \infty$ such that $s' \geqq [1/2 - \Delta\{\nu - \mathrm{Re}(\kappa)\} - \mathrm{Re}(\mu)]^{-1}$ with $1/s + 1/s' = 1$.*

(b) *If $1 < r \leqq 2$, then the transform $\boldsymbol{H}^1_{\sigma,\kappa}$ is one-to-one on $\mathfrak{L}_{\nu,r}$ and there holds the equality (5.1.14) for $f \in \mathfrak{L}_{\nu,r}$ and $\mathrm{Re}(s) = \nu - \mathrm{Re}(\kappa+\sigma)$.*

(c) *If $f \in \mathfrak{L}_{\nu,r}$ and $g \in \mathfrak{L}_{1-\nu+\mathrm{Re}(\kappa+\sigma),s}$ with $1 < s < \infty, 1/r + 1/s \geqq 1$ and $\Delta[\nu - \mathrm{Re}(\kappa)] + \mathrm{Re}(\mu) \leqq 1/2 - \max[\gamma(r), \gamma(s)]$, then the relation (5.1.19) holds.*

(d) *If $1 - \nu + \mathrm{Re}(\kappa) \notin \mathcal{E}_{\mathcal{H}}$, then the transform $\boldsymbol{H}^1_{\sigma,\kappa}$ is one-to-one on $\mathfrak{L}_{\nu,r}$. If we set $\eta = -\Delta\alpha - \mu - 1$ for $\Delta > 0$ and $\eta = -\Delta\beta - \mu - 1$ for $\Delta < 0$, then $\mathrm{Re}(\eta) > -1$ and there hold*

$$\boldsymbol{H}^1_{\sigma,\kappa}(\mathfrak{L}_{\nu,r}) = \left(M_{\sigma+\mu/\Delta+1/2}\mathbb{H}_{\Delta,\eta}\right)\left(\mathfrak{L}_{1/2-\nu-\mathrm{Re}(\mu)/\Delta-\mathrm{Re}(\kappa),r}\right). \tag{5.5.6}$$

When $1 - \nu + \mathrm{Re}(\kappa) \in \mathcal{E}_{\mathcal{H}}$, $\boldsymbol{H}^1_{\sigma,\kappa}(\mathfrak{L}_{\nu,r})$ is a subset of the right-hand side of (5.5.6).

(e) *If $f \in \mathfrak{L}_{\nu,r}$, $\lambda \in \mathbb{C}, h > 0$ and $\Delta[\nu - \mathrm{Re}(\kappa)] + \mathrm{Re}(\mu) \leqq 1/2 - \gamma(r)$, then $\boldsymbol{H}^1_{\sigma,\kappa}f$ is given in (5.5.1) for $\mathrm{Re}(\lambda) > [\nu - \mathrm{Re}(\kappa)]h - 1$, while in (5.5.2) for $\mathrm{Re}(\lambda) < [\nu - \mathrm{Re}(\kappa)]h - 1$. If $\Delta[\nu - \mathrm{Re}(\kappa)] + \mathrm{Re}(\mu) < 0$, $\boldsymbol{H}^1_{\sigma,\kappa}f$ is given in (5.1.4).*

From (5.1.9), (5.4.1) and Theorem 5.5, the $\mathfrak{L}_{\nu,r}$-theory of the transform $\boldsymbol{H}^1_{\sigma,\kappa}$ in (5.1.4) with $a^* > 0$ in $\mathfrak{L}_{\nu,r}$-space can be established for any $\nu \in \mathbb{R}$ and $1 \leqq r \leqq \infty$.

Theorem 5.41. *Let $a^* > 0$, $\alpha < \nu - \mathrm{Re}(\kappa) < \beta$ and $1 \leqq r \leqq s \leqq \infty$.*

(a) *The transform $\boldsymbol{H}^1_{\sigma,\kappa}$ defined on $\mathfrak{L}_{\nu,2}$ can be extended to $\mathfrak{L}_{\nu,r}$ as an element of $[\mathfrak{L}_{\nu,r}, \mathfrak{L}_{\nu-\mathrm{Re}(\kappa+\sigma),s}]$. If $1 \leqq r \leqq 2$, then $\boldsymbol{H}^1_{\sigma,\kappa}$ is a one-to-one transform from $\mathfrak{L}_{\nu,r}$ onto $\mathfrak{L}_{\nu-\mathrm{Re}(\kappa+\sigma),s}$.*

(b) *If $f \in \mathfrak{L}_{\nu,r}$ and $g \in \mathfrak{L}_{1-\nu+\mathrm{Re}(\kappa+\sigma),s'}$ with $1/s + 1/s' = 1$, then the relation (5.1.19) holds.*

By virtue of (5.1.9), (5.4.1), Theorems 5.6–5.10 and Lemma 3.2 we characterize the boundedness and the range of $\boldsymbol{H}^1_{\sigma,\kappa}$ on $\mathfrak{L}_{\nu,r}$ for $a^* > 0$ in five cases.

Theorem 5.42. *Let $a_1^* > 0, a_2^* > 0, m > 0, n > 0, \alpha < \nu - \mathrm{Re}(\kappa) < \beta$ and $\omega = \mu + a_1^*\alpha - a_2^*\beta + 1$ and let $1 < r < \infty$.*

(a) *If $1-\nu+\mathrm{Re}(\kappa) \notin \mathcal{E}_{\mathcal{H}}$, or if $1 \leqq r \leqq 2$, then the transform $\boldsymbol{H}^1_{\sigma,\kappa}$ is one-to-one on $\mathfrak{L}_{\nu,r}$.*

(b) *If $\mathrm{Re}(\omega) \geqq 0$ and $1-\nu+\mathrm{Re}(\kappa) \notin \mathcal{E}_{\mathcal{H}}$, then*

$$\boldsymbol{H}^1_{\sigma,\kappa}(\mathfrak{L}_{\nu,r}) = \left(\mathbb{L}_{a_1^*,\alpha-\sigma}\mathbb{L}_{a_2^*,1-\beta+\sigma-\omega/a_2^*}\right)\left(\mathfrak{L}_{\nu-\mathrm{Re}(\kappa+\sigma),r}\right). \tag{5.5.7}$$

When $1-\nu+\mathrm{Re}(\kappa) \in \mathcal{E}_{\mathcal{H}}$, $\boldsymbol{H}^1_{\sigma,\kappa}(\mathfrak{L}_{\nu,r})$ is a subset of the right-hand side of (5.5.7).

(c) *If $\mathrm{Re}(\omega) < 0$ and $1-\nu+\mathrm{Re}(\kappa) \notin \mathcal{E}_{\mathcal{H}}$, then*

$$\boldsymbol{H}^1_{\sigma,\kappa}(\mathfrak{L}_{\nu,r}) = \left(I^{-\omega}_{-;1/a_1^*,a_1^*(\sigma-\alpha)}\mathbb{L}_{a_1^*,\alpha-\sigma}\mathbb{L}_{a_2^*,1-\beta+\sigma}\right)\left(\mathfrak{L}_{\nu-\mathrm{Re}(\kappa+\sigma),r}\right). \tag{5.5.8}$$

When $1-\nu+\mathrm{Re}(\kappa) \in \mathcal{E}_{\mathcal{H}}$, $\boldsymbol{H}^1_{\sigma,\kappa}(\mathfrak{L}_{\nu,r})$ is a subset of the right-hand side of (5.5.8).

Theorem 5.43. *Let $m > 0, a_1^* > 0, a_2^* = 0, \alpha < \nu - \mathrm{Re}(\kappa) < \beta$ and $\omega = \mu + a_1^*\alpha + 1/2$ and let $1 < r < \infty$.*

(a) *If $1-\nu+\mathrm{Re}(\kappa) \notin \mathcal{E}_{\mathcal{H}}$, or if $1 < r \leqq 2$, then the transform $\boldsymbol{H}^1_{\sigma,\kappa}$ is one-to-one on $\mathfrak{L}_{\nu,r}$.*

(b) *If $\mathrm{Re}(\omega) \geqq 0$ and $1-\nu+\mathrm{Re}(\kappa) \notin \mathcal{E}_{\mathcal{H}}$, then*

$$\boldsymbol{H}^1_{\sigma,\kappa}(\mathfrak{L}_{\nu,r}) = \mathbb{L}_{a_1^*,\alpha-\sigma-\omega/a_1^*}\left(\mathfrak{L}_{1-\nu+\mathrm{Re}(\kappa+\sigma),r}\right). \tag{5.5.9}$$

When $1-\nu+\mathrm{Re}(\kappa) \in \mathcal{E}_{\mathcal{H}}$, $\boldsymbol{H}^1_{\sigma,\kappa}(\mathfrak{L}_{\nu,r})$ is a subset of the right-hand side of (5.5.9).

(c) *If $\mathrm{Re}(\omega) < 0$ and $1-\nu+\mathrm{Re}(\kappa) \notin \mathcal{E}_{\mathcal{H}}$, then*

$$\boldsymbol{H}^1_{\sigma,\kappa}(\mathfrak{L}_{\nu,r}) = \left(I^{-\omega}_{-;1/a_1^*,a_1^*(\sigma-\alpha)}\mathbb{L}_{a_1^*,\alpha-\sigma}\right)\left(\mathfrak{L}_{1-\nu+\mathrm{Re}(\kappa+\sigma),r}\right). \tag{5.5.10}$$

When $1-\nu+\mathrm{Re}(\kappa) \in \mathcal{E}_{\mathcal{H}}$, $\boldsymbol{H}^1_{\sigma,\kappa}(\mathfrak{L}_{\nu,r})$ is a subset of the right-hand side of (5.5.10).

Theorem 5.44. *Let $a_1^* = 0, a_2^* > 0, n > 0, \alpha < \nu - \mathrm{Re}(\kappa) < \beta$ and $\omega = \mu - a_2^*\beta + 1/2$ and let $1 < r < \infty$.*

(a) *If $1-\nu+\mathrm{Re}(\kappa) \notin \mathcal{E}_{\mathcal{H}}$, or if $1 < r \leqq 2$, then the transform $\boldsymbol{H}^1_{\sigma,\kappa}$ is one-to-one on $\mathfrak{L}_{\nu,r}$.*

(b) *If $\mathrm{Re}(\omega) \geqq 0$ and $1-\nu+\mathrm{Re}(\kappa) \notin \mathcal{E}_{\mathcal{H}}$, then*

$$\boldsymbol{H}^1_{\sigma,\kappa}(\mathfrak{L}_{\nu,r}) = \mathbb{L}_{-a_2^*,\beta-\sigma+\omega/a_2^*}\left(\mathfrak{L}_{1-\nu+\mathrm{Re}(\kappa+\sigma),r}\right). \tag{5.5.11}$$

When $1-\nu+\mathrm{Re}(\kappa) \in \mathcal{E}_{\mathcal{H}}$, $\boldsymbol{H}^1_{\sigma,\kappa}(\mathfrak{L}_{\nu,r})$ is a subset of the right-hand side of (5.5.11).

(c) *If $\mathrm{Re}(\omega) < 0$ and $1-\nu+\mathrm{Re}(\kappa) \notin \mathcal{E}_{\mathcal{H}}$, then*

$$\boldsymbol{H}^1_{\sigma,\kappa}(\mathfrak{L}_{\nu,r}) = \left(I^{-\omega}_{0+;1/a_2^*,a_2^*(\beta-\sigma)-1}\mathbb{L}_{-a_2^*,\beta-\sigma}\right)\left(\mathfrak{L}_{1-\nu+\mathrm{Re}(\kappa+\sigma),r}\right). \tag{5.5.12}$$

When $1-\nu+\mathrm{Re}(\kappa) \in \mathcal{E}_{\mathcal{H}}$, $\boldsymbol{H}^1_{\sigma,\kappa}(\mathfrak{L}_{\nu,r})$ is a subset of the right-hand side of (5.5.12).

Theorem 5.45. *Let $a^* > 0$, $a_1^* > 0$, $a_2^* < 0$, $\alpha < \nu - \mathrm{Re}(\kappa) < \beta$ and let $1 < r < \infty$.*

(a) *If $1-\nu+\mathrm{Re}(\kappa) \notin \mathcal{E}_{\mathcal{H}}$, or if $1 < r \leqq 2$, then the transform $\boldsymbol{H}^1_{\sigma,\kappa}$ is one-to-one on $\mathfrak{L}_{\nu,r}$.*

(b) *Let* $\omega, \eta, \zeta \in \mathbb{C}$ *be chosen as*

$$\omega = a^*\eta - \mu - \frac{1}{2}; \tag{5.5.13}$$

$$a^*\mathrm{Re}(\eta) \geqq \gamma(r) + 2a_2^*[\nu - \mathrm{Re}(\kappa)] + \mathrm{Re}(\mu); \tag{5.5.14}$$

$$\mathrm{Re}(\eta) > -\nu + \mathrm{Re}(\kappa); \tag{5.5.15}$$

$$\mathrm{Re}(\zeta) < \nu - \mathrm{Re}(\kappa). \tag{5.5.16}$$

If $1 - \nu + \mathrm{Re}(\kappa) \notin \mathcal{E}_{\mathcal{H}}$, *then*

$$\boldsymbol{H}^1_{\sigma,\kappa}(\mathfrak{L}_{\nu,r}) = \left(M_{\sigma+1/2+\omega/(2a_2^*)}\mathbb{H}_{-2a_2^*,2a_2^*\zeta+\omega-1}\mathbb{L}_{-a^*,1/2+\eta-\omega/(2a_2^*)}\right)$$
$$\left(\mathfrak{L}_{\nu+1/2+\mathrm{Re}(\omega)/(2a_2^*)-\mathrm{Re}(\kappa),r}\right). \tag{5.5.17}$$

When $1 - \nu + \mathrm{Re}(\kappa) \in \mathcal{E}_{\mathcal{H}}$, $\boldsymbol{H}^1_{\sigma,\kappa}(\mathfrak{L}_{\nu,r})$ *is a subset of the right-hand side of* (5.5.17).

Theorem 5.46. *Let* $a^* > 0$, $a_1^* < 0$, $a_2^* > 0$, $\alpha < \nu - \mathrm{Re}(\kappa) < \beta$ *and let* $1 < r < \infty$.
(a) *If* $1-\nu+\mathrm{Re}(\kappa) \notin \mathcal{E}_{\mathcal{H}}$, *or if* $1 < r \leqq 2$, *then the transform* $\boldsymbol{H}^1_{\sigma,\kappa}$ *is one-to-one on* $\mathfrak{L}_{\nu,r}$.
(b) *Let* $\omega, \eta, \zeta \in \mathbb{C}$ *be chosen as*

$$\omega = a^*\eta - \Delta - \mu - \frac{1}{2}; \tag{5.5.18}$$

$$a^*\mathrm{Re}(\eta) \geqq \gamma(r) - 2a_1^*[\nu - \mathrm{Re}(\kappa)] + \Delta + \mathrm{Re}(\mu); \tag{5.5.19}$$

$$\mathrm{Re}(\eta) > -\nu + \mathrm{Re}(\kappa); \tag{5.5.20}$$

$$\mathrm{Re}(\zeta) < \nu - \mathrm{Re}(\kappa). \tag{5.5.21}$$

If $1 - \nu + \mathrm{Re}(\kappa) \notin \mathcal{E}_{\mathcal{H}}$, *then*

$$\boldsymbol{H}^1_{\sigma,\kappa}(\mathfrak{L}_{\nu,r}) = \left(M_{\sigma-1/2-\omega/(2a_1^*)}\mathbb{H}_{2a_1^*,2a_1^*\zeta+\omega-1}\mathbb{L}_{a^*,1/2-\eta+\omega/(2a_1^*)}\right)$$
$$\left(\mathfrak{L}_{\nu-1/2-\mathrm{Re}(\omega)/(2a_1^*)-\mathrm{Re}(\kappa),r}\right). \tag{5.5.22}$$

When $1 - \nu + \mathrm{Re}(\kappa) \in \mathcal{E}_{\mathcal{H}}$, $\boldsymbol{H}^1_{\sigma,\kappa}(\mathfrak{L}_{\nu,r})$ *is a subset of the right-hand side of* (5.5.22).

The inversion formulas for the transform $\boldsymbol{H}^1_{\sigma,\kappa}$ can be found in a similar manner as before by only noting that (5.1.4) is equaivalent to

$$\boldsymbol{H}^1 M_\kappa f = M_{-\sigma}\boldsymbol{H}^1_{\sigma,\kappa} f$$

for $f \in \mathfrak{L}_{\nu,r}$. Thus by virtue of (5.2.27) and (5.2.28) we have formally that

$$f(x) = -hx^{(\lambda+1)/h-\kappa}\frac{d}{dx}x^{-(\lambda+1)/h}$$
$$\cdot \int_0^\infty H^{q-m,p-n+1}_{p+1,q+1}\left[\frac{t}{x}\middle| \begin{matrix}(-\lambda,h),(1-a_i-\alpha_i,\alpha_i)_{n+1,p},(1-a_i-\alpha_i,\alpha_i)_{1,n}\\(1-b_j-\beta_j,\beta_j)_{m+1,q},(1-b_j-\beta_j,\beta_j)_{1,m},(-\lambda-1,h)\end{matrix}\right]$$
$$\cdot\, t^{-\sigma}\left(\boldsymbol{H}^1_{\sigma,\kappa}f\right)(t)dt \tag{5.5.23}$$

or

$$f(x) = hx^{(\lambda+1)/h-\kappa}\frac{d}{dx}x^{-(\lambda+1)/h}$$

$$\cdot \int_0^\infty H^{q-m+1,p-n}_{p+1,q+1}\left[\frac{t}{x}\,\middle|\,\begin{matrix}(1-a_i-\alpha_i,\alpha_i)_{n+1,p},(1-a_i-\alpha_i,\alpha_i)_{1,n},(-\lambda,h)\\(-\lambda-1,h),(1-b_j-\beta_j,\beta_j)_{m+1,q},(1-b_j-\beta_j,\beta_j)_{1,m}\end{matrix}\right]$$

$$\cdot t^{-\sigma}\left(\boldsymbol{H}^1_{\sigma,\kappa}f\right)(t)dt. \tag{5.5.24}$$

The conditions for the validity of (5.5.23) and (5.5.24) follow from Theorems 5.11–5.12.

Theorem 5.47. *Let* $a^* = 0$, $\alpha < \nu - \mathrm{Re}(\kappa) < \beta$ *and* $\alpha_0 < 1 - \nu + \mathrm{Re}(\kappa) < \beta_0$, *and let* $\lambda \in \mathbb{C}, h > 0$.

(a) *If* $\Delta[\nu - \mathrm{Re}(\kappa)] + \mathrm{Re}(\mu) = 0$ *and* $f \in \mathfrak{L}_{\nu,2}$, *then the inversion formula* (5.5.23) *holds for* $\mathrm{Re}(\lambda) > [1 - \nu + \mathrm{Re}(\kappa)]h - 1$ *and* (5.5.24) *for* $\mathrm{Re}(\lambda) < [1 - \nu + \mathrm{Re}(\kappa)]h - 1$.

(b) *If* $\Delta = \mathrm{Re}(\mu) = 0$ *and* $f \in \mathfrak{L}_{\nu,r}$ $(1 < r < \infty)$, *then the inversion formula* (5.5.23) *holds for* $\mathrm{Re}(\lambda) > [1 - \nu + \mathrm{Re}(\kappa)]h - 1$ *and* (5.5.24) *for* $\mathrm{Re}(\lambda) < [1 - \nu + \mathrm{Re}(\kappa)]h - 1$.

Theorem 5.48. *Let* $a^* = 0$, $1 < r < \infty$ *and* $\Delta[\nu - \mathrm{Re}(\kappa)] + \mathrm{Re}(\mu) \leqq 1/2 - \gamma(r)$, *and let* $\lambda \in \mathbb{C}, h > 0$.

(a) *If* $\Delta > 0, m > 0, \alpha < \nu - \mathrm{Re}(\kappa) < \beta, \alpha_0 < 1-\nu+\mathrm{Re}(\kappa) < \min[\beta_0, \{\mathrm{Re}(\mu+1/2)/\Delta\}+1]$ *and if* $f \in \mathfrak{L}_{\nu,r}$, *then the inversion formulas* (5.5.23) *and* (5.5.24) *hold for* $\mathrm{Re}(\lambda) > [1 - \nu + \mathrm{Re}(\kappa)]h - 1$ *and for* $\mathrm{Re}(\lambda) < [1 - \nu + \mathrm{Re}(\kappa)]h - 1$, *respectively.*

(b) *If* $\Delta < 0, n > 0, \alpha < \nu - \mathrm{Re}(\kappa) < \beta, \max[\alpha_0, \{\mathrm{Re}(\mu+1/2)/\Delta\}+1] < 1-\nu+\mathrm{Re}(\kappa) < \beta_0$ *and if* $f \in \mathfrak{L}_{\nu,r}$, *then the inversion formulas* (5.5.23) *and* (5.5.24) *hold for* $\mathrm{Re}(\lambda) > [1 - \nu + \mathrm{Re}(\kappa)]h - 1$ *and for* $\mathrm{Re}(\lambda) < [1 - \nu + \mathrm{Re}(\kappa)]h - 1$, *respectively.*

Remark 5.3. The results in this section generalize those in Section 5.2. Namely, Theorems 5.1–5.12 follow from Theorems 5.37–5.48 when $\sigma = \kappa = 0$.

5.6. $H^2_{\sigma,\kappa}$-Transform on the Space $\mathfrak{L}_{\nu,r}$

Finally, let us study the transform $\boldsymbol{H}^2_{\sigma,\kappa}$ defined in (5.1.5). According to (5.1.10) this transform is connected with the transform $\boldsymbol{H}^2$ via the elementary operators M_σ and M_κ in the same way as the $\boldsymbol{H}^1_{\sigma,\kappa}$ in (5.1.4) is connected with the $\boldsymbol{H}^1$. Therefore by using (5.1.10) and the isometric property (5.4.1), we can apply the results in Section 5.3 for the transform $\boldsymbol{H}^2 f$ by replacing ν by $\nu - \mathrm{Re}(\kappa)$.

Theorem 5.49. *Suppose that* **(a)** $\alpha < 1 - \nu + \mathrm{Re}(\kappa) < \beta$ *and that either of the conditions* **(b)** $a^* > 0$, *or* **(c)** $a^* = 0$, $\Delta[1 - \nu + \mathrm{Re}(\kappa)] + \mathrm{Re}(\mu) \leqq 0$ *holds. Then we have the following results:*

(i) *There is a one-to-one transform* $\boldsymbol{H}^2_{\sigma,\kappa} \in [\mathfrak{L}_{\nu,2}, \mathfrak{L}_{\nu-\mathrm{Re}(\kappa+\sigma),2}]$ *such that* (5.1.15) *holds for* $\mathrm{Re}(s) = \nu - \mathrm{Re}(\kappa+\sigma)$ *and* $f \in \mathfrak{L}_{\nu,2}$. *If* $a^* = 0$, $\Delta[1-\nu+\mathrm{Re}(\kappa)] + \mathrm{Re}(\mu) = 0$ *and* $\nu - \mathrm{Re}(\kappa) \notin \mathcal{E}_{\mathcal{H}}$, *then the transform* $\boldsymbol{H}^2_{\sigma,\kappa}$ *maps* $\mathfrak{L}_{\nu,2}$ *onto* $\mathfrak{L}_{\nu-\mathrm{Re}(\kappa+\sigma),2}$.

(ii) *If* $f \in \mathfrak{L}_{\nu,2}$ *and* $g \in \mathfrak{L}_{1-\nu+\mathrm{Re}(\kappa+\sigma),2}$, *then the relation* (5.1.20) *holds for* $\boldsymbol{H}^2_{\sigma,\kappa}$.

(iii) *Let* $\lambda \in \mathbb{C}, h > 0$ *and* $f \in \mathfrak{L}_{\nu,2}$. *When* $\mathrm{Re}(\lambda) > [1-\nu+\mathrm{Re}(\kappa)]h - 1$, $\boldsymbol{H}^2_{\sigma,\kappa}f$ *is given by*

$$\left(\boldsymbol{H}^2_{\sigma,\kappa}f\right)(x) = -hx^{\sigma+(\lambda+1)/h}\frac{d}{dx}x^{-(\lambda+1)h}$$
$$\cdot \int_0^\infty H^{m,n+1}_{p+1,q+1}\left[\frac{t}{x}\,\middle|\,\begin{matrix}(-\lambda,h),(a_i,\alpha_i)_{1,p}\\(b_j,\beta_j)_{1,q},(-\lambda-1,h)\end{matrix}\right]t^\kappa f(t)dt. \tag{5.6.1}$$

When $\mathrm{Re}(\lambda) < [1-\nu+\mathrm{Re}(\kappa)]h - 1$,

$$\left(\boldsymbol{H}^2_{\sigma,\kappa}f\right)(x) = hx^{\sigma+(\lambda+1)/h}\frac{d}{dx}x^{-(\lambda+1)/h}$$
$$\cdot \int_0^\infty H^{m+1,n}_{p+1,q+1}\left[\frac{t}{x}\,\middle|\,\begin{matrix}(a_i,\alpha_i)_{1,p},(-\lambda,h)\\(-\lambda-1,h),(b_j,\beta_j)_{1,q}\end{matrix}\right]t^\kappa f(t)dt. \tag{5.6.2}$$

(iv) *The transform* $\boldsymbol{H}^2_{\sigma,\kappa}$ *is independent of* ν *in the sense that, if* ν *and* $\widetilde{\nu}$ *satisfy* (a), *and either* (b) *or* (c), *and if the transforms* $\boldsymbol{H}^2_{\sigma,\kappa}$ *and* $\widetilde{\boldsymbol{H}^2}_{\sigma,\kappa}$ *are defined in* $\mathfrak{L}_{\nu,2}$ *and* $\mathfrak{L}_{\widetilde{\nu},2}$ *respectively by* (5.1.15), *then* $\boldsymbol{H}^2_{\sigma,\kappa}f = \widetilde{\boldsymbol{H}^2}_{\sigma,\kappa}f$ *for* $f \in \mathfrak{L}_{\nu,2} \cap \mathfrak{L}_{\widetilde{\nu},2}$.

(v) *If* $a^* > 0$ *or if* $a^* = 0$, $\Delta[1-\nu+\mathrm{Re}(\kappa)] + \mathrm{Re}(\mu) < 0$, *then for* $f \in \mathfrak{L}_{\nu,2}$, $\boldsymbol{H}^2_{\sigma,\kappa}f$ *is given in* (5.1.5).

The results in Section 5.3 deduce the $\mathfrak{L}_{\nu,r}$-theory of the transform $\boldsymbol{H}^2_{\sigma,\kappa}$ when $a^* = 0$ by virtue of (5.1.10), (5.4.1) and Lemmas 3.1(i) and 3.2.

Theorem 5.50. *Let* $a^* = \Delta = 0, \mathrm{Re}(\mu) = 0, \alpha < 1-\nu+\mathrm{Re}(\kappa) < \beta$ *and let* $1 < r < \infty$.

(a) *The transform* $\boldsymbol{H}^2_{\sigma,\kappa}$ *defined on* $\mathfrak{L}_{\nu,2}$ *can be extended to* $\mathfrak{L}_{\nu,r}$ *as an element of* $[\mathfrak{L}_{\nu,r}, \mathfrak{L}_{\nu-\mathrm{Re}(\kappa+\sigma),r}]$.

(b) *If* $1 < r \leqq 2$, *then the transform* $\boldsymbol{H}^2_{\sigma,\kappa}$ *is one-to-one on* $\mathfrak{L}_{\nu,r}$ *and there holds the equality* (5.1.15) *for* $f \in \mathfrak{L}_{\nu,r}$ *and* $\mathrm{Re}(s) = \nu - \mathrm{Re}(\kappa+\sigma)$.

(c) *If* $f \in \mathfrak{L}_{\nu,r}$ *and* $g \in \mathfrak{L}_{1-\nu+\mathrm{Re}(\kappa+\sigma),r'}$ *with* $r' = r/(r-1)$, *then the relation* (5.1.20) *holds.*

(d) *If* $\nu - \mathrm{Re}(\kappa) \notin \mathcal{E}_{\mathcal{H}}$, *then the transform* $\boldsymbol{H}^2_{\sigma,\kappa}$ *is one-to-one on* $\mathfrak{L}_{\nu,r}$ *and there holds*

$$\boldsymbol{H}^2_{\sigma,\kappa}(\mathfrak{L}_{\nu,r}) = \mathfrak{L}_{\nu-\mathrm{Re}(\kappa+\sigma),r}. \tag{5.6.3}$$

(e) *If* $f \in \mathfrak{L}_{\nu,r}, \lambda \in \mathbb{C}$ *and* $h > 0$, *then* $\boldsymbol{H}^2_{\sigma,\kappa}f$ *is given in* (5.6.1) *for* $\mathrm{Re}(\lambda) > [1-\nu+\mathrm{Re}(\kappa)]h - 1$, *while in* (5.6.2) *for* $\mathrm{Re}(\lambda) < [1-\nu+\mathrm{Re}(\kappa)]h - 1$.

Theorem 5.51. *Let* $a^* = \Delta = 0, \mathrm{Re}(\mu) < 0$ *and* $\alpha < 1-\nu+\mathrm{Re}(\kappa) < \beta$, *and let either* $m > 0$ *or* $n > 0$. *Let* $1 < r < \infty$.

(a) *The transform* $\boldsymbol{H}^2_{\sigma,\kappa}$ *defined on* $\mathfrak{L}_{\nu,2}$ *can be extended to* $\mathfrak{L}_{\nu,r}$ *as an element of* $[\mathfrak{L}_{\nu,r}, \mathfrak{L}_{\nu-\mathrm{Re}(\kappa+\sigma),s}]$ *for all* $s \geqq r$ *such that* $1/s > 1/r + \mathrm{Re}(\mu)$.

(b) *If* $1 < r \leqq 2$, *then the transform* $\boldsymbol{H}^2_{\sigma,\kappa}$ *is one-to-one on* $\mathfrak{L}_{\nu,r}$ *and there holds the equality* (5.1.15) *for* $f \in \mathfrak{L}_{\nu,r}$ *and* $\text{Re}(s) = \nu - \text{Re}(\kappa + \sigma)$.

(c) *If* $f \in \mathfrak{L}_{\nu,r}$ *and* $g \in \mathfrak{L}_{1-\nu+\text{Re}(\kappa+\sigma),s}$ *with* $1 < s < \infty$ *and* $1 \leqq 1/r + 1/s < 1 - \text{Re}(\mu)$, *then the relation* (5.1.20) *holds.*

(d) *Let* $k > 0$. *If* $\nu - \text{Re}(\kappa) \notin \mathcal{E}_{\mathcal{H}}$, *then the transform* $\boldsymbol{H}^2_{\sigma,\kappa}$ *is one-to-one on* $\mathfrak{L}_{\nu,r}$ *and there hold*

$$\boldsymbol{H}^2_{\sigma,\kappa}\left(\mathfrak{L}_{\nu,r}\right) = I^{-\mu}_{0+;k,(1-\alpha-\sigma)/k-1}\left(\mathfrak{L}_{\nu-\text{Re}(\kappa+\sigma),r}\right) \tag{5.6.4}$$

for $m > 0$, *and*

$$\boldsymbol{H}^2_{\sigma,\kappa}\left(\mathfrak{L}_{\nu,r}\right) = I^{-\mu}_{-;k,(\beta+\sigma-1)/k}\left(\mathfrak{L}_{\nu-\text{Re}(\kappa+\sigma),r}\right) \tag{5.6.5}$$

for $n > 0$. *When* $\nu - \text{Re}(\kappa) \in \mathcal{E}_{\mathcal{H}}$, $\boldsymbol{H}^2_{\sigma,\kappa}\left(\mathfrak{L}_{\nu,r}\right)$ *is a subset of the right-hand sides of* (5.6.4) *and* (5.6.5) *in the respective cases.*

(e) *If* $f \in \mathfrak{L}_{\nu,r}, \lambda \in \mathbb{C}$ *and* $h > 0$, *then* $\boldsymbol{H}^2_{\sigma,\kappa}f$ *is given in* (5.6.1) *for* $\text{Re}(\lambda) > [1 - \nu + \text{Re}(\kappa)]h - 1$, *while in* (5.6.2) *for* $\text{Re}(\lambda) < [1 - \nu + \text{Re}(\kappa)]h - 1$. *Furthermore* $\boldsymbol{H}^2_{\sigma,\kappa}f$ *is given in* (5.1.5).

Theorem 5.52. *Let* $a^* = 0, \Delta \neq 0, m > 0, \alpha < 1 - \nu + \text{Re}(\kappa) < \beta$, $1 < r < \infty$ *and* $\Delta[1 - \nu + \text{Re}(\kappa)] + \text{Re}(\mu) \leqq 1/2 - \gamma(r)$. *Assume that* $m > 0$ *if* $\Delta > 0$ *and* $n > 0$ *if* $\Delta < 0$.

(a) *The transform* $\boldsymbol{H}^2_{\sigma,\kappa}$ *defined on* $\mathfrak{L}_{\nu,2}$ *can be extended to* $\mathfrak{L}_{\nu,r}$ *as an element of* $[\mathfrak{L}_{\nu,r}, \mathfrak{L}_{\nu-\text{Re}(\kappa+\sigma),s}]$ *for all* s *with* $r \leqq s < \infty$ *such that* $s' \geqq [1/2 - \Delta\{1-\nu+\text{Re}(\kappa)\} - \text{Re}(\mu)]^{-1}$ *with* $1/s + 1/s' = 1$.

(b) *If* $1 < r \leqq 2$, *the transform* $\boldsymbol{H}^2_{\sigma,\kappa}$ *is one-to-one on* $\mathfrak{L}_{\nu,r}$ *and there holds the equality* (5.1.15) *for* $f \in \mathfrak{L}_{\nu,r}$ *and* $\text{Re}(s) = \nu - \text{Re}(\kappa + \sigma)$.

(c) *If* $f \in \mathfrak{L}_{\nu,r}$ *and* $g \in \mathfrak{L}_{1-\nu+\text{Re}(\kappa+\sigma),s}$ *with* $1 < s < \infty, 1/r + 1/s \geqq 1$ *and* $\Delta[\nu - \text{Re}(\kappa)] + \text{Re}(\mu) \leqq 1/2 - \max[\gamma(r), \gamma(s)]$, *then the relation* (5.1.20) *holds.*

(d) *If* $\nu - \text{Re}(\kappa) \notin \mathcal{E}_{\mathcal{H}}$, *then the transform* $\boldsymbol{H}^2_{\sigma,\kappa}$ *is one-to-one on* $\mathfrak{L}_{\nu,r}$. *If we set* $\eta = -\Delta\alpha - \mu - 1$ *for* $\Delta > 0$ *and* $\eta = -\Delta\beta - \mu - 1$ *for* $\Delta < 0$, *then* $\text{Re}(\eta) > -1$ *and there holds*

$$\boldsymbol{H}^2_{\sigma,\kappa}(\mathfrak{L}_{\nu,r}) = \left(M_{\sigma-\mu/\Delta-1/2}H_{-\Delta,\eta}\right)\left(\mathfrak{L}_{3/2-\nu+\text{Re}(\mu)/\Delta+\text{Re}(\kappa),r}\right). \tag{5.6.6}$$

When $\nu - \text{Re}(\kappa) \in \mathcal{E}_{\mathcal{H}}$, $\boldsymbol{H}^2_{\sigma,\kappa}(\mathfrak{L}_{\nu,r})$ *is a subset of the right-hand side of* (5.6.6).

(e) *If* $f \in \mathfrak{L}_{\nu,r}, \lambda \in \mathbb{C}, h > 0$ *and* $\Delta[1 - \nu + \text{Re}(\kappa)] + \text{Re}(\mu) \leqq 1/2 - \gamma(r)$, *then* $\boldsymbol{H}^2_{\sigma,\kappa}f$ *is given in* (5.6.1) *for* $\text{Re}(\lambda) > [1-\nu+\text{Re}(\kappa)]h-1$, *while in* (5.6.2) *for* $\text{Re}(\lambda) < [1-\nu+\text{Re}(\kappa)]h-1$. *If* $\Delta[1 - \nu + \text{Re}(\kappa)] + \text{Re}(\mu) < 0$, $\boldsymbol{H}^2_{\sigma,\kappa}f$ *is given in* (5.1.5).

From (5.1.10), (5.4.1) and Theorem 5.5, the $\mathfrak{L}_{\nu,r}$-theory of the transform $\boldsymbol{H}^2_{\sigma,\kappa}$ in (5.1.5) with $a^* > 0$ in $\mathfrak{L}_{\nu,r}$-space can be established for any $\nu \in \mathbb{R}$ and $1 \leqq r \leqq \infty$.

Theorem 5.53. *Let* $a^* > 0$, $\alpha < 1 - \nu + \text{Re}(\kappa) < \beta$ *and* $1 \leqq r \leqq s \leqq \infty$.

(a) *The transform* $\boldsymbol{H}^2_{\sigma,\kappa}$ *defined on* $\mathfrak{L}_{\nu,2}$ *can be extended to* $\mathfrak{L}_{\nu,r}$ *as an element of* $[\mathfrak{L}_{\nu,r}, \mathfrak{L}_{\nu-\text{Re}(\kappa+\sigma),s}]$. *If* $1 \leqq r \leqq 2$, *then* $\boldsymbol{H}^2_{\sigma,\kappa}$ *is a one-to-one transform from* $\mathfrak{L}_{\nu,r}$ *onto* $\mathfrak{L}_{1-\nu+\text{Re}(\kappa+\sigma),s}$.

(b) *If $f \in \mathfrak{L}_{\nu,r}$ and $g \in \mathfrak{L}_{1-\nu+\mathrm{Re}(\kappa+\sigma),s'}$ with $1/s+1/s'=1$, then the relation (5.1.20) holds.*

By virtue of (5.1.10), (5.4.1), Theorems 5.18–5.22 and Lemma 3.2 we characterize the boundedness and the range of $\boldsymbol{H}^2_{\sigma,\kappa}$ on $\mathfrak{L}_{\nu,r}$ for $a^*>0$.

Theorem 5.54. *Let $a_1^*>0, a_2^*>0, m>0, n>0, \alpha<1-\nu+\mathrm{Re}(\kappa)<\beta$ and $\omega=\mu+a_1^*\alpha-a_2^*\beta+1$ and let $1<r<\infty$.*

(a) *If $\nu-\mathrm{Re}(\kappa)\notin\mathcal{E}_{\mathcal{H}}$, or if $1\leqq r\leqq 2$, then the transform $\boldsymbol{H}^2_{\sigma,\kappa}$ is one-to-one on $\mathfrak{L}_{\nu,r}$.*

(b) *If $\mathrm{Re}(\omega)\geqq 0$ and $\nu-\mathrm{Re}(\kappa)\notin\mathcal{E}_{\mathcal{H}}$, then*

$$\boldsymbol{H}^2_{\sigma,\kappa}(\mathfrak{L}_{\nu,r})=\left(\mathbb{L}_{-a_1^*,1-\alpha-\sigma}\mathbb{L}_{-a_2^*,\beta+\sigma+\omega/a_2^*}\right)\left(\mathfrak{L}_{\nu-\mathrm{Re}(\kappa+\sigma),r}\right). \tag{5.6.7}$$

When $\nu-\mathrm{Re}(\kappa)\in\mathcal{E}_{\mathcal{H}}$, $\boldsymbol{H}^1_{\sigma,\kappa}(\mathfrak{L}_{\nu,r})$ is a subset of the right-hand side of (5.6.7).

(c) *If $\mathrm{Re}(\omega)<0$ and $\nu-\mathrm{Re}(\kappa)\notin\mathcal{E}_{\mathcal{H}}$, then*

$$\boldsymbol{H}^2_{\sigma,\kappa}(\mathfrak{L}_{\nu,r})=\left(I^{-\omega}_{0+;1/a_1^*,a_1^*(1-\sigma-\alpha)-1}\mathbb{L}_{-a_1^*,1-\alpha-\sigma}\mathbb{L}_{-a_2^*,\beta+\sigma}\right)\left(\mathfrak{L}_{\nu-\mathrm{Re}(\kappa+\sigma),r}\right). \tag{5.6.8}$$

When $\nu-\mathrm{Re}(\kappa)\in\mathcal{E}_{\mathcal{H}}$, $\boldsymbol{H}^2_{\sigma,\kappa}(\mathfrak{L}_{\nu,r})$ is a subset of the right-hand side of (5.6.8).

Theorem 5.55. *Let $a_1^*>0, a_2^*=0, m>0, \alpha<1-\nu+\mathrm{Re}(\kappa)<\beta$ and $\omega=\mu+a_1^*\alpha+1/2$ and let $1<r<\infty$.*

(a) *If $\nu-\mathrm{Re}(\kappa)\notin\mathcal{E}_{\mathcal{H}}$, or if $1<r\leqq 2$, then the transform $\boldsymbol{H}^2_{\sigma,\kappa}$ is one-to-one on $\mathfrak{L}_{\nu,r}$.*

(b) *If $\mathrm{Re}(\omega)\geqq 0$ and $\nu-\mathrm{Re}(\kappa)\notin\mathcal{E}_{\mathcal{H}}$, then*

$$\boldsymbol{H}^2_{\sigma,\kappa}(\mathfrak{L}_{\nu,r})=\mathbb{L}_{-a_1^*,1-\alpha-\sigma+\omega/a_1^*}\left(\mathfrak{L}_{1-\nu+\mathrm{Re}(\kappa+\sigma),r}\right). \tag{5.6.9}$$

When $\nu-\mathrm{Re}(\kappa)\in\mathcal{E}_{\mathcal{H}}$, $\boldsymbol{H}^2_{\sigma,\kappa}(\mathfrak{L}_{\nu,r})$ is a subset of the right-hand side of (5.6.9).

(c) *If $\mathrm{Re}(\omega)<0$ and $\nu-\mathrm{Re}(\kappa)\notin\mathcal{E}_{\mathcal{H}}$, then*

$$\boldsymbol{H}^2_{\sigma,\kappa}(\mathfrak{L}_{\nu,r})=\left(I^{-\omega}_{0+;1/a_1^*,a_1^*(1-\sigma-\alpha)-1}\mathbb{L}_{-a_1^*,1-\alpha-\sigma}\right)\left(\mathfrak{L}_{1-\nu+\mathrm{Re}(\kappa+\sigma),r}\right). \tag{5.6.10}$$

When $\nu-\mathrm{Re}(\kappa)\in\mathcal{E}_{\mathcal{H}}$, $\boldsymbol{H}^2_{\sigma,\kappa}(\mathfrak{L}_{\nu,r})$ is a subset of the right-hand side of (5.6.10).

Theorem 5.56. *Let $a_1^*=0, a_2^*>0, n>0, \alpha<1-\nu+\mathrm{Re}(\kappa)<\beta$ and $\omega=\mu-a_2^*\beta+1/2$ and let $1<r<\infty$.*

(a) *If $\nu-\mathrm{Re}(\kappa)\notin\mathcal{E}_{\mathcal{H}}$, or if $1<r\leqq 2$, then the transform $\boldsymbol{H}^2_{\sigma,\kappa}$ is one-to-one on $\mathfrak{L}_{\nu,r}$.*

(b) *If $\mathrm{Re}(\omega)\geqq 0$ and $\nu-\mathrm{Re}(\kappa)\notin\mathcal{E}_{\mathcal{H}}$, then*

$$\boldsymbol{H}^2_{\sigma,\kappa}(\mathfrak{L}_{\nu,r})=\left(\mathbb{L}_{a_2^*,1-\beta-\sigma-\omega/a_2^*}\right)\left(\mathfrak{L}_{1-\nu+\mathrm{Re}(\kappa+\sigma),r}\right). \tag{5.6.11}$$

When $\nu-\mathrm{Re}(\kappa)\in\mathcal{E}_{\mathcal{H}}$, $\boldsymbol{H}^2_{\sigma,\kappa}(\mathfrak{L}_{\nu,r})$ is a subset of the right-hand side of (5.6.11).

(c) *If $\mathrm{Re}(\omega)<0$ and $\nu-\mathrm{Re}(\kappa)\notin\mathcal{E}_{\mathcal{H}}$, then*

$$\boldsymbol{H}^2_{\sigma,\kappa}(\mathfrak{L}_{\nu,r})=\left(I^{-\omega}_{-;1/a_2^*,a_2^*(\beta+\sigma-1)}\mathbb{L}_{a_2^*,1-\beta-\sigma}\right)\left(\mathfrak{L}_{1-\nu+\mathrm{Re}(\kappa+\sigma),r}\right). \tag{5.6.12}$$

When $\nu-\mathrm{Re}(\kappa)\in\mathcal{E}_{\mathcal{H}}$, $\boldsymbol{H}^2_{\sigma,\kappa}(\mathfrak{L}_{\nu,r})$ is a subset of the right-hand side of (5.6.12).

Theorem 5.57. *Let $a^* > 0$, $a_1^* > 0$, $a_2^* < 0$, $\alpha < 1 - \nu + \mathrm{Re}(\kappa) < \beta$ and let $1 < r < \infty$.*
(a) *If $\nu - \mathrm{Re}(\kappa) \notin \mathcal{E}_{\mathcal{H}}$, or if $1 < r \leqq 2$, then the transform $\boldsymbol{H}^2_{\sigma,\kappa}$ is one-to-one on $\mathfrak{L}_{\nu,r}$.*
(b) *Let $\omega, \eta, \zeta \in \mathbb{C}$ be chosen as*

$$\omega = a^*\eta - \mu - \frac{1}{2}; \tag{5.6.13}$$

$$a^*\mathrm{Re}(\eta) \geqq \gamma(r) + 2a_2^*[\nu - \mathrm{Re}(\kappa) - 1] + \mathrm{Re}(\mu); \tag{5.6.14}$$

$$\mathrm{Re}(\eta) > \nu - \mathrm{Re}(\kappa) - 1; \tag{5.6.15}$$

$$\mathrm{Re}(\zeta) < 1 - \nu + \mathrm{Re}(\kappa). \tag{5.6.16}$$

If $\nu - \mathrm{Re}(\kappa) \notin \mathcal{E}_{\mathcal{H}}$, then

$$\boldsymbol{H}^2_{\sigma,\kappa}(\mathfrak{L}_{\nu,r}) = \left(M_{\sigma-1/2-\omega/(2a_2^*)}\mathbb{H}_{2a_2^*,2a_2^*\zeta+\omega-1}\mathbb{L}_{a^*,1/2-\eta+\omega/(2a_2^*)}\right)$$
$$\left(\mathfrak{L}_{\nu-1/2-\mathrm{Re}(\omega)/(2a_2^*)-\mathrm{Re}(\kappa),r}\right). \tag{5.6.17}$$

When $\nu - \mathrm{Re}(\kappa) \in \mathcal{E}_{\mathcal{H}}$, $\boldsymbol{H}^2_{\sigma,\kappa}(\mathfrak{L}_{\nu,r})$ is a subset of the right-hand side of (5.6.17).

Theorem 5.58. *Let $a^* > 0$, $a_1^* < 0$, $a_2^* > 0$, $\alpha < 1 - \nu + \mathrm{Re}(\kappa) < \beta$ and let $1 < r < \infty$.*
(a) *If $\nu - \mathrm{Re}(\kappa) \notin \mathcal{E}_{\mathcal{H}}$, or if $1 < r \leqq 2$, then the transform $\boldsymbol{H}^2_{\sigma,\kappa}$ is one-to-one on $\mathfrak{L}_{\nu,r}$.*
(b) *Let $\omega, \eta, \zeta \in \mathbb{C}$ be chosen as*

$$\omega = a^*\eta - \Delta - \mu - \frac{1}{2}; \tag{5.6.18}$$

$$a^*\mathrm{Re}(\eta) \geqq \gamma(r) - 2a_1^*[\nu - \mathrm{Re}(\kappa)] + \Delta + \mathrm{Re}(\mu); \tag{5.6.19}$$

$$\mathrm{Re}(\eta) > -\nu + \mathrm{Re}(\kappa); \tag{5.6.20}$$

$$\mathrm{Re}(\zeta) < \nu - \mathrm{Re}(\kappa). \tag{5.6.21}$$

If $\nu - \mathrm{Re}(\kappa) \notin \mathcal{E}_{\mathcal{H}}$, then

$$\boldsymbol{H}^2_{\sigma,\kappa}(\mathfrak{L}_{\nu,r}) = \left(M_{\sigma+1/2+\omega/(2a_1^*)}\mathbb{H}_{-2a_1^*,2a_1^*\zeta+\omega-1}\mathbb{L}_{-a^*,1/2+\eta-\omega/(2a_1^*)}\right)$$
$$\left(\mathfrak{L}_{\nu+1/2+\mathrm{Re}(\omega)/(2a_1^*)-\mathrm{Re}(\kappa),r}\right). \tag{5.6.22}$$

When $\nu - \mathrm{Re}(\kappa) \in \mathcal{E}_{\mathcal{H}}$, $\boldsymbol{H}^2_{\sigma,\kappa}(\mathfrak{L}_{\nu,r})$ is a subset of the right-hand side of (5.6.22).

The inversion formulas for the transform $\boldsymbol{H}^2_{\sigma,\kappa}$ follow by noting that (5.1.5) is equaivalent to

$$\boldsymbol{H}^2 M_\kappa f = M_{-\sigma}\boldsymbol{H}^2_{\sigma,\kappa}$$

from (5.1.10) and Lemma 3.1(iii), where the isometric property (5.4.1) and the relation (5.4.24) for the operator $(M_\zeta)^{-1}$ inverse to M_ζ are used, and we have formally

$$f(x) = hx^{1-\kappa-(\lambda+1)/h}\frac{d}{dx}x^{(\lambda+1)/h}$$

$$\cdot \int_0^\infty H^{q-m,p-n+1}_{p+1,q+1}\left[\frac{x}{t}\,\middle|\,\begin{matrix}(-\lambda,h),(1-a_i-\alpha_i,\alpha_i)_{n+1,p},(1-a_i-\alpha_i,\alpha_i)_{1,n}\\(1-b_j-\beta_j,\beta_j)_{m+1,q},(1-b_j-\beta_j,\beta_j)_{1,m},(-\lambda-1,h)\end{matrix}\right]$$
$$\cdot\, t^{-\sigma-1}\left(\boldsymbol{H}^2_{\sigma,\kappa}f\right)(t)dt \tag{5.6.23}$$

or

$$f(x)=-hx^{1-\kappa-(\lambda+1)/h}\frac{d}{dx}x^{-(\lambda+1)/h}$$
$$\cdot \int_0^\infty H^{q-m+1,p-n}_{p+1,q+1}\left[\frac{x}{t}\,\middle|\,\begin{matrix}(1-a_i-\alpha_i,\alpha_i)_{n+1,p},(1-a_i-\alpha_i,\alpha_i)_{1,n},(-\lambda,h)\\(-\lambda-1,h),(1-b_j-\beta_j,\beta_j)_{m+1,q},(1-b_j-\beta_j,\beta_j)_{1,m}\end{matrix}\right]$$
$$\cdot\, t^{-\sigma-1}\left(\boldsymbol{H}^2_{\sigma,\kappa}f\right)(t)dt. \tag{5.6.24}$$

The validity of (5.6.23) and (5.6.24) can be proved by virtue of Theorems 5.23–5.24.

Theorem 5.59. *Let $a^*=0$, $\alpha<1-\nu+\mathrm{Re}(\kappa)<\beta$ and $\alpha_0<\nu-\mathrm{Re}(\kappa)<\beta_0$, and let $\lambda\in\mathbb{C},h>0$.*

(a) *If $\Delta[1-\nu+\mathrm{Re}(\kappa)]+\mathrm{Re}(\mu)=0$ and $f\in\mathfrak{L}_{\nu,2}$, then the inversion formula* (5.6.23) *holds for* $\mathrm{Re}(\lambda)>[\nu-\mathrm{Re}(\kappa)]h-1$ *and* (5.6.24) *for* $\mathrm{Re}(\lambda)<[\nu-\mathrm{Re}(\kappa)]h-1$.

(b) *If $\Delta=\mathrm{Re}(\mu)=0$ and $f\in\mathfrak{L}_{\nu,r}$ $(1<r<\infty)$, then the inversion formula* (5.6.23) *holds for* $\mathrm{Re}(\lambda)>[\nu-\mathrm{Re}(\kappa)]h-1$ *and* (5.6.24) *for* $\mathrm{Re}(\lambda)<[\nu-\mathrm{Re}(\kappa)]h-1$.

Theorem 5.60. *Let $a^*=0$, $1<r<\infty$ and $\Delta[1-\nu+\mathrm{Re}(\kappa)]+\mathrm{Re}(\mu)\leqq 1/2-\gamma(r)$, and let $\lambda\in\mathbb{C},h>0$.*

(a) *If $\Delta>0,m>0,\alpha<1-\nu+\mathrm{Re}(\kappa)<\beta$, $\alpha_0<\nu-\mathrm{Re}(\kappa)<\min[\beta_0,\{\mathrm{Re}(\mu+1/2)/\Delta\}+1]$ and if $f\in\mathfrak{L}_{\nu,r}$, then the inversion formulas* (5.6.23) *and* (5.6.24) *hold for* $\mathrm{Re}(\lambda)>[\nu-\mathrm{Re}(\kappa)]h-1$ *and for* $\mathrm{Re}(\lambda)<[\nu-\mathrm{Re}(\kappa)]h-1$*, respectively.*

(b) *If $\Delta<0,n>0,\alpha<1-\nu+\mathrm{Re}(\kappa)<\beta$, $\max[\alpha_0,\{\mathrm{Re}(\mu+1/2)/\Delta\}+1]<\nu-\mathrm{Re}(\kappa)<\beta_0$ and if $f\in\mathfrak{L}_{\nu,r}$, then the inversion formulas* (5.6.23) *and* (5.6.24) *hold for* $\mathrm{Re}(\lambda)>[\nu-\mathrm{Re}(\kappa)]h-1$ *and for* $\mathrm{Re}(\lambda)<[\nu-\mathrm{Re}(\kappa)]h-1$*, respectively.*

Remark 5.4. The results in this section generalize those in Section 5.3 by putting $\sigma=\kappa=0$.

5.7. Bibliographical Remarks and Additional Information on Chapter 5

For Section 5.1. The integral transforms with H-function kernels (5.1.1)–(5.1.5) are also investigated with a variable x in the upper or lower limit of the integral. These transformations, generalizing the Riemann–Liouville and Erdélyi–Kober fractional integrals (2.7.1), (3.3.1) and (2.7.2), (3.3.2), are defined for $x>0$ by

$$\left(\boldsymbol{H}_{U,0+}f\right)(x)=\sigma x^{-\eta-\sigma\gamma-1}\int_0^x(x^\sigma-t^\sigma)^\gamma H^{m,n}_{p,q}\left[kU\left(\frac{t}{x}\right)\,\middle|\,\begin{matrix}(a_i,\alpha_i)_{1,p}\\(b_j,\beta_j)_{1,q}\end{matrix}\right]t^\eta f(t)dt \tag{5.7.1}$$

and

$$\left(\boldsymbol{H}_{U,-}f\right)(x)=\sigma x^{\delta}\int_x^\infty(t^\sigma-x^\sigma)^\gamma H^{m,n}_{p,q}\left[kU\left(\frac{x}{t}\right)\,\middle|\,\begin{matrix}(a_i,\alpha_i)_{1,p}\\(b_j,\beta_j)_{1,q}\end{matrix}\right]t^{-\delta-\sigma\gamma-1}f(t)dt, \tag{5.7.2}$$

respectively, where

$$U(z) = (1 - z^\sigma)^\alpha z^{\beta\sigma} \tag{5.7.3}$$

and $\sigma > 0$, α, β, γ, η and k are complex numbers. Such generalized operators were introduced by R.K. Saxena and Kumbhat [2] who investigated these transforms in the space $L_r(\mathbb{R}_+)$ $(r \geqq 1)$, and proved their Mellin transforms for $f(x) \in L_r(\mathbb{R}_+)$ $(1 \leqq r \leqq 2)$ and the relation of integration by parts

$$\int_0^\infty g(x)\left(\boldsymbol{H}_{U,0+}f\right)(x)dx = \int_0^\infty f(x)\left(\boldsymbol{H}_{U,-}g\right)(x)dx \tag{5.7.4}$$

$$\left(f \in L_r(\mathbb{R}_+),\ g \in L_{r'}(\mathbb{R}_+),\ \frac{1}{r} + \frac{1}{r'} = 1\right),$$

provided that $a^* = 0$, $|\arg(k)| < a^*\pi/2$ and some other conditions are satisfied.

It should be noted that such results for the integral operators (5.7.1) and (5.7.2) with $\gamma = \alpha = 0$ were first obtained by Kalla [4]. In [5] Kalla established the inversion formulas for such operators on the basis of their representations in the form of the Mellin convolution

$$(k * f)(x) \equiv \int_0^\infty k\left(\frac{x}{t}\right) f(t)\frac{dt}{t} \tag{5.7.5}$$

with the kernel k involving the H-functions considered. Srivastava and Buschman [1] showed that the product of two operators of the form (5.7.1) and (5.7.2), studied by Kalla in [4] and [5], is the integral transform the kernel of which contains the H-function of two variables (see, for example, Srivastava, Gupta and Goyal [1]). One may find the results above in Kalla [4], [5], [10, Section 4] and Srivastava and Buschman [1]. We also mention the paper by Galué, Kalla and Srivastava [1] where results similar to the above were established for the multiplier of Erdélyi–Kober type operators of the forms (3.3.1) and (3.3.2) involving the H-functions in their kernels.

Dighe and Bhise [1] derived the composition of fractional integral operators with H-function kernels more general than those in (5.7.1) and (5.7.2), which led to an integral operator having a Fourier type kernel and generalizing the operator studied by Srivastava and Buschman [1]. Bhise and Dighe [1] also gave the conditions for the operator obtained to be bounded from $L_r(\mathbb{R}_+)$ to $L_q(\mathbb{R}_+)$.

R.K. Saxena and Singh [1] studied the existence, the Mellin transform, composition properties and the inversion relations for more general integral transforms than (5.7.1) and (5.7.2) in which the H-function $H^{m,n}_{p,q}(z)$ is replaced by a more general construction.

A method based on the Laplace transform was used by Srivastava and Buschman [2] and Srivastava [7] (see also Srivastava and Buschman [4]) to obtain the inversion formulas for the $\boldsymbol{H}$-transforms

$$\left(\widehat{\boldsymbol{H}}_{0+}f\right)(x) = \int_0^x (x-t)^\gamma H^{1,n}_{p,q}\left[x-t \,\middle|\, \begin{matrix}(a_i,\alpha_i)_{1,p}\\ (0,1),(b_j,\beta_j)_{2,q}\end{matrix}\right] t^\eta f(t)dt \tag{5.7.6}$$

and

$$\left(\widehat{\boldsymbol{H}}_{-}f\right)(x) = \int_x^\infty (t-x)^\gamma H^{m,n}_{p,q}\left[t-x \,\middle|\, \begin{matrix}(a_i,\alpha_i)_{1,p}\\ (b_j,\beta_j)_{1,q}\end{matrix}\right] t^{-\delta-\sigma\gamma-1} f(t)dt, \tag{5.7.7}$$

respectively, which are the modifications of the transforms (5.7.1) and (5.7.2) with $\sigma = 1$, $\alpha = 1$ and $\beta = -1$.

For Sections 5.2 and 5.3. The results presented in these sections were proved by the authors in Kilbas and Saigo [7].

McBride and Spratt [2] (see also McBride [4], [5]) investigated the transform $\boldsymbol{H}^1$ in (5.1.1) in the subspace $\mathsf{F}_{r,\nu}$ of the space $\mathfrak{L}_{\nu,r}$ defined for $\nu \in \mathbb{C}$ and $1 \leqq r \leqq \infty$ by

$$\mathsf{F}_{r,\nu} = \left\{\varphi \in C^\infty(\mathbb{R}_+) :\ x^n\frac{d^n f}{dx^n} \in \mathfrak{L}_{-\nu,r},\ n \in \mathbb{N}_0\right\}. \tag{5.7.8}$$

This space, studied by McBride in [2], is a Fréchet space with respect to the topology generated by the seminorms $\{\gamma_n^{r,\nu}\}_0^\infty$, where

$$\gamma_n^{r,\nu}(f) = \left\| x^n \frac{d^n f}{dx^n} \right\|_{-\nu,r} \quad (f \in F_{r,\nu},\ n \in \mathbb{N}_0). \tag{5.7.9}$$

McBride and Spratt defined the transform $\boldsymbol{H}^1$ by the relation (5.1.11) in the space $\mathsf{F}_{r,\nu}$ with $1 < r < \infty$ and $\alpha < \nu < \beta$, where α and β are any real numbers such that $\mathcal{H}(s) = \mathcal{H}_{p,q}^{m,n}(s)$ given by (1.1.2) is analytic in the strip $\alpha < \mathrm{Re}(s) < \beta$. They proved the following results [2, Theorems 5.4 amd 5.6]:

1) If $a^* = 0$ and if $1-\nu$ does not belong to the exceptional set $\mathcal{E}_{\mathcal{H}}$ of $\mathcal{H}(s)$ (see Definition 3.4), then $\boldsymbol{H}^1$ is a continuous linear mapping from $\mathsf{F}_{r,\nu}$ into $\mathsf{F}_{r,\nu}$, and if, in addition, $1-\nu$ does not belong to the exceptional set $1/\mathcal{E}_{\mathcal{H}}$, that $\boldsymbol{H}^1$ is a homeomorhism of $\mathsf{F}_{r,\nu}$ onto $\mathsf{F}_{r,\nu}$ whose inverse is the transform of the form (5.1.11) with $\mathcal{H}(s)$ replaced by $1/\mathcal{H}(s)$.

2) If $a^* > 0$ and if $1-\nu \notin \mathcal{E}_{\mathcal{H}}$, then $\boldsymbol{H}^1$ is a continuous linear mapping from $\mathsf{F}_{r,\nu}$ into the space F_{r,ν,a^*}, and if, in addition, $1-\nu \notin \mathcal{E}_{1/\mathcal{H}}$, that $\boldsymbol{H}^1$ is a homeomorhism of $\mathsf{F}_{r,\nu}$ onto F_{r,ν,a^*}.

Here the space F_{r,ν,a^*} is a certain subspace of $\mathsf{F}_{r,\nu}$ which is defined in terms of the simplest transform $\boldsymbol{H}^1$ in (5.1.11) with $\mathcal{H}_{0,1}^{1,0}(s) = \Gamma(\eta + s/m)$ with $\eta \in \mathbb{C}$ and $m > 0$, which for $1 \leqq p < \infty$ and $\mathrm{Re}(\eta - \nu/m) > 0$ is given by

$$\left(\mathbb{N}_m^\eta f\right)(x) = m\int_0^\infty t^{m\eta} \exp\left(-t^m\right) f\left(\frac{x}{t}\right)\frac{dt}{t} \quad (x > 0;\ \eta \in \mathbb{C},\ m > 0) \tag{5.7.10}$$

for $f \in \mathfrak{L}_{\nu,r}$. The substitution $t = x\tau$ show that this transform is the modified Laplace type transform of the form (7.1.10) with $\gamma = m\eta$, $k = m$, $\sigma = \kappa = 0$ and f being replaced by Rf:

$$\left(\mathbb{N}_m^\eta f\right)(x) = m\left(\mathbb{L}^*_{m\eta,m;0,0} Rf\right)(x), \tag{5.7.11}$$

where R is the elementary transform (3.3.13).

Several authors have studied modified transforms $\boldsymbol{H}^1$ and $\boldsymbol{H}^2$ in (5.1.1) and (5.1.2) in the form (5.7.1) and (5.7.2). Kiryakova [3] and Kalla and Kiryakova [1] investigated the operators

$$\left(I_{\beta;m}^{\gamma,\delta} f\right)(x) = \frac{1}{x}\int_0^x H_{m,m}^{m,0}\left[\frac{t}{x}\,\middle|\,\begin{matrix}\left(\gamma_i+\delta_i+1-\frac{1}{\beta_i},\frac{1}{\beta_i}\right)_{1,m}\\ \left(\gamma_j+1-\frac{1}{\beta_j},\frac{1}{\beta_j}\right)_{1,m}\end{matrix}\right] f(t)dt; \tag{5.7.12}$$

$$\left(K_{\beta;m}^{\gamma,\delta} f\right)(x) = \frac{1}{x}\int_x^\infty H_{m,m}^{m,0}\left[\frac{x}{t}\,\middle|\,\begin{matrix}\left(\gamma_i+\delta_i+\frac{1}{\beta_i},\frac{1}{\beta_i}\right)_{1,m}\\ \left(\gamma_j+\frac{1}{\beta_j},\frac{1}{\beta_j}\right)_{1,m}\end{matrix}\right] f(t)dt \tag{5.7.13}$$

in the space $L^p(\mathbb{R}_+)$ $(p \geqq 1)$. They obtained the conditions for the boundedness of the operators $I_{\beta;m}^{\gamma,\delta}$ and $K_{\beta;m}^{\gamma,\delta}$ in $L^p(\mathbb{R}_+)$, and proved some properties of them including the relations for their Mellin transforms, inversion formulas and the representations of the operators in (5.7.11) and (5.7.12) as compositions of m commuting Erdélyi–Kober type fractional integral operators (3.3.1) and (3.3.2), respectively. Kiryakova [4] established some of these properties for the operator (5.7.12) in some space of analytic functions. See also the book by Kiryakova [5] in this connection.

Raina and Saigo [1] studied the generalized fractional integral operators $I_{\beta;m}^{\gamma,\delta}$ and $K_{\beta;m}^{\gamma,\delta}$ in the spaces $\mathsf{F}_{r,\nu}$ and proved the conditions for their boundedeness, and some properties including their compositions with the differential operator δ defined by

$$\left(\delta f\right)(x) = x\frac{df}{dx}(x). \tag{5.7.14}$$

These investigations were continued by Saigo, Raina and Kilbas [1] who proved some properties of the operators (5.7.12) and (5.7.13), their Mellin transforms, the relation of fractional integration by parts

$$\int_0^\infty \left(I_{\beta;m}^{\gamma,\delta} g\right)(x) f(x)dx = \int_0^\infty \left(K_{\beta;m}^{\gamma,\delta} f\right)(x) g(x)dx \tag{5.7.15}$$

$$\left(f \in \mathsf{F}_{r,\nu},\ g \in \mathsf{F}_{r',-\nu},\ 1 \leqq r < \infty,\ \frac{1}{r} + \frac{1}{r'} = 1\right),$$

the decomposition relations in terms of m commuting Erdélyi–Kober type fractional integral operators (3.3.1) and (3.3.2), extensions of the range of parameteres and the compositions of $I_{\beta;m}^{\gamma,\delta}$ and $K_{\beta;m}^{\gamma,\delta}$ with the axisymmetric differential operator of potential theory L_σ defined by

$$\left(L_\sigma f\right)(x) = x^{-2\sigma-1}\frac{d}{dx}x^{2\sigma+1}\frac{d}{dx}f(x) = \frac{d^2 f(x)}{dx^2} + \frac{2\sigma+1}{x}\frac{df(x)}{dx}. \tag{5.7.16}$$

Using (5.7.15), Saigo, Raina and Kilbas [1] also defined the operators $I_{\beta;m}^{\gamma,\delta}$ and $K_{\beta;m}^{\gamma,\delta}$ in the spaces $\mathsf{F}'_{r,\nu}$ of generalized functions (space of continuous linear functionals on $\mathsf{F}_{r,\nu}$ equipped with the weak topology for which see McBride [2]), and proved in such a space $\mathsf{F}'_{r,\nu}$ the properties of the integral operators (5.7.12) and (5.7.13) similarly to those obtained in the space $\mathsf{F}_{r,\nu}$. Raina and Saigo [2] investigated in $\mathsf{F}'_{r,\nu}$ the compositions of such $\boldsymbol{H}$-transforms with the fractional integration operators involving the Gauss hypergeometric function (2.9.2) in the kernel defined by Saigo in [1] (see (7.12.45) and (7.12.46) in Section 7.12).

More general than (5.7.12) and (5.7.13) the modified $\boldsymbol{H}$-transforms

$$\left(\mathbf{I}_{p,q}^{m,n} f\right)(x) = \frac{1}{x}\int_0^x H_{p,q}^{m,n}\left[\frac{t}{x}\,\middle|\,\begin{matrix}(a_i,\alpha_i)_{1,p}\\(b_j,\beta_j)_{1,q}\end{matrix}\right] f(t)dt \tag{5.7.17}$$

and

$$\left(\mathbf{K}_{p,q}^{m,n} f\right)(x) = \frac{1}{x}\int_x^\infty H_{p,q}^{m,n}\left[\frac{x}{t}\,\middle|\,\begin{matrix}(a_i,\alpha_i)_{1,p}\\(b_j,\beta_j)_{1,q}\end{matrix}\right] f(t)dt \tag{5.7.18}$$

in the spaces $\mathsf{F}_{r,\nu}$ and $\mathsf{F}'_{r,\nu}$ were investigated by Kilbas and Saigo. The mapping properties were proved in Kilbas and Saigo [2], [3], while the compositions of these transforms with the differential operator of axisymmetric theory L_σ in (5.7.16) were given in Saigo and Kilbas [2], [3].

For Sections 5.4–5.6. As was indicated in Section 4.11 (For Sections 4.9 and 4.10), the papers by R. Singh [1], K.C. Gupta and P.K. Mittal [1], [2] and Nasim [2] are devoted to the inversion of the tramsforms $\boldsymbol{H}_{1,0}$ and $\boldsymbol{H}_{0,\kappa}$ given by (5.1.3). Mittal [1] showed that the composition of the $\boldsymbol{H}$-transform considered by K.C. Gupta and P.K. Mittal [1], with the Varma transform (see Section 7.2) also gives the $\boldsymbol{H}$-transform with another H-function in the kernel.

Using the Mellin transform, Mehra [2] proved the inversion formula for the generalized $\boldsymbol{H}^1$-transform (5.1.5) of the form

$$\left(\boldsymbol{H}^1 f\right)(x) = \int_0^\infty H_{p+1,q}^{m,n+1}\left[a\left(\frac{x}{t}\right)^\lambda\,\middle|\,\begin{matrix}(0,\lambda),(a_i,\alpha_i)_{1,p}\\(b_j,\beta_j)_{1,q}\end{matrix}\right] f(t)\frac{dt}{t} \quad (x>0) \tag{5.7.19}$$

with real a and λ.

de Amin and Kalla [1] derived the relation betweeen the Hardy transform (8.12.6) and the $\boldsymbol{H}$-transform of the form (4.11.17) considered by K.C. Gupta and P.K. Mittal [1], and pointed out that in special cases their result reduces to the relations between the Hardy transform and the Mejer, Varma, Hankel and Laplace transforms (see Chapter 7).

Treating the H-function transform of the form $\boldsymbol{H}_{0,1}$ given by (5.1.3), Srivastava and Buschman [3] established an expression for the H-function transform of the Mellin convolution (5.7.4) of two functions in terms of the Mellin convolution of H-function transforms of a function.

Dange and Chaudhary [1] proved the Abelian theorems for the modified $\boldsymbol{H}^2$-transform of the form

$$\left(\boldsymbol{H}^2_{0,0}f\right)(x) = \frac{1}{x}\int_0^\infty H^{1,2}_{2,2}\left[\left(\frac{x}{t}\right)^\lambda \middle| \begin{matrix}(a_1,\alpha_1),(1-a_2-\alpha_2,\lambda\alpha_2)\\(b_1,\beta_1),(b_2,\beta_2)\end{matrix}\right] f(t)dt \quad (x>0) \tag{5.7.20}$$

with real λ and showed that such Abelian theorems also hold for the special space of distributions f discussed by Zemanian [6].

Malgonde [1] extended the modified $\boldsymbol{H}$-transform (5.1.4) of the form

$$(\boldsymbol{H}^1_{0,\eta}f)(x) = \int_0^\infty \frac{t^{-\eta}}{\Gamma(\eta)} H^{m,n+1}_{p+1,q}\left[\frac{x}{t} \middle| \begin{matrix}(1-\eta,1),(a_i,\alpha_i)_{1,p}\\(b_j,\beta_j)_{1,q}\end{matrix}\right] f(t)dt \quad (x>0) \tag{5.7.21}$$

with $\eta > 0$ to a class of Banach space valued distributions, following the procedure by Zemanian [6], and derived a complex inversion formula for such a transform.

Virchenko and Haidey [1] considered the modified $\boldsymbol{H}$-transform

$$\int_0^\infty H^{0,1}_{2m,0}\left[\frac{x}{t} \middle| \begin{matrix}\overline{\qquad\qquad}\\ \left(1-\frac{\nu}{2},1\right)_{m\ \text{times}}, \left(\nu+1-\frac{\nu\lambda}{2},\lambda\right)_{m\ \text{times}}\end{matrix}\right] f(2\sqrt{t})dt \quad (x>0) \tag{5.7.22}$$

and proved its existence and the inversion relation in the space $\mathfrak{M}^{-1}_{c,\gamma}$ (see Samko, Kilbas and Marichev [1, Section 36]).

The results in Sections 5.4–5.6 have not been published before.

Chapter 6

G-TRANSFORM AND MODIFIED G-TRANSFORMS ON THE SPACE $\mathfrak{L}_{\nu,r}$

6.1. G-Transform on the Space $\mathfrak{L}_{\nu,r}$

This section deals with the $\boldsymbol{G}$-transforms, namely, the integral transforms of the form of (3.1.2):

$$\left(\boldsymbol{G}f\right)(x)=\int_0^\infty G_{p,q}^{m,n}\left[xt\,\middle|\,\begin{matrix}(a_i)_{1,p}\\(b_j)_{1,q}\end{matrix}\right]f(t)dt \tag{6.1.1}$$

with the Meijer G-function $G_{p,q}^{m,n}\left[z\,\middle|\,\begin{matrix}(a_i)_{1,p}\\(b_j)_{1,q}\end{matrix}\right]$ defined in (2.9.1) as kernel. A formal Mellin transform $\mathfrak{M}$, defined in (2.5.1), of (6.1.1) gives a similar relation to (3.1.5)

$$\left(\mathfrak{M}\boldsymbol{G}f\right)(s)=\mathcal{G}_{p,q}^{m,n}\left[\begin{matrix}(a_i)_{1,p}\\(b_j)_{1,q}\end{matrix}\,\middle|\,s\right]\left(\mathfrak{M}f\right)(1-s), \tag{6.1.2}$$

where

$$\mathcal{G}_{p,q}^{m,n}\left[\begin{matrix}(a_i)_{1,p}\\(b_j)_{1,q}\end{matrix}\,\middle|\,s\right]=\mathcal{G}_{p,q}^{m,n}\left[\begin{matrix}(a)_p\\(b)_q\end{matrix}\,\middle|\,s\right]=\mathcal{G}_{p,q}^{m,n}\left[\begin{matrix}a_1,\cdots,a_p\\b_1,\cdots,b_q\end{matrix}\,\middle|\,s\right]$$
$$=\frac{\prod_{j=1}^{m}\Gamma(b_j+s)\prod_{i=1}^{n}\Gamma(1-a_i-s)}{\prod_{i=n+1}^{p}\Gamma(a_i+s)\prod_{j=m+1}^{q}\Gamma(1-b_j-s)}. \tag{6.1.3}$$

In this section on the basis of the results in Chapters 3 and 4 we charactrerize the mapping properties such as the existence, boundedness and representative properties of the $\boldsymbol{G}$-transform (6.1.1) on the spaces $\mathfrak{L}_{\nu,r}$ and also give inversion formulas for this transform. As indicated in Section 3.1, (6.1.1) is a particular case of the $\boldsymbol{H}$-transform in (3.1.1) when

$$\alpha_1=\cdots=\alpha_p=\beta_1=\cdots=\beta_q=1. \tag{6.1.4}$$

The numbers $a^*,\Delta,a_1^*,a_2^*,\alpha$ and β given in (1.1.7), (1.1.8), (1.1.11), (1.1.12), (3.4.1) and (3.4.2) for the H-function (1.1.1), are simplified for the Meijer G-function (6.1.2) and take the forms:

$$a^* = 2(m+n) - p - q; \tag{6.1.5}$$

$$\Delta = q - p; \tag{6.1.6}$$

$$a_1^* = m + n - p; \tag{6.1.7}$$

$$a_2^* = m + n - q; \tag{6.1.8}$$

$$\alpha = \begin{cases} -\min\limits_{1 \leqq j \leqq m} [\mathrm{Re}(b_j)] & \text{if } m > 0, \\ -\infty & \text{if } m = 0; \end{cases} \tag{6.1.9}$$

and

$$\beta = \begin{cases} 1 - \max\limits_{1 \leqq i \leqq n} [\mathrm{Re}(a_i)] & \text{if } n > 0, \\ \infty & \text{if } n = 0, \end{cases} \tag{6.1.10}$$

while μ in (1.1.10) remains the same:

$$\mu = \sum_{j=1}^{q} b_j - \sum_{i=1}^{p} a_i + \frac{p-q}{2}. \tag{6.1.11}$$

Definition 3.4 on the exceptional set takes the following form.

Definition 6.1. Let the function $\mathcal{G}(s) = G_{p,q}^{m,n}\left[z \,\middle|\, \begin{matrix} (a_i)_{1,p} \\ (b_j)_{1,q} \end{matrix}\right]$ be given in (6.1.3) and let the real numbers α and β be defined in (6.1.9) and (6.1.10), respectively. We call the exceptional set of $\mathcal{G}$ the set $\mathcal{E}_{\mathcal{G}}$ of real numbers ν such that $\alpha < 1 - \nu < \beta$ and $\mathcal{G}(s)$ has a zero on the line $\mathrm{Re}(s) = 1 - \nu$.

From Theorems 3.6 and 3.7 we obtain the $\mathfrak{L}_{\nu,2}$-theory of the $\boldsymbol{G}$-transform (6.1.1).

Theorem 6.1. *We suppose that* **(a)** $\alpha < 1 - \nu < \beta$ *and either of the conditions* **(b)** $a^* > 0$, *or* **(c)** $a^* = 0$, $\Delta(1-\nu) + \mathrm{Re}(\mu) \leqq 0$ *holds. Then we have the results:*

(i) *There is a one-to-one transform* $\boldsymbol{G} \in [\mathfrak{L}_{\nu,2}, \mathfrak{L}_{1-\nu,2}]$ *such that* (6.1.2) *holds for* $\mathrm{Re}(s) = 1 - \nu$ *and* $f \in \mathfrak{L}_{\nu,2}$. *If* $a^* = 0$, $\Delta(1-\nu) + \mathrm{Re}(\mu) = 0$ *and* $\nu \notin \mathcal{E}_{\mathcal{G}}$, *then the transform* $\boldsymbol{G}$ *maps* $\mathfrak{L}_{\nu,2}$ *onto* $\mathfrak{L}_{1-\nu,2}$.

(ii) *If* $f, g \in \mathfrak{L}_{\nu,2}$, *then the relation*

$$\int_0^\infty f(x)\big(\boldsymbol{G}g\big)(x)dx = \int_0^\infty \big(\boldsymbol{G}f\big)(x)g(x)dx \tag{6.1.12}$$

holds.

(iii) *Let* $f \in \mathfrak{L}_{\nu,2}$ *and* $\lambda \in \mathbb{C}$. *If* $\mathrm{Re}(\lambda) > -\nu$, *then* $\boldsymbol{G}f$ *is given by*

$$\big(\boldsymbol{G}f\big)(x) = x^{-\lambda}\frac{d}{dx}x^{\lambda+1}\int_0^\infty G_{p+1,q+1}^{m,n+1}\left[xt \,\middle|\, \begin{matrix} -\lambda, a_1, \cdots, a_p \\ b_1, \cdots, b_q, -\lambda-1 \end{matrix}\right] f(t)dt. \tag{6.1.13}$$

When $\mathrm{Re}(\lambda) < -\nu$,

$$\left(\boldsymbol{G}f\right)(x) = -x^{-\lambda}\frac{d}{dx}x^{\lambda+1}\int_0^\infty G^{m+1,n}_{p+1,q+1}\left[xt \,\middle|\, \begin{matrix} a_1, \cdots, a_p, -\lambda \\ -\lambda-1, b_1, \cdots, b_q \end{matrix}\right] f(t)dt. \tag{6.1.14}$$

(iv) *The transform* $\boldsymbol{G}$ *is independent of* ν *in the sense that, if* ν *and* $\widetilde{\nu}$ *satisfy* (a), *and either* (b) *or* (c), *and if the transforms* $\boldsymbol{G}$ *and* $\widetilde{\boldsymbol{G}}$ *are defined in* $\mathfrak{L}_{\nu,2}$ *and* $\mathfrak{L}_{\widetilde{\nu},2}$, *respectively, by* (6.1.2), *then* $\boldsymbol{G}f = \widetilde{\boldsymbol{G}}f$ *for* $f \in \mathfrak{L}_{\nu,2} \cap \mathfrak{L}_{\widetilde{\nu},2}$.

(v) *If* $a^* > 0$ *or if* $a^* = 0$, $\Delta(1-\nu) + \mathrm{Re}(\mu) < 0$, *then* $\boldsymbol{G}f$ *is given in* (6.1.1) *for* $f \in \mathfrak{L}_{\nu,2}$.

Corollary 6.1.1. *Let* $\alpha < \beta$ *and let one of the following conditions hold:*

(b) $a^* > 0$,

(e) $a^* = 0$, $\Delta > 0$ and $\alpha < -\dfrac{\mathrm{Re}(\mu)}{\Delta}$,

(f) $a^* = 0$, $\Delta < 0$ and $\beta > -\dfrac{\mathrm{Re}(\mu)}{\Delta}$,

(g) $a^* = 0$, $\Delta = 0$ and $\mathrm{Re}(\mu) \leqq 0$.

Then the $\boldsymbol{G}$*-transform can be defined on* $\mathfrak{L}_{\nu,2}$ *with* $\alpha < \nu < \beta$.

Theorem 6.2. *Let* $\alpha < 1 - \nu < \beta$ *and either of the following conditions hold:*

(b) $a^* > 0$,

(d) $a^* = 0$, $\Delta(1-\nu) + \mathrm{Re}(\mu) < 0$.

Then for $f \in \mathfrak{L}_{\nu,2}$ *and* $x > 0$, $\left(\boldsymbol{G}f\right)(x)$ *is given in* (6.1.1).

Corollary 6.2.1. *Let* $\alpha < \beta$ *and let one of the following conditions hold:*

(b) $a^* > 0$,

(h) $a^* = 0$, $\Delta > 0$ and $\alpha < -\dfrac{\mathrm{Re}(\mu)+1}{\Delta}$,

(i) $a^* = 0$, $\Delta < 0$ and $\beta > -\dfrac{\mathrm{Re}(\mu)+1}{\Delta}$,

(j) $a^* = 0$, $\Delta = 0$ and $\mathrm{Re}(\mu) < 0$.

Then the $\boldsymbol{G}$*-transform can be defined by* (6.1.1) *on* $\mathfrak{L}_{\nu,2}$ *with* $\alpha < \nu < \beta$.

Using the results in Sections 4.1–4.4, we present the $\mathfrak{L}_{\nu,r}$-theory of the $\boldsymbol{G}$-transform (6.1.1) when $a^* = 0$. From Theorems 4.1–4.4 we obtain the mapping properties and the range of $\boldsymbol{G}$ on $\mathfrak{L}_{\nu,r}$ in three different cases when either $\Delta = \mathrm{Re}(\mu) = 0$ or $\Delta = 0$, $\mathrm{Re}(\mu) < 0$ or $\Delta \neq 0$. Here $a^*, \Delta, a_1^*, a_2^*, \alpha, \beta$ and μ are taken as in (6.1.5), (6.1.6), (6.1.7), (6.1.8), (6.1.9), (6.1.10) and (6.1.11), respectively.

Theorem 6.3. *Let $a^* = \Delta = 0, \mathrm{Re}(\mu) = 0$ and $\alpha < 1 - \nu < \beta$. Let $1 < r < \infty$.*

(a) *The transform G defined on $\mathfrak{L}_{\nu,2}$ can be extended to $\mathfrak{L}_{\nu,r}$ as an element of $[\mathfrak{L}_{\nu,r}, \mathfrak{L}_{1-\nu,r}]$.*

(b) *If $1 < r \leqq 2$, then the transform G is one-to-one on $\mathfrak{L}_{\nu,r}$ and there holds the equality (6.1.2) for $f \in \mathfrak{L}_{\nu,r}$ and $\mathrm{Re}(s) = 1 - \nu$.*

(c) *If $f \in \mathfrak{L}_{\nu,r}$ and $g \in \mathfrak{L}_{\nu,r'}$ with $r' = r/(r-1)$, then the relation (6.1.12) holds.*

(d) *If $\nu \notin \mathcal{E}_{\mathcal{G}}$, then the transform G is one-to-one on $\mathfrak{L}_{\nu,r}$ and there holds*

$$G(\mathfrak{L}_{\nu,r}) = \mathfrak{L}_{1-\nu,r}. \tag{6.1.15}$$

(e) *If $f \in \mathfrak{L}_{\nu,r}$ and $\lambda \in \mathbb{C}$, then Gf is given in (6.1.13) for $\mathrm{Re}(\lambda) > -\nu$, while in (6.1.14) for $\mathrm{Re}(\lambda) < -\nu$.*

Theorem 6.4. *Let $a^* = \Delta = 0, \mathrm{Re}(\mu) < 0$ and $\alpha < 1 - \nu < \beta$, and let either $m > 0$ or $n > 0$. Let $1 < r < \infty$.*

(a) *The transform G defined on $\mathfrak{L}_{\nu,2}$ can be extended to $\mathfrak{L}_{\nu,r}$ as an element of $[\mathfrak{L}_{\nu,r}, \mathfrak{L}_{1-\nu,s}]$ for all $s \geqq r$ such that $1/s > 1/r + \mathrm{Re}(\mu)$.*

(b) *If $1 < r \leqq 2$, then the transform G is one-to-one on $\mathfrak{L}_{\nu,r}$ and there holds the equality (6.1.2) for $f \in \mathfrak{L}_{\nu,r}$ and $\mathrm{Re}(s) = 1 - \nu$.*

(c) *If $f \in \mathfrak{L}_{\nu,r}$ and $g \in \mathfrak{L}_{\nu,s}$ with $1 < s < \infty$ and $1 \leqq 1/r + 1/s < 1 - \mathrm{Re}(\mu)$, then the relation (6.1.12) holds.*

(d) *Let $k > 0$. If $\nu \notin \mathcal{E}_{\mathcal{G}}$, then the transform G is one-to-one on $\mathfrak{L}_{\nu,r}$ and there hold*

$$G(\mathfrak{L}_{\nu,r}) = I^{-\mu}_{-;k,-\alpha/k}(\mathfrak{L}_{1-\nu,r}) \tag{6.1.16}$$

for $m > 0$, and

$$G(\mathfrak{L}_{\nu,r}) = I^{-\mu}_{0+;k,\beta/k-1}(\mathfrak{L}_{1-\nu,r}) \tag{6.1.17}$$

for $n > 0$. If $\nu \in \mathcal{E}_{\mathcal{G}}$, $G(\mathfrak{L}_{\nu,r})$ is a subset of the right-hand sides of (6.1.16) and (6.1.17) in the respective cases.

(e) *If $f \in \mathfrak{L}_{\nu,r}$ and $\lambda \in \mathbb{C}$, then Gf is given in (6.1.13) for $\mathrm{Re}(\lambda) > -\nu$, while in (6.1.14) for $\mathrm{Re}(\lambda) < -\nu$. Furthermore Gf is given in (6.1.1).*

Theorem 6.5. *Let $a^* = 0, \Delta \neq 0, \alpha < 1-\nu < \beta, 1 < r < \infty$ and $\Delta(1-\nu) + \mathrm{Re}(\mu) \leqq 1/2 - \gamma(r)$, where $\gamma(r)$ is defined in (3.3.9). Assume that $m > 0$ if $\Delta > 0$ and $n > 0$ if $\Delta < 0$.*

(a) *The transform G defined on $\mathfrak{L}_{\nu,2}$ can be extended to $\mathfrak{L}_{\nu,r}$ as an element of $[\mathfrak{L}_{\nu,r}, \mathfrak{L}_{1-\nu,s}]$ for all s with $r \leqq s < \infty$ such that $s' \geqq [1/2 - \Delta(1-\nu) - \mathrm{Re}(\mu)]^{-1}$ with $1/s + 1/s' = 1$.*

(b) *If $1 < r \leqq 2$, then the transform G is one-to-one on $\mathfrak{L}_{\nu,r}$ and there holds the equality (6.1.2) for $f \in \mathfrak{L}_{\nu,r}$ and $\mathrm{Re}(s) = 1 - \nu$.*

(c) *If $f \in \mathfrak{L}_{\nu,r}$ and $g \in \mathfrak{L}_{\nu,s}$ with $1 < s < \infty, 1/r + 1/s \geqq 1$ and $\Delta(1-\nu) + \mathrm{Re}(\mu) \leqq 1/2 - \max[\gamma(r), \gamma(s)]$, then the relation (6.1.12) holds.*

(d) *If $\nu \notin \mathcal{E}_{\mathcal{G}}$, then the transform G is one-to-one on $\mathfrak{L}_{\nu,r}$. If we set $\eta = -\Delta\alpha - \mu - 1$ for $\Delta > 0$ and $\eta = -\Delta\beta - \mu - 1$ for $\Delta < 0$, then $\mathrm{Re}(\eta) > -1$ and there holds*

$$G(\mathfrak{L}_{\nu,r}) = \left(M_{\mu/\Delta+1/2}\mathbb{H}_{\Delta,\eta}\right)\left(\mathfrak{L}_{\nu-1/2-\mathrm{Re}(\mu)/\Delta,r}\right). \tag{6.1.18}$$

When $\nu \in \mathcal{E}_{\mathcal{G}}$, $G(\mathfrak{L}_{\nu,r})$ is a subset of the right-hand side of (6.1.18).

(e) *If $f \in \mathfrak{L}_{\nu,r}, \lambda \in \mathbb{C}$ and $\Delta(1-\nu) + \mathrm{Re}(\mu) \leqq 1/2 - \gamma(r)$, then $\boldsymbol{G}f$ is given in* (6.1.13) *for* $\mathrm{Re}(\lambda) > -\nu$, *while in* (6.1.14) *for* $\mathrm{Re}(\lambda) < -\nu$. *If* $\Delta(1-\nu) + \mathrm{Re}(\mu) < 0$, $\boldsymbol{G}f$ *is given in* (6.1.1).

Corollary 6.5.1. *Let $1 < r < \infty$, $\alpha < \beta$, $a^* = 0$ and let one of the following conditions hold:*

(a) $\Delta > 0,\ \alpha < \dfrac{1}{\Delta}\left[\dfrac{1}{2} - \mathrm{Re}(\mu) - \gamma(r)\right]$,

(b) $\Delta < 0,\ \beta > \dfrac{1}{\Delta}\left[\dfrac{1}{2} - \mathrm{Re}(\mu) - \gamma(r)\right]$,

(c) $\Delta = 0,\ \mathrm{Re}(\mu) \leqq 0$.

Then the transform $\boldsymbol{G}$ can be defined on $\mathfrak{L}_{\nu,r}$ with $\alpha < 1 - \nu < \beta$.

From Theorem 4.5 in Section 4.5 we obtain the $\mathfrak{L}_{\nu,r}$-theory of the $\boldsymbol{G}$-transform (6.1.1) with $a^* > 0$ in $\mathfrak{L}_{\nu,r}$-spaces for any $\nu \in \mathbb{C}$ and $1 \leqq r \leqq \infty$.

Theorem 6.6. *Let $a^* > 0, \alpha < 1 - \nu < \beta$ and $1 \leqq r \leqq s \leqq \infty$.*

(a) *The $\boldsymbol{G}$-transform defined on $\mathfrak{L}_{\nu,2}$ can be extended to $\mathfrak{L}_{\nu,r}$ as an element of $[\mathfrak{L}_{\nu,r}, \mathfrak{L}_{1-\nu,s}]$. When $1 \leqq r \leqq 2$, $\boldsymbol{G}$ is a one-to-one transform from $\mathfrak{L}_{\nu,r}$ onto $\mathfrak{L}_{1-\nu,s}$.*

(b) *If $f \in \mathfrak{L}_{\nu,r}$ and $g \in \mathfrak{L}_{\nu,s'}$ with $1/s + 1/s' = 1$, then the relation* (6.1.12) *holds.*

According to Theorems 4.6–4.10 we characterize the boundedness and the range of $\boldsymbol{G}$ on $\mathfrak{L}_{\nu,r}$ for $a^* > 0$ but with various combinations of the signs of a_1^* and a_2^* which are classified in five cases: $a_1^* > 0, a_2^* > 0$; $a_1^* > 0, a_2^* = 0$; $a_1^* = 0, a_2^* > 0$; $a^* > 0, a_1^* > 0, a_2^* < 0$; and $a^* > 0, a_1^* < 0, a_2^* > 0$.

Theorem 6.7. *Let $a_1^* > 0, a_2^* > 0, m > 0, n > 0, \alpha < 1-\nu < \beta$ and $\omega = \mu + a_1^*\alpha - a_2^*\beta + 1$ and let $1 < r < \infty$.*

(a) *If $\nu \notin \mathcal{E}_{\mathcal{G}}$, or if $1 \leqq r \leqq 2$, then the transform $\boldsymbol{G}$ is one-to-one on $\mathfrak{L}_{\nu,r}$.*

(b) *If $\mathrm{Re}(\omega) \geqq 0$ and $\nu \notin \mathcal{E}_{\mathcal{G}}$, then*

$$\boldsymbol{G}(\mathfrak{L}_{\nu,r}) = \left(\mathbb{L}_{a_1^*,\alpha}\mathbb{L}_{a_2^*,1-\beta-\omega/a_2^*}\right)(\mathfrak{L}_{1-\nu,r}). \tag{6.1.19}$$

When $\nu \in \mathcal{E}_{\mathcal{G}}$, $\boldsymbol{G}(\mathfrak{L}_{\nu,r})$ is a subset of the right-hand side of (6.1.19).

(c) *If $\mathrm{Re}(\omega) < 0$ and $\nu \notin \mathcal{E}_{\mathcal{G}}$, then*

$$\boldsymbol{G}(\mathfrak{L}_{\nu,r}) = \left(I^{-\omega}_{-;1/a_1^*,-a_1^*\alpha}\mathbb{L}_{a_1^*,\alpha}\mathbb{L}_{a_2^*,1-\beta}\right)(\mathfrak{L}_{1-\nu,r}). \tag{6.1.20}$$

When $\nu \in \mathcal{E}_{\mathcal{G}}$, $\boldsymbol{G}(\mathfrak{L}_{\nu,r})$ is a subset of the right-hand side of (6.1.20).

Theorem 6.8. *Let $a_1^* > 0, a_2^* = 0, m > 0, \alpha < 1 - \nu < \beta$ and $\omega = \mu + a_1^*\alpha + 1/2$ and let $1 < r < \infty$.*

(a) *If $\nu \notin \mathcal{E}_{\mathcal{G}}$, or if $1 < r \leqq 2$, then the transform $\boldsymbol{G}$ is one-to-one on $\mathfrak{L}_{\nu,r}$.*

(b) *If $\mathrm{Re}(\omega) \geqq 0$ and $\nu \notin \mathcal{E}_{\mathcal{G}}$, then*

$$\boldsymbol{G}(\mathfrak{L}_{\nu,r}) = \mathbb{L}_{a_1^*,\alpha-\omega/a_1^*}(\mathfrak{L}_{\nu,r}). \tag{6.1.21}$$

When $\nu \in \mathcal{E}_{\mathcal{G}}$, $\boldsymbol{G}(\mathfrak{L}_{\nu,r})$ is a subset of the right-hand side of (6.1.21).

(c) *If* $\mathrm{Re}(\omega) < 0$ *and* $\nu \notin \mathcal{E}_{\mathcal{G}}$, *then*

$$\boldsymbol{G}(\mathfrak{L}_{\nu,r}) = \left(I^{-\omega}_{-;1/a_1^*,-a_1^*\alpha}\mathbb{L}_{a_1^*,\alpha}\right)(\mathfrak{L}_{\nu,r}). \tag{6.1.22}$$

When $\nu \in \mathcal{E}_{\mathcal{G}}$, $\boldsymbol{G}(\mathfrak{L}_{\nu,r})$ is a subset of the right-hand side of (6.1.22).

Theorem 6.9. *Let* $a_1^* = 0, a_2^* > 0, n > 0, \alpha < 1-\nu < \beta$ *and* $\omega = \mu - a_2^*\beta + 1/2$ *and let* $1 < r < \infty$.

(a) *If* $\nu \notin \mathcal{E}_{\mathcal{G}}$, *or if* $1 < r \leqq 2$, *then the transform* $\boldsymbol{G}$ *is one-to-one on* $\mathfrak{L}_{\nu,r}$.

(b) *If* $\mathrm{Re}(\omega) \geqq 0$ *and* $\nu \notin \mathcal{E}_{\mathcal{G}}$, *then*

$$\boldsymbol{G}(\mathfrak{L}_{\nu,r}) = \mathbb{L}_{-a_2^*,\beta+\omega/a_2^*}(\mathfrak{L}_{\nu,r}). \tag{6.1.23}$$

When $\nu \in \mathcal{E}_{\mathcal{G}}$, $\boldsymbol{G}(\mathfrak{L}_{\nu,r})$ is a subset of the right-hand side of (6.1.23).

(c) *If* $\mathrm{Re}(\omega) < 0$ *and* $\nu \notin \mathcal{E}_{\mathcal{G}}$, *then*

$$\boldsymbol{G}(\mathfrak{L}_{\nu,r}) = \left(I^{-\omega}_{0+;1/a_2^*,a_2^*\beta-1}\mathbb{L}_{-a_2^*,\beta}\right)(\mathfrak{L}_{\nu,r}). \tag{6.1.24}$$

When $\nu \in \mathcal{E}_{\mathcal{G}}$, $\boldsymbol{G}(\mathfrak{L}_{\nu,r})$ is a subset of the right-hand side of (6.1.24).

Theorem 6.10. *Let* $a^* > 0, a_1^* > 0, a_2^* < 0, \alpha < 1-\nu < \beta$ *and let* $1 < r < \infty$.

(a) *If* $\nu \notin \mathcal{E}_{\mathcal{G}}$, *or if* $1 < r \leqq 2$, *then the transform* $\boldsymbol{G}$ *is one-to-one on* $\mathfrak{L}_{\nu,r}$.

(b) *Let* $\omega, \eta, \zeta \in \mathbb{C}$ *be chosen as*

$$\omega = a^*\eta - \mu - \frac{1}{2}; \tag{6.1.25}$$

$$a^*\mathrm{Re}(\eta) \geqq \gamma(r) + 2a_2^*(\nu-1) + \mathrm{Re}(\mu); \tag{6.1.26}$$

$$\mathrm{Re}(\eta) > \nu - 1; \tag{6.1.27}$$

$$\mathrm{Re}(\zeta) < 1 - \nu. \tag{6.1.28}$$

If $\nu \notin \mathcal{E}_{\mathcal{G}}$, *then*

$$\boldsymbol{G}(\mathfrak{L}_{\nu,r}) = \left(M_{1/2+\omega/(2a_2^*)}\mathbb{H}_{-2a_2^*,2a_2^*\zeta+\omega-1}\mathbb{L}_{-a^*,1/2+\eta-\omega/(2a_2^*)}\right)\left(\mathfrak{L}_{3/2-\nu+\mathrm{Re}(\omega)/(2a_2^*),r}\right). \tag{6.1.29}$$

When $\nu \in \mathcal{E}_{\mathcal{G}}$, $\boldsymbol{G}(\mathfrak{L}_{\nu,r})$ is a subset of the right-hand side of (6.1.29).

Corollary 6.10.1. *Let* $a^* > 0, a_1^* > 0, a_2^* < 0, \alpha < 1-\nu < \beta$ *and let* $1 < r < \infty$.

(a) *If* $\nu \notin \mathcal{E}_{\mathcal{G}}$, *or if* $1 < r \leqq 2$, *then the transform* $\boldsymbol{G}$ *is one-to-one on* $\mathfrak{L}_{\nu,r}$.

(b) *Let* $\omega = a^*\eta - \mu - 1/2$ *and let* η *and* ζ *be chosen such that any of the following conditions holds:*

(i) $a^*\mathrm{Re}(\eta) \geqq \gamma(r) - 2a_2^*\beta + \mathrm{Re}(\mu),\ \mathrm{Re}(\eta) \geqq -\alpha,\ \mathrm{Re}(\zeta) \leqq \alpha$ *if* $m > 0, n > 0$;

(ii) $a^*\mathrm{Re}(\eta) \leqq \gamma(r) - 2a_2^*\beta + \mathrm{Re}(\mu),\ \mathrm{Re}(\eta) > \nu - 1,\ \mathrm{Re}(\zeta) < 1 - \nu$ *if* $m = 0, n > 0$;

(iii) $a^*\mathrm{Re}(\eta) \geqq \gamma(r) + 2a_2^*(\nu - 1) + \mathrm{Re}(\mu)$, $\mathrm{Re}(\eta) \geqq -\alpha$, $\mathrm{Re}(\zeta) \leqq \alpha$ *if* $m > 0, n = 0$.

Then, if $\nu \notin \mathcal{E}_{\mathcal{G}}$, $\boldsymbol{G}(\mathfrak{L}_{\nu,r})$ *can be represented by the relation* (6.1.29). *When* $\nu \in \mathcal{E}_{\mathcal{G}}$, $\boldsymbol{G}(\mathfrak{L}_{\nu,r})$ *is a subset of the right-hand side of* (6.1.29).

Corollary 6.10.2. *Let* $a^* > 0, a_1^* > 0, a_2^* < 0, m > 0, n > 0, \alpha < 1 - \nu < \beta$ *and* $1 < r < \infty$.

(a) *If* $\nu \notin \mathcal{E}_{\mathcal{G}}$, *or if* $1 < r \leqq 2$, *then the transform* $\boldsymbol{G}$ *is one-to-one on* $\mathfrak{L}_{\nu,r}$.

(b) *Let* $a^*\alpha - 2a_2^*\beta + \mathrm{Re}(\mu) + \gamma(r) \leqq 0$, $\omega = -a^*\alpha - \mu - 1/2$ *and let* ζ *be chosen such that* $\mathrm{Re}(\zeta) \leqq \alpha$. *Then if* $\nu \notin \mathcal{E}_{\mathcal{G}}$, $\boldsymbol{G}(\mathfrak{L}_{\nu,r})$ *can be represented in the form* (6.1.29). *When* $\nu \in \mathcal{E}_{\mathcal{G}}$, $\boldsymbol{G}(\mathfrak{L}_{\nu,r})$ *is a subset of the right-hand side of* (6.1.29).

Theorem 6.11. *Let* $a^* > 0, a_1^* < 0, a_2^* > 0, \alpha < 1 - \nu < \beta$ *and let* $1 < r < \infty$.

(a) *If* $\nu \notin \mathcal{E}_{\mathcal{G}}$, *or if* $1 < r \leqq 2$, *then the transform* $\boldsymbol{G}$ *is one-to-one on* $\mathfrak{L}_{\nu,r}$.

(b) *Let* $\omega, \eta, \zeta \in \mathbb{C}$ *be chosen as*

$$\omega = a^*\eta - \Delta - \mu - \frac{1}{2}; \tag{6.1.30}$$

$$a^*\mathrm{Re}(\eta) \geqq \gamma(r) - 2a_1^*\nu + \Delta + \mathrm{Re}(\mu); \tag{6.1.31}$$

$$\mathrm{Re}(\eta) > -\nu; \tag{6.1.32}$$

$$\mathrm{Re}(\zeta) < \nu. \tag{6.1.33}$$

If $\nu \notin \mathcal{E}_{\mathcal{G}}$, *then*

$$\boldsymbol{G}(\mathfrak{L}_{\nu,r}) = \left(M_{-1/2-\omega/(2a_1^*)}\mathbb{H}_{2a_1^*,2a_1^*\zeta+\omega-1}\mathbb{L}_{a^*,1/2-\eta+\omega/(2a_1^*)}\right) \left(\mathfrak{L}_{1/2-\nu-\mathrm{Re}(\omega)/(2a_1^*),r}\right). \tag{6.1.34}$$

When $\nu \in \mathcal{E}_{\mathcal{G}}$, $\boldsymbol{G}(\mathfrak{L}_{\nu,r})$ *is a subset of the right-hand side of* (6.1.34).

Corollary 6.11.1. *Let* $a^* > 0, a_1^* < 0, a_2^* > 0, \alpha < 1 - \nu < \beta$ *and let* $1 < r < \infty$.

(a) *If* $\nu \notin \mathcal{E}_{\mathcal{G}}$, *or if* $1 < r \leqq 2$, *then the transform* $\boldsymbol{G}$ *is one-to-one on* $\mathfrak{L}_{\nu,r}$.

(b) *Let* $\omega = a^*\eta - \Delta - \mu - 1/2$, *and let* η *and* ζ *be chosen such that either of the following conditions holds:*

(i) $a^*\mathrm{Re}(\eta) \geqq \gamma(r) - 2a_1^*(1-\nu) + \Delta + \mathrm{Re}(\mu)$, $\mathrm{Re}(\eta) \geqq \beta - 1$, $\mathrm{Re}(\zeta) \leqq 1 - \beta$ *if* $m > 0, n > 0$;

(ii) $a^*\mathrm{Re}(\eta) \geqq \gamma(r) - 2a_1^*\beta + \Delta + \mathrm{Re}(\mu)$, $\mathrm{Re}(\eta) \geqq \beta - 1$, $\mathrm{Re}(\zeta) \leqq 1 - \beta$ *if* $m = 0, n > 0$;

(iii) $a^*\mathrm{Re}(\eta) \geqq \gamma(r) - 2a_1^*(1 - \alpha) + \Delta + \mathrm{Re}(\mu)$, $\mathrm{Re}(\eta) > -\nu$, $\mathrm{Re}(\zeta) < \nu$ *if* $m > 0, n = 0$.

Then, if $\nu \notin \mathcal{E}_{\mathcal{G}}$, $\boldsymbol{G}(\mathfrak{L}_{\nu,r})$ *can be represented by the relation* (6.1.34). *When* $\nu \in \mathcal{E}_{\mathcal{G}}$, $\boldsymbol{G}(\mathfrak{L}_{\nu,r})$ *is a subset of the right-hand side of* (6.1.34).

Corollary 6.11.2. *Let* $a^* > 0, a_1^* < 0, a_2^* > 0, m > 0, n > 0, \alpha < 1 - \nu < \beta$, *and let* $1 < r < \infty$.

(a) *If* $\nu \notin \mathcal{E}_{\mathcal{G}}$, *or if* $1 < r \leqq 2$, *then the transform* $\boldsymbol{G}$ *is one-to-one on* $\mathfrak{L}_{\nu,r}$.

(b) *Let $2a_1^*\alpha - a_2^*\beta + \mathrm{Re}(\mu) + \gamma(r) \leqq 0$, and let ζ be chosen such that $\mathrm{Re}(\zeta) \leqq 1-\beta$. Then, if $\nu \notin \mathcal{E}_{\mathcal{G}}$, $G(\mathfrak{L}_{\nu,r})$ can be represented by the relation (6.1.34). When $\nu \in \mathcal{E}_{\mathcal{G}}$, $G(\mathfrak{L}_{\nu,r})$ is a subset of the right-hand side of (6.1.34).*

It follows from the results in Sections 4.9 and 4.10 that the inversion formulas for the G-transform (6.1.1) have the respective forms:

$$f(x) = x^{-\lambda}\frac{d}{dx}x^{\lambda+1}$$

$$\cdot \int_0^\infty G^{q-m,p-n+1}_{p+1,q+1}\left[xt \,\middle|\, \begin{matrix} -\lambda, -a_{n+1}, \cdots, -a_p, -a_1, \cdots, -a_n \\ -b_{m+1}, \cdots, -b_q, -b_1, \cdots, -b_m, -\lambda-1 \end{matrix}\right](Gf)(t)dt; \quad (6.1.35)$$

$$f(x) = -x^{-\lambda}\frac{d}{dx}x^{\lambda+1}$$

$$\cdot \int_0^\infty G^{q-m+1,p-n}_{p+1,q+1}\left[xt \,\middle|\, \begin{matrix} -a_{n+1}, \cdots, -a_p, -a_1, \cdots, -a_n, -\lambda \\ -\lambda-1, -b_{m+1}, \cdots, -b_q, -b_1, \cdots, -b_m \end{matrix}\right](Gf)(t)dt \quad (6.1.36)$$

corresponding to (6.1.13) and (6.1.14), provided that $a^* = 0$. The conditions for the validity of (6.1.35) and (6.1.36) follow from Theorems 4.11–4.14, if we take into account that the numbers α_0 and β_0 in (4.9.6) and (4.9.7) take the forms:

$$\alpha_0 = \begin{cases} \max\limits_{m+1\leqq j\leqq q}[\mathrm{Re}(b_j)] & \text{if } q > m; \\ -\infty & \text{if } q = m \end{cases} \quad (6.1.37)$$

and

$$\beta_0 = \begin{cases} \min\limits_{n+1\leqq i\leqq p}[\mathrm{Re}(a_i)] + 1 & \text{if } p > n; \\ \infty & \text{if } p = n. \end{cases} \quad (6.1.38)$$

We note that if $\alpha_0 < \nu < \beta_0$, ν is not in the exceptional set of $\mathcal{G}$ (see Definition 6.1).

From Theorems 4.11 and 4.12 we obtain the conditions for the inversion formulas (6.1.35) and (6.1.36) in the space $\mathfrak{L}_{\nu,2}$.

Theorem 6.12. *Let $a^* = 0, \alpha < 1-\nu < \beta$ and $\alpha_0 < \nu < \beta_0$, and let $\lambda \in \mathbb{C}$.*

(a) *If $\Delta(1-\nu) + \mathrm{Re}(\mu) = 0$ and if $f \in \mathfrak{L}_{\nu,2}$, then the inversion formula (6.1.35) holds for $\mathrm{Re}(\lambda) > \nu - 1$ and (6.1.36) for $\mathrm{Re}(\lambda) < \nu - 1$.*

(b) *If $\Delta = \mathrm{Re}(\mu) = 0$ and if $f \in \mathfrak{L}_{\nu,r}$ $(1 < r < \infty)$, then the inversion formula (6.1.35) holds for $\mathrm{Re}(\lambda) > \nu - 1$ and (6.1.36) for $\mathrm{Re}(\lambda) < \nu - 1$.*

Theorems 4.13 and 4.14 give the conditions for the inversion formulas (6.1.35) and (6.1.36) in $\mathfrak{L}_{\nu,r}$-space when $\Delta \neq 0$.

Theorem 6.13. *Let $a^* = 0$, $1 < r < \infty$ and $\Delta(1-\nu) + \mathrm{Re}(\mu) \leqq 1/2 - \gamma(r)$, and let $\lambda \in \mathbb{C}$.*

(a) *If* $\Delta > 0, m > 0, \alpha < 1-\nu < \beta, \alpha_0 < \nu < \min[\beta_0, \{\mathrm{Re}(\mu+1/2)/\Delta\}+1]$ *and* $f \in \mathfrak{L}_{\nu,r}$, *then the inversion formulas* (6.1.35) *and* (6.1.36) *hold for* $\mathrm{Re}(\lambda) > \nu-1$ *and for* $\mathrm{Re}(\lambda) < \nu-1$, *respectively.*

(b) *If* $\Delta < 0, n > 0, \alpha < 1-\nu < \beta, \max[\alpha_0, \{\mathrm{Re}(\mu+1/2)/\Delta\}+1] < \nu < \beta_0$ *and if* $f \in \mathfrak{L}_{\nu,r}$, *then the inversion formulas* (6.1.35) *and* (6.1.36) *hold for* $\mathrm{Re}(\lambda) > \nu-1$ *and for* $\mathrm{Re}(\lambda) < \nu-1$, *respectively.*

6.2. Modified G-Transforms

Let us consider the following modifications of the G-transform (6.1.1):

$$\left(G^1 f\right)(x) = \int_0^\infty G^{m,n}_{p,q}\left[\frac{x}{t}\,\middle|\,\begin{matrix}(a_i)_{1,p}\\(b_j)_{1,q}\end{matrix}\right] f(t)\frac{dt}{t}; \tag{6.2.1}$$

$$\left(G^2 f\right)(x) = \int_0^\infty G^{m,n}_{p,q}\left[\frac{t}{x}\,\middle|\,\begin{matrix}(a_i)_{1,p}\\(b_j)_{1,q}\end{matrix}\right] f(t)\frac{dt}{x}; \tag{6.2.2}$$

$$\left(G_{\sigma,\kappa} f\right)(x) = x^\sigma \int_0^\infty G^{m,n}_{p,q}\left[xt\,\middle|\,\begin{matrix}(a_i)_{1,p}\\(b_j)_{1,q}\end{matrix}\right] t^\kappa f(t)dt; \tag{6.2.3}$$

$$\left(G^1_{\sigma,\kappa} f\right)(x) = x^\sigma \int_0^\infty G^{m,n}_{p,q}\left[\frac{x}{t}\,\middle|\,\begin{matrix}(a_i)_{1,p}\\(b_j)_{1,q}\end{matrix}\right] t^\kappa f(t)\frac{dt}{t} \tag{6.2.4}$$

and

$$\left(G^2_{\sigma,\kappa} f\right)(x) = x^\sigma \int_0^\infty G^{m,n}_{p,q}\left[\frac{t}{x}\,\middle|\,\begin{matrix}(a_i)_{1,p}\\(b_j)_{1,q}\end{matrix}\right] t^\kappa f(t)\frac{dt}{x}, \tag{6.2.5}$$

where $\sigma, \kappa \in \mathbb{C}$.

These transforms are connected with the G-transform (6.1.1) by the relations

$$\left(G^1 f\right)(x) = \left(G\mathrm{R}f\right)(x); \tag{6.2.6}$$

$$\left(G^2 f\right)(x) = \left(RGf\right)(x); \tag{6.2.7}$$

$$\left(G_{\sigma,\kappa} f\right)(x) = \left(M_\sigma G M_\kappa f\right)(x); \tag{6.2.8}$$

$$\left(G^1_{\sigma,\kappa} f\right)(x) = \left(M_\sigma G R M_\kappa f\right)(x) = \left(M_\sigma G^1 M_\kappa f\right)(x) \tag{6.2.9}$$

and

$$\left(G^2_{\sigma,\kappa} f\right)(x) = \left(M_\sigma R G M_\kappa f\right)(x) = \left(M_\sigma G^2 M_\kappa f\right)(x), \tag{6.2.10}$$

respectively, where R and M_η are the operators given in (3.3.13) and (3.3.11). Due to (6.2.6)–(6.2.10), (6.1.2) and Lemma 3.1 the Mellin transforms of the modified G-transforms (6.2.1)–

(6.2.5) for "sufficiently good" function f are given by the relations

$$\left(\mathfrak{M}\boldsymbol{G}^1 f\right)(s) = \mathcal{G}_{p,q}^{m,n}\left[\begin{array}{c}(a_i)_{1,p}\\(b_j)_{1,q}\end{array}\middle| s\right]\left(\mathfrak{M}f\right)(s); \tag{6.2.11}$$

$$\left(\mathfrak{M}\boldsymbol{G}^2 f\right)(s) = \mathcal{G}_{p,q}^{m,n}\left[\begin{array}{c}(a_i)_{1,p}\\(b_j)_{1,q}\end{array}\middle| 1-s\right]\left(\mathfrak{M}f\right)(s); \tag{6.2.12}$$

$$\left(\mathfrak{M}\boldsymbol{G}_{\sigma,\kappa} f\right)(s) = \mathcal{G}_{p,q}^{m,n}\left[\begin{array}{c}(a_i)_{1,p}\\(b_j)_{1,q}\end{array}\middle| s+\sigma\right]\left(\mathfrak{M}f\right)(1-s-\sigma+\kappa); \tag{6.2.13}$$

$$\left(\mathfrak{M}\boldsymbol{G}_{\sigma,\kappa}^1 f\right)(s) = \mathcal{G}_{p,q}^{m,n}\left[\begin{array}{c}(a_i)_{1,p}\\(b_j)_{1,q}\end{array}\middle| s+\sigma\right]\left(\mathfrak{M}f\right)(s+\sigma+\kappa) \tag{6.2.14}$$

and

$$\left(\mathfrak{M}\boldsymbol{G}_{\sigma,\kappa}^2 f\right)(s) = \mathcal{G}_{p,q}^{m,n}\left[\begin{array}{c}(a_i)_{1,p}\\(b_j)_{1,q}\end{array}\middle| 1-s-\sigma\right]\left(\mathfrak{M}f\right)(s+\sigma+\kappa) \tag{6.2.15}$$

in terms of the function $\mathcal{G}_{p,q}^{m,n}\left[\begin{array}{c}(a_i)_{1,p}\\(b_j)_{1,q}\end{array}\middle| s\right]$ defined in (6.1.3).

It is directly verified that for the "sufficiently good" functions f and g the following formulas hold

$$\int_0^\infty f(x)\left(\boldsymbol{G}^1 g\right)(x)dx = \int_0^\infty \left(\boldsymbol{G}^2 f\right)(x)g(x)dx; \tag{6.2.16}$$

$$\int_0^\infty f(x)\left(\boldsymbol{G}^2 g\right)(x)dx = \int_0^\infty \left(\boldsymbol{G}^1 f\right)(x)g(x)dx; \tag{6.2.17}$$

$$\int_0^\infty f(x)\left(\boldsymbol{G}_{\sigma,\kappa} g\right)(x)dx = \int_0^\infty \left(\boldsymbol{G}_{\kappa,\sigma} f\right)(x)g(x)dx; \tag{6.2.18}$$

$$\int_0^\infty f(x)\left(\boldsymbol{G}_{\sigma,\kappa}^1 g\right)(x)dx = \int_0^\infty \left(\boldsymbol{G}_{\kappa,\sigma}^2 f\right)(x)g(x)dx \tag{6.2.19}$$

and

$$\int_0^\infty f(x)\left(\boldsymbol{G}_{\sigma,\kappa}^2 g\right)(x)dx = \int_0^\infty \left(\boldsymbol{G}_{\kappa,\sigma}^1 f\right)(x)g(x)dx. \tag{6.2.20}$$

The modified $\boldsymbol{G}$-transforms (6.2.1)–(6.2.5) are particular cases of the modified $\boldsymbol{H}$-transforms (5.1.1)–(5.1.5) when the condition (6.1.4) is satisfied. Therefore the properties in $\mathfrak{L}_{\nu,r}$-space for the modified $\boldsymbol{G}$-transforms follow from the corresponding results for the modified $\boldsymbol{H}$-transforms proved in Chapter 5.

We suppose in Sections 6.2–6.6 below that $a^*, \Delta, a_1^*, a_2^*, \alpha, \beta$ and μ are given in (6.1.5), (6.1.6), (6.1.7), (6.1.8), (6.1.9), (6.1.10) and (6.1.11), respectively

6.3. G^1-Transform on the Space $\mathfrak{L}_{\nu,r}$

We first investigate the transform G^1 defined in (6.2.1). Let $\mathcal{E}_{\mathcal{G}}$ be the exceptional set of the function $\mathcal{G}_{p,q}^{m,n}(s)$ given in Definition 6.1. From the results in Section 5.2 we obtain the $\mathfrak{L}_{\nu,2}$- and $\mathfrak{L}_{\nu,r}$-theory of the modified G^1-tranform (6.2.1). The first result follows from Theorem 5.1.

Theorem 6.14. *We suppose that* **(a)** $\alpha < \nu < \beta$ *and that either of the conditions* **(b)** $a^* > 0$, *or* **(c)** $a^* = 0$, $\Delta\nu + \mathrm{Re}(\mu) \leqq 0$ *holds. Then we have the following results:*

(i) *There is a one-to-one transform* $G^1 \in [\mathfrak{L}_{\nu,2}, \mathfrak{L}_{\nu,2}]$ *such that* (6.2.11) *holds for* $\mathrm{Re}(s) = \nu$ *and* $f \in \mathfrak{L}_{\nu,2}$. *If* $a^* = 0$, $\Delta\nu + \mathrm{Re}(\mu) = 0$ *and* $1 - \nu \notin \mathcal{E}_{\mathcal{G}}$, *then the transform* G^1 *maps* $\mathfrak{L}_{\nu,2}$ *onto* $\mathfrak{L}_{\nu,2}$.

(ii) *If* $f \in \mathfrak{L}_{\nu,2}$ *and* $g \in \mathfrak{L}_{1-\nu,2}$, *then the relation* (6.2.16) *holds for* G^1.

(iii) *Let* $f \in \mathfrak{L}_{\nu,2}$ *and* $\lambda \in \mathbb{C}$. *If* $\mathrm{Re}(\lambda) > \nu - 1$, *then* $G^1 f$ *is given by*

$$\left(G^1 f\right)(x) = x^{-\lambda}\frac{d}{dx}x^{\lambda+1}\int_0^\infty G_{p+1,q+1}^{m,n+1}\left[\frac{x}{t}\,\middle|\,\begin{matrix}-\lambda, a_1, \cdots, a_p\\ b_1, \cdots, b_q, -\lambda-1\end{matrix}\right]\frac{f(t)}{t}dt. \tag{6.3.1}$$

When $\mathrm{Re}(\lambda) < \nu - 1$,

$$\left(G^1 f\right)(x) = -x^{-\lambda}\frac{d}{dx}x^{\lambda+1}\int_0^\infty G_{p+1,q+1}^{m+1,n}\left[\frac{x}{t}\,\middle|\,\begin{matrix}a_1, \cdots, a_p, -\lambda\\ -\lambda-1, b_1, \cdots, b_q\end{matrix}\right]\frac{f(t)}{t}dt. \tag{6.3.2}$$

(iv) *The transform* G^1 *is independent of* ν *in the sense that, if* ν *and* $\widetilde{\nu}$ *satisfy* (a), *and* (b) *or* (c), *and if the transforms* G^1 *and* $\widetilde{G^1}$ *are defined in* $\mathfrak{L}_{\nu,2}$ *and* $\mathfrak{L}_{\widetilde{\nu},2}$, *respectively, by* (6.2.11), *then* $G^1 f = \widetilde{G^1} f$ *for* $f \in \mathfrak{L}_{\nu,2} \cap \mathfrak{L}_{\widetilde{\nu},2}$.

(v) *If* $a^* > 0$ *or if* $a^* = 0$, $\Delta\nu + \mathrm{Re}(\mu) < 0$, *then for* $f \in \mathfrak{L}_{\nu,2}$, $G^1 f$ *is given in* (6.2.1).

When $a^* = 0$, from Theorems 5.2–5.4 we obtain the mapping properties and the range of G^1 on $\mathfrak{L}_{\nu,r}$ in three different cases when either $\Delta = \mathrm{Re}(\mu) = 0$ or $\Delta = 0$, $\mathrm{Re}(\mu) < 0$ or $\Delta \neq 0$.

Theorem 6.15. *Let* $a^* = \Delta = 0, \mathrm{Re}(\mu) = 0, \alpha < \nu < \beta$ *and let* $1 < r < \infty$.

(a) *The transform* G^1 *defined on* $\mathfrak{L}_{\nu,2}$ *can be extended to* $\mathfrak{L}_{\nu,r}$ *as an element of* $[\mathfrak{L}_{\nu,r}, \mathfrak{L}_{\nu,r}]$.

(b) *If* $1 < r \leqq 2$, *then the transform* G^1 *is one-to-one on* $\mathfrak{L}_{\nu,r}$ *and there holds the equality* (6.2.11) *for* $f \in \mathfrak{L}_{\nu,r}$ *and* $\mathrm{Re}(s) = \nu$.

(c) *If* $f \in \mathfrak{L}_{\nu,r}$ *and* $g \in \mathfrak{L}_{1-\nu,r'}$ *with* $r' = r/(r-1)$, *then the relation* (6.2.16) *holds.*

(d) *If* $1 - \nu \notin \mathcal{E}_{\mathcal{G}}$, *then the transform* G^1 *is one-to-one on* $\mathfrak{L}_{\nu,r}$ *and there holds*

$$G^1(\mathfrak{L}_{\nu,r}) = \mathfrak{L}_{\nu,r}. \tag{6.3.3}$$

(e) *If* $f \in \mathfrak{L}_{\nu,r}$ *and* $\lambda \in \mathbb{C}$, *then* $G^1 f$ *is given in* (6.3.1) *for* $\mathrm{Re}(\lambda) > \nu - 1$, *while in* (6.3.2) *for* $\mathrm{Re}(\lambda) < \nu - 1$.

Theorem 6.16. *Let* $a^* = \Delta = 0, \mathrm{Re}(\mu) < 0$ *and* $\alpha < \nu < \beta$, *and let either* $m > 0$ *or* $n > 0$. *Let* $1 < r < \infty$.

(a) *The transform* G^1 *defined on* $\mathfrak{L}_{\nu,2}$ *can be extended to* $\mathfrak{L}_{\nu,r}$ *as an element of* $[\mathfrak{L}_{\nu,r}, \mathfrak{L}_{\nu,s}]$ *for all* $s \geqq r$ *such that* $1/s > 1/r + \mathrm{Re}(\mu)$.

(b) *If* $1 < r \leqq 2$, *then the transform* $\boldsymbol{G}^1$ *is one-to-one on* $\mathfrak{L}_{\nu,r}$ *and there holds the equality* (6.2.11) *for* $f \in \mathfrak{L}_{\nu,2}$ *and* $\mathrm{Re}(s) = \nu$.

(c) *If* $f \in \mathfrak{L}_{\nu,r}$ *and* $g \in \mathfrak{L}_{1-\nu,s}$ *with* $1 < s < \infty$ *and* $1 \leqq 1/r + 1/s < 1 - \mathrm{Re}(\mu)$, *then the relation* (6.2.16) *holds.*

(d) *Let* $k > 0$. *If* $1 - \nu \notin \mathcal{E}_{\mathcal{G}}$, *then the transform* $\boldsymbol{G}^1$ *is one-to-one on* $\mathfrak{L}_{\nu,r}$ *and there hold*

$$\boldsymbol{G}^1\left(\mathfrak{L}_{\nu,r}\right) = I^{-\mu}_{-;k,-\alpha/k}\left(\mathfrak{L}_{\nu,r}\right) \tag{6.3.4}$$

for $m > 0$, *and*

$$\boldsymbol{G}^1\left(\mathfrak{L}_{\nu,r}\right) = I^{-\mu}_{0+;k,\beta/k-1}\left(\mathfrak{L}_{\nu,r}\right) \tag{6.3.5}$$

for $n > 0$. *When* $1 - \nu \in \mathcal{E}_{\mathcal{G}}$, $\boldsymbol{G}^1\left(\mathfrak{L}_{\nu,r}\right)$ *is a subset of the right-hand sides of* (6.3.4) *and* (6.3.5) *in the respective cases.*

(e) *If* $f \in \mathfrak{L}_{\nu,r}$ *and* $\lambda \in \mathbb{C}$, *then* $\boldsymbol{G}^1 f$ *is given in* (6.3.1) *for* $\mathrm{Re}(\lambda) > \nu - 1$, *while in* (6.3.2) *for* $\mathrm{Re}(\lambda) < \nu - 1$. *Furthermore* $\boldsymbol{G}^1 f$ *is given in* (6.2.1).

Theorem 6.17. *Let* $a^* = 0, \Delta \neq 0, \alpha < \nu < \beta, 1 < r < \infty$ *and* $\Delta\nu + \mathrm{Re}(\mu) \leqq 1/2 - \gamma(r)$. *Assume that* $m > 0$ *if* $\Delta > 0$ *and* $n > 0$ *if* $\Delta < 0$.

(a) *The transform* $\boldsymbol{G}^1$ *defined on* $\mathfrak{L}_{\nu,2}$ *can be extended to* $\mathfrak{L}_{\nu,r}$ *as an element of* $[\mathfrak{L}_{\nu,r}, \mathfrak{L}_{\nu,s}]$ *for all* s *with* $r \leqq s < \infty$ *such that* $s' \geqq [1/2 - \Delta\nu - \mathrm{Re}(\mu)]^{-1}$ *with* $1/s + 1/s' = 1$.

(b) *If* $1 < r \leqq 2$, *then the transform* $\boldsymbol{G}^1$ *is one-to-one on* $\mathfrak{L}_{\nu,r}$ *and there holds the equality* (6.2.11) *for* $f \in \mathfrak{L}_{\nu,2}$ *and* $\mathrm{Re}(s) = \nu$.

(c) *If* $f \in \mathfrak{L}_{\nu,r}$ *and* $g \in \mathfrak{L}_{1-\nu,s}$ *with* $1 < s < \infty, 1/r + 1/s \geqq 1$ *and* $\Delta\nu + \mathrm{Re}(\mu) \leqq 1/2 - \max[\gamma(r), \gamma(s)]$, *then the relation* (6.2.16) *holds.*

(d) *If* $1 - \nu \notin \mathcal{E}_{\mathcal{G}}$, *then the transform* $\boldsymbol{G}^1$ *is one-to-one on* $\mathfrak{L}_{\nu,r}$. *If we set* $\eta = -\Delta\alpha - \mu - 1$ *for* $\Delta > 0$ *and* $\eta = -\Delta\beta - \mu - 1$ *for* $\Delta < 0$, *then* $\mathrm{Re}(\eta) > -1$ *and there holds*

$$\boldsymbol{G}^1(\mathfrak{L}_{\nu,r}) = \left(M_{\mu/\Delta+1/2}\mathbb{H}_{\Delta,\eta}\right)\left(\mathfrak{L}_{1/2-\nu-\mathrm{Re}(\mu)/\Delta,r}\right). \tag{6.3.6}$$

When $1 - \nu \in \mathcal{E}_{\mathcal{G}}$, $\boldsymbol{G}^1(\mathfrak{L}_{\nu,r})$ *is a subset of the right-hand side of* (6.3.6).

(e) *If* $f \in \mathfrak{L}_{\nu,r}, \lambda \in \mathbb{C}$ *and* $\Delta\nu + \mathrm{Re}(\mu) \leqq 1/2 - \gamma(r)$, *then* $\boldsymbol{G}^1 f$ *is given in* (6.3.1) *for* $\mathrm{Re}(\lambda) > \nu - 1$, *while in* (6.3.2) *for* $\mathrm{Re}(\lambda) < \nu - 1$. *If* $\Delta\nu + \mathrm{Re}(\mu) < 0$, $\boldsymbol{G}^1 f$ *is given in* (6.2.1).

From Theorem 5.5 we obtain the $\mathfrak{L}_{\nu,r}$-theory of the transform $\boldsymbol{G}^1$ in (6.2.1) for $a^* > 0$.

Theorem 6.18. *Let* $a^* > 0, \alpha < \nu < \beta$ *and* $1 \leqq r \leqq s \leqq \infty$.

(a) *The transform* $\boldsymbol{G}^1$ *defined on* $\mathfrak{L}_{\nu,2}$ *can be extended to* $\mathfrak{L}_{\nu,r}$ *as an element of* $[\mathfrak{L}_{\nu,r}, \mathfrak{L}_{\nu,s}]$. *If* $1 \leqq r \leqq 2$, *then* $\boldsymbol{G}^1$ *is a one-to-one transform from* $\mathfrak{L}_{\nu,r}$ *onto* $\mathfrak{L}_{\nu,s}$.

(b) *If* $f \in \mathfrak{L}_{\nu,r}$ *and* $g \in \mathfrak{L}_{1-\nu,s'}$ *with* $1/s + 1/s' = 1$, *then the relation* (6.2.16) *holds.*

When $a^* > 0$, the boundedness and the range of $\boldsymbol{G}^1$ on $\mathfrak{L}_{\nu,r}$ can be obtained in five cases as in Theorems 5.6–5.10.

Theorem 6.19. *Let* $a_1^* > 0, a_2^* > 0, m > 0, n > 0, \alpha < \nu < \beta$ *and* $\omega = \mu + a_1^*\alpha - a_2^*\beta + 1$ *and let* $1 < r < \infty$.

(a) *If* $1-\nu \notin \mathcal{E}_{\mathcal{G}}$, *or if* $1 \leqq r \leqq 2$, *then the transform* G^1 *is one-to-one on* $\mathfrak{L}_{\nu,r}$.

(b) *If* $\mathrm{Re}(\omega) \geqq 0$ *and* $1-\nu \notin \mathcal{E}_{\mathcal{G}}$, *then*

$$G^1(\mathfrak{L}_{\nu,r}) = \left(\mathbb{L}_{a_1^*,\alpha}\mathbb{L}_{a_2^*,1-\beta-\omega/a_2^*}\right)(\mathfrak{L}_{\nu,r}). \tag{6.3.7}$$

When $1-\nu \in \mathcal{E}_{\mathcal{G}}$, $G^1(\mathfrak{L}_{\nu,r})$ *is a subset of the right-hand side of* (6.3.7).

(c) *If* $\mathrm{Re}(\omega) < 0$ *and* $1-\nu \notin \mathcal{E}_{\mathcal{G}}$, *then*

$$G^1(\mathfrak{L}_{\nu,r}) = \left(I^{-\omega}_{-;1/a_1^*,-a_1^*\alpha}\mathbb{L}_{a_1^*,\alpha}\mathbb{L}_{a_2^*,1-\beta}\right)(\mathfrak{L}_{\nu,r}). \tag{6.3.8}$$

When $1-\nu \in \mathcal{E}_{\mathcal{G}}$, $G^1(\mathfrak{L}_{\nu,r})$ *is a subset of the right-hand side of* (6.3.8).

Theorem 6.20. *Let* $a_1^* > 0, a_2^* = 0, m > 0, \alpha < \nu < \beta$ *and* $\omega = \mu + a_1^*\alpha + 1/2$ *and let* $1 < r < \infty$.

(a) *If* $1-\nu \notin \mathcal{E}_{\mathcal{G}}$, *or if* $1 < r \leqq 2$, *then the transform* G^1 *is one-to-one on* $\mathfrak{L}_{\nu,r}$.

(b) *If* $\mathrm{Re}(\omega) \geqq 0$ *and* $1-\nu \notin \mathcal{E}_{\mathcal{G}}$, *then*

$$G^1(\mathfrak{L}_{\nu,r}) = \mathbb{L}_{a_1^*,\alpha-\omega/a_1^*}(\mathfrak{L}_{1-\nu,r}). \tag{6.3.9}$$

When $1-\nu \in \mathcal{E}_{\mathcal{G}}$, $G^1(\mathfrak{L}_{\nu,r})$ *is a subset of the right-hand side of* (6.3.9).

(c) *If* $\mathrm{Re}(\omega) < 0$ *and* $1-\nu \notin \mathcal{E}_{\mathcal{G}}$, *then*

$$G^1(\mathfrak{L}_{\nu,r}) = \left(I^{-\omega}_{-;1/a_1^*,-a_1^*\alpha}\mathbb{L}_{a_1^*,\alpha}\right)(\mathfrak{L}_{1-\nu,r}). \tag{6.3.10}$$

When $1-\nu \in \mathcal{E}_{\mathcal{G}}$, $G^1(\mathfrak{L}_{\nu,r})$ *is a subset of the right-hand side of* (6.3.10).

Theorem 6.21. *Let* $a_1^* = 0, a_2^* > 0, n > 0, \alpha < \nu < \beta$ *and* $\omega = \mu - a_2^*\beta + 1/2$ *and let* $1 < r < \infty$.

(a) *If* $1-\nu \notin \mathcal{E}_{\mathcal{G}}$, *or if* $1 < r \leqq 2$, *then the transform* G^1 *is one-to-one on* $\mathfrak{L}_{\nu,r}$.

(b) *If* $\mathrm{Re}(\omega) \geqq 0$ *and* $1-\nu \notin \mathcal{E}_{\mathcal{G}}$, *then*

$$G^1(\mathfrak{L}_{\nu,r}) = \mathbb{L}_{-a_2^*,\beta+\omega/a_2^*}(\mathfrak{L}_{1-\nu,r}). \tag{6.3.11}$$

When $1-\nu \in \mathcal{E}_{\mathcal{G}}$, $G^1(\mathfrak{L}_{\nu,r})$ *is a subset of the right-hand side of* (6.3.11).

(c) *If* $\mathrm{Re}(\omega) < 0$ *and* $1-\nu \notin \mathcal{E}_{\mathcal{G}}$, *then*

$$G^1(\mathfrak{L}_{\nu,r}) = \left(I^{-\omega}_{0+;1/a_2^*,a_2^*\beta-1}\mathbb{L}_{-a_2^*,\beta}\right)(\mathfrak{L}_{1-\nu,r}). \tag{6.3.12}$$

When $1-\nu \in \mathcal{E}_{\mathcal{G}}$, $G^1(\mathfrak{L}_{\nu,r})$ *is a subset of the right-hand side of* (6.3.12).

Theorem 6.22. *Let* $a^* > 0, a_1^* > 0, a_2^* < 0, \alpha < \nu < \beta$ *and let* $1 < r < \infty$.

(a) *If* $1-\nu \notin \mathcal{E}_{\mathcal{G}}$, *or if* $1 < r \leqq 2$, *then the transform* G^1 *is one-to-one on* $\mathfrak{L}_{\nu,r}$.

(b) *Let* $\omega, \eta, \zeta \in \mathbb{C}$ *be chosen as*

$$\omega = a^*\eta - \mu - \frac{1}{2}; \tag{6.3.13}$$

$$a^*\mathrm{Re}(\eta) \geqq \gamma(r) - 2a_2^*\nu + \mathrm{Re}(\mu); \tag{6.3.14}$$

$$\mathrm{Re}(\eta) > -\nu; \tag{6.3.15}$$

$$\mathrm{Re}(\zeta) < \nu. \tag{6.3.16}$$

If $1-\nu \notin \mathcal{E}_{\mathcal{G}}$, *then*

$$G^1(\mathfrak{L}_{\nu,r}) = \left(M_{1/2+\omega/(2a_2^*)}\mathbb{H}_{-2a_2^*,2a_2^*\zeta+\omega-1}\mathbb{L}_{-a^*,1/2+\eta-\omega/(2a_2^*)}\right)$$
$$\left(\mathfrak{L}_{\nu+1/2+\mathrm{Re}(\omega)/(2a_2^*),r}\right). \tag{6.3.17}$$

When $1-\nu \in \mathcal{E}_{\mathcal{G}}$, $G^1(\mathfrak{L}_{\nu,r})$ *is a subset of the right-hand side of* (6.3.17).

Theorem 6.23. *Let* $a^* > 0, a_1^* < 0, a_2^* > 0, \alpha < \nu < \beta$ *and let* $1 < r < \infty$.
(a) *If* $1-\nu \notin \mathcal{E}_{\mathcal{G}}$, *or if* $1 < r \leqq 2$, *then the transform* G^1 *is one-to-one on* $\mathfrak{L}_{\nu,r}$.
(b) *Let* $\omega, \eta, \zeta \in \mathbb{C}$ *be chosen as*

$$\omega = a^*\eta - \Delta - \mu - \frac{1}{2}; \tag{6.3.18}$$

$$a^*\mathrm{Re}(\eta) \geqq \gamma(r) + 2a_1^*(\nu-1) + \Delta + \mathrm{Re}(\mu); \tag{6.3.19}$$

$$\mathrm{Re}(\eta) > \nu - 1; \tag{6.3.20}$$

$$\mathrm{Re}(\zeta) < 1 - \nu. \tag{6.3.21}$$

If $1-\nu \notin \mathcal{E}_{\mathcal{G}}$, *then*

$$G^1(\mathfrak{L}_{\nu,r}) = \left(M_{-1/2-\omega/(2a_1^*)}\mathbb{H}_{2a_1^*,2a_1^*\zeta+\omega-1}\mathbb{L}_{a^*,1/2-\eta+\omega/(2a_1^*)}\right)$$
$$\left(\mathfrak{L}_{\nu-1/2-\mathrm{Re}(\omega)/(2a_1^*),r}\right). \tag{6.3.22}$$

When $1-\nu \in \mathcal{E}_{\mathcal{G}}$, $G^1(\mathfrak{L}_{\nu,r})$ *is a subset of the right-hand side of* (6.3.22).

The inversion formulas for the transform G^1 in (6.2.1) follow from those for (5.2.27) and (5.2.28). They take the forms

$$f(x) = -x^{\lambda+1}\frac{d}{dx}x^{-(\lambda+1)}$$
$$\cdot \int_0^\infty G^{q-m,p-n+1}_{p+1,q+1}\left[\frac{t}{x}\,\middle|\,\begin{matrix}-\lambda, -a_{n+1}, \cdots, -a_p, -a_1, \cdots, -a_n \\ -b_{m+1}, \cdots, -b_q, -b_1, \cdots, -b_m, -\lambda-1\end{matrix}\right](G^1 f)(t)dt \tag{6.3.23}$$

and

$$f(x) = x^{\lambda+1}\frac{d}{dx}x^{-(\lambda+1)}$$
$$\cdot \int_0^\infty G^{q-m+1,p-n}_{p+1,q+1}\left[\frac{t}{x}\,\middle|\,\begin{matrix}-a_{n+1}, \cdots, -a_p, -a_1, \cdots, -a_n, -\lambda \\ -\lambda-1, -b_{m+1}, \cdots, -b_q, -b_1, \cdots, -b_m\end{matrix}\right](G^1 f)(t)dt, \tag{6.3.24}$$

respectively.

The conditions for the validity of (6.3.23) and (6.3.24) follow from Theorems 5.11 and 5.12, if we take into account that the numbers α_0 and β_0 are given in (6.1.37) and (6.1.38).

Theorem 6.24. *Let* $a^* = 0, \alpha < \nu < \beta$ *and* $\alpha_0 < 1-\nu < \beta_0$, *and let* $\lambda \in \mathbb{C}$.

(a) *If* $\Delta\nu + \mathrm{Re}(\mu) = 0$ *and* $f \in \mathfrak{L}_{\nu,2}$, *then the inversion formula* (6.3.23) *holds for* $\mathrm{Re}(\lambda) > -\nu$ *and* (6.3.24) *for* $\mathrm{Re}(\lambda) < -\nu$.

(b) *If* $\Delta = \mathrm{Re}(\mu) = 0$ *and* $f \in \mathfrak{L}_{\nu,r}$ $(1 < r < \infty)$, *then the inversion formula* (6.3.23) *holds for* $\mathrm{Re}(\lambda) > -\nu$ *and* (6.3.24) *for* $\mathrm{Re}(\lambda) < -\nu$.

Theorem 6.25. *Let* $a^* = 0, 1 < r < \infty$ and $\Delta\nu + \mathrm{Re}(\mu) \leqq 1/2 - \gamma(r)$, and *let* $\lambda \in \mathbb{C}$.

(a) *If* $\Delta > 0, m > 0, \alpha < \nu < \beta$, $\alpha_0 < 1 - \nu < \min[\beta_0, \{\mathrm{Re}(\mu + 1/2)/\Delta\} + 1]$ *and if* $f \in \mathfrak{L}_{\nu,r}$, *then the inversion formulas* (6.3.23) *and* (6.3.24) *hold for* $\mathrm{Re}(\lambda) > \nu$ *and for* $\mathrm{Re}(\lambda) < -\nu$, *respectively.*

(b) *If* $\Delta < 0, n > 0, \alpha < \nu < \beta$, $\max[\alpha_0, \{\mathrm{Re}(\mu + 1/2)/\Delta\} + 1] < 1 - \nu < \beta_0$ *and if* $f \in \mathfrak{L}_{\nu,r}$, *then the inversion formulas* (6.3.23) *and* (6.3.24) *hold for* $\mathrm{Re}(\lambda) > \nu$ *and for* $\mathrm{Re}(\lambda) < -\nu$, *respectively.*

6.4. G^2-Transform on the Space $\mathfrak{L}_{\nu,r}$

Now we treat the transform G^2 defined in (6.2.2). From the results in Section 5.3 we obtain $\mathfrak{L}_{\nu,2}$- and $\mathfrak{L}_{\nu,r}$-theory for this transform. The first result comes from Theorem 5.13.

Theorem 6.26. *We suppose that* **(a)** $\alpha < 1 - \nu < \beta$ *and that either of the conditions* **(b)** $a^* > 0$, or **(c)** $a^* = 0$, $\Delta(1-\nu) + \mathrm{Re}(\mu) \leqq 0$ *holds. Then we have the following results:*

(i) *There is a one-to-one transform* $G^2 \in [\mathfrak{L}_{\nu,2}, \mathfrak{L}_{\nu,2}]$ *such that* (6.2.12) *holds for* $\mathrm{Re}(s) = \nu$ *and* $f \in \mathfrak{L}_{\nu,2}$. *If* $a^* = 0$, $\Delta(1-\nu) + \mathrm{Re}(\mu) = 0$ *and* $\nu \notin \mathcal{E}_{\mathcal{G}}$, *then the transform* G^2 *maps* $\mathfrak{L}_{\nu,2}$ *onto* $\mathfrak{L}_{\nu,2}$.

(ii) *If* $f \in \mathfrak{L}_{\nu,2}$ *and* $g \in \mathfrak{L}_{1-\nu,2}$, *then the relation* (6.2.17) *holds.*

(iii) *Let* $f \in \mathfrak{L}_{\nu,2}$ *and* $\lambda \in \mathbb{C}$. *If* $\mathrm{Re}(\lambda) > -\nu$, *then* $G^2 f$ *is given by*

$$\left(G^2 f\right)(x) = -x^{\lambda+1}\frac{d}{dx}x^{-\lambda-1}\int_0^\infty G^{m,n+1}_{p+1,q+1}\left[\frac{t}{x}\,\middle|\,\begin{matrix}-\lambda, a_1, \cdots, a_p\\ b_1, \cdots, b_q, -\lambda-1\end{matrix}\right]f(t)dt. \tag{6.4.1}$$

When $\mathrm{Re}(\lambda) < -\nu$,

$$\left(G^2 f\right)(x) = x^{\lambda+1}\frac{d}{dx}x^{-\lambda-1}\int_0^\infty G^{m+1,n}_{p+1,q+1}\left[\frac{t}{x}\,\middle|\,\begin{matrix}a_1, \cdots, a_p, -\lambda\\ -\lambda-1, b_1, \cdots, b_q\end{matrix}\right]f(t)dt. \tag{6.4.2}$$

(iv) *The transform* G^2 *is independent of* ν *in the sense that, if* ν *and* $\widetilde{\nu}$ *satisfy* (a), *and* (b) *or* (c), *and if the transforms* G^2 *and* $\widetilde{G^2}$ *are defined in* $\mathfrak{L}_{\nu,2}$ *and* $\mathfrak{L}_{\widetilde{\nu},2}$, *respectively, by* (6.2.12), *then* $G^2 f = \widetilde{G^2} f$ *for* $f \in \mathfrak{L}_{\nu,2} \cap \mathfrak{L}_{\widetilde{\nu},2}$.

(v) *If* $a^* > 0$ *or if* $a^* = 0$, $\Delta(1-\nu) + \mathrm{Re}(\mu) < 0$, *then for* $f \in \mathfrak{L}_{\nu,2}$, $G^1 f$ *is given in* (6.2.2).

When $a^* = 0$, from Theorems 5.14–5.16 we obtain the mapping properties and the range of G^2 on $\mathfrak{L}_{\nu,r}$ in three different cases when either $\Delta = \mathrm{Re}(\mu) = 0$ or $\Delta = 0, \mathrm{Re}(\mu) < 0$ or $\Delta \neq 0$.

Theorem 6.27. *Let* $a^* = \Delta = 0, \mathrm{Re}(\mu) = 0$ *and* $\alpha < 1 - \nu < \beta$, *and let* $1 < r < \infty$.

(a) *The transform* G^2 *defined on* $\mathfrak{L}_{\nu,2}$ *can be extended to* $\mathfrak{L}_{\nu,r}$ *as an element of* $[\mathfrak{L}_{\nu,r}, \mathfrak{L}_{\nu,r}]$.

(b) *If* $1 < r \leqq 2$, *then the transform* $\boldsymbol{G}^2$ *is one-to-one on* $\mathfrak{L}_{\nu,r}$ *and there holds the equality* (6.2.12) *for* $f \in \mathfrak{L}_{\nu,2}$ *and* $\mathrm{Re}(s) = \nu$.

(c) *If* $f \in \mathfrak{L}_{\nu,r}$ *and* $g \in \mathfrak{L}_{1-\nu,r'}$ *with* $r' = r/(r-1)$, *then the relation* (6.2.17) *holds.*

(d) *If* $\nu \notin \mathcal{E}_{\mathcal{G}}$, *then the transform* $\boldsymbol{G}^2$ *is one-to-one on* $\mathfrak{L}_{\nu,r}$ *and there holds*

$$\boldsymbol{G}^2(\mathfrak{L}_{\nu,r}) = \mathfrak{L}_{\nu,r}. \tag{6.4.3}$$

(e) *If* $f \in \mathfrak{L}_{\nu,r}$ *and* $\lambda \in \mathbb{C}$, *then* $\boldsymbol{G}^2 f$ *is given in* (6.4.1) *for* $\mathrm{Re}(\lambda) > -\nu$, *while in* (6.4.2) *for* $\mathrm{Re}(\lambda) < -\nu$.

Theorem 6.28. *Let* $a^* = \Delta = 0, \mathrm{Re}(\mu) < 0$ *and* $\alpha < 1-\nu < \beta$, *and let either* $m > 0$ *or* $n > 0$. *Let* $1 < r < \infty$.

(a) *The transform* $\boldsymbol{G}^2$ *defined on* $\mathfrak{L}_{\nu,2}$ *can be extended to* $\mathfrak{L}_{\nu,r}$ *as an element of* $[\mathfrak{L}_{\nu,r}, \mathfrak{L}_{\nu,s}]$ *for all* $s \geqq r$ *such that* $1/s > 1/r + \mathrm{Re}(\mu)$.

(b) *If* $1 < r \leqq 2$, *then the transform* $\boldsymbol{G}^2$ *is one-to-one on* $\mathfrak{L}_{\nu,r}$ *and there holds the equality* (6.2.12) *for* $f \in \mathfrak{L}_{\nu,2}$ *and* $\mathrm{Re}(s) = \nu$.

(c) *If* $f \in \mathfrak{L}_{\nu,r}$ *and* $g \in \mathfrak{L}_{1-\nu,s}$ *with* $1 < s < \infty$ *and* $1 \leqq 1/r + 1/s < 1 - \mathrm{Re}(\mu)$, *then the relation* (6.2.17) *holds.*

(d) *Let* $k > 0$. *If* $\nu \notin \mathcal{E}_{\mathcal{G}}$, *then the transform* $\boldsymbol{G}^2$ *is one-to-one on* $\mathfrak{L}_{\nu,r}$ *and there hold*

$$\boldsymbol{G}^2\left(\mathfrak{L}_{\nu,r}\right) = I^{-\mu}_{0+;k,(1-\alpha)/k-1}\left(\mathfrak{L}_{\nu,r}\right) \tag{6.4.4}$$

for $m > 0$, *and*

$$\boldsymbol{G}^2\left(\mathfrak{L}_{\nu,r}\right) = I^{-\mu}_{-;k,(\beta-1)/k}\left(\mathfrak{L}_{\nu,r}\right) \tag{6.4.5}$$

for $n > 0$. *If* $\nu \in \mathcal{E}_{\mathcal{G}}$, *then* $\boldsymbol{G}^2\left(\mathfrak{L}_{\nu,r}\right)$ *is a subset of the right-hand sides of* (6.4.4) *and* (6.4.5) *in the respective cases.*

(e) *If* $f \in \mathfrak{L}_{\nu,r}$ *and* $\lambda \in \mathbb{C}$, *then* $\boldsymbol{G}^2 f$ *is given in* (6.4.1) *for* $\mathrm{Re}(\lambda) > -\nu$, *while in* (6.4.2) *for* $\mathrm{Re}(\lambda) < -\nu$. *Furthermore* $\boldsymbol{G}^2 f$ *is given in* (6.2.2).

Theorem 6.29. *Let* $a^* = 0, \Delta \neq 0, \alpha < 1-\nu < \beta, 1 < r < \infty$ *and* $\Delta(1-\nu) + \mathrm{Re}(\mu) \leqq 1/2 - \gamma(r)$. *Assume that* $m > 0$ *if* $\Delta > 0$ *and* $n > 0$ *if* $\Delta < 0$.

(a) *The transform* $\boldsymbol{G}^2$ *defined on* $\mathfrak{L}_{\nu,2}$ *can be extended to* $\mathfrak{L}_{\nu,r}$ *as an element of* $[\mathfrak{L}_{\nu,r}, \mathfrak{L}_{\nu,s}]$ *for all* s *with* $r \leqq s < \infty$ *such that* $s' \geqq [1/2 - \Delta(1-\nu) - \mathrm{Re}(\mu)]^{-1}$ *with* $1/s + 1/s' = 1$.

(b) *If* $1 < r \leqq 2$, *then the transform* $\boldsymbol{G}^2$ *is one-to-one on* $\mathfrak{L}_{\nu,r}$ *and there holds the equality* (6.2.12) *for* $f \in \mathfrak{L}_{\nu,2}$ *and* $\mathrm{Re}(s) = \nu$.

(c) *If* $f \in \mathfrak{L}_{\nu,r}$ *and* $g \in \mathfrak{L}_{1-\nu,s}$ *with* $1 < s < \infty, 1/r + 1/s \geqq 1$ *and* $\Delta(1-\nu) + \mathrm{Re}(\mu) \leqq 1/2 - \max[\gamma(r), \gamma(s)]$, *then the relation* (6.2.17) *holds.*

(d) *If* $\nu \notin \mathcal{E}_{\mathcal{G}}$, *then the transform* $\boldsymbol{G}^2$ *is one-to-one on* $\mathfrak{L}_{\nu,r}$. *If we set* $\eta = -\Delta\alpha - \mu - 1$ *for* $\Delta > 0$ *and* $\eta = -\Delta\beta - \mu - 1$ *for* $\Delta < 0$, *then* $\mathrm{Re}(\eta) > -1$ *and there holds*

$$\boldsymbol{G}^2(\mathfrak{L}_{\nu,r}) = \left(M_{-\mu/\Delta-1/2}\mathbb{H}_{-\Delta,\eta}\right)\left(\mathfrak{L}_{3/2-\nu+\mathrm{Re}(\mu)/\Delta,r}\right). \tag{6.4.6}$$

When $\nu \in \mathcal{E}_{\mathcal{G}}$, $\boldsymbol{G}^2(\mathfrak{L}_{\nu,r})$ *is a subset of the right-hand side of* (6.4.6).

(e) *If* $f \in \mathfrak{L}_{\nu,r}, \lambda \in \mathbb{C}$ *and* $\Delta(1-\nu) + \mathrm{Re}(\mu) \leqq 1/2 - \gamma(r)$, *then* $\boldsymbol{G}^2 f$ *is given in* (6.4.1) *for* $\mathrm{Re}(\lambda) > -\nu$, *while in* (6.4.2) *for* $\mathrm{Re}(\lambda) < -\nu$. *If* $\Delta(1-\nu) + \mathrm{Re}(\mu) < 0$, $\boldsymbol{G}^2 f$ *is given in*

(6.2.2).

From Theorem 5.17 we have the $\mathfrak{L}_{\nu,r}$-theory of the transform G^2 in (6.2.2) with $a^* > 0$.

Theorem 6.30. *Let $a^* > 0, \alpha < 1-\nu < \beta$ and $1 \leqq r \leqq s \leqq \infty$.*

(a) *The transform G^2 defined on $\mathfrak{L}_{\nu,2}$ can be extended to $\mathfrak{L}_{\nu,r}$ as an element of $[\mathfrak{L}_{\nu,r}, \mathfrak{L}_{\nu,s}]$. If $1 \leqq r \leqq 2$, then G^2 is a one-to-one transform from $\mathfrak{L}_{\nu,r}$ onto $\mathfrak{L}_{\nu,s}$.*

(b) *If $f \in \mathfrak{L}_{\nu,r}$ and $g \in \mathfrak{L}_{1-\nu,s'}$ with $1/s+1/s'=1$, then the relation (6.2.17) holds.*

When $a^* > 0$, from Theorems 5.18–5.22 we obtain the characterization of the boundedness and the range of G^2 on $\mathfrak{L}_{\nu,r}$ in five different cases.

Theorem 6.31. *Let $a_1^* > 0, a_2^* > 0, m > 0, n > 0, \alpha < 1-\nu < \beta$ and $\omega = \mu + a_1^*\alpha - a_2^*\beta + 1$ and let $1 < r < \infty$.*

(a) *If $\nu \notin \mathcal{E}_{\mathcal{G}}$, or if $1 \leqq r \leqq 2$, then the transform G^2 is one-to-one on $\mathfrak{L}_{\nu,r}$.*

(b) *If $\mathrm{Re}(\omega) \geqq 0$ and $\nu \notin \mathcal{E}_{\mathcal{G}}$, then*

$$G^2(\mathfrak{L}_{\nu,r}) = \left(\mathbb{L}_{-a_1^*,1-\alpha}\mathbb{L}_{-a_2^*,\beta+\omega/a_2^*}\right)(\mathfrak{L}_{\nu,r}). \tag{6.4.7}$$

When $\nu \in \mathcal{E}_{\mathcal{G}}$, $G^2(\mathfrak{L}_{\nu,r})$ is a subset of the right-hand side of (6.4.7).

(c) *If $\mathrm{Re}(\omega) < 0$ and $\nu \notin \mathcal{E}_{\mathcal{G}}$, then*

$$G^2(\mathfrak{L}_{\nu,r}) = \left(I^{-\omega}_{0+;1/a_1^*,(1-\alpha)a_1^*-1}\mathbb{L}_{-a_1^*,1-\alpha}\mathbb{L}_{-a_2^*,\beta}\right)(\mathfrak{L}_{\nu,r}). \tag{6.4.8}$$

When $\nu \in \mathcal{E}_{\mathcal{G}}$, $G^2(\mathfrak{L}_{\nu,r})$ is a subset of the right-hand side of (6.4.8).

Theorem 6.32. *Let $a_1^* > 0, a_2^* = 0, m > 0, \alpha < 1-\nu < \beta$ and $\omega = \mu + a_1^*\alpha + 1/2$ and let $1 < r < \infty$.*

(a) *If $\nu \notin \mathcal{E}_{\mathcal{G}}$, or if $1 < r \leqq 2$, then the transform G^2 is one-to-one on $\mathfrak{L}_{\nu,r}$.*

(b) *If $\mathrm{Re}(\omega) \geqq 0$ and $\nu \notin \mathcal{E}_{\mathcal{G}}$, then*

$$G^2(\mathfrak{L}_{\nu,r}) = \mathbb{L}_{-a_1^*,1-\alpha+\omega/a_1^*}(\mathfrak{L}_{1-\nu,r}). \tag{6.4.9}$$

When $\nu \in \mathcal{E}_{\mathcal{G}}$, $G^1(\mathfrak{L}_{\nu,r})$ is a subset of the right-hand side of (6.4.9).

(c) *If $\mathrm{Re}(\omega) < 0$ and $\nu \notin \mathcal{E}_{\mathcal{G}}$, then*

$$G^2(\mathfrak{L}_{\nu,r}) = \left(I^{-\omega}_{0+;1/a_1^*,(1-\alpha)a_1^*-1}\mathbb{L}_{-a_1^*,1-\alpha}\right)(\mathfrak{L}_{1-\nu,r}). \tag{6.4.10}$$

When $\nu \in \mathcal{E}_{\mathcal{G}}$, $G^2(\mathfrak{L}_{\nu,r})$ is a subset of the right-hand side of (6.4.10).

Theorem 6.33. *Let $a_1^* = 0, a_2^* > 0, n > 0, \alpha < 1-\nu < \beta$ and $\omega = \mu - a_2^*\beta + 1/2$ and let $1 < r < \infty$.*

(a) *If $\nu \notin \mathcal{E}_{\mathcal{G}}$, or if $1 < r \leqq 2$, then the transform G^2 is one-to-one on $\mathfrak{L}_{\nu,r}$.*

(b) *If $\mathrm{Re}(\omega) \geqq 0$ and $\nu \notin \mathcal{E}_{\mathcal{G}}$, then*

$$G^2(\mathfrak{L}_{\nu,r}) = \mathbb{L}_{a_2^*,1-\beta-\omega/a_2^*}(\mathfrak{L}_{1-\nu,r}). \tag{6.4.11}$$

When $\nu \in \mathcal{E}_{\mathcal{G}}$, $G^2(\mathfrak{L}_{\nu,r})$ is a subset of the right-hand side of (6.4.11).

(c) *If* $\mathrm{Re}(\omega) < 0$ *and* $\nu \notin \mathcal{E}_{\mathcal{G}}$, *then*

$$G^2(\mathfrak{L}_{\nu,r}) = \left(I^{-\omega}_{-;1/a_2^*,a_2^*(\beta-1)}\mathbb{L}_{a_2^*,1-\beta}\right)(\mathfrak{L}_{1-\nu,r}). \tag{6.4.12}$$

When $\nu \in \mathcal{E}_{\mathcal{G}}$, $G^2(\mathfrak{L}_{\nu,r})$ *is a subset of the right-hand side of* (6.4.12).

Theorem 6.34. *Let* $a^* > 0, a_1^* > 0, a_2^* < 0, \alpha < 1-\nu < \beta$ *and let* $1 < r < \infty$.
(a) *If* $\nu \notin \mathcal{E}_{\mathcal{G}}$, *or if* $1 < r \leqq 2$, *then the transform* G^2 *is one-to-one on* $\mathfrak{L}_{\nu,r}$.
(b) *Let* $\omega, \eta, \zeta \in \mathbb{C}$ *be chosen as*

$$\omega = a^*\eta - \mu - \frac{1}{2}; \tag{6.4.13}$$

$$a^*\mathrm{Re}(\eta) \geqq \gamma(r) + 2a_2^*(\nu - 1) + \mathrm{Re}(\mu); \tag{6.4.14}$$

$$\mathrm{Re}(\eta) > \nu - 1; \tag{6.4.15}$$

$$\mathrm{Re}(\zeta) < 1 - \nu. \tag{6.4.16}$$

If $\nu \notin \mathcal{E}_{\mathcal{H}}$, *then*

$$G^2(\mathfrak{L}_{\nu,r}) = \left(M_{-1/2-\omega/(2a_2^*)}\mathbb{H}_{2a_2^*,2a_2^*\zeta+\omega-1}\mathbb{L}_{a^*,1/2-\eta+\omega/(2a_2^*)}\right)\left(\mathfrak{L}_{\nu-1/2-\mathrm{Re}(\omega)/(2a_2^*),r}\right). \tag{6.4.17}$$

When $\nu \in \mathcal{E}_{\mathcal{G}}$, $G^2(\mathfrak{L}_{\nu,r})$ *is a subset of the right-hand side of* (6.4.17).

Theorem 6.35. *Let* $a^* > 0, a_1^* < 0, a_2^* > 0, \alpha < 1-\nu < \beta$ *and let* $1 < r < \infty$.
(a) *If* $\nu \notin \mathcal{E}_{\mathcal{G}}$, *or if* $1 < r \leqq 2$, *then the transform* G^2 *is one-to-one on* $\mathfrak{L}_{\nu,r}$.
(b) *Let* $\omega, \eta, \zeta \in \mathbb{C}$ *be chosen as*

$$\omega = a^*\eta - \Delta - \mu - \frac{1}{2}; \tag{6.4.18}$$

$$a^*\mathrm{Re}(\eta) \geqq \gamma(r) - 2a_1^*\nu + \Delta + \mathrm{Re}(\mu); \tag{6.4.19}$$

$$\mathrm{Re}(\eta) > -\nu; \tag{6.4.20}$$

$$\mathrm{Re}(\zeta) < \nu. \tag{6.4.21}$$

If $\nu \notin \mathcal{E}_{\mathcal{G}}$, *then*

$$G^2(\mathfrak{L}_{\nu,r}) = \left(M_{1/2+\omega/(2a_1^*)}\mathbb{H}_{-2a_1^*,2a_1^*\zeta+\omega-1}\mathbb{L}_{-a^*,1/2+\eta-\omega/(2a_1^*)}\right)\left(\mathfrak{L}_{\nu+1/2+\mathrm{Re}(\omega)/(2a_1^*),r}\right). \tag{6.4.22}$$

When $\nu \in \mathcal{E}_{\mathcal{G}}$, $G^2(\mathfrak{L}_{\nu,r})$ *is a subset of the right-hand side of* (6.4.22).

The inversion formulas for the transform G^2 in (6.2.2) on $\mathfrak{L}_{\nu,r}$ when $a^* = 0$, are obtained from (5.3.25) and (5.3.26) under the condition (6.1.4), and take the forms

$$f(x) = x^{-\lambda}\frac{d}{dx}x^{\lambda+1} \cdot \int_0^\infty G^{q-m,p-n+1}_{p+1,q+1}\left[\frac{x}{t}\,\middle|\,\begin{matrix}-\lambda, -a_{n+1}, \cdots, -a_p, -a_1, \cdots, -a_n \\ -b_{m+1}, \cdots, -b_q, -b_1, \cdots, -b_m, -\lambda-1\end{matrix}\right]\frac{1}{t}(G^2 f)(t)dt \tag{6.4.23}$$

and

$$f(x) = -x^{-\lambda}\frac{d}{dx}x^{\lambda+1}$$

$$\cdot \int_0^\infty G^{q-m+1,p-n}_{p+1,q+1}\left[\frac{x}{t}\,\middle|\,\begin{matrix} -a_{n+1},\cdots,-a_p,-a_1,\cdots,-a_n,-\lambda \\ -\lambda-1,-b_{m+1},\cdots,-b_q,-b_1,\cdots,-b_m \end{matrix}\right]\frac{1}{t}(G^2 f)(t)dt. \quad (6.4.24)$$

The validity of (6.4.23) and (6.4.24) are deduced from Theorems 5.23 and 5.24, where $a^*, \Delta, \alpha, \beta, \mu, \alpha_0$ and β_0 are given in (6.1.5), (6.1.6), (6.1.9), (6.1.10), (6.1.11), (6.1.37) and (6.1.38), respectively.

Theorem 6.36. *Let $a^* = 0, \alpha < 1-\nu < \beta$ and $\alpha_0 < \nu < \beta_0$, and let $\lambda \in \mathbb{C}$.*

(a) *If $\Delta(1-\nu) + \mathrm{Re}(\mu) = 0$ and $f \in \mathfrak{L}_{\nu,2}$, then the inversion formula* (6.4.23) *holds for* $\mathrm{Re}(\lambda) > \nu - 1$ and (6.4.24) for $\mathrm{Re}(\lambda) < \nu - 1$.

(b) *If $\Delta = \mathrm{Re}(\mu) = 0$ and $f \in \mathfrak{L}_{\nu,r}$ $(1 < r < \infty)$, then the inversion formula* (6.4.23) *holds for* $\mathrm{Re}(\lambda) > \nu - 1$ *and* (6.4.24) *for* $\mathrm{Re}(\lambda) < \nu - 1$.

Theorem 6.37. *Let $a^* = 0, 1 < r < \infty$ and $\Delta(1-\nu) + \mathrm{Re}(\mu) \leqq 1/2 - \gamma(r)$, and let $\lambda \in \mathbb{C}$.*

(a) *If $\Delta > 0, m > 0, \alpha < 1-\nu < \beta$, $\alpha_0 < \nu < \min[\beta_0, \{\mathrm{Re}(\mu + 1/2)/\Delta\} + 1]$ and if $f \in \mathfrak{L}_{\nu,r}$, then the inversion formulas* (6.4.23) *and* (6.4.24) *hold for* $\mathrm{Re}(\lambda) > \nu - 1$ and for $\mathrm{Re}(\lambda) < \nu - 1$, *respectively.*

(b) *If $\Delta < 0, n > 0, \alpha < 1-\nu < \beta$, $\max[\alpha_0, \{\mathrm{Re}(\mu + 1/2)/\Delta\} + 1] < \nu < \beta_0$ and if $f \in \mathfrak{L}_{\nu,r}$, then the inversion formulas* (6.4.23) *and* (6.4.24) *hold for* $\mathrm{Re}(\lambda) > \nu - 1$ and for $\mathrm{Re}(\lambda) < \nu - 1$, *respectively.*

6.5. $G_{\sigma,\kappa}$-Transform on the Space $\mathfrak{L}_{\nu,r}$

We consider the transform $G_{\sigma,\kappa}$ defined in (6.2.3). From the results in Section 5.4 we obtain $\mathfrak{L}_{\nu,2}$- and $\mathfrak{L}_{\nu,r}$-theory for the transform $G_{\sigma,\kappa}$.The first result follows from Theorem 5.25.

Theorem 6.38. *We suppose that* **(a)** *$\alpha < 1 - \nu + \mathrm{Re}(\kappa) < \beta$ and that either of the conditions* **(b)** *$a^* > 0$, or* **(c)** *$a^* = 0$, $\Delta[1 - \nu + \mathrm{Re}(\kappa)] + \mathrm{Re}(\mu) \leqq 0$ holds. Then we have the following results:*

(i) *There is a one-to-one transform $G_{\sigma,\kappa} \in [\mathfrak{L}_{\nu,2}, \mathfrak{L}_{1-\nu+\mathrm{Re}(\kappa-\sigma),2}]$ such that* (6.2.13) *holds for* $\mathrm{Re}(s) = 1 - \nu + \mathrm{Re}(\kappa - \sigma)$ *and $f \in \mathfrak{L}_{\nu,2}$. If $a^* = 0$, $\Delta[1 - \nu + \mathrm{Re}(\kappa)] + \mathrm{Re}(\mu) = 0$ and $\nu - \mathrm{Re}(\kappa) \notin \mathcal{E}_{\mathcal{G}}$, then the transform $G_{\sigma,\kappa}$ maps $\mathfrak{L}_{\nu,2}$ onto $\mathfrak{L}_{1-\nu+\mathrm{Re}(\kappa-\sigma),2}$.*

(ii) *If $f \in \mathfrak{L}_{\nu,2}$ and $g \in \mathfrak{L}_{\nu+\mathrm{Re}(\kappa-\sigma),2}$, then the relation* (6.2.18) *holds for $G_{\sigma,\kappa}$.*

(iii) *Let $\lambda \in \mathbb{C}$ and $f \in \mathfrak{L}_{\nu,2}$. If $\mathrm{Re}(\lambda) > -\nu + \mathrm{Re}(\kappa)$, then $G_{\sigma,\kappa} f$ is given by*

$$\left(G_{\sigma,\kappa} f\right)(x) = x^{\sigma-\lambda}\frac{d}{dx}x^{\lambda+1}\int_0^\infty G^{m,n+1}_{p+1,q+1}\left[xt\,\middle|\,\begin{matrix} -\lambda,,a_1,\cdots,a_p \\ b_1,\cdots,b_q,-\lambda-1 \end{matrix}\right]t^\kappa f(t)dt. \quad (6.5.1)$$

When $\mathrm{Re}(\lambda) < -\nu + \mathrm{Re}(\kappa)$,

$$\left(\boldsymbol{G}_{\sigma,\kappa}f\right)(x) = -x^{\sigma-\lambda}\frac{d}{dx}x^{\lambda+1}\int_0^\infty G^{m+1,n}_{p+1,q+1}\left[xt \,\middle|\, \begin{matrix} a_1,\cdots,a_p,-\lambda \\ -\lambda-1,b_1,\cdots,b_q \end{matrix}\right] t^\kappa f(t)dt. \tag{6.5.2}$$

(iv) *The transform $\boldsymbol{G}_{\sigma,\kappa}$ is independent of ν in the sense that, if ν and $\widetilde{\nu}$ satisfy* (a), *and* (b) *or* (c), *and if the transforms $\boldsymbol{G}_{\sigma,\kappa}$ and $\widetilde{\boldsymbol{G}}_{\sigma,\kappa}$ are defined in $\mathfrak{L}_{\nu,2}$ and $\mathfrak{L}_{\widetilde{\nu},2}$, respectively, by* (6.2.13), *then $\boldsymbol{G}_{\sigma,\kappa}f = \widetilde{\boldsymbol{G}}_{\sigma,\kappa}f$ for $f \in \mathfrak{L}_{\nu,2}\cap\mathfrak{L}_{\widetilde{\nu},2}$.*

(v) *If $a^* > 0$ or if $a^* = 0$, $\Delta[1-\nu+\text{Re}(\kappa)]+\text{Re}(\mu) < 0$, then for $f \in \mathfrak{L}_{\nu,2}$, $\boldsymbol{G}_{\sigma,\kappa}f$ is given in* (6.2.3).

If $a^* = 0$, Theorems 5.26–5.28 give the mapping properties and the range of $\boldsymbol{G}_{\sigma,\kappa}$ on $\mathfrak{L}_{\nu,r}$.

Theorem 6.39. *Let $a^* = \Delta = 0, \text{Re}(\mu) = 0$ and $\alpha < 1-\nu+\text{Re}(\kappa) < \beta$. Let $1 < r < \infty$.*

(a) *The transform $\boldsymbol{G}_{\sigma,\kappa}$ defined on $\mathfrak{L}_{\nu,2}$ can be extended to $\mathfrak{L}_{\nu,r}$ as an element of $[\mathfrak{L}_{\nu,r}, \mathfrak{L}_{1-\nu+\text{Re}(\kappa-\sigma),r}]$.*

(b) *If $1 < r \leqq 2$, then the transform $\boldsymbol{G}_{\sigma,\kappa}$ is one-to-one on $\mathfrak{L}_{\nu,r}$ and there holds the equality* (6.2.13) *for $f \in \mathfrak{L}_{\nu,r}$ and $\text{Re}(s) = 1-\nu+\text{Re}(\kappa-\sigma)$.*

(c) *If $f \in \mathfrak{L}_{\nu,r}$ and $g \in \mathfrak{L}_{\nu+\text{Re}(\kappa-\sigma),r'}$ with $r' = r/(r-1)$, then the relation* (6.2.18) *holds.*

(d) *If $\nu - \text{Re}(\kappa) \notin \mathcal{E}_{\mathcal{G}}$, then the transform $\boldsymbol{G}_{\sigma,\kappa}$ is one-to-one on $\mathfrak{L}_{\nu,r}$ and there holds*

$$\boldsymbol{G}_{\sigma,\kappa}(\mathfrak{L}_{\nu,r}) = \mathfrak{L}_{1-\nu+\text{Re}(\kappa-\sigma),r}. \tag{6.5.3}$$

(e) *If $f \in \mathfrak{L}_{\nu,r}$ and $\lambda \in \mathbb{C}$, then $\boldsymbol{G}_{\sigma,\kappa}f$ is given in* (6.5.1) *for $\text{Re}(\lambda) > -\nu+\text{Re}(\kappa)$, while in* (6.5.2) *for $\text{Re}(\lambda) < -\nu+\text{Re}(\kappa)$.*

Theorem 6.40. *Let $a^* = \Delta = 0, \text{Re}(\mu) < 0$ and $\alpha < 1-\nu+\text{Re}(\kappa) < \beta$, and let either $m > 0$ or $n > 0$. Let $1 < r < \infty$.*

(a) *The transform $\boldsymbol{G}_{\sigma,\kappa}$ defined on $\mathfrak{L}_{\nu,2}$ can be extended to $\mathfrak{L}_{\nu,r}$ as an element of $[\mathfrak{L}_{\nu,r}, \mathfrak{L}_{1-\nu+\text{Re}(\kappa-\sigma),s}]$ for all $s \geqq r$ such that $1/s > 1/r+\text{Re}(\mu)$.*

(b) *If $1 < r \leqq 2$, then the transform $\boldsymbol{G}_{\sigma,\kappa}$ is one-to-one on $\mathfrak{L}_{\nu,r}$ and there holds the equality* (6.2.13) *for $f \in \mathfrak{L}_{\nu,r}$ and $\text{Re}(s) = 1-\nu+\text{Re}(\kappa-\sigma)$.*

(c) *If $f \in \mathfrak{L}_{\nu,r}$ and $g \in \mathfrak{L}_{\nu+\text{Re}(\kappa-\sigma),s}$ with $1 < s < \infty$ and $1 \leqq 1/r+1/s < 1-\text{Re}(\mu)$, then the relation* (6.2.18) *holds.*

(d) *Let $k > 0$. If $\nu - \text{Re}(\kappa) \notin \mathcal{E}_{\mathcal{G}}$, then the transform $\boldsymbol{G}_{\sigma,\kappa}$ is one-to-one on $\mathfrak{L}_{\nu,r}$ and there hold*

$$\boldsymbol{G}_{\sigma,\kappa}\left(\mathfrak{L}_{\nu,r}\right) = I^{-\mu}_{-;k,(\sigma-\alpha)/k}\left(\mathfrak{L}_{1-\nu+\text{Re}(\kappa-\sigma),r}\right) \tag{6.5.4}$$

for $m > 0$, and

$$\boldsymbol{G}_{\sigma,\kappa}\left(\mathfrak{L}_{\nu,r}\right) = I^{-\mu}_{0+;k,(\beta-\sigma)/k-1}\left(\mathfrak{L}_{1-\nu+\text{Re}(\kappa-\sigma),r}\right) \tag{6.5.5}$$

for $n > 0$. When $\nu - \text{Re}(\kappa) \in \mathcal{E}_{\mathcal{G}}$, $\boldsymbol{G}_{\sigma,\kappa}\left(\mathfrak{L}_{\nu,r}\right)$ is a subset of the right-hand sides of (6.5.4) *and* (6.5.5) *in the respective cases.*

(e) *If $f \in \mathfrak{L}_{\nu,r}$ and $\lambda \in \mathbb{C}$, then $\boldsymbol{G}_{\sigma,\kappa}f$ is given in* (6.5.1) *for $\text{Re}(\lambda) > -\nu+\text{Re}(\kappa)$, while in* (6.5.2) *for $\text{Re}(\lambda) < -\nu+\text{Re}(\kappa)$. Furthermore $\boldsymbol{G}_{\sigma,\kappa}f$ is given in* (6.2.3).

Theorem 6.41. *Let $a^* = 0, \Delta \neq 0, \alpha < 1 - \nu + \mathrm{Re}(\kappa) < \beta$, $1 < r < \infty$ and $\Delta[1 - \nu + \mathrm{Re}(\kappa)] + \mathrm{Re}(\mu) \leqq 1/2 - \gamma(r)$. Assume that $m > 0$ if $\Delta > 0$ and $n > 0$ if $\Delta < 0$.*

(a) *The transform $\boldsymbol{G}_{\sigma,\kappa}$ defined on $\mathfrak{L}_{\nu,2}$ can be extended to $\mathfrak{L}_{\nu,r}$ as an element of $[\mathfrak{L}_{\nu,r}, \mathfrak{L}_{1-\nu+\mathrm{Re}(\kappa-\sigma),s}]$ for all s with $r \leqq s < \infty$ such that $s' \geqq \{1/2 - \Delta[1-\nu+\mathrm{Re}(\kappa)] - \mathrm{Re}(\mu)\}^{-1}$ with $1/s + 1/s' = 1$.*

(b) *If $1 < r \leqq 2$, then the transform $\boldsymbol{G}_{\sigma,\kappa}$ is one-to-one on $\mathfrak{L}_{\nu,r}$ and there holds the equality (6.2.13) for $f \in \mathfrak{L}_{\nu,r}$ and $\mathrm{Re}(s) = 1 - \nu + \mathrm{Re}(\kappa - \sigma)$.*

(c) *If $f \in \mathfrak{L}_{\nu,r}$ and $g \in \mathfrak{L}_{\nu+\mathrm{Re}(\kappa-\sigma),s}$ with $1 < s < \infty, 1/r + 1/s \geqq 1$ and $\Delta[1-\nu+\mathrm{Re}(\kappa)] + \mathrm{Re}(\mu) \leqq 1/2 - \max[\gamma(r), \gamma(s)]$, then the relation (6.2.18) holds.*

(d) *If $\nu - \mathrm{Re}(\kappa) \notin \mathcal{E}_{\mathcal{G}}$, then the transform $\boldsymbol{G}_{\sigma,\kappa}$ is one-to-one on $\mathfrak{L}_{\nu,r}$. If we set $\eta = -\Delta\alpha - \mu - 1$ for $\Delta > 0$ and $\eta = -\Delta\beta - \mu - 1$ for $\Delta < 0$, then $\mathrm{Re}(\eta) > -1$ and there holds*

$$\boldsymbol{G}_{\sigma,\kappa}(\mathfrak{L}_{\nu,r}) = \left(M_{\sigma+\mu/\Delta+1/2}\mathbb{H}_{\Delta,\eta}\right)\left(\mathfrak{L}_{\nu-1/2-\mathrm{Re}(\mu)/\Delta-\mathrm{Re}(\kappa),r}\right). \tag{6.5.6}$$

When $\nu - \mathrm{Re}(\kappa) \in \mathcal{E}_{\mathcal{G}}$, $\boldsymbol{G}_{\sigma,\kappa}(\mathfrak{L}_{\nu,r})$ is a subset of the right-hand side of (6.5.6).

(e) *If $f \in \mathfrak{L}_{\nu,r}, \lambda \in \mathbb{C}$ and $\Delta[1 - \nu + \mathrm{Re}(\kappa)] + \mathrm{Re}(\mu) \leqq 1/2 - \gamma(r)$, then $\boldsymbol{G}_{\sigma,\kappa}f$ is given in (6.5.1) for $\mathrm{Re}(\lambda) > -\nu + \mathrm{Re}(\kappa)$, while in (6.5.2) for $\mathrm{Re}(\lambda) < -\nu + \mathrm{Re}(\kappa)$. If $\Delta[1 - \nu + \mathrm{Re}(\kappa)] + \mathrm{Re}(\mu) < 0$, $\boldsymbol{G}_{\sigma,\kappa}f$ is given in (6.2.3).*

From Theorem 5.29 we obtain the $\mathfrak{L}_{\nu,r}$-theory of the transform $\boldsymbol{G}_{\sigma,\kappa}$ in (6.2.3) with $a^* > 0$.

Theorem 6.42. *Let $a^* > 0, \alpha < 1 - \nu + \mathrm{Re}(\kappa) < \beta$ and $1 \leqq r \leqq s \leqq \infty$.*

(a) *The transform $\boldsymbol{G}_{\sigma,\kappa}$ defined on $\mathfrak{L}_{\nu,2}$ can be extended to $\mathfrak{L}_{\nu,r}$ as an element of $[\mathfrak{L}_{\nu,r}, \mathfrak{L}_{1-\nu+\mathrm{Re}(\kappa-\sigma),s}]$. If $1 \leqq r \leqq 2$, then $\boldsymbol{G}_{\sigma,\kappa}$ is a one-to-one transform from $\mathfrak{L}_{\nu,r}$ onto $\mathfrak{L}_{1-\nu+\mathrm{Re}(\kappa-\sigma),s}$.*

(b) *If $f \in \mathfrak{L}_{\nu,r}$ and $g \in \mathfrak{L}_{\nu-\mathrm{Re}(\kappa-\sigma),s'}$ with $1/s + 1/s' = 1$, then the relation (6.2.18) holds.*

For $a^* > 0$, Theorems 5.30–5.34 imply the characterization theorems of the boundedness and the range of $\boldsymbol{G}_{\sigma,\kappa}$ on $\mathfrak{L}_{\nu,r}$.

Theorem 6.43. *Let $a_1^* > 0, a_2^* > 0, m > 0, n > 0, \alpha < 1 - \nu + \mathrm{Re}(\kappa) < \beta$ and $\omega = \mu + a_1^*\alpha - a_2^*\beta + 1$ and let $1 < r < \infty$.*

(a) *If $\nu - \mathrm{Re}(\kappa) \notin \mathcal{E}_{\mathcal{G}}$, or if $1 \leqq r \leqq 2$, then the transform $\boldsymbol{G}_{\sigma,\kappa}$ is one-to-one on $\mathfrak{L}_{\nu,r}$.*

(b) *If $\mathrm{Re}(\omega) \geqq 0$ and $\nu - \mathrm{Re}(\kappa) \notin \mathcal{E}_{\mathcal{G}}$, then*

$$\boldsymbol{G}_{\sigma,\kappa}(\mathfrak{L}_{\nu,r}) = \left(\mathbb{L}_{a_1^*,\alpha-\sigma}\mathbb{L}_{a_2^*,1-\beta+\sigma-\omega/a_2^*}\right)\left(\mathfrak{L}_{1-\nu+\mathrm{Re}(\kappa-\sigma),r}\right). \tag{6.5.7}$$

When $\nu - \mathrm{Re}(\kappa) \in \mathcal{E}_{\mathcal{G}}$, $\boldsymbol{G}_{\sigma,\kappa}(\mathfrak{L}_{\nu,r})$ is a subset of the right-hand side of (6.5.7).

(c) *If $\mathrm{Re}(\omega) < 0$ and $\nu - \mathrm{Re}(\kappa) \notin \mathcal{E}_{\mathcal{G}}$, then*

$$\boldsymbol{G}_{\sigma,\kappa}(\mathfrak{L}_{\nu,r}) = \left(I^{-\omega}_{-;1/a_1^*,a_1^*(\sigma-\alpha)}\mathbb{L}_{a_1^*,\alpha-\sigma}\mathbb{L}_{a_2^*,1-\beta+\sigma}\right)\left(\mathfrak{L}_{1-\nu+\mathrm{Re}(\kappa-\sigma),r}\right). \tag{6.5.8}$$

When $\nu - \mathrm{Re}(\kappa) \in \mathcal{E}_{\mathcal{G}}$, $\boldsymbol{G}_{\sigma,\kappa}(\mathfrak{L}_{\nu,r})$ is a subset of the right-hand side of (6.5.8).

Theorem 6.44. *Let $a_1^* > 0, a_2^* = 0, m > 0, \alpha < 1 - \nu + \mathrm{Re}(\kappa) < \beta$ and $\omega = \mu + a_1^*\alpha + 1/2$ and let $1 < r < \infty$.*

(a) *If* $\nu - \mathrm{Re}(\kappa) \notin \mathcal{E}_{\mathcal{G}}$, *or if* $1 < r \leqq 2$, *then the transform* $\boldsymbol{G}_{\sigma,\kappa}$ *is one-to-one on* $\mathfrak{L}_{\nu,r}$.

(b) *If* $\mathrm{Re}(\omega) \geqq 0$ *and* $\nu - \mathrm{Re}(\kappa) \notin \mathcal{E}_{\mathcal{G}}$, *then*

$$\boldsymbol{G}_{\sigma,\kappa}(\mathfrak{L}_{\nu,r}) = \mathbb{L}_{a_1^*,\alpha-\sigma-\omega/a_1^*}\left(\mathfrak{L}_{\nu-\mathrm{Re}(\kappa-\sigma),r}\right). \tag{6.5.9}$$

When $\nu - \mathrm{Re}(\kappa) \in \mathcal{E}_{\mathcal{G}}$, $\boldsymbol{G}_{\sigma,\kappa}(\mathfrak{L}_{\nu,r})$ *is a subset of the right-hand side of* (6.5.9).

(c) *If* $\mathrm{Re}(\omega) < 0$ *and* $\nu - \mathrm{Re}(\kappa) \notin \mathcal{E}_{\mathcal{G}}$, *then*

$$\boldsymbol{G}_{\sigma,\kappa}(\mathfrak{L}_{\nu,r}) = \left(I_{-;1/a_1^*,a_1^*(\sigma-\alpha)}^{-\omega}\mathbb{L}_{a_1^*,\alpha-\sigma}\right)\left(\mathfrak{L}_{\nu-\mathrm{Re}(\kappa-\sigma),r}\right). \tag{6.5.10}$$

When $\nu - \mathrm{Re}(\kappa) \in \mathcal{E}_{\mathcal{G};\omega}$, $\boldsymbol{G}_{\sigma,\kappa}(\mathfrak{L}_{\nu,r})$ *is a subset of the right-hand side of* (6.5.10).

Theorem 6.45. *Let* $a_1^* = 0, a_2^* > 0, n > 0, \alpha < 1-\nu+\mathrm{Re}(\kappa) < \beta$ *and* $\omega = \mu - a_2^*\beta + 1/2$ *and let* $1 < r < \infty$.

(a) *If* $\nu - \mathrm{Re}(\kappa) \notin \mathcal{E}_{\mathcal{G}}$, *or if* $1 < r \leqq 2$, *then the transform* $\boldsymbol{G}_{\sigma,\kappa}$ *is one-to-one on* $\mathfrak{L}_{\nu,r}$.

(b) *If* $\mathrm{Re}(\omega) \geqq 0$ *and* $\nu - \mathrm{Re}(\kappa) \notin \mathcal{E}_{\mathcal{G}}$, *then*

$$\boldsymbol{G}_{\sigma,\kappa}(\mathfrak{L}_{\nu,r}) = \mathbb{L}_{-a_2^*,\beta-\sigma+\omega/a_2^*}\left(\mathfrak{L}_{\nu-\mathrm{Re}(\kappa-\sigma),r}\right). \tag{6.5.11}$$

When $\nu - \mathrm{Re}(\kappa) \in \mathcal{E}_{\mathcal{G}}$, $\boldsymbol{G}_{\sigma,\kappa}(\mathfrak{L}_{\nu,r})$ *is a subset of the right-hand side of* (6.5.11).

(c) *If* $\mathrm{Re}(\omega) < 0$ *and* $\nu - \mathrm{Re}(\kappa) \notin \mathcal{E}_{\mathcal{G}}$, *then*

$$\boldsymbol{G}_{\sigma,\kappa}(\mathfrak{L}_{\nu,r}) = \left(I_{0+;1/a_2^*,a_2^*(\beta-\sigma)-1}^{-\omega}\mathbb{L}_{-a_2^*,\beta-\sigma}\right)\left(\mathfrak{L}_{\nu-\mathrm{Re}(\kappa-\sigma),r}\right). \tag{6.5.12}$$

When $\nu - \mathrm{Re}(\kappa) \in \mathcal{E}_{\mathcal{G}}$, $\boldsymbol{G}_{\sigma,\kappa}(\mathfrak{L}_{\nu,r})$ *is a subset of the right-hand side of* (6.5.12).

Theorem 6.46. *Let* $a^* > 0, a_1^* > 0, a_2^* < 0, \alpha < 1-\nu+\mathrm{Re}(\kappa) < \beta$ *and let* $1 < r < \infty$.

(a) *If* $\nu - \mathrm{Re}(\kappa) \notin \mathcal{E}_{\mathcal{G}}$, *or if* $1 < r \leqq 2$, *then the transform* $\boldsymbol{G}_{\sigma,\kappa}$ *is one-to-one on* $\mathfrak{L}_{\nu,r}$.

(b) *Let* $\omega, \eta, \zeta \in \mathbb{C}$ *be chosen as*

$$\omega = a^*\eta - \mu - \frac{1}{2}; \tag{6.5.13}$$

$$a^*\mathrm{Re}(\eta) \geqq \gamma(r) + 2a_2^*[\nu - \mathrm{Re}(\kappa) - 1] + \mathrm{Re}(\mu); \tag{6.5.14}$$

$$\mathrm{Re}(\eta) > \nu - \mathrm{Re}(\kappa) - 1; \tag{6.5.15}$$

$$\mathrm{Re}(\zeta) < 1 - \nu + \mathrm{Re}(\kappa). \tag{6.5.16}$$

If $\nu - \mathrm{Re}(\kappa) \notin \mathcal{E}_{\mathcal{G}}$, *then*

$$\boldsymbol{G}_{\sigma,\kappa}(\mathfrak{L}_{\nu,r}) = \left(M_{\sigma+1/2+\omega/(2a_2^*)}\mathbb{H}_{-2a_2^*,2a_2^*\zeta+\omega-1}\mathbb{L}_{-a^*,1/2+\eta-\omega/(2a_2^*)}\right)$$
$$\left(\mathfrak{L}_{3/2-\nu+\mathrm{Re}(\omega)/(2a_2^*)+\mathrm{Re}(\kappa),r}\right). \tag{6.5.17}$$

When $\nu - \mathrm{Re}(\kappa) \in \mathcal{E}_{\mathcal{G}}$, $\boldsymbol{G}_{\sigma,\kappa}(\mathfrak{L}_{\nu,r})$ *is a subset of the right-hand side of* (6.5.17).

Theorem 6.47. *Let* $a^* > 0, a_1^* < 0, a_2^* > 0, \alpha < 1-\nu+\mathrm{Re}(\kappa) < \beta$ *and let* $1 < r < \infty$.

(a) *If* $\nu - \mathrm{Re}(\kappa) \notin \mathcal{E}_{\mathcal{G}}$, *or if* $1 < r \leqq 2$, *then the transform* $\boldsymbol{G}_{\sigma,\kappa}$ *is one-to-one on* $\mathfrak{L}_{\nu,r}$.

(b) *Let $\omega, \eta, \zeta \in \mathbb{C}$ be chosen as*

$$\omega = a^*\eta - \Delta - \mu - \frac{1}{2}; \tag{6.5.18}$$

$$a^*\mathrm{Re}(\eta) \geqq \gamma(r) - 2a_1^*[\nu - \mathrm{Re}(\kappa)] + \Delta + \mathrm{Re}(\mu); \tag{6.5.19}$$

$$\mathrm{Re}(\eta) > -\nu + \mathrm{Re}(\kappa); \tag{6.5.20}$$

$$\mathrm{Re}(\zeta) < \nu - \mathrm{Re}(\kappa). \tag{6.5.21}$$

If $\nu - \mathrm{Re}(\kappa) \notin \mathcal{E}_{\mathcal{G}}$, then

$$\boldsymbol{G}_{\sigma,\kappa}(\mathfrak{L}_{\nu,r}) = \left(M_{\sigma-1/2-\omega/(2a_1^*)}\mathbb{H}_{2a_1^*,2a_1^*\zeta+\omega-1}\mathbb{L}_{a^*,1/2-\eta+\omega/(2a_1^*)}\right)$$
$$\left(\mathfrak{L}_{1/2-\nu-\mathrm{Re}(\omega)/(2a_1^*)-\mathrm{Re}(\kappa),r}\right). \tag{6.5.22}$$

When $\nu - \mathrm{Re}(\kappa) \in \mathcal{E}_{\mathcal{G}}$, $\boldsymbol{G}_{\sigma,\kappa}(\mathfrak{L}_{\nu,r})$ is a subset of the right-hand side of (6.5.22).

The inversion formulas for the transform $\boldsymbol{G}_{\sigma,\kappa}$ in (6.2.3) on $\mathfrak{L}_{\nu,r}$ when $a^* = 0$, are obtained from (5.4.25) and (5.4.26) under the relation (6.1.4), and take the form

$$f(x) = x^{-\kappa-\lambda}\frac{d}{dx}x^{\lambda+1}$$
$$\cdot \int_0^\infty H_{p+1,q+1}^{q-m,p-n+1}\left[xt \,\middle|\, \begin{matrix} -\lambda, -a_{n+1}, \cdots, -a_p, -a_1, \cdots, -a_n \\ -b_{m+1}, \cdots, -b_q, -b_1, \cdots, -b_m, -\lambda-1 \end{matrix}\right]$$
$$\cdot\, t^{-\sigma}\left(\boldsymbol{G}_{\sigma,\kappa}f\right)(t)dt \tag{6.5.23}$$

or

$$f(x) = -x^{-\kappa-\lambda}\frac{d}{dx}x^{\lambda+1}$$
$$\cdot \int_0^\infty H_{p+1,q+1}^{q-m+1,p-n}\left[xt \,\middle|\, \begin{matrix} -a_{n+1}, \cdots, -a_p, -a_1, \cdots, -a_n, -\lambda \\ -\lambda-1, -b_{m+1}, \cdots, -b_q, -b_1, \cdots, -b_m \end{matrix}\right]$$
$$\cdot\, t^{-\sigma}\left(\boldsymbol{G}_{\sigma,\kappa}f\right)(t)dt. \tag{6.5.24}$$

The conditions for the validity of (6.5.23) and (6.5.24) follow from Theorems 5.35 and 5.36, where $a^*, \Delta, \alpha, \beta, \mu, \alpha_0$ and β_0 are given in (6.1.5), (6.1.6), (6.1.9)–(6.1.11), (6.1.37) and (6.1.38), respectively.

Theorem 6.48. *Let $a^* = 0, \alpha < 1 - \nu + \mathrm{Re}(\kappa) < \beta$ and $\alpha_0 < \nu - \mathrm{Re}(\kappa) < \beta_0$, and let $\lambda \in \mathbb{C}$.*

(a) *If $\Delta[1 - \nu + \mathrm{Re}(\kappa)] + \mathrm{Re}(\mu) = 0$ and $f \in \mathfrak{L}_{\nu,2}$, then the inversion formula* (6.5.23) *holds for $\mathrm{Re}(\lambda) > \nu - \mathrm{Re}(\kappa) - 1$ and* (6.5.24) *for $\mathrm{Re}(\lambda) < \nu - \mathrm{Re}(\kappa) - 1$.*

(b) *If $\Delta = \mathrm{Re}(\mu) = 0$ and $f \in \mathfrak{L}_{\nu,r}$ $(1 < r < \infty)$, then the inversion formula* (6.5.23) *holds for $\mathrm{Re}(\lambda) > \nu - \mathrm{Re}(\kappa) - 1$ and* (6.5.24) *for $\mathrm{Re}(\lambda) < \nu - \mathrm{Re}(\kappa) - 1$.*

Theorem 6.49. *Let $a^* = 0, 1 < r < \infty$ and $\Delta[1-\nu+\text{Re}(\kappa)]+\text{Re}(\mu) \leqq 1/2-\gamma(r)$, and let $\lambda \in \mathbb{C}$.*

(a) *If $\Delta > 0, m > 0, \alpha < 1-\nu+\text{Re}(\kappa) < \beta, \alpha_0 < \nu-\text{Re}(\kappa) < \min[\beta_0, \{\text{Re}(\mu+1/2)/\Delta\}+1]$ and if $f \in \mathfrak{L}_{\nu,r}$, then the inversion formulas (6.5.23) and (6.5.24) hold for $\text{Re}(\lambda) > \nu-\text{Re}(\kappa)-1$ and for $\text{Re}(\lambda) < \nu-\text{Re}(\kappa)-1$, respectively.*

(b) *If $\Delta < 0, n > 0, \alpha < 1-\nu+\text{Re}(\kappa) < \beta, \max[\alpha_0, \{\text{Re}(\mu+1/2)/\Delta\}+1] < \nu-\text{Re}(\kappa) < \beta_0$ and if $f \in \mathfrak{L}_{\nu,r}$, then the inversion formulas (6.5.23) and (6.5.24) hold for $\text{Re}(\lambda) > \nu-\text{Re}(\kappa)-1$ and for $\text{Re}(\lambda) < \nu-\text{Re}(\kappa)-1$, respectively.*

Remark 6.1. The results in this section generalize those in Section 6.1. Namely, Theorems 6.1–6.12 follow from Theorems 6.37 and 6.38 when $\sigma = \kappa = 0$.

6.6. $G^1_{\sigma,\kappa}$-Transform on the Space $\mathfrak{L}_{\nu,r}$

Let us study the transform $G^1_{\sigma,\kappa}$ defined in (6.2.4). The theory of the transform $G^1_{\sigma,\kappa}$ on the spaces $\mathfrak{L}_{\nu,2}$ and $\mathfrak{L}_{\nu,r}$ is obtained from that in Section 5.5, for which the parameters are defined in (6.1.5)–(6.1.11).

Theorem 6.50. *We suppose that* **(a)** $\alpha < \nu-\text{Re}(\kappa) < \beta$ *and that either of the conditions* **(b)** $a^* > 0$, *or* **(c)** $a^* = 0$, $\Delta[\nu-\text{Re}(\kappa)]+\text{Re}(\mu) \leqq 0$ *holds. Then we have the following results:*

(i) *There is a one-to-one transform $G^1_{\sigma,\kappa} \in [\mathfrak{L}_{\nu,2}, \mathfrak{L}_{\nu-\text{Re}(\kappa+\sigma),2}]$ such that (6.2.14) holds for $\text{Re}(s) = \nu-\text{Re}(\kappa+\sigma)$ and $f \in \mathfrak{L}_{\nu,2}$. If $a^* = 0$, $\Delta[\nu-\text{Re}(\kappa)]+\text{Re}(\mu) = 0$ and $1-\nu+\text{Re}(\kappa) \notin \mathcal{E}_{\mathcal{G}}$, then the transform $G^1_{\sigma,\kappa}$ maps $\mathfrak{L}_{\nu,2}$ onto $\mathfrak{L}_{\nu-\text{Re}(\kappa+\sigma),2}$.*

(ii) *If $f \in \mathfrak{L}_{\nu,2}$ and $g \in \mathfrak{L}_{1-\nu+\text{Re}(\kappa+\sigma),2}$, then the relation (6.2.19) holds for $G^1_{\sigma,\kappa}$.*

(iii) *Let $\lambda \in \mathbb{C}$ and $f \in \mathfrak{L}_{\nu,2}$. If $\text{Re}(\lambda) > \nu-\text{Re}(\kappa)-1$, then $G^1_{\sigma,\kappa}f$ is given by*

$$\left(G^1_{\sigma,\kappa}f\right)(x) = x^{\sigma-\lambda}\frac{d}{dx}x^{\lambda+1}\int_0^\infty G^{m,n+1}_{p+1,q+1}\left[\frac{x}{t}\,\middle|\,\begin{matrix}-\lambda, a_1, \cdots, a_p\\ b_1, \cdots, b_q, -\lambda-1\end{matrix}\right]t^{\kappa-1}f(t)dt. \tag{6.6.1}$$

When $\text{Re}(\lambda) < \nu-\text{Re}(\kappa)-1$,

$$\left(G^1_{\sigma,\kappa}f\right)(x) = -x^{\sigma-\lambda}\frac{d}{dx}x^{\lambda+1}\int_0^\infty G^{m+1,n}_{p+1,q+1}\left[\frac{x}{t}\,\middle|\,\begin{matrix}a_1, \cdots, a_p, -\lambda\\ -\lambda-1, b_1, \cdots, b_q\end{matrix}\right]t^{\kappa-1}f(t)dt. \tag{6.6.2}$$

(iv) *The transform $G^1_{\sigma,\kappa}$ is independent of ν in the sense that, if ν and $\widetilde{\nu}$ satisfy* (a), *and either* (b) *or* (c), *and if the transforms $G^1_{\sigma,\kappa}$ and $\widetilde{G^1}_{\sigma,\kappa}$ are defined in $\mathfrak{L}_{\nu,2}$ and $\mathfrak{L}_{\widetilde{\nu},2}$, respectively, by (6.2.14), then $G^1_{\sigma,\kappa}f = \widetilde{G^1}_{\sigma,\kappa}f$ for $f \in \mathfrak{L}_{\nu,2} \cap \mathfrak{L}_{\widetilde{\nu},2}$.*

(v) *If $a^* > 0$ or if $a^* = 0$, $\Delta[\nu-\text{Re}(\kappa)]+\text{Re}(\mu) < 0$, then for $f \in \mathfrak{L}_{\nu,2}$, $G^1_{\sigma,\kappa}f$ is given in (6.2.4).*

If $a^* = 0$, Theorems 5.38–5.40 yield the mapping properties and the range of $G^1_{\sigma,\kappa}$ on $\mathfrak{L}_{\nu,r}$.

Theorem 6.51. *Let $a^* = \Delta = 0, \text{Re}(\mu) = 0$ and $\alpha < \nu-\text{Re}(\kappa) < \beta$. Let $1 < r < \infty$.*

(a) *The transform* $G^1_{\sigma,\kappa}$ *defined on* $\mathfrak{L}_{\nu,2}$ *can be extended to* $\mathfrak{L}_{\nu,r}$ *as an element of* $[\mathfrak{L}_{\nu,r}, \mathfrak{L}_{\nu-\mathrm{Re}(\kappa+\sigma),r}]$.

(b) *If* $1 < r \leqq 2$, *then the transform* $G^1_{\sigma,\kappa}$ *is one-to-one on* $\mathfrak{L}_{\nu,r}$ *and there holds the equality* (6.2.14) *for* $f \in \mathfrak{L}_{\nu,r}$ *and* $\mathrm{Re}(s) = \nu - \mathrm{Re}(\kappa+\sigma)$.

(c) *If* $f \in \mathfrak{L}_{\nu,r}$ *and* $g \in \mathfrak{L}_{1-\nu+\mathrm{Re}(\kappa+\sigma),r'}$ *with* $r' = r/(r-1)$, *then the relation* (6.2.19) *holds.*

(d) *If* $1-\nu+\mathrm{Re}(\kappa) \notin \mathcal{E}_{\mathcal{G}}$, *then the transform* $G^1_{\sigma,\kappa}$ *is one-to-one on* $\mathfrak{L}_{\nu,r}$ *and there holds*

$$G^1_{\sigma,\kappa}(\mathfrak{L}_{\nu,r}) = \mathfrak{L}_{\nu-\mathrm{Re}(\kappa+\sigma),r}. \tag{6.6.3}$$

(e) *If* $f \in \mathfrak{L}_{\nu,r}$ *and* $\lambda \in \mathbb{C}$, *then* $G^1_{\sigma,\kappa}f$ *is given in* (6.6.1) *for* $\mathrm{Re}(\lambda) > \nu - \mathrm{Re}(\kappa) - 1$, *while in* (6.6.2) *for* $\mathrm{Re}(\lambda) < \nu - \mathrm{Re}(\kappa) - 1$.

Theorem 6.52. *Let* $a^* = \Delta = 0, \mathrm{Re}(\mu) < 0$ *and* $\alpha < \nu - \mathrm{Re}(\kappa) < \beta$, *and let either* $m > 0$ *or* $n > 0$. *Let* $1 < r < \infty$.

(a) *The transform* $G^1_{\sigma,\kappa}$ *defined on* $\mathfrak{L}_{\nu,2}$ *can be extended to* $\mathfrak{L}_{\nu,r}$ *as an element of* $[\mathfrak{L}_{\nu,r}, \mathfrak{L}_{\nu-\mathrm{Re}(\kappa+\sigma),s}]$ *for all* $s \geqq r$ *such that* $1/s > 1/r + \mathrm{Re}(\mu)$.

(b) *If* $1 < r \leqq 2$, *then the transform* $G^1_{\sigma,\kappa}$ *is one-to-one on* $\mathfrak{L}_{\nu,r}$ *and there holds the equality* (6.2.14) *for* $f \in \mathfrak{L}_{\nu,r}$ *and* $\mathrm{Re}(s) = \nu - \mathrm{Re}(\kappa+\sigma)$.

(c) *If* $f \in \mathfrak{L}_{\nu,r}$ *and* $g \in \mathfrak{L}_{1-\nu+\mathrm{Re}(\kappa+\sigma),s}$ *with* $1 < s < \infty$ *and* $1 \leqq 1/r + 1/s < 1 - \mathrm{Re}(\mu)$, *then the relation* (6.2.19) *holds.*

(d) *Let* $k > 0$. *If* $1-\nu+\mathrm{Re}(\kappa) \notin \mathcal{E}_{\mathcal{G}}$, *then the transform* $G^1_{\sigma,\kappa}$ *is one-to-one on* $\mathfrak{L}_{\nu,r}$ *and there hold*

$$G^1_{\sigma,\kappa}(\mathfrak{L}_{\nu,r}) = I^{-\mu}_{-;k,(\sigma-\alpha)/k}\left(\mathfrak{L}_{\nu-\mathrm{Re}(\kappa+\sigma),r}\right) \tag{6.6.4}$$

for $m > 0$, *and*

$$G^1_{\sigma,\kappa}(\mathfrak{L}_{\nu,r}) = I^{-\mu}_{0+;k,(\beta-\sigma)/k-1}\left(\mathfrak{L}_{\nu-\mathrm{Re}(\kappa+\sigma),r}\right) \tag{6.6.5}$$

for $n > 0$. *If* $1-\nu+\mathrm{Re}(\kappa) \in \mathcal{E}_{\mathcal{G}}$, *then* $G^1_{\sigma,\kappa}(\mathfrak{L}_{\nu,r})$ *is a subset of the right-hand sides of* (6.6.4) *and* (6.6.5) *in the respective cases.*

(e) *If* $f \in \mathfrak{L}_{\nu,r}$ *and* $\lambda \in \mathbb{C}$, *then* $G^1_{\sigma,\kappa}f$ *is given in* (6.6.1) *for* $\mathrm{Re}(\lambda) > \nu - \mathrm{Re}(\kappa) - 1$, *while in* (6.6.2) *for* $\mathrm{Re}(\lambda) < \nu - \mathrm{Re}(\kappa) - 1$. *Furthermore* $G^1_{\sigma,\kappa}f$ *is given in* (6.2.4).

Theorem 6.53. *Let* $a^* = 0, \Delta \neq 0, m > 0, \alpha < \nu - \mathrm{Re}(\kappa) < \beta, 1 < r < \infty$ *and* $\Delta[\nu - \mathrm{Re}(\kappa)] + \mathrm{Re}(\mu) \leqq 1/2 - \gamma(r)$. *Assume that* $m > 0$ *if* $\Delta > 0$ *and* $n > 0$ *if* $\Delta < 0$.

(a) *The transform* $G^1_{\sigma,\kappa}$ *defined on* $\mathfrak{L}_{\nu,2}$ *can be extended to* $\mathfrak{L}_{\nu,r}$ *as an element of* $[\mathfrak{L}_{\nu,r}, \mathfrak{L}_{\nu-\mathrm{Re}(\kappa+\sigma),s}]$ *for all* s *with* $r \leqq s < \infty$ *such that* $s' \geqq [1/2 - \Delta\{\nu - \mathrm{Re}(\kappa)\} - \mathrm{Re}(\mu)]^{-1}$ *with* $1/s + 1/s' = 1$.

(b) *If* $1 < r \leqq 2$, *then the transform* $G^1_{\sigma,\kappa}$ *is one-to-one on* $\mathfrak{L}_{\nu,r}$ *and there holds the equality* (6.2.14) *for* $f \in \mathfrak{L}_{\nu,r}$ *and* $\mathrm{Re}(s) = \nu - \mathrm{Re}(\kappa+\sigma)$.

(c) *If* $f \in \mathfrak{L}_{\nu,r}$ *and* $g \in \mathfrak{L}_{1-\nu+\mathrm{Re}(\kappa+\sigma),s}$ *with* $1 < s < \infty, 1/r + 1/s \geqq 1$ *and* $\Delta[\nu - \mathrm{Re}(\kappa)] + \mathrm{Re}(\mu) \leqq 1/2 - \max[\gamma(r), \gamma(s)]$, *then the relation* (6.2.19) *holds.*

(d) *If* $1-\nu+\mathrm{Re}(\kappa) \notin \mathcal{E}_{\mathcal{G}}$, *then the transform* $G^1_{\sigma,\kappa}$ *is one-to-one on* $\mathfrak{L}_{\nu,r}$. *If we set* $\eta = -\Delta\alpha - \mu - 1$ *for* $\Delta > 0$ *and* $\eta = -\Delta\beta - \mu - 1$ *for* $\Delta < 0$, *then* $\mathrm{Re}(\eta) > -1$ *and there* *holds*

$$\boldsymbol{G}^1_{\sigma,\kappa}(\mathfrak{L}_{\nu,r}) = \left(M_{\sigma+\mu/\Delta+1/2}\mathbb{H}_{\Delta,\eta}\right)\left(\mathfrak{L}_{1/2-\nu-\mathrm{Re}(\mu)/\Delta-\mathrm{Re}(\kappa),r}\right). \tag{6.6.6}$$

When $1-\nu+\mathrm{Re}(\kappa) \in \mathcal{E}_{\mathcal{G}}$, $\boldsymbol{G}^1_{\sigma,\kappa}(\mathfrak{L}_{\nu,r})$ *is a subset of the set on the right-hand side of* (6.6.6).

(e) *If* $f \in \mathfrak{L}_{\nu,r}$, $\lambda \in \mathbb{C}$ *and* $\Delta[\nu-\mathrm{Re}(\kappa)]+\mathrm{Re}(\mu) \leqq 1/2-\gamma(r)$, *then* $\boldsymbol{G}^1_{\sigma,\kappa}f$ *is given in* (6.6.1) *for* $\mathrm{Re}(\lambda) > \nu-\mathrm{Re}(\kappa)-1$, *while in* (6.6.2) *for* $\mathrm{Re}(\lambda) < \nu-\mathrm{Re}(\kappa)-1$. *If* $\Delta[\nu-\mathrm{Re}(\kappa)]+\mathrm{Re}(\mu) < 0$, $\boldsymbol{G}^1_{\sigma,\kappa}f$ *is given in* (6.2.4).

From Theorem 5.41 we obtain the $\mathfrak{L}_{\nu,r}$-theory of the transform $\boldsymbol{G}^1_{\sigma,\kappa}$ in (6.2.4) with $a^* > 0$.

Theorem 6.54. *Let* $a^* > 0, \alpha < \nu - \mathrm{Re}(\kappa) < \beta$ *and* $1 \leqq r \leqq s \leqq \infty$.

(a) *The transform* $\boldsymbol{G}^1_{\sigma,\kappa}$ *defined on* $\mathfrak{L}_{\nu,2}$ *can be extended to* $\mathfrak{L}_{\nu,r}$ *as an element of* $[\mathfrak{L}_{\nu,r}, \mathfrak{L}_{\nu-\mathrm{Re}(\kappa+\sigma),s}]$. *If* $1 \leqq r \leqq 2$, *then* $\boldsymbol{G}^1_{\sigma,\kappa}$ *is a one-to-one transform from* $\mathfrak{L}_{\nu,r}$ *onto* $\mathfrak{L}_{\nu-\mathrm{Re}(\kappa+\sigma),s}$.

(b) *If* $f \in \mathfrak{L}_{\nu,r}$ *and* $g \in \mathfrak{L}_{1-\nu+\mathrm{Re}(\kappa+\sigma),s'}$ *with* $1/s+1/s' = 1$, *then the relation* (6.2.19) *holds.*

By virtue of Theorems 5.42–5.44 we characterize the boundedness and the range of $\boldsymbol{G}^1_{\sigma,\kappa}$ on $\mathfrak{L}_{\nu,r}$ in five cases.

Theorem 6.55. *Let* $a_1^* > 0, a_2^* > 0, m > 0, n > 0, \alpha < \nu - \mathrm{Re}(\kappa) < \beta$ *and* $\omega = \mu + a_1^*\alpha - a_2^*\beta + 1$ *and let* $1 < r < \infty$.

(a) *If* $1-\nu+\mathrm{Re}(\kappa) \notin \mathcal{E}_{\mathcal{G}}$, *or if* $1 \leqq r \leqq 2$, *then the transform* $\boldsymbol{G}^1_{\sigma,\kappa}$ *is one-to-one on* $\mathfrak{L}_{\nu,r}$.

(b) *If* $\mathrm{Re}(\omega) \geqq 0$ *and* $1-\nu+\mathrm{Re}(\kappa) \notin \mathcal{E}_{\mathcal{G}}$, *then*

$$\boldsymbol{G}^1_{\sigma,\kappa}(\mathfrak{L}_{\nu,r}) = \left(\mathbb{L}_{a_1^*,\alpha-\sigma}\mathbb{L}_{a_2^*,1-\beta+\sigma-\omega/a_2^*}\right)\left(\mathfrak{L}_{\nu-\mathrm{Re}(\kappa+\sigma),r}\right). \tag{6.6.7}$$

When $1-\nu+\mathrm{Re}(\kappa) \in \mathcal{E}_{\mathcal{G}}$, $\boldsymbol{G}^1_{\sigma,\kappa}(\mathfrak{L}_{\nu,r})$ *is a subset of the right-hand side of* (6.6.7).

(c) *If* $\mathrm{Re}(\omega) < 0$ *and* $1-\nu+\mathrm{Re}(\kappa) \notin \mathcal{E}_{\mathcal{G}}$, *then*

$$\boldsymbol{G}^1_{\sigma,\kappa}(\mathfrak{L}_{\nu,r}) = \left(I^{-\omega}_{-;1/a_1^*,a_1^*(\sigma-\alpha)}\mathbb{L}_{a_1^*,\alpha-\sigma}\mathbb{L}_{a_2^*,1-\beta+\sigma}\right)\left(\mathfrak{L}_{\nu-\mathrm{Re}(\kappa+\sigma),r}\right). \tag{6.6.8}$$

When $1-\nu+\mathrm{Re}(\kappa) \in \mathcal{E}_{\mathcal{G}}$, $\boldsymbol{G}^1_{\sigma,\kappa}(\mathfrak{L}_{\nu,r})$ *is a subset of the right-hand side of* (6.6.8).

Theorem 6.56. *Let* $a_1^* > 0, a_2^* = 0, m > 0, \alpha < \nu - \mathrm{Re}(\kappa) < \beta$ *and* $\omega = \mu + a_1^*\alpha + 1/2$ *and let* $1 < r < \infty$.

(a) *If* $1-\nu+\mathrm{Re}(\kappa) \notin \mathcal{E}_{\mathcal{G}}$, *or if* $1 < r \leqq 2$, *then the transform* $\boldsymbol{G}^1_{\sigma,\kappa}$ *is one-to-one on* $\mathfrak{L}_{\nu,r}$.

(b) *If* $\mathrm{Re}(\omega) \geqq 0$ *and* $1-\nu+\mathrm{Re}(\kappa) \notin \mathcal{E}_{\mathcal{G}}$, *then*

$$\boldsymbol{G}^1_{\sigma,\kappa}(\mathfrak{L}_{\nu,r}) = \mathbb{L}_{a_1^*,\alpha-\sigma-\omega/a_1^*}\left(\mathfrak{L}_{1-\nu+\mathrm{Re}(\kappa+\sigma),r}\right). \tag{6.6.9}$$

When $1-\nu+\mathrm{Re}(\kappa) \in \mathcal{E}_{\mathcal{G}}$, $\boldsymbol{G}^1_{\sigma,\kappa}(\mathfrak{L}_{\nu,r})$ *is a subset of the right-hand side of* (6.6.9).

(c) *If* $\mathrm{Re}(\omega) < 0$ *and* $1-\nu+\mathrm{Re}(\kappa) \notin \mathcal{E}_{\mathcal{G}}$, *then*

$$\boldsymbol{G}^1_{\sigma,\kappa}(\mathfrak{L}_{\nu,r}) = \left(I^{-\omega}_{-;1/a_1^*,a_1^*(\sigma-\alpha)}\mathbb{L}_{a_1^*,\alpha-\sigma}\right)\left(\mathfrak{L}_{1-\nu+\mathrm{Re}(\kappa+\sigma),r}\right). \tag{6.6.10}$$

When $1-\nu+\mathrm{Re}(\kappa) \in \mathcal{E}_{\mathcal{G}}$, $\boldsymbol{G}^1_{\sigma,\kappa}(\mathfrak{L}_{\nu,r})$ *is a subset of the right-hand side of* (6.6.10).

Theorem 6.57. *Let $a_1^* = 0, a_2^* > 0, n > 0, \alpha < \nu - \text{Re}(\kappa) < \beta$ and $\omega = \mu - a_2^*\beta + 1/2$ and let $1 < r < \infty$.*

(a) *If $1 - \nu + \text{Re}(\kappa) \notin \mathcal{E}_{\mathcal{G}}$, or if $1 < r \leqq 2$, then the transform $\boldsymbol{G}^1_{\sigma,\kappa}$ is one-to-one on $\mathfrak{L}_{\nu,r}$.*

(b) *If $\text{Re}(\omega) \geqq 0$ and $1 - \nu + \text{Re}(\kappa) \notin \mathcal{E}_{\mathcal{G}}$, then*

$$\boldsymbol{G}^1_{\sigma,\kappa}(\mathfrak{L}_{\nu,r}) = \mathbb{L}_{-a_2^*,\beta-\sigma+\omega/a_2^*}\left(\mathfrak{L}_{1-\nu+\text{Re}(\kappa+\sigma),r}\right). \tag{6.6.11}$$

When $1 - \nu + \text{Re}(\kappa) \in \mathcal{E}_{\mathcal{G}}$, $\boldsymbol{G}^1_{\sigma,\kappa}(\mathfrak{L}_{\nu,r})$ is a subset of the right-hand side of (6.6.11).

(c) *If $\text{Re}(\omega) < 0$ and $1 - \nu + \text{Re}(\kappa) \notin \mathcal{E}_{\mathcal{G}}$, then*

$$\boldsymbol{G}^1_{\sigma,\kappa}(\mathfrak{L}_{\nu,r}) = \left(I^{-\omega}_{0+;1/a_2^*,a_2^*(\beta-\sigma)-1}\mathbb{L}_{-a_2^*,\beta-\sigma}\right)\left(\mathfrak{L}_{1-\nu+\text{Re}(\kappa+\sigma),r}\right). \tag{6.6.12}$$

If $1 - \nu + \text{Re}(\kappa) \in \mathcal{E}_{\mathcal{G}}$, $\boldsymbol{G}^1_{\sigma,\kappa}(\mathfrak{L}_{\nu,r})$ is a subset of the right-hand side of (6.6.12).

Theorem 6.58. *Let $a^* > 0, a_1^* > 0, a_2^* < 0, \alpha < \nu - \text{Re}(\kappa)) < \beta$ and let $1 < r < \infty$.*

(a) *If $1 - \nu + \text{Re}(\kappa) \notin \mathcal{E}_{\mathcal{G}}$, or if $1 < r \leqq 2$, then the transform $\boldsymbol{G}^1_{\sigma,\kappa}$ is one-to-one on $\mathfrak{L}_{\nu,r}$.*

(b) *Let $\omega, \eta, \zeta \in \mathbb{C}$ be chosen as*

$$\omega = a^*\eta - \mu - \frac{1}{2}; \tag{6.6.13}$$

$$a^*\text{Re}(\eta) \geqq \gamma(r) + 2a_2^*[\nu - \text{Re}(\kappa)] + \text{Re}(\mu); \tag{6.6.14}$$

$$\text{Re}(\eta) > -\nu + \text{Re}(\kappa); \tag{6.6.15}$$

$$\text{Re}(\zeta) < \nu - \text{Re}(\kappa). \tag{6.6.16}$$

If $1 - \nu + \text{Re}(\kappa) \notin \mathcal{E}_{\mathcal{G}}$, then

$$\boldsymbol{G}^1_{\sigma,\kappa}(\mathfrak{L}_{\nu,r}) = \left(M_{\sigma+1/2+\omega/(2a_2^*)}\mathbb{H}_{-2a_2^*,2a_2^*\zeta+\omega-1}\mathbb{L}_{-a^*,1/2+\eta-\omega/(2a_2^*)}\right)$$
$$\left(\mathfrak{L}_{\nu+1/2+\text{Re}(\omega)/(2a_2^*)-\text{Re}(\kappa),r}\right). \tag{6.6.17}$$

When $1 - \nu + \text{Re}(\kappa) \in \mathcal{E}_{\mathcal{G}}$, $\boldsymbol{G}^1_{\sigma,\kappa}(\mathfrak{L}_{\nu,r})$ is a subset of the right-hand side of (6.6.17).

Theorem 6.59. *Let $a^* > 0, a_1^* < 0, a_2^* > 0, \alpha < \nu - \text{Re}(\kappa) < \beta$ and let $1 < r < \infty$.*

(a) *If $1 - \nu + \text{Re}(\kappa) \notin \mathcal{E}_{\mathcal{G}}$, or if $1 < r \leqq 2$, then the transform $\boldsymbol{G}^1_{\sigma,\kappa}$ is one-to-one on $\mathfrak{L}_{\nu,r}$.*

(b) *Let $\omega, \eta, \zeta \in \mathbb{C}$ be chosen as*

$$\omega = a^*\eta - \Delta - \mu - \frac{1}{2}; \tag{6.6.18}$$

$$a^*\text{Re}(\eta) \geqq \gamma(r) - 2a_1^*[\nu - \text{Re}(\kappa)] + \Delta + \text{Re}(\mu); \tag{6.6.19}$$

$$\text{Re}(\eta) > -\nu + \text{Re}(\kappa); \tag{6.6.20}$$

$$\text{Re}(\zeta) < \nu - \text{Re}(\kappa). \tag{6.6.21}$$

If $1 - \nu + \text{Re}(\kappa) \notin \mathcal{E}_{\mathcal{G}}$, then

$$\boldsymbol{G}^1_{\sigma,\kappa}(\mathfrak{L}_{\nu,r}) = \left(M_{\sigma-1/2-\omega/(2a_1^*)}\mathbb{H}_{2a_1^*,2a_1^*\zeta+\omega-1}\mathbb{L}_{a^*,1/2-\eta+\omega/(2a_1^*)}\right)$$
$$\left(\mathfrak{L}_{\nu-1/2-\text{Re}(\omega)/(2a_1^*)-\text{Re}(\kappa),r}\right). \tag{6.6.22}$$

When $1-\nu+\mathrm{Re}(\kappa)\in\mathcal{E}_{\mathcal{G}}$, $\boldsymbol{G}^1_{\sigma,\kappa}(\mathfrak{L}_{\nu,r})$ *is a subset of the right-hand side of* (6.6.22).

The inversion formulas for the transform $\boldsymbol{G}^1_{\sigma,\kappa}$ in (6.2.4) on $\mathfrak{L}_{\nu,r}$, when $a^*=0$, are obtained from (5.5.23) and (5.5.24) under the condition (6.1.4):

$$f(x)=-x^{\lambda+1-\kappa}\frac{d}{dx}x^{-\lambda-1}$$
$$\cdot\int_0^\infty G^{q-m,p-n+1}_{p+1,q+1}\left[\frac{t}{x}\,\middle|\,\begin{matrix}-\lambda,-a_{n+1},\cdots,-a_p,-a_1,\cdots,-a_n\\ -b_{m+1},\cdots,-b_q,-b_1,\cdots,-b_m,-\lambda-1\end{matrix}\right]$$
$$\cdot\, t^{-\sigma}\left(\boldsymbol{G}^1_{\sigma,\kappa}f\right)(t)dt \tag{6.6.23}$$

or

$$f(x)=x^{\lambda+1-\kappa}\frac{d}{dx}x^{-\lambda-1}$$
$$\cdot\int_0^\infty G^{q-m+1,p-n}_{p+1,q+1}\left[\frac{t}{x}\,\middle|\,\begin{matrix}-a_{n+1},\cdots,-a_p,-a_1,\cdots,-a_n,-\lambda\\ -\lambda-1,-b_{m+1},\cdots,-b_q,-b_1,\cdots,-b_m\end{matrix}\right]$$
$$\cdot\, t^{-\sigma}\left(\boldsymbol{G}^1_{\sigma,\kappa}f\right)(t)dt. \tag{6.6.24}$$

Theorems 5.47 and 5.48 deduce the validity of the inversion formulas (6.6.23) and (6.6.24).

Theorem 6.60. *Let* $a^*=0,\alpha<\nu-\mathrm{Re}(\kappa)<\beta$ *and* $\alpha_0<1-\nu+\mathrm{Re}(\kappa)<\beta_0$, *and let* $\lambda\in\mathbb{C}$.

(a) *If* $\Delta[\nu-\mathrm{Re}(\kappa)]+\mathrm{Re}(\mu)=0$ *and* $f\in\mathfrak{L}_{\nu,2}$, *then the inversion formula* (6.6.23) *holds for* $\mathrm{Re}(\lambda)>-\nu+\mathrm{Re}(\kappa)$ *and* (6.6.24) *for* $\mathrm{Re}(\lambda)<-\nu+\mathrm{Re}(\kappa)$.

(b) *If* $\Delta=\mathrm{Re}(\mu)=0$ *and* $f\in\mathfrak{L}_{\nu,r}$ $(1<r<\infty)$, *then the inversion formula* (6.6.23) *holds for* $\mathrm{Re}(\lambda)>-\nu+\mathrm{Re}(\kappa)$ *and* (6.6.24) *for* $\mathrm{Re}(\lambda)<-\nu+\mathrm{Re}(\kappa)$.

Theorem 6.61. *Let* $a^*=0,1<r<\infty$ *and* $\Delta[\nu-\mathrm{Re}(\kappa)]+\mathrm{Re}(\mu)\leqq 1/2-\gamma(r)$, *and let* $\lambda\in\mathbb{C}$.

(a) *If* $\Delta>0,m>0,\alpha<\nu-\mathrm{Re}(\kappa)<\beta$, $\alpha_0<1-\nu+\mathrm{Re}(\kappa)<\min[\beta_0,\{\mathrm{Re}(\mu+1/2)/\Delta\}+1]$ *and if* $f\in\mathfrak{L}_{\nu,r}$, *then the inversion formulas* (6.6.23) *and* (6.6.24) *hold for* $\mathrm{Re}(\lambda)>-\nu+\mathrm{Re}(\kappa)$ *and for* $\mathrm{Re}(\lambda)<-\nu+\mathrm{Re}(\kappa)$, *respectively.*

(b) *If* $\Delta<0,n>0,\alpha<\nu-\mathrm{Re}(\kappa)<\beta$, $\max[\alpha_0,\{\mathrm{Re}(\mu+1/2)/\Delta\}+1]<1-\nu+\mathrm{Re}(\kappa)<\beta_0$ *and if* $f\in\mathfrak{L}_{\nu,r}$, *then the inversion formulas* (6.6.23) *and* (6.6.24) *hold for* $\mathrm{Re}(\lambda)>-\nu+\mathrm{Re}(\kappa)$ *and for* $\mathrm{Re}(\lambda)<-\nu+\mathrm{Re}(\kappa)$, *respectively.*

Remark 6.2. The results in this section generalize those in Section 6.3. That is, Theorems 6.13–6.24 follow from Theorems 6.48–6.60 when $\sigma=\kappa=0$.

6.7. $G^2_{\sigma,\kappa}$-Transform on the Space $\mathfrak{L}_{\nu,r}$

Lastly we treat the transform $G^2_{\sigma,\kappa}$ defined in (6.2.5) by using the results in Section 5.6.

Theorem 6.62. *We suppose that* **(a)** $\alpha < 1-\nu+\mathrm{Re}(\kappa) < \beta$ *and that either of the conditions* **(b)** $a^* > 0$, *or* **(c)** $a^* = 0,\ \Delta[1-\nu+\mathrm{Re}(\kappa)] + \mathrm{Re}(\mu) \leqq 0$ *holds. Then we have the following results:*

(i) *There is a one-to-one transform* $G^2_{\sigma,\kappa} \in [\mathfrak{L}_{\nu,2}, \mathfrak{L}_{\nu-\mathrm{Re}(\kappa+\sigma),2}]$ *such that* (6.2.15) *holds for* $\mathrm{Re}(s) = \nu - \mathrm{Re}(\kappa+\sigma)$ *and* $f \in \mathfrak{L}_{\nu,2}$. *If* $a^* = 0$, $\Delta[1-\nu+\mathrm{Re}(\kappa)] + \mathrm{Re}(\mu) = 0$ *and* $\nu - \mathrm{Re}(\kappa) \notin \mathcal{E}_{\mathcal{G}}$, *then the transform* $G^2_{\sigma,\kappa}$ *maps* $\mathfrak{L}_{\nu,2}$ *onto* $\mathfrak{L}_{\nu-\mathrm{Re}(\kappa+\sigma),2}$.

(ii) *If* $f \in \mathfrak{L}_{\nu,2}$ *and* $g \in \mathfrak{L}_{1-\nu+\mathrm{Re}(\kappa+\sigma),2}$, *then the relation* (6.2.20) *holds for* $G^2_{\sigma,\kappa}$.

(iii) *Let* $\lambda \in \mathbb{C}$ *and* $f \in \mathfrak{L}_{\nu,2}$. *If* $\mathrm{Re}(\lambda) > -\nu + \mathrm{Re}(\kappa)$, *then* $G^2_{\sigma,\kappa}f$ *is given by*

$$\left(G^2_{\sigma,\kappa}f\right)(x) = -x^{\sigma+\lambda+1}\frac{d}{dx}x^{-\lambda-1}\int_0^\infty G^{m,n+1}_{p+1,q+1}\left[\frac{t}{x}\,\middle|\,\begin{matrix}-\lambda, a_1,\cdots,a_p\\ b_1,\cdots,b_q,-\lambda-1\end{matrix}\right]t^\kappa f(t)dt. \quad (6.7.1)$$

When $\mathrm{Re}(\lambda) < -\nu + \mathrm{Re}(\kappa)$,

$$\left(G^2_{\sigma,\kappa}f\right)(x) = x^{\sigma+\lambda+1}\frac{d}{dx}x^{-\lambda-1}\int_0^\infty G^{m+1,n}_{p+1,q+1}\left[\frac{t}{x}\,\middle|\,\begin{matrix}a_1,\cdots,a_p,-\lambda\\ -\lambda-1, b_1,\cdots,b_q\end{matrix}\right]t^\kappa f(t)dt. \quad (6.7.2)$$

(iv) *The transform* $G^2_{\sigma,\kappa}$ *is independent of* ν *in the sense that, if* ν *and* $\widetilde{\nu}$ *satisfy* (a), *and either* (b) *or* (c), *and if the transforms* $G^2_{\sigma,\kappa}$ *and* $\widetilde{G^2}_{\sigma,\kappa}$ *are defined in* $\mathfrak{L}_{\nu,2}$ *and* $\mathfrak{L}_{\widetilde{\nu},2}$, *respectively, by* (6.2.15), *then* $G^2_{\sigma,\kappa}f = \widetilde{G^2}_{\sigma,\kappa}f$ *for* $f \in \mathfrak{L}_{\nu,2} \cap \mathfrak{L}_{\widetilde{\nu},2}$.

(v) *If* $a^* > 0$ *or if* $a^* = 0$, $\Delta[1-\nu+\mathrm{Re}(\kappa)] + \mathrm{Re}(\mu) < 0$, *then for* $f \in \mathfrak{L}_{\nu,2}$, $G^2_{\sigma,\kappa}f$ *is given in* (6.2.5).

Theorems 5.50–5.52 lead to the mapping properties and the range of $G^2_{\sigma,\kappa}$ on $\mathfrak{L}_{\nu,r}$ for $a^* > 0$.

Theorem 6.63. *Let* $a^* = \Delta = 0, \mathrm{Re}(\mu) = 0$ *and* $\alpha < 1-\nu+\mathrm{Re}(\kappa) < \beta$. *Let* $1 < r < \infty$.

(a) *The transform* $G^2_{\sigma,\kappa}$ *defined on* $\mathfrak{L}_{\nu,2}$ *can be extended to* $\mathfrak{L}_{\nu,r}$ *as an element of* $[\mathfrak{L}_{\nu,r}, \mathfrak{L}_{\nu-\mathrm{Re}(\kappa+\sigma),r}]$.

(b) *If* $1 < r \leqq 2$, *then the transform* $G^2_{\sigma,\kappa}$ *is one-to-one on* $\mathfrak{L}_{\nu,r}$ *and there holds the equality* (6.2.15) *for* $f \in \mathfrak{L}_{\nu,r}$ *and* $\mathrm{Re}(s) = \nu - \mathrm{Re}(\kappa+\sigma)$.

(c) *If* $f \in \mathfrak{L}_{\nu,r}$ *and* $g \in \mathfrak{L}_{1-\nu+\mathrm{Re}(\kappa+\sigma),r'}$ *with* $r' = r/(r-1)$, *then the relation* (6.2.20) *holds.*

(d) *If* $\nu - \mathrm{Re}(\kappa) \notin \mathcal{E}_{\mathcal{G}}$, *then the transform* $G^2_{\sigma,\kappa}$ *is one-to-one on* $\mathfrak{L}_{\nu,r}$ *and there holds*

$$G^2_{\sigma,\kappa}(\mathfrak{L}_{\nu,r}) = \mathfrak{L}_{\nu-\mathrm{Re}(\kappa+\sigma),r}. \quad (6.7.3)$$

(e) *If* $f \in \mathfrak{L}_{\nu,r}$ *and* $\lambda \in \mathbb{C}$, *then* $G^2_{\sigma,\kappa}f$ *is given in* (6.7.1) *for* $\mathrm{Re}(\lambda) > -\nu+\mathrm{Re}(\kappa)$, *while in* (6.7.2) *for* $\mathrm{Re}(\lambda) < -\nu+\mathrm{Re}(\kappa)$.

Theorem 6.64. *Let* $a^* = \Delta = 0, \mathrm{Re}(\mu) < 0$ *and* $\alpha < 1-\nu+\mathrm{Re}(\kappa) < \beta$, *and let either* $m > 0$ *or* $n > 0$. *Let* $1 < r < \infty$.

(a) *The transform $G^2_{\sigma,\kappa}$ defined on $\mathfrak{L}_{\nu,2}$ can be extended to $\mathfrak{L}_{\nu,r}$ as an element of $[\mathfrak{L}_{\nu,r}, \mathfrak{L}_{\nu-\mathrm{Re}(\kappa+\sigma),s}]$ for all $s \geqq r$ such that $1/s > 1/r + \mathrm{Re}(\mu)$.*

(b) *If $1 < r \leqq 2$, then the transform $G^2_{\sigma,\kappa}$ is one-to-one on $\mathfrak{L}_{\nu,r}$ and there holds the equality (6.2.15) for $f \in \mathfrak{L}_{\nu,r}$ and $\mathrm{Re}(s) = \nu - \mathrm{Re}(\kappa+\sigma)$.*

(c) *If $f \in \mathfrak{L}_{\nu,r}$ and $g \in \mathfrak{L}_{1-\nu+\mathrm{Re}(\kappa+\sigma),s}$ with $1 < s < \infty$ and $1 \leqq 1/r + 1/s < 1 - \mathrm{Re}(\mu)$, then the relation (6.2.20) holds.*

(d) *Let $k > 0$. If $\nu - \mathrm{Re}(\kappa) \notin \mathcal{E}_{\mathcal{G}}$, then the transform $G^2_{\sigma,\kappa}$ is one-to-one on $\mathfrak{L}_{\nu,r}$ and there hold*

$$G^2_{\sigma,\kappa}(\mathfrak{L}_{\nu,r}) = I^{-\mu}_{0+;k,(1-\alpha-\sigma)/k-1}\left(\mathfrak{L}_{\nu-\mathrm{Re}(\kappa+\sigma),r}\right) \tag{6.7.4}$$

for $m > 0$, and

$$G^2_{\sigma,\kappa}(\mathfrak{L}_{\nu,r}) = I^{-\mu}_{-;k,(\beta+\sigma-1)/k}\left(\mathfrak{L}_{\nu-\mathrm{Re}(\kappa+\sigma),r}\right) \tag{6.7.5}$$

for $n > 0$. When $\nu - \mathrm{Re}(\kappa) \in \mathcal{E}_{\mathcal{G}}$, $G^2_{\sigma,\kappa}(\mathfrak{L}_{\nu,r})$ is a subset of the right-hand sides of (6.7.4) and (6.7.5) in the respective cases.

(e) *If $f \in \mathfrak{L}_{\nu,r}$ and $\lambda \in \mathbb{C}$, then $G^2_{\sigma,\kappa}f$ is given in (6.7.1) for $\mathrm{Re}(\lambda) > -\nu + \mathrm{Re}(\kappa)$, while in (6.7.2) for $\mathrm{Re}(\lambda) < -\nu + \mathrm{Re}(\kappa)$. Furthermore $G^2_{\sigma,\kappa}f$ is given in (6.2.5).*

Theorem 6.65. *Let $a^* = 0, \Delta \neq 0, \alpha < 1 - \nu + \mathrm{Re}(\kappa) < \beta$, $1 < r < \infty$ and $\Delta[1 - \nu + \mathrm{Re}(\kappa)] + \mathrm{Re}(\mu) \leqq 1/2 - \gamma(r)$. Assume that $m > 0$ if $\Delta > 0$ and $n > 0$ if $\Delta < 0$.*

(a) *The transform $G^2_{\sigma,\kappa}$ defined on $\mathfrak{L}_{\nu,2}$ can be extended to $\mathfrak{L}_{\nu,r}$ as an element of $[\mathfrak{L}_{\nu,r}, \mathfrak{L}_{\nu-\mathrm{Re}(\kappa+\sigma),s}]$ for all s with $r \leqq s < \infty$ such that $s' \geqq [1/2 - \Delta\{1-\nu+\mathrm{Re}(\kappa)\} - \mathrm{Re}(\mu)]^{-1}$ with $1/s + 1/s' = 1$.*

(b) *If $1 < r \leqq 2$, then the transform $G^2_{\sigma,\kappa}$ is one-to-one on $\mathfrak{L}_{\nu,r}$ and there holds the equality (6.2.15) for $f \in \mathfrak{L}_{\nu,r}$ and $\mathrm{Re}(s) = \nu - \mathrm{Re}(\kappa+\sigma)$.*

(c) *If $f \in \mathfrak{L}_{\nu,r}$ and $g \in \mathfrak{L}_{1-\nu+\mathrm{Re}(\kappa+\sigma),s}$ with $1 < s < \infty, 1/r + 1/s \geqq 1$ and $\Delta[\nu - \mathrm{Re}(\kappa)] + \mathrm{Re}(\mu) \leqq 1/2 - \max[\gamma(r), \gamma(s)]$, then the relation (5.1.20) holds.*

(d) *If $\nu - \mathrm{Re}(\kappa) \notin \mathcal{E}_{\mathcal{G}}$, then the transform $G^2_{\sigma,\kappa}$ is one-to-one on $\mathfrak{L}_{\nu,r}$. If we set $\eta = -\Delta\alpha - \mu - 1$ for $\Delta > 0$ and $\eta = -\Delta\beta - \mu - 1$ for $\Delta < 0$, then $\mathrm{Re}(\eta) > -1$ and there holds*

$$G^2(\mathfrak{L}_{\nu,r}) = \left(M_{\sigma-\mu/\Delta-1/2}\mathbb{H}_{-\Delta,\eta}\right)\left(\mathfrak{L}_{3/2-\nu+\mathrm{Re}(\mu)/\Delta+\mathrm{Re}(\kappa),r}\right). \tag{6.7.6}$$

When $\nu - \mathrm{Re}(\kappa) \in \mathcal{E}_{\mathcal{G}}$, $G^2_{\sigma,\kappa}(\mathfrak{L}_{\nu,r})$ is a subset of the set on the right-hand side of (6.7.6).

(e) *If $f \in \mathfrak{L}_{\nu,r}, \lambda \in \mathbb{C}$ and $\Delta[1 - \nu + \mathrm{Re}(\kappa)] + \mathrm{Re}(\mu) \leqq 1/2 - \gamma(r)$, then $G^2_{\sigma,\kappa}f$ is given in (6.7.1) for $\mathrm{Re}(\lambda) > -\nu + \mathrm{Re}(\kappa)$, while in (6.7.2) for $\mathrm{Re}(\lambda) < -\nu + \mathrm{Re}(\kappa)$. If $\Delta[1 - \nu + \mathrm{Re}(\kappa)] + \mathrm{Re}(\mu) < 0$, $G^2_{\sigma,\kappa}f$ is given in (6.2.5).*

From Theorem 5.53 we have the $\mathfrak{L}_{\nu,r}$-theory of the transform $G^2_{\sigma,\kappa}$ in (5.2.5) with $a^* > 0$ in $\mathfrak{L}_{\nu,r}$-spaces.

Theorem 6.66. *Let $a^* > 0, \alpha < 1 - \nu + \mathrm{Re}(\kappa) < \beta$ and $1 \leqq r \leqq s \leqq \infty$.*

(a) *The transform $G^2_{\sigma,\kappa}$ defined on $\mathfrak{L}_{\nu,2}$ can be extended to $\mathfrak{L}_{\nu,r}$ as an element of $[\mathfrak{L}_{\nu,r}, \mathfrak{L}_{\nu-\mathrm{Re}(\kappa+\sigma),s}]$. If $1 \leqq r \leqq 2$, then $G^2_{\sigma,\kappa}$ is a one-to-one transform from $\mathfrak{L}_{\nu,r}$ onto $\mathfrak{L}_{\nu-\mathrm{Re}(\kappa+\sigma),s}$.*

(b) *If $f \in \mathfrak{L}_{\nu,r}$ and $g \in \mathfrak{L}_{1-\nu+\mathrm{Re}(\kappa+\sigma),s'}$ with $1/s+1/s'=1$, then the relation (6.2.20) holds.*

Now we have the characterization theorems from Theorems 5.54–5.58 of the boundedness and the range of $G^2_{\sigma,\kappa}$ on $\mathfrak{L}_{\nu,r}$.

Theorem 6.67. *Let $a_1^* > 0, a_2^* > 0, m > 0, n > 0, \alpha < 1-\nu+\mathrm{Re}(\kappa) < \beta$ and $\omega = \mu + a_1^*\alpha - a_2^*\beta + 1$ and let $1 < r < \infty$.*

(a) *If $\nu - \mathrm{Re}(\kappa) \notin \mathcal{E}_{\mathcal{G}}$, or if $1 \leqq r \leqq 2$, then the transform $G^2_{\sigma,\kappa}$ is one-to-one on $\mathfrak{L}_{\nu,r}$.*

(b) *If $\mathrm{Re}(\omega) \geqq 0$ and $\nu - \mathrm{Re}(\kappa) \notin \mathcal{E}_{\mathcal{G}}$, then*

$$G^2_{\sigma,\kappa}(\mathfrak{L}_{\nu,r}) = \left(\mathbb{L}_{-a_1^*,1-\alpha-\sigma}\mathbb{L}_{-a_2^*,\beta+\sigma+\omega/a_2^*}\right)\left(\mathfrak{L}_{\nu-\mathrm{Re}(\kappa+\sigma),r}\right). \tag{6.7.7}$$

When $\nu - \mathrm{Re}(\kappa) \in \mathcal{E}_{\mathcal{G}}$, $G^1_{\sigma,\kappa}(\mathfrak{L}_{\nu,r})$ is a subset of the right-hand side of (6.7.7).

(c) *If $\mathrm{Re}(\omega) < 0$ and $\nu - \mathrm{Re}(\kappa) \notin \mathcal{E}_{\mathcal{G}}$, then*

$$G^2_{\sigma,\kappa}(\mathfrak{L}_{\nu,r}) = \left(I^{-\omega}_{0+;1/a_1^*,a_1^*(1-\sigma-\alpha)-1}\mathbb{L}_{-a_1^*,1-\alpha-\sigma}\mathbb{L}_{-a_2^*,\beta+\sigma}\right)\left(\mathfrak{L}_{\nu-\mathrm{Re}(\kappa+\sigma),r}\right). \tag{6.7.8}$$

When $\nu - \mathrm{Re}(\kappa) \in \mathcal{E}_{\mathcal{G}}$, $G^2_{\sigma,\kappa}(\mathfrak{L}_{\nu,r})$ is a subset of the right-hand side of (6.7.8).

Theorem 6.68. *Let $a_1^* > 0, a_2^* = 0, m > 0, \alpha < 1-\nu+\mathrm{Re}(\kappa) < \beta$ and $\omega = \mu + a_1^*\alpha + 1/2$ and let $1 < r < \infty$.*

(a) *If $\nu - \mathrm{Re}(\kappa) \notin \mathcal{E}_{\mathcal{G}}$, or if $1 < r \leqq 2$, then the transform $G^2_{\sigma,\kappa}$ is one-to-one on $\mathfrak{L}_{\nu,r}$.*

(b) *If $\mathrm{Re}(\omega) \geqq 0$ and $\nu - \mathrm{Re}(\kappa) \notin \mathcal{E}_{\mathcal{G}}$, then*

$$G^2_{\sigma,\kappa}(\mathfrak{L}_{\nu,r}) = \mathbb{L}_{-a_1^*,1-\alpha-\sigma+\omega/a_1^*}\left(\mathfrak{L}_{1-\nu+\mathrm{Re}(\kappa+\sigma),r}\right). \tag{6.7.9}$$

When $\nu - \mathrm{Re}(\kappa) \in \mathcal{E}_{\mathcal{G}}$, $G^2_{\sigma,\kappa}(\mathfrak{L}_{\nu,r})$ is a subset of the right-hand side of (6.7.9).

(c) *If $\mathrm{Re}(\omega) < 0$ and $\nu - \mathrm{Re}(\kappa) \notin \mathcal{E}_{\mathcal{G}}$, then*

$$G^2_{\sigma,\kappa}(\mathfrak{L}_{\nu,r}) = \left(I^{-\omega}_{0+;1/a_1^*,a_1^*(1-\alpha-\sigma)-1}\mathbb{L}_{-a_1^*,1-\alpha-\sigma}\right)\left(\mathfrak{L}_{1-\nu+\mathrm{Re}(\kappa+\sigma),r}\right). \tag{6.7.10}$$

When $\nu - \mathrm{Re}(\kappa) \in \mathcal{E}_{\mathcal{G}}$, $G^2_{\sigma,\kappa}(\mathfrak{L}_{\nu,r})$ is a subset of the right-hand side of (6.7.10).

Theorem 6.69. *Let $a_1^* = 0, a_2^* > 0, n > 0, \alpha < 1-\nu+\mathrm{Re}(\kappa) < \beta$ and $\omega = \mu - a_2^*\beta + 1/2$ and let $1 < r < \infty$.*

(a) *If $\nu - \mathrm{Re}(\kappa) \notin \mathcal{E}_{\mathcal{G}}$, or if $1 < r \leqq 2$, then the transform $G^2_{\sigma,\kappa}$ is one-to-one on $\mathfrak{L}_{\nu,r}$.*

(b) *If $\mathrm{Re}(\omega) \geqq 0$ and $\nu - \mathrm{Re}(\kappa) \notin \mathcal{E}_{\mathcal{G}}$, then*

$$G^2_{\sigma,\kappa}(\mathfrak{L}_{\nu,r}) = \mathbb{L}_{a_2^*,1-\beta-\sigma-\omega/a_2^*}\left(\mathfrak{L}_{1-\nu+\mathrm{Re}(\kappa+\sigma),r}\right). \tag{6.7.11}$$

When $\nu - \mathrm{Re}(\kappa) \in \mathcal{E}_{\mathcal{G}}$, $G^2_{\sigma,\kappa}(\mathfrak{L}_{\nu,r})$ is a subset of the right-hand side of (6.7.11).

(c) *If $\mathrm{Re}(\omega) < 0$ and $\nu - \mathrm{Re}(\kappa) \notin \mathcal{E}_{\mathcal{G}}$, then*

$$G^2_{\sigma,\kappa}(\mathfrak{L}_{\nu,r}) = \left(I^{-\omega}_{-;1/a_2^*,a_2^*(\beta+\sigma-1)}\mathbb{L}_{a_2^*,1-\beta-\sigma}\right)\left(\mathfrak{L}_{1-\nu+\mathrm{Re}(\kappa+\sigma),r}\right). \tag{6.7.12}$$

When $\nu - \mathrm{Re}(\kappa) \in \mathcal{E}_{\mathcal{G}}$, $G^2_{\sigma,\kappa}(\mathfrak{L}_{\nu,r})$ is a subset of the right-hand side of (6.7.12).

Theorem 6.70. *Let $a^* > 0, a_1^* > 0, a_2^* < 0, \alpha < 1 - \nu + \mathrm{Re}(\kappa) < \beta$ and let $1 < r < \infty$.*
(a) *If $\nu - \mathrm{Re}(\kappa) \notin \mathcal{E}_{\mathcal{G}}$, or if $1 < r \leqq 2$, then the transform $\boldsymbol{G}^2_{\sigma,\kappa}$ is one-to-one on $\mathfrak{L}_{\nu,r}$.*
(b) *Let $\omega, \eta, \zeta \in \mathbb{C}$ be chosen as*

$$\omega = a^*\eta - \mu - \frac{1}{2}; \tag{6.7.13}$$

$$a^*\mathrm{Re}(\eta) \geqq \gamma(r) + 2a_2^*[\nu - \mathrm{Re}(\kappa) - 1] + \mathrm{Re}(\mu); \tag{6.7.14}$$

$$\mathrm{Re}(\eta) > \nu - \mathrm{Re}(\kappa) - 1; \tag{6.7.15}$$

$$\mathrm{Re}(\zeta) < 1 - \nu + \mathrm{Re}(\kappa). \tag{6.7.16}$$

If $\nu - \mathrm{Re}(\kappa) \notin \mathcal{E}_{\mathcal{G}}$, then

$$\boldsymbol{G}^2_{\sigma,\kappa}(\mathfrak{L}_{\nu,r}) = \left(M_{\sigma-1/2-\omega/(2a_2^*)}\mathbb{H}_{2a_2^*,2a_2^*\zeta+\omega-1}\mathbb{L}_{a^*,1/2-\eta+\omega/(2a_2^*)}\right)$$
$$\left(\mathfrak{L}_{\nu-1/2-\mathrm{Re}(\omega)/(2a_2^*)-\mathrm{Re}(\kappa),r}\right). \tag{6.7.17}$$

When $\nu - \mathrm{Re}(\kappa) \in \mathcal{E}_{\mathcal{G}}$, then $\boldsymbol{G}^2_{\sigma,\kappa}(\mathfrak{L}_{\nu,r})$ is a subset of the right-hand side of (6.7.17).

Theorem 6.71. *Let $a^* > 0, a_1^* < 0, a_2^* > 0, \alpha < 1 - \nu + \mathrm{Re}(\kappa) < \beta$ and let $1 < r < \infty$.*
(a) *If $\nu - \mathrm{Re}(\kappa) \notin \mathcal{E}_{\mathcal{G}}$, or if $1 < r \leqq 2$, then the transform $\boldsymbol{G}^2_{\sigma,\kappa}$ is one-to-one on $\mathfrak{L}_{\nu,r}$.*
(b) *Let $\omega, \eta, \zeta \in \mathbb{C}$ be chosen as*

$$\omega = a^*\eta - \Delta - \mu - \frac{1}{2}; \tag{6.7.18}$$

$$a^*\mathrm{Re}(\eta) \geqq \gamma(r) - 2a_1^*[\nu - \mathrm{Re}(\kappa)] + \Delta + \mathrm{Re}(\mu); \tag{6.7.19}$$

$$\mathrm{Re}(\eta) > -\nu + \mathrm{Re}(\kappa); \tag{6.7.20}$$

$$\mathrm{Re}(\zeta) < \nu - \mathrm{Re}(\kappa). \tag{6.7.21}$$

If $\nu - \mathrm{Re}(\kappa) \notin \mathcal{E}_{\mathcal{G}}$, then

$$\boldsymbol{G}^2_{\sigma,\kappa}(\mathfrak{L}_{\nu,r}) = \left(M_{\sigma+1/2+\omega/(2a_1^*)}\mathbb{H}_{-2a_1^*,2a_1^*\zeta+\omega-1}\mathbb{L}_{-a^*,1/2+\eta-\omega/(2a_1^*)}\right)$$
$$\left(\mathfrak{L}_{\nu+1/2+\mathrm{Re}(\omega)/(2a_1^*)-\mathrm{Re}(\kappa),r}\right). \tag{6.7.22}$$

When $\nu - \mathrm{Re}(\kappa) \in \mathcal{E}_{\mathcal{G}}$, $\boldsymbol{G}^2_{\sigma,\kappa}(\mathfrak{L}_{\nu,r})$ is a subset of the right-hand side of (6.7.22).

Now we state the inversion formulas for the transform $\boldsymbol{G}^2_{\sigma,\kappa}$ in (6.2.5) on $\mathfrak{L}_{\nu,r}$ when $a^* = 0$. From (5.6.23) and (5.6.24), we find

$$f(x) = x^{-\kappa-\lambda}\frac{d}{dx}x^{\lambda+1}$$
$$\cdot \int_0^\infty G^{q-m,p-n+1}_{p+1,q+1}\left[\frac{x}{t}\,\middle|\,\begin{matrix}-\lambda, -a_{n+1}, \cdots, -a_p, -a_1, \cdots, -a_n \\ -b_{m+1}, \cdots, -b_q, -b_1, \cdots, -b_m, -\lambda - 1\end{matrix}\right]$$
$$\cdot\, t^{-\sigma-1}\left(\boldsymbol{G}^2_{\sigma,\kappa}f\right)(t)dt \tag{6.7.23}$$

or

$$f(x) = -x^{-\kappa-\lambda}\frac{d}{dx}x^{\lambda+1}$$

$$\cdot \int_0^\infty G^{q-m+1,p-n}_{p+1,q+1}\left[\frac{x}{t}\,\middle|\,\begin{matrix}-a_{n+1},\cdots,-a_p,-a_1,\cdots,-a_n,-\lambda\\ -\lambda-1,-b_{m+1},\cdots,-b_q,-b_1,\cdots,-b_m\end{matrix}\right]$$

$$\cdot t^{-\sigma-1}\left(\boldsymbol{G}^2_{\sigma,\kappa}f\right)(t)dt. \tag{6.7.24}$$

The conditions for the validity of (6.7.23) and (6.7.24) follow from Theorems 5.59 and 5.60, where $a^*, \Delta, \alpha, \beta, \mu, \alpha_0$ and β_0 are taken as in (6.1.5), (6.1.6), (6.1.9)–(6.1.11), (6.1.37) and (6.1.38).

Theorem 6.72. *Let $a^* = 0, \alpha < 1 - \nu + \mathrm{Re}(\kappa) < \beta$ and $\alpha_0 < \nu - \mathrm{Re}(\kappa) < \beta_0$, and let $\lambda \in \mathbb{C}$.*

(a) *If $\Delta[1 - \nu + \mathrm{Re}(\kappa)] + \mathrm{Re}(\mu) = 0$ and $f \in \mathfrak{L}_{\nu,2}$, then the inversion formula (6.7.23) holds for $\mathrm{Re}(\lambda) > \nu - \mathrm{Re}(\kappa) - 1$ and (6.7.24) for $\mathrm{Re}(\lambda) < \nu - \mathrm{Re}(\kappa) - 1$.*

(b) *If $\Delta = \mathrm{Re}(\mu) = 0$ and $f \in \mathfrak{L}_{\nu,r}$ $(1 < r < \infty)$, then the inversion formula (6.7.23) holds for $\mathrm{Re}(\lambda) > \nu - \mathrm{Re}(\kappa) - 1$ and (6.7.24) for $\mathrm{Re}(\lambda) < \nu - \mathrm{Re}(\kappa) - 1$.*

Theorem 6.73. *Let $a^* = 0, 1 < r < \infty$ and $\Delta[1 - \nu + \mathrm{Re}(\kappa)] + \mathrm{Re}(\mu) \leqq 1/2 - \gamma(r)$, and let $\lambda \in \mathbb{C}$.*

(a) *If $\Delta > 0, m > 0, \alpha < 1-\nu+\mathrm{Re}(\kappa) < \beta, \alpha_0 < \nu-\mathrm{Re}(\kappa) < \min[\beta_0, \{\mathrm{Re}(\mu+1/2)/\Delta\}+1]$ and if $f \in \mathfrak{L}_{\nu,r}$, then the inversion formulas (6.7.23) and (6.7.24) hold for $\mathrm{Re}(\lambda) > \nu-\mathrm{Re}(\kappa)-1$ and for $\mathrm{Re}(\lambda) < \nu - \mathrm{Re}(\kappa) - 1$, respectively.*

(b) *If $\Delta < 0, n > 0, \alpha < 1-\nu+\mathrm{Re}(\kappa) < \beta, \max[\alpha_0, \{\mathrm{Re}(\mu+1/2)/\Delta\}+1] < \nu-\mathrm{Re}(\kappa) < \beta_0$ and if $f \in \mathfrak{L}_{\nu,r}$, then the inversion formulas (6.7.23) and (6.7.24) hold for $\mathrm{Re}(\lambda) > \nu-\mathrm{Re}(\kappa)-1$ and for $\mathrm{Re}(\lambda) < \nu - \mathrm{Re}(\kappa) - 1$, respectively.*

Remark 6.3. The results in this section generalize those in Section 5.3. Namely, Theorems 6.25–6.36 follow from Theorems 6.61–6.72 when $\sigma = \kappa = 0$.

6.8. Bibliographical Remarks and Additional Information on Chapter 6

For Section 6.1. The integral transforms with the general Meijer G-function (2.9.1) as kernel were first considered by Kesarwani [8]–[10] while investigating the G-function as an unsymmetrical Fourier kernel, though earlier he treated in [2] the particular case

$$\left(\boldsymbol{G}f\right)(x) = \int_0^\infty G^{4,0}_{2,4}\left[xt\,\middle|\,\begin{matrix}m-k+\frac{1}{2},\eta-\lambda\\ 2m,0,\eta+\mu-\frac{1}{2},\eta-\mu-\frac{1}{2}\end{matrix}\right]f(t)dt. \tag{6.8.1}$$

Kesarwani [8] showed that the functions

$$K^{m,p}_{p+q,m+n}(x) = 2\gamma x^{\gamma-1/2}G^{m,p}_{p+q,m+n}\left[x^{2\gamma}\,\middle|\,\begin{matrix}a_1,\cdots,a_p,b_1\cdots,b_q\\ c_1,\cdots,c_m,d_1,\cdots,d_n\end{matrix}\right] \tag{6.8.2}$$

and

$$H^{n,q}_{q+p,n+m}(x) = 2\gamma x^{\gamma-1/2} G^{n,q}_{q+p,n+m}\left[x^{2\gamma} \middle| \begin{matrix} -b_1,\cdots,-b_q,-a_1\cdots,-a_p \\ -d_1,\cdots,-d_n,-c_1,\cdots,-c_m \end{matrix}\right] \tag{6.8.3}$$

are a pair of unsymmetrical kernels, since their Mellin transforms $(\mathfrak{M}G_1)(s)$ and $(\mathfrak{M}G_2)(s)$ satisfy the relation $(\mathfrak{M}G_1)(s)(\mathfrak{M}G_2)(1-s)=1$, where

$$G_1(x) = \int_0^x K^{m,p}_{p+q,m+n}(t)dt, \qquad G_2(x) = \int_0^x H^{n,q}_{q+p,m+n}(t)dt. \tag{6.8.4}$$

In [9] Kesarwani considered the transform $\boldsymbol{G}$

$$\left(\boldsymbol{G}f\right)(x) = 2\gamma \int_0^\infty (xt)^{\gamma-1/2} G^{m,p}_{p+q,m+n}\left[(xt)^{2\gamma} \middle| \begin{matrix} a_1,\cdots,a_p,b_1\cdots,b_q \\ c_1,\cdots,c_m,d_1,\cdots,d_n \end{matrix}\right] f(t)dt \tag{6.8.5}$$

and proved its inversion formula

$$\frac{f(x+0)+f(x-0)}{2}$$
$$= 2\gamma \int_0^\infty (xt)^{\gamma-1/2} G^{n,q}_{q+p,n+m}\left[x^{2\gamma} \middle| \begin{matrix} -b_1,\cdots,-b_q,-a_1\cdots,-a_p \\ -d_1,\cdots,-d_n,-c_1,\cdots,-c_m \end{matrix}\right] (\boldsymbol{G}f)(t)dt, \tag{6.8.6}$$

provided that

$$\begin{gathered} \gamma>0, \quad n-p=m-q=\frac{\Delta}{2}>0, \quad \sum_{i=1}^p a_i - \sum_{i=1}^q b_i = \sum_{j=1}^m c_j - \sum_{j=1}^n d_j, \\ \mathrm{Re}(a_i) < \frac{\Delta-1}{2\Delta} \quad (1 \leqq i \leqq p), \quad \mathrm{Re}(b_i) > \frac{1-\Delta}{2\Delta} \quad (1 \leqq i \leqq q), \\ \mathrm{Re}(c_j) > \frac{1-\Delta}{2\Delta} \quad (1 \leqq j \leqq m), \quad \mathrm{Re}(d_j) < \frac{\Delta-1}{2\Delta} \quad (1 \leqq j \leqq n), \end{gathered} \tag{6.8.7}$$

$f(t)t^{\Delta/\gamma-1/2} \in L_1(\mathbb{R}_+)$ and $f(t)$ is of bounded variation near $t=x$ $(x>0)$.

In [10] Kesarwani established that if the conditions in (6.8.7) are satisfied, and further, $\mathrm{Re}(a_i)<1/2$ $(1 \leqq i \leqq p)$, $\mathrm{Re}(b_i) > -1/2$ $(1 \leqq i \leqq q)$, $\mathrm{Re}(c_j) > -1/2$ $(1 \leqq j \leqq m)$, $\mathrm{Re}(d_j) < 1/2$ $(1 \leqq j \leqq n)$ and $f(x) \in L_2(\mathbb{R}_+)$ are assumed, then the formula

$$g(x) = \frac{d}{dx}\int_0^\infty G_1(xt)f(t)\frac{dt}{t} \tag{6.8.8}$$

is defined as a function in $L_2(\mathbb{R}_+)$, and the reciprocal formula

$$f(x) = \frac{d}{dx}\int_0^\infty G_2(xt)g(t)\frac{dt}{t} \tag{6.8.9}$$

holds almost everywhere, where $G_1(x), G_2(x)$ are given in (6.8.4).

The special case of the above results by Kesarwani [9], [10] in which $q=p$, $n=m$ and $b_i=-a_i$, $d_j=-c_j$ for each $1 \leqq i \leqq p$, $1 \leqq j \leqq m$ was earlier studied by Fox [2, Theorems 1 and 2]. Fox [2] also indicated that the function

$$G(x) = \lambda G^{q,p}_{2p,2q}\left[x^{1/\lambda} \middle| \begin{matrix} a_1,\cdots,a_p,1-\lambda-a_1,\cdots,1-\lambda-a_p \\ b_1,\cdots,b_q,1-\lambda-b_1,\cdots,1-\lambda-b_q \end{matrix}\right] \tag{6.8.10}$$

with $\lambda>0$ is a symmetric Fourier kernel. Such a result was also given by Masood and Kapoor [1] in a different way.

Kesarwani [12] and Wong and Kesarwani [1] derived necessary and sufficient conditions for the pair of functions $f(x) \in L_2(\mathbb{R}_+)$ and $g(x) \in L_2(\mathbb{R}_+)$ to be the $\boldsymbol{G}$-transform of each other:

$$f(x) = \int_0^\infty G(xt)g(t)dt, \quad g(x) = \int_0^\infty G(xt)f(t)dt, \tag{6.8.11}$$

where $G(x)$ is given by

$$G(x) = \gamma\mu^{\gamma/2}x^{(\gamma-1)/2}G_{2p,2q}^{q,p}\left[(\mu x)^\gamma \,\middle|\, \begin{matrix} a_1,\cdots,a_p,-a_1\cdots,-a_p \\ b_1,\cdots,b_q,-b_1,\cdots,-b_q \end{matrix}\right] \tag{6.8.12}$$

with positive $\gamma > 0$ and $\mu > 0$.

Bhise [5] established theorems on functions connected by the $\boldsymbol{G}$-transform of the form

$$\left(\boldsymbol{G}f\right)(x) = \int_0^\infty G_{2,4}^{2,1}\left[xt \,\middle|\, \begin{matrix} k-m-\frac{\eta}{2}-\frac{1}{2}, -k+m+\frac{\eta}{2}+\frac{1}{2} \\ \frac{\eta}{2}, \frac{\eta}{2}+2m, -\frac{\eta}{2}, -2m-\frac{\eta}{2} \end{matrix}\right] f(t)dt. \tag{6.8.13}$$

Raj.K. Saxena [1] expressed via this transform the transform containing the Whittaker function $W_{\rho,\gamma}(x)$ (see (7.2.2)) in the kernel. R.U. Verma [2] proved Parseval's theorem and some properties for a more general integral transform than (6.8.13) with the $G_{2,4}^{2,1}$-function as kernel. Using the technique of the Laplace transform $\mathbb{L}$ and its inverse $\mathbb{L}^{-1}$ developed by Fox in [4] and [5], R.U. Verma in [6], [7] and [8] proved the inversion formulas for the $\boldsymbol{G}$-transforms (6.1.1) with the functions $G_{2p,q}^{0,p}(x)$, $G_{2p+n,2q+m}^{q+m,p}(x)$ and $G_{p,q}^{q,0}(x)$ as the kernels, respectively.

Using the method developed by Zemanian [6], Misra [3] proved that the real inversion formula for the general $\boldsymbol{G}$-transform (6.1.1) can be extended to certain spaces of generalized functions. In [4] Misra obtained such a result for the particular case of the $\boldsymbol{G}$-transform with $m = q = 2$, $n = 1$, $p = 1$

$$\left(\boldsymbol{G}f\right)(x) = \int_0^\infty G_{1,2}^{2,0}\left[xt \,\middle|\, \begin{matrix} a \\ b_1, b_2 \end{matrix}\right] f(t)dt, \tag{6.8.14}$$

which is reduced to the Laplace transform (2.5.2) when $a = b_1$ and $b_2 = 0$. Pathak and J.N. Pandey [4] extended the $\boldsymbol{G}$-transform (6.1.1) to another class of generalized functions. In particular, they established that the inversion formula (6.8.6) due to Kesarwani [9] can be extended to distributions in the sense of weak convergence. The above results were presented in Brychkov and Prudnikov [1, Section 8.1].

O.P. Sharma [2] used the transform (6.8.5) to obtain the formula for the Meijer transform (8.9.1) of the product of two functions $f(x)$ and $g(x)$. Pathak [9] gave two Abelian theorems for the transform (6.8.4) and showed that Abelian theorems for integral transforms such as the Hankel transform $\mathbb{H}_\eta$ in (8.1.1), the $\mathbb{Y}_\eta$-transform in (8.7.1), the Struve transform $\mathcal{H}_\eta$ in (8.8.1), the Meijer transform $\mathbf{K}_\eta$ in (8.9.1) and the Hardy transform $\mathbf{J}_{\eta,\sigma}$ in (8.12.6) follow by suitably specializing the parameters in the kernel $K_{p+q,m+n}^{m,p}(x)$ given in (6.8.2) (see also Section 8.14 in this connection).

The technique of factorization of the $\boldsymbol{G}$-transform (6.1.1) to representations via simpler $\boldsymbol{G}$-transforms, together with special tables and notation, were first developed by Brychkov, Glaeske and Marichev [1], without characterization of conditions and spaces of functions (see also their [2]). The factorization of the $\boldsymbol{G}$-transform as well as its mapping properties and inversion formulas in the special spaces of functions $\mathfrak{M}_{c,\gamma}^{-1}$ and $\mathfrak{L}_2^{(c,\gamma)}$ were proved by Marichev and Vu Kim Tuan [1], [2], Vu Kim Tuan, Marichev and Yakubovich [1] and Vu Kim Tuan [1] (see the results in the bibliography in Sections 36 and 39.1 of the book by Samko, Kilbas and Marichev [1], and in the book by Yakubovich and Luchko [1].).

The results of these sections concerning the boundedness, representation, range and inversion of the $\boldsymbol{G}$-transforms (6.1.1) in $\mathfrak{L}_{\nu,r}$-spaces (3.1.3) were proved by Rooney [6]. We obtained these results as the particular cases of the corresponding statements for the H-transforms (3.1.1) presented in Chapters 3 and 4. They basically coincide with those in Rooney [6] if we replace μ by ν and take into account the relations

$$\delta = \frac{a^*}{2}, \quad k = a_1^*, \quad l = a_2^*, \quad \nu = \mu + \frac{\Delta}{2}, \tag{6.8.15}$$

comparing Rooney's notation with ours.

We only note that in Theorem 6.4 the representations (6.1.16) and (6.1.17) for any $k>0$ generalize those given by Rooney [6, Theorem 6.2] for $k=1$, and in Theorems 6.9 and 6.10 the conditions (6.1.26)–(6.1.28) and (6.1.31)–(6.1.33) for the validity of the relations (6.1.29) and (6.1.34), respectively are more general and simpler than those proved by Rooney [6, Theorems 7.5 and 7.6]. We also indicate that Corollaries 6.10.1, 6.10.2 and 6.11.1, 6.11.2 present the sufficient conditions for (6.1.26)–(6.1.28) and (6.1.31)–(6.1.33) to hold.

For Section 6.2. A series of papers was devoted to investigating the generalizations and the modifications of the G-transform (6.1.1) which are different from those in (6.2.1)–(6.2.5). Kesarwani [1] first considered the modified Meijer G-transform of the form

$$\left(\mathbb{G}f\right)(x)=2^{-\eta}\int_0^\infty (xt)^{\eta+1/2}\chi_{\eta,k,m}\left(\frac{x^2t^2}{4}\right)f(t)dt,$$

$$\chi_{\eta,k,m}(x)=x^{-\eta}G^{2,1}_{2,4}\left[x\,\middle|\,\begin{matrix}k-m-\frac{1}{2},\eta-k+m+\frac{1}{2}\\ \eta,\eta+2m,-2m,0\end{matrix}\right],\tag{6.8.16}$$

which reduces to the Hankel transform $\mathbb{H}_\eta f$ in (8.1.1) when $k+m=1/2$. Kesarwani [1] proved the inversion formula for such a transform and discussed self-reciprocal functions connected with the kernel of this transform.

K.C. Sharma [3] defined the integral transform

$$\left(\mathbb{G}f\right)(x)=\int_0^\infty e^{-nxt/4}G^{4,n}_{m+n,m+n+2}\left[axt\,\middle|\,\begin{matrix}a_1,\cdots,a_n,c_1,\cdots,c_m\\ b_1,\cdots,b_4,d_1,\cdots,d_{m+n-2}\end{matrix}\right]f(t)dt\tag{6.8.17}$$

with $0\leqq m\leqq 3$, $n\geqq 0$, $m+n\geqq 2$, $n\in\mathbb{N}_0$ and proved the inversion formula for such a transform by using the Mellin transform $\mathfrak{M}$ and its inverse $\mathfrak{M}^{-1}$. For such a transform with $m=2$ and $n=0$ S.P. Goyal [2] obtained three chains connecting the originals and their images.

R.P. Goyal [2] and Golas [1] gave the conditions of convergence and uniform convergence for the modified G-transform

$$\left(\mathbb{G}f\right)(x)=\int_0^\infty e^{-bxt}G^{m,n}_{p+1,q}\left[axt\,\middle|\,\begin{matrix}a_1,\cdots,a_p,\sigma\\ b_1,\cdots,b_q\end{matrix}\right]d\alpha(t).\tag{6.8.18}$$

Misra [2] generalized part of the Abelian theorems for the Laplace transform given in Widder [1, p.181] to the G-transform

$$\left(\mathbb{G}f\right)(x)=\int_0^\infty e^{-bxt}G^{m,n+1}_{p+1,q}\left[\frac{2}{xt}\,\middle|\,\begin{matrix}-\gamma,a_1,\cdots,a_p\\ b_1,\cdots,b_q\end{matrix}\right]f(t)dt\tag{6.8.19}$$

for functions and for generalized functions that are equivalent to Lebesgue integrable functions in the neighborhood of the origin.

R.U. Verma [1] gave an inversion formula for the transform

$$\left(\mathbb{G}f\right)(x)=\int_0^\infty e^{-xt/2}G^{m,n+1}_{p+1,q}\left[(xt)^2\,\middle|\,\begin{matrix}(a_i)_{1,p}\\ (b_j)_{1,q}\end{matrix}\right]f(t)dt,\tag{6.8.20}$$

provided that $x^cf(x)\in L(\mathbb{R}_+)$.

Kapoor and Masood [1] obtained the inversion formula for the generalized G-transform (6.1.1) of the form

$$\left(\widetilde{G}f\right)(x)=\int_0^\infty G^{m,n}_{p,q}\left[a(xt)^\lambda\,\middle|\,\begin{matrix}a_1,\cdots,a_p,\sigma\\ b_1,\cdots,b_q\end{matrix}\right]f(t)dt\tag{6.8.21}$$

in terms of the Mellin transform and of another G-function.

For Sections 6.3 and 6.4. The results presented in these sections were proved by the authors in Saigo and Kilbas [5].

Kapoor [1] introduced the $\boldsymbol{G}^1$-transform (6.2.1) in the form

$$\left(\boldsymbol{G}^1 f\right)(x) = \int_0^\infty G_{p+1,q}^{m,n+1}\left[\frac{x}{t}\,\middle|\,\begin{matrix}0, a_1, \cdots, a_p\\ b_1, \cdots, b_q\end{matrix}\right]\frac{f(t)}{t}dt \tag{6.8.22}$$

as a generalization of the Stieltjes transform, and proved its inversion formula in terms of the Mellin transform and of another G-function. In [3] Kapoor iterated this transform with the Laplace transform $\mathbb{L}$ in (2.5.2) to obtain generalizations of the third and fourth iterates of Laplace transforms and to establish some operational results.

R.K. Saxena and N. Gupta [1] studied the asymptotic expansion of the transform $\boldsymbol{G}^1$

$$\left(\boldsymbol{G}^1 f\right)(x) = \int_0^\infty G_{3,3}^{3,1}\left[\frac{x}{t}\,\middle|\,\begin{matrix}-\mu, \lambda+\nu, \gamma+\delta\\ 0, \gamma, \delta\end{matrix}\right]\frac{f(t)}{t}dt. \tag{6.8.23}$$

For Sections 6.5–6.7. Bhise [1] first considered the modified Meijer $\boldsymbol{G}$-transform of the form (6.2.3) with $\sigma = 1$ and $\kappa = 0$:

$$\left(\boldsymbol{G}_{1,0} f\right)(x) = x\int_0^\infty G_{m,m+1}^{m+1,0}\left[xt\,\middle|\,\begin{matrix}\eta_1+\alpha_1, \cdots, \eta_m+\alpha_m\\ \eta_1, \cdots, \eta_m, \sigma\end{matrix}\right]f(t)dt \tag{6.8.24}$$

and proved its inversion formula. Further, Bhise studied properties of such a transform, namely, certain rules and recurrence relations [3], certain finite and infinite series [4] and composition relations [6]. Mittal [2] established two theorems involving H-functions for the G-transform which generalize the result given by Bhise [1], while K.C. Gupta and P.K. Mittal [3] developed a chain rule for the transform (6.8.24).

Kapoor [2] obtained the relations of a more general transform

$$\left(\boldsymbol{G}_{0,0} f\right)(x) = x\int_0^\infty G_{p,q}^{m,n}\left[xt\,\middle|\,\begin{matrix}(a_i)_{1,p}\\ (b_j)_{1,q}\end{matrix}\right]f(t)dt \tag{6.8.25}$$

than (6.8.24) with the Hankel transform $\mathbb{H}_\eta$ and the $\mathbb{Y}_\eta$-transform given in (8.1.1) and (8.7.1), respectively.

Using the Laplace transform $\mathbb{L}$ and its inverse $\mathbb{L}^{-1}$, R.U. Verma [10] proved the inversion formula for the modified transform in (6.2.3)

$$\left(\boldsymbol{G}_{\mu,-\mu} f\right)(x) = \int_0^\infty \left(\frac{x}{t}\right)^\mu G_{1,2}^{2,0}\left[xt\,\middle|\,\begin{matrix}a\\ b, c\end{matrix}\right]f(t)dt \tag{6.8.26}$$

in the form

$$f = \boldsymbol{M}_{\mu-b}\mathbb{L}^{-1}\boldsymbol{M}_{a-b}\mathbb{L}\boldsymbol{M}_{a-c}\mathbb{L}^{-1}\boldsymbol{M}_{-c-\mu}\boldsymbol{G}f, \tag{6.8.27}$$

where the operator $\boldsymbol{M}_\zeta$ is given by (3.3.11).

A series of papers was devoted to investigating the integral transforms with the Meijer G-functions as kernels of the form (5.7.1) and (5.7.2) which generalize the Riemann–Liouville and Erdélyi–Kober type fractional integrals (2.7.1), (3.3.1) and (2.7.2), (3.3.2). Parashar [1] first defined such generalizations in the form

$$\left(\boldsymbol{G}_{0+} f\right)(x) = x^{-\eta-1}\int_0^x G_{p+2,q+2}^{m,n}\left[\frac{at}{x}\,\middle|\,\begin{matrix}\alpha, 1-\beta+l, a_1, \cdots, a_p\\ \nu, 1-\beta, b_1, \cdots, b_q\end{matrix}\right]t^\eta f(t)dt \tag{6.8.28}$$

and

$$\left(\boldsymbol{G}_{-}f\right)(x)=x^{\eta}\int_x^{\infty}G_{p+2,q+2}^{m,n}\left[\frac{ax}{t}\,\middle|\,\begin{matrix}\alpha,1-\beta+l,a_1,\cdots,a_p\\ \nu,1-\beta,b_1,\cdots,b_q\end{matrix}\right]t^{-\eta-1}f(t)dt \tag{6.8.29}$$

with $a>0$ and investigated some properties of these transforms including their domains and ranges.

Kalla [6] studied the operators

$$\left(\boldsymbol{G}_{\eta;0+}f\right)(x)=x^{-\eta-1}\int_0^{x}G_{p,q}^{m,n}\left[a\left(\frac{t}{x}\right)^{\sigma}\,\middle|\,\begin{matrix}a_1,\cdots,a_p\\ \nu,1-\beta,b_1,\cdots,b_q\end{matrix}\right]t^{\eta}f(t)dt \tag{6.8.30}$$

and

$$\left(\boldsymbol{G}_{\delta;-}f\right)(x)=x^{\delta}\int_x^{\infty}G_{p,q}^{m,n}\left[a\left(\frac{t}{x}\right)^{\sigma}\,\middle|\,\begin{matrix}a_1,\cdots,a_p\\ b_1,\cdots,b_q\end{matrix}\right]t^{-\delta-1}f(t)dt \tag{6.8.31}$$

with $\sigma>0$ and complex η and δ in the space $L_r(\mathbb{R}_+)$ $(r\geqq 1)$ and proved their existence for $\mathrm{Re}(\eta)>1/r-1$ and $\mathrm{Re}(\eta)>-1/r$, respectively. Kalla [6] also proved the formulas for the Mellin transforms of these integrals for $f(x)\in L_r(\mathbb{R}_+)$ $(1\leqq r\leqq 2)$ and the relation of integration by parts

$$\int_0^{\infty}\left(\boldsymbol{G}_{\delta;0+}f\right)(x)g(x)dx=\int_0^{\infty}f(x)\left(\boldsymbol{G}_{\delta;-}g\right)(x)dx \tag{6.8.32}$$

$$\left(f\in L_r(\mathbb{R}_+),\ g\in L_{r'}(\mathbb{R}_+),\ \frac{1}{r}+\frac{1}{r'}=1\right),$$

provided that $2(m+n)>p+q$ and $\mathrm{Re}(\delta)>\gamma(r)$, where $\gamma(r)$ is given by (3.3.9). Further, in [7] Kalla obtained inversion formulas for the integral transforms (6.8.30) and (6.8.31).

Kiryakova [1], [2] introduced the modified G-transform of the form (5.7.12) and (5.7.13) in the form

$$\left(\mathbf{I}_{p,q}^{m,n}f\right)(x)\equiv\left(\mathbf{K}_{p,q}^{m,n}f\right)(x)=\int_0^{1}G_{m,m}^{m,0}\left[t\,\middle|\,\begin{matrix}(\gamma_i+\delta_i)_{1,m}\\ (\gamma_j)_{1,m}\end{matrix}\right]f(xt^{1/\beta})dt, \tag{6.8.33}$$

and studied some of its properties in inversion formulas provided that $f(x)$ is continuous on $\mathbb{R}_+$ or analytic in a starlike complex domain, considering an associated power weight (see also Kalla and Kiryakova [1] and Kiryakova [5] in this connection). Kiryakova, Raina and Saigo [1] expressed the operator $\mathbf{I}_{p,q}^{m,n}$ in terms of the Laplace transform $\mathbb{L}$ and its inverse $\mathbb{L}^{-1}$ in the space $L_r(\mathbb{R}_+)$ $(r\geqq 1)$.

Luchko and Kiryakova [1] considered the modified G-transforms

$$\left(\mathbf{G}_{\beta}^{1}f\right)=\frac{\beta}{\sqrt{\eta}}\int_0^{\infty}G_{0,2m}^{m,0}\left[\left(\frac{x}{\eta}\right)^{\beta}\,\middle|\,\begin{matrix}\overline{}\\ \left(\gamma_j+1-\frac{1}{\beta}\right)_{1,2m}\end{matrix}\right]f(t)dt\quad(x>0) \tag{6.8.34}$$

and

$$\left(\mathbf{G}_{\beta}^{2}f\right)=\frac{\beta}{\sqrt{\eta}}\int_0^{\infty}G_{0,2m}^{m,0}\left[\left(\frac{x}{\eta}\right)^{\beta}\,\middle|\,\begin{matrix}\overline{}\\ \left(\gamma_j+1-\frac{1}{\beta}\right)_{1,m},(-\gamma_j)_{1,m}\end{matrix}\right]f(t)dt$$

$$(x>0) \tag{6.8.35}$$

with $\beta>0,\eta>0$ and $\gamma_j\in\mathbb{R}$ $(j=1,\cdots,2m)$, proved the isomorphism of $\mathbf{G}_{\beta}^{1}$ and $\mathbf{G}_{\beta}^{2}$ in a special space, found the inversion relations, and established compositions of (6.8.34) and (6.8.35) with hyper-Bessel differential operators.

The results in Sections 6.5–6.7 have not been published before.

Chapter 7

HYPERGEOMETRIC TYPE INTEGRAL TRANSFORMS ON THE SPACE $\mathfrak{L}_{\nu,r}$

7.1. Laplace Type Transforms

We consider the integral transform

$$\left(\mathbb{L}^*_{\gamma,k}f\right)(x) = \int_0^\infty (xt)^\gamma e^{-(xt)^k} f(t)dt \tag{7.1.1}$$

with $\gamma \in \mathbb{C}$ and $k > 0$. This is a modification of the generalized Laplace transform $\mathbb{L}_{k,\alpha}$ given in (3.3.3):

$$\left(\mathbb{L}^*_{\gamma,k}f\right)(x) = k^{-\gamma/k}\left(W_{k^{-1/k}}\mathbb{L}_{1/k,-\gamma}f\right)(x) = k^{-\gamma/k}\left(\mathbb{L}_{1/k,-\gamma}f\right)\left(k^{1/k}x\right), \tag{7.1.2}$$

where the operator W_δ is defined by (3.3.12).

According to (7.1.2), (3.3.15) and (3.3.8) the Mellin transform (2.5.1) of (7.1.1) for a "sufficiently good" function f is given by the relation

$$\left(\mathfrak{M}\mathbb{L}^*_{\gamma,k}f\right)(s) = \frac{1}{k}\mathcal{H}^{1,0}_{0,1}(s)\left(\mathfrak{M}f\right)(1-s), \quad \mathcal{H}^{1,0}_{0,1}(s) = \Gamma\left(\frac{\gamma+s}{k}\right) \tag{7.1.3}$$

for $\text{Re}(\gamma + s) > 0$. This relation shows that (7.1.1) is the $\boldsymbol{H}$-transform (3.1.1) of the form

$$\left(\mathbb{L}^*_{\gamma,k}f\right)(x) = \frac{1}{k}\int_0^\infty H^{1,0}_{0,1}\left[xt \,\middle|\, \begin{matrix} \overline{} \\ \left(\frac{\gamma}{k}, \frac{1}{k}\right) \end{matrix}\right] f(t)dt. \tag{7.1.4}$$

The constants $a^*, \Delta, a_1^*, a_2^*, \alpha, \beta$ and μ in (1.1.7), (1.1.8), (1.1.11), (1.1.12), (3.4.1), (3.4.2) and (1.1.10) take the forms

$$a^* = \frac{1}{k}, \ \Delta = \frac{1}{k}, \ a_1^* = \frac{1}{k}, \ a_2^* = 0, \ \alpha = -\text{Re}(\gamma), \ \beta = \infty, \ \mu = \frac{\gamma}{k} - \frac{1}{2}. \tag{7.1.5}$$

Let $\mathcal{E}_1$ be the exceptional set of the function $\mathcal{H}^{1,0}_{0,1}(s)$ in (7.1.3) given in Definition 3.4. Since the gamma function $\Gamma(z)$ does not have zeros, then $\mathcal{E}_1$ is empty.

From Theorems 3.6, 3.7 and 4.5, 4.7 we obtain the $\mathfrak{L}_{\nu,2}$- and $\mathfrak{L}_{\nu,r}$-theory of the transform (7.1.1).

Theorem 7.1. *Let* $1 - \nu > -\text{Re}(\gamma)$.

(i) *There is a one-to-one transform* $\mathbb{L}^*_{\gamma,k} \in [\mathfrak{L}_{\nu,2}, \mathfrak{L}_{1-\nu,2}]$ *such that* (7.1.3) *holds for* $\text{Re}(s) = 1 - \nu$ *and* $f \in \mathfrak{L}_{\nu,2}$.

(ii) *If $f, g \in \mathfrak{L}_{\nu,2}$, then the relation*

$$\int_0^\infty f(x)\left(\mathbb{L}^*_{\gamma,k}g\right)(x)dx = \int_0^\infty \left(\mathbb{L}^*_{\gamma,k}f\right)(x)g(x)dx \tag{7.1.6}$$

holds for $\mathbb{L}^*_{\gamma,k}$.

(iii) *Let $f \in \mathfrak{L}_{\nu,2}$, $\lambda \in \mathbb{C}$ and $h > 0$. When $\mathrm{Re}(\lambda) > (1-\nu)h - 1$, $\mathbb{L}^*_{\gamma,k}f$ is given by*

$$\left(\mathbb{L}^*_{\gamma,k}f\right)(x) = hx^{1-(\lambda+1)/h}\frac{d}{dx}x^{(\lambda+1)/h}$$

$$\cdot \int_0^\infty H^{1,1}_{1,2}\left[xt \,\middle|\, \begin{matrix}(-\lambda, h)\\ \left(\frac{\gamma}{k}, \frac{1}{k}\right), (-\lambda-1, h)\end{matrix}\right] f(t)dt. \tag{7.1.7}$$

When $\mathrm{Re}(\lambda) < (1-\nu)h - 1$,

$$\left(\mathbb{L}^*_{\gamma,k}f\right)(x) = -hx^{1-(\lambda+1)/h}\frac{d}{dx}x^{(\lambda+1)/h}$$

$$\cdot \int_0^\infty H^{2,0}_{1,2}\left[xt \,\middle|\, \begin{matrix}(-\lambda, h)\\ (-\lambda-1, h), \left(\frac{\gamma}{k}, \frac{1}{k}\right)\end{matrix}\right] f(t)dt. \tag{7.1.8}$$

(iv) *The transform $\mathbb{L}^*_{\gamma,k}$ is independent of ν in the sense that if $1-\nu > -\mathrm{Re}(\gamma)$ and $1-\widetilde{\nu} > -\mathrm{Re}(\gamma)$ and if the transforms $\mathbb{L}^*_{\gamma,k}$ and $\widetilde{\mathbb{L}}^*_{\gamma,k}$ on $\mathfrak{L}_{\nu,2}$ and $\mathfrak{L}_{\widetilde{\nu},2}$, respectively, are given in (7.1.3), then $\mathbb{L}^*_{\gamma,k}f = \widetilde{\mathbb{L}}^*_{\gamma,k}f$ for $f \in \mathfrak{L}_{\nu,2} \cap \mathfrak{L}_{\widetilde{\nu},2}$.*

(v) *For $f \in \mathfrak{L}_{\nu,2}$ and $x > 0$, $\left(\mathbb{L}^*_{\gamma,k}f\right)(x)$ is given in (7.1.1) and (7.1.4).*

Theorem 7.2. *Let $1-\nu > -\mathrm{Re}(\gamma)$ and $1 \leqq r \leqq s \leqq \infty$.*

(a) *The transform $\mathbb{L}^*_{\gamma,k}$ defined on $\mathfrak{L}_{\nu,2}$ can be extended to $\mathfrak{L}_{\nu,r}$ as an element of $[\mathfrak{L}_{\nu,r}, \mathfrak{L}_{1-\nu,s}]$. If $1 \leqq r \leqq 2$, then $\mathbb{L}^*_{\gamma,k}$ is a one-to-one transform from $\mathfrak{L}_{\nu,r}$ onto $\mathfrak{L}_{1-\nu,s}$.*

(b) *If $f \in \mathfrak{L}_{\nu,r}$ and $g \in \mathfrak{L}_{\nu,s'}$ with $1/s + 1/s' = 1$, then the relation (7.1.6) holds.*

(c) *If $1 < r < \infty$, then there holds the representation*

$$\mathbb{L}^*_{\gamma,k}(\mathfrak{L}_{\nu,r}) = \mathbb{L}_{1/k,-\gamma}(\mathfrak{L}_{\nu,r}), \tag{7.1.9}$$

where the operator $\mathbb{L}_{-\gamma,1/k}$ is given in (3.3.3).

Next we consider a further modification of the integral transform (7.1.1) in the form (5.1.3):

$$\left(\mathbb{L}^*_{\gamma,k;\sigma,\kappa}f\right)(x) = x^\sigma \int_0^\infty (xt)^\gamma e^{-(xt)^k} t^\kappa f(t)dt \tag{7.1.10}$$

with $\gamma, \sigma, \kappa \in \mathbb{C}$ and $k > 0$. For this transform the relations (7.1.3) and (7.1.6) take the forms (5.1.13) and (5.1.18):

$$\left(\mathfrak{M}\mathbb{L}^*_{\gamma,k;\sigma,\kappa}f\right)(s) = \frac{1}{k}\Gamma\left(\frac{\gamma+s+\sigma}{k}\right)\left(\mathfrak{M}f\right)(1-s-\sigma+\kappa) \tag{7.1.11}$$

and

$$\int_0^\infty f(x)\left(\mathbb{L}^*_{\gamma,k;\sigma,\kappa}g\right)(x)dx = \int_0^\infty \left(\mathbb{L}^*_{\gamma,k;\kappa,\sigma}f\right)(x)g(x)dx, \tag{7.1.12}$$

respectively. It follows from (7.1.11) that (7.1.10) is a kind of $\boldsymbol{H}_{\sigma,\kappa}$-transform of the form

$$\left(\mathbb{L}^*_{\gamma,k;\sigma,\kappa}f\right)(x)=\frac{x^\sigma}{k}\int_0^\infty H^{1,0}_{0,1}\left[xt\left|\begin{matrix}\overline{\qquad}\\ \left(\frac{\gamma+\sigma}{k},\frac{\sigma}{k}\right)\end{matrix}\right.\right]t^\kappa f(t)dt. \tag{7.1.13}$$

Theorems 5.25, 5.29 and 5.31 lead to the $\mathfrak{L}_{\nu,2}$- and $\mathfrak{L}_{\nu,r}$-theory of the transform (7.1.10).

Theorem 7.3. *Let* $1-\nu+\mathrm{Re}(\kappa)>-\mathrm{Re}(\gamma)$.

(i) *There is a one-to-one transform* $\mathbb{L}^*_{\gamma,k;\sigma,\kappa}\in[\mathfrak{L}_{\nu,2},\mathfrak{L}_{1-\nu+\mathrm{Re}(\kappa-\sigma),2}]$ *such that* (7.1.11) *holds for* $\mathrm{Re}(s)=1-\nu+\mathrm{Re}(\kappa-\sigma)$ *and* $f\in\mathfrak{L}_{\nu,2}$.

(ii) *If* $f\in\mathfrak{L}_{\nu,2}$ *and* $g\in\mathfrak{L}_{\nu+\mathrm{Re}(\kappa-\sigma),2}$, *then the relation* (7.1.12) *holds for* $\mathbb{L}^*_{\gamma,k;\sigma,\kappa}$.

(iii) *Let* $\lambda\in\mathbb{C}$, $h>0$ *and* $f\in\mathfrak{L}_{\nu,2}$. *When* $\mathrm{Re}(\lambda)>[1-\nu+\mathrm{Re}(\kappa)]h-1$, $\mathbb{L}^*_{\gamma,k;\sigma,\kappa}f$ *is given by*

$$\left(\mathbb{L}^*_{\gamma,k;\sigma,\kappa}f\right)(x)=hx^{\sigma+1-(\lambda+1)/h}\frac{d}{dx}x^{(\lambda+1)/h}$$

$$\cdot\int_0^\infty H^{1,1}_{1,2}\left[xt\left|\begin{matrix}(-\lambda,h)\\ \left(\frac{\gamma+\sigma}{k},\frac{\sigma}{k}\right),(-\lambda-1,h)\end{matrix}\right.\right]t^\kappa f(t)dt. \tag{7.1.14}$$

When $\mathrm{Re}(\lambda)<[1-\nu+\mathrm{Re}(\kappa)]h-1$,

$$\left(\mathbb{L}^*_{\gamma,k;\sigma,\kappa}f\right)(x)=-hx^{\sigma+1-(\lambda+1)/h}\frac{d}{dx}x^{(\lambda+1)/h}$$

$$\cdot\int_0^\infty H^{2,0}_{1,2}\left[xt\left|\begin{matrix}(-\lambda,h)\\ (-\lambda-1,h),\left(\frac{\gamma+\sigma}{k},\frac{\sigma}{k}\right)\end{matrix}\right.\right]t^\kappa f(t)dt. \tag{7.1.15}$$

(iv) *The transform* $\mathbb{L}^*_{\gamma,k;\sigma,\kappa}$ *is independent of* ν *in the sense that if* $1-\nu+\kappa>-\mathrm{Re}(\gamma)$ *and* $1-\widetilde{\nu}+\kappa>-\mathrm{Re}(\gamma)$ *and if the transforms* $\mathbb{L}^*_{\gamma,k;\sigma,\kappa}$ *and* $\widetilde{\mathbb{L}}^*_{\gamma,k;\sigma,\kappa}$ *on* $\mathfrak{L}_{\nu,2}$ *and* $\mathfrak{L}_{\widetilde{\nu},2}$, *respectively, are given in* (7.1.11), *then* $\mathbb{L}^*_{\gamma,k;\sigma,\kappa}f=\widetilde{\mathbb{L}}^*_{\gamma,k;\sigma,\kappa}f$ *for* $f\in\mathfrak{L}_{\nu,2}\cap\mathfrak{L}_{\widetilde{\nu},2}$.

(v) *For* $f\in\mathfrak{L}_{\nu,2}$ *and* $x>0$, $(\mathbb{L}^*_{\gamma,k;\sigma,\kappa}f)(x)$ *is given in* (7.1.10) *and* (7.1.13).

Theorem 7.4. *Let* $1-\nu+\mathrm{Re}(\kappa)>-\mathrm{Re}(\gamma)$ *and* $1\leqq r\leqq s\leqq\infty$.

(a) *The transform* $\mathbb{L}^*_{\gamma,k;\sigma,\kappa}$ *defined on* $\mathfrak{L}_{\nu,2}$ *can be extended to* $\mathfrak{L}_{\nu,r}$ *as an element of* $[\mathfrak{L}_{\nu,r},\mathfrak{L}_{1-\nu+\mathrm{Re}(\kappa-\sigma),s}]$. *If* $1\leqq r\leqq 2$, *then* $\mathbb{L}^*_{\gamma,k}$ *is a one-to-one transform from* $\mathfrak{L}_{\nu,r}$ *onto* $\mathfrak{L}_{1-\nu+\mathrm{Re}(\kappa-\sigma),s}$.

(b) *If* $f\in\mathfrak{L}_{\nu,r}$ *and* $g\in\mathfrak{L}_{\nu+\mathrm{Re}(\kappa-\sigma),s'}$ *with* $1/s+1/s'=1$, *then the relation* (7.1.12) *holds.*

(c) *If* $1<r<\infty$,

$$\mathbb{L}^*_{\gamma,k;\sigma,\kappa}(\mathfrak{L}_{\nu,r})=\mathbb{L}_{1/k,-\gamma-\sigma}(\mathfrak{L}_{\nu-\mathrm{Re}(\kappa-\sigma),r}), \tag{7.1.16}$$

where the operators $\mathbb{L}_{1/k,-\gamma-\sigma}$ *is given in* (3.3.3).

When $\gamma=0$ and $k=1$, (7.1.4) is the Laplace transform (2.5.2), and (7.1.3) and (7.1.4) take the forms

$$\left(\mathfrak{M}\mathbb{L}f\right)(s)=\mathcal{G}^{1,0}_{0,1}(s)\left(\mathfrak{M}f\right)(1-s),\quad \mathcal{G}^{1,0}_{0,1}(s)=\Gamma(s) \tag{7.1.17}$$

and

$$\left(\mathbb{L}f\right)(x) = \int_0^\infty G_{0,1}^{1,0}\left[xt \,\middle|\, \begin{matrix} - \\ 0 \end{matrix}\right] f(t)dt, \tag{7.1.18}$$

respectively. Then from Theorems 6.1, 6.2, 6.6 and 6.8 we have

Corollary 7.4.1. *Let $\nu < 1$ and $1 \leqq r \leqq s \leqq \infty$.*

(a) *There is a one-to-one transform $\mathbb{L} \in [\mathfrak{L}_{\nu,2}, \mathfrak{L}_{1-\nu,2}]$ such that the relation (7.1.17) holds for $\mathrm{Re}(s) = 1 - \nu$ and $f \in \mathfrak{L}_{\nu,2}$.*

(b) *Let $\lambda \in \mathbb{C}$ and $f \in \mathfrak{L}_{\nu,2}$. If $\mathrm{Re}(\lambda) > -\nu$, then $\mathbb{L}f$ is given by*

$$\left(\mathbb{L}f\right)(x) = x^{-\lambda}\frac{d}{dx}x^{\lambda+1}\int_0^\infty G_{1,2}^{1,1}\left[xt \,\middle|\, \begin{matrix} -\lambda \\ 0, -\lambda-1 \end{matrix}\right] f(t)dt, \tag{7.1.19}$$

while if $\mathrm{Re}(\lambda) < -\nu$,

$$\left(\mathbb{L}f\right)(x) = -x^{-\lambda}\frac{d}{dx}x^{\lambda+1}\int_0^\infty G_{1,2}^{2,0}\left[xt \,\middle|\, \begin{matrix} -\lambda \\ -\lambda-1, 0 \end{matrix}\right] f(t)dt. \tag{7.1.20}$$

(c) *$\mathbb{L}$ is independent of ν in the sense that if $\nu < 1$ and $\widetilde{\nu} < 1$ and if the transforms $\mathbb{L}$ and $\widetilde{\mathbb{L}}$ on $\mathfrak{L}_{\nu,2}$ and $\mathfrak{L}_{\widetilde{\nu},2}$, respectively, are given in (7.1.17), then $\mathbb{L}f = \widetilde{\mathbb{L}}f$ for $f \in \mathfrak{L}_{\nu,2} \cap \mathfrak{L}_{\widetilde{\nu},2}$.*

(d) *For $x > 0$ and $f \in \mathfrak{L}_{\nu,2}$, $\mathbb{L}f$ is given in (2.5.2) and (7.1.18).*

(e) *The Laplace transform $\mathbb{L}$ defined on $\mathfrak{L}_{\nu,2}$ can be extended to $\mathfrak{L}_{\nu,r}$ as an element of $[\mathfrak{L}_{\nu,r}, \mathfrak{L}_{1-\nu,s}]$. If $1 \leqq r \leqq 2$, then $\mathbb{L}$ is a one-to-one transform from $\mathfrak{L}_{\nu,r}$ onto $\mathfrak{L}_{1-\nu,s}$.*

(f) *If $f \in \mathfrak{L}_{\nu,r}$ and $g \in \mathfrak{L}_{\nu,s'}$ with $1/s + 1/s' = 1$, then*

$$\int_0^\infty f(x)\left(\mathbb{L}g\right)(x)dx = \int_0^\infty \left(\mathbb{L}f\right)(x)g(x)dx. \tag{7.1.21}$$

7.2. Meijer and Varma Integral Transforms

We consider the Meijer transform $\boldsymbol{M}_{k,m}$ defined by

$$\left(\boldsymbol{M}_{k,m}f\right)(x) = \int_0^\infty (xt)^{-k-1/2}e^{-xt/2}W_{k+1/2,m}(xt)f(t)dt \tag{7.2.1}$$

with $k, m \in \mathbb{R}$ containing the Whittaker function $W_{l,m}(z)$ in the kernel. This function is given by

$$W_{l,m}(z) = e^{-z/2}z^{m+1/2}\,\Psi\left(m - l + \frac{1}{2}, 2m+1; z\right), \tag{7.2.2}$$

where $\Psi(a, c; x)$ is the confluent hypergeometric function of Tricomi:

$$\Psi(a, c; x) = \frac{1}{\Gamma(a)\Gamma(a-c+1)}G_{1,2}^{2,1}\left[xt \,\middle|\, \begin{matrix} 1-a \\ 0, 1-c \end{matrix}\right], \tag{7.2.3}$$

which has the integral representation

$$\Psi(a,c;x)=\frac{1}{\Gamma(a)}\int_0^\infty e^{-xt}t^{a-1}(1+t)^{c-a-1}dt \qquad (\mathrm{Re}(a)>0) \tag{7.2.4}$$

(see the formulas in Erdélyi, Magnus, Oberhettinger and Tricomi [1, 6.9(4) and 6.5(2)] and in Prudnikov, Brychkov and Marichev [3, 8.4.46.1]).

When $k=-m$, since $\Psi(a,a+1;z)=z^{-a}$ from (7.2.4) and $W_{k+1/2,m}(x)=x^{-m+1/2}e^{-x/2}$ by (7.2.2), we find that the transform $\boldsymbol{M}_{-m,m}$ (7.2.1) coincides with the Laplace transform (2.5.2).

By (3.3.11), (3.3.14) and the formula in Prudnikov, Brychkov and Marichev [3, 8.4.44.1] we have the following relation for the Mellin transform of $\boldsymbol{M}_{k,m}f$ for a "sufficiently good" function f:

$$\left(\mathfrak{M}\boldsymbol{M}_{k,m}f\right)(s)=\mathcal{G}_{1,2}^{2,0}(s)\left(\mathfrak{M}f\right)(1-s) \tag{7.2.5}$$

with

$$\mathcal{G}_{1,2}^{2,0}(s)=\frac{\Gamma(m-k+s)\Gamma(-m-k+s)}{\Gamma(-2k+s)},\quad \mathrm{Re}(s)>|m|+k. \tag{7.2.6}$$

By (7.2.6) $\boldsymbol{M}_{k,m}$ is the $\boldsymbol{G}$-transform (5.1.1) in the form

$$\left(\boldsymbol{M}_{k,m}f\right)(x)=\int_0^\infty G_{1,2}^{2,0}\left[xt\,\middle|\,\begin{matrix}-2k\\ m-k,-m-k\end{matrix}\right]f(t)dt \tag{7.2.7}$$

and the constants (6.1.5)–(6.1.11) take the forms

$$a^*=\Delta=a_1^*=1,\ a_2^*=0,\ \alpha=k+|m|,\ \beta=\infty,\ \mu=-\frac{1}{2}, \tag{7.2.8}$$

respectively.

Let $\mathcal{E}_2$ be the exceptional set of the function $\mathcal{G}_{1,2}^{2,0}(s)$ in (7.2.6), which is the set of $s\in\mathbb{C}$ of zero points of the function $\mathcal{G}_{1,2}^{2,0}(s)$ having zero only at the poles of the gamma function $\Gamma(-2k+s)$. Then, in accordance with Definition 6.1, $\nu\notin\mathcal{E}_2$ means that

$$s\neq 2k-i\quad (i=0,1,2,\cdots)\quad \text{for}\quad \mathrm{Re}(s)=1-\nu. \tag{7.2.9}$$

From Theorems 6.1, 6.2 and 6.6, 6.8 we obtain the $\mathfrak{L}_{\nu,2}$- and $\mathfrak{L}_{\nu,r}$-theory of the integral transform given in (7.2.1).

Theorem 7.5. *Let $\alpha=k+|m|$ and $1-\nu>\alpha$.*

(i) *There is a one-to-one transform $\boldsymbol{M}_{k,m}\in[\mathfrak{L}_{\nu,2},\mathfrak{L}_{1-\nu,2}]$ such that (7.2.5) holds for $\mathrm{Re}(s)=1-\nu$ and $f\in\mathfrak{L}_{\nu,2}$.*

(ii) *If $f,g\in\mathfrak{L}_{\nu,2}$, then the relation*

$$\int_0^\infty f(x)\left(\boldsymbol{M}_{k,m}g\right)(x)dx=\int_0^\infty\left(\boldsymbol{M}_{k,m}f\right)(x)g(x)dx \tag{7.2.10}$$

holds.

(iii) *Let $f \in \mathfrak{L}_{\nu,2}$ and $\lambda \in \mathbb{C}$. If $\mathrm{Re}(\lambda) > -\nu$, then $\boldsymbol{M}_{k,m}f$ is given by*

$$\left(\boldsymbol{M}_{k,m}f\right)(x) = x^{-\lambda}\frac{d}{dx}x^{\lambda+1}\int_0^\infty G_{2,3}^{2,1}\left[xt \,\middle|\, \begin{matrix} -\lambda, -2k \\ m-k, -m-k, -\lambda-1 \end{matrix}\right] f(t)dt. \tag{7.2.11}$$

When $\mathrm{Re}(\lambda) < -\nu$,

$$\left(\boldsymbol{M}_{k,m}f\right)(x) = -x^{-\lambda}\frac{d}{dx}x^{\lambda+1}\int_0^\infty G_{2,3}^{3,0}\left[xt \,\middle|\, \begin{matrix} -2k, -\lambda \\ -\lambda-1, m-k, -m-k \end{matrix}\right] f(t)dt. \tag{7.2.12}$$

(iv) *The transform $\boldsymbol{M}_{k,m}$ is independent of ν in the sense that if $1-\nu > \alpha$ and $1-\widetilde{\nu} > \alpha$ and if the transforms $\boldsymbol{M}_{k,m}$ and $\widetilde{\boldsymbol{M}}_{k,m}$ on $\mathfrak{L}_{\nu,2}$ and $\mathfrak{L}_{\widetilde{\nu},2}$, respectively, are given in (7.2.5), then $\boldsymbol{M}_{k,m}f = \widetilde{\boldsymbol{M}}_{k,m}f$ for $f \in \mathfrak{L}_{\nu,2} \cap \mathfrak{L}_{\widetilde{\nu},2}$.*

(v) *For $f \in \mathfrak{L}_{\nu,2}$ and $x > 0$, $\left(\boldsymbol{M}_{k,m}f\right)(x)$ is given in (7.2.1) and (7.2.7).*

Theorem 7.6. *Let $\alpha = k + |m|$ and $1 - \nu > \alpha$.*

(a) *Let $1 \leqq r \leqq s \leqq \infty$. The transform $\boldsymbol{M}_{k,m}$ defined on $\mathfrak{L}_{\nu,2}$ can be extended to $\mathfrak{L}_{\nu,r}$ as an element of $[\mathfrak{L}_{\nu,r}, \mathfrak{L}_{1-\nu,s}]$. If $1 \leqq r \leqq 2$, then $\boldsymbol{M}_{k,m}$ is a one-to-one transform from $\mathfrak{L}_{\nu,r}$ onto $\mathfrak{L}_{1-\nu,s}$.*

(b) *Let $1 \leqq r \leqq s \leqq \infty$. If $f \in \mathfrak{L}_{\nu,r}$ and $g \in \mathfrak{L}_{\nu,s'}$ with $1/s + 1/s' = 1$, then the relation (7.2.10) holds.*

(c) *Let $1 < r < \infty$ and the condition (7.2.9) holds. Then $\boldsymbol{M}_{k,m}$ is one-to-one on $\mathfrak{L}_{\nu,r}$.*

(d) *Let $1 < r < \infty$ and $\alpha \geqq 0$. If (7.2.9) holds, then*

$$\boldsymbol{M}_{k,m}(\mathfrak{L}_{\nu,r}) = \mathbb{L}(\mathfrak{L}_{\nu,r}), \tag{7.2.13}$$

where the Laplace operator $\mathbb{L}$ is given in (2.5.2). If (7.2.9) is not valid, then $\boldsymbol{M}_{k,m}(\mathfrak{L}_{\nu,r})$ is a subset of the right-hand side of (7.2.13).

(e) *Let $1 < r < \infty$ and $\alpha < 0$. If (7.2.9) holds, then*

$$\boldsymbol{M}_{k,m}(\mathfrak{L}_{\nu,r}) = \left(I_{-;1,-\alpha}^{-\alpha}\mathbb{L}_{1,\alpha}\right)(\mathfrak{L}_{\nu,r}), \tag{7.2.14}$$

where the operators $I_{-;1,-\alpha}^{-\alpha}$ and $\mathbb{L}_{1,\alpha}$ are given in (3.3.2) and (3.3.3). If (7.2.9) is not valid, then $\boldsymbol{M}_{k,m}(\mathfrak{L}_{\nu,r})$ is a subset of the right-hand side of (7.2.14).

We also consider the Varma transform defined by

$$\left(\boldsymbol{V}_{k,m}f\right)(x) = \int_0^\infty (xt)^{m-1/2}e^{-xt/2}W_{k,m}(xt)f(t)dt. \tag{7.2.15}$$

According to (3.3.11) and (3.3.14) and Prudnikov, Brychkov and Marichev [3, 8.4.44.1], the Mellin transform (2.5.1) of (7.2.1) for a "sufficiently good" function f is given by the relation

$$\left(\mathfrak{M}\boldsymbol{V}_{k,m}f\right)(s) = \mathcal{G}_{1,2}^{2,0}(s)\left(\mathfrak{M}f\right)(1-s) \tag{7.2.16}$$

with

$$\mathcal{G}_{1,2}^{2,0}(s) = \frac{\Gamma(2m+s)\Gamma(s)}{\Gamma\left(m-k+\dfrac{1}{2}+s\right)}, \qquad \mathrm{Re}(s) > -m+|m|. \tag{7.2.17}$$

The relation (7.2.17) shows that (7.2.1) is the $\boldsymbol{G}$-transform (5.1.1) of the form

$$\left(\boldsymbol{V}_{k,m}f\right)(x) = \int_0^\infty G_{1,2}^{2,0}\left[xt \,\middle|\, \begin{matrix} \frac{1}{2}+m-k \\ 2m, 0 \end{matrix}\right] f(t)dt. \tag{7.2.18}$$

When $m+k=1/2$, from (7.2.2) and (7.2.4) we know $W_{1/2-m,m} = x^{1/2-m}e^{-x/2}$ and we find the transform $\boldsymbol{V}_{k,m}$ in (7.2.15) coincides with the Laplace transform (2.5.2).

The constants $a^*, \Delta, a_1^*, a_2^*, \alpha, \beta$ and μ in (6.1.5)–(6.1.11) take the forms

$$a^* = \Delta = a_1^* = 1,\ a_2^* = 0,\ \alpha = -m+|m|,\ \beta = \infty,\ \mu = m+k-1. \tag{7.2.19}$$

We note that if $\mathrm{Re}(s) = 1-\nu$, then the condition (7.2.17) is equivalent to $1-\nu > \alpha$.

Let $\mathcal{E}_3$ be the exceptional set of the function $\mathcal{G}_{1,2}^{2,0}(s)$ in (7.2.17). Since $\mathcal{G}_{1,2}^{2,0}(s)$ has zero only in the poles of the gamma function $\Gamma(m-k+1/2+s)$, then in accordance with Definition 6.1 in Section 6.1 the condition $\nu \notin \mathcal{E}_3$ means that

$$s \neq k-m-\frac{1}{2}-i \quad (i=0,1,2,\cdots) \quad \text{for} \quad \mathrm{Re}(s) = 1-\nu. \tag{7.2.20}$$

Similarly to Theorems 7.5 and 7.6, we obtain the $\mathfrak{L}_{\nu,2}$- and $\mathfrak{L}_{\nu,r}$-theory of the transform (7.1.1) from Theorems 6.1, 6.2 and 6.6, 6.8.

Theorem 7.7. *Let $\alpha = -m+|m|$ and $1-\nu > \alpha$.*

(i) *There is a one-to-one transform $\boldsymbol{V}_{k,m} \in [\mathfrak{L}_{\nu,2}, \mathfrak{L}_{1-\nu,2}]$ such that (7.2.16) holds for $\mathrm{Re}(s) = 1-\nu$ and $f \in \mathfrak{L}_{\nu,2}$.*

(ii) *If $f, g \in \mathfrak{L}_{\nu,2}$, then the relation*

$$\int_0^\infty f(x)\left(\boldsymbol{V}_{k,m}g\right)(x)dx = \int_0^\infty \left(\boldsymbol{V}_{k,m}f\right)(x)g(x)dx \tag{7.2.21}$$

holds.

(iii) *Let $f \in \mathfrak{L}_{\nu,2}$ and $\lambda \in \mathbb{C}$. If $\mathrm{Re}(\lambda) > -\nu$, then $\boldsymbol{V}_{k,m}f$ is given by*

$$\left(\boldsymbol{V}_{k,m}f\right)(x) = x^{-\lambda}\frac{d}{dx}x^{\lambda+1}\int_0^\infty G_{2,3}^{2,1}\left[xt \,\middle|\, \begin{matrix} -\lambda, m-k+\frac{1}{2} \\ 2m, 0, -\lambda-1 \end{matrix}\right] f(t)dt. \tag{7.2.22}$$

When $\mathrm{Re}(\lambda) < -\nu$,

$$\left(\boldsymbol{V}_{k,m}f\right)(x) = -x^{-\lambda}\frac{d}{dx}x^{\lambda+1}\int_0^\infty G_{2,3}^{3,0}\left[xt \,\middle|\, \begin{matrix} m-k+\frac{1}{2}, -\lambda \\ -\lambda-1, 2m, 0 \end{matrix}\right] f(t)dt. \tag{7.2.23}$$

(iv) *The transform $\boldsymbol{V}_{k,m}$ is independent of ν in the sense that if $1-\nu > \alpha$ and $1-\widetilde{\nu} > \alpha$ and if the transforms $\boldsymbol{V}_{k,m}$ and $\widetilde{\boldsymbol{V}}_{k,m}$ on $\mathfrak{L}_{\nu,2}$ and $\mathfrak{L}_{\widetilde{\nu},2}$, respectively, are given in (7.2.16), then $\boldsymbol{V}_{k,m}f = \widetilde{\boldsymbol{V}}_{k,m}f$ for $f \in \mathfrak{L}_{\nu,2} \cap \mathfrak{L}_{\widetilde{\nu},2}$.*

(v) *For $f \in \mathfrak{L}_{\nu,2}$ and $x > 0$, $\left(\boldsymbol{V}_{k,m}f\right)(x)$ is given in (7.2.15) and (7.2.18).*

Theorem 7.8. *Let $\alpha = -m+|m|$ and $1-\nu > \alpha$.*

(a) *Let $1 \leqq r \leqq s \leqq \infty$. The transform $\boldsymbol{V}_{k,m}$ defined on $\mathfrak{L}_{\nu,2}$ can be extended to $\mathfrak{L}_{\nu,r}$ as an element of $[\mathfrak{L}_{\nu,r}, \mathfrak{L}_{1-\nu,s}]$. If $1 \leqq r \leqq 2$, then $\boldsymbol{V}_{k,m}$ is a one-to-one transform from $\mathfrak{L}_{\nu,r}$ onto $\mathfrak{L}_{1-\nu,s}$.*

(b) *Let $1 \leqq r \leqq s \leqq \infty$. If $f \in \mathfrak{L}_{\nu,r}$ and $g \in \mathfrak{L}_{\nu,s'}$ with $1/s + 1/s' = 1$, then the relation (7.2.21) holds.*

(c) *Let $1 < r < \infty$. If the condition (7.2.20) holds, then $\boldsymbol{V}_{k,m}$ is one-to-one on $\mathfrak{L}_{\nu,r}$.*

(d) *Let $1 < r < \infty$ and $|m| + k - 1/2 \geqq 0$. If (7.2.20) holds, then*

$$\boldsymbol{V}_{k,m}(\mathfrak{L}_{\nu,r}) = \mathbb{L}_{1,1/2-k-m}(\mathfrak{L}_{\nu,r}), \tag{7.2.24}$$

where the operator $\mathbb{L}_{1,1/2-k-m}$ is given in (3.3.3). If (7.2.20) is not valid, then $\boldsymbol{V}_{k,m}(\mathfrak{L}_{\nu,r})$ is a subset of the right-hand side of (7.2.24).

(e) *Let $1 < r < \infty$ and $|m| + k - 1/2 < 0$. If (7.2.20) holds, then*

$$\boldsymbol{V}_{k,m}(\mathfrak{L}_{\nu,r}) = \left(I_{-;1,-\alpha}^{-|m|-k+1/2}\mathbb{L}_{1,\alpha}\right)(\mathfrak{L}_{\nu,r}), \tag{7.2.25}$$

where the operators $I_{-;1,-\alpha}^{-|m|-k+1/2}$ and $\mathbb{L}_{1,\alpha}$ are given in (3.3.2) and (3.3.3). If (7.2.20) is not valid, then $\boldsymbol{V}_{k,m}(\mathfrak{L}_{\nu,r})$ is a subset of the right-hand side of (7.2.25).

In conclusion of this section we note that the relations (7.2.13), (7.2.14), (7.2.24) and (7.2.25) exhibit the ranges of the transforms $\boldsymbol{M}_{k,m}$ and $\boldsymbol{V}_{k,m}$ in (7.2.1) and (7.1.15) on the spaces $\mathfrak{L}_{\nu,r}$ for any $1 < r < \infty$. They were obtained from the results in Section 6.1 for the $\boldsymbol{G}$-transform (6.1.1). But such ranges of the $\boldsymbol{G}$-transform can be estimated more precisely for certain integral transforms. For example, using the technique of the Mellin transform $\mathfrak{M}$ and taking into acount the formulas (3.3.6)–(3.3.8) and (3.3.14), we directly prove the following relations:

$$\left(\mathfrak{M}\mathbb{L}x^{\beta}I_{0+}^{\alpha}f\right)(s) = \frac{\Gamma(-\alpha-\beta+s)\Gamma(s)}{\Gamma(-\beta+s)}\left(\mathfrak{M}f\right)(1-s+\alpha+\beta) \tag{7.2.26}$$

$$(\mathrm{Re}(s-\beta) = 1-\nu+\alpha)$$

and

$$\left(\mathfrak{M}I_{-}^{\alpha}x^{\beta}\mathbb{L}f\right)(s) = \frac{\Gamma(\alpha+\beta+s)\Gamma(s)}{\Gamma(\alpha+s)}\left(\mathfrak{M}f\right)(1-s-\alpha-\beta) \tag{7.2.27}$$

$$(\mathrm{Re}(s+\beta) = 1-\nu-\alpha)$$

for the Riemann–Liouville fractional integration operators I_{0+}^{α} and I_{-}^{α} in (2.6.1) and (2.6.2) and the Laplace transform $\mathbb{L}$ in (2.5.2).

On the other hand, by using (7.2.16) and (3.3.14) we have

$$\left(\mathfrak{M}\boldsymbol{V}_{k,m}x^{\alpha+\beta}f\right)(s) = \frac{\Gamma(2m+s)\Gamma(s)}{\Gamma\left(m-k+\dfrac{1}{2}+s\right)}\left(\mathfrak{M}f\right)(1-s+\alpha+\beta) \tag{7.2.28}$$

$$(\mathrm{Re}(s-\beta) = 1-\nu+\alpha)$$

and

$$\left(\mathfrak{M}\boldsymbol{V}_{k,m}x^{-\alpha-\beta}f\right)(s)=\frac{\Gamma(2m+s)\Gamma(s)}{\Gamma\left(m-k+\frac{1}{2}+s\right)}\left(\mathfrak{M}f\right)(1-s-\alpha-\beta) \tag{7.2.29}$$

$$(\mathrm{Re}(s+\beta)=1-\nu-\alpha).$$

(7.2.26) coincides with (7.2.28) when $\alpha=1/2-m-k$, $\beta=k-m-1/2$, and

$$\left(\mathfrak{M}\mathbb{L}x^{k-m-1/2}I_{0+}^{1/2-m-k}f\right)(s)=\left(\mathfrak{M}\boldsymbol{V}_{k,m}x^{-2m}f\right)(s)\quad(\mathrm{Re}(s)=1-\nu-2m)$$

or

$$\left(\mathfrak{M}\mathbb{L}x^{k-m-1/2}I_{0+}^{1/2-m-k}x^{2m}f\right)(s)=\left(\mathfrak{M}\boldsymbol{V}_{k,m}f\right)(s)\quad(\mathrm{Re}(s)=1-\nu). \tag{7.2.30}$$

Similarly, (7.2.27) coincides with (7.2.29) when $\alpha=m-k+1/2$, $\beta=m+k-1/2$, and

$$\left(\mathfrak{M}I_{-}^{m-k+1/2}x^{m+k-1/2}\mathbb{L}f\right)(s)=\left(\mathfrak{M}\boldsymbol{V}_{k,m}x^{-2m}f\right)(s)\quad(\mathrm{Re}(s)=1-\nu-2m)$$

or

$$\left(\mathfrak{M}I_{-}^{m-k+1/2}x^{m+k-1/2}\mathbb{L}x^{2m}f\right)(s)=\left(\mathfrak{M}\boldsymbol{V}_{k,m}f\right)(s)\quad(\mathrm{Re}(s)=1-\nu). \tag{7.2.31}$$

Motivated by the relations (7.2.30) and (7.2.31), we define the transforms $\boldsymbol{V}^1_{k,m}$ and $\boldsymbol{V}^2_{k,m}$ by

$$\boldsymbol{V}^1_{k,m}=\mathbb{L}M_{k-m-1/2}I_{0+}^{1/2-m-k}M_{2m} \tag{7.2.32}$$

and

$$\boldsymbol{V}^2_{k,m}=I_{-}^{m-k+1/2}M_{m+k-1/2}\mathbb{L}M_{2m}. \tag{7.2.33}$$

Applying the relations

$$\left(I^{\alpha}_{0+}f\right)(x)=x^{\alpha}\left(I^{\alpha}_{0+;1,0}f\right)(x),\quad\left(I^{\alpha}_{-}f\right)(x)=\left(I^{\alpha}_{-;1,0}t^{\alpha}f\right)(x),\quad\mathbb{L}=\mathbb{L}_{1,0} \tag{7.2.34}$$

connecting the Riemann–Liouville fractional integrals (2.7.1) and (2.7.2) with the Erdélyi–Kober type integrals (3.3.1) and (3.3.2), and the Laplace transform (2.5.2) and the generalized Laplace transform (3.3.3), as well as (3.3.23) and (3.3.22), we can rewrite (7.2.32) and (7.2.33) as

$$\boldsymbol{V}^1_{k,m}=\mathbb{L}I^{1/2-m-k}_{0+:1,2m} \tag{7.2.35}$$

and

$$\boldsymbol{V}^2_{k,m}=I^{m-k+1/2}_{-;1,0}\mathbb{L}_{1,-2m}, \tag{7.2.36}$$

respectively.

On the basis of (7.2.35) and (7.2.36) we obtain the following results from Theorem 3.2(a)–(c).

Theorem 7.9. *Let $1 \leqq r \leqq \infty$ and $\nu < 1 + 2m$.*

(a) *If $m + k < 1/2$ and $\nu < 1$, then for all $s \geqq r$ such that $1/s > 1/r + m + k - 1/2$, the operator $\boldsymbol{V}^1_{k,m}$ belongs to $[\mathfrak{L}_{\nu,r}, \mathfrak{L}_{1-\nu,s}]$ and is a one-to-one transform from $\mathfrak{L}_{\nu,r}$ onto $\mathfrak{L}_{1-\nu,s}$. For $1 \leqq r \leqq 2$ and $f \in \mathfrak{L}_{\nu,r}$*

$$\left(\mathfrak{M}\boldsymbol{V}^1_{k,m} f\right)(s) = \frac{\Gamma(2m+s)\Gamma(s)}{\Gamma\left(m - k + \dfrac{1}{2} + s\right)} \left(\mathfrak{M}f\right)(1-s) \quad \text{for} \quad \text{Re}(s) = 1 - \nu. \tag{7.2.37}$$

(b) *If $m - k > -1/2$ and $\nu > 0$, then for all $s \geqq r$ such that $1/s > 1/r + k - m - 1/2$, the operator $\boldsymbol{V}^2_{k,m}$ belongs to $[\mathfrak{L}_{\nu,r}, \mathfrak{L}_{1-\nu,s}]$ and is a one-to-one transform from $\mathfrak{L}_{\nu,r}$ onto $\mathfrak{L}_{1-\nu,s}$. For $1 \leqq r \leqq 2$ and $f \in \mathfrak{L}_{\nu,r}$*

$$\left(\mathfrak{M}\boldsymbol{V}^2_{k,m} f\right)(s) = \frac{\Gamma(2m+s)\Gamma(s)}{\Gamma\left(m - k + \dfrac{1}{2} + s\right)} \left(\mathfrak{M}f\right)(1-s) \quad \text{for} \quad \text{Re}(s) = 1 - \nu. \tag{7.2.38}$$

According to this theorem and (7.2.15)–(7.2.17) the Varma transform can be interpreted by the transforms $\boldsymbol{V}^1_{k,m}$ and $\boldsymbol{V}^2_{k,m}$ as $\boldsymbol{V}_{k,m} = \boldsymbol{V}^1_{k,m}$ and $\boldsymbol{V}_{k,m} = \boldsymbol{V}^2_{k,m}$ for $m + k < 1/2$ and $m - k > -1/2$, respectively. Taking the same arguments as in Chapter 4, we obtain the range of the Varma transform.

Theorem 7.10. *Let $1 < r < \infty$ and $\nu < 1 + 2m$.*

(a) *Let $k + m < 1/2$ and $\nu < 1$. If (7.2.9) holds, then*

$$\boldsymbol{V}_{k,m}(\mathfrak{L}_{\nu,r}) = \left(\mathbb{L} I^{1/2-m-k}_{0+;1,2m}\right)(\mathfrak{L}_{\nu,r}). \tag{7.2.39}$$

When (7.2.9) is not valid, then $\boldsymbol{V}_{k,m}(\mathfrak{L}_{\nu,r})$ is a subset of the right-hand side of (7.2.39).

(b) *Let $m - k > -1/2$ and $\nu > 0$. If (7.2.9) holds, then*

$$\boldsymbol{V}_{k,m}(\mathfrak{L}_{\nu,r}) = \left(I^{m-k+1/2}_{-;1,0} \mathbb{L}_{1,-2m}\right)(\mathfrak{L}_{\nu,r}). \tag{7.2.40}$$

When (7.2.9) is not valid, then $\boldsymbol{V}_{k,m}(\mathfrak{L}_{\nu,r})$ is a subset of the right-hand side of (7.2.40).

Remark 7.1. The relations (7.2.32) and (7.2.33) can be used to find the inverse formulas for the Varma transform $\boldsymbol{V}_{k,m}$ in $\mathfrak{L}_{\nu,r}$-spaces.

7.3. Generalized Whittaker Transforms

We consider the integral transform

$$\left(\boldsymbol{W}^k_{\rho,\gamma} f\right)(x) = \int_0^\infty (xt)^k e^{-xt/2} W_{\rho,\gamma}(xt) f(t) dt \tag{7.3.1}$$

with ρ, $\gamma \in \mathbb{C}$ and $k \in \mathbb{R}$, containing the Whittaker function (7.2.2) in the kernel. It is known that the Meijer transform (7.2.1) and the Varma transform (7.2.15) are particular cases of this generalized Whittaker transform $\boldsymbol{W}^k_{\rho,\gamma}$:

$$\left(\boldsymbol{M}_{k,m} f\right)(x) = \left(\boldsymbol{W}^{-k-1/2}_{k+1/2,m} f\right)(x), \qquad \left(\boldsymbol{V}_{k,m} f\right)(x) = \left(\boldsymbol{W}^{m-1/2}_{k,m} f\right)(x). \tag{7.3.2}$$

Due to Prudnikov, Brychkov and Marichev [3, 8.4.44.1], the Mellin transform (2.5.1) of (7.3.1) for a "sufficiently good" function f is given by the relation

$$\left(\mathfrak{M}\boldsymbol{W}^k_{\rho,\gamma}f\right)(s) = \mathcal{G}^{2,0}_{1,2}(s)\left(\mathfrak{M}f\right)(1-s), \tag{7.3.3}$$

$$\mathcal{G}^{2,0}_{1,2}(s) = \frac{\Gamma\left(k+\gamma+\frac{1}{2}+s\right)\Gamma\left(k-\gamma+\frac{1}{2}+s\right)}{\Gamma(k-\rho+1+s)},$$

provided that

$$\mathrm{Re}(s) > |\mathrm{Re}(\gamma)| - k - \frac{1}{2}. \tag{7.3.4}$$

The relation (7.3.3) shows that (7.3.1) is the $\boldsymbol{G}$-transform (5.1.1) of the form

$$\left(\boldsymbol{W}^k_{\rho,\gamma}f\right)(x) = \int_0^\infty G^{2,0}_{1,2}\left[xt \,\middle|\, \begin{matrix} 1-\rho+k \\ k+\gamma+\frac{1}{2}, k-\gamma+\frac{1}{2} \end{matrix}\right] f(t)dt. \tag{7.3.5}$$

The constants $a^*, \Delta, a_1^*, a_2^*, \alpha, \beta$ and μ in (6.1.5)–(6.1.11) take the forms

$$a^* = \Delta = a_1^* = 1,\ a_2^* = 0,\ \alpha = |\mathrm{Re}(\gamma)| - k - \frac{1}{2},\ \beta = \infty,\ \mu = k + \varrho - \frac{1}{2}. \tag{7.3.6}$$

We note that for $\mathrm{Re}(s) = 1-\nu$ the condition (7.3.4) is equivalent to $1-\nu > \alpha$.

Lat $\mathcal{E}_4$ be the exceptional set of the function $\mathcal{G}^{2,0}_{1,2}(s)$ in (7.3.3). Since $\mathcal{G}^{2,0}_{1,2}(s)$ has a zero only at the poles of the gamma function $\Gamma(k-\rho+1+s)$, Definition 6.1 implies that the condition $\nu \notin \mathcal{E}_4$ means that

$$s \neq \rho - k - 1 - i \quad (i = 0, 1, 2, \cdots) \quad \text{for} \quad \mathrm{Re}(s) = 1 - \nu. \tag{7.3.7}$$

From Theorems 6.1, 6.2 and 6.6, 6.8 we obtain the $\mathfrak{L}_{\nu,2}$- and $\mathfrak{L}_{\nu,r}$-theory of the transform (7.3.1).

Theorem 7.11. *Let $\alpha = |\mathrm{Re}(\gamma)| - k - 1/2$ and $1-\nu > \alpha$.*

(i) *There is a one-to-one transform $\boldsymbol{W}^k_{\rho,\gamma} \in [\mathfrak{L}_{\nu,2}, \mathfrak{L}_{1-\nu,2}]$ such that (7.3.3) holds for $\mathrm{Re}(s) = 1-\nu$ and $f \in \mathfrak{L}_{\nu,2}$.*

(ii) *If $f, g \in \mathfrak{L}_{\nu,2}$, then the relation*

$$\int_0^\infty f(x)\left(\boldsymbol{W}^k_{\rho,\gamma}g\right)(x)dx = \int_0^\infty \left(\boldsymbol{W}^k_{\rho,\gamma}f\right)(x)g(x)dx \tag{7.3.8}$$

holds.

(iii) *Let $f \in \mathfrak{L}_{\nu,2}$ and $\lambda \in \mathbb{C}$. If $\mathrm{Re}(\lambda) > -\nu$, then $\boldsymbol{W}^k_{\rho,\gamma}f$ is given by*

$$\left(\boldsymbol{W}^k_{\rho,\gamma}f\right)(x)$$

$$= x^{-\lambda}\frac{d}{dx}x^{\lambda+1}\int_0^\infty G^{2,1}_{2,3}\left[xt \,\middle|\, \begin{matrix} -\lambda, 1+k-\rho \\ k+\gamma+\frac{1}{2}, k-\gamma+\frac{1}{2}, -\lambda-1 \end{matrix}\right] f(t)dt. \tag{7.3.9}$$

When $\mathrm{Re}(\lambda) < -\nu$,

$$\left(\boldsymbol{W}^k_{\rho,\gamma}f\right)(x)$$
$$= -x^{-\lambda}\frac{d}{dx}x^{\lambda+1}\int_0^\infty G^{3,0}_{2,3}\left[xt \,\middle|\, \begin{matrix} 1+k-\rho, -\lambda \\ -\lambda-1, k+\gamma+\frac{1}{2}, k-\gamma+\frac{1}{2} \end{matrix}\right] f(t)dt. \tag{7.3.10}$$

(iv) *Moreover, $\boldsymbol{W}^k_{\rho,\gamma}$ is independent of ν in the sense that if $1-\nu > \alpha$ and $1-\widetilde{\nu} > \alpha$ and if the transforms $\boldsymbol{W}^k_{\rho,\gamma}$ and $\widetilde{\boldsymbol{W}}^k_{\rho,\gamma}$ on $\mathfrak{L}_{\nu,2}$ and $\mathfrak{L}_{\widetilde{\nu},2}$, respectively, are given in (7.3.3), then $\boldsymbol{W}^k_{\rho,\gamma}f = \widetilde{\boldsymbol{W}}^k_{\rho,\gamma}f$ for $f \in \mathfrak{L}_{\nu,2} \cap \mathfrak{L}_{\widetilde{\nu},2}$.*

(v) *For $f \in \mathfrak{L}_{\nu,2}$ and $x > 0$, $\left(\boldsymbol{W}^k_{\rho,\gamma}f\right)(x)$ is given in (7.3.1) and (7.3.5).*

Theorem 7.12. *Let $\alpha = |\mathrm{Re}(\gamma)| - k - 1/2$ and $1-\nu > \alpha$.*

(a) *Let $1 \leqq r \leqq s \leqq \infty$. The transform $\boldsymbol{W}^k_{\rho,\gamma}$ defined on $\mathfrak{L}_{\nu,2}$ can be extended to $\mathfrak{L}_{\nu,r}$ as an element of $[\mathfrak{L}_{\nu,r}, \mathfrak{L}_{1-\nu,s}]$. If $1 \leqq r \leqq 2$, then $\boldsymbol{W}^k_{\rho,\gamma}$ is a one-to-one transform from $\mathfrak{L}_{\nu,r}$ onto $\mathfrak{L}_{1-\nu,s}$.*

(b) *Let $1 \leqq r \leqq s \leqq \infty$, $f \in \mathfrak{L}_{\nu,r}$ and $g \in \mathfrak{L}_{\nu,s'}$ with $1/s + 1/s' = 1$. Then the relation (7.3.8) holds.*

(c) *Let $1 < r < \infty$. If the condition (7.3.7) holds, then $\boldsymbol{W}^k_{\rho,\gamma}$ is one-to-one on $\mathfrak{L}_{\nu,r}$.*

(d) *Let $1 < r < \infty$ and $\mathrm{Re}(\rho) + |\mathrm{Re}(\gamma)| - 1/2 \geqq 0$. If (7.3.7) holds, then*

$$\boldsymbol{W}^k_{\rho,\gamma}(\mathfrak{L}_{\nu,r}) = \mathbb{L}_{1,-k-\rho}(\mathfrak{L}_{\nu,r}), \tag{7.3.11}$$

where the operator $\mathbb{L}_{1,-k-\rho}$ is given in (3.3.3). If (7.3.7) is not valid, then $\boldsymbol{W}^k_{\rho,\gamma}(\mathfrak{L}_{\nu,r})$ is a subset of the right-hand side of (7.3.11).

(e) *Let $1 < r < \infty$ and $\mathrm{Re}(\rho) + |\mathrm{Re}(\gamma)| - 1/2 < 0$. If (7.3.7) holds, then*

$$\boldsymbol{W}^k_{\rho,\gamma}(\mathfrak{L}_{\nu,r}) = \left(I^{-(\rho+\alpha+k)}_{-;1,-\alpha}\mathbb{L}_{1,\alpha}\right)(\mathfrak{L}_{\nu,r}), \tag{7.3.12}$$

where the operators $I^{-(\rho+\alpha+k)}_{-;1,-\alpha}$ and $\mathbb{L}_{1,\alpha}$ are given in (3.3.1) and (3.3.3). If (7.3.7) is not valid, then $\boldsymbol{W}^k_{\rho,\gamma}(\mathfrak{L}_{\nu,r})$ is a subset of the right-hand side of (7.3.12).

Remark 7.2. In view of (7.3.2), Theorems 7.5, 7.7 and 7.6, 7.8 in the previous section are particular cases of Theorems 7.11 and 7.12.

Now we consider the modification of the transform (7.3.1) in the form (6.2.3):

$$\left(\boldsymbol{W}^k_{\rho,\gamma;\sigma,\kappa}f\right)(x) = x^\sigma\int_0^\infty (xt)^k e^{-xt/2}W_{\rho,\gamma}(xt)t^\kappa f(t)dt \tag{7.3.13}$$

with $\rho,\gamma,\sigma,\kappa \in \mathbb{C}$ and $k \in \mathbb{R}$. For this transform the relations (7.3.3), (7.3.5) and (7.3.8) take the forms, by consulting (6.2.13), (6.2.3) and (6.2.18):

$$\left(\mathfrak{M}\boldsymbol{W}^k_{\rho,\gamma;\sigma,\kappa}f\right)(s) = \mathcal{G}^{2,0}_{1,2}(s+\sigma)\left(\mathfrak{M}f\right)(1-s-\sigma+\kappa), \tag{7.3.14}$$

where $\mathcal{G}^{2,0}_{1,2}(s)$ is given in (7.3.3);

$$\left(\boldsymbol{W}^k_{\rho,\gamma;\sigma,\kappa}f\right)(x) = x^\sigma\int_0^\infty G^{2,0}_{1,2}\left[xt \,\middle|\, \begin{matrix} 1+k-\rho \\ k+\gamma+\frac{1}{2}, k-\gamma+\frac{1}{2} \end{matrix}\right] t^\kappa f(t)dt \tag{7.3.15}$$

and

$$\int_0^\infty f(x)\left(\boldsymbol{W}^k_{\rho,\gamma;\sigma,\kappa}g\right)(x)dx = \int_0^\infty \left(\boldsymbol{W}^k_{\rho,\gamma;\kappa,\sigma}f\right)(x)g(x)dx. \tag{7.3.16}$$

The exceptional set for this kernel $\mathcal{G}^{2,0}_{1,2}(s)$ is the same as $\mathcal{E}_4$ for (7.3.3). Then (7.3.7) shows that the condition $\nu - \mathrm{Re}(\kappa) \notin \mathcal{E}_4$ means that

$$s \neq \rho - k - 1 - i \quad (i = 0, 1, 2, \cdots) \quad \text{for} \quad \mathrm{Re}(s) = 1 - \nu + \mathrm{Re}(\kappa). \tag{7.3.17}$$

From Theorems 6.38, 6.42 and 6.44 we have the $\mathfrak{L}_{\nu,2}$- and $\mathfrak{L}_{\nu,r}$-theory of the transform (7.3.13).

Theorem 7.13. *Let $\alpha = |\mathrm{Re}(\gamma)| - k - 1/2$ and $1 - \nu + \mathrm{Re}(\kappa) > \alpha$.*

(i) *There is a one-to-one transform $\boldsymbol{W}^k_{\rho,\gamma;\sigma,\kappa} \in [\mathfrak{L}_{\nu,2}, \mathfrak{L}_{1-\nu+\mathrm{Re}(\kappa-\sigma),2}]$ such that (7.3.14) holds for $\mathrm{Re}(s) = 1 - \nu + \mathrm{Re}(\kappa - \sigma)$ and $f \in \mathfrak{L}_{\nu,2}$.*

(ii) *If $f \in \mathfrak{L}_{\nu,2}$ and $g \in \mathfrak{L}_{\nu+\mathrm{Re}(\kappa-\sigma),2}$, then the relation (7.3.16) holds for $\boldsymbol{W}^k_{\rho,\gamma;\sigma,\kappa}$.*

(iii) *Let $f \in \mathfrak{L}_{\nu,2}$ and $\lambda \in \mathbb{C}$. If $\mathrm{Re}(\lambda) > -\nu + \mathrm{Re}(\kappa)$, then $\boldsymbol{W}^k_{\rho,\gamma;\sigma,\kappa}f$ is given by*

$$\left(\boldsymbol{W}^k_{\rho,\gamma;\sigma,\kappa}f\right)(x)$$
$$= x^{\sigma-\lambda}\frac{d}{dx}x^{\lambda+1}\int_0^\infty G^{2,1}_{2,3}\left[xt \,\middle|\, \begin{matrix} -\lambda, 1+k-\rho \\ k+\gamma+\frac{1}{2}, k-\gamma+\frac{1}{2}, -\lambda-1 \end{matrix}\right] t^\kappa f(t)dt. \tag{7.3.18}$$

When $\mathrm{Re}(\lambda) < -\nu + \mathrm{Re}(\kappa)$,

$$\left(\boldsymbol{W}^k_{\rho,\gamma;\sigma,\kappa}f\right)(x)$$
$$= -x^{\sigma-\lambda}\frac{d}{dx}x^{\lambda+1}\int_0^\infty G^{3,0}_{2,3}\left[xt \,\middle|\, \begin{matrix} 1+k-\rho, -\lambda \\ -\lambda-1, k+\frac{1}{2}, k-\gamma+\frac{1}{2} \end{matrix}\right] t^\kappa f(t)dt. \tag{7.3.19}$$

(iv) *The transform $\boldsymbol{W}^k_{\rho,\gamma;\sigma,\kappa}$ is independent of ν in the sense that if $1 - \nu + \kappa > \alpha$ and $1 - \widetilde{\nu} + \kappa > \alpha$ and if the transforms $\boldsymbol{W}^k_{\rho,\gamma;\sigma,\kappa}$ and $\widetilde{\boldsymbol{W}}^k_{\rho,\gamma;\sigma,\kappa}$ on $\mathfrak{L}_{\nu,2}$ and $\mathfrak{L}_{\widetilde{\nu},2}$, respectively, are given in (7.3.14), then $\boldsymbol{W}^k_{\rho,\gamma;\sigma,\kappa}f = \widetilde{\boldsymbol{W}}^k_{\rho,\gamma;\sigma,\kappa}f$ for $f \in \mathfrak{L}_{\nu,2} \cap \mathfrak{L}_{\widetilde{\nu},2}$.*

(v) *For $f \in \mathfrak{L}_{\nu,2}$ and $x > 0$, $\left(\boldsymbol{W}^k_{\rho,\gamma;\sigma,\kappa}f\right)(x)$ is given in (7.3.13) and (7.3.15).*

Theorem 7.14. *Let $\alpha = |\mathrm{Re}(\gamma)| - k - 1/2$ and $1 - \nu + \mathrm{Re}(\kappa) > \alpha$.*

(a) *If $1 \leqq r \leqq s \leqq \infty$, then the transform $\boldsymbol{W}^k_{\rho,\gamma;\sigma,\kappa}$ defined on $\mathfrak{L}_{\nu,2}$ can be extended to $\mathfrak{L}_{\nu,r}$ as an element of $[\mathfrak{L}_{\nu,r}, \mathfrak{L}_{1-\nu+\mathrm{Re}(\kappa-\sigma),s}]$. If $1 \leqq r \leqq 2$, then $\boldsymbol{W}^k_{\rho,\gamma;\sigma,\kappa}$ is a one-to-one transform from $\mathfrak{L}_{\nu,r}$ onto $\mathfrak{L}_{1-\nu+\mathrm{Re}(\kappa-\sigma),s}$.*

(b) *If $1 \leqq r \leqq s \leqq \infty$, $f \in \mathfrak{L}_{\nu,r}$ and $g \in \mathfrak{L}_{\nu+\mathrm{Re}(\kappa-\sigma),s'}$ with $1/s + 1/s' = 1$, then the relation (7.3.16) holds.*

(c) *If $1 < r < \infty$ and the condition (7.3.17) holds, then $\boldsymbol{W}^k_{\rho,\gamma;\sigma,\kappa}$ is one-to-one on $\mathfrak{L}_{\nu,r}$.*

(d) *Let $1 < r < \infty$ and $\mathrm{Re}(\rho) + |\mathrm{Re}(\gamma)| - 1/2 \geqq 0$. If (7.3.17) holds, then*

$$\boldsymbol{W}^k_{\rho,\gamma;\sigma,\kappa}(\mathfrak{L}_{\nu,r}) = \mathbb{L}_{1,-k-\rho-\sigma}\left(\mathfrak{L}_{\nu-\mathrm{Re}(\kappa-\sigma),r}\right), \tag{7.3.20}$$

where the operator $\mathbb{L}_{1,-k-\rho-\sigma}$ *is given in* (3.3.3). *When* (7.3.17) *is not valid,* $\boldsymbol{W}^k_{\rho,\gamma;\sigma,\kappa}(\mathfrak{L}_{\nu,r})$ *is a subset of the right-hand side of* (7.3.20).

(e) *Let* $1 < r < \infty$ *and* $\mathrm{Re}(\rho) + |\mathrm{Re}(\gamma)| - 1/2 < 0$. *If* (7.3.17) *holds, then*

$$\boldsymbol{W}^k_{\rho,\gamma;\sigma,\kappa}(\mathfrak{L}_{\nu,r}) = \left(I^{-(\rho+\alpha+k)}_{-;1,\sigma-\alpha}\mathbb{L}_{1,\alpha-\sigma}\right)\left(\mathfrak{L}_{\nu-\mathrm{Re}(\kappa-\sigma),r}\right). \tag{7.3.21}$$

If (7.3.17) *is not valid, then* $\boldsymbol{W}^k_{\rho,\gamma;\sigma,\kappa}(\mathfrak{L}_{\nu,r})$ *is a subset of the right-hand side of* (7.3.21).

Remark 7.3. The results in Theorems 7.11 and 7.12 follow from those in Theorems 7.13 and 7.14 when $\sigma = \kappa = 0$.

7.4. D_γ-Transforms

We consider the integral transform

$$\left(\boldsymbol{D}_\gamma f\right)(x) = 2^{-\gamma/2}\int_0^\infty e^{-xt/2}D_\gamma\left((2xt)^{1/2}\right)f(t)dt \tag{7.4.1}$$

with $\gamma \in \mathbb{C}$, containing the parabolic cylinder function $D_\gamma(z)$ defined in terms of the confluent hypergeometric function of Tricomi (7.2.3) by

$$D_\gamma(z) = 2^{\gamma/2}e^{-z^2/4}\Psi\left(\frac{-\gamma}{2},\frac{1}{2};\frac{z^2}{2}\right) \tag{7.4.2}$$

(see Erdélyi, Magnus, Oberhettinger and Tricomi [2, 8.2]). According to Prudnikov, Brychkov and Marichev [3, 8.4.18.1] the Mellin transform (7.4.1) for a "sufficiently good" function f is given by

$$\left(\mathfrak{M}\boldsymbol{D}_\gamma f\right)(s) = \mathcal{G}^{2,0}_{1,2}(s)\left(\mathfrak{M}f\right)(1-s), \qquad \mathcal{G}^{2,0}_{1,2}(s) = \frac{\Gamma(s)\Gamma\left(\dfrac{1}{2}+s\right)}{\Gamma\left(\dfrac{1-\gamma}{2}+s\right)}, \tag{7.4.3}$$

provided that $\mathrm{Re}(s) > 0$. This relation shows that (7.4.1) is the $\boldsymbol{G}$-transform (6.1.1) of the form

$$\left(\boldsymbol{D}_\gamma f\right)(x) = \int_0^\infty G^{2,0}_{1,2}\left[xt\,\middle|\,\begin{matrix}\dfrac{1-\gamma}{2}\\ 0,\dfrac{1}{2}\end{matrix}\right]f(t)dt. \tag{7.4.4}$$

The constants $a^*, \Delta, a_1^*, a_2^*, \alpha, \beta$ and μ in (6.1.5)–(6.1.11) take the forms

$$a^* = \Delta = a_1^* = 1,\ a_2^* = 0,\ \alpha = 0,\ \beta = \infty,\ \mu = \frac{\gamma-1}{2}. \tag{7.4.5}$$

Let $\mathcal{E}_5$ be the exceptional set of the function $\mathcal{G}^{2,0}_{1,2}(s)$ in (7.4.3). Since $\mathcal{G}^{2,0}_{1,2}(s)$ has zero only at the poles of the gamma function $\Gamma(s+(1-\gamma)/2)$, Definition 6.1 implies that the condition $\nu \notin \mathcal{E}_5$ means that

$$s \neq \frac{\gamma-1}{2} - i \quad (i = 0,1,2,\cdots) \quad \text{for} \quad \mathrm{Re}(s) = 1-\nu. \tag{7.4.6}$$

From Theorems 6.1, 6.2 and 6.6, 6.8 we obtain the $\mathfrak{L}_{\nu,2}$- and $\mathfrak{L}_{\nu,r}$-theory of the transform (7.4.1).

Theorem 7.15. *Let $\nu < 1$.*

(i) *There is a one-to-one transform $\boldsymbol{D}_\gamma \in [\mathfrak{L}_{\nu,2}, \mathfrak{L}_{1-\nu,2}]$ such that (7.4.3) holds for $\mathrm{Re}(s) = 1-\nu$ and $f \in \mathfrak{L}_{\nu,2}$.*

(ii) *If $f, g \in \mathfrak{L}_{\nu,2}$, then the relation*

$$\int_0^\infty f(x)\big(\boldsymbol{D}_\gamma g\big)(x)dx = \int_0^\infty \big(\boldsymbol{D}_\gamma f\big)(x)g(x)dx \tag{7.4.7}$$

holds.

(iii) *Let $f \in \mathfrak{L}_{\nu,2}$ and $\lambda \in \mathbb{C}$. If $\mathrm{Re}(\lambda) > -\nu$, then $\boldsymbol{D}_\gamma f$ is given by*

$$\big(\boldsymbol{D}_\gamma f\big)(x) = x^{-\lambda}\frac{d}{dx}x^{\lambda+1}\int_0^\infty G_{2,3}^{2,1}\left[xt \,\middle|\, \begin{matrix} -\lambda, \dfrac{1-\gamma}{2} \\ 0, \dfrac{1}{2}, -\lambda-1 \end{matrix}\right] f(t)dt. \tag{7.4.8}$$

When $\mathrm{Re}(\lambda) < -\nu$,

$$\big(\boldsymbol{D}_\gamma f\big)(x) = -x^{-\lambda}\frac{d}{dx}x^{\lambda+1}\int_0^\infty G_{2,3}^{3,0}\left[xt \,\middle|\, \begin{matrix} \dfrac{1-\gamma}{2}, -\lambda \\ -\lambda-1, 0, \dfrac{1}{2} \end{matrix}\right] f(t)dt. \tag{7.4.9}$$

(iv) *The transform $\boldsymbol{D}_\gamma$ is independent of ν in the sense that if $\nu < 1$ and $\widetilde{\nu} < 1$ and if the transforms $\boldsymbol{D}_\gamma$ and $\widetilde{\boldsymbol{D}}_\gamma$ on $\mathfrak{L}_{\nu,2}$ and $\mathfrak{L}_{\widetilde{\nu},2}$, respectively, are given in (7.4.3), then $\boldsymbol{D}_\gamma f = \widetilde{\boldsymbol{D}}_\gamma f$ for $f \in \mathfrak{L}_{\nu,2} \cap \mathfrak{L}_{\widetilde{\nu},2}$.*

(v) *For $f \in \mathfrak{L}_{\nu,2}$ and $x > 0$, $\big(\boldsymbol{D}_\gamma f\big)(x)$ is given in (7.4.1) and (7.4.4).*

Theorem 7.16. *Let $\nu < 1$ and $1 \leqq r \leqq s \leqq \infty$.*

(a) *The transform $\boldsymbol{D}_\gamma$ defined on $\mathfrak{L}_{\nu,2}$ can be extended to $\mathfrak{L}_{\nu,r}$ as an element of $[\mathfrak{L}_{\nu,r}, \mathfrak{L}_{1-\nu,s}]$. If $1 \leqq r \leqq 2$, $\boldsymbol{D}_\gamma$ is a one-to-one transform from $\mathfrak{L}_{\nu,r}$ onto $\mathfrak{L}_{1-\nu,s}$.*

(b) *If $f \in \mathfrak{L}_{\nu,r}$ and $g \in \mathfrak{L}_{\nu,s'}$ with $1/s + 1/s' = 1$, then the relation (7.4.7) holds.*

(c) *If $1 < r < \infty$ and the condition (7.4.6) holds, then $\boldsymbol{D}_\gamma$ is one-to-one on $\mathfrak{L}_{\nu,r}$.*

(d) *If $1 < r < \infty$, $\mathrm{Re}(\gamma) \geqq 0$ and the condition (7.4.6) holds, then*

$$\boldsymbol{D}_\gamma(\mathfrak{L}_{\nu,r}) = \mathbb{L}_{1,-\gamma/2}(\mathfrak{L}_{\nu,r}), \tag{7.4.10}$$

where the operator $\mathbb{L}_{1,-\gamma/2}$ is given in (3.3.3). When (7.4.6) is not valid, $\boldsymbol{D}_\gamma(\mathfrak{L}_{\nu,r})$ is a subset of the right-hand side of (7.4.10).

(e) *If $1 < r < \infty$, $\mathrm{Re}(\gamma) < 0$ and (7.4.6) holds, then*

$$\boldsymbol{D}_\gamma(\mathfrak{L}_{\nu,r}) = \left(I_{-;1,0}^{-\gamma/2}\mathbb{L}\right)(\mathfrak{L}_{\nu,r}), \tag{7.4.11}$$

where the operators $I_{-;1,0}^{-\gamma/2}$ and $\mathbb{L}$ are given in (3.3.2) and (2.5.2). When (7.4.6) is not valid, $\boldsymbol{D}_\gamma(\mathfrak{L}_{\nu,r})$ is a subset of the right-hand side of (7.4.11).

Now we consider a generalization of (7.4.1) in the form

$$\left(\boldsymbol{D}_{\gamma;\sigma,\kappa}f\right)(x) = 2^{-\gamma/2}x^{\sigma}\int_0^\infty e^{-xt/2}D_\gamma\left((2xt)^{1/2}\right)t^\kappa f(t)dt \tag{7.4.12}$$

with $\gamma, \sigma, \kappa \in \mathbb{C}$. In view of (6.2.8) and (7.4.4) this is a modified transform (6.2.3) of the form

$$\left(\boldsymbol{D}_{\gamma;\sigma,\kappa}f\right)(x) = x^{\sigma}\int_0^\infty G_{1,2}^{2,0}\left[xt \,\middle|\, \begin{matrix} \frac{1-\gamma}{2} \\ 0, \frac{1}{2} \end{matrix}\right] t^\kappa f(t)dt. \tag{7.4.13}$$

The Mellin transform (2.5.1) of (7.4.12) for a "sufficiently good" function f is given by

$$\left(\mathfrak{M}\boldsymbol{D}_{\gamma;\sigma,\kappa}f\right)(s) = \mathcal{G}_{1,2}^{2,0}(s+\sigma)\left(\mathfrak{M}f\right)(1-s-\sigma+\kappa), \tag{7.4.14}$$

where $\mathcal{G}_{1,2}^{2,0}(s)$ is the function in (7.4.3).

For the exceptional set $\mathcal{E}_5$ of the function $\mathcal{G}_{1,2}^{2,0}(s)$ in (7.4.3), the condition $\nu - \mathrm{Re}(\kappa) \notin \mathcal{E}_{\mathcal{G}}$ means that

$$s \neq \frac{\gamma-1}{2} - i \ (i = 0, 1, 2, \cdots) \ \text{ for } \ \mathrm{Re}(s) = 1 - \nu + \mathrm{Re}(\kappa). \tag{7.4.15}$$

From Theorems 6.38, 6.42 and 6.44 we give the $\mathfrak{L}_{\nu,2}$- and $\mathfrak{L}_{\nu,r}$-theory of the transform (7.4.12).

Theorem 7.17. *Let* $\nu - \mathrm{Re}(\kappa) < 1$.

(i) *There is a one-to-one transform* $\boldsymbol{D}_{\gamma;\sigma,\kappa} \in [\mathfrak{L}_{\nu,2}, \mathfrak{L}_{1-\nu+\mathrm{Re}(\kappa-\sigma),2}]$ *such that* (7.4.14) *holds for* $\mathrm{Re}(s) = 1 - \nu + \mathrm{Re}(\kappa - \sigma)$ *and* $f \in \mathfrak{L}_{\nu,2}$.

(ii) *If* $f \in \mathfrak{L}_{\nu,2}$ *and* $g \in \mathfrak{L}_{\nu+\mathrm{Re}(\kappa-\sigma),2}$, *then the relation*

$$\int_0^\infty f(x)\left(\boldsymbol{D}_{\gamma;\sigma,\kappa}g\right)(x)dx = \int_0^\infty \left(\boldsymbol{D}_{\gamma;\kappa,\sigma}f\right)(x)g(x)dx \tag{7.4.16}$$

holds.

(iii) *Let* $f \in \mathfrak{L}_{\nu,2}$ *and* $\lambda \in \mathbb{C}$. *If* $\mathrm{Re}(\lambda) > -\nu + \mathrm{Re}(\kappa)$, *then* $\boldsymbol{D}_{\gamma;\sigma,\kappa}f$ *is given by*

$$\left(\boldsymbol{D}_{\gamma;\sigma,\kappa}f\right)(x) = x^{\sigma-\lambda}\frac{d}{dx}x^{\lambda+1}\int_0^\infty G_{2,3}^{2,1}\left[xt \,\middle|\, \begin{matrix} -\lambda, \frac{1-\gamma}{2} \\ 0, \frac{1}{2}, -\lambda-1 \end{matrix}\right] t^\kappa f(t)dt. \tag{7.4.17}$$

When $f \in \mathfrak{L}_{\nu,2}$ *and* $\mathrm{Re}(\lambda) < -\nu + \mathrm{Re}(\kappa)$,

$$\left(\boldsymbol{D}_{\gamma;\sigma,\kappa}f\right)(x) = -x^{\sigma-\lambda}\frac{d}{dx}x^{\lambda+1}\int_0^\infty G_{2,3}^{3,0}\left[xt \,\middle|\, \begin{matrix} \frac{1-\gamma}{2}, -\lambda \\ -\lambda-1, 0, \frac{1}{2} \end{matrix}\right] t^\kappa f(t)dt. \tag{7.4.18}$$

(iv) *The transform* $\boldsymbol{D}_{\gamma;\sigma,\kappa}$ *is independent of* ν *in the sense that if* $1 - \nu + \mathrm{Re}(\kappa) > 0$ *and* $1 - \widetilde{\nu} + \mathrm{Re}(\kappa) > 0$ *and if the transforms* $\boldsymbol{D}_{\gamma;\sigma,\kappa}$ *and* $\widetilde{\boldsymbol{D}}_{\gamma;\sigma,\kappa}$ *on* $\mathfrak{L}_{\nu,2}$ *and* $\mathfrak{L}_{\widetilde{\nu},2}$, *respectively, are given in* (7.4.14), *then* $\boldsymbol{D}_{\gamma;\sigma,\kappa}f = \widetilde{\boldsymbol{D}}_{\gamma;\sigma,\kappa}f$ *for* $f \in \mathfrak{L}_{\nu,2} \cap \mathfrak{L}_{\widetilde{\nu},2}$.

(v) *For* $f \in \mathfrak{L}_{\nu,2}$ *and* $x > 0$, $\left(\boldsymbol{D}_{\gamma;\sigma,\kappa}f\right)(x)$ *is given in* (7.4.12) *and* (7.4.13).

Theorem 7.18. *Let $\nu - \mathrm{Re}(\kappa) < 1$ and $1 \leqq r \leqq s \leqq \infty$.*

(a) *The transform $\boldsymbol{D}_{\gamma;\sigma,\kappa}$ defined on $\mathfrak{L}_{\nu,2}$ can be extended to $\mathfrak{L}_{\nu,r}$ as an element of $[\mathfrak{L}_{\nu,r}, \mathfrak{L}_{1-\nu+\mathrm{Re}(\kappa-\sigma),s}]$. If $1 \leqq r \leqq 2$, then $\boldsymbol{D}_{\gamma;\sigma,\kappa}$ is a one-to-one transform from $\mathfrak{L}_{\nu,r}$ onto $\mathfrak{L}_{1-\nu+\mathrm{Re}(\kappa-\sigma),s}$.*

(b) *If $f \in \mathfrak{L}_{\nu,r}$ and $g \in \mathfrak{L}_{\nu+\mathrm{Re}(\kappa-\sigma),s'}$ with $1/s + 1/s' = 1$, then the relation (7.4.16) holds.*

(c) *If $1 < r < \infty$ and the condition (7.4.15) holds, then $\boldsymbol{D}_{\gamma;\sigma,\kappa}$ is a one-to-one transform on $\mathfrak{L}_{\nu,r}$.*

(d) *If $1 < r < \infty$, $\mathrm{Re}(\gamma) \geqq 0$ and the condition (7.4.15) holds, then*

$$\boldsymbol{D}_{\gamma;\sigma,\kappa}(\mathfrak{L}_{\nu,r}) = \mathbb{L}_{1,-\sigma-\gamma/2}\left(\mathfrak{L}_{\nu-\mathrm{Re}(\kappa-\sigma),r}\right), \tag{7.4.19}$$

where the operator $\mathbb{L}_{1,-\sigma-\gamma/2}$ is given in (3.3.3). If (7.4.15) is not valid, then $\boldsymbol{D}_{\gamma;\sigma,\kappa}(\mathfrak{L}_{\nu,r})$ is a subset of the right-hand side of (7.4.19).

(e) *If $1 < r < \infty$, $\mathrm{Re}(\gamma) < 0$ and (7.4.15) holds, then*

$$\boldsymbol{D}_{\gamma;\sigma,\kappa}(\mathfrak{L}_{\nu,r}) = \left(I_{-;1,\sigma}^{-\gamma/2}\mathbb{L}_{1,-\sigma}\right)\left(\mathfrak{L}_{\nu-\mathrm{Re}(\kappa-\sigma),r}\right), \tag{7.4.20}$$

where the operators $I_{-;1,\sigma}^{-\gamma/2}$ and $\mathbb{L}_{1,-\sigma}$ are given in (3.3.2) and (3.3.3). If (7.4.15) is not valid, then $\boldsymbol{D}_{\gamma;\sigma,\kappa}(\mathfrak{L}_{\nu,r})$ is a subset of the right-hand side of (7.4.20).

7.5. ${}_1\boldsymbol{F}_1$-Transforms

We now proceed to consider the integral transform

$$\left({}_1\boldsymbol{F}_1{}^k f\right)(x) = \frac{\Gamma(a)}{\Gamma(c)}\int_0^\infty (xt)^k \, {}_1F_1(a;c;-xt)f(t)dt \tag{7.5.1}$$

with $a, c \in \mathbb{C}$ ($\mathrm{Re}(a) > 0$) and $k \in \mathbb{R}$ containing the confluent hypergeometric or Kummer function ${}_1F_1(a;c;z)$ in the kernel (see (2.9.3) and (2.9.14)). The transform (7.5.1), first considered by Erdélyi [3], is called the ${}_1F_1$-transform. According to Prudnikov, Brychkov and Marichev [3, 8.4.45.1], (3.3.11) and (3.3.14), the Mellin transform (2.5.1) of (7.5.1) for a "sufficiently good" function f is given by

$$\left(\mathfrak{M}\,{}_1\boldsymbol{F}_1^k f\right)(s) = \mathcal{G}_{1,2}^{1,1}(s)\left(\mathfrak{M}f\right)(1-s), \qquad \mathcal{G}_{1,2}^{1,1}(s) = \frac{\Gamma(k+s)\Gamma(a-k-s)}{\Gamma(c-k-s)}, \tag{7.5.2}$$

provided that $0 < \mathrm{Re}(s+k) < \mathrm{Re}(a)$. This relation shows that (7.5.1) is the $\boldsymbol{G}$-transform (6.1.1) of the form

$$\left({}_1\boldsymbol{F}_1{}^k f\right)(x) = \int_0^\infty G_{1,2}^{1,1}\left[xt \,\middle|\, \begin{matrix} 1-a+k \\ k, 1-c+k \end{matrix}\right] f(t)dt. \tag{7.5.3}$$

The constants $a^*, \Delta, a_1^*, a_2^*, \alpha, \beta$ and μ in (6.1.5)–(6.1.11) take the forms

$$a^* = \Delta = a_1^* = 1, \; a_2^* = 0, \; \alpha = -k, \; \beta = \mathrm{Re}(a-k), \; \mu = a - c + k - \frac{1}{2}. \tag{7.5.4}$$

Let $\mathcal{E}_6$ be the exceptional set of the function $\mathcal{G}_{1,2}^{1,1}(s)$ in (7.5.2). It is composed of the poles of the gamma function $\Gamma(c-k-s)$, so Definition 6.1 implies the condition $\nu \notin \mathcal{E}_6$ means that

$$s \neq c - k + i \quad (i = 0, 1, 2, \cdots) \quad \text{for} \quad \mathrm{Re}(s) = 1 - \nu. \tag{7.5.5}$$

From Theorems 6.1, 6.2 and 6.6, 6.8 we obtain the $\mathfrak{L}_{\nu,2}$- and $\mathfrak{L}_{\nu,r}$-theory of the transform (7.5.1).

Theorem 7.19. *Let $-\mathrm{Re}(k) < 1-\nu < \mathrm{Re}(a-k)$.*

(i) *There is a one-to-one transform ${}_1\boldsymbol{F}_1{}^k \in [\mathfrak{L}_{\nu,2}, \mathfrak{L}_{1-\nu,2}]$ such that (7.5.2) holds for $\mathrm{Re}(s) = 1-\nu$ and $f \in \mathfrak{L}_{\nu,2}$.*

(ii) *If $f, g \in \mathfrak{L}_{\nu,2}$, then the relation*

$$\int_0^\infty f(x)\left({}_1\boldsymbol{F}_1{}^k g\right)(x)dx = \int_0^\infty \left({}_1\boldsymbol{F}_1{}^k f\right)(x)g(x)dx \tag{7.5.6}$$

holds.

(iii) *Let $f \in \mathfrak{L}_{\nu,2}$ and $\lambda \in \mathbb{C}$. If $\mathrm{Re}(\lambda) > -\nu$, then ${}_1\boldsymbol{F}_1{}^k f$ is given by*

$$\left({}_1\boldsymbol{F}_1{}^k f\right)(x) = x^{-\lambda}\frac{d}{dx}x^{\lambda+1}\int_0^\infty G_{2,3}^{1,2}\left[xt \,\middle|\, \begin{matrix} -\lambda, 1-a+k \\ k, 1-c+k, -\lambda-1 \end{matrix}\right] f(t)dt. \tag{7.5.7}$$

When $\mathrm{Re}(\lambda) < -\nu$,

$$\left({}_1\boldsymbol{F}_1{}^k f\right)(x) = -x^{-\lambda}\frac{d}{dx}x^{\lambda+1}\int_0^\infty G_{2,3}^{2,1}\left[xt \,\middle|\, \begin{matrix} 1-a+k, -\lambda \\ -\lambda-1, k, 1-c+k \end{matrix}\right] f(t)dt. \tag{7.5.8}$$

(iv) *The transform ${}_1\boldsymbol{F}_1{}^k$ is independent of ν in the sense that if $-\mathrm{Re}(k) < 1-\nu < \mathrm{Re}(a-k)$ and $-\mathrm{Re}(k) < 1-\tilde{\nu} < \mathrm{Re}(a-k)$ and if the transforms ${}_1\boldsymbol{F}_1{}^k$ and ${}_1\widetilde{\boldsymbol{F}_1}{}^k$ on $\mathfrak{L}_{\nu,2}$ and $\mathfrak{L}_{\tilde{\nu},2}$, respectively, are given in (7.5.2), then ${}_1\boldsymbol{F}_1{}^k f = {}_1\widetilde{\boldsymbol{F}_1}{}^k f$ for $f \in \mathfrak{L}_{\nu,2} \cap \mathfrak{L}_{\tilde{\nu},2}$.*

(v) *For $f \in \mathfrak{L}_{\nu,2}$ and $x > 0$, $\left({}_1\boldsymbol{F}_1{}^k f\right)(x)$ is given in (7.5.1) and (7.5.3).*

Theorem 7.20. *Let $-\mathrm{Re}(k) < 1-\nu < \mathrm{Re}(a-k)$ and $1 \leqq r \leqq s \leqq \infty$.*

(a) *The transform ${}_1\boldsymbol{F}_1{}^k$ defined on $\mathfrak{L}_{\nu,2}$ can be extended to $\mathfrak{L}_{\nu,r}$ as an element of $[\mathfrak{L}_{\nu,r}, \mathfrak{L}_{1-\nu,s}]$. When $1 \leqq r \leqq 2$, ${}_1\boldsymbol{F}_1{}^k$ is a one-to-one transform from $\mathfrak{L}_{\nu,r}$ onto $\mathfrak{L}_{1-\nu,s}$.*

(b) *If $f \in \mathfrak{L}_{\nu,r}$ and $g \in \mathfrak{L}_{\nu,s'}$ with $1/s + 1/s' = 1$, then the relation (7.5.6) holds.*

(c) *If $1 < r < \infty$ and the condition (7.5.5) holds, then ${}_1\boldsymbol{F}_1{}^k$ is a one-to-one transform on $\mathfrak{L}_{\nu,r}$.*

(d) *If $1 < r < \infty$, $\mathrm{Re}(a-c) \geqq 0$ and the condition (7.5.5) holds, then*

$${}_1\boldsymbol{F}_1{}^k(\mathfrak{L}_{\nu,r}) = \mathbb{L}_{1,c-a-k}(\mathfrak{L}_{\nu,r}), \tag{7.5.9}$$

where the operator $\mathbb{L}_{1,c-a-k}$ is given in (3.3.3). When (7.5.5) is not valid, ${}_1\boldsymbol{F}_1{}^k(\mathfrak{L}_{\nu,r})$ is a subset of the right-hand side of (7.5.9).

(e) *If $1 < r < \infty$, $\mathrm{Re}(a-c) < 0$ and (7.5.5) holds, then*

$${}_1\boldsymbol{F}_1{}^k(\mathfrak{L}_{\nu,r}) = \left(I_{-;1,k}^{c-a}\mathbb{L}_{1,-k}\right)(\mathfrak{L}_{\nu,r}), \tag{7.5.10}$$

where the operators $I_{-;1,k}^{c-a}$ and $\mathbb{L}_{1,-k}$ are given in (3.3.2) and (3.3.3). When (7.5.5) is not valid, ${}_1\boldsymbol{F}_1{}^k(\mathfrak{L}_{\nu,r})$ is a subset of the right-hand side of (7.5.10).

Now we treat a generalization of (7.5.1) in the form

$$\left({}_1\boldsymbol{F}_1{}^{\sigma,\kappa} f\right)(x) = \frac{\Gamma(a)}{\Gamma(c)} x^\sigma \int_0^\infty {}_1F_1(a;c;-xt)t^\kappa f(t)dt \tag{7.5.11}$$

with $a, c, \sigma, \kappa \in \mathbb{C}$ ($\text{Re}(a) > 0; c \neq 0, -1, -2, \cdots$). This is a modified transform (6.2.3) of the form

$$\left({}_1\boldsymbol{F}_1{}^{\sigma,\kappa} f\right)(x) = x^\sigma \int_0^\infty G_{1,2}^{1,1}\left[xt \,\middle|\, \begin{matrix} 1-a \\ 0, 1-c \end{matrix}\right] t^\kappa f(t)dt. \tag{7.5.12}$$

The Mellin transform (2.5.1) of (7.5.11) for a "sufficiently good" function f is given by

$$\left(\mathfrak{M}\,{}_1\boldsymbol{F}_1{}^{\sigma,\kappa} f\right)(s) = \mathcal{G}_{1,2}^{1,1}(s+\sigma)\left(\mathfrak{M}f\right)(1-s-\sigma+\kappa), \quad \mathcal{G}_{1,2}^{1,1}(s) = \frac{\Gamma(s)\Gamma(a-s)}{\Gamma(c-s)}, \tag{7.5.13}$$

provided that $0 < \text{Re}(s+\sigma) < \text{Re}(a)$. The constants $a^*, \Delta, a_1^*, a_2^*, \alpha, \beta$ and μ in (6.1.5)–(6.1.11) take the forms

$$a^* = \Delta = a_1^* = 1,\ a_2^* = 0,\ \alpha = 0,\ \beta = \text{Re}(a),\ \mu = a - c - \frac{1}{2}. \tag{7.5.14}$$

Let $\mathcal{E}_7$ be the exceptional set of the function $\mathcal{G}_{1,2}^{1,1}(s)$ in (7.5.12). Then similarly to (7.5.5) the condition $\nu - \text{Re}(\kappa) \notin \mathcal{E}_7$ means that

$$s \neq c + i \quad (i = 0, 1, 2, \cdots) \quad \text{for} \quad \text{Re}(s) = 1 - \nu + \text{Re}(\kappa). \tag{7.5.15}$$

Thus Theorems 6.38, 6.42 and 6.44 lead to the $\mathfrak{L}_{\nu,2}$- and $\mathfrak{L}_{\nu,r}$-theory of the transform (7.5.11).

Theorem 7.21. *Let* $0 < 1 - \nu + \text{Re}(\kappa) < \text{Re}(a)$.

(i) *There is a one-to-one transform* ${}_1\boldsymbol{F}_1{}^{\sigma,\kappa} \in [\mathfrak{L}_{\nu,2}, \mathfrak{L}_{1-\nu+\text{Re}(\kappa-\sigma),2}]$ *such that* (7.5.13) *holds for* $\text{Re}(s) = 1 - \nu + \text{Re}(\kappa - \sigma)$ *and* $f \in \mathfrak{L}_{\nu,2}$.

(ii) *If* $f \in \mathfrak{L}_{\nu,2}$ *and* $g \in \mathfrak{L}_{\nu+\text{Re}(\kappa-\sigma),2}$, *then the relation*

$$\int_0^\infty f(x)\left({}_1\boldsymbol{F}_1{}^{\sigma,\kappa} g\right)(x)dx = \int_0^\infty \left({}_1\boldsymbol{F}_1{}^{\kappa,\sigma} f\right)(x)g(x)dx \tag{7.5.16}$$

holds.

(iii) *Let* $f \in \mathfrak{L}_{\nu,2}$ *and* $\lambda \in \mathbb{C}$. *If* $\text{Re}(\lambda) > -\nu + \text{Re}(\kappa)$, *then* ${}_1\boldsymbol{F}_1{}^{\sigma,\kappa} f$ *is given by*

$$\left({}_1\boldsymbol{F}_1{}^{\sigma,\kappa} f\right)(x) = x^{\sigma-\lambda}\frac{d}{dx}x^{\lambda+1}\int_0^\infty G_{2,3}^{1,2}\left[xt \,\middle|\, \begin{matrix} -\lambda, 1-a \\ 0, 1-c, -\lambda-1 \end{matrix}\right] t^\kappa f(t)dt. \tag{7.5.17}$$

When $\text{Re}(\lambda) < -\nu + \text{Re}(\kappa)$,

$$\left({}_1\boldsymbol{F}_1{}^{\sigma,\kappa} f\right)(x) = -x^{\sigma-\lambda}\frac{d}{dx}x^{\lambda+1}\int_0^\infty G_{2,3}^{2,1}\left[xt \,\middle|\, \begin{matrix} 1-a, -\lambda \\ -\lambda-1, 0, 1-c \end{matrix}\right] t^\kappa f(t)dt. \tag{7.5.18}$$

(iv) *The transform* ${}_1\boldsymbol{F}_1{}^{\sigma,\kappa}$ *is independent of* ν *in the sense that if* $0 < 1-\nu+\text{Re}(\kappa) < \text{Re}(a)$ *and* $0 < 1 - \tilde{\nu} + \text{Re}(\kappa) < \text{Re}(a)$ *and if the transforms* ${}_1\boldsymbol{F}_1{}^{\sigma,\kappa}$ *and* ${}_1\widetilde{\boldsymbol{F}}_1{}^{\sigma,\kappa}$ *on* $\mathfrak{L}_{\nu,2}$ *and* $\mathfrak{L}_{\tilde{\nu},2}$, *respectively, are given in* (7.5.13), *then* ${}_1\boldsymbol{F}_1{}^{\sigma,\kappa} f = {}_1\widetilde{\boldsymbol{F}}_1{}^{\sigma,\kappa} f$ *for* $f \in \mathfrak{L}_{\nu,2} \cap \mathfrak{L}_{\tilde{\nu},2}$.

(v) For $f \in \mathfrak{L}_{\nu,2}$ and $x > 0$, $\left({}_1F_1^{\sigma,\kappa} f\right)(x)$ is given in (7.5.11) and (7.5.12).

Let $\sigma = \kappa = k \in \mathbb{C}$. We have

Corollary 7.21.1. *Let $0 < 1 - \nu + \mathrm{Re}(k) < \mathrm{Re}(a)$.*
(iii′) *Let $f \in \mathfrak{L}_{\nu,2}$ and $\lambda \in \mathbb{C}$. If $\mathrm{Re}(\lambda) > -\nu + \mathrm{Re}(k)$, then ${}_1F_1^{k} f$ is given by*

$$\left({}_1F_1^{k} f\right)(x) = x^{k-\lambda}\frac{d}{dx}x^{\lambda+1}\int_0^{\infty} G_{2,3}^{1,2}\left[xt \,\middle|\, \begin{matrix} -\lambda, 1-a \\ 0, 1-c, -\lambda-1 \end{matrix}\right] t^k f(t)dt. \tag{7.5.19}$$

When $\mathrm{Re}(\lambda) < -\nu + \mathrm{Re}(k)$,

$$\left({}_1F_1^{k} f\right)(x) = -x^{k-\lambda}\frac{d}{dx}x^{\lambda+1}\int_0^{\infty} G_{2,3}^{2,1}\left[xt \,\middle|\, \begin{matrix} 1-a, -\lambda \\ -\lambda-1, 0, 1-c \end{matrix}\right] t^k f(t)dt. \tag{7.5.20}$$

Theorem 7.22. *Let $0 < 1 - \nu + \mathrm{Re}(\kappa) < \mathrm{Re}(a)$ and $1 \leqq r \leqq s \leqq \infty$.*

(a) *The transform ${}_1F_1^{\sigma,\kappa}$ defined on $\mathfrak{L}_{\nu,2}$ can be extended to $\mathfrak{L}_{\nu,r}$ as an element of $[\mathfrak{L}_{\nu,r}, \mathfrak{L}_{1-\nu+\mathrm{Re}(\kappa-\sigma),s}]$. If $1 \leqq r \leqq 2$, then ${}_1F_1^{\sigma,\kappa}$ is a one-to-one transform from $\mathfrak{L}_{\nu,r}$ onto $\mathfrak{L}_{1-\nu+\mathrm{Re}(\kappa-\sigma),s}$.*

(b) *If $f \in \mathfrak{L}_{\nu,r}$ and $g \in \mathfrak{L}_{\nu+\mathrm{Re}(\kappa-\sigma),s'}$ with $1/s + 1/s' = 1$, then the relation (7.5.16) holds.*

(c) *If $1 < r < \infty$ and the condition (7.5.15) holds, then ${}_1F_1^{\sigma,\kappa}$ is a one-to-one transform on $\mathfrak{L}_{\nu,r}$.*

(d) *If $1 < r < \infty$, $\mathrm{Re}(a - c) \geqq 0$ and the condition (7.5.15) holds, then*

$${}_1F_1^{\sigma,\kappa}(\mathfrak{L}_{\nu,r}) = \mathbb{L}_{1,c-a-\sigma}\left(\mathfrak{L}_{\nu-\mathrm{Re}(\kappa-\sigma),r}\right), \tag{7.5.21}$$

where the operator $\mathbb{L}_{1,c-a-\sigma}$ is given in (3.3.3). When (7.5.15) is not valid, ${}_1F_1^{\sigma,\kappa}(\mathfrak{L}_{\nu,r})$ is a subset of the right-hand side of (7.5.21).

(e) *If $1 < r < \infty$, $\mathrm{Re}(a - c) < 0$ and (7.5.15) holds, then*

$${}_1F_1^{\sigma,\kappa}(\mathfrak{L}_{\nu,r}) = \left(I_{-;1,\sigma}^{c-a}\mathbb{L}_{1,-\sigma}\right)\left(\mathfrak{L}_{\nu-\mathrm{Re}(\kappa-\sigma),r}\right), \tag{7.5.22}$$

where the operators $I_{-;1,k}^{c-a}$ and $\mathbb{L}$ are given in (3.3.2) and (2.5.2). When (7.5.15) is not valid, ${}_1F_1^{\sigma,\kappa}(\mathfrak{L}_{\nu,r})$ is a subset of the right-hand side of (7.5.22).

Corollary 7.22.1. *Let $0 < 1 - \nu + \mathrm{Re}(k) < \mathrm{Re}(a)$.*
(d′) *If $1 < r < \infty$, $\mathrm{Re}(a - c) \geqq 0$ and the condition (7.5.5) holds, then*

$${}_1F_1^{k}(\mathfrak{L}_{\nu,r}) = \mathbb{L}_{1,c-a-k}(\mathfrak{L}_{\nu,r}), \tag{7.5.23}$$

where the operator $\mathbb{L}_{1,c-a-k}$ is given in (3.3.3). When (7.5.5) is not valid, ${}_1F_1^{k}(\mathfrak{L}_{\nu,r})$ is a subset of the right-hand side of (7.5.23).

(e′) *If $1 < r < \infty$, $\mathrm{Re}(a - c) < 0$ and (7.5.5) holds, then*

$${}_1F_1^{k}(\mathfrak{L}_{\nu,r}) = \left(I_{-;1,k}^{c-a}\mathbb{L}_{1,-k}\right)(\mathfrak{L}_{\nu,r}), \tag{7.5.24}$$

where the operators $I_{-;1,k}^{c-a}$ and $\mathbb{L}_{1,-k}$ are given in (3.3.2) *and* (3.3.3). *When* (7.5.5) *is not valid,* ${}_1\boldsymbol{F}_1{}^k(\mathfrak{L}_{\nu,r})$ *is a subset of the right-hand side of* (7.5.24).

Remark 7.4. When $\sigma = \kappa = k$, the transform ${}_1\boldsymbol{F}_1{}^k$ in (7.5.13) coincides with the transform ${}_1\boldsymbol{F}_1{}^k$ in (7.5.1). From Theorems 7.21 and 7.22 we obtain the statements (i), (ii), (iv), (v) of Theorem 7.19 and the statements (a), (b), (c) of Theorem 7.20, respectively. As for the statements given in Theorem 7.19(iii) and Theorem 7.20(d)–(e), characterizing the representation and the range of the ${}_1\boldsymbol{F}_1{}^k$-transform (7.5.1), Corollaries 7.21.1 and 7.22.1 show that the latter in (7.5.23) and (7.5.24) are the same as in (7.5.9) and (7.5.10), while the former are changed. This is caused by the fact that these representations are valid for different conditions: the representations in (7.5.19) and (7.5.20) hold for $\text{Re}(\lambda) > -\nu + \text{Re}(k)$ and $\text{Re}(\lambda) < -\nu + \text{Re}(k)$, while those in (7.5.7) and (7.5.8) hold for $\text{Re}(\lambda) > -\nu$ and $\text{Re}(\lambda) < -\nu$, respectively.

7.6. ${}_1\boldsymbol{F}_2$-Transforms

We consider the integral transform

$$\left({}_1\boldsymbol{F}_2 f\right)(x) = \frac{\Gamma(a)}{\Gamma(c)\Gamma(d)} \int_0^\infty {}_1F_2(a; c, d; -xt) f(t) dt \tag{7.6.1}$$

with $a, c, d \in \mathbb{C}$ $(\text{Re}(a) > 0)$ containing the hypergeometric function ${}_1F_2(a; c, d; z)$ in the kernel. According to Prudnikov, Brychkov and Marichev [3, 8.4.48.1] the Mellin transform (2.5.1) of (7.6.1) for a "sufficiently good" function f is given by

$$\left(\mathfrak{M}_1 \boldsymbol{F}_2 f\right)(s) = \mathcal{G}_{1,3}^{1,1}(s)\left(\mathfrak{M}f\right)(1-s), \qquad \mathcal{G}_{1,3}^{1,1}(s) = \frac{\Gamma(s)\Gamma(a-s)}{\Gamma(c-s)\Gamma(d-s)}, \tag{7.6.2}$$

provided that $0 < \text{Re}(s) < \min[\text{Re}(a), 1/4 + \text{Re}(\{c + d - a\}/2)]$. This relation shows that (7.6.1) is the $\boldsymbol{G}$-transform (6.1.1) of the form

$$\left({}_1\boldsymbol{F}_2 f\right)(x) = \int_0^\infty G_{1,3}^{1,1}\left[xt \,\middle|\, \begin{matrix} 1-a \\ 0, 1-c, 1-d \end{matrix}\right] f(t) dt. \tag{7.6.3}$$

The constants $a^*, \Delta, a_1^*, a_2^*, \alpha, \beta$ and μ in (6.1.5)–(6.1.11) take the forms

$$a^* = 0, \ \Delta = 2, \ a_1^* = 1, \ a_2^* = -1, \ \alpha = 0, \ \beta = \text{Re}(a), \ \mu = a - c - d. \tag{7.6.4}$$

Let $\mathcal{E}_8$ be the exceptional set of the function $\mathcal{G}_{1,3}^{1,1}(s)$ in (7.6.2). Then the condition $\nu \notin \mathcal{E}_8$ means that

$$s \neq c + i, \ s \neq d + j \quad (i, j = 0, 1, 2, \cdots) \quad \text{for} \quad \text{Re}(s) = 1 - \nu. \tag{7.6.5}$$

Thus from Theorems 6.1, 6.2 and 6.5 we have the $\mathfrak{L}_{\nu,2}$- and $\mathfrak{L}_{\nu,r}$-theory of the transform (7.6.1).

Theorem 7.23. *Let* $0 < 1 - \nu < \text{Re}(a)$ *and* $2(1-\nu) + \text{Re}(a - c - d) \leqq 0$.

(i) *There is a one-to-one transform* ${}_1\boldsymbol{F}_2 \in [\mathfrak{L}_{\nu,2}, \mathfrak{L}_{1-\nu,2}]$ *such that* (7.6.2) *holds for* $\mathrm{Re}(s) = 1-\nu$ *and* $f \in \mathfrak{L}_{\nu,2}$. *If* $2(1-\nu)+\mathrm{Re}(a-c-d)=0$ *and if the condition* (7.6.5) *holds, then the transform* ${}_1\boldsymbol{F}_2$ *maps* $\mathfrak{L}_{\nu,2}$ *onto* $\mathfrak{L}_{1-\nu,2}$.

(ii) *If* $f, g \in \mathfrak{L}_{\nu,2}$, *then the relation*

$$\int_0^\infty f(x)\left({}_1\boldsymbol{F}_2\, g\right)(x)dx = \int_0^\infty \left({}_1\boldsymbol{F}_2\, f\right)(x)g(x)dx \tag{7.6.6}$$

holds for ${}_1\boldsymbol{F}_2$.

(iii) *Let* $f \in \mathfrak{L}_{\nu,2}$ *and* $\lambda \in \mathbb{C}$. *If* $\mathrm{Re}(\lambda) > -\nu$, *then* ${}_1\boldsymbol{F}_2\, f$ *is given by*

$$\left({}_1\boldsymbol{F}_2\, f\right)(x) = x^{-\lambda}\frac{d}{dx}x^{\lambda+1}\int_0^\infty G_{2,4}^{1,2}\left[xt \,\middle|\, \begin{matrix} -\lambda, 1-a \\ 0, 1-c, 1-d, -\lambda-1 \end{matrix}\right] f(t)dt. \tag{7.6.7}$$

When $\mathrm{Re}(\lambda) < -\nu$,

$$\left({}_1\boldsymbol{F}_2\, f\right)(x) = -x^{-\lambda}\frac{d}{dx}x^{\lambda+1}\int_0^\infty G_{2,4}^{2,1}\left[xt \,\middle|\, \begin{matrix} 1-a, -\lambda \\ -\lambda-1, 0, 1-c, 1-d \end{matrix}\right] f(t)dt. \tag{7.6.8}$$

(iv) *The transform* ${}_1\boldsymbol{F}_2$ *is independent of* ν *in the sense that if* $0 < 1-\nu < \mathrm{Re}(a)$, $2(1-\nu)+\mathrm{Re}(a-c-d) \leqq 0$, $0 < 1-\widetilde{\nu} < \mathrm{Re}(a)$ *and* $2(1-\widetilde{\nu})+\mathrm{Re}(a-c-d) \leqq 0$, *and if the transforms* ${}_1\boldsymbol{F}_2$ *and* ${}_1\widetilde{\boldsymbol{F}}_2$ *on* $\mathfrak{L}_{\nu,2}$ *and* $\mathfrak{L}_{\widetilde{\nu},2}$, *respectively, are given in* (7.6.2), *then* ${}_1\boldsymbol{F}_2\, f = {}_1\widetilde{\boldsymbol{F}}_2\, f$ *for* $f \in \mathfrak{L}_{\nu,2} \cap \mathfrak{L}_{\widetilde{\nu},2}$.

(v) *If* $2(1-\nu)+\mathrm{Re}(a-c-d) < 0$, *then for* $x > 0$ *and* $f \in \mathfrak{L}_{\nu,2}$, $\left({}_1\boldsymbol{F}_2\, f\right)(x)$ *is given in* (7.6.1) *and* (7.6.3).

Theorem 7.24. *Let* $0 < 1-\nu < \mathrm{Re}(a), 1 < r < \infty$ *and* $2(1-\nu)+\mathrm{Re}(a-c-d) \leqq 1/2-\gamma(r)$, *where* $\gamma(r)$ *is defined in* (3.3.9).

(a) *The transform* ${}_1\boldsymbol{F}_2$ *defined on* $\mathfrak{L}_{\nu,2}$ *can be extended to* $\mathfrak{L}_{\nu,r}$ *as an element of* $[\mathfrak{L}_{\nu,r}, \mathfrak{L}_{1-\nu,s}]$ *for all* s *with* $r \leqq s < \infty$ *such that* $s' \geqq [1/2-2(1-\nu)-\mathrm{Re}(a-c-d)]^{-1}$ *with* $1/s+1/s'=1$.

(b) *If* $1 < r \leqq 2$, *the transform* ${}_1\boldsymbol{F}_2$ *is one-to-one on* $\mathfrak{L}_{\nu,r}$ *and there holds the equality* (7.6.2) *for* $f \in \mathfrak{L}_{\nu,r}$ *and* $\mathrm{Re}(s) = 1-\nu$.

(c) *If* $f \in \mathfrak{L}_{\nu,r}$ *and* $g \in \mathfrak{L}_{\nu,s}$ *with* $1 < s < \infty, 1/r+1/s \geqq 1$ *and* $2(1-\nu)+\mathrm{Re}(a-c-d) \leqq 1/2-\max[\gamma(r),\gamma(s)]$, *then the relation* (7.6.6) *holds.*

(d) *If the conditions in* (7.6.5) *hold, the transform* ${}_1\boldsymbol{F}_2$ *is one-to-one on* $\mathfrak{L}_{\nu,r}$ *and there holds the relation*

$${}_1\boldsymbol{F}_2(\mathfrak{L}_{\nu,r}) = \left(M_{(a-c-d+1)/2}H_{2,c+d-a-1}\right)\left(\mathfrak{L}_{\nu-\mathrm{Re}(a-c-d)/2-1/2,r}\right), \tag{7.6.9}$$

where M_ζ *and* $H_{2,\eta}$ *are given in* (3.3.11) *and* (3.3.4). *When* (7.6.5) *is not valid, then* ${}_1\boldsymbol{F}_2(\mathfrak{L}_{\nu,r})$ *is a subset of the right-hand side of* (7.6.9).

(e) *If* $f \in \mathfrak{L}_{\nu,r}, \lambda \in \mathbb{C}$ *and* $2(1-\nu)+\mathrm{Re}(a-c-d) \leqq 1/2-\gamma(r)$, *then* ${}_1\boldsymbol{F}_2\, f$ *is given in* (7.6.7) *for* $\mathrm{Re}(\lambda) > -\nu$, *while in* (7.6.8) *for* $\mathrm{Re}(\lambda) < -\nu$. *If* $2(1-\nu)+\mathrm{Re}(a-c-d) < 0$, ${}_1\boldsymbol{F}_2\, f$ *is given in* (7.6.1) *and* (7.6.3) *for* $f \in \mathfrak{L}_{\nu,r}$.

Theorems 6.12 and 6.13 give the inversion formulas for the transform (7.6.1).

Theorem 7.25. *Let $0 < 1-\nu < \mathrm{Re}(a)$ and $\alpha_0 = \max[1-\mathrm{Re}(c), 1-\mathrm{Re}(d)]$.*

(a) *If $\nu > \alpha_0$, $2(1-\nu)+\mathrm{Re}(a-c-d) = 0$ and $f \in \mathfrak{L}_{\nu,2}$, then the inversion formula*

$$f(x) = x^{-\lambda}\frac{d}{dx}x^{\lambda+1}\int_0^\infty G_{2,4}^{2,1}\left[xt \,\middle|\, \begin{matrix} -\lambda, a-1 \\ c-1, d-1, 0, -\lambda-1 \end{matrix}\right]\left({}_1\boldsymbol{F}_2 f\right)(t)dt \tag{7.6.10}$$

holds for $\mathrm{Re}(\lambda) > \nu - 1$ and

$$f(x) = -x^{-\lambda}\frac{d}{dx}x^{\lambda+1}\int_0^\infty G_{2,4}^{3,0}\left[xt \,\middle|\, \begin{matrix} a-1, -\lambda \\ -\lambda-1, c-1, d-1, 0 \end{matrix}\right]\left({}_1\boldsymbol{F}_2 f\right)(t)dt \tag{7.6.11}$$

for $\mathrm{Re}(\lambda) < \nu - 1$.

(b) *Let $1 < r < \infty, \alpha_0 < \nu < \mathrm{Re}(a-c-d)/2+5/4$ and $2(1-\nu)+\mathrm{Re}(a-c-d) \leqq 1/2-\gamma(r)$, where $\gamma(r)$ is given in (3.3.9). If $f \in \mathfrak{L}_{\nu,r}$, then the inversion formulas (7.6.10) and (7.6.11) hold for $\mathrm{Re}(\lambda) > \nu-1$ and for $\mathrm{Re}(\lambda) < \nu-1$, respectively.*

Now we consider a modification of (7.6.1) in the form

$$\left({}_1\boldsymbol{F}_2{}^{\sigma,\kappa} f\right)(x) = \frac{\Gamma(a)}{\Gamma(c)\Gamma(d)}x^\sigma \int_0^\infty {}_1F_2(a;c,d;-xt)t^\kappa f(t)dt \tag{7.6.12}$$

with $a, c, d, \sigma, \kappa \in \mathbb{C}$ ($\mathrm{Re}(a) > 0$). From (7.6.2) and (3.3.14) the Mellin transform of (7.6.12) for a "sufficiently good" function f is given by

$$\left(\mathfrak{M}\,{}_1\boldsymbol{F}_2{}^{\sigma,\kappa} f\right)(s) = \mathcal{G}_{1,3}^{1,1}(s+\sigma)\left(\mathfrak{M}f\right)(1-s-\sigma+\kappa) \tag{7.6.13}$$

for $0 < \mathrm{Re}(s+\sigma) < \mathrm{Re}(a)$, $\mathrm{Re}(s+\sigma) < 1/4+\mathrm{Re}(c+d-a)/2$, $c, d \neq 0, -1, -2, \cdots$ and $\mathcal{G}_{1,3}^{1,1}(s)$ in (7.6.2). Hence the transform (7.6.12) is a modified transform (6.2.3) of the form

$$\left({}_1\boldsymbol{F}_2{}^{\sigma,\kappa} f\right)(x) = x^\sigma \int_0^\infty G_{1,3}^{1,1}\left[xt \,\middle|\, \begin{matrix} 1-a \\ 0, 1-c, 1-d \end{matrix}\right] t^\kappa f(t)dt. \tag{7.6.14}$$

If $\mathcal{E}_8$ is the exceptional set of the function $\mathcal{G}_{1,3}^{1,1}(s)$ in (7.6.2), then the condition $\nu \notin \mathcal{E}_8$ means that

$$s \neq c+i,\ s \neq d+j \quad (i,j = 0,1,2,\cdots) \quad \text{for} \quad \mathrm{Re}(s) = 1-\nu+\mathrm{Re}(\kappa). \tag{7.6.15}$$

From Theorems 6.38 and 6.41 we obtain the $\mathfrak{L}_{\nu,2}$- and $\mathfrak{L}_{\nu,r}$-theory of the ${}_1\boldsymbol{F}_2{}^{\sigma,\kappa}$-transform (7.6.12).

Theorem 7.26. *Let $0 < 1-\nu+\mathrm{Re}(\kappa) < \mathrm{Re}(a)$ and $2[1-\nu+\mathrm{Re}(\kappa)]+\mathrm{Re}(a-c-d) \leqq 0$.*

(i) *There is a one-to-one transform ${}_1\boldsymbol{F}_2{}^{\sigma,\kappa} \in [\mathfrak{L}_{\nu,2}, \mathfrak{L}_{1-\nu,2}]$ such that (7.6.13) holds for $\mathrm{Re}(s) = 1-\nu+\mathrm{Re}(\kappa-\sigma)$ and $f \in \mathfrak{L}_{\nu,2}$. If $2[1-\nu+\mathrm{Re}(\kappa)]+\mathrm{Re}(a-c-d) = 0$ and if the condition (7.6.15) holds, then the transform ${}_1\boldsymbol{F}_2{}^{\sigma,\kappa}$ maps $\mathfrak{L}_{\nu,2}$ onto $\mathfrak{L}_{1-\nu+\mathrm{Re}(\kappa-\sigma),2}$.*

(ii) *If $f \in \mathfrak{L}_{\nu,2}$ and $g \in \mathfrak{L}_{\nu+\mathrm{Re}(\kappa-\sigma),2}$, then the relation*

$$\int_0^\infty f(x)\left({}_1\boldsymbol{F}_2{}^{\sigma,\kappa} g\right)(x)dx = \int_0^\infty g(x)\left({}_1\boldsymbol{F}_2{}^{\kappa,\sigma} f\right)(x)dx \tag{7.6.16}$$

holds.

(iii) *Let $\lambda \in \mathbb{C}$ and $f \in \mathfrak{L}_{\nu,2}$. If $\mathrm{Re}(\lambda) > -\nu + \mathrm{Re}(\kappa)$, then ${}_1\boldsymbol{F}_2^{\sigma,\kappa} f$ is given by*

$$\left({}_1\boldsymbol{F}_2^{\sigma,\kappa} f\right)(x)$$
$$= x^{\sigma-\lambda}\frac{d}{dx}x^{\lambda+1}\int_0^\infty G_{2,4}^{1,2}\left[xt \,\middle|\, \begin{matrix} -\lambda, 1-a \\ 0, 1-c, 1-d, -\lambda-1 \end{matrix}\right] t^{\kappa} f(t)dt. \tag{7.6.17}$$

When $\mathrm{Re}(\lambda) < -\nu + \mathrm{Re}(\kappa)$,

$$\left({}_1\boldsymbol{F}_2^{\sigma,\kappa} f\right)(x)$$
$$= -x^{\sigma-\lambda}\frac{d}{dx}x^{\lambda+1}\int_0^\infty G_{2,4}^{2,1}\left[xt \,\middle|\, \begin{matrix} 1-a, -\lambda \\ -\lambda-1, 0, 1-c, 1-d \end{matrix}\right] t^{\kappa} f(t)dt. \tag{7.6.18}$$

(iv) *The transform ${}_1\boldsymbol{F}_2^{\sigma,\kappa}$ is independent of ν in the sense that if $0 < 1 - \nu + \mathrm{Re}(\kappa) < \mathrm{Re}(a)$, $2[1-\nu+\mathrm{Re}(\kappa)]+\mathrm{Re}(a-c-d) \leqq 0$ and $0 < 1-\widetilde{\nu}+\mathrm{Re}(\kappa) < \mathrm{Re}(a)$ and $2[1-\widetilde{\nu}+\mathrm{Re}(\kappa)]+\mathrm{Re}(a-c-d) \leqq 0$. and if the transforms ${}_1\boldsymbol{F}_2^{\sigma,\kappa}$ and ${}_1\widetilde{\boldsymbol{F}}_2^{\sigma,\kappa}$ on $\mathfrak{L}_{\nu,2}$ and $\mathfrak{L}_{\widetilde{\nu},2}$, respectively, are given in (7.6.13), then ${}_1\boldsymbol{F}_2^{\sigma,\kappa} f = {}_1\widetilde{\boldsymbol{F}}_2^{\sigma,\kappa} f$ for $f \in \mathfrak{L}_{\nu,2} \cap \mathfrak{L}_{\widetilde{\nu},2}$.*

(v) *If $2[1-\nu+\mathrm{Re}(\kappa)]+\mathrm{Re}(a-c-d) < 0$, then for $f \in \mathfrak{L}_{\nu,2}$ and $x > 0$, $\left({}_1\boldsymbol{F}_2^{\sigma,\kappa} f\right)(x)$ is given in (7.6.12) and (7.6.14).*

Theorem 7.27. *Let $0 < 1 - \nu + \mathrm{Re}(\kappa) < \mathrm{Re}(a)$, $1 < r < \infty$ and $2[1-\nu+\mathrm{Re}(\kappa)] + \mathrm{Re}(a-c-d) \leqq 1/2 - \gamma(r)$, where $\gamma(r)$ is defined in (3.3.9).*

(a) *The transform ${}_1\boldsymbol{F}_2^{\sigma,\kappa}$ defined on $\mathfrak{L}_{\nu,2}$ can be extended to $\mathfrak{L}_{\nu,r}$ as an element of $[\mathfrak{L}_{\nu,r}, \mathfrak{L}_{1-\nu+\mathrm{Re}(\kappa-\sigma),s}]$ for all s with $r \leqq s < \infty$ such that $s' \geqq [1/2 - 2\{1-\nu+\mathrm{Re}(\kappa)\} - \mathrm{Re}(a-c-d)]^{-1}$ with $1/s + 1/s' = 1$.*

(b) *If $1 < r \leqq 2$, the transform ${}_1\boldsymbol{F}_2^{\sigma,\kappa}$ is one-to-one on $\mathfrak{L}_{\nu,r}$ and there holds the equality (7.6.13) for $f \in \mathfrak{L}_{\nu,r}$ and $\mathrm{Re}(s) = 1 - \nu + \mathrm{Re}(\kappa - \sigma)$.*

(c) *If $f \in \mathfrak{L}_{\nu,r}$ and $g \in \mathfrak{L}_{\nu+\mathrm{Re}(\kappa-\sigma),s}$ with $1 < s < \infty$, $1/r + 1/s \geqq 1$ and $2[1-\nu+\mathrm{Re}(\kappa)]+\mathrm{Re}(a-c-d) \leqq 1/2 - \max[\gamma(r), \gamma(s)]$, then the relation (7.6.16) holds.*

(d) *If the conditions in (7.6.15) hold, then the transform ${}_1\boldsymbol{F}_2^{\sigma,\kappa}$ is one-to-one on $\mathfrak{L}_{\nu,r}$ and there holds the relation*

$${}_1\boldsymbol{F}_2^{\sigma,\kappa}(\mathfrak{L}_{\nu,r}) = \left(M_{\sigma+(a-c-d+1)/2}\mathbb{H}_{2,c+d-a-1}\right)\left(\mathfrak{L}_{\nu-\mathrm{Re}(a-c-d+1)/2-\mathrm{Re}(\kappa),r}\right), \tag{7.6.19}$$

where M_ζ and $\mathbb{H}_{2,\eta}$ are given in (3.3.11) and (3.3.4). When (7.6.15) is not valid, then ${}_1\boldsymbol{F}_2^{\sigma,\kappa}(\mathfrak{L}_{\nu,r})$ is a subset of the right-hand side of (7.6.19).

(e) *If $f \in \mathfrak{L}_{\nu,r}$ with $1 < r < \infty$, $\lambda \in \mathbb{C}$ and $2[1-\nu+\mathrm{Re}(\kappa)]+\mathrm{Re}(a-c-d) \leqq 1/2 - \gamma(r)$, then ${}_1\boldsymbol{F}_2^{\sigma,\kappa} f$ is given in (7.6.17) for $\mathrm{Re}(\lambda) > -\nu + \mathrm{Re}(\kappa)$, while ${}_1\boldsymbol{F}_2^{\sigma,\kappa} f$ is given in (7.6.18) for $\mathrm{Re}(\lambda) < -\nu + \mathrm{Re}(\kappa)$. If $2[1-\nu+\mathrm{Re}(\kappa)]+\mathrm{Re}(a-c-d) < 0$, ${}_1\boldsymbol{F}_2^{\sigma,\kappa} f$ is given in (7.6.12).*

From Theorems 6.48 and 6.49 we have the inversion formulas for the transform (7.6.12).

Theorem 7.28. *Let $0 < 1 - \nu + \mathrm{Re}(\kappa) < \mathrm{Re}(a)$ and $\alpha_0 = \max[1 - \mathrm{Re}(c), 1 - \mathrm{Re}(d)]$.*

(a) *If* $\nu > \alpha_0$, $2[1-\nu+\mathrm{Re}(\kappa)] + \mathrm{Re}(a-c-d) = 0$ *and* $f \in \mathfrak{L}_{\nu,2}$, *then the inversion formulas*

$$f(x) = x^{-\kappa-\lambda}\frac{d}{dx}x^{\lambda+1} \cdot \int_0^\infty G_{2,4}^{2,1}\left[xt \,\middle|\, \begin{matrix} -\lambda, a-1 \\ c-1, d-1, 0, -\lambda-1 \end{matrix}\right] t^{-\sigma}\left({}_1\boldsymbol{F}_2{}^{\sigma,\kappa} f\right)(t)dt \tag{7.6.20}$$

hold for $\mathrm{Re}(\lambda) > \nu - \mathrm{Re}(\kappa) - 1$ *and*

$$f(x) = -x^{-\kappa-\lambda}\frac{d}{dx}x^{\lambda+1} \cdot \int_0^\infty G_{2,4}^{3,0}\left[xt \,\middle|\, \begin{matrix} a-1, -\lambda \\ -\lambda-1, c-1, d-1, 0 \end{matrix}\right] t^{-\sigma}\left({}_1\boldsymbol{F}_2{}^{\sigma,\kappa} f\right)(t)dt \tag{7.6.21}$$

for $\mathrm{Re}(\lambda) < \nu - \mathrm{Re}(\kappa) - 1$.

(b) *Let* $1 < r < \infty$, $\alpha_0 < \nu - \mathrm{Re}(\kappa) < \mathrm{Re}(a-c-d)/2 + 5/4$ *and* $2[1-\nu+\mathrm{Re}(\kappa)] + \mathrm{Re}(a-c-d) \leqq 1/2 - \gamma(r)$, *where* $\gamma(r)$ *is given in* (3.3.9). *If* $f \in \mathfrak{L}_{\nu,r}$, *then the relations* (7.6.20) *and* (7.6.21) *hold for* $\mathrm{Re}(\lambda) > \nu - \mathrm{Re}(\kappa) - 1$ *and for* $\mathrm{Re}(\lambda) < \nu - \mathrm{Re}(\kappa) - 1$, *respectively.*

7.7. ${}_2\boldsymbol{F}_1$-Transforms

Let us consider the integral transforms

$$\left({}_2\boldsymbol{F}_1 f\right)(x) = \frac{\Gamma(a)\Gamma(b)}{\Gamma(c)}\int_0^\infty {}_2F_1(a,b;c;-xt)f(t)dt \tag{7.7.1}$$

and

$$\left({}_2\boldsymbol{F}_1{}^* f\right)(x) = \frac{\Gamma(a)\Gamma(b)}{\Gamma(c)}\int_0^\infty {}_2F_1\left(a,b;c;-\frac{1}{xt}\right)f(t)dt \tag{7.7.2}$$

with $a, b, c \in \mathbb{C}$ $(\mathrm{Re}(a) > 0, \mathrm{Re}(b) > 0)$ involving the Gauss hypergeometric function ${}_2F_1(a,b;c;z)$ in the kernels. According to Prudnikov, Brychkov and Marichev [3, 8.4.49.13 and 8.4.49.14] the Mellin transforms of (7.7.1) and (7.7.2) for a "sufficiently good" function f are given by

$$\left(\mathfrak{M}\,{}_2\boldsymbol{F}_1 f\right)(s) = \mathcal{G}_{2,2}^{1,2}(s)\left(\mathfrak{M}f\right)(1-s), \quad \mathcal{G}_{2,2}^{1,2}(s) = \frac{\Gamma(s)\Gamma(a-s)\Gamma(b-s)}{\Gamma(c-s)} \tag{7.7.3}$$

and

$$\left(\mathfrak{M}\,{}_2\boldsymbol{F}_1{}^* f\right)(s) = \mathcal{G}_{2,2}^{2,1}(s)\left(\mathfrak{M}f\right)(1-s), \quad \mathcal{G}_{2,2}^{2,1}(s) = \frac{\Gamma(-s)\Gamma(a+s)\Gamma(b+s)}{\Gamma(c+s)} \tag{7.7.4}$$

under the restrictions $0 < \mathrm{Re}(s) < \min[\mathrm{Re}(a), \mathrm{Re}(b)]$ and $-\min[\mathrm{Re}(a), \mathrm{Re}(b)] < \mathrm{Re}(s) < 0$, respectively. These relations show that (7.7.1) and (7.7.2) are $\boldsymbol{G}$-transforms (6.1.1) of the form

$$\left({}_2\boldsymbol{F}_1 f\right)(x) = \int_0^\infty G_{2,2}^{1,2}\left[xt \,\middle|\, \begin{matrix} 1-a, 1-b \\ 0, 1-c \end{matrix}\right] f(t)dt \tag{7.7.5}$$

and

$$\left({}_2\boldsymbol{F}_1{}^* f\right)(x) = \int_0^\infty G_{2,2}^{2,1}\left[xt \,\middle|\, \begin{matrix} 1, c \\ a, b \end{matrix}\right] f(t)dt. \tag{7.7.6}$$

The constants $a^*, \Delta, a_1^*, a_2^*, \alpha, \beta$ and μ in (6.1.5)–(6.1.11) have the forms

$$\begin{gathered} a^* = 2,\ \Delta = 0,\ a_1^* = 1,\ a_2^* = 1, \\ \alpha = 0,\ \beta = \min[\mathrm{Re}(a), \mathrm{Re}(b)],\ \mu = a + b - c - 1 \end{gathered} \tag{7.7.7}$$

and

$$\begin{gathered} a^* = 2,\ \Delta = 0,\ a_1^* = 1,\ a_2^* = 1, \\ \alpha = -\min[\mathrm{Re}(a), \mathrm{Re}(b)],\ \beta = 0,\ \mu = a + b - c - 1. \end{gathered} \tag{7.7.8}$$

Let $\mathcal{E}_9$ and $\mathcal{E}_{10}$ be the exceptional sets of the functions $\mathcal{G}_{2,2}^{1,2}(s)$ and $\mathcal{G}_{2,2}^{2,1}(s)$ in (7.7.3) and (7.7.4). Due to Definition 6.1 the conditions $\nu \notin \mathcal{E}_9$ and $\nu \notin \mathcal{E}_{10}$ mean that

$$s \neq c + i \quad (i = 0, 1, 2, \cdots) \quad \text{for} \quad \mathrm{Re}(s) = 1 - \nu \tag{7.7.9}$$

and

$$s \neq -c - i \quad (i = 0, 1, 2, \cdots) \quad \text{for} \quad \mathrm{Re}(s) = 1 - \nu, \tag{7.7.10}$$

respectively.

Theorems 6.1 and 6.2 deduce the following $\mathfrak{L}_{\nu,2}$-theory of the transforms ${}_2\boldsymbol{F}_1$ and ${}_2\boldsymbol{F}_1{}^*$.

Theorem 7.29. *Let* $0 < 1 - \nu < \min[\mathrm{Re}(a), \mathrm{Re}(b)]$.

(i) *There is a one-to-one transform* ${}_2\boldsymbol{F}_1 \in [\mathfrak{L}_{\nu,2}, \mathfrak{L}_{1-\nu,2}]$ *such that* (7.7.3) *holds for* $\mathrm{Re}(s) = 1 - \nu$ *and* $f \in \mathfrak{L}_{\nu,2}$.

(ii) *If* $f, g \in \mathfrak{L}_{\nu,2}$, *then the relation*

$$\int_0^\infty f(x)\left({}_2\boldsymbol{F}_1 g\right)(x)dx = \int_0^\infty \left({}_2\boldsymbol{F}_1 f\right)(x)g(x)dx \tag{7.7.11}$$

holds.

(iii) *Let* $f \in \mathfrak{L}_{\nu,2}$ *and* $\lambda \in \mathbb{C}$. *If* $\mathrm{Re}(\lambda) > -\nu$, *then* ${}_2\boldsymbol{F}_1 f$ *is given by*

$$\left({}_2\boldsymbol{F}_1 f\right)(x) = x^{-\lambda}\frac{d}{dx}x^{\lambda+1}\int_0^\infty G_{3,3}^{1,3}\left[xt \,\middle|\, \begin{matrix} -\lambda, 1 - a, 1 - b \\ 0, 1 - c, -\lambda - 1 \end{matrix}\right] f(t)dt. \tag{7.7.12}$$

When $\mathrm{Re}(\lambda) < -\nu$,

$$\left({}_2\boldsymbol{F}_1 f\right)(x) = -x^{-\lambda}\frac{d}{dx}x^{\lambda+1}\int_0^\infty G_{3,3}^{2,2}\left[xt \,\middle|\, \begin{matrix} 1 - a, 1 - b, -\lambda \\ -\lambda - 1, 0, 1 - c \end{matrix}\right] f(t)dt. \tag{7.7.13}$$

(iv) *The transform* ${}_2\boldsymbol{F}_1$ *is independent of* ν *in the sense that if* $0 < 1 - \nu < \min[\mathrm{Re}(a), \mathrm{Re}(b)]$ *and* $0 < 1 - \widetilde{\nu} < \min[\mathrm{Re}(a), \mathrm{Re}(b)]$ *and if the transforms* ${}_2\boldsymbol{F}_1$ *and* ${}_2\widetilde{\boldsymbol{F}}_1$ *on* $\mathfrak{L}_{\nu,2}$ *and* $\mathfrak{L}_{\widetilde{\nu},2}$, *respectively, are given in* (7.7.3), *then* ${}_2\boldsymbol{F}_1 f = {}_2\widetilde{\boldsymbol{F}}_1 f$ *for* $f \in \mathfrak{L}_{\nu,2} \cap \mathfrak{L}_{\widetilde{\nu},2}$.

(v) *For $f \in \mathfrak{L}_{\nu,2}$ and $x > 0$, $\left({}_2\boldsymbol{F}_1 f\right)(x)$ is given in (7.7.1) and (7.7.5).*

Theorem 7.30. *Let $-\min[\mathrm{Re}(a), \mathrm{Re}(b)] < 1 - \nu < 0$.*

(i) *There is a one-to-one transform ${}_2\boldsymbol{F}_1{}^* \in [\mathfrak{L}_{\nu,2}, \mathfrak{L}_{1-\nu,2}]$ such that (7.7.4) holds for $\mathrm{Re}(s) = 1 - \nu$ and $f \in \mathfrak{L}_{\nu,2}$.*

(ii) *If $f, g \in \mathfrak{L}_{\nu,2}$, then the relation*

$$\int_0^\infty f(x)\left({}_2\boldsymbol{F}_1{}^* g\right)(x)dx = \int_0^\infty \left({}_2\boldsymbol{F}_1{}^* f\right)(x)g(x)dx \tag{7.7.14}$$

holds.

(iii) *Let $f \in \mathfrak{L}_{\nu,2}$ and $\lambda \in \mathbb{C}$. If $\mathrm{Re}(\lambda) > -\nu$, then ${}_2\boldsymbol{F}_1{}^* f$ is given by*

$$\left({}_2\boldsymbol{F}_1{}^* f\right)(x) = x^{-\lambda}\frac{d}{dx}x^{\lambda+1}\int_0^\infty G_{3,3}^{2,2}\left[xt \,\middle|\, \begin{matrix} -\lambda, 1, c \\ a, b, -\lambda-1 \end{matrix}\right] f(t)dt. \tag{7.7.15}$$

When $\mathrm{Re}(\lambda) < -\nu$,

$$\left({}_2\boldsymbol{F}_1{}^* f\right)(x) = -x^{-\lambda}\frac{d}{dx}x^{\lambda+1}\int_0^\infty G_{3,3}^{3,1}\left[xt \,\middle|\, \begin{matrix} 1, c, -\lambda \\ -\lambda-1, a, b \end{matrix}\right] f(t)dt. \tag{7.7.16}$$

(iv) *The transform ${}_2\boldsymbol{F}_1{}^*$ is independent of ν in the sense that if $-\min[\mathrm{Re}(a), \mathrm{Re}(b)] < 1 - \nu < 0$ and $-\min[\mathrm{Re}(a), \mathrm{Re}(b)] < 1 - \widetilde{\nu} < 0$ and if the transforms ${}_2\boldsymbol{F}_1{}^*$ and ${}_2\widetilde{\boldsymbol{F}}_1{}^*$ on $\mathfrak{L}_{\nu,2}$ and $\mathfrak{L}_{\widetilde{\nu},2}$, respectively, are given in (7.7.4), then ${}_2\boldsymbol{F}_1{}^* f = {}_2\widetilde{\boldsymbol{F}}_1{}^* f$ for $f \in \mathfrak{L}_{\nu,2} \cap \mathfrak{L}_{\widetilde{\nu},2}$.*

(v) *For $f \in \mathfrak{L}_{\nu,2}$ and $x > 0$, $\left({}_2\boldsymbol{F}_1{}^* f\right)(x)$ is given in (7.7.2) and (7.7.6).*

From Theorems 6.6 and 6.7 we obtain the $\mathfrak{L}_{\nu,r}$-theory of the transforms ${}_2\boldsymbol{F}_1$ and ${}_2\boldsymbol{F}_1{}^*$.

Theorem 7.31. *Let $\beta = \min[\mathrm{Re}(a), \mathrm{Re}(b)]$, $0 < 1 - \nu < \beta$ and $\omega = a + b - c - \beta$.*

(a) *If $1 \leqq r \leqq s \leqq \infty$, then the transform ${}_2\boldsymbol{F}_1$ defined on $\mathfrak{L}_{\nu,2}$ can be extended to $\mathfrak{L}_{\nu,r}$ as an element of $[\mathfrak{L}_{\nu,r}, \mathfrak{L}_{1-\nu,s}]$. If $1 \leqq r \leqq 2$, then ${}_2\boldsymbol{F}_1$ is a one-to-one transform from $\mathfrak{L}_{\nu,r}$ onto $\mathfrak{L}_{\nu,s}$.*

(b) *If $1 \leqq r \leqq s \leqq \infty$, $f \in \mathfrak{L}_{\nu,r}$ and $g \in \mathfrak{L}_{\nu,s'}$ with $1/s + 1/s' = 1$, then the relation (7.7.11) holds.*

(c) *If $1 < r < \infty$ and the condition (7.7.9) holds, then ${}_2\boldsymbol{F}_1$ is one-to-one on $\mathfrak{L}_{\nu,r}$.*

(d) *Let $1 < r < \infty$ and $\mathrm{Re}(\omega) \geqq 0$. If the condition (7.7.9) holds, then*

$${}_2\boldsymbol{F}_1(\mathfrak{L}_{\nu,r}) = \left(\mathbb{L}\mathbb{L}_{1,1+c-a-b}\right)(\mathfrak{L}_{1-\nu,r}), \tag{7.7.17}$$

where $\mathbb{L}$ and $\mathbb{L}_{1,1+c-a-b}$ are given in (2.5.2) and (3.3.3). If (7.7.9) is not valid, then ${}_2\boldsymbol{F}_1(\mathfrak{L}_{\nu,r})$ is a subset of the right-hand side of (7.7.17).

(e) *Let $1 < r < \infty$ and $\mathrm{Re}(\omega) < 0$. If the condition (7.7.9) holds, then*

$${}_2\boldsymbol{F}_1(\mathfrak{L}_{\nu,r}) = \left(I_{-;1,0}^{-\omega}\mathbb{L}\mathbb{L}_{1,1-\beta}\right)(\mathfrak{L}_{1-\nu,r}), \tag{7.7.18}$$

where $I_{-;1,0}^{-\omega}$ is given in (3.3.2). If (7.7.9) is not valid, then ${}_2\boldsymbol{F}_1(\mathfrak{L}_{\nu,r})$ is a subset of the right-hand side of (7.7.18).

Theorem 7.32. *Let $\alpha = -\min[\mathrm{Re}(a), \mathrm{Re}(b)]$, $\alpha < 1-\nu < 0$ and $\omega = a+b-c+\alpha$.*

(a) *If $1 \leqq r \leqq s \leqq \infty$, then the transform ${}_2\boldsymbol{F}_1{}^*$ defined on $\mathfrak{L}_{\nu,2}$ can be extended to $\mathfrak{L}_{\nu,r}$ as an element of $[\mathfrak{L}_{\nu,r}, \mathfrak{L}_{1-\nu,s}]$. If $1 \leqq r \leqq 2$, then ${}_2\boldsymbol{F}_1{}^*$ is a one-to-one transform from $\mathfrak{L}_{\nu,r}$ onto $\mathfrak{L}_{1-\nu,s}$.*

(b) *If $1 \leqq r \leqq s \leqq \infty$, $f \in \mathfrak{L}_{\nu,r}$ and $g \in \mathfrak{L}_{\nu,s'}$ with $1/s + 1/s' = 1$, then the relation (7.7.14) holds.*

(c) *If $1 < r < \infty$ and the condition (7.7.10) holds, then ${}_2\boldsymbol{F}_1{}^*$ is one-to-one on $\mathfrak{L}_{\nu,r}$.*

(d) *Let $1 < r < \infty$ and $\mathrm{Re}(\omega) \geqq 0$. If the condition (7.7.10) holds, then*

$$ {}_2\boldsymbol{F}_1{}^*(\mathfrak{L}_{\nu,r}) = \left(\mathbb{L}_{1,\alpha}\mathbb{L}_{1,1-\omega}\right)(\mathfrak{L}_{1-\nu,r}), \tag{7.7.19} $$

where $\mathbb{L}_{1,\alpha}$ is given in (3.3.3). If (7.7.10) is not valid, then ${}_2\boldsymbol{F}_1{}^*(\mathfrak{L}_{\nu,r})$ is a subset of the right-hand side of (7.7.19).

(e) *Let $1 < r < \infty$ and $\mathrm{Re}(\omega) < 0$. If the condition (7.7.10) holds, then*

$$ {}_2\boldsymbol{F}_1{}^*(\mathfrak{L}_{\nu,r}) = \left(I^{-\omega}_{-;1,-\alpha}\mathbb{L}_{1,\alpha}\mathbb{L}_{1,1}\right)(\mathfrak{L}_{1-\nu,r}), \tag{7.7.20} $$

where $I^{-\omega}_{-;1,-\alpha}$ is given in (3.3.2). If (7.7.10) is not valid, then ${}_2\boldsymbol{F}_1{}^*(\mathfrak{L}_{\nu,r})$ is a subset of the right-hand side of (7.7.20).

Now we treat generalizations of (7.7.1) and (7.7.2) in the forms

$$ \left({}_2\boldsymbol{F}_1{}^{\sigma,\kappa} f\right)(x) = \frac{\Gamma(a)\Gamma(b)}{\Gamma(c)} x^{\sigma} \int_0^{\infty} {}_2F_1(a,b;c;-xt)t^{\kappa} f(t)dt \tag{7.7.21} $$

and

$$ \left({}_2\boldsymbol{F}_1{}^{*;\sigma,\kappa} f\right)(x) = \frac{\Gamma(a)\Gamma(b)}{\Gamma(c)} x^{\sigma} \int_0^{\infty} {}_2F_1\left(a,b;c;-\frac{1}{xt}\right) t^{\kappa} f(t)dt \tag{7.7.22} $$

with $a, b, c, \sigma, \kappa \in \mathbb{C}$ ($\mathrm{Re}(a) > 0, \mathrm{Re}(b) > 0$). By virtue of (7.7.3), (7.7.4) and (3.3.14) the Mellin transforms of (7.7.21) and (7.7.22) for a "sufficiently good" function f are given by

$$ \left(\mathfrak{M}\,{}_2\boldsymbol{F}_1{}^{\sigma,\kappa} f\right)(s) = \mathcal{G}^{1,2}_{2,2}(s+\sigma)\left(\mathfrak{M}f\right)(1-s-\sigma+\kappa) \tag{7.7.23} $$

and

$$ \left(\mathfrak{M}\,{}_2\boldsymbol{F}_1{}^{*;\sigma,\kappa} f\right)(s) = \mathcal{G}^{2,1}_{2,2}(s+\sigma)\left(\mathfrak{M}f\right)(1-s-\sigma+\kappa), \tag{7.7.24} $$

where $\mathcal{G}^{1,2}_{2,2}(s)$ and $\mathcal{G}^{2,1}_{2,2}(s)$ are defined in (7.7.3) and (7.7.4). Thus the transforms (7.7.21) and (7.7.22) are modified transforms (6.2.3) of the forms

$$ \left({}_2\boldsymbol{F}_1{}^{\sigma,\kappa} f\right)(x) = x^{\sigma} \int_0^{\infty} G^{1,2}_{2,2}\left[xt \,\middle|\, \begin{matrix} 1-a, 1-b \\ 0, 1-c \end{matrix}\right] t^{\kappa} f(t)dt \tag{7.7.25} $$

and

$$ \left({}_2\boldsymbol{F}_1{}^{*;\sigma,\kappa} f\right)(x) = x^{\sigma} \int_0^{\infty} G^{2,1}_{2,2}\left[xt \,\middle|\, \begin{matrix} 1, c \\ a, b \end{matrix}\right] t^{\kappa} f(t)dt, \tag{7.7.26} $$

respectively.

For the exceptional sets $\mathcal{E}_9$ and $\mathcal{E}_{10}$ of the functions $\mathcal{G}^{1,2}_{2,2}(s)$ and $\mathcal{G}^{2,1}_{2,2}(s)$ in (7.7.3) and (7.7.4), $\nu - \mathrm{Re}(\kappa) \notin \mathcal{E}_9$ and $\nu - \mathrm{Re}(\kappa) \notin \mathcal{E}_{10}$ mean that

$$s \neq c + i \quad (i = 0, 1, 2, \cdots) \quad \text{for} \quad \mathrm{Re}(s) = 1 - \nu + \mathrm{Re}(\kappa) \tag{7.7.27}$$

and

$$s \neq -c - i \quad (i = 0, 1, 2, \cdots) \quad \text{for} \quad \mathrm{Re}(s) = 1 - \nu + \mathrm{Re}(\kappa). \tag{7.7.28}$$

Theorem 6.38 yields the $\mathfrak{L}_{\nu,2}$-theory of the transforms ${}_2\boldsymbol{F}_1{}^{\sigma,\kappa}$ and ${}_2\boldsymbol{F}_1{}^{*;\sigma,\kappa}$.

Theorem 7.33. *Let* $0 < 1 - \nu + \mathrm{Re}(\kappa) < \min[\mathrm{Re}(a), \mathrm{Re}(b)]$.

(i) *There is a one-to-one transform* ${}_2\boldsymbol{F}_1{}^{\sigma,\kappa} \in [\mathfrak{L}_{\nu,2}, \mathfrak{L}_{1-\nu+\mathrm{Re}(\kappa-\sigma),2}]$ *such that* (7.7.23) *holds for* $\mathrm{Re}(s) = 1 - \nu + \mathrm{Re}(\kappa - \sigma)$ *and* $f \in \mathfrak{L}_{\nu,2}$.

(ii) *If* $f \in \mathfrak{L}_{\nu,2}$ *and* $g \in \mathfrak{L}_{\nu+\mathrm{Re}(\kappa-\sigma),2}$, *then the relation*

$$\int_0^\infty f(x)\left({}_2\boldsymbol{F}_1{}^{\sigma,\kappa} g\right)(x)dx = \int_0^\infty \left({}_2\boldsymbol{F}_1{}^{\kappa,\sigma} f\right)(x)g(x)dx \tag{7.7.29}$$

holds.

(iii) *Let* $f \in \mathfrak{L}_{\nu,2}$ *and* $\lambda \in \mathbb{C}$. *If* $\mathrm{Re}(\lambda) > -\nu + \mathrm{Re}(\kappa)$, *then* ${}_2\boldsymbol{F}_1{}^{\sigma,\kappa} f$ *is given by*

$$\left({}_2\boldsymbol{F}_1{}^{\sigma,\kappa} f\right)(x) = x^{\sigma-\lambda}\frac{d}{dx}x^{\lambda+1}\int_0^\infty G^{1,3}_{3,3}\left[xt \,\middle|\, \begin{matrix} -\lambda, 1-a, 1-b \\ 0, 1-c, -\lambda-1 \end{matrix}\right] t^\kappa f(t)dt. \tag{7.7.30}$$

When $\mathrm{Re}(\lambda) < -\nu + \mathrm{Re}(\kappa)$,

$$\left({}_2\boldsymbol{F}_1{}^{\sigma,\kappa} f\right)(x) = -x^{\sigma-\lambda}\frac{d}{dx}x^{\lambda+1}\int_0^\infty G^{2,2}_{3,3}\left[xt \,\middle|\, \begin{matrix} 1-a, 1-b, -\lambda \\ -\lambda-1, 0, 1-c \end{matrix}\right] t^\kappa f(t)dt. \tag{7.7.31}$$

(iv) *The transform* ${}_2\boldsymbol{F}_1{}^{\sigma,\kappa}$ *is independent of* ν *in the sense that if* $0 < 1 - \nu < \min[\mathrm{Re}(a), \mathrm{Re}(b)]$ *and* $0 < 1 - \tilde{\nu} < \min[\mathrm{Re}(a), \mathrm{Re}(b)]$ *and if the transforms* ${}_2\boldsymbol{F}_1{}^{\sigma,\kappa}$ *and* ${}_2\widetilde{\boldsymbol{F}}_1{}^{\sigma,\kappa}$ *on* $\mathfrak{L}_{\nu,2}$ *and* $\mathfrak{L}_{\tilde{\nu},2}$, *respectively, are given in* (7.7.23), *then* ${}_2\boldsymbol{F}_1{}^{\sigma,\kappa} f = {}_2\widetilde{\boldsymbol{F}}_1{}^{\sigma,\kappa} f$ *for* $f \in \mathfrak{L}_{\nu,2} \cap \mathfrak{L}_{\tilde{\nu},2}$.

(v) *For* $f \in \mathfrak{L}_{\nu,2}$ *and* $x > 0$, $\left({}_2\boldsymbol{F}_1{}^{\sigma,\kappa} f\right)(x)$ *is given in* (7.7.21) *and* (7.7.25).

Theorem 7.34. *Let* $-\min[\mathrm{Re}(a), \mathrm{Re}(b)] < 1 - \nu + \mathrm{Re}(\kappa) < 0$.

(i) *There is a one-to-one transform* ${}_2\boldsymbol{F}_1{}^{*;\sigma,\kappa} \in [\mathfrak{L}_{\nu,2}, \mathfrak{L}_{1-\nu+\mathrm{Re}(\kappa-\sigma),2}]$ *such that* (7.7.24) *holds for* $\mathrm{Re}(s) = 1 - \nu + \mathrm{Re}(\kappa - \sigma)$ *and* $f \in \mathfrak{L}_{\nu,2}$.

(ii) *If* $f \in \mathfrak{L}_{\nu,2}$ *and* $g \in \mathfrak{L}_{\nu+\mathrm{Re}(\kappa-\sigma),2}$, *then the relation*

$$\int_0^\infty f(x)\left({}_2\boldsymbol{F}_1{}^{*;\sigma,\kappa} g\right)(x)dx = \int_0^\infty \left({}_2\boldsymbol{F}_1{}^{*;\kappa,\sigma} f\right)(x)g(x)dx \tag{7.7.32}$$

holds.

(iii) *Let* $f \in \mathfrak{L}_{\nu,2}$ *and* $\lambda \in \mathbb{C}$. *If* $\mathrm{Re}(\lambda) > -\nu + \mathrm{Re}(\kappa)$, *then* ${}_2\boldsymbol{F}_1{}^{*;\sigma,\kappa} f$ *is given by*

$$\left({}_2\boldsymbol{F}_1{}^{*;\sigma,\kappa} f\right)(x) = x^{\sigma-\lambda}\frac{d}{dx}x^{\lambda+1}\int_0^\infty G^{2,2}_{3,3}\left[xt \,\middle|\, \begin{matrix} -\lambda, 1, c \\ a, b, -\lambda-1 \end{matrix}\right] t^\kappa f(t)dt. \tag{7.7.33}$$

When $\mathrm{Re}(\lambda) < -\nu + \mathrm{Re}(\kappa)$,

$$\left({}_2\boldsymbol{F}_1{}^{*;\sigma,\kappa} f\right)(x) = -x^{\sigma-\lambda}\frac{d}{dx}x^{\lambda+1}\int_0^\infty G_{3,3}^{3,1}\left[xt \,\middle|\, \begin{matrix} 1, c, -\lambda \\ -\lambda-1, a, b \end{matrix}\right] t^\kappa f(t)dt. \tag{7.7.34}$$

(iv) *The transform ${}_2\boldsymbol{F}_1{}^{*;\sigma,\kappa}$ is independent of ν in the sense that if* $-\min[\mathrm{Re}(a),\mathrm{Re}(b)] < 1-\nu < 0$ *and* $-\min[\mathrm{Re}(a),\mathrm{Re}(b)] < 1-\widetilde{\nu} < 0$ *and if the transforms ${}_2\boldsymbol{F}_1{}^{*;\sigma,\kappa}$ and ${}_2\widetilde{\boldsymbol{F}}_1{}^{*;\sigma,\kappa}$ on $\mathfrak{L}_{\nu,2}$ and $\mathfrak{L}_{\widetilde{\nu},2}$, respectively, are given in (7.7.24), then ${}_2\boldsymbol{F}_1{}^{*;\sigma,\kappa} f = {}_2\widetilde{\boldsymbol{F}}_1{}^{*;\sigma,\kappa} f$ for $f \in \mathfrak{L}_{\nu,2}\cap\mathfrak{L}_{\widetilde{\nu},2}$.*

(v) *For $f \in \mathfrak{L}_{\nu,2}$ and $x > 0$, $\left({}_2\boldsymbol{F}_1{}^{*;\sigma,\kappa} f\right)(x)$ is given in (7.7.22) and (7.7.26).*

Theorems 6.42 and 6.43 produce the $\mathfrak{L}_{\nu,r}$-theory of the transforms ${}_2\boldsymbol{F}_1{}^{\sigma,\kappa}$ and ${}_2\boldsymbol{F}_1{}^{*;\sigma,\kappa}$.

Theorem 7.35. *Let $\beta = \min[\mathrm{Re}(a),\mathrm{Re}(b)]$, $0 < 1-\nu+\mathrm{Re}(\kappa) < \beta$ and $\omega = a+b-c-\beta$.*

(a) *If $1 \leqq r \leqq s \leqq \infty$, then the transform ${}_2\boldsymbol{F}_1{}^{\sigma,\kappa}$ defined on $\mathfrak{L}_{\nu,2}$ can be extended to $\mathfrak{L}_{\nu,r}$ as an element of $[\mathfrak{L}_{\nu,r},\mathfrak{L}_{1-\nu+\mathrm{Re}(\kappa-\sigma),s}]$. If $1 \leqq r \leqq 2$, then ${}_2\boldsymbol{F}_1{}^{\sigma,\kappa}$ is a one-to-one transform from $\mathfrak{L}_{\nu,r}$ onto $\mathfrak{L}_{1-\nu+\mathrm{Re}(\kappa-\sigma),s}$.*

(b) *If $1 \leqq r \leqq s \leqq \infty$, $f \in \mathfrak{L}_{\nu,r}$ and $g \in \mathfrak{L}_{\nu+\mathrm{Re}(\kappa-\sigma),s'}$ with $1/s+1/s' = 1$, then the relation (7.7.29) holds.*

(c) *If $1 < r < \infty$ and the condition (7.7.27) holds, then the transform ${}_2\boldsymbol{F}_1{}^{\sigma,\kappa}$ is one-to-one on $\mathfrak{L}_{\nu,r}$.*

(d) *Let $1 < r < \infty$ and $\mathrm{Re}(\omega) \geqq 0$. If the condition (7.7.27) holds, then*

$${}_2\boldsymbol{F}_1{}^{\sigma,\kappa}(\mathfrak{L}_{\nu,r}) = \left(\mathbb{L}_{1,-\sigma}\mathbb{L}_{1,1+c-a-b+\sigma}\right)(\mathfrak{L}_{1-\nu+\mathrm{Re}(\kappa-\sigma),r}), \tag{7.7.35}$$

where $\mathbb{L}_{1,-\sigma}$ and $\mathbb{L}_{1,1+c-a-b}$ are given in (3.3.3). If (7.7.27) is not valid, then ${}_2\boldsymbol{F}_1{}^{\sigma,\kappa}(\mathfrak{L}_{\nu,r})$ is a subset of the right-hand side of (7.7.35).

(e) *Let $1 < r < \infty$ and $\mathrm{Re}(\omega) < 0$. If the condition (7.7.27) holds, then*

$${}_2\boldsymbol{F}_1{}^{\sigma,\kappa}(\mathfrak{L}_{\nu,r}) = \left(I_{-;1,\sigma}^{-\omega}\mathbb{L}_{1,-\sigma}\mathbb{L}_{1,1-\beta+\sigma}\right)(\mathfrak{L}_{1-\nu+\mathrm{Re}(\kappa-\sigma),r}), \tag{7.7.36}$$

where $I_{-;1,\sigma}^{-\omega}$ is given in (3.3.2). If (7.7.27) is not valid, then ${}_2\boldsymbol{F}_1{}^{\sigma,\kappa}\,\mathfrak{L}_{\nu,r})$ is a subset of the right-hand side of (7.7.36).

Theorem 7.36. *Let $\alpha = -\min[\mathrm{Re}(a),\mathrm{Re}(b)]$, $\alpha < 1-\nu+\mathrm{Re}(\kappa) < 0$ and $\omega = a+b-c+\alpha$.*

(a) *If $1 \leqq r \leqq s \leqq \infty$, then the transform ${}_2\boldsymbol{F}_1{}^{*;\sigma,\kappa}$ defined on $\mathfrak{L}_{\nu,2}$ can be extended to $\mathfrak{L}_{\nu,r}$ as element of $[\mathfrak{L}_{\nu,r},\mathfrak{L}_{1-\nu+\mathrm{Re}(\kappa-\sigma),s}]$. If $1 \leqq r \leqq 2$, then ${}_2\boldsymbol{F}_1{}^{*;\sigma,\kappa}$ is a one-to-one transform from $\mathfrak{L}_{\nu,r}$ onto $\mathfrak{L}_{1-\nu+\mathrm{Re}(\kappa-\sigma),s}$.*

(b) *If $1 \leqq r \leqq s \leqq \infty$, $f \in \mathfrak{L}_{\nu,r}$ and $g \in \mathfrak{L}_{\nu+\mathrm{Re}(\kappa-\sigma),s'}$ with $1/s+1/s' = 1$, then the relation (7.7.32) holds.*

(c) *If $1 < r < \infty$ and the condition (7.7.28) holds, then the transform ${}_2\boldsymbol{F}_1{}^{*;\sigma,\kappa}$ is one-to-one on $\mathfrak{L}_{\nu,r}$.*

(d) *Let $1 < r < \infty$ and $\mathrm{Re}(\omega) \geqq 0$. If the condition (7.7.28) holds, then*

$${}_2\boldsymbol{F}_1{}^{*;\sigma,\kappa}(\mathfrak{L}_{\nu,r}) = \left(\mathbb{L}_{1,\alpha-\sigma}\mathbb{L}_{1,1+\sigma-\omega}\right)(\mathfrak{L}_{1-\nu+\mathrm{Re}(\kappa-\sigma),r}), \tag{7.7.37}$$

where $\mathbb{L}_{1,\alpha-\sigma}$ and $\mathbb{L}_{1,1+\sigma-\omega}$ are given in (3.3.3). If (7.7.28) is not valid, then ${}_2\boldsymbol{F}_1{}^{*;\sigma,\kappa}(\mathfrak{L}_{\nu,r})$ is a subset of the right-hand side of (7.7.37).

(e) *Let* $1 < r < \infty$ *and* $\mathrm{Re}(\omega) < 0$. *If the condition* (7.7.28) *holds, then*

$$ {}_2\boldsymbol{F}_1{}^{*;\sigma,\kappa}(\mathfrak{L}_{\nu,r}) = \left(I^{-\omega}_{-;1,\sigma-\alpha}\mathbb{L}_{1,\alpha-\sigma}\mathbb{L}_{1,1+\sigma}\right)(\mathfrak{L}_{1-\nu+\mathrm{Re}(\kappa-\sigma),r}), \tag{7.7.38} $$

where $I^{-\omega}_{-;1,\sigma-\alpha}$ *is given in* (3.3.2). *If* (7.7.28) *is not valid, then* ${}_2\boldsymbol{F}_1{}^{*;\sigma,\kappa}(\mathfrak{L}_{\nu,r})$ *is a subset of the right-hand side of* (7.7.38).

7.8. Modified ${}_2F_1$-Transforms

Let us study the transforms

$$ \left({}_2\boldsymbol{F}_1{}^1 f\right)(x) = \frac{\Gamma(a)\Gamma(b)}{\Gamma(c)}\int_0^\infty {}_2F_1\left(a,b;c;-\frac{x}{t}\right)\frac{f(t)}{t}dt \tag{7.8.1} $$

and

$$ \left({}_2\boldsymbol{F}_1{}^2 f\right)(x) = \frac{\Gamma(a)\Gamma(b)}{\Gamma(c)}\int_0^\infty {}_2F_1\left(a,b;c;-\frac{t}{x}\right)\frac{f(t)}{x}dt \tag{7.8.2} $$

with $a,b,c \in \mathbb{C}$ ($\mathrm{Re}(a) > 0, \mathrm{Re}(b) > 0$) which are modifications of the transforms (7.7.1) and (7.7.2) in the forms (6.2.1) and (6.2.2). Comparing with (6.2.11) and (6.2.12), we find that the Mellin transforms of (7.8.1) and (7.8.2) have the forms

$$ \left(\mathfrak{M}\,{}_2\boldsymbol{F}_1{}^1 f\right)(s) = \mathcal{G}^{1,2}_{2,2}(s)\left(\mathfrak{M}f\right)(s) \tag{7.8.3} $$

and

$$ \left(\mathfrak{M}\,{}_2\boldsymbol{F}_1{}^2 f\right)(s) = \mathcal{G}^{1,2}_{2,2}(1-s)\left(\mathfrak{M}f\right)(s), \tag{7.8.4} $$

where $\mathcal{G}^{1,2}_{2,2}(s)$ is given in (7.7.3). These relations show that (7.8.1) and (7.8.2) can be regarded as the modified $\boldsymbol{G}$-transforms (6.2.1) and (6.2.2) in the forms

$$ \left({}_2\boldsymbol{F}_1{}^1 f\right)(x) = \int_0^\infty G^{1,2}_{2,2}\left[\frac{x}{t}\,\middle|\,\begin{matrix}1-a,1-b\\0,1-c\end{matrix}\right]\frac{f(t)}{t}dt \tag{7.8.5} $$

and

$$ \left({}_2\boldsymbol{F}_1{}^2 f\right)(x) = \int_0^\infty G^{1,2}_{2,2}\left[\frac{t}{x}\,\middle|\,\begin{matrix}1-a,1-b\\0,1-c\end{matrix}\right]\frac{f(t)}{x}dt. \tag{7.8.6} $$

Now we can apply the results in Sections 6.3 and 6.4 to obtain the $\mathfrak{L}_{\nu,2}$- and $\mathfrak{L}_{\nu,r}$-theory of the transforms (7.8.1) and (7.8.2) if we take into account that by (7.7.9) the conditions $1-\nu \notin \mathcal{E}_9$ and $\nu \notin \mathcal{E}_9$ for the exceptional set $\mathcal{E}_9$ of the function $\mathcal{G}^{1,2}_{2,2}(s)$ in (7.7.3) are equivalent to

$$ s \neq c+i \quad (i = 0,1,2,\cdots) \quad \text{for} \quad \mathrm{Re}(s) = \nu \tag{7.8.7} $$

and

$$ s \neq c+i \quad (i = 0,1,2,\cdots) \quad \text{for} \quad \mathrm{Re}(s) = 1-\nu, \tag{7.8.8} $$

respectively.

From Theorems 6.14 and 6.26 we obtain the $\mathfrak{L}_{\nu,2}$-theory of the transforms (7.8.1) and (7.8.2).

Theorem 7.37. *Let $0 < \nu < \min[\mathrm{Re}(a), \mathrm{Re}(b)]$.*

(i) *There is a one-to-one transform ${}_2\boldsymbol{F}_1{}^1 \in [\mathfrak{L}_{\nu,2}]$ such that (7.8.3) holds for $\mathrm{Re}(s) = \nu$ and $f \in \mathfrak{L}_{\nu,2}$.*

(ii) *If $f \in \mathfrak{L}_{\nu,2}$ and $g \in \mathfrak{L}_{1-\nu,2}$, then the relation*

$$\int_0^\infty f(x)\left({}_2\boldsymbol{F}_1{}^1 g\right)(x)dx = \int_0^\infty \left({}_2\boldsymbol{F}_1{}^2 f\right)(x)g(x)dx \tag{7.8.9}$$

holds.

(iii) *Let $f \in \mathfrak{L}_{\nu,2}$ and $\lambda \in \mathbb{C}$. If $\mathrm{Re}(\lambda) > \nu - 1$, then ${}_2\boldsymbol{F}_1{}^1 f$ is given by*

$$\left({}_2\boldsymbol{F}_1{}^1 f\right)(x) = x^{-\lambda}\frac{d}{dx}x^{\lambda+1}\int_0^\infty G_{3,3}^{1,3}\left[\frac{x}{t}\,\middle|\,\begin{matrix}-\lambda, 1-a, 1-b\\ 0, 1-c, -\lambda-1\end{matrix}\right]\frac{f(t)}{t}dt. \tag{7.8.10}$$

When $\mathrm{Re}(\lambda) < \nu - 1$,

$$\left({}_2\boldsymbol{F}_1{}^1 f\right)(x) = -x^{-\lambda}\frac{d}{dx}x^{\lambda+1}\int_0^\infty G_{3,3}^{2,2}\left[\frac{x}{t}\,\middle|\,\begin{matrix}1-a, 1-b, -\lambda\\ -\lambda-1, 0, 1-c\end{matrix}\right]\frac{f(t)}{t}dt. \tag{7.8.11}$$

(iv) *The transform ${}_2\boldsymbol{F}_1{}^1$ is independent of ν in the sense that if $0 < \nu < \min[\mathrm{Re}(a), \mathrm{Re}(b)]$ and $0 < \widetilde{\nu} < \min[\mathrm{Re}(a), \mathrm{Re}(b)]$ and if the transforms ${}_2\boldsymbol{F}_1{}^1$ and $\widetilde{{}_2\boldsymbol{F}_1{}^1}$ are defined on $\mathfrak{L}_{\nu,2}$ and $\mathfrak{L}_{\widetilde{\nu},2}$, respectively by (7.8.3), then ${}_2\boldsymbol{F}_1{}^1 f = \widetilde{{}_2\boldsymbol{F}_1{}^1} f$ for $f \in \mathfrak{L}_{\nu,2} \cap \mathfrak{L}_{\widetilde{\nu},2}$.*

(v) *For $f \in \mathfrak{L}_{\nu,2}$ and $x > 0$, $\left({}_2\boldsymbol{F}_1{}^1 f\right)(x)$ is given in (7.8.1) and (7.8.5).*

Theorem 7.38. *Let $0 < 1 - \nu < \min[\mathrm{Re}(a), \mathrm{Re}(b)]$.*

(i) *There is a one-to-one transform ${}_2\boldsymbol{F}_1{}^2 \in [\mathfrak{L}_{\nu,2}]$ such that (7.8.4) holds for $\mathrm{Re}(s) = \nu$ and $f \in \mathfrak{L}_{\nu,2}$.*

(ii) *If $f \in \mathfrak{L}_{\nu,2}$ and $g \in \mathfrak{L}_{1-\nu,2}$, then the relation*

$$\int_0^\infty f(x)\left({}_2\boldsymbol{F}_1{}^2 g\right)(x)dx = \int_0^\infty \left({}_2\boldsymbol{F}_1{}^1 f\right)(x)g(x)dx \tag{7.8.12}$$

holds.

(iii) *Let $f \in \mathfrak{L}_{\nu,2}$ and $\lambda \in \mathbb{C}$. If $\mathrm{Re}(\lambda) > -\nu$, then ${}_2\boldsymbol{F}_1{}^2 f$ is given by*

$$\left({}_2\boldsymbol{F}_1{}^2 f\right)(x) = -x^{\lambda+1}\frac{d}{dx}x^{-\lambda}\int_0^\infty G_{3,3}^{1,3}\left[\frac{t}{x}\,\middle|\,\begin{matrix}-\lambda, 1-a, 1-b\\ 0, 1-c, -\lambda-1\end{matrix}\right]\frac{f(t)}{x}dt. \tag{7.8.13}$$

When $\mathrm{Re}(\lambda) < -\nu$,

$$\left({}_2\boldsymbol{F}_1{}^2 f\right)(x) = x^{\lambda+1}\frac{d}{dx}x^{-\lambda}\int_0^\infty G_{3,3}^{2,2}\left[\frac{t}{x}\,\middle|\,\begin{matrix}1-a, 1-b, -\lambda\\ -\lambda-1, 0, 1-c\end{matrix}\right]\frac{f(t)}{x}dt. \tag{7.8.14}$$

(iv) *The transform ${}_2\boldsymbol{F}_1{}^2$ is independent of ν in the sense that if $0 < \nu < \min[\mathrm{Re}(a), \mathrm{Re}(b)]$ and $0 < \widetilde{\nu} < \min[\mathrm{Re}(a), \mathrm{Re}(b)]$ and if the transforms ${}_2\boldsymbol{F}_1{}^2$ and $\widetilde{{}_2\boldsymbol{F}_1{}^2}$ are defined on $\mathfrak{L}_{\nu,2}$ and $\mathfrak{L}_{\widetilde{\nu},2}$, respectively by (7.8.4), then ${}_2\boldsymbol{F}_1{}^2 f = \widetilde{{}_2\boldsymbol{F}_1{}^2} f$ for $f \in \mathfrak{L}_{\nu,2} \cap \mathfrak{L}_{\widetilde{\nu},2}$.*

(v) For $f \in \mathfrak{L}_{\nu,2}$ and $x > 0$, $\left({}_2\boldsymbol{F}_1{}^2 f\right)(x)$ *is given in* (7.8.2) *and* (7.8.6).

From Theorems 6.18, 6.19 and 6.30, 6.31 we obtain the $\mathfrak{L}_{\nu,r}$-theory of the transform (7.8.1) and (7.8.2).

Theorem 7.39. *Let* $\beta = \min[\mathrm{Re}(a), \mathrm{Re}(b)]$, $0 < \nu < \beta$ *and* $\omega = a + b - c - \beta$.

(a) *If* $1 \leqq r \leqq s \leqq \infty$, *then the transform* ${}_2\boldsymbol{F}_1{}^1$ *defined on* $\mathfrak{L}_{\nu,2}$ *can be extended to* $\mathfrak{L}_{\nu,r}$ *as an element of* $[\mathfrak{L}_{\nu,r}, \mathfrak{L}_{\nu,s}]$. *If* $1 \leqq r \leqq 2$, *then* ${}_2\boldsymbol{F}_1{}^1$ *is a one-to-one transform from* $\mathfrak{L}_{\nu,r}$ *onto* $\mathfrak{L}_{\nu,s}$.

(b) *If* $1 \leqq r \leqq s \leqq \infty$, $f \in \mathfrak{L}_{\nu,r}$ *and* $g \in \mathfrak{L}_{1-\nu,s'}$ *with* $1/s + 1/s' = 1$, *then the relation* (7.8.9) *holds.*

(c) *If* $1 < r < \infty$ *and the condition* (7.8.7) *holds, then* ${}_2\boldsymbol{F}_1{}^1$ *is one-to-one on* $\mathfrak{L}_{\nu,r}$.

(d) *Let* $1 < r < \infty$ *and* $\mathrm{Re}(\omega) \geqq 0$. *If the condition* (7.8.7) *holds, then*

$$ {}_2\boldsymbol{F}_1{}^1(\mathfrak{L}_{\nu,r}) = \left(\mathbb{L}\mathbb{L}_{1,1+c-a-b}\right)(\mathfrak{L}_{\nu,r}), \tag{7.8.15} $$

where $\mathbb{L}$ *and* $\mathbb{L}_{1,1+c-a-b}$ *are given in* (2.5.2) *and* (3.3.3). *If* (7.8.7) *is not valid, then* ${}_2\boldsymbol{F}_1{}^1(\mathfrak{L}_{\nu,r})$ *is a subset of the right-hand side of* (7.8.15).

(e) *Let* $1 < r < \infty$ *and* $\mathrm{Re}(\omega) < 0$. *If the condition* (7.8.7) *holds, then*

$$ {}_2\boldsymbol{F}_1{}^1(\mathfrak{L}_{\nu,r}) = \left(I^{-\omega}_{-;1,0}\mathbb{L}\mathbb{L}_{1,1-\beta}\right)(\mathfrak{L}_{\nu,r}), \tag{7.8.16} $$

where $I^{-\omega}_{-;1,0}$ *is given in* (3.3.2). *If* (7.8.7) *is not valid, then* ${}_2\boldsymbol{F}_1{}^1(\mathfrak{L}_{\nu,r})$ *is a subset of the right-hand side of* (7.8.16).

Theorem 7.40. *Let* $\beta = \min[\mathrm{Re}(a), \mathrm{Re}(b)]$, $0 < 1 - \nu < \beta$ *and* $\omega = a + b - c - \beta$.

(a) *If* $1 \leqq r \leqq s \leqq \infty$, *then the transform* ${}_2\boldsymbol{F}_1{}^2$ *defined on* $\mathfrak{L}_{\nu,2}$ *can be extended to* $\mathfrak{L}_{\nu,r}$ *as an element of* $[\mathfrak{L}_{\nu,r}, \mathfrak{L}_{\nu,s}]$. *If* $1 \leqq r \leqq 2$, *then* ${}_2\boldsymbol{F}_1{}^2$ *is a one-to-one transform from* $\mathfrak{L}_{\nu,r}$ *onto* $\mathfrak{L}_{\nu,s}$.

(b) *If* $1 \leqq r \leqq s \leqq \infty$, $f \in \mathfrak{L}_{\nu,r}$ *and* $g \in \mathfrak{L}_{1-\nu,s'}$ *with* $1/s + 1/s' = 1$, *then the relation* (7.8.12) *holds.*

(c) *If* $1 < r < \infty$ *and the condition* (7.8.8) *holds, then* ${}_2\boldsymbol{F}_1{}^2$ *is one-to-one on* $\mathfrak{L}_{\nu,r}$.

(d) *Let* $1 < r < \infty$ *and* $\mathrm{Re}(\omega) \geqq 0$. *If the condition* (7.8.8) *holds, then*

$$ {}_2\boldsymbol{F}_1{}^2(\mathfrak{L}_{\nu,r}) = \left(\mathbb{L}_{-1,1}\mathbb{L}_{-1,\beta+\omega}\right)(\mathfrak{L}_{\nu,r}), \tag{7.8.17} $$

where $\mathbb{L}_{1,\alpha}$ *is given in* (3.3.3). *If* (7.8.8) *is not valid, then* ${}_2\boldsymbol{F}_1{}^2(\mathfrak{L}_{\nu,r})$ *is a subset of the right-hand side of* (7.8.17).

(e) *Let* $1 < r < \infty$ *and* $\mathrm{Re}(\omega) < 0$. *If the condition* (7.8.8) *holds, then*

$$ {}_2\boldsymbol{F}_1{}^2(\mathfrak{L}_{\nu,r}) = \left(I^{-\omega}_{0+;1,0}\mathbb{L}_{-1,1}\mathbb{L}_{-1,\beta}\right)(\mathfrak{L}_{\nu,r}), \tag{7.8.18} $$

where $I^{-\omega}_{-;1,-\alpha}$ *is given in* (3.3.2). *If* (7.8.8) *is not valid, then* ${}_2\boldsymbol{F}_1{}^2(\mathfrak{L}_{\nu,r})$ *is a subset of the right-hand side of* (7.8.18).

Now we discuss further generalizations of (7.8.1) and (7.8.2) in the forms

$$\left({}_2\boldsymbol{F}_1{}^{1;\sigma,\kappa} f\right)(x) = \frac{\Gamma(a)\Gamma(b)}{\Gamma(c)} x^{\sigma} \int_0^{\infty} {}_2F_1\left(a, b; c; -\frac{x}{t}\right) t^{\kappa-1} f(t)dt \tag{7.8.19}$$

and

$$\left({}_2\boldsymbol{F}_1{}^{2;\sigma,\kappa} f\right)(x) = \frac{\Gamma(a)\Gamma(b)}{\Gamma(c)} x^{\sigma-1} \int_0^{\infty} {}_2F_1\left(a, b; c; -\frac{t}{x}\right) t^{\kappa} f(t)dt \tag{7.8.20}$$

with $a, b, c, \sigma, \kappa \in \mathbb{C}$ ($\mathrm{Re}(a) > 0, \mathrm{Re}(b) > 0$). The Mellin transforms of (7.8.19) and (7.8.20) for a "sufficiently good" function f are given by

$$\left(\mathfrak{M}\, {}_2\boldsymbol{F}_1{}^{1;\sigma,\kappa} f\right)(s) = \mathcal{G}_{2,2}^{1,2}(s+\sigma)\left(\mathfrak{M}f\right)(s+\sigma+\kappa) \tag{7.8.21}$$

and

$$\left(\mathfrak{M}\, {}_2\boldsymbol{F}_1{}^{2;\sigma,\kappa} f\right)(s) = \mathcal{G}_{2,2}^{1,2}(1-s-\sigma)\left(\mathfrak{M}f\right)(s+\sigma+\kappa) \tag{7.8.22}$$

with $\mathcal{G}_{2,2}^{1,2}(s)$ being defined in (7.7.3). Then they are modified $\boldsymbol{G}$-transforms (6.2.4) and (6.2.5) of the forms

$$\left({}_2\boldsymbol{F}_1{}^{1;\sigma,\kappa} f\right)(x) = x^{\sigma} \int_0^{\infty} G_{2,2}^{1,2}\left[\frac{x}{t} \,\middle|\, \begin{matrix} 1-a, 1-b \\ 0, 1-c \end{matrix}\right] t^{\kappa-1} f(t)dt \tag{7.8.23}$$

and

$$\left({}_2\boldsymbol{F}_1{}^{2;\sigma,\kappa} f\right)(x) = x^{\sigma-1} \int_0^{\infty} G_{2,2}^{1,2}\left[\frac{t}{x} \,\middle|\, \begin{matrix} 1-a, 1-b \\ 0, 1-c \end{matrix}\right] t^{\kappa} f(t)dt. \tag{7.8.24}$$

Now we apply the results in Section 6.6 and 6.7 to obtain the $\mathfrak{L}_{\nu,2}$- and $\mathfrak{L}_{\nu,r}$-theory of the transforms (7.8.19) and (7.8.20), by taking into account that from (7.8.7) and (7.8.8) the condition $1-\nu+\mathrm{Re}(\kappa) \notin \mathcal{E}_9$ and $\nu - \mathrm{Re}(\kappa) \notin \mathcal{E}_9$ for the exceptional set $\mathcal{E}_9$ are equivalent to

$$s \neq c+i \quad (i = 0, 1, 2, \cdots) \quad \text{for} \quad \mathrm{Re}(s) = \nu - \mathrm{Re}(\kappa) \tag{7.8.25}$$

and

$$s \neq c+i \quad (i = 0, 1, 2, \cdots) \quad \text{for} \quad \mathrm{Re}(s) = 1-\nu+\mathrm{Re}(\kappa). \tag{7.8.26}$$

Theorems 6.50 and 6.62 yield the $\mathfrak{L}_{\nu,2}$-theory of the transforms ${}_2\boldsymbol{F}_1{}^{1;\sigma,\kappa}$ and ${}_2\boldsymbol{F}_1{}^{2;\sigma,\kappa}$.

Theorem 7.41. *Let* $0 < \nu - \mathrm{Re}(\kappa) < \min[\mathrm{Re}(a), \mathrm{Re}(b)]$.

(i) *There is a one-to-one transform* ${}_2\boldsymbol{F}_1{}^{1;\sigma,\kappa} \in [\mathfrak{L}_{\nu,2}, \mathfrak{L}_{\nu-\mathrm{Re}(\kappa+\sigma),2}]$ *such that* (7.8.21) *holds for* $\mathrm{Re}(s) = \nu - \mathrm{Re}(\kappa+\sigma)$ *and* $f \in \mathfrak{L}_{\nu,2}$.

(ii) *If* $f \in \mathfrak{L}_{\nu,2}$ *and* $g \in \mathfrak{L}_{1-\nu+\mathrm{Re}(\kappa+\sigma),2}$, *then the relation*

$$\int_0^{\infty} f(x)\left({}_2\boldsymbol{F}_1{}^{1;\sigma,\kappa} g\right)(x)dx = \int_0^{\infty} \left({}_2\boldsymbol{F}_1{}^{2;\kappa,\sigma} f\right)(x) g(x)dx \tag{7.8.27}$$

holds.

(iii) *Let $f \in \mathfrak{L}_{\nu,2}$ and $\lambda \in \mathbb{C}$. If $\mathrm{Re}(\lambda) > \nu - \mathrm{Re}(\kappa) - 1$, then ${}_2F_1{}^{1;\sigma,\kappa} f$ is given by*

$$\left({}_2F_1{}^{1;\sigma,\kappa} f\right)(x) = x^{\sigma-\lambda}\frac{d}{dx}x^{\lambda+1}\int_0^\infty G_{3,3}^{1,3}\left[\frac{x}{t}\,\middle|\,\begin{matrix}-\lambda, 1-a, 1-b\\ 0, 1-c, -\lambda-1\end{matrix}\right] t^{\kappa-1}f(t)dt. \quad (7.8.28)$$

When $\mathrm{Re}(\lambda) < \nu - \mathrm{Re}(\kappa) - 1$,

$$\left({}_2F_1{}^{1;\sigma,\kappa} f\right)(x) = -x^{\sigma-\lambda}\frac{d}{dx}x^{\lambda+1}\int_0^\infty G_{3,3}^{2,2}\left[\frac{x}{t}\,\middle|\,\begin{matrix}1-a, 1-b, -\lambda\\ -\lambda-1, 0, 1-c\end{matrix}\right] t^{\kappa-1}f(t)dt. \quad (7.8.29)$$

(iv) *The transform ${}_2F_1{}^{1;\sigma,\kappa}$ is independent of ν in the sense that if $0 < \nu - \mathrm{Re}(\kappa) < \min[\mathrm{Re}(a), \mathrm{Re}(b)]$ and $0 < \widetilde{\nu} - \mathrm{Re}(\kappa) < \min[\mathrm{Re}(a), \mathrm{Re}(b)]$ and if the transforms ${}_2F_1{}^{1;\sigma,\kappa}$ and ${}_2\widetilde{F}_1{}^{\sigma,\kappa}$ are defined on $\mathfrak{L}_{\nu,2}$ an $\mathfrak{L}_{\widetilde{\nu},2}$, respectively by (7.8.21), then ${}_2F_1{}^{1;\sigma,\kappa} f = {}_2\widetilde{F}_1{}^{\sigma,\kappa} f$ for $f \in \mathfrak{L}_{\nu,2} \cap \mathfrak{L}_{\widetilde{\nu},2}$.*

(v) *For $f \in \mathfrak{L}_{\nu,2}$ and $x > 0$, $\left({}_2F_1{}^{1;\sigma,\kappa} f\right)(x)$ is given in (7.8.19) and (7.8.23).*

Theorem 7.42. *Let $0 < 1 - \nu + \mathrm{Re}(\kappa) < \min[\mathrm{Re}(a), \mathrm{Re}(b)]$.*

(i) *There is a one-to-one transform ${}_2F_1{}^{2;\sigma,\kappa} \in [\mathfrak{L}_{\nu,2}, \mathfrak{L}_{\nu-\mathrm{Re}(\kappa+\sigma),2}]$ such that (7.8.22) holds for $\mathrm{Re}(s) = \nu - \mathrm{Re}(\kappa + \sigma)$ and $f \in \mathfrak{L}_{\nu,2}$.*

(ii) *If $f \in \mathfrak{L}_{\nu,2}$ and $g \in \mathfrak{L}_{1-\nu+\mathrm{Re}(\kappa+\sigma),2}$, then the relation*

$$\int_0^\infty f(x)\left({}_2F_1{}^{2;\sigma,\kappa} g\right)(x)dx = \int_0^\infty \left({}_2F_1{}^{1;\sigma,\kappa} f\right)(x)g(x)dx \quad (7.8.30)$$

holds.

(iii) *Let $f \in \mathfrak{L}_{\nu,2}$ and $\lambda \in \mathbb{C}$. If $\mathrm{Re}(\lambda) > -\nu + \mathrm{Re}(\kappa)$, then ${}_2F_1{}^{2;\sigma,\kappa} f$ is given by*

$$\left({}_2F_1{}^{2;\sigma,\kappa} f\right)(x) = -x^{\sigma+\lambda+1}\frac{d}{dx}x^{-\lambda-1}\int_0^\infty G_{3,3}^{1,3}\left[\frac{t}{x}\,\middle|\,\begin{matrix}-\lambda, 1-a, 1-b\\ 0, 1-c, -\lambda-1\end{matrix}\right] t^{\kappa}f(t)dt. \quad (7.8.31)$$

When $\mathrm{Re}(\lambda) < -\nu + \mathrm{Re}(\kappa)$,

$$\left({}_2F_1{}^{2;\sigma,\kappa} f\right)(x) = x^{\sigma+\lambda+1}\frac{d}{dx}x^{-\lambda-1}\int_0^\infty G_{3,3}^{2,2}\left[\frac{t}{x}\,\middle|\,\begin{matrix}1-a, 1-b, -\lambda\\ -\lambda-1, 0, 1-c\end{matrix}\right] t^{\kappa}f(t)dt. \quad (7.8.32)$$

(iv) *The transform ${}_2F_1{}^{2;\sigma,\kappa}$ is independent of ν in the sense that if $0 < 1 - \nu + \mathrm{Re}(\kappa) < \min[\mathrm{Re}(a), \mathrm{Re}(b)]$ and $0 < 1 - \widetilde{\nu} + \mathrm{Re}(\kappa) < \min[\mathrm{Re}(a), \mathrm{Re}(b)]$ and if the transforms ${}_2F_1{}^{2;\sigma,\kappa}$ and ${}_2\widetilde{F}_1{}^{2;\sigma,\kappa}$ are defined on $\mathfrak{L}_{\nu,2}$ an $\mathfrak{L}_{\widetilde{\nu},2}$, respectively by (7.8.22), then ${}_2F_1{}^{2;\sigma,\kappa} f = {}_2\widetilde{F}_1{}^{2;\sigma,\kappa} f$ for $f \in \mathfrak{L}_{\nu,2} \cap \mathfrak{L}_{\widetilde{\nu},2}$.*

(v) *For $f \in \mathfrak{L}_{\nu,2}$ and $x > 0$, $\left({}_2F_1{}^{2;\sigma,\kappa} f\right)(x)$ is given in (7.8.20) and (7.8.24).*

Theorems 6.54, 6.55 and 6.66, 6.67 give the $\mathfrak{L}_{\nu,r}$-theory of the transforms ${}_2F_1{}^{1;\sigma,\kappa}$ and ${}_2F_1{}^{2;\sigma,\kappa}$.

Theorem 7.43. *Let $\beta = \min[\mathrm{Re}(a), \mathrm{Re}(b)]$, $0 < \nu - \mathrm{Re}(\kappa) < \beta$ and $\omega = a + b - c - \beta$.*

(a) *If $1 \leqq r \leqq s \leqq \infty$, then the transform ${}_2F_1{}^{1;\sigma,\kappa}$ defined on $\mathfrak{L}_{\nu,2}$ can be extended to $\mathfrak{L}_{\nu,r}$ as an element of $[\mathfrak{L}_{\nu,r}, \mathfrak{L}_{\nu-\mathrm{Re}(\kappa+\sigma),s}]$. If $1 \leqq r \leqq 2$, then ${}_2F_1{}^{1;\sigma,\kappa}$ is a one-to-one transform from $\mathfrak{L}_{\nu,r}$ onto $\mathfrak{L}_{\nu-\mathrm{Re}(\kappa+\sigma),s}$.*

(b) *If* $1 \leqq r \leqq s \leqq \infty$, $f \in \mathfrak{L}_{\nu,r}$ and $g \in \mathfrak{L}_{1-\nu+\mathrm{Re}(\kappa+\sigma),s'}$ *with* $1/s + 1/s' = 1$, *then the relation* (7.8.27) *holds.*

(c) *If* $1 < r < \infty$ *and the condition* (7.8.25) *holds, then* ${}_2\boldsymbol{F}_1{}^{1;\sigma,\kappa}$ *is one-to-one on* $\mathfrak{L}_{\nu,r}$.

(d) *Let* $1 < r < \infty$ *and* $\mathrm{Re}(\omega) \geqq 0$. *If the condition* (7.8.25) *holds, then*

$$ {}_2\boldsymbol{F}_1{}^{1;\sigma,\kappa}(\mathfrak{L}_{\nu,r}) = \left(\mathbb{L}_{1,-\sigma}\mathbb{L}_{1,1+c-a-b+\sigma}\right)\left(\mathfrak{L}_{\nu-\mathrm{Re}(\kappa+\sigma),r}\right), \tag{7.8.33} $$

where $\mathbb{L}_{1,-\sigma}$ and $\mathbb{L}_{1,1+c-a-b+\sigma}$ *are given in* (3.3.3). *If* (7.8.25) *is not valid, then* ${}_2\boldsymbol{F}_1{}^{1;\sigma,\kappa}(\mathfrak{L}_{\nu,r})$ *is a subset of the right-hand side of* (7.8.33).

(e) *Let* $1 < r < \infty$ and $\mathrm{Re}(\omega) < 0$. *If the condition* (7.8.25) *holds, then*

$$ {}_2\boldsymbol{F}_1{}^{1;\sigma,\kappa}(\mathfrak{L}_{\nu,r}) = \left(I_{-;1,\sigma}^{-\omega}\mathbb{L}_{1,-\sigma}\mathbb{L}_{1,1-\beta+\sigma}\right)\left(\mathfrak{L}_{\nu-\mathrm{Re}(\kappa+\sigma),r}\right), \tag{7.8.34} $$

where $I_{-;1,\sigma}^{-\omega}$ *is given in* (3.3.2). *If* (7.8.25) *is not valid, then* ${}_2\boldsymbol{F}_1{}^{1;\sigma,\kappa}(\mathfrak{L}_{\nu,r})$ *is a subset of the right-hand side of* (7.8.34).

Theorem 7.44. *Let* $\beta = \min[\mathrm{Re}(a), \mathrm{Re}(b)]$, $0 < 1 - \nu + \mathrm{Re}(\kappa) < \beta$ and $\omega = a + b - c - \beta$

(a) *If* $1 \leqq r \leqq s \leqq \infty$, *then the transform* ${}_2\boldsymbol{F}_1{}^{2;\sigma,\kappa}$ *defined on* $\mathfrak{L}_{\nu,2}$ *can be extended to* $\mathfrak{L}_{\nu,r}$ *as an element of* $[\mathfrak{L}_{\nu,r}, \mathfrak{L}_{\nu-\mathrm{Re}(\kappa+\sigma),s}]$. *If* $1 \leqq r \leqq 2$, *then* ${}_2\boldsymbol{F}_1{}^{2;\sigma,\kappa}$ is a one-to-one transform *from* $\mathfrak{L}_{\nu,r}$ *onto* $\mathfrak{L}_{\nu-\mathrm{Re}(\kappa+\sigma),s}$.

(b) *If* $1 \leqq r \leqq s \leqq \infty$, $f \in \mathfrak{L}_{\nu,r}$ and $g \in \mathfrak{L}_{1-\nu+\mathrm{Re}(\kappa+\sigma),s'}$ *with* $1/s + 1/s' = 1$, *then the relation* (7.8.30) *holds.*

(c) *If* $1 < r < \infty$ *and the condition* (7.8.26) *holds, then* ${}_2\boldsymbol{F}_1{}^{2;\sigma,\kappa}$ *is one-to-one on* $\mathfrak{L}_{\nu,r}$.

(d) *Let* $1 < r < \infty$ and $\mathrm{Re}(\omega) \geqq 0$. *If the condition* (7.8.26) *holds, then*

$$ {}_2\boldsymbol{F}_1{}^{2;\sigma,\kappa}(\mathfrak{L}_{\nu,r}) = \left(\mathbb{L}_{-1,1-\sigma}\mathbb{L}_{-1,a+b-c+\sigma}\right)\left(\mathfrak{L}_{\nu-\mathrm{Re}(\kappa+\sigma),r}\right), \tag{7.8.35} $$

where $\mathbb{L}_{-1,1-\sigma}$ and $\mathbb{L}_{-1,a+b-c+\sigma}$ *are given in* (3.3.3). *If* (7.8.26) *is not valid, then* ${}_2\boldsymbol{F}_1{}^{2;\sigma,\kappa}(\mathfrak{L}_{\nu,r})$ *is a subset of the right-hand side of* (7.8.35).

(e) *Let* $1 < r < \infty$ and $\mathrm{Re}(\omega) < 0$. *If the condition* (7.8.26) *holds, then*

$$ {}_2\boldsymbol{F}_1{}^{2;\sigma,\kappa}(\mathfrak{L}_{\nu,r}) = \left(I_{0+;1,-\sigma}^{-\omega}\mathbb{L}_{-1,1-\sigma}\mathbb{L}_{-1,\beta+\sigma}\right)\left(\mathfrak{L}_{\nu-\mathrm{Re}(\kappa+\sigma),r}\right), \tag{7.8.36} $$

where $I_{0+;1,-\sigma}^{-\omega}$ *is given in* (3.3.1). *If* (7.8.26) *is not valid, then* ${}_2\boldsymbol{F}_1{}^{2;\sigma,\kappa}(\mathfrak{L}_{\nu,r})$ *is a subset of the right-hand side of* (7.8.36).

7.9. The Generalized Stieltjes Transform

As an application of the results in the previous section, we consider the integral transform

$$ \left(\mathfrak{S}_{\beta,\eta,\alpha}f\right)(x) = \frac{\Gamma(\beta+\eta+1)\Gamma(\beta+1)}{\Gamma(\alpha+\beta+\eta+1)} \cdot \int_0^\infty \left(\frac{t}{x}\right)^\beta {}_2F_1\left(\beta+\eta+1, \beta+1; \alpha+\beta+\eta+1; -\frac{t}{x}\right)\frac{f(t)}{x}dt \tag{7.9.1} $$

with $\alpha, \beta, \eta \in \mathbb{C}$ ($\mathrm{Re}(\beta+\eta)+1>0, \mathrm{Re}(\beta)+1>0$). When $\alpha=0$, by the known relation ${}_2F_1(a,b;a;z) = {}_1F_0(b;z) = (1-z)^{-b}$, (7.9.1) takes the form

$$\left(\mathfrak{S}_{\beta,\eta,0}f\right)(x) = \Gamma(\beta+1)\int_0^\infty \left(\frac{t}{x}\right)^\beta {}_1F_0\left(\beta+1;-\frac{t}{x}\right)\frac{f(t)}{x}dt$$

$$= \Gamma(\beta+1)\int_0^\infty \frac{t^\beta}{(t+x)^{\beta+1}}f(t)dt \tag{7.9.2}$$

and becomes the Stieltjes transform for $\beta=0$:

$$(\mathfrak{S}f)(x) = \int_0^\infty \frac{1}{t+x}f(t)dt. \tag{7.9.3}$$

The transform (7.9.1) is a particular case of the modified transform ${}_2\boldsymbol{F}_1{}^{2;\sigma,\kappa}$ (7.8.20) for which

$$\sigma=-\beta, \quad \kappa=\beta, \quad a=\beta+\eta+1, \quad b=\beta+1, \quad c=\alpha+\beta+\eta+1. \tag{7.9.4}$$

The relations (7.8.22), (7.8.24), (7.8.30) and (7.8.26) have the forms

$$\left(\mathfrak{M}\mathfrak{S}_{\beta,\eta,\alpha}f\right)(s) = \frac{\Gamma(1+\beta-s)\Gamma(\eta+s)\Gamma(s)}{\Gamma(\alpha+\eta+s)}\left(\mathfrak{M}f\right)(s), \tag{7.9.5}$$

$$\left(\mathfrak{S}_{\beta,\eta,\alpha}f\right)(x) = \int_0^\infty \left(\frac{t}{x}\right)^\beta G_{2,2}^{1,2}\left[\frac{t}{x}\,\middle|\,\begin{matrix}-\beta-\eta,-\beta\\ 0,-\alpha-\beta-\eta\end{matrix}\right]\frac{f(t)}{x}dt, \tag{7.9.6}$$

$$\int_0^\infty f(x)\left(\mathfrak{S}_{\beta,\eta,\alpha}g\right)(x)dx = \int_0^\infty \left(\mathfrak{S}^*_{\beta,\eta,\alpha}f\right)(x)g(x)dx \tag{7.9.7}$$

with

$$\left(\mathfrak{S}^*_{\beta,\eta,\alpha}f\right)(x) = \frac{\Gamma(\beta+\eta+1)\Gamma(\beta+1)}{\Gamma(\alpha+\beta+\eta+1)}$$

$$\cdot\int_0^\infty \left(\frac{x}{t}\right)^\beta {}_2F_1\left(\beta+\eta+1,\beta+1;\alpha+\beta+\eta+1;-\frac{x}{t}\right)\frac{f(t)}{t}dt \tag{7.9.8}$$

and

$$s \neq \alpha+\beta+\eta+1+i \quad (i=0,1,2,\cdots) \quad \text{for} \quad \mathrm{Re}(s) = 1-\nu+\mathrm{Re}(\beta). \tag{7.9.9}$$

Theorems 7.42 and 7.44 lead to the $\mathfrak{L}_{\nu,2}$- and $\mathfrak{L}_{\nu,r}$-theory of the generalized Stieltjes transform $\mathfrak{S}_{\beta,\eta,\alpha}$.

Theorem 7.45. *Let* $-\mathrm{Re}(\beta) < 1-\nu < 1+\min[0,\mathrm{Re}(\eta)]$.

(i) *There is a one-to-one transform* $\mathfrak{S}_{\beta,\eta,\alpha} \in [\mathfrak{L}_{\nu,2}]$ *such that the relation* (7.9.5) *holds for* $\mathrm{Re}(s)=\nu$ *and* $f \in \mathfrak{L}_{\nu,2}$.

(ii) *If* $f \in \mathfrak{L}_{\nu,2}$ *and* $g \in \mathfrak{L}_{1-\nu,2}$, *then the relation* (7.9.7) *holds for* $\mathfrak{S}_{\beta,\eta,\alpha}$.

(iii) *Let* $f \in \mathfrak{L}_{\nu,2}$ *and* $\lambda \in \mathbb{C}$. *If* $\mathrm{Re}(\lambda) > -\nu+\mathrm{Re}(\beta)$, *then* $\mathfrak{S}_{\beta,\eta,\alpha}f$ *is given by*

$$\left(\mathfrak{S}_{\beta,\eta,\alpha}f\right)(x)$$

$$= -x^{-\beta+\lambda+1}\frac{d}{dx}x^{-\lambda-1}\int_0^\infty G_{3,3}^{1,3}\left[\frac{t}{x}\,\middle|\,\begin{matrix}-\lambda,-\beta-\eta,-\beta\\ 0,-\alpha-\beta-\eta,-\lambda-1\end{matrix}\right]t^\beta f(t)dt. \tag{7.9.10}$$

When $\text{Re}(\lambda) < -\nu + \text{Re}(\beta)$,

$$\left(\mathfrak{S}_{\beta,\eta,\alpha}f\right)(x)$$

$$= x^{-\beta+\lambda+1}\frac{d}{dx}x^{-\lambda-1}\int_0^\infty G_{3,3}^{2,2}\left[\frac{t}{x}\,\middle|\,\begin{matrix}-\beta-\eta,-\beta,-\lambda\\ -\lambda-1,0,-\alpha-\beta-\eta\end{matrix}\right]t^{\beta}f(t)dt. \tag{7.9.11}$$

(iv) *The transform* $\mathfrak{S}_{\beta,\eta,\alpha}$ *is independent of* ν *in the sense that if* $-\text{Re}(\beta) < 1-\nu < 1+\min[0,\text{Re}(\eta)]$ *and* $-\text{Re}(\beta) < 1-\widetilde{\nu} < 1+\min[0,\text{Re}(\eta)]$ *and if the transforms* $\mathfrak{S}_{\beta,\eta,\alpha}$ *and* $\widetilde{\mathfrak{S}}_{\beta,\eta,\alpha}$ *are defined on* $\mathfrak{L}_{\nu,2}$ *and* $\mathfrak{L}_{\widetilde{\nu},2}$, *respectively by* (7.9.5), *then* $\mathfrak{S}_{\beta,\eta,\alpha}f = \widetilde{\mathfrak{S}}_{\beta,\eta,\alpha}f$ *for* $f \in \mathfrak{L}_{\nu,2}\cap\mathfrak{L}_{\widetilde{\nu},2}$.

(v) *For* $f \in \mathfrak{L}_{\nu,2}$ *and* $x > 0$, $\left(\mathfrak{S}_{\beta,\eta,\alpha}f\right)(x)$ *is given in* (7.9.1) *and* (7.9.6).

Theorem 7.46. *Let* $-\text{Re}(\beta) < 1-\nu < 1+\min[0,\text{Re}(\eta)]$.

(a) *If* $1 \leqq r \leqq s \leqq \infty$, *then the transform* $\mathfrak{S}_{\beta,\eta,\alpha}$ *defined on* $\mathfrak{L}_{\nu,2}$ *can be extended to* $\mathfrak{L}_{\nu,r}$ *as an element of* $[\mathfrak{L}_{\nu,r},\mathfrak{L}_{\nu,s}]$. *If* $1 \leqq r \leqq 2$, *then* $\mathfrak{S}_{\beta,\eta,\alpha}$ *is a one-to-one transform from* $\mathfrak{L}_{\nu,r}$ *onto* $\mathfrak{L}_{\nu,s}$.

(b) *If* $1 \leqq r \leqq s \leqq \infty$, $f \in \mathfrak{L}_{\nu,r}$ *and* $g \in \mathfrak{L}_{1-\nu,s'}$ *with* $1/s+1/s'=1$, *then the relation* (7.9.7) *holds.*

(c) *If* $1 < r < \infty$ *and the condition* (7.9.9) *holds, then* $\mathfrak{S}_{\beta,\eta,\alpha}$ *is one-to-one on* $\mathfrak{L}_{\nu,r}$.

(d) *Let* $1 < r < \infty$ *and* $\text{Re}(\alpha)+\min[0,\text{Re}(\eta)] \leqq 0$. *If the condition* (7.9.9) *holds, then*

$$\mathfrak{S}_{\beta,\eta,\alpha}(\mathfrak{L}_{\nu,r}) = \left(\mathbb{L}_{-1,1+\beta}\mathbb{L}_{-1,1-\alpha}\right)(\mathfrak{L}_{\nu,r}), \tag{7.9.12}$$

where $\mathbb{L}_{-1,1+\beta}$ *and* $\mathbb{L}_{-1,1-\alpha}$ *are given in* (3.3.3). *If* (7.9.9) *is not valid, then* $\mathfrak{S}_{\beta,\eta,\alpha}(\mathfrak{L}_{\nu,r})$ *is a subset of the right-hand side of* (7.9.13).

(e) *Let* $1 < r < \infty$ *and* $\text{Re}(\alpha)+\min[0,\text{Re}(\eta)] > 0$. *If the condition* (7.9.9) *holds, then*

$$\mathfrak{S}_{\beta,\eta,\alpha}(\mathfrak{L}_{\nu,r}) = \left(I_{0+;1,\beta}^{-\omega}\mathbb{L}_{-1,1+\beta}\mathbb{L}_{-1,0}\right)(\mathfrak{L}_{\nu,r}), \tag{7.9.13}$$

where $\omega = -\alpha-\min[0,\text{Re}(\eta)]$ *and* $I_{0+;1,\beta}^{-\omega}$ *is given in* (3.3.1). *If* (7.9.9) *is not valid, then* $\mathfrak{S}_{\beta,\eta,\alpha}(\mathfrak{L}_{\nu,r})$ *is a subset of the right-hand side of* (7.9.14).

7.10. ${}_p\boldsymbol{F}_q$-Transform

Let us investigate the integral transform of the type

$$\left({}_p\boldsymbol{F}_q\,f\right)(x) = \frac{\prod_{i=1}^{p}\Gamma(a_i)}{\prod_{j=1}^{q}\Gamma(b_j)}\int_0^\infty {}_pF_q(a_1,\cdots,a_p;b_1,\cdots,b_q;-xt)f(t)dt \tag{7.10.1}$$

with $a_i, b_j \in \mathbb{C}$ ($\text{Re}(a_i) > 0$; $i = 1,2,\cdots,p$; $j = 1,2,\cdots q$), for which we take $p \in \mathbb{N}$ and either $q = p$, $q = p+1$ or $q = p-1$. These transforms, containing the generalized hypergeometric function (2.9.2) in the kernel, generalize the transforms ${}_1\boldsymbol{F}_1$, ${}_1\boldsymbol{F}_2$ and ${}_2\boldsymbol{F}_1$ given in (7.5.1), (7.6.1) and (7.7.1), respectively.

Due to Prudnikov, Brychkov and Marichev [3, 8.4.51.1], the Mellin transform of (7.10.1) for a "sufficiently good" function f is given by the relation

$$\left(\mathfrak{M}\, {}_p\boldsymbol{F}_q f\right)(s) = \mathcal{G}^{1,p}_{p,q+1}(s)\left(\mathfrak{M}f\right)(1-s) \tag{7.10.2}$$

with

$$\mathcal{G}^{1,p}_{p,q+1}\left[\begin{array}{c}(1-a_i)_{1,p}\\ 0,(1-b_j)_{1,q}\end{array}\middle| s\right] = \frac{\Gamma(s)\prod_{i=1}^{p}\Gamma(a_i-s)}{\prod_{j=1}^{q}\Gamma(b_j-s)}, \tag{7.10.3}$$

provided that

$$0 < \mathrm{Re}(s) < \min_{1\leqq i\leqq p}\mathrm{Re}(a_i) \tag{7.10.4}$$

for $q = p$ and $q = p-1$, and

$$0 < \mathrm{Re}(s) < \min_{1\leqq i\leqq p}\mathrm{Re}(a_i), \quad \mathrm{Re}(s) < \frac{1}{4} - \frac{1}{2}\mathrm{Re}\left(\sum_{i=1}^{p} a_i - \sum_{i=1}^{p+1} b_i\right) \tag{7.10.5}$$

for $q = p+1$.

The relations (7.10.2) and (7.10.3) show that, for the cases $q = p$, $q = p+1$ and $q = p-1$, (7.10.1) is the $\boldsymbol{G}$-transform (1.5.1) of the forms

$$\left({}_p\boldsymbol{F}_p f\right)(x) = \int_0^\infty G^{1,p}_{p,p+1}\left[xt\,\middle|\begin{array}{c}1-a_1,\cdots,1-a_p\\ 0,1-b_1,\cdots,1-b_p\end{array}\right] f(t)dt, \tag{7.10.6}$$

$$\left({}_p\boldsymbol{F}_{p+1} f\right)(x) = \int_0^\infty G^{1,p}_{p,p+2}\left[xt\,\middle|\begin{array}{c}1-a_1,\cdots,1-a_p\\ 0,1-b_1,\cdots,1-b_{p+1}\end{array}\right] f(t)dt \tag{7.10.7}$$

and

$$\left({}_p\boldsymbol{F}_{p-1} f\right)(x) = \int_0^\infty G^{1,p}_{p,p}\left[xt\,\middle|\begin{array}{c}1-a_1,\cdots,1-a_p\\ 0,1-b_1,\cdots,1-b_{p-1}\end{array}\right] f(t)dt. \tag{7.10.8}$$

For the transforms (7.10.6), (7.10.7) and (7.10.8), the constants $a^*, \Delta, a_1^*, a_2^*, \alpha, \beta$ and μ in (6.1.5)–(6.1.11) take the forms

$$a^* = 1,\ \Delta = 1,\ a_1^* = 1,\ a_2^* = 0,\ \alpha = 0,\ \beta = \min_{1\leqq i\leqq p}[\mathrm{Re}(a_i)],\quad \mu = \sum_{i=1}^{p}(a_i - b_i) - \frac{1}{2}, \tag{7.10.9}$$

$$a^* = 0,\ \Delta = 2,\ a_1^* = 1,\ a_2^* = -1,\ \alpha = 0,\ \beta = \min_{1\leqq i\leqq p}\mathrm{Re}(a_i),\quad \mu = \sum_{i=1}^{p} a_i - \sum_{i=1}^{p+1} b_i - 1 \tag{7.10.10}$$

and

$$a^* = 2, \quad \Delta = 0, \quad a_1^* = 1, \quad a_2^* = 1,$$
$$\alpha = 0, \quad \beta = \min_{1 \leqq i \leqq p} \mathrm{Re}(a_i), \quad \mu = \sum_{i=1}^{p} a_i - \sum_{i=1}^{p-1} b_i, \tag{7.10.11}$$

respectively. According to these relations, the conditions in (7.10.4) and (7.10.5) take the forms

$$0 < \mathrm{Re}(s) < \beta \tag{7.10.12}$$

for $q = p$ and $q = p - 1$, and

$$0 < \mathrm{Re}(s) < \beta, \ \mathrm{Re}(s) < -\frac{1}{4} - \frac{1}{2}\mathrm{Re}(\mu) \tag{7.10.13}$$

for $q = p + 1$.

Let $\mathcal{E}_{11}$ be the exceptional set of the function $\mathcal{G}_{p,q+1}^{1,p}(s)$ in (7.10.3). Considering the poles of the gamma functions $\Gamma(b_j - s)$ $(1 \leqq j \leqq q)$ in the function $\mathcal{G}_{p,q+1}^{1,p}(s)$, we find that $\nu \notin \mathcal{E}_{11}$ means that

$$s \neq b_j + i \quad (j = 1, 2, \cdots, q; \ i = 0, 1, 2, \cdots) \quad \text{for} \quad \mathrm{Re}(s) = 1 - \nu. \tag{7.10.14}$$

First we state the $\mathfrak{L}_{\nu,2}$- and $\mathfrak{L}_{\nu,r}$-theory of the transform (7.10.1) in the case when $q = p$ by using Theorems 6.1, 6.2 and 6.6, 6.8.

Theorem 7.47. *Let β is be given in (7.10.9) and $0 < 1 - \nu < \beta$.*

(i) *There is a one-to-one transform ${}_p\boldsymbol{F}_p \in [\mathfrak{L}_{\nu,2}, \mathfrak{L}_{1-\nu,2}]$ such that (7.10.2) holds for $q = p$, $\mathrm{Re}(s) = 1 - \nu$ and $f \in \mathfrak{L}_{\nu,2}$.*

(ii) *If $f, g \in \mathfrak{L}_{\nu,2}$, then the relation*

$$\int_0^\infty f(x)\left({}_p\boldsymbol{F}_p g\right)(x)dx = \int_0^\infty \left({}_p\boldsymbol{F}_p f\right)(x)g(x)dx \tag{7.10.15}$$

holds.

(iii) *Let $f \in \mathfrak{L}_{\nu,2}$ and $\lambda \in \mathbb{C}$. If $\mathrm{Re}(\lambda) > -\nu$, then ${}_p\boldsymbol{F}_p f$ is given by*

$$\left({}_p\boldsymbol{F}_p f\right)(x)$$
$$= x^{-\lambda}\frac{d}{dx}x^{\lambda+1}\int_0^\infty G_{p+1,p+2}^{1,p+1}\left[xt \,\middle|\, \begin{matrix} -\lambda, 1 - a_1, \cdots, 1 - a_p \\ 0, 1 - b_1, \cdots, 1 - b_p, -\lambda - 1 \end{matrix}\right] f(t)dt. \tag{7.10.16}$$

When $\mathrm{Re}(\lambda) < -\nu$, then ${}_p\boldsymbol{F}_p f$ is given by

$$\left({}_p\boldsymbol{F}_p f\right)(x)$$
$$= -x^{-\lambda}\frac{d}{dx}x^{\lambda+1}\int_0^\infty G_{p+1,p+2}^{2,p}\left[xt \,\middle|\, \begin{matrix} 1 - a_1, \cdots, 1 - a_p, -\lambda \\ -\lambda - 1, 0, 1 - b_1, \cdots, 1 - b_p \end{matrix}\right] f(t)dt. \tag{7.10.17}$$

(iv) *The transform ${}_p\boldsymbol{F}_p$ is independent of ν in the sense that if $0 < 1-\nu < \beta$ and $0 < 1-\tilde{\nu} < \beta$ and if the transforms ${}_p\boldsymbol{F}_p$ and ${}_p\widetilde{\boldsymbol{F}}_p$ are defined on $\mathfrak{L}_{\nu,2}$ an $\mathfrak{L}_{\tilde{\nu},2}$, respectively by* (7.10.2) *with $q=p$, then ${}_p\boldsymbol{F}_pf = {}_p\widetilde{\boldsymbol{F}}_pf$ for $f \in \mathfrak{L}_{\nu,2} \cap \mathfrak{L}_{\tilde{\nu},2}$.*

(v) *For $f \in \mathfrak{L}_{\nu,2}$ and $x > 0$, $\left({}_p\boldsymbol{F}_p f\right)(x)$ is given in* (7.10.6) *as well as* (7.10.1) *with $q = p$.*

Theorem 7.48. *Let β and μ be given in* (7.10.9), *$0 < 1-\nu < \beta$, $\omega = \mu + 1/2$ and $1 \leqq r \leqq s \leqq \infty$.*

(a) *The transform ${}_p\boldsymbol{F}_p$ defined on $\mathfrak{L}_{\nu,2}$ can be extended to $\mathfrak{L}_{\nu,r}$ as an element of $[\mathfrak{L}_{\nu,r}, \mathfrak{L}_{1-\nu,s}]$. When $1 \leqq r \leqq 2$, ${}_p\boldsymbol{F}_p$ is a one-to-one transform from $\mathfrak{L}_{\nu,r}$ onto $\mathfrak{L}_{1-\nu,s}$.*

(b) *If $f \in \mathfrak{L}_{\nu,r}$ and $g \in \mathfrak{L}_{\nu,s'}$ with $1/s + 1/s' = 1$, then the relation* (7.10.15) *holds.*

(c) *If $1 < r < \infty$ and the condition* (7.10.14) *for $q = p$ holds, then the transform ${}_p\boldsymbol{F}_p$ is one-to-one on $\mathfrak{L}_{\nu,r}$.*

(d) *If $1 < r < \infty$, $\mathrm{Re}(\omega) \geqq 0$ and the condition* (7.10.14) *for $q = p$ holds, then*

$$ {}_p\boldsymbol{F}_p(\mathfrak{L}_{\nu,r}) = \mathbb{L}_{1,-\omega}(\mathfrak{L}_{\nu,r}), \tag{7.10.18} $$

where the operator $\mathbb{L}_{1,-\omega}$ is given in (3.3.3). *If the condition* (7.10.14) *for $q = p$ is not valid, then ${}_p\boldsymbol{F}_p(\mathfrak{L}_{\nu,r})$ is a subset of the right-hand side of* (7.10.18).

(e) *If $1 < r < \infty$, $\mathrm{Re}(\omega) < 0$ and the condition* (7.10.14) *for $q = p$ holds, then*

$$ {}_p\boldsymbol{F}_p(\mathfrak{L}_{\nu,r}) = \left(I_{-;1,0}^{-\omega}\mathbb{L}\right)(\mathfrak{L}_{\nu,r}), \tag{7.10.19} $$

where the operators $I_{-;1,0}^{-\omega}$ and $\mathbb{L}$ are given in (3.3.2) *and* (2.5.2). *If the condition* (7.10.14) *for $q = p$ is not valid, then ${}_p\boldsymbol{F}_p(\mathfrak{L}_{\nu,r})$ is a subset of the right-hand side of* (7.10.19).

Theorems 6.1, 6.2 and 6.5 lead to the $\mathfrak{L}_{\nu,2}$- and $\mathfrak{L}_{\nu,r}$-theory of the transform (7.10.1) in the case when $q = p+1$.

Theorem 7.49. *Let β and μ be given in* (7.10.10), *$0 < 1-\nu < \beta$ and $2(1-\nu) + \mathrm{Re}(\mu) \leqq 0$.*

(i) *There is a one-to-one transform ${}_p\boldsymbol{F}_{p+1} \in [\mathfrak{L}_{\nu,2}, \mathfrak{L}_{1-\nu,2}]$ such that* (7.10.2) *holds for $q = p+1$, $\mathrm{Re}(s) = 1-\nu$ and $f \in \mathfrak{L}_{\nu,2}$. If $2(1-\nu) + \mathrm{Re}(\mu) = 0$ and if the condition* (7.10.14) *with $q = p+1$ holds, then the transform ${}_p\boldsymbol{F}_{p+1}$ maps $\mathfrak{L}_{\nu,2}$ onto $\mathfrak{L}_{1-\nu,2}$.*

(ii) *If $f, g \in \mathfrak{L}_{\nu,2}$, then the relation*

$$ \int_0^\infty f(x)\left({}_p\boldsymbol{F}_{p+1}\, g\right)(x)dx = \int_0^\infty \left({}_p\boldsymbol{F}_{p+1}\, f\right)(x)g(x)dx \tag{7.10.20} $$

holds.

(iii) *Let $f \in \mathfrak{L}_{\nu,2}$ and $\lambda \in \mathbb{C}$. If $\mathrm{Re}(\lambda) > -\nu$, then ${}_p\boldsymbol{F}_{p+1}f$ is given by*

$$ \left({}_p\boldsymbol{F}_{p+1}\, f\right)(x) $$

$$ = x^{-\lambda}\frac{d}{dx}x^{\lambda+1}\int_0^\infty G_{p+1,p+3}^{1,p+1}\left[xt \,\middle|\, \begin{matrix} -\lambda, 1-a_1, \cdots, 1-a_p \\ 0, 1-b_1, \cdots, 1-b_{p+1}, -(\lambda+1) \end{matrix}\right] f(t)dt. \tag{7.10.21} $$

When $\mathrm{Re}(\lambda) < -\nu$,

$$ \left({}_p\boldsymbol{F}_{p+1}\, f\right)(x) $$

$$ = -x^{-\lambda}\frac{d}{dx}x^{\lambda+1}\int_0^\infty G_{p+1,p+3}^{2,p}\left[xt \,\middle|\, \begin{matrix} 1-a_1, \cdots, 1-a_p, -\lambda \\ -\lambda-1, 0, 1-b_1, \cdots, 1-b_{p+1} \end{matrix}\right] f(t)dt. \tag{7.10.22} $$

(iv) *The transform* ${}_p\boldsymbol{F}_{p+1}$ *is independent of* ν *in the sense that if* $0 < 1-\nu < \beta$, $2(1-\nu)+\text{Re}(\mu) \leqq 0$ *and* $0 < 1-\tilde{\nu} < \beta$, $2(1-\tilde{\nu})+\text{Re}(\mu) \leqq 0$ *and if the transforms* ${}_p\boldsymbol{F}_{p+1}$ *and* ${}_p\widetilde{\boldsymbol{F}}_{p+1}$ *are defined on* $\mathfrak{L}_{\nu,2}$ *and* $\mathfrak{L}_{\tilde{\nu},2}$, *respectively by* (7.10.2) *with* $q = p+1$, *then* ${}_p\boldsymbol{F}_{p+1}f = {}_p\widetilde{\boldsymbol{F}}_{p+1}f$ *for* $f \in \mathfrak{L}_{\nu,2} \cap \mathfrak{L}_{\tilde{\nu},2}$.

(v) *If* $2(1-\nu)+\text{Re}(\mu) < 0$, *then for* $f \in \mathfrak{L}_{\nu,2}$ *and* $x > 0$, $\left({}_p\boldsymbol{F}_{p+1}f\right)(x)$ *is given in* (7.10.7) *as well as* (7.10.1) *with* $q = p+1$.

Theorem 7.50. *Let* β *and* μ *be given in* (7.10.10), $0 < 1-\nu < \beta$, $1 < r < \infty$ *and* $2(1-\nu)+\text{Re}(\mu) \leqq 1/2-\gamma(r)$, *where* $\gamma(r)$ *is defined in* (3.3.9).

(a) *The transform* ${}_p\boldsymbol{F}_{p+1}$ *defined on* $\mathfrak{L}_{\nu,2}$ *can be extended to* $\mathfrak{L}_{\nu,r}$ *as an element of* $[\mathfrak{L}_{\nu,r}, \mathfrak{L}_{1-\nu,s}]$ *for all* s *with* $r \leqq s < \infty$ *such that* $s' \geqq [1/2-2(1-\nu)-\text{Re}(\mu)]^{-1}$ *with* $1/s + 1/s' = 1$.

(b) *If* $1 < r \leqq 2$, *the transform* ${}_p\boldsymbol{F}_{p+1}$ *is one-to-one on* $\mathfrak{L}_{\nu,r}$ *and there holds the equality* (7.10.2) *with* $q = p+1$ *and* $\text{Re}(s) = 1-\nu$.

(c) *If* $f \in \mathfrak{L}_{\nu,r}$ *and* $g \in \mathfrak{L}_{\nu,s}$ *with* $1 < s < \infty, 1/r+1/s \geqq 1$ *and* $2(1-\nu)+\text{Re}(\mu) \leqq 1/2 - \max[\gamma(r), \gamma(s)]$, *then the relation* (7.10.20) *holds.*

(d) *If the condition* (7.10.14) *with* $q = p+1$ *holds, then the transform* ${}_p\boldsymbol{F}_{p+1}$ *is one-to-one on* $\mathfrak{L}_{\nu,r}$. *If we set* $\eta = -\mu-1$, *then* $\text{Re}(\eta) > -1$ *and there holds*

$$ {}_p\boldsymbol{F}_{p+1}(\mathfrak{L}_{\nu,r}) = \left(M_{(\mu+1)/2}\mathbb{H}_{2,\eta}\right)\left(\mathfrak{L}_{\nu-\text{Re}(\mu)/2-1/2,r}\right), \tag{7.10.23} $$

where M_ζ *and* $\mathbb{H}_{2,\eta}$ *are given in* (3.3.11) *and* (3.3.4). *When the condition* (7.10.14) *with* $q = p+1$ *is not valid, then* ${}_p\boldsymbol{F}_{p+1}$ *is a subset of the right-hand side of* (7.10.23).

(e) *If* $f \in \mathfrak{L}_{\nu,r}$ *with* $\lambda \in \mathbb{C}$ *and* $2(1-\nu)+\text{Re}(\mu) \leqq 1/2-\gamma(r)$, *then* ${}_p\boldsymbol{F}_{p+1}f$ *is given in* (7.10.21) *for* $\text{Re}(\lambda) > -\nu$, *while in* (7.10.22) *for* $\text{Re}(\lambda) < -\nu$. *If* $2(1-\nu)+\text{Re}(\mu) < 0$, ${}_p\boldsymbol{F}_{p+1}f$ *is given in* (7.10.7) *as well as* (7.10.1) *with* $q = p+1$.

For the case $q = p+1$, we can apply the inversion formulas in Theorems 6.12 and 6.13.

Theorem 7.51. *Let* β *and* μ *be given in* (7.10.10), $0 < 1-\nu < \beta$, $\lambda \in \mathbb{C}$ *and* $\alpha_0 = 1 - \min\limits_{1 \leqq j \leqq p+1}[\text{Re}(b_j)]$.

(a) *If* $\nu > \alpha_0$ *and* $2(1-\nu)+\text{Re}(\mu) = 0$, *then for* $f \in \mathfrak{L}_{\nu,2}$ *the inversion formulas*

$$ f(x) = x^{-\lambda}\frac{d}{dx}x^{\lambda+1} $$

$$ \cdot \int_0^\infty G^{p+1,1}_{p+1,p+3}\left[xt \,\middle|\, \begin{matrix} -\lambda, a_1-1, \cdots, a_p-1 \\ b_1-1, \cdots, b_{p+1}-1, 0, -\lambda-1 \end{matrix}\right]\left({}_p\boldsymbol{F}_{p+1}f\right)(t)dt \tag{7.10.24} $$

and

$$ f(x) = -x^{-\lambda}\frac{d}{dx}x^{\lambda+1} $$

$$ \cdot \int_0^\infty G^{p+2,0}_{p+1,p+3}\left[xt \,\middle|\, \begin{matrix} a_1-1, \cdots, a_p-1, -\lambda \\ -\lambda-1, b_1-1, \cdots, b_{p+1}-1, 0 \end{matrix}\right]\left({}_p\boldsymbol{F}_{p+1}f\right)(t)dt \tag{7.10.25} $$

hold for $\text{Re}(\lambda) > \nu-1$ *and* $\text{Re}(\lambda) < \nu-1$, *respectively.*

(b) *Let $1 < r < \infty$, $\alpha_0 < \nu < \mathrm{Re}(\mu + 1/2)/2 + 1$ and $2(1-\nu) + \mathrm{Re}(\mu) \leqq 1/2 - \gamma(r)$, where $\gamma(r)$ is given in (3.3.9). If $f \in \mathfrak{L}_{\nu,r}$, then the relations (7.10.24) and (7.10.25) hold for $\mathrm{Re}(\lambda) > \nu - 1$ and for $\mathrm{Re}(\lambda) < \nu - 1$, respectively.*

Lastly we obtain the $\mathfrak{L}_{\nu,2}$- and $\mathfrak{L}_{\nu,r}$-theory of the transform (7.10.1) in the case when $q = p - 1$ from Theorems 6.1, 6.2 and 6.6, 6.7.

Theorem 7.52. *Let $p \geqq 1$, β be given in (7.10.11) and $0 < 1 - \nu < \beta$.*

(i) *There is a one-to-one transform ${}_p\boldsymbol{F}_{p-1} \in [\mathfrak{L}_{\nu,2}, \mathfrak{L}_{1-\nu,2}]$ such that (7.10.2) holds for $q = p - 1$, $\mathrm{Re}(s) = 1 - \nu$ and $f \in \mathfrak{L}_{\nu,2}$.*

(ii) *If $f, g \in \mathfrak{L}_{\nu,2}$, then the relation*

$$\int_0^\infty f(x)\left({}_p\boldsymbol{F}_{p-1}\, g\right)(x)dx = \int_0^\infty \left({}_p\boldsymbol{F}_{p-1}\, f\right)(x)g(x)dx \tag{7.10.26}$$

holds.

(iii) *Let $f \in \mathfrak{L}_{\nu,2}$ and $\lambda \in \mathbb{C}$. If $\mathrm{Re}(\lambda) > -\nu$, then ${}_p\boldsymbol{F}_{p-1} f$ is given by*

$$\left({}_p\boldsymbol{F}_{p-1}\, f\right)(x)$$

$$= x^{-\lambda}\frac{d}{dx}x^{\lambda+1}\int_0^\infty G^{1,p+1}_{p+1,p+1}\left[xt \,\middle|\, \begin{matrix} -\lambda, 1-a_1, \cdots, 1-a_p \\ 0, 1-b_1, \cdots, 1-b_{p-1}, -\lambda-1 \end{matrix}\right] f(t)dt. \tag{7.10.27}$$

When $\mathrm{Re}(\lambda) < -\nu$,

$$\left({}_p\boldsymbol{F}_{p-1}\, f\right)(x)$$

$$= -x^{-\lambda}\frac{d}{dx}x^{\lambda+1}\int_0^\infty G^{2,p}_{p+1,p+1}\left[xt \,\middle|\, \begin{matrix} 1-a_1, \cdots, 1-a_p, -\lambda \\ -\lambda-1, 0, 1-b_1, \cdots, 1-b_{p-1} \end{matrix}\right] f(t)dt. \tag{7.10.28}$$

(iv) *The transform ${}_p\boldsymbol{F}_{p-1}$ is independent of ν in the sense that if $0 < 1 - \nu < \beta$ and $0 < 1 - \tilde{\nu} < \beta$ and if the transforms ${}_p\boldsymbol{F}_{p-1}$ and ${}_p\widetilde{\boldsymbol{F}}_{p-1}$ are defined on $\mathfrak{L}_{\nu,2}$ and $\mathfrak{L}_{\tilde{\nu},2}$, respectively by (7.10.2) with $q = p - 1$, then ${}_p\boldsymbol{F}_{p-1} f = {}_p\widetilde{\boldsymbol{F}}_{p-1} f$ for $f \in \mathfrak{L}_{\nu,2} \cap \mathfrak{L}_{\tilde{\nu},2}$.*

(v) *For $f \in \mathfrak{L}_{\nu,2}$ and $x > 0$, $\left({}_p\boldsymbol{F}_{p-1}\, f\right)(x)$ is given in (7.10.8) as well as (7.10.1) with $q = p - 1$.*

Theorem 7.53. *Let $p \geqq 1$, β and μ be given in (7.10.11), $0 < 1 - \nu < \beta$, $\omega = \mu - \beta + 1$ and $1 \leqq r \leqq s \leqq \infty$.*

(a) *The transform ${}_p\boldsymbol{F}_{p-1}$ defined on $\mathfrak{L}_{\nu,2}$ can be extended to $\mathfrak{L}_{\nu,r}$ as an element of $[\mathfrak{L}_{\nu,r}, \mathfrak{L}_{1-\nu,s}]$. If $1 \leqq r \leqq 2$, then ${}_p\boldsymbol{F}_{p-1}$ is a one-to-one transform from $\mathfrak{L}_{\nu,r}$ onto $\mathfrak{L}_{1-\nu,s}$.*

(b) *If $f \in \mathfrak{L}_{\nu,r}$ and $g \in \mathfrak{L}_{\nu,s'}$ with $1/s + 1/s' = 1$, then the relation (7.10.26) holds.*

(c) *If $1 < r < \infty$ and the condition (7.10.14) for $q = p - 1$ holds, then ${}_p\boldsymbol{F}_{p-1}$ is a one-to-one transform on $\mathfrak{L}_{\nu,r}$.*

(d) *If $1 < r < \infty$, $\mathrm{Re}(\omega) \geqq 0$ and the condition (7.10.14) for $q = p - 1$ holds, then*

$${}_p\boldsymbol{F}_{p-1}(\mathfrak{L}_{\nu,r}) = \left(\mathbb{L}\mathbb{L}_{1,-\mu}\right)(\mathfrak{L}_{1-\nu,r}), \tag{7.10.29}$$

where the operators $\mathbb{L}$ *and* $\mathbb{L}_{1,-\mu}$ *are given in* (2.5.2) *and* (3.3.3). *If the condition* (7.10.14) *for* $q = p-1$ *is not valid, then* ${}_p\boldsymbol{F}_{p-1}(\mathfrak{L}_{\nu,r})$ *is a subset of the right-hand side of* (7.10.29).

(e) *If* $1 < r < \infty$, $\text{Re}(\omega) < 0$ *and the condition* (7.10.14) *for* $q = p-1$ *holds, then*

$${}_p\boldsymbol{F}_{p-1}(\mathfrak{L}_{\nu,r}) = \left(I^{-\omega}_{-;1,0}\mathbb{L}\mathbb{L}_{1,1-\beta}\right)(\mathfrak{L}_{1-\nu,r}), \tag{7.10.30}$$

where the operator $I^{-\omega}_{-;1,0}$ *is given in* (3.3.2). *If the condition* (7.10.14) *for* $q = p-1$ *is not valid, then* ${}_p\boldsymbol{F}_{p-1}(\mathfrak{L}_{\nu,r})$ *is a subset of the right-hand side of* (7.10.30).

7.11. The Wright Transform

We consider the integral transform of the form

$$\left({}_p\Psi_q f\right)(x) = \int_0^\infty {}_p\Psi_q\left[\begin{array}{c|} (a_i,\alpha_i)_{1,p} \\ (b_j,\beta_j)_{1,q}\end{array}\, xt\right] f(t)dt \tag{7.11.1}$$

containing the Wright function ${}_p\Psi_q(z)$ defined in (2.9.30) in the kernel. Due to (2.9.29) this transform is the $\boldsymbol{H}$-transform (3.1.1) of the form

$$\left({}_p\Psi_q f\right)(x) = \int_0^\infty H^{1,p}_{p,q+1}\left[xt \,\middle|\, \begin{array}{c} (1-a_i,\alpha_i)_{1,p} \\ (0,1),(1-b_j,\beta_j)_{1,q}\end{array}\right] f(t)dt. \tag{7.11.2}$$

Therefore the Mellin transform of (7.11.1) is given by

$$\left(\mathfrak{M}\,{}_p\Psi_q f\right)(s) = \mathcal{H}^{1,p}_{p,q+1}(s)\left(\mathfrak{M}f\right)(1-s), \tag{7.11.3}$$

where

$$\mathcal{H}^{1,p}_{p,q+1}(s) = \mathcal{H}^{1,p}_{p,q+1}\left[\begin{array}{c|} (1-a_i,\alpha_i)_{1,p} \\ (0,1),(1-b_j,\beta_j)_{1,q}\end{array}\, s\right] = \frac{\Gamma(s)\prod_{i=1}^{p}\Gamma(a_i-\alpha_i s)}{\prod_{j=1}^{q}\Gamma(b_j-\beta_j s)}. \tag{7.11.4}$$

Applying the results in Sections 3.6 and 4.1–4.8, we can construct the $\mathfrak{L}_{\nu,2}$- and $\mathfrak{L}_{\nu,r}$-theory of the transform (7.11.1) by taking into account that the constants a^*, Δ, δ, μ, a_1^*, a_2^*, α and β in (1.1.7)–(1.1.13), (3.4.1) and (3.4.2) have the forms

$$a^* = \sum_{i=1}^{p}\alpha_i - \sum_{j=1}^{q}\beta_j + 1; \tag{7.11.5}$$

$$\Delta = \sum_{j=1}^{q}\beta_j - \sum_{i=1}^{p}\alpha_i + 1; \tag{7.11.6}$$

$$\delta = \prod_{i=1}^{p}\alpha_i^{-\alpha_i}\prod_{j=1}^{q}\beta_j^{\beta_j}; \tag{7.11.7}$$

$$\mu = \sum_{i=1}^{p}a_i - \sum_{j=1}^{q}b_j - \frac{p-q+1}{2}; \tag{7.11.8}$$

$$a_1^* = 1; \quad a_2^* = \sum_{i=1}^{p}\alpha_i - \sum_{j=1}^{q}\beta_j \tag{7.11.9}$$

and

$$\alpha = 0, \quad \beta = \min_{1 \leqq i \leqq p} \left[\frac{\mathrm{Re}(a_i)}{\alpha_i} \right]. \tag{7.11.10}$$

Let $\mathcal{E}_{12}$ be the exceptional set of the function $\mathcal{H}^{1,p}_{p,q+1}(s)$ in (7.11.4) and $s \notin \mathcal{E}_{12}$ means that

$$s \neq \frac{b_j + i}{\beta_j} \quad (i = 0, 1, 2, \cdots; \ j = 1, \cdots, q) \quad \text{for} \quad \mathrm{Re}(s) = 1 - \nu. \tag{7.11.11}$$

Now Theorems 3.6 and 3.7 lead to the $\mathfrak{L}_{\nu,2}$-theory of the transform (7.11.1).

Theorem 7.54. *Let a^*, Δ, μ and β be given in* (7.11.5), (7.11.6), (7.11.8) *and* (7.11.10). *We suppose that* **(a)** $0 < 1 - \nu < \beta$ *and that either of conditions* **(b)** $a^* > 0$, *or* **(c)** $a^* = 0, \Delta(1-\nu) + \mathrm{Re}(\mu) \leqq 0$ *holds. Then we have the following results:*

(i) *There is a one-to-one transform ${}_p\Psi_q \in [\mathfrak{L}_{\nu,2}, \mathfrak{L}_{1-\nu,2}]$ such that* (7.11.3) *holds for* $\mathrm{Re}(s) = 1 - \nu$ *and* $f \in \mathfrak{L}_{\nu,2}$. *If* $a^* = 0$, $\Delta(1-\nu) + \mathrm{Re}(\mu) = 0$ *and the condition* (7.11.11) *holds, then the transform ${}_p\Psi_q$ maps $\mathfrak{L}_{\nu,2}$ onto $\mathfrak{L}_{1-\nu,2}$.*

(ii) *If $f, g \in \mathfrak{L}_{\nu,2}$, then the relation*

$$\int_0^\infty f(x)\left({}_p\Psi_q g\right)(x)dx = \int_0^\infty \left({}_p\Psi_q f\right)(x)g(x)dx \tag{7.11.12}$$

holds.

(iii) *Let $f \in \mathfrak{L}_{\nu,2}$, $\lambda \in \mathbb{C}$ and $h > 0$. If* $\mathrm{Re}(\lambda) > (1-\nu)h - 1$, *then ${}_p\Psi_q f$ is given by*

$$\left({}_p\Psi_q f\right)(x) = hx^{1-(\lambda+1)/h}\frac{d}{dx}x^{(\lambda+1)/h}$$

$$\cdot \int_0^\infty H^{1,p+1}_{p+1,q+2}\left[xt \,\middle|\, \begin{matrix} (-\lambda, h), (1-a_i, \alpha_i)_{1,p} \\ (0,1), (1-b_j, \beta_j)_{1,q}, (-\lambda-1, h) \end{matrix}\right] f(t)dt. \tag{7.11.13}$$

When $\mathrm{Re}(\lambda) < (1-\nu)h - 1$,

$$\left({}_p\Psi_q f\right)(x) = -hx^{1-(\lambda+1)/h}\frac{d}{dx}x^{(\lambda+1)/h}$$

$$\cdot \int_0^\infty H^{2,p}_{p+1,q+2}\left[xt \,\middle|\, \begin{matrix} (1-a_i, \alpha_i)_{1,p}, (-\lambda, h) \\ (-\lambda-1, h), (0,1), (1-b_j, \beta_j)_{1,q} \end{matrix}\right] f(t)dt. \tag{7.11.14}$$

(iv) *The transform ${}_p\Psi_q$ is independent of ν in the sense that if ν and $\widetilde{\nu}$ satisfy* (a), *and* (b) *or* (c), *and if the transforms ${}_p\Psi_q$ and ${}_p\widetilde{\Psi}_q$ are defined on $\mathfrak{L}_{\nu,2}$ and $\mathfrak{L}_{\widetilde{\nu},2}$, respectively by* (7.11.3), *then ${}_p\Psi_q f = {}_p\widetilde{\Psi}_q f$ for $f \in \mathfrak{L}_{\nu,2} \cap \mathfrak{L}_{\widetilde{\nu},2}$.*

(v) *If either* **(b)** $a^* > 0$ *or* **(d)** $a^* = 0$, $\Delta(1-\nu) + \mathrm{Re}(\mu) < 0$, *then for $f \in \mathfrak{L}_{\nu,2}$ and $x > 0$, $\left({}_p\Psi_q f\right)(x)$ is given in* (7.11.1) *and* (7.11.2).

From Theorems 4.1, 4.2 and 4.3, 4.4 we obtain the $\mathfrak{L}_{\nu,r}$-theory of the transform (7.11.1) in the case when $a^* = 0$.

Theorem 7.55. *Let a^*, Δ, μ and β given in* (7.11.5), (7.11.6), (7.11.8) *and* (7.11.10) *be such that* $a^* = \Delta = 0, \mathrm{Re}(\mu) = 0$ *and* $0 < 1 - \nu < \beta$.

(a) *The transform ${}_p\Psi_q$ defined on $\mathfrak{L}_{\nu,2}$ can be extended to $\mathfrak{L}_{\nu,r}$ as an element of $[\mathfrak{L}_{\nu,r}, \mathfrak{L}_{1-\nu,r}]$ for $1 < r < \infty$.*

(b) *If $1 < r \leqq 2$, the transform ${}_p\Psi_q$ is one-to-one on $\mathfrak{L}_{\nu,r}$ and there holds the equality (7.11.3) for $f \in \mathfrak{L}_{\nu,r}$.*

(c) *If the condition (7.11.11) is satisfied, then ${}_p\Psi_q$ is one-to-one on $\mathfrak{L}_{\nu,r}$ and there holds*

$$ {}_p\Psi_q(\mathfrak{L}_{\nu,r}) = \mathfrak{L}_{1-\nu,r}. \tag{7.11.15} $$

(d) *If $f \in \mathfrak{L}_{\nu,r}$ and $g \in \mathfrak{L}_{\nu,r'}$ with $1 < r < \infty$ and $r' = r/(r-1)$, then the relation (7.11.12) holds.*

(e) *If $1 < r < \infty, f \in \mathfrak{L}_{\nu,r}$, $\lambda \in \mathbb{C}$ and $h > 0$, then ${}_p\Psi_q f$ is given in (7.11.13) for $\mathrm{Re}(\lambda) > (1-\nu)h - 1$, while ${}_p\Psi_q f$ in (7.11.14) for $\mathrm{Re}(\lambda) < (1-\nu)h - 1$.*

Theorem 7.56. *Let a^*, Δ, μ and β given in (7.11.5), (7.11.6), (7.11.8) and (7.11.10) be such that $a^* = \Delta = 0, \mathrm{Re}(\mu) < 0$ and $0 < 1 - \nu < \beta$.*

(a) *Let $1 < r < \infty$. The transform ${}_p\Psi_q$ defined on $\mathfrak{L}_{\nu,2}$ can be extended to $\mathfrak{L}_{\nu,r}$ as an element of $[\mathfrak{L}_{\nu,r}, \mathfrak{L}_{1-\nu,s}]$ for all $s \geqq r$ such that $1/s > 1/r + \mathrm{Re}(\mu)$.*

(b) *If $1 < r \leqq 2$, then ${}_p\Psi_q$ is a one-to-one transform on $\mathfrak{L}_{\nu,r}$ and there holds the equality (7.11.3).*

(c) *If the condition (7.11.11) is satisfied, then the transform ${}_p\Psi_q$ is one-to-one on $\mathfrak{L}_{\nu,r}$ and there hold*

$$ {}_p\Psi_q\left(\mathfrak{L}_{\nu,r}\right) = I^{-\mu}_{-;k,0}\left(\mathfrak{L}_{1-\nu,r}\right) \tag{7.11.16} $$

for $k \geqq 1$, and

$$ {}_p\Psi_q\left(\mathfrak{L}_{\nu,r}\right) = I^{-\mu}_{0+;k,\beta/k-1}\left(\mathfrak{L}_{1-\nu,r}\right) \tag{7.11.17} $$

for $0 < k \leqq 1$ and $p > 0$, where the operators $I^{-\mu}_{-;k,0}$ and $I^{-\mu}_{0+;k,\beta/k-1}$ are defined in (3.3.1) and (3.3.2). If the condition (7.11.11) is not satisfied, ${}_p\Psi_q\left(\mathfrak{L}_{\nu,r}\right)$ is a subset of the right-hand sides of (7.11.16) and (7.11.17) in the respective cases.

(d) *If $f \in \mathfrak{L}_{\nu,r}$ and $g \in \mathfrak{L}_{\nu,s}$ with $1 < r < \infty, 1 < s < \infty$ and $1 \leqq 1/r + 1/s < 1 - \mathrm{Re}(\mu)$, then the relation (7.11.12) holds.*

(e) *If $1 < r < \infty$, $f \in \mathfrak{L}_{\nu,r}$, $\lambda \in \mathbb{C}$ and $h > 0$, then ${}_p\Psi_q f$ is given in (7.11.13) for $\mathrm{Re}(\lambda) > (1-\nu)h - 1$, while ${}_p\Psi_q f$ in (7.11.14) for $\mathrm{Re}(\lambda) < (1-\nu)h - 1$. Furthermore ${}_p\Psi_q f$ is given in (7.11.1) and (7.11.2).*

Theorem 7.57. *Let a^*, Δ, μ and β given in (7.11.5), (7.11.6), (7.11.8) and (7.11.10) be such that $a^* = 0$, $\Delta \neq 0$, $0 < 1 - \nu < \beta, 1 < r < \infty$ and $\Delta(1-\nu) + \mathrm{Re}(\mu) \leqq 1/2 - \gamma(r)$, where $\gamma(r)$ is defined in (3.3.9). Further assume that $p > 0$ if $\Delta < 0$.*

(a) *The transform ${}_p\Psi_q$ defined on $\mathfrak{L}_{\nu,2}$ can be extended to $\mathfrak{L}_{\nu,r}$ as an element of $[\mathfrak{L}_{\nu,r}, \mathfrak{L}_{1-\nu,s}]$ for all s with $r \leqq s < \infty$ such that $s' \geqq [1/2 - \Delta(1-\nu) - \mathrm{Re}(\mu)]^{-1}$ with $1/s + 1/s' = 1$.*

(b) *If $1 < r \leqq 2$, the transform ${}_p\Psi_q$ is one-to-one on $\mathfrak{L}_{\nu,r}$ and there holds the equality (7.11.3).*

(c) *If the condition (7.11.11) is satisfied, then the transform ${}_p\Psi_q$ is one-to-one on $\mathfrak{L}_{\nu,r}$. If we set $\eta = -\mu - 1$ for $\Delta > 0$ and $\eta = -\Delta\beta - \mu - 1$ for $\Delta < 0$, then $\mathrm{Re}(\eta) > -1$ and there* *holds*

$$ {}_p\Psi_q(\mathfrak{L}_{\nu,r}) = \left(M_{\mu/\Delta+1/2}H_{\Delta,\eta}\right)\left(\mathfrak{L}_{\nu-\mathrm{Re}(\mu)/\Delta-1/2,r}\right). \tag{7.11.18}$$

If the condition (7.11.11) *is not satisfied,* ${}_p\Psi_q(\mathfrak{L}_{\nu,r})$ *is a subset of the right-hand side of* (7.11.18).

(d) *If* $f \in \mathfrak{L}_{\nu,r}$ *and* $g \in \mathfrak{L}_{\nu,s}$ *with* $1 < s < \infty, 1/r+1/s \geqq 1$ *and* $\Delta(1-\nu)+\mathrm{Re}(\mu) \leqq 1/2 - \max[\gamma(r),\gamma(s)]$, *then the relation* (7.11.12) *holds.*

(e) *If* $f \in \mathfrak{L}_{\nu,r}, \lambda \in \mathbb{C}, h > 0$ *and* $\Delta(1-\nu)+\mathrm{Re}(\mu) \leqq 1/2 - \gamma(r)$, *then* ${}_p\Psi_q f$ *is given in* (7.11.13) *for* $\mathrm{Re}(\lambda) > (1-\nu)h-1$, *while* ${}_p\Psi_q f$ *in* (7.11.14) *for* $\mathrm{Re}(\lambda) < (1-\nu)h-1$. *If* $\Delta(1-\nu)+\mathrm{Re}(\mu) < 0$, ${}_p\Psi_q f$ *is given in* (7.11.1) *and* (7.11.2).

From Theorems 4.5–4.7 and 4.9 we obtain the $\mathfrak{L}_{\nu,r}$-theory of the transform (7.11.1) in the case when $a^* \neq 0$.

Theorem 7.58. *Let* a^* *and* β *given in* (7.11.5) *and* (7.11.10) *be such that* $a^* > 0$, $0 < 1-\nu < \beta$ *and* $1 \leqq r \leqq s \leqq \infty$.

(a) *The transform* ${}_p\Psi_q$ *defined on* $\mathfrak{L}_{\nu,2}$ *can be extended to* $\mathfrak{L}_{\nu,r}$ *as an element of* $[\mathfrak{L}_{\nu,r}, \mathfrak{L}_{1-\nu,s}]$. *When* $1 \leqq r \leqq 2$, *the transform* ${}_p\Psi_q$ *is one-to-one from* $\mathfrak{L}_{\nu,r}$ *onto* $\mathfrak{L}_{1-\nu,s}$.

(b) *If* $f \in \mathfrak{L}_{\nu,r}$ *and* $g \in \mathfrak{L}_{\nu,s'}$ *with* $1/s+1/s' = 1$, *then the relation* (7.11.12) *holds.*

Theorem 7.59. *Let* μ, a_2^* *and* β *given in* (7.11.8), (7.11.9) *and* (7.11.10) *be such that* $a_2^* > 0$, $0 < 1-\nu < \beta$ *and* $\omega = \mu - a_2^*\beta + 1$ *and let* $1 < r < \infty$. *Let further* $p > 0$.

(a) *Let the condition* (7.11.11) *be satisfied, or if* $1 \leqq r \leqq 2$, *then the transform* ${}_p\Psi_q$ *is one-to-one on* $\mathfrak{L}_{\nu,r}$.

(b) *If* $\mathrm{Re}(\omega) \geqq 0$ *and the condition* (7.11.11) *is satisfied, then*

$$ {}_p\Psi_q(\mathfrak{L}_{\nu,r}) = \left(\mathbb{L}\mathbb{L}_{a_2^*,1-\beta-\omega/a_2^*}\right)(\mathfrak{L}_{1-\nu,r}), \tag{7.11.19}$$

where $\mathbb{L}$ *and* $\mathbb{L}_{k,\alpha}$ *are given in* (2.5.2) *and* (3.3.3). *If the condition* (7.11.11) *is not satisfied, then* ${}_p\Psi_q(\mathfrak{L}_{\nu,r})$ *is a subset of the right-hand side of* (7.11.19).

(c) *If* $\mathrm{Re}(\omega) < 0$ *and the condition* (7.11.11) *is satisfied, then*

$$ {}_p\Psi_q(\mathfrak{L}_{\nu,r}) = \left(I_{-;1,0}^{-\omega}\mathbb{L}\mathbb{L}_{a_2^*,1-\beta}\right)(\mathfrak{L}_{1-\nu,r}), \tag{7.11.20}$$

where $I_{-;1,0}^{-\omega}$ *is given in* (3.3.2). *If the condition* (7.11.11) *is not satisfied, then* ${}_p\Psi_q(\mathfrak{L}_{\nu,r})$ *is a subset of the right-hand side of* (7.11.20).

Theorem 7.60. *Let* μ, a_2^* *and* β *given in* (7.11.8), (7.11.9) *and* (7.11.10) *be such that* $a_2^* = 0$, $0 < 1-\nu < \beta$ *and* $\omega = \mu + 1/2$ *and let* $1 < r < \infty$.

(a) *Let the condition* (7.11.11) *be satisfied, or if* $1 < r \leqq 2$, *then* ${}_p\Psi_q$ *is a one-to-one transform on* $\mathfrak{L}_{\nu,r}$.

(b) *If* $\mathrm{Re}(\omega) \geqq 0$ *and the condition* (7.11.11) *is satisfied, then*

$$ {}_p\Psi_q(\mathfrak{L}_{\nu,r}) = \mathbb{L}_{1,-\omega}(\mathfrak{L}_{\nu,r}), \tag{7.11.21}$$

where $\mathbb{L}_{1,-\omega}$ *is given in* (3.3.3). *If the condition* (7.11.11) *is not satisfied, then* ${}_p\Psi_q(\mathfrak{L}_{\nu,r})$ *is a subset of the right-hand side of* (7.11.21).

(c) *If* $\mathrm{Re}(\omega) < 0$ *and the condition* (7.11.11) *is satisfied, then*

$$ {}_p\Psi_q(\mathfrak{L}_{\nu,r}) = \left(I_{-;1,0}^{-\omega}\mathbb{L}\right)(\mathfrak{L}_{\nu,r}), \tag{7.11.22}$$

where $I_{-;1,0}^{-\omega}$ *and* $\mathbb{L}$ *are given in* (3.3.2) *and* (2.5.2). *If the condition* (7.11.11) *is not satisfied, then* ${}_p\Psi_q(\mathfrak{L}_{\nu,r})$ *is a subset of the right-hand side of* (7.11.22).

Theorem 7.61. *Let* a^*, μ, a_2^* *and* β *given in* (7.11.5), (7.11.8), (7.11.9) *and* (7.11.10) *be such that* $a^* > 0$, $a_2^* < 0$, $0 < 1 - \nu < \beta$ *and let* $1 < r < \infty$.

(a) *Let the condition* (7.11.11) *be satisfied, or if* $1 < r \leqq 2$, *then the transform* ${}_p\Psi_q$ *is one-to-one on* $\mathfrak{L}_{\nu,r}$.

(b) *Let* ω, η, ζ *be chosen as*

$$\omega = a^*\eta - \mu - \frac{1}{2}, \tag{7.11.23}$$

$$a^*\mathrm{Re}(\eta) \geqq \gamma(r) + 2a_2^*(\nu - 1) + \mathrm{Re}(\mu), \quad \mathrm{Re}(\eta) > \nu - 1, \tag{7.11.24}$$

$$\mathrm{Re}(\zeta) < 1 - \nu, \tag{7.11.25}$$

where $\gamma(r)$ *is given in* (3.3.9). *If the condition* (7.11.11) *is satisfied, then*

$$ {}_p\Psi_q(\mathfrak{L}_{\nu,r}) = \left(M_{1/2+\omega/(2a_2^*)}H_{-2a_2^*,2a_2^*\zeta+\omega-1}\mathbb{L}_{-a^*,1/2+\eta-\omega/(2a_2^*)}\right) \left(\mathfrak{L}_{3/2+\mathrm{Re}(\omega)/(2a_2^*)-\nu,r}\right). \tag{7.11.26}$$

If the condition (7.11.11) *is not satisfied, then* ${}_p\Psi_q(\mathfrak{L}_{\nu,r})$ *is a subset of the right-hand side of* (7.11.26).

If $a^* = 0$, from (4.9.1) and (4.9.2) we get the inversion formulas for the transform (7.11.1):

$$f(x) = hx^{1-(\lambda+1)/h}\frac{d}{dx}x^{(\lambda+1)/h} \cdot \int_0^\infty H_{p+1,q+2}^{q,1}\left[xt \,\middle|\, \begin{matrix}(-\lambda,h),(a_i-\alpha_i,\alpha_i)_{1,p}, \\ (b_j-\beta_j,\beta_j)_{1,q},(0,1),(-\lambda-1,h)\end{matrix}\right]\left({}_p\Psi_q f\right)(t)dt \tag{7.11.27}$$

and

$$f(x) = -hx^{1-(\lambda+1)/h}\frac{d}{dx}x^{(\lambda+1)/h} \cdot \int_0^\infty H_{p+1,q+2}^{q+1,0}\left[xt \,\middle|\, \begin{matrix}(a_i-\alpha_i,\alpha_i)_{1,p},(-\lambda,h) \\ (-\lambda-1,h),(b_j-\beta_j,\beta_j)_{1,q},(0,1),\end{matrix}\right]\left({}_p\Psi_q f\right)(t)dt, \tag{7.11.28}$$

respectively.

Then Theorems 4.11, 4.12 and 4.13, 4,14 imply the conditions of the inversion formulas (7.11.27) and (7.11.28) for the transform (7.11.1) in the cases when $\Delta = 0$ and $\Delta \neq 0$, respectively. The constants α_0 and β_0 in (4.9.6) and (4.9.7) take the forms

$$\alpha_0 = 1 - \min_{1\leqq j\leqq q}\left[\frac{\mathrm{Re}(b_j)}{\beta_j}\right], \quad \beta_0 = \infty. \tag{7.11.29}$$

Theorem 7.62. *Let* a^*, Δ, μ, β *and* α_0 *be given in* (7.11.5), (7.11.6), (7.11.8), (7.11.10) *and* (7.11.29). *Let* $a^* = 0, 0 < 1 - \nu < \beta, \alpha_0 < \nu < 0, \lambda \in \mathbb{C}$ *and* $h > 0$.

(a) *If* $\Delta(1-\nu) + \mathrm{Re}(\mu) = 0$ *and if* $f \in \mathfrak{L}_{\nu,2}$, *the inversion formula* (7.11.27) *holds for* $\mathrm{Re}(\lambda) > \nu h - 1$ and (7.11.28) for $\mathrm{Re}(\lambda) < \nu h - 1$.

(b) *If* $\Delta = \mathrm{Re}(\mu) = 0$ *and if* $f \in \mathfrak{L}_{\nu,r}$ $(1 < r < \infty)$, *then the inversion formula* (7.11.27) *holds for* $\mathrm{Re}(\lambda) > \nu h - 1$ and (7.11.28) for $\mathrm{Re}(\lambda) < \nu h - 1$.

Theorem 6.63. *Let* a^*, Δ, μ, β *and* α_0 *be given in* (7.11.5), (7.11.6), (7.11.8), (7.11.10) *and* (7.11.29). *Let* $a^* = 0$ *and* $\Delta(1-\nu) + \mathrm{Re}(\mu) \leqq 1/2 - \gamma(r)$, *where* $\gamma(r)$ *is given in* (3.3.9), *and let* $1 < r < \infty, \lambda \in \mathbb{C}$ *and* $h > 0$.

(a) *If* $\Delta > 0, 0 < 1 - \nu < \beta, \alpha_0 < \nu < \mathrm{Re}(\mu + 1/2)/\Delta + 1$ *and if* $f \in \mathfrak{L}_{\nu,r}$, *then the inversion formulas* (7.11.27) *and* (7.11.28) *hold for* $\mathrm{Re}(\lambda) > \nu h - 1$ *and for* $\mathrm{Re}(\lambda) < \nu h - 1$, *respectively.*

(b) *If* $p > 0, \Delta < 0, 0 < 1 - \nu < \beta, \max[\alpha_0, \{\mathrm{Re}(\mu + 1/2)/\Delta\} + 1] < \nu < 0$ *and if* $f \in \mathfrak{L}_{\nu,r}$, *then the inversion formulas* (7.11.27) *and* (7.11.28) *hold for* $\mathrm{Re}(\lambda) > \nu h - 1$ *and for* $\mathrm{Re}(\lambda) < \nu h - 1$, *respectively.*

7.12. Bibliographical Remarks and Additional Information on Chapter 7

The results presented in Sections 7.1–7.11 were obtained by the authors and have not been published before except those in Section 7.9. Below we give historical comments and a review of other investigations connected with the integral transforms of hypergeometric type.

For Section 7.1. The classical theory of the Laplace transform $\mathbb{L}$ in (2.5.2), in particular its properties in the space $L_2(\mathbb{R}_+)$, is well known, for which see the books by Doetsch [1], [2], Widder [1], Titchmarsh [3], Sneddon [1] and Ditkin and Prudnikov [1]. The Laplace transform of generalized functions was considered by Zemanian [6] and by Brychkov and Prudnikov [1, Section 3.4].

The existence and the range of the generalized Laplace transform $\mathbb{L}_{k,\alpha}$ given in (3.3.3) in the space $\mathfrak{L}_{\nu,r}$ presented in Theorem 3.2(c) was proved by Rooney [6, Theorem 5.1(d)]. McBride and Spratt [1] studied the range and invertibiliy of the generalized Laplace type transform N_m^η given in (5.7.10) in $\mathfrak{L}_{\nu,r}$. In particular, they proved in [1, Theorem 5.4] that, if $\nu \in \mathbb{C}$, $\eta \in \mathbb{C}$ and $m > 0$ are such that $\mathrm{Re}(\eta - \nu/m) > 0$, then for the transform N_m^η on $\mathfrak{L}_{\nu,r}$ $(1 \leqq r < \infty)$ there holds the following inversion formula

$$f(x) = \lim_{n\to\infty} \left(L_m^{n,\eta,\alpha} N_m^\eta f\right)(x) \quad (f \in \mathfrak{L}_{\nu,r}), \tag{7.12.1}$$

where (for sufficiently large n)

$$\left(L_m^{n,\eta,\alpha} g\right)(x) = \frac{n^{-(\eta+\alpha)}}{\Gamma(n)} \left(W_{n^{-1/m}} K_m^{\eta+\alpha+n,-\alpha-n} g\right)(x) \quad (g \in F_{r,\nu}). \tag{7.12.2}$$

Here W_δ is given in (3.3.12) and $K_m^{\eta,\alpha} = I_{-;m,\eta}^\alpha$ is the Erdélyi–Kober type operator (3.3.2) defined in terms of the Mellin transform by (3.3.7) for $f \in F_{r,\nu}$ $(1 \leqq r \leqq 2)$, where the space $F_{r,\nu}$ is defined in (5.7.8), $\alpha \in \mathbb{C}$, $\nu \in \mathbb{C}$ and $m > 0$ such that $\mathrm{Re}(\eta + \nu/m) \neq -j$ $(j = 0, 1, 2, \cdots)$. By (5.7.10) one may deduce from (7.12.1) and (7.12.2) the corresponding inversion formula for the generalized Laplace transform $L_{\gamma,k;\sigma,\kappa}^*$ given in (7.1.10).

For Section 7.2. The transform $M_{k,m}$ defined by (7.2.1) was introduced by Meijer [3] (1941) and is given his name. Meijer [3], [4] has proved the inversion formula for the transform $M_{k,m}$ in the form

$$f(x) = \lim_{R\to\infty} \frac{1}{2\pi i} \frac{\Gamma(1-k+m)}{\Gamma(1+2m)} \int_{\beta-iR}^{\beta+iR} (xt)^{k-1/2} e^{xt/2} M_{k-1/2,m}(xt) \left(\boldsymbol{M}_{k,m} f\right)(t) dt, \tag{7.12.3}$$

where $M_{k-1/2,m}(z)$ is the Whittaker function defined via the confluent hypergeometric function of Tricomi (7.2.3) by

$$M_{k-1/2,m}(z) = e^{-z/2} z^{2m+1} \Psi(1-k+m, 2m+1; z) \tag{7.12.4}$$

(see Erdélyi, Magnus, Oberhettinger and Tricomi [1, 6.9(1)]). Meijer [3], [4] gave sufficient conditions on $f(x)$ under which (7.12.3) follows from (7.2.1) in the case $k \leqq m \leqq -k$, and under which (7.2.1) is a consequence of (7.12.3), when $\mathrm{Re}(k) \leqq -\mathrm{Re}(m) < 1/2$.

It should be noted that, when $k = -m$, the relation (7.2.1) is reduced to the Laplace transform $\mathbb{L}$ in (2.5.2), while (7.12.3) is reduced to the inverse Laplace transform $\mathbb{L}^{-1}$ known as a complex inversion formula. Saksena [3] constructed a real inversion formula for the Meijer transform (7.2.1), analogous to that constructed by Widder [1] for the Laplace integral, with the aid of the operator $K^{-}_{\eta,\alpha}$, which is one of the so-called Kober fractional integration operators defined for $\alpha > 0$ by

$$\left(I^{+}_{\eta,\alpha} f\right)(x) \equiv \left(I^{\alpha}_{0+;1,\eta} f\right)(x) = \frac{x^{-\alpha-\eta}}{\Gamma(\alpha)} \int_0^x \frac{t^\eta}{(x-t)^{1-\alpha}} f(t) dt; \tag{7.12.5}$$

$$\left(K^{-}_{\eta,\alpha} f\right)(x) \equiv \left(I^{\alpha}_{-;1,\eta} f\right)(x) = \frac{x^\eta}{\Gamma(\alpha)} \int_x^\infty \frac{t^{-\eta-\alpha}}{(t-x)^{1-\alpha}} f(t) dt \tag{7.12.6}$$

(see Samko, Kilbas and Marichev [1, (18.5)–(18.6)]). Saksena [3] defined sequences of differential operators

$$\left(\boldsymbol{U}_n f\right)(x) = (-1)^n x^{n-k-m} \left(\frac{d}{dx}\right)^n [x^{k+m} f(x)] \tag{7.12.7}$$

and proved the inversion formula in the form

$$f(x) = \lim_{n\to\infty} \frac{1}{t\Gamma(n)} \left(\boldsymbol{U}_n \boldsymbol{M}_{k,m} f\right)(t)\Big|_{t=n/x} \qquad (x > 0), \tag{7.12.8}$$

provided that $f(t)$ is bounded for $0 < t < \infty$. He also deduced necessary and sufficient conditions for a given function $g(x)$ to be represented by the integral $g(x) = (\boldsymbol{M}_{k,m} f)(x)$ with bounded $f(x)$. K.J. Srivastava [1] applied the Kober operators (7.12.5) and (7.12.6) to the investigation of the Meijer transform $\boldsymbol{M}_{k,m}$ as the operator acting from $L_r(\mathbb{R}_+)$ into $L_{r'}(\mathbb{R}_+)$ for $1 \leqq r \leqq 2$ $(1/r + 1/r' = 1)$.

We also indicate some other investigations of the Meijer transform $\boldsymbol{M}_{k,m}$ in (7.2.1). Arya [3] studied the convergence and some properties of the Meijer type transform of the form (7.2.1) in which $f(t)dt$ is replaced by $da(t)$ with a special function $a(t)$. Misra [1] gave Abelian and Tauberian theorems for the transform $\boldsymbol{M}_{k,m}$. Mehra [1], [3] studied some operational properties of the transform (7.2.1). In particular, he considered the images of the Meijer transforms which are self-reciprocal in the Hankel transform (8.1.1). Pathak and Rai [1] proved certain Abelian (initial and final value) theorems for the Meijer transform $\boldsymbol{M}_{k,m}$.

The Varma transform $\boldsymbol{V}_{k,m}$ in (7.2.15) is a generalization of the Laplace integral transform $\mathbb{L}$ in (2.5.2), when $k+m = 1/2$. Saksena [1] (1953) probably first called this transform the Varma transform. Varma himself in [3] (1951) studied some properties of a more general transform than (7.2.15) of the form

$$\int_0^\infty (xt)^{m-1/2} e^{-xt/2} W_{k,m}(xt) da(t) \tag{7.12.9}$$

with a certain function $a(x)$ and, in particular, he gave several real inversion formulas for such a transform. Earlier Varma in [1] (1947) obtained for the transform

$$x \int_0^\infty (2xt)^{-1/4} W_{k,m}(2xt) f(t) dt \tag{7.12.10}$$

theorems which are analogous to certain results for the Laplace transform (2.5.2), while in [2] (1949) he derived a complex inversion formula for the transform (7.12.10).

Saksena [1] established three inversion formulas for the transform $\boldsymbol{V}_{k,m}$ which correspond to an inversion for the Laplace transform $\mathbb{L}$ in (2.5.2). In [2] and [4] Saksena proved representations for the Varma transform (7.2.15) as compositions of the Laplace operator $\mathbb{L}$ and the Kober operators (7.12.5) and (7.12.6). The representations by Saksena can be obtained from (7.2.30) and (7.2.31) and have the forms

$$\left(\boldsymbol{V}_{k,m}f\right)(x) = \left(\mathbb{L}I^{+}_{2m,1/2-k-m}f\right)(x) \tag{7.12.11}$$

and

$$\left(\boldsymbol{V}_{k,m}f\right)(x) = \left(K^{-}_{0,m-k-1/2}x^{2m}\mathbb{L}x^{2m}f\right)(x). \tag{7.12.12}$$

By using such representations Saksena proved a real inversion formula for the Varma transform. Similar results were also obtained by R.K. Saxena [7], Kalla [1], [3], Pathak [6] and Habibullah [2] for the Varma transform (7.2.15) and by Habibullah [3] for the transform of the form (7.2.15), in which the Whittaker function $W_{k,m}(xt)$ given in (7.2.2) is replaced by the Whittaker function $M_{k,m}(xt)$ given in (7.12.4). We also indicated that Fox [5] established the inversion formula for the Varma transform $\boldsymbol{V}_{k,m}$ in terms of the direct $\mathbb{L}$ and inverse $\mathbb{L}^{-1}$ Laplace transforms.

Gupta [3] established connections between the Meijer and Varma transforms given in (7.2.1) and (7.2.15). Connections of the Varma transform $\boldsymbol{V}_{k,m}$ with the Hankel transform $\mathbb{H}_\eta$ in (8.1.1) and the Meijer transform $\mathfrak{K}_\eta$ in (8.9.1) were studied by R.K. Saxena [4] and Gupta [1], respectively. R.K. Saxena and K.C. Gupta [1] and Kalla and Munot [1] obtained several theorems involving the Laplace transform (2.5.2) and the Varma transform (7.2.15). Saigo, Goyal and S. Saxena [1] proved a relationship between $\mathbb{L}$, $\boldsymbol{V}_{k,m}$ and the fractional integral, more general than the Riemann–Liouville fractional integral I^{α}_{-} in (2.7.2), involving the general class of polynomials and a generalized polynomial set.

Kesarwani [2]–[7] and R.K. Saxena [1]–[3], [5]–[6] studied the properties of the simple modification of the Varma transform (7.2.15) in the form

$$x\int_0^{\infty}(xt)^{m-1/2}e^{-xt/2}W_{k,m}(xt)f(t)dt. \tag{7.12.13}$$

Pathak in [7] and [8] extended the Varma transform (7.2.15) and the Meijer transform $\boldsymbol{M}_{k,m}$ in (7.2.1) to certain spaces of generalized functions and investigated properties such as analyticity, the inversion relation, uniqueness, characterization and the structure formulas. Tiwari [1] obtained similar results for the transform $\boldsymbol{V}_{k,m}$ in another space of tested functions (see Brychkov and Prudnikov [1, Sections 7.1 and 7.2] in this connection).

For Section 7.3. H.M. Srivastava [5] (1968) first considered the generalized Whittaker transform $\boldsymbol{W}^{k}_{\rho,\gamma}$ in (7.3.1) and proved that if $f(x) \in L_2(\mathbb{R}_+)$ and $-k \leqq \rho \leqq k + 1/2$, then an inversion formula for the transform $\boldsymbol{W}^{k}_{\rho,\gamma}$ is given via the inverse Laplace transform $\mathbb{L}^{-1}$ and the Kober fractional integration operator $K^{-}_{1+k-\rho,k+\rho}$ in (7.12.6) by

$$f(x) = \left(\mathbb{L}^{-1}K^{-}_{1+k-\rho,k+\rho}\boldsymbol{W}^{k}_{\rho,\gamma}f\right)(x). \tag{7.12.14}$$

Pathak [5] gave two other inversion relations for the generalized Whittaker transform (7.3.1). The first one has the form

$$f(x) = \left(\boldsymbol{M}_{-(k+\gamma+1/2)}\mathbb{L}^{-1}\boldsymbol{M}_{-(k+\gamma+1/2)}K^{-}_{1+k-\rho,\rho+\gamma-1/2}\boldsymbol{W}^{k}_{\rho,\gamma}f\right)(x), \tag{7.12.15}$$

where $\boldsymbol{M}_{-(k+\gamma+1/2)}$ is the elementatry operator (3.3.11), provided that $1 + k > \max[|\gamma|, \rho - 1/2]$, $\gamma + \rho > 1/2$ and $f(x) \in L_2(\mathbb{R}_+)$, $x^{k+\gamma+1/2}f(x) \in L_2(\mathbb{R}_+)$. The second inversion relation is given by

$$f(x) = \left(\mathbb{L}^{-1}(K^{-}_{k+\gamma+1/2,1/2-\gamma-\rho})^{-1}\boldsymbol{W}^{k}_{\rho,\gamma}f\right)(x), \tag{7.12.16}$$

provided that $-k-1<\gamma<1/2-\rho$, $f(x)\in L_2(\mathbb{R}_+)$ and the right-hand side of (7.12.16) exists.

Nasim [3] derived the inversion operators in terms of certain differential operators, for the operator $\boldsymbol{W}^k_{\rho,\gamma}$ and for the operator of the form (7.3.1) with $W_{\rho,\gamma}(xt)$ being replaced by $M_{\rho,\gamma}(xt)$ given in (7.12.4).

Srivastava and Vyas [1] obtained the relationship between the transform $\boldsymbol{W}^k_{\rho,\gamma}$ and the generalized Hankel transform given in (8.14.57).

Srivastava [6] (1968) introduced the generalization of the Whittaker transform (7.3.1) in the form

$$\left(\boldsymbol{S}^{\rho,\sigma}_{q,k,m}f\right)(x)=\int_0^\infty (xt)^{\sigma-1/2}e^{-qxt/2}W_{k,m}(pxt)f(t)dt, \tag{7.12.17}$$

investigated its inversion formula and obtained the relation of this transform with the Hardy transform (8.14.51). Srivastava, Goyal and Jain [1] studied the relationship between the transform (7.12.17) and the fractional integral, more general than the Riemann–Liouville fractional integral I^α_- in (2.7.2), involving the general class of polynomials.

Tiwari [4], [5], Tiwari and Ko [1] and Rao [9] extended the generalized Whittaker transform $\boldsymbol{S}^{\rho,\sigma}_{q,k,m}$ in (7.12.17) to certain spaces of generalized functions and discussed such properties as smoothness, the inversion formula and uniqueness. Tiwari [4] also proved some Abelian theorems for this transform. Carmichael and Pathak [1] considered the transform $\left(\boldsymbol{S}^{\rho,\sigma}_{q,k,m}f\right)(x)$ for complex $x\in\mathbb{C}$, studied its behavior as $x\to\infty$ and as $x\to 0$ provided that $f(t)$ has a certain behavior when $t\to 0+$ or $t\to\infty$, respectively, and extended the results obtained to a certain space of generalized functions described previously by Pathak [7].

We also mention C.K. Sharma [1] who established some properties of the transform, more general than the transform (7.3.1) involving the H-function.

Banerjee [1] (1961) and Mainra [1] (1961) probably first considered further modifications of the generalized Whittaker transform (7.3.13) in the forms

$$\int_0^\infty (2xt)^{m-1/2}e^{-pxt/2}\Psi(a,c;2xt)f(t)dt \tag{7.12.18}$$

and

$$\left(\mathbf{W}^{-\lambda-1/2}_{k+1/2,m;1,0}f\right)(x)=x\int_0^\infty (xt)^{-\lambda-1/2}e^{-xt/2}W_{k+1/2,m}(xt)f(t)dt, \tag{7.12.19}$$

respectively. We note that, in accordance with (7.2.2), the transform in (7.12.18) with the Tricomi hypergeometric function (7.2.3) in the kernel is the generalized Whittaker transform of the form (7.3.13). Banerjee [1] also introduced the transform of the form (7.12.18) in which $\Psi(a,c;2xt)$ is replaced by another confluent hypergeometric function $\Phi(a,c;2xt)$ (see Erdélyi, Magnus, Oberhettinger and Tricomi [2, Chapter VI]) and proved the inversion formulas for these transforms. Bora [1] proved two results for the transform (7.12.18). Mainra [1] studied (7.12.19) with complex $x\in\mathbb{C}$ ($\mathrm{Re}(x)>0$) and proved several properties, in particular an analogy of the relation (7.3.16) and an inversion formula. Gupta [3] proved a theorem connecting the transform (7.12.19) and a generalized Stieltjes transform.

U.C. Jain [2] and Parashar [2] established some other properties for the transform (7.12.19). R.P. Goyal [1] investigated the convergence of the generalized Whittaker transform of the form

$$\int_0^\infty (2xt)^\lambda W_{k,m}(2xt)da(t), \tag{7.12.20}$$

where $a(t)$ is a normalized function of bounded variation. Habibullah [3] considered the integral transform of the form (7.12.18)

$$x^\lambda\int_0^\infty (xt)^{a-1}\Psi(a;c;-xt)f(t)dt, \tag{7.12.21}$$

proved its representation as a composition of the transforms generalizing Stieltjes' and Laplace's, (7.9.3) and (2.5.2), and apply this result to investigate the boundedness of the transform (7.12.21) from $L_p(\mathbb{R}_+)$ $(p \geqq 1)$ into $L_q(\mathbb{R}_+)$ $(1/q = 1 - 1/p - \lambda \geqq 0)$. We also mention Arya [4] who proved the Abelian theorem for the integral transform

$$\int_0^\infty (xt)^{2m} e^{-xt} \Psi(-k+m; m+1; xt) da(t). \tag{7.12.22}$$

For Section 7.4. The transform $\boldsymbol{D}_\gamma$ in (7.4.1) was defined and studied by Saksena [6] in a special space of generalized functions, in which the properties such as analyticity, the inversion formula and characterization were investigated (see the book by Brychkov and Prudnikov [1, Sections 7.8]). Mahato and Saksena [1]–[3] studied the transform $\boldsymbol{D}_\gamma$ in other spaces of tested and generalized functions and indicated the application to solve a special boundary value problem.

We also indicate that earlier Saksena in [5] obtained in terms of the transform $\boldsymbol{D}_\gamma$ the following inversion formula for the Laplace transform (2.5.2):

$$\left(\mathbb{L}^{-1} f\right)(x)$$
$$= \lim_{k\to\infty} \frac{k^{k+1} 2^{-\nu-(k+1)/2}}{\sqrt{\pi}\Gamma(k+1)} \int_0^\infty (xt)^{-\nu-k/2} e^{-k^2/(8xt)} D_{k+2\nu-1}\left(\frac{k}{\sqrt{2xt}}\right) \left(\mathbb{L} f\right)(t) dt \tag{7.12.23}$$

under certain conditions on $f(x)$.

Marichev and Vu Kim Tuan [1], [2] (see also Samko, Kilbas and Marichev [1, Theorem 36.14]) studied the isomorphic property of the modification of the integral transform (7.4.1) with $e^{-xt/2} D_\gamma\left(\sqrt{2xt}\right)$ being replaced by $e^{-x/(2t)} D_\gamma\left(\sqrt{2x/t}\right)$. They proved two composition representations for such a transform in terms of the Riemann–Liouville fractional integral operator I_-^α in (2.7.2) and a modification of the Laplace transform (7.1.1) with $k = 1$ in which xt is replaced by x/t and the multiplier $1/t$ is added.

For Section 7.5. The transform ${}_1\boldsymbol{F}_1^k$ in (7.5.1) was first introduced by Erdélyi [3] in the form

$$\left({}_1\boldsymbol{F}_1^\beta f\right)(x) = \frac{\Gamma(\beta+\eta+1)}{\Gamma(\alpha+\beta+\eta+1)} \int_0^\infty (xt)^\beta \, {}_1F_1(\beta+\eta+1; \alpha+\beta+\eta+1; -xt) f(t) dt \tag{7.12.24}$$

with complex parameters α, β and η.

Joshi [1], [3] investigated this transform in the space $L_p(\mathbb{R}_+)$ $(p \geqq 1)$ and proved two real inversion formulas and a representation theorem for this transform. The first inversion relation is based on the representation of (7.12.24) as the composition of the Kober fractional integral operator $I_{\eta,\alpha}^+$ in (7.12.5) and the generalized Laplace transform $\mathbb{L}_{1,-\beta}$ in (3.3.3):

$${}_1\boldsymbol{F}_1^\beta = I_{\eta,\alpha}^+ \mathbb{L}_{1,-\beta}. \tag{7.12.25}$$

The second inversion formula contains a differential operator which is an inverse to the Stieltjes transform (7.9.3) and has a similar form to the one given in (7.12.7) and (7.12.8) for the Meijer transform (7.2.1). Joshi used the first inversion formula to give in a representation theorem necessary and sufficient conditions for the function $g(x)$ to have the representation in the form (7.12.24): $g = {}_1\boldsymbol{F}_1^\beta f$ with $f \in L_p(\mathbb{R}_+)$ $(p \geqq 1)$.

In [2] Joshi showed that the Laplace transform (2.5.2) of the transform (7.12.24) yields the generalized transform ${}_2\boldsymbol{F}_1^\beta$ of the form (7.8.20):

$$\left(\mathbb{L}\, {}_2\boldsymbol{F}_1^\beta f\right)(x) = \frac{\Gamma(\beta+\eta+1)\Gamma(\beta+1)}{\Gamma(\alpha+\beta+\eta+1)}$$
$$\cdot \frac{1}{x}\int_0^\infty \left(\frac{t}{x}\right)^\beta {}_2F_1\left(\eta+\beta+1, \beta+1; \alpha+\beta+\eta+1; -\frac{t}{x}\right) f(t) dt. \tag{7.12.26}$$

Joshi in [4] also proved certain Abelian theorems for the transform ${}_1\boldsymbol{F}_1^\beta$ in (7.12.24), and in [6], [7] gave a representation theorem for such a transform on Lorentz spaces.

Habibullah [3] and Love, Prabhakar and Kashyap [1] investigated the generalized transform ${}_1\boldsymbol{F}_1^{\sigma,\kappa}$ in (7.5.11) with $\sigma = \lambda + a - 1$, $\kappa = a - 1$ and $\sigma = 0$, $\kappa = c - 1$, respectively. They obtained the representations for such transforms as the compositions of the Kober fractional integral operator $K_{\eta,\alpha}^{-}$ in (7.12.6) and a certain generalized Laplace transform $\mathbb{L}_{k,\beta}$ in (3.3.3) and applied these results to find the inversion formulas of the transforms considered.

Rao [1], [3]–[7] extended the transform ${}_1\boldsymbol{F}_1^{\beta}$ in (7.12.24) of the form (7.5.1) to a certain space of generalized functions and established analyticity, representation, inversion and uniqueness. In [8] Rao proved the Abelian theorems for such a transform and extended the results in the distributional sense (see Brychkov and Prudnikov [1, Sections 7.5] in this conection).

We also mention that Wimp [1], Prabhakar [1], [2] and Habibullah [1] obtained the inversion formulas for the integral transform containing ${}_1F_1(a;c;z)$ in the kernel by generalizing the Riemann–Liouville fractional integration operators to the form

$$\frac{1}{\Gamma(\alpha)}\int_a^x (x-t)^{\alpha-1}\, {}_1F_1\left(\beta;\alpha;\lambda(x-t)\right) f(t)dt \qquad (\alpha > 0), \tag{7.12.27}$$

and the corresponding integral transform with integration over (x, b) under certain conditions on $f(t)$ and the integral in (7.12.27), where $-\infty < a < x < b < \infty$. The boundedness and inversion of these operators in the space $L_p(a,b)$ $(1 \leqq p < \infty)$ were given by Marichev (see Samko, Kilbas and Marichev [1, Section 37.1]). In Section 10.4 of the book, Marichev also proved the representation and boundedness theorem for such operators with $a = 0$ and $b = \infty$ in $L_p(\mathbb{R}_+)$ $(1 \leqq p < \infty)$.

For Sections 7.7 and 7.8. The modified transform ${}_2\boldsymbol{F}_1{}^2$ in (7.8.2) was first considered by Swaroop [1] (1964). Using the inverse Mellin transform (see, for example, the book by Marichev [1, Theorem 25]), he gave several complex inversion formulas for this transform, proved the uniqueness theorem and the relation of integration by parts (7.8.12) also called a Parseval type property. K.C. Gupta and S.S. Mittal [1]–[3] established several theorems involving the transform ${}_2\boldsymbol{F}_1{}^2$ and the modification of the Whittaker transform (7.12.19). Kalla [2] showed that the composition of the transform (7.8.2) and a modification of the Varma transform (7.2.15) yields the integral transform containing the Meijer G-function (2.9.1) as a kernel. Marichev [1, Section 8.2] showed that the transform ${}_2\boldsymbol{F}_1{}^2$ can be represented in the form

$${}_2\boldsymbol{F}_1{}^2 = \boldsymbol{M}_{a-1}\mathbb{L}\boldsymbol{M}_{a-1}\mathbb{L}x^{b-1}I_{-}^{c-b}\boldsymbol{M}_{1-c}, \tag{7.12.28}$$

where the operators $\boldsymbol{M}_{\zeta}$, $\mathbb{L}$ and I_{-}^{c-b} are given in (3.3.11), (2.5.2) and (2.7.2), and proved the relation (7.8.12).

Several authors used the compositions of the fractional integration operators of Riemann–Liouville (2.7.1), (2.7.2) and of Erdélyi–Kober type (3.3.1), (3.3.2) with the generalized Stieltjes transform (7.9.1) to obtain the inversion formulas for the generalized hypergemetric transforms. Love [3] first proved such an inversion formula for the transform ${}_2\boldsymbol{F}_1^{1;0,b}$ in (7.8.19). Prabhakar and Kashyap [1] and Habibullah [3] established the inversion relations for the transforms ${}_2\boldsymbol{F}_1^{2;0,c-1}$ in (7.8.20) and ${}_2\boldsymbol{F}_1^{\lambda+b,b}$ in (7.7.21), respectively. We also mention that Erdélyi [4] considered the transform ${}_2\boldsymbol{F}_1^{1;0,-b}$ of the form (7.8.19) in a certain space of generalized functions.

Many authors investigated the integral transforms involving the Gauss hypergeometric function ${}_2F_1(a,b;c;z)$ in kernels, which generalize the fractional integration operators of Riemann–Liouville (2.7.1), (2.7.2), Kober (7.12.5), (7.12.6) and Erdélyi–Kober type (3.3.1), (3.3.2). Love [1] (1967), [2] (1967) first considered the integral transforms of the forms for $x > 0$

$$\left({}_1\boldsymbol{I}_{0+}^{c}(a,b)f\right)(x) = \frac{1}{\Gamma(c)}\int_0^x (x-t)^{c-1}\, {}_2F_1\left(a,b;c;1-\frac{x}{t}\right) f(t)dt; \tag{7.12.29}$$

$$\left({}_2\boldsymbol{I}_{0+}^{c}(a,b)f\right)(x) = \frac{1}{\Gamma(c)}\int_0^x (x-t)^{c-1}\, {}_2F_1\left(a,b;c;1-\frac{t}{x}\right) f(t)dt \tag{7.12.30}$$

and

$$\left({}_1\boldsymbol{I}_{-}^{c}(a,b)f\right)(x) = \frac{1}{\Gamma(c)}\int_x^\infty (t-x)^{c-1}\, {}_2F_1\left(a,b;c;1-\frac{x}{t}\right) f(t)dt, \tag{7.12.31}$$

$$\left({}_2I^c_-(a,b)f\right)(x) = \frac{1}{\Gamma(c)}\int_x^\infty (t-x)^{c-1}\, {}_2F_1\left(a,b;c;1-\frac{t}{x}\right) f(t)dt \tag{7.12.32}$$

with $a, b, c \in \mathbb{C}$ ($\mathrm{Re}(c) > 0$). Love proved the representations of the above transforms as compositions of two Riemann–Liouville fractional integrals (2.7.1) and (2.7.2) with power weights. These representations for the integrals (7.12.29) and (7.12.30) have the forms

$${}_1I^c_{0+}(a,b) = I^{c-b}_{0+}M_{-a}I^b_{0+}M_a = M_{c-a-b}I^b_{0+}M_{a-c}I^{c-b}_{0+}M_b; \tag{7.12.33}$$

$${}_2I^c_{0+}(a,b) = M_aI^b_{0+}M_{-a}I^{c-b}_{0+}M_{c-b} = M_bI^{c-b}_{0+}M_{a-c}I^b_{0+}M_{c-b}, \tag{7.12.34}$$

where M_ξ is the elementary operator (3.3.11). The representations for the integrals in (7.12.31) and (7.12.32) are obtained from (7.12.33) and (7.12.34) by replacing ${}_1I^c_{0+}(a,b)$, ${}_2I^c_{0+}(a,b)$, I^{c-b}_{0+} and I^b_{0+} by ${}_1I^c_-(a,b)$, ${}_2I^c_-(a,b)$, I^{c-b}_- and I^b_-, respectively.

Love [1], [2] established the boundedness of the operators ${}_1I^c_{0+}(a,b)$, ${}_2I^c_{0+}(a,b)$ in the space $L^q_1(0,d)$ and ${}_1I^c_-(a,b)$, ${}_2I^c_-(a,b)$ in $L^r_1(e,\infty)$, where $0 < d, e < \infty$, $1 \leqq p < \infty$, $q, r \in \mathbb{C}$,

$$L^q_p(0,d) = \{f(x):\ x^q f(x) \in L_p(0,d)\} \tag{7.12.35}$$

and

$$L^r_p(e,\infty) = \{f(x):\ x^r f(x) \in L_p(e,\infty)\}. \tag{7.12.36}$$

He applied the results obtained to find the explicit inversion formulas for these transforms. Marichev extended the results by Love to more general weighted spaces $L^q_p(0,d)$ and $L^r_p(e,\infty)$ with any p ($1 \leqq p < \infty$) (see Samko, Kilbas and Marichev [1, Sections 10.1 and 35.1]).

It should be noted that earlier Wimp [1] (1964) and Higgins [2] (1964) proved the inversion formulas for the transforms of the forms (7.12.31) and (7.12.32) in which the upper limit ∞ is replaced by 1 (see Samko, Kilbas and Marichev [1, Section 39.2, Note 35.1]).

R.K. Saxena [9] (1967) introduced the operators

$$\frac{\sigma x^{-\eta-1}}{\Gamma(1-\alpha)}\int_0^x {}_2F_1\left(\alpha,\beta+m;\gamma;\frac{at^\sigma}{x^\sigma}\right) t^\eta f(t)dt \tag{7.12.37}$$

and

$$\frac{\sigma x^\delta}{\Gamma(1-\alpha)}\int_x^\infty {}_2F_1\left(\alpha,\beta+m;\gamma;\frac{ax^\sigma}{t^\sigma}\right) t^{-\delta-1} f(t)dt \tag{7.12.38}$$

with complex parameters $\alpha, \beta, \gamma \in \mathbb{C}$ satisfying certain conditions and $\sigma = a = 1$. He gave the formulas of their Mellin transform for $f(x) \in L_p(\mathbb{R}_+)$ if $1 \leqq p \leqq 2$ and for $f(x)$ in a special space $\mathcal{M}_p$ if $p > 2$, and the relation of integration by parts being an analog of (7.7.11). Kalla and R.K. Saxena in [1] extended these results to the general transforms (7.12.37) and (7.12.38) with $\sigma > 0$ and complex $a \in \mathbb{C}$, and in [3] obtained the inversion formulas for these transforms (see also Kalla [10, Section 3]).

We note that such a transform of the form (7.12.37) was first studied by Erdélyi [1] (1940) while investigating the composition of the Kober operators (7.12.5) and (7.12.6) with the elementary operator R in (3.3.13):

$$\left(T_\alpha f\right)(x) = \left(I^+_{\alpha+\eta/2,-\alpha}K^-_{\eta/2,\alpha}Rf\right)(x) \tag{7.12.39}$$

with $f(x) \in L_2(\mathbb{R})$. In particular, when $\mathrm{Re}(\eta) > -1$ and $n \in \mathbb{N}$, he obtained the relation

$$\left(T_n f\right)(x) = (-1)^n (Rf)(x)$$
$$+\frac{\Gamma(\eta+1)}{\Gamma(n)\Gamma(\eta+1)}\int_0^{1/x} (xt)^{\eta/2}\, {}_2F_1(1-n,\eta+n-1;\eta+1;xt)f(t)dt. \tag{7.12.40}$$

If we replace x by $1/x$, then the second term is reduced to the transform of the form (7.12.37).

R.K. Saxena and Kumbhat [1], [3] studied the generalizations of the transforms (7.12.5) and (7.12.6) in the forms

$$\frac{x^{-\alpha-\eta}}{\Gamma(\alpha)}\int_0^x (x-t)^{\alpha-1}\,{}_2F_1\left(\alpha+\beta,-\eta;\alpha;1-\frac{t}{x}\right)t^{\eta}f(t)dt \tag{7.12.41}$$

and

$$\frac{x^{\eta}}{\Gamma(\alpha)}\int_x^\infty (t-x)^{\alpha-1}\,{}_2F_1\left(\alpha+\beta,-\eta;\alpha;1-\frac{x}{t}\right)t^{-\alpha-\eta}f(t)dt \tag{7.12.42}$$

with $\alpha>0$, $\beta,\eta\in\mathbb{C}$, $f(x)\in L_p(\mathbb{R}_+)$ $(1\leqq p\leqq 2)$. In [1] they investigated the relations between these transforms and the Kober transforms $I^+_{\eta,\alpha}$ and $K^-_{\eta,\alpha}$ given in (7.12.5) and (7.12.6), while in [3] they established the uniqueness theorem and the inversion formulas and gave the relations between the operators in (7.12.41) and (7.12.42) and the Hankel transform $\mathbb{H}_\eta$ in (8.1.1) and the Laplace transform $\mathbb{L}$ in (2.5.2).

Using the approach by Love [1], [2], McBride [1] obtained the inversion formulas for the transform

$$\int_0^x \frac{(x^m-t^m)^{c-1}}{\Gamma(c)}\,{}_2F_1\left(a,b;c;1-\frac{x^m}{t^m}\right)mt^{m-1}f(t)dt \qquad (x>0) \tag{7.12.43}$$

with $m>0$, $a,b,c\in\mathbb{C}$ $(\mathrm{Re}(c)>0)$, more general than (7.12.29), and for similar generalizations of (7.12.30)–(7.12.32) in spaces of generalized functions $\mathsf{F}'_{r,\nu}$ (see For Section 5.7 (Sections 5.2 and 5.3)), while Prabhakar [3] gave the inversion relation for the transform

$$\int_x^d \frac{(t^m-x^m)^{c-1}}{\Gamma(c)}\,{}_2F_1\left(a,b;c;1-\frac{x^m}{t^m}\right)mt^{m-1}f(t)dt \quad (a<x<d,\ m>0) \tag{7.12.44}$$

with $m>0$, $a,b,c\in\mathbb{C}$ $(\mathrm{Re}(c)>0)$ in the space $L_1(a,b)$. We also mention that Braaksma and Schuitman [1] obtained the inversion relations for the integral transform of the form (7.12.32) with $(t-x)^{c-1}$ being replaced by $(1-x/t)^{c-1}$ in certain spaces of tested and generalized functions.

Saigo [1] (1979) introduced the transforms

$$\left(I^{\alpha,\beta,\eta}_{0+}f\right)(x)=\frac{x^{-\alpha-\beta}}{\Gamma(\alpha)}\int_0^x (x-t)^{\alpha-1}\,{}_2F_1\left(\alpha+\beta,-\eta;\alpha;1-\frac{t}{x}\right)f(t)dt \tag{7.12.45}$$

and

$$\left(I^{\alpha,\beta,\eta}_{-}f\right)(x)=\frac{1}{\Gamma(\alpha)}\int_x^\infty (t-x)^{\alpha-1}\,{}_2F_1\left(\alpha+\beta,-\eta;\alpha;1-\frac{x}{t}\right)t^{-\alpha-\beta}f(t)dt \tag{7.12.46}$$

with $\alpha,\beta,\eta\in\mathbb{C}$ $(\mathrm{Re}(\alpha)>0)$, which are modifications of the transforms (7.12.30) and (7.12.31), and investigated composition properties, the relation of integration by parts and inversion formulas in the space L_p with any $1\leqq p\leqq\infty$ (see also Samko, Kilbas and Marichev [1, Section 23.2, Note 18.6] in this connection). Srivastava and Saigo [1] evaluated multiplications of the operators $I^{\alpha,\beta,\eta}_{0+}$ and $I^{\alpha,\beta,\eta}_{-}$. Saigo and Glaeske [1] extended these and other properties of the transforms (7.12.45) and (7.12.46) to McBride spaces of tested and generalized functions $\mathsf{F}_{\nu,r}$ and $\mathsf{F}'_{\nu,r}$. Kilbas, Repin and Saigo [1] constructed $\mathfrak{L}_{\nu,r}$-theory for the operators $I^{\alpha,\beta,\eta}_{0+}$ and $I^{\alpha,\beta,\eta}_{-}$ as special cases of the modified $G^2_{-\beta,0}$- and $G^1_{0,\beta}$-transforms (6.2.5) and (6.2.4) with $m=p=q=2$ and $n=0$. Glaeske and Saigo [2] evaluated the Laplace transform (2.5.2) of the generlaized fractional integrals (7.12.45) and (7.12.46) and conversely these fractional integrals of the Laplace transform in $\mathsf{F}'_{\nu,r}$. Saigo and Raina [1] obtained the images of some elementary functions under the operators $I^{\alpha,\beta,\eta}_{0+}$ and $I^{\alpha,\beta,\eta}_{-}$ and gave applications to certain statistical distributions. Saigo, R.K. Saxena and Ram [2] established the relations between the Mellin transform $\mathfrak{M}$ of $x^\beta I^{\alpha,\beta,\eta}_{0+}$, $x^\beta I^{\alpha,\beta,\eta}_{0+}$ and $\mathfrak{M}f$, and proved the representations of the transforms (7.12.45) and (7.12.46) via the Laplace transform $\mathbb{L}$ and the inverse $\mathbb{L}^{-1}$. Compositions of $I^{\alpha,\beta,\eta}_{0+}$ and $I^{\alpha,\beta,\eta}_{-}$ with the axisymmetric differential operator of potential theory L_σ in (5.7.16) was studied by Kilbas, Saigo and Zhuk [1].

Grin'ko and Kilbas [1] investigated the mapping properties of the operators ${}_1I^c_{0+}(a,b)$, ${}_2I^c_{0+}(a,b)$, ${}_1I^c_{-}(a,b)$ and ${}_2I^c_{-}(a,b)$ given in (7.12.29)–(7.12.32) in weighted Hölder spaces with power weight on the half-axis $\mathbb{R}_+$, established the conditions for these operators to realize an isomorphism between two such weighted Hölder spaces and gave applications to the integral transforms (7.12.45) and (7.12.46). In [2] Grin'ko and Kilbas found the conditions on the parameters a, b and c in (7.12.29)–(7.12.32) sufficient for all compositions of these operators with special power weights to be operators of the same form. See in this connection Samko, Kilbas and Marichev [1, Section 17.2, Note 10.1]. The authors in Saigo and Kilbas [1] proved mapping properties of the operator (7.12.45) defined on a finite interval $(0,d)$ $(0 < d < \infty)$ in the space of Hölder functions.

A series of papers was devoted to studying the inversion for special cases of the integral transform of the form (7.12.32) on a finite interval $(0,1)$, integration over (x,∞) being replaced by integration over (x,∞), with the Chebyshev polynomial $T_n(x)$ and the Legendre polynomial $P_k(x)$ in the kernels, and for modifications of these transforms with the Legendre function $P^\nu_\mu(x)$, the Jacobi polynomial $P^{(\alpha,\beta)}_n(x)$, the Gegenbauer polynomial $C^\lambda_k(x)$ and other generalized polynomials in the kernels (see Samko, Kilbas and Marichev [1, Section 35.1 and Section 39.2, Notes 35.1 and 35.2]). Heywood and Rooney [5] investigated in the space $\mathfrak{L}_{\nu,r}$ $(1 \leqq r \leqq \infty,\ \nu \in \mathbb{C})$ the integral transform of the form

$$\left(G^\lambda_k f\right)(x) = \frac{2}{\Gamma(\lambda+1/2)} \int_x^\infty \left(1-\frac{x^2}{t^2}\right)^{\lambda-1/2} \mathcal{G}^\lambda_k\left(\frac{x}{t}\right) f(t)\frac{dt}{t} \qquad (k \in \mathbb{N}_0), \tag{7.12.47}$$

where $\mathcal{G}^\lambda_k(x) = C^\lambda_k(x)/C^\lambda_k(1)$ $(\lambda > -1/2,\ \lambda \neq 0)$, $C^\lambda_k(x)$ is the Gegenbauer polynomial of index λ and order k defined by

$$C^\lambda_k(x) = \frac{(2\lambda)_k}{k!}\, {}_2F_1\left(-k, k+2\lambda; \lambda+\frac{1}{2}; \frac{1-x}{2}\right) \qquad (k \in \mathbb{N}, \lambda \in \mathbb{C}) \tag{7.12.48}$$

and $\mathcal{G}^0_k(x) = T_k(x)$ is the Chebyshev polynomial of degree k (see Erdélyi, Magnus, Oberhettinger and Tricomi [2, Section 10.9]). Heywood and Rooney [5] established the boundedness and the range of the transform (7.12.47) in $\mathfrak{L}_{\nu,r}$ and proved its inversion formulas. We also mention that earlier the inversion relations for the integral transforms of the form (7.12.47) and similar transforms with integration over (x,d) $(0 < d \leqq \infty)$ were given by K.N. Srivastava [1] (1961/62), Buschman [1] (1962) and Higgins [1] (1963). We also mention Saigo and Maeda [1] who investigated on the McBride space $\mathsf{F}_{\nu,r}$ properties of integral transforms generalizing (7.12.45) and (7.12.46) and involving the Appell function F_3 in (2.10.4) as the kernels.

For Section 7.9. The results presented in this section were obtained by the authors in Saigo and Kilbas [5].

The generalized Stieltjes transform $\mathfrak{S}_{\beta,\eta,\alpha}$ (7.9.1) in the particular case when $\beta = 0$, $\eta = 2m$ and $\alpha = -m-k+1/2$

$$\left(\mathfrak{S}_{0,2m,-m-k+1/2} f\right)(x) = \frac{\Gamma(2m+1)}{\Gamma(m-k+3/2)} \int_0^\infty {}_2F_1\left(2m+1, 1; m-k+\frac{3}{2}; -\frac{t}{x}\right) f(t)\frac{dt}{x} \tag{7.12.49}$$

with $\mathrm{Re}(m) > -1/2$ was first considered by Arya. In [1] (1958) he proved Abelian type theorems for the transform which generalize the corresponding Abelian theorems for the Laplace transform (2.5.2) proved by Widder [1]. In [2] (1958) Arya used the Mellin transform $\mathfrak{M}$ in (2.5.1) to establish the real inversion formula for the transform (7.12.49) in the form

$$\frac{1}{2}[f(t+0)+f(t-0)] = \frac{1}{2\pi i}\int_{c-i\infty}^{c+i\infty} \frac{\Gamma(m-k+s+1/2)}{\Gamma(2m+s)\Gamma(s)\Gamma(1-s)} t^{-s} \left(\mathfrak{M}\mathfrak{S}_{0,2m,-m-k+1/2} f\right)(s)ds \tag{7.12.50}$$

under certain conditions on $f(x)$ as well as the result that $f(x) \in \mathfrak{L}_{c,1}$: $x^{c-1}f(x) \in L_1(\mathbb{R}_+)$. In [5] (1960) Arya proved the complex inversion formula for (7.12.49) on the basis of the corresponding complex inverse formula for the Meijer transform K_η in (8.9.1). Rao [2], [7] extended the transform $\mathfrak{S}_{0,2m,-m-k+1/2}$ to a certain space of generalized functions and proved a number of Abelian (an initial-value and a final-value) theorems for this transform.

Golas [2] investigated the invertibility of the transform (7.9.1) and its generalization

$$\frac{\Gamma(\beta+\eta+1)\Gamma(\beta+1)}{\Gamma(\alpha+\beta+\eta+1)}\frac{1}{x}$$
$$\cdot\int_0^\infty\left(\frac{t}{x}\right)^\beta {}_2F_1\left(\beta+\eta+1,\beta+1;\alpha+\beta+\eta+1;-\frac{t}{x}\right)da(t), \tag{7.12.51}$$

where $a(t)$ is a function of bounded variation in every finite interval. He established the real inversion formulas for these transforms via a certain integro-differential operator. In [3] Golas investigated some other properties of the transform (7.9.1). Joshi [5] proved the complex inversion relation for the generalized Stieltjes transform $\mathfrak{S}_{\beta,\eta,\alpha}$ in (7.9.1). Tiwari [2], [3], [6] extended this transform to a certain space of generalized functions. Tiwari and Koranne [1] proved Abelian theorems of the initial and final value type for the transform (7.9.1) and extended the results to a certain class of generalized functions. See in this connection also Brychkov and Prudnikov [1, Section 7.11].

We also note that Glaeske and Saigo [2] investigated the left- and right-sided compositions of the generalized Stieltjes transform

$$\left(\mathfrak{S}_\rho f\right)(x)=\int_0^\infty\frac{f(t)}{(x+t)^\rho}dt \tag{7.12.52}$$

with the generalized fractional integrals (7.12.45) and (7.12.46) in the McBride space $\mathsf{F}'_{p,\mu}$ of generalized functions and showed that such compositions are expressed in terms of the Kampé de Fériet series (see the book by Srivastava and Karlsson [1]).

For Section 7.10. Mehra and R.K. Saxena [1] (1967), by considering the transform ${}_pF_p$ in (7.10.1), proved two theorems and derived some results in particular cases when $p=1$ and $p=2$.

Goyal and Jain [1] studied the generalizations of the integral transforms (7.12.41) and (7.12.42) in the forms

$$\frac{x^{-\alpha-\eta}}{\Gamma(\alpha)}\int_0^x(x-t)^{\alpha-1}\,{}_pF_q\left(a_1,\cdots,a_p;b_1,\cdots,b_p;\lambda\left[1-\frac{t}{x}\right]\right)t^\eta f(t)dt \tag{7.12.53}$$

and

$$\frac{x^\eta}{\Gamma(\alpha)}\int_x^\infty(t-x)^{\alpha-1}\,{}_pF_q\left(a_1,\cdots,a_p;b_1,\cdots,b_p;\lambda\left[1-\frac{x}{t}\right]\right)t^{-\alpha-\eta}f(t)dt \tag{7.12.54}$$

with complex parameters α $(\mathrm{Re}(\alpha)>0), \eta, a_i$ $(1\leqq i\leqq p)$, b_j $(1\leqq j\leqq q)$ and λ. They showed that the composition of two transforms (7.12.53) (or (7.12.54)) with different parameters α, η, a_i, b_j and λ becomes the integral transform of the form (7.12.53) (or (7.12.54)) with ${}_pF_q$ being replaced by a special triple hypergeometric series. Goyal and Jain [1] also proved formally the Mellin transforms of (7.12.53) and (7.12.54) and applied the relations to give their inversion formulas and then they established certain relations of the generalized Hankel transform (8.14.62). Goyal, Jain and Gaur [1], [2] extended some of these results to more general integral transforms involving additionally in the kernels of (7.12.53) and (7.12.54) the terms with a general class of polynomials.

Srivastava, Saigo and Raina [1] investigated compositions of the Laplace transform (2.5.1) with the generalized fractional integral operators (7.12.45) and (7.12.46). They proved that the compositions $\mathbb{L}x^\lambda I_-^{\alpha,\beta,\eta}f$ and $I_{0+}^{\alpha,\beta,\eta}x^\lambda\mathbb{L}f$ are the transform ${}_2F_2$ of the form

$$\frac{\Gamma(\lambda+1)\Gamma(\lambda-\beta+\eta+1)}{\Gamma(\lambda-\beta+1)\Gamma(\lambda+\alpha+\eta+1)}x^{\lambda-\beta}$$
$$\cdot\int_0^\infty {}_2F_2\left(\lambda+1,\lambda-\beta+\eta+1;\lambda-\beta+1,\lambda+\alpha+\eta+1;-xt\right)f(t)dt, \tag{7.12.55}$$

while $\mathbb{L}x^\lambda I_{0+}^{\alpha,\beta,\eta}f$ and $I_{-}^{\alpha,\beta,\eta}x^\lambda\mathbb{L}f$ take more complicated shapes and are represented as sums of three kinds of transform ${}_2\boldsymbol{F}_2$.

We also indicate that Prabhakar [4] studied the inversion of the integral transforms of the form (7.12.43) and (17.12.44) with ${}_2F_1$ being replaced by the confluent hypergeometric function of two variables in the space $L_1(a,b)$. He proved that such an integral transform can be represented as the composition of two Riemann–Liouville fractional integrals (2.7.1) with power-exponential weights and applied this result to prove the inversion fomula. In [5] Prabhakar extended these assertions to more general integral transforms (in this connection see Samko, Kilbas and Marichev [1, (10.63), (10.64) and Section 17.1, Notes to §10.4]).

Chapter 8

BESSEL TYPE INTEGRAL TRANSFORMS ON THE SPACE $\mathfrak{L}_{\nu,r}$

8.1. The Hankel Transform

We consider the Hankel integral transform defined in (2.6.1) by

$$\left(\mathbb{H}_\eta f\right)(x) = \int_0^\infty (xt)^{1/2} J_\eta(xt) f(t)dt \quad (x > 0) \tag{8.1.1}$$

for $\mathrm{Re}(\eta) > -1$. This transform (8.1.1) is clearly defined for a continuous function $f \in C_0$ with compact support on $\mathbb{R}_+$ for the range of parameters indicated.

We note that the Hankel transform $\mathbb{H}_\eta$ generalizes the Fourier cosine transform

$$\left(\mathfrak{F}_c f\right)(x) = \left(\mathbb{H}_{-1/2} f\right)(x) = \left(\frac{2}{\pi}\right)^{1/2} \int_0^\infty \cos(xt) f(t)dt \quad (x > 0) \tag{8.1.2}$$

and the Fourier sine transform

$$\left(\mathfrak{F}_s f\right)(x) = \left(\mathbb{H}_{1/2} f\right)(x) = \left(\frac{2}{\pi}\right)^{1/2} \int_0^\infty \sin(xt) f(t)dt \quad (x > 0) \tag{8.1.3}$$

in view of the well-known formulas (Erdélyi, Magnus, Oberhettinger and Tricomi [2, 7.11(14), (15)])

$$J_{-1/2}(z) = \left(\frac{2}{\pi z}\right)^{1/2} \cos z, \qquad J_{1/2}(z) = \left(\frac{2}{\pi z}\right)^{1/2} \sin z.$$

First we present the results characterizing the boundedness and the representation of the Hankel transform (8.1.1) in $\mathfrak{L}_{\nu,r}$-space.

Theorem 8.1. *Let* $1 \leqq r \leqq \infty$ *and let* $\gamma(r)$ *be given in* (3.3.9).

(a) *If* $1 < r < \infty$ *and* $\gamma(r) \leqq \nu < \mathrm{Re}(\eta) + 3/2$, *then for all* $s \geqq r$ *such that* $s' \geqq 1/\nu$ *and* $1/s + 1/s' = 1$, *the operator* $\mathbb{H}_\eta$ *belongs to* $[\mathfrak{L}_{\nu,r}, \mathfrak{L}_{1-\nu,s}]$ *and is a one-to-one transform from* $\mathfrak{L}_{\nu,r}$ *onto* $\mathfrak{L}_{1-\nu,s}$. *If* $1 < r \leqq 2$ *and* $f \in \mathfrak{L}_{\nu,r}$, *then the Mellin transform of* (8.1.1) *for* $\mathrm{Re}(s) = 1 - \nu$ *is given by*

$$\left(\mathfrak{M}\mathbb{H}_\eta f\right)(s) = 2^{s-1/2} \frac{\Gamma\left(\frac{1}{2}\left\{\eta + s + \frac{1}{2}\right\}\right)}{\Gamma\left(\frac{1}{2}\left\{\eta - s + \frac{3}{2}\right\}\right)} \left(\mathfrak{M}f\right)(1 - s). \tag{8.1.4}$$

(b) *If* $1 \leqq \nu \leqq \mathrm{Re}(\eta) + 3/2$, *then* $\mathbb{H}_\eta \in [\mathfrak{L}_{\nu,1}, \mathfrak{L}_{1-\nu,\infty}]$. *If* $1 < \nu < \mathrm{Re}(\eta) + 3/2$, *then for all* r $(1 \leqq r \leqq \infty)$ $\mathbb{H}_\eta \in [\mathfrak{L}_{\nu,1}, \mathfrak{L}_{1-\nu,r}]$.

(c) *If* $f \in \mathfrak{L}_{\nu,r}$ *and* $g \in \mathfrak{L}_{\nu,s}$, *where* $1 < r < \infty$ *and* $1 < s < \infty$ *are such that* $1/r + 1/s \geqq 1$ *and* $\max[\gamma(r), \gamma(s)] \leqq \nu < \mathrm{Re}(\eta) + 3/2$, *then the following relation holds:*

$$\int_0^\infty f(x)\left(\mathbb{H}_\eta g\right)(x)dx = \int_0^\infty \left(\mathbb{H}_\eta f\right)(x)g(x)dx. \tag{8.1.5}$$

(d) *If* $f \in \mathfrak{L}_{\nu,r}$, *where* $1 < r < \infty$ *and* $\gamma(r) \leqq \nu < \mathrm{Re}(\eta) + 3/2$, *then for almost all* $x > 0$

$$\left(\mathbb{H}_\eta f\right)(x) = x^{-(\eta+1/2)}\frac{d}{dx}x^{\eta+1/2}\int_0^\infty (xt)^{1/2}J_{\eta+1}(xt)f(t)\frac{dt}{t}. \tag{8.1.6}$$

Proof. The assertion (a) coincides with those in Theorem 8.50(a),(b) which will be proved in Section 8.12. To prove (b) we first note that by (2.6.2) and Erdélyi, Magnus, Oberhettinger and Tricomi [2, 7.13(3)] the asymptotic behavior of $J_\eta(z)$ near zero and infinity have the forms

$$J_\eta(z) \sim \frac{2^{-\eta}}{\Gamma(\eta+1)}z^\eta \quad (z \to 0), \tag{8.1.7}$$

$$J_\eta(z) \sim \left(\frac{2}{\pi z}\right)^{1/2}\cos\left[z - \frac{1}{2}\left(\eta + \frac{1}{2}\right)\pi\right] \quad (z \to \infty). \tag{8.1.8}$$

Hence there is a constant K_η such that, for all $x > 0$,

$$|J_\eta(x)| \leqq K_\eta x^{\nu-3/2},$$

if $1 \leqq \nu \leqq \mathrm{Re}(\eta) + 3/2$. Then for $f \in C_0$ and $1 \leqq \nu \leqq \mathrm{Re}(\eta) + 3/2$, we have

$$\left|\left(\mathbb{H}_\eta f\right)(x)\right| \leqq \int_0^\infty (xt)^{1/2}|J_\eta(xt)||f(t)|dt \leqq K_\eta x^{\nu-1}\int_0^\infty t^{\nu-1}|f(t)|dt = K_\eta x^{\nu-1}||f||_{\nu,1}$$

and

$$\operatorname*{ess\,sup}_{x\in\mathbb{R}}\left[x^{1-\nu}\left|\left(\mathbb{H}_\eta f\right)(x)\right|\right] = \left\|\mathbb{H}_\eta f\right\|_{1-\nu,\infty} \leqq K_\eta||f||_{\nu,1}.$$

Thus $\mathbb{H}_\eta$ can be extended to $\mathfrak{L}_{\nu,1}$ for $1 \leqq \nu \leqq \mathrm{Re}(\eta) + 3/2$, and

$$\mathbb{H}_\eta \in [\mathfrak{L}_{\nu,1}, \mathfrak{L}_{1-\nu,\infty}] \tag{8.1.9}$$

and $\mathbb{H}_\eta$ is clearly given in (8.1.1) on $\mathfrak{L}_{\nu,1}$.

If $1 < \nu < \mathrm{Re}(\eta) + 3/2$, we have

$$\begin{aligned}\int_0^\infty x^{1-\nu}\left|\left(\mathbb{H}_\eta f\right)(x)\right|\frac{dx}{x} &\leqq \int_0^\infty x^{-\nu}dx\int_0^\infty (xt)^{1/2}|J_\eta(xt)||f(t)|dt\\ &= \int_0^\infty t^{1/2}|f(t)|dt\int_0^\infty x^{1/2-\nu}|J_\eta(xt)|dx\\ &= M\int_0^\infty t^{\nu-1}|f(t)|dt,\end{aligned}$$

where

$$M = \int_0^\infty x^{1/2-\nu}|J_\eta(x)|dx < \infty, \tag{8.1.10}$$

if and only if $1 < \nu < \mathrm{Re}(\eta) + 3/2$ in view of (8.1.7) and (8.1.8). Hence

$$\mathbb{H}_\eta \in [\mathfrak{L}_{\nu,1}, \mathfrak{L}_{1-\nu,1}]. \tag{8.1.11}$$

By the interpolation of Banach spaces (see Stein [1, Theorem 2]), if $1 < \nu < \mathrm{Re}(\eta) + 3/2$ and $1 \leqq r \leqq \infty$, we obtain $\mathbb{H}_\eta \in [\mathfrak{L}_{\nu,1}, \mathfrak{L}_{1-\nu,r}]$, which completes the proof of the assertion (b).

To prove (c) we note that for $f, g \in C_0$, the relation (8.1.5) is proved directly by using Fubini's theorem. Hence, since C_0 is dense in $\mathfrak{L}_{\nu,r}$ and $\mathfrak{L}_{\nu,s}$ (see Rooney [1, Lemma 2.2]), (8.1.5) will be true if we show that both sides of (8.1.5) represent bounded bilinear functionals on $\mathfrak{L}_{\nu,r} \times \mathfrak{L}_{\nu,s}$. Now the assumption $1/r + 1/s \geqq 1$ implies $r' \geqq s$ and the fact that $1/r \leqq \gamma(r) \leqq \nu$ yields $(r')' = r \geqq 1/\nu$. Thus the assumption (a) deduces $\mathbb{H}_\eta \in [\mathfrak{L}_{\nu,s}, \mathfrak{L}_{1-\nu,r'}]$. Applying the Hölder inequality (4.1.13) we have

$$\left|\int_0^\infty f(x)\left(\mathbb{H}_\eta g\right)(x)dx\right| \leqq \int_0^\infty |x^\nu f(x)| \left|x^{1-\nu}\left(\mathbb{H}_\eta g\right)(x)\right| \frac{dx}{x}$$
$$\leqq K\|f\|_{\nu,r}\left\|\mathbb{H}_\eta g\right\|_{1-\nu,r'} \leqq K_\nu \|f\|_{\nu,r}\|g\|_{\nu,s},$$

where K_ν is a bound for $\mathbb{H}_\eta$ as an element of $[\mathfrak{L}_{\nu,s}, \mathfrak{L}_{1-\nu,r'}]$. So the left-hand side of (8.1.5) is a bounded bilinear functional on $\mathfrak{L}_{\nu,r} \times \mathfrak{L}_{\nu,s}$ as the right-hand side of (8.1.5) is by a similar calculation, which shows (c).

To prove (d) we need in the function of the form (3.5.5):

$$g_{\eta,x}(t) = \begin{cases} t^{\eta+1/2}, & \text{if } 0 < t < x; \\ 0, & \text{if } t > x \end{cases} \tag{8.1.12}$$

and the function

$$h_{\eta,x}(t) = x^{\eta+1} t^{-1/2} J_{\eta+1}(xt), \tag{8.1.13}$$

for which we prove the following auxiliary result.

Lemma 8.1. *Let $1 < r < \infty$. The following assertions hold:*
(a) $g_{\eta,x} \in \mathfrak{L}_{\nu,r}$ *if and only if* $\nu > -\mathrm{Re}(\eta) - 1/2$.
(b) *If* $\mathrm{Re}(\eta) > -3/2$, *then* $h_{\eta,x} \in \mathfrak{L}_{\nu,r}$ *if and only if* $-\mathrm{Re}(\eta) - 1/2 < \nu < 1$.
(c) *If* $\mathrm{Re}(\eta) > -1$, *then*

$$\mathbb{H}_\eta g_{\eta,x} = h_{\eta,x} \tag{8.1.14}$$

and

$$\mathbb{H}_\eta h_{\eta,x} = g_{\eta,x}. \tag{8.1.15}$$

Proof. According to (3.1.3)

$$\|g_{\eta,x}\|_{\nu,r} = \left(\int_0^x t^{r(\nu+\mathrm{Re}(\eta)+1/2)-1} dt\right)^{1/r} < \infty,$$

if and only if $\nu > -\mathrm{Re}(\eta) - 1/2$, which proves (a).

Since from (8.1.7), if $\text{Re}(\eta) > -2$

$$h_{\eta,x}(t) \sim \frac{1}{2^{\eta+1}\Gamma(\eta+2)} x^{2\eta+2} t^{\eta+1/2} \quad (t \to +0) \tag{8.1.16}$$

and from (8.1.8)

$$h_{\eta,x}(t) \sim \left(\frac{2}{\pi}\right)^{1/2} x^{\eta+1/2} t^{-1} \cos\left[xt - \frac{1}{2}\left(\eta + \frac{1}{2}\right)\pi\right] \quad (t \to \infty), \tag{8.1.17}$$

then $h_{\eta,x} \in \mathfrak{L}_{\nu,r}$, if and only if

$$\int_0^\delta t^{r(\nu+\eta+1/2)-1} dt < \infty, \quad \int_R^\infty t^{r(\nu-1)-1} dt < \infty$$

for some positive δ and R, and thus, for $-\text{Re}(\eta) - 1/2 < \nu < 1$ guaranteed by the assumption $\text{Re}(\eta) > -3/2$, (b) is proved.

Putting $\nu = 1/2, r = 2$ in (a), we have $g_{\eta,x} \in \mathfrak{L}_{1/2,2} = L_2(\mathbb{R}_+)$ if $\text{Re}(\eta) > -1$. By noting that

$$(ut)^{1/2} J_\eta(ut) = \frac{1}{u}\frac{d}{dt} \int_0^{ut} v^{1/2} J_\eta(v) dv,$$

when $\text{Re}(\eta) > -1/2$, and by taking into account (8.1.12) and the relation

$$\int z^{\eta+1} J_\eta(z) dz = z^{\eta+1} J_{\eta+1}(z) \tag{8.1.18}$$

(see Erdélyi, Magnus, Oberhettinger and Tricomi [2, 7.7(1)]), we have

$$\begin{aligned}
\left(\mathbb{H}_\eta g_{\eta,x}\right)(t) &= \frac{d}{dt} \int_0^\infty g_{\eta,x}(u) \frac{du}{u} \int_0^{ut} v^{1/2} J_\eta(v) dv = \frac{d}{dt} \int_0^x u^{\eta-1/2} du \int_0^{ut} v^{1/2} J_\eta(v) dv \\
&= \frac{d}{dt} \int_0^x u^{\eta+1} du \int_0^t v^{1/2} J_\eta(uv) dv = \frac{d}{dt} \int_0^t v^{1/2} dv \int_0^x u^{\eta+1} J_\eta(uv) du \\
&= t^{1/2} \int_0^x u^{\eta+1} J_\eta(ut) du = t^{-(\eta+3/2)} \int_0^{xt} u^{\eta+1} J_\eta(u) du = x^{\eta+1} t^{-1/2} J_{\eta+1}(xt) \\
&= h_{\eta,x}(t).
\end{aligned}$$

It follows from the Titchmarsh theorem

$$f(x) = \frac{d}{dx} \int_0^\infty k(xt) g(t) \frac{dt}{t}, \quad g(x) = \frac{d}{dx} \int_0^\infty k(xt) f(t) \frac{dt}{t} \tag{8.1.19}$$

in $L_2(\mathbb{R}_+)$ (see Titchmarsh [3, Theorem 129]) and (8.1.4), that on $\mathfrak{L}_{1/2,2}$ the operator $\mathbb{H}_\eta^2$ is an identity: $\mathbb{H}_\eta^2 = I$. Since $h_{\eta,x} \in \mathfrak{L}_{1/2,2} = L_2(\mathbb{R}_+)$ by the result of Theorem 8.1(a),

$$\mathbb{H}_\eta h_{\eta,x} = \mathbb{H}_\eta^2 g_{\eta,x} = g_{\eta,x},$$

which completes the proof of Lemma 8.1.

Now we can prove the assertion (d) of Theorem 8.1. By Lemma 8.1(a), $g_{\eta,x} \in \mathfrak{L}_{\nu,r'}$. Applying (8.1.12), Theorem 8.1(c), (8.1.14) and (8.1.13), we have for $x > 0$

$$\int_0^x t^{\eta+1/2}\left(\mathbb{H}_\eta f\right)(t)dt = \int_0^\infty g_{\eta,x}(t)\left(\mathbb{H}_\eta f\right)(t)dt = \int_0^\infty \left(\mathbb{H}_\eta g_{\eta,x}\right)(t)f(t)dt$$

$$= \int_0^\infty h_{\eta,x}(t)f(t)dt = x^{\eta+1/2}\int_0^\infty (xt)^{1/2}J_{\eta+1}(xt)f(t)\frac{dt}{t}$$

and the result in (8.1.6) follows on differentiating. Thus Theorem 8.1 is proved.

Next we present the results characterizing the range of the Hankel transform (8.1.1) in the space $\mathfrak{L}_{\nu,r}$. First we indicate the constancy of the range of $\mathbb{H}_\eta$:

$$\mathbb{H}_\eta(\mathfrak{L}_{\nu,r}) = \mathbb{H}_\zeta(\mathfrak{L}_{\nu,r}), \tag{8.1.20}$$

if $1 < r < \infty, \gamma(r) \leqq \nu < \min[\mathrm{Re}(\eta), \mathrm{Re}(\zeta)] + 3/2$. This relation for real η and ζ was proved by Rooney [3, Theorem 1] by using a technique based on Mellin multipliers, and can be clearly extended to complex η and ζ.

To give a complete description of $\mathbb{H}_\eta(\mathfrak{L}_{\nu,r})$ we prove the representation of $\mathbb{H}_\eta f$ in terms of the shift operator M_ζ and the Erdélyi–Kober operator

$$\left(I_{\alpha,\xi}f\right)(x) = \frac{2x^{-2(\alpha+\xi)}}{\Gamma(\alpha)}\int_0^x (x^2-t^2)^{\alpha-1}t^{2\xi+1}f(t)dt \quad (\alpha,\ \xi \in \mathbb{C};\ \mathrm{Re}(\alpha) > 0) \tag{8.1.21}$$

(see Samko, Kilbas and Marichev [1, (18.8)]) with

$$\left(I_{0,\xi}f\right)(x) = f(x), \tag{8.1.22}$$

where (8.1.21) is the special case of the operator $I^\alpha_{0+;\sigma,\xi}$ defined in (3.3.1) as $I_{\alpha,\xi} = I^\alpha_{0+;2,\xi}$.

Lemma 8.2. *Let $1 < r < \infty$, $\gamma(r) \leqq \nu < \mathrm{Re}(\eta) + 3/2$, where $\gamma = \gamma(r)$ is given in* (3.3.9). *If $f \in \mathfrak{L}_{\nu,r}$, then the representation*

$$\mathbb{H}_\eta f = 2^{\gamma-\nu}M_{\nu-\gamma}I_{\nu-\gamma,(\eta-\nu+\gamma-1/2)/2}\mathbb{H}_{\eta-\nu+\gamma}M_{\nu-\gamma}f \tag{8.1.23}$$

holds.

Proof. By putting $\alpha = \eta - \nu + \gamma$ and $\beta = \nu - \gamma$, we rewrite (8.1.23) in the simpler form

$$\mathbb{H}_{\alpha+\beta}f = 2^{-\beta}M_\beta I_{\beta,(\alpha-1/2)/2}\mathbb{H}_\alpha M_\beta f, \tag{8.1.24}$$

which we are going to prove. Let $f \in C_0$. Note that $\mathrm{Re}(\alpha) > -1$ and $\beta \geqq 0$ from the assumptions. The result is obvious if $\beta = 0$ in view of (8.1.22). Let $\beta > 0$. Using the relation

$$J_{\alpha+\beta}(xt) = \frac{1}{2^{\beta-1}\Gamma(\beta)}x^{-(\alpha+\beta)}t^\beta\int_0^x (x^2-u^2)^{\beta-1}u^{\alpha+1}J_\alpha(tu)du \tag{8.1.25}$$

in Prudnikov, Brychkov and Marichev [2, (2.12.4.6)] and Fubini's theorem, we have

$$\left(\mathbb{H}_{\alpha+\beta}f\right)(x) = \int_0^\infty (xt)^{1/2}J_{\alpha+\beta}(xt)f(t)dt$$

$$= \frac{1}{2^{\beta-1}\Gamma(\beta)}x^{1/2-(\alpha+\beta)}\int_0^\infty t^{\beta+1/2}f(t)dt\int_0^x (x^2-u^2)^{\beta-1}u^{\alpha+1}J_\alpha(tu)du$$

$$= \frac{1}{2^{\beta-1}\Gamma(\beta)}x^{1/2-(\alpha+\beta)}\int_0^x (x^2-u^2)^{\beta-1}u^{\alpha+1/2}du\int_0^\infty (tu)^{1/2}J_\alpha(tu)t^\beta f(t)dt$$

$$= 2^{-\beta}\left(M_\beta I_{\beta,(\alpha-1/2)/2}\mathbb{H}_\alpha M_\beta f\right)(x)$$

and (8.1.24) is proved for $f \in C_0$.

Due to Rooney [2, Lemma 2.2], C_0 is dense in $\mathfrak{L}_{\nu,r}$ and by Theorem 8.1(a), we have $\mathbb{H}_\eta \in [\mathfrak{L}_{\nu,r}, \mathfrak{L}_{1-\nu,r}]$. Thus to complete the proof we must show that

$$M_\beta I_{\beta,(\alpha-1/2)/2}\mathbb{H}_\alpha M_\beta \in [\mathfrak{L}_{\nu,r}, \mathfrak{L}_{1-\nu,r}]. \tag{8.1.26}$$

But by Lemma 3.1(i) M_β maps $\mathfrak{L}_{\nu,r}$ isometrically onto $\mathfrak{L}_{\nu-\beta,r} = \mathfrak{L}_{\gamma,r}$ and by Theorem 8.1(a) $\mathbb{H}_\alpha$ maps $\mathfrak{L}_{\gamma,r}$ boundedly into $\mathfrak{L}_{1-\gamma,r}$ under $\gamma < \mathrm{Re}(\alpha) + 3/2$. Then according to Theorem 3.2(a) $I_{\beta,(\alpha-1/2)/2}$ maps $\mathfrak{L}_{1-\gamma,r}$ boundedly into itself, if $1-\gamma < \mathrm{Re}(\alpha) + 3/2$ which is valid from the assumption. Finally M_β maps $\mathfrak{L}_{1-\gamma,r}$ isometrically onto $\mathfrak{L}_{1-\gamma-\beta,r} = \mathfrak{L}_{1-\nu,r}$, and thus the result in (8.1.24) follows. This completes the proof of Lemma 8.2.

Now we are ready to prove the range of the Hankel transform (8.1.1) in terms of the shift operator M_ζ in (3.3.11), the Erdélyi–Kober operator $I_{\alpha,\xi}$ in (8.1.21) and the cosine transform $\mathfrak{F}_c$ in (8.1.2).

Theorem 8.2. *Let $1 < r < \infty$ and $\gamma \leqq \nu < \mathrm{Re}(\eta) + 3/2$, where $\gamma = \gamma(r)$ is given in (3.3.9). Then there holds*

$$\mathbb{H}_\eta(\mathfrak{L}_{\nu,r}) = \left(M_{\nu-\gamma}I_{\nu-\gamma,-1/2}\mathfrak{F}_c\right)(\mathfrak{L}_{\gamma,r}). \tag{8.1.27}$$

In particular, if $r = 2$ and $1/2 \leqq \nu < \mathrm{Re}(\eta) + 3/2$, then

$$\mathbb{H}_\eta(\mathfrak{L}_{\nu,2}) = \left(M_{\nu-1/2}I_{\nu-1/2,-1/2}\mathfrak{F}_c\right)\left(L_2(\mathbb{R}_+)\right). \tag{8.1.28}$$

Proof. Let $\zeta = \nu-\gamma-1/2$. Then $\zeta+3/2 = \nu-\gamma+1 > \nu$, since $\gamma < 1$. Hence by (8.1.20),

$$\mathbb{H}_\eta(\mathfrak{L}_{\nu,r}) = \mathbb{H}_\zeta(\mathfrak{L}_{\nu,r}) = \mathbb{H}_{\nu-\gamma-1/2}(\mathfrak{L}_{\nu,r}). \tag{8.1.29}$$

According to Lemma 8.2 and (8.1.2), if $f \in \mathfrak{L}_{\nu,r}$, we have

$$\begin{aligned}\mathbb{H}_{\nu-\gamma-1/2}f &= 2^{\gamma-\nu}M_{\nu-\gamma}I_{\nu-\gamma,-1/2}\mathbb{H}_{-1/2}M_{\nu-\gamma}f \\ &= 2^{\gamma-\nu}M_{\nu-\gamma}I_{\nu-\gamma,-1/2}\mathfrak{F}_cM_{\nu-\gamma}f.\end{aligned} \tag{8.1.30}$$

Let $g \in \mathbb{H}_\eta(\mathfrak{L}_{\nu,r})$. Then by (8.1.29) and (8.1.30) there exists $f \in \mathfrak{L}_{\nu,r}$ such that

$$g = 2^{\gamma-\nu}M_{\nu-\gamma}I_{\nu-\gamma,-1/2}\mathfrak{F}_cM_{\nu-\gamma}f.$$

Since $M_{\nu-\gamma}f \in \mathfrak{L}_{\gamma,r}$, $g \in (M_{\nu-\gamma}I_{\nu-\gamma,-1/2}\mathfrak{F}_c)(\mathfrak{L}_{\gamma,r})$ which means that

$$\mathbb{H}_\eta(\mathfrak{L}_{\nu,r}) \subset \left(M_{\nu-\gamma}I_{\nu-\gamma,-1/2}\mathfrak{F}_c\right)(\mathfrak{L}_{\gamma,r}). \tag{8.1.31}$$

Conversely, if $g \in \left(M_{\nu-\gamma}I_{\nu-\gamma,-1/2}\mathfrak{F}_c\right)(\mathfrak{L}_{\gamma,r})$, there exists $f \in \mathfrak{L}_{\gamma,r}$ such that $g = M_{\nu-\gamma}I_{\nu-\gamma,-1/2}\mathfrak{F}_c f$. Let $f_1 = M_{\nu-\gamma}f$. Then $f_1 \in \mathfrak{L}_{\nu,r}$ and $g = 2^{\nu-\gamma}\mathbb{H}_{\nu-\gamma-1/2}f_1$ from (8.1.30), which imply that $g \in \mathbb{H}_{\nu-\gamma-1/2}(\mathfrak{L}_{\nu,r})$. Thus by virtue of (8.1.29) we find

$$\left(M_{\nu-\gamma}I_{\nu-\gamma,-1/2}\mathfrak{F}_c\right)(\mathfrak{L}_{\gamma,r}) \subset \mathbb{H}_\eta(\mathfrak{L}_{\nu,r}). \tag{8.1.32}$$

From (8.1.31) and (8.1.32) we obtain (8.1.27), which completes the proof of Theorem 8.2.

Theorem 8.1(c) and Lemma 8.1 allow us to find the inversion formula for the Hankel transform (8.1.1) in the space $\mathfrak{L}_{\nu,r}$.

Theorem 8.3. *Let* $1 < r < \infty$ *and* $\gamma(r) \leqq \nu < \min[1, \mathrm{Re}(\eta) + 3/2]$. *If* $f \in \mathfrak{L}_{\nu,r}$, *then the inversion relation*

$$f(x) = x^{-(\eta+1/2)}\frac{d}{dx}x^{\eta+1/2}\int_0^\infty (xt)^{1/2}J_{\eta+1}(xt)\left(\mathbb{H}_\eta f\right)(t)\frac{dt}{t} \tag{8.1.33}$$

holds for almost all $x > 0$.

Proof. Since $\mathrm{Re}(\eta) > -1$ and $-\mathrm{Re}(\eta) - 1/2 < 1/2 \leqq \gamma(r) \leqq \nu < 1$, Lemma 8.1(b) yields that $h_{\eta,x}$ defined in (8.1.13) is in the space $\mathfrak{L}_{\nu,r'}$. Then by Theorem 8.1(c) and Lemma 8.1(c) we have

$$x^{\eta+1/2}\int_0^\infty (xt)^{1/2}J_{\eta+1}(xt)\left(\mathbb{H}_\eta f\right)(t)\frac{dt}{t} = \int_0^\infty h_{\eta,x}(t)\left(\mathbb{H}_\eta f\right)(t)dt$$
$$= \int_0^\infty \left(\mathbb{H}_\eta h_{\eta,x}\right)(t)f(t)dt = \int_0^x t^{\eta+1/2}f(t)dt$$

and the result in (8.1.33) follows on differentiating.

Corollary 8.3.1. *If* $1 < r < \infty$ *and* $\gamma(r) \leqq \nu < \mathrm{Re}(\eta) + 3/2$, *then the transform* $\mathbb{H}_\eta$ *is one-to-one on* $\mathfrak{L}_{\nu,r}$.

Proof. Let $f \in \mathfrak{L}_{\nu,r}$ and $\mathbb{H}_\eta f = 0$. If $\gamma(r) \leqq \nu < \min[1, \mathrm{Re}(\eta) + 3/2]$, then $f = 0$ according to Theorem 8.3. When $\gamma(r) \leqq \nu < \mathrm{Re}(\eta) + 3/2$, then by Lemma 8.2

$$M_{\nu-\gamma}I_{\nu-\gamma,(\eta-\nu+\gamma-1/2)/2}\mathbb{H}_{\eta-\nu+\gamma}M_{\nu-\gamma}f = 0$$

and hence $\mathbb{H}_{\eta-\nu+\gamma}M_{\nu-\gamma}f = 0$ taking into account the isomorphism of the operator M_ξ and $I_{\alpha,\xi}$ (see Lemma 3.1(i) and Rooney [1, Lemma 3.4]). Thus $M_{\nu-\gamma}f \in \mathfrak{L}_{\gamma,r}$. If we note $\gamma < 1$ and $\gamma < \mathrm{Re}(\eta) - \nu + \gamma + 3/2$, then by the previous result with η being replaced by $\eta - \nu + \gamma$, we have $M_{\nu-\gamma}f = 0$ and $f = 0$.

This result will be used in Section 8.7 while studying the transform $\mathbb{Y}_\eta$.

Remark 8.1. The right-hand sides of (8.1.6) and (8.1.33) are the same except that in (8.1.33) f is replaced by $\mathbb{H}_\eta f$ which means formally $\mathbb{H}_\eta^{-1} = \mathbb{H}_\eta$ or $\mathbb{H}_\eta^2$ is an identity: $\mathbb{H}_\eta^2 = I$. However, except for the case $\mathfrak{L}_{1/2,2} = L_2(\mathbb{R}_+)$ this is purely formal. If $f \in \mathfrak{L}_{\nu,r}$ $(1 < r < \infty)$ and $\gamma(r) \leqq \nu < \mathrm{Re}(\eta) + 3/2$, then $\mathbb{H}_\eta f \in \mathfrak{L}_{1-\nu,r}$ and thus for $\mathbb{H}_\eta^2$ to be defined we require that $\gamma(r) \leqq 1 - \nu < \mathrm{Re}(\eta) + 3/2$. But since $\gamma(r) \geqq 1/2$ with the equality only if $r = 2$, then

$1-\nu \leqq 1/2 \leqq \gamma(r)$ with the equality only if $r=2$, and thus $\nu = 1/2$ and $r=2$.

In Theorem 8.3 we found the inversion for the Hankel transform $\mathbb{H}_\eta$ on $\mathfrak{L}_{\nu,r}$, but with the restriction that $\nu < 1$. The result below is the exception of this restriction.

Theorem 8.4. *Let $1 < r < \infty$ and $\gamma(r) \leqq \nu < \mathrm{Re}(\eta)+3/2$, or $r=1$ and $1 \leqq \nu \leqq \mathrm{Re}(\eta)+3/2$ where $\gamma(r)$ is given in (3.3.9). If we choose the integer $l > \nu$ and if $f \in \mathfrak{L}_{\nu,r}$, then the inversion relation*

$$f(x) = x^{-\eta+1/2}\left(\frac{1}{x}\frac{d}{dx}\right)^l x^{\eta+l-1/2}\int_0^\infty (xt)^{1/2}J_{\eta+l}(xt)\left(\mathbb{H}_\eta f\right)(t)\frac{dt}{t^l} \tag{8.1.34}$$

holds for almost all $x > 0$,

Proof. This is proved similarly to Theorem 8.3 on the basis of Theorem 8.1(c) and Lemma 8.1 if we choose the function

$$h_{\eta,x}(t) = x^{\eta+l}t^{-l+1/2}J_{\eta+l}(xt) \quad (l \in \mathbb{N}) \tag{8.1.35}$$

instead of that in (8.1.13) and take into account Theorem 8.1(b) and the property

$$\left(\frac{1}{z}\frac{d}{dz}\right)^l [z^\eta J_\eta(z)] = z^{\eta-l}J_{\eta-l}(z) \tag{8.1.36}$$

(see Erdélyi, Magnus, Oberhettinger and Tricomi [2, 7.2(52)]).

8.2. Fourier Cosine and Sine Transforms

We consider the Fourier cosine and sine transforms $\mathfrak{F}_c$ and $\mathfrak{F}_s$ defined in (8.1.2) and (8.1.3). As indicated in Section 8.1, these transforms are particular cases of the Hankel transform (8.1.1) when $\eta = -1/2$ and $\eta = 1/2$, respectively. Therefore from the results in Section 8.1 we obtain the $\mathfrak{L}_{\nu,r}$-theory of them.

The boundedness and the representation of the cosine and sine transforms in the space $\mathfrak{L}_{\nu,r}$ follow from Theorem 8.1.

Theorem 8.5. *Let $1 \leqq r < \infty$ and let $\gamma(r)$ be given in (3.3.9).*

(a) *If $1 < r < \infty$ and $\gamma(r) \leqq \nu < 1$, then for all $s \geqq r$ such that $s' \geqq 1/\nu$ and $1/s+1/s' = 1$, the operator $\mathfrak{F}_c$ belongs to $[\mathfrak{L}_{\nu,r}, \mathfrak{L}_{1-\nu,s}]$ and is a one-to-one transform from $\mathfrak{L}_{\nu,r}$ onto $\mathfrak{L}_{1-\nu,s}$. If $1 < r \leqq 2$ and $f \in \mathfrak{L}_{\nu,r}$, then the Mellin transform of (8.1.2) for $\mathrm{Re}(s) = 1-\nu$ is given by*

$$\left(\mathfrak{M}\mathfrak{F}_c f\right)(s) = 2^{s-1/2}\frac{\Gamma\left(\frac{s}{2}\right)}{\Gamma\left(\frac{1-s}{2}\right)}\left(\mathfrak{M}f\right)(1-s). \tag{8.2.1}$$

(b) $\mathfrak{F}_c \in [L_1(\mathbb{R}_+), L_\infty(\mathbb{R}_+)]$.

(c) *If $f \in \mathfrak{L}_{\nu,r}$ and $g \in \mathfrak{L}_{\nu,s}$, where $1 < r < \infty$ and $1 < s < \infty$ are such that $1/r+1/s \geqq 1$ and $\max[\gamma(r),\gamma(s)] \leqq \nu < 1$, then the following relation holds:*

$$\int_0^\infty f(x)\left(\mathfrak{F}_c g\right)(x)dx = \int_0^\infty \left(\mathfrak{F}_c f\right)(x)g(x)dx. \tag{8.2.2}$$

(d) *If $f \in \mathfrak{L}_{\nu,r}$, where $1 < r < \infty$ and $\gamma(r) \leqq \nu < 1$, then for almost all $x > 0$,*

$$\left(\mathfrak{F}_c f\right)(x) = \frac{d}{dx}\int_0^\infty (xt)^{1/2} J_{1/2}(xt) f(t) \frac{dt}{t} = \left(\frac{2}{\pi}\right)^{1/2} \frac{d}{dx}\int_0^\infty \sin(xt) f(t) \frac{dt}{t}. \tag{8.2.3}$$

Theorem 8.6. *Let $1 \leqq r < \infty$ and let $\gamma(r)$ be given in (3.3.9).*

(a) *If $1 < r < \infty$ and $\gamma(r) \leqq \nu < 2$, then for all $s \geqq r$ such that $s' \geqq 1/\nu$ and $1/s + 1/s' = 1$, the operator $\mathfrak{F}_s$ belongs to $[\mathfrak{L}_{\nu,r}, \mathfrak{L}_{1-\nu,s}]$ and is a one-to-one transform from $\mathfrak{L}_{\nu,r}$ onto $\mathfrak{L}_{1-\nu,s}$. If $1 < r \leqq 2$ and $f \in \mathfrak{L}_{\nu,r}$, then the Mellin transform of (8.1.3) for $\mathrm{Re}(s) = 1 - \nu$ is given by*

$$\left(\mathfrak{M}\mathfrak{F}_s f\right)(s) = 2^{s-1/2} \frac{\Gamma\left(\dfrac{1+s}{2}\right)}{\Gamma\left(\dfrac{2-s}{2}\right)} \left(\mathfrak{M} f\right)(1-s). \tag{8.2.4}$$

(b) *If $1 \leqq \nu \leqq 2$, then $\mathfrak{F}_s \in [\mathfrak{L}_{\nu,1}, \mathfrak{L}_{1-\nu,\infty}]$. If $1 < \nu < 2$, then for all r $(1 \leqq r \leqq \infty)$, $\mathfrak{F}_s \in [\mathfrak{L}_{\nu,1}, \mathfrak{L}_{1-\nu,r}]$.*

(c) *If $f \in \mathfrak{L}_{\nu,r}$ and $g \in \mathfrak{L}_{\nu,s}$, where $1 < r < \infty$ and $1 < s < \infty$ are such that $1/r + 1/s \geqq 1$ and $\max[\gamma(r), \gamma(s)] \leqq \nu < 2$, then the following relation holds:*

$$\int_0^\infty f(x)\left(\mathfrak{F}_s g\right)(x) dx = \int_0^\infty \left(\mathfrak{F}_s f\right)(x) g(x) dx. \tag{8.2.5}$$

(d) *If $f \in \mathfrak{L}_{\nu,r}$, where $1 < r < \infty$ and $\gamma(r) \leqq \nu < 2$, then for almost all $x > 0$,*

$$\left(\mathfrak{F}_s f\right)(x) = \frac{1}{x}\frac{d}{dx} x \int_0^\infty (xt)^{1/2} J_{3/2}(xt) f(t) \frac{dt}{t}. \tag{8.2.6}$$

Theorem 8.2 gives the range of the Fourier sine transform in the space $\mathfrak{L}_{\nu,r}$.

Theorem 8.7. *Let $1 < r < \infty$ and $\gamma(r)$ be given in (3.3.9). If $\gamma \leqq \nu < 2$, then*

$$\mathfrak{F}_s(\mathfrak{L}_{\nu,r}) = \left(M_{\nu-\gamma} I_{\nu-\gamma,-1/2} \mathfrak{F}_c\right)\left(\mathfrak{L}_{\gamma,r}\right), \tag{8.2.7}$$

where $I_{\alpha,\xi}$ is the Erdélyi–Kober operator (8.1.21). In particular if $r = 2$ and $1/2 \leqq \nu < 2$,

$$\mathfrak{F}_s(\mathfrak{L}_{\nu,2}) = \left(M_{\nu-1/2} I_{\nu-1/2,-1/2} \mathfrak{F}_c\right)\left(L_2(\mathbb{R}_+)\right). \tag{8.2.8}$$

From Theorem 8.4 we come to the inversion formulas for the Fourier cosine and sine transforms in the space $\mathfrak{L}_{\nu,r}$

Theorem 8.8. *Let $1 < r < \infty$, and let l be an integer such that $l > \nu$.*

(a) *Let $\gamma(r) \leqq \nu < 1$, or $r = 1$ and $\nu = 1$. If $f \in \mathfrak{L}_{\nu,r}$, then the inversion relation*

$$f(x) = x\left(\frac{1}{x}\frac{d}{dx}\right)^l x^{l-1} \int_0^\infty (xt)^{1/2} J_{l-1/2}(xt) \left(\mathfrak{F}_c f\right)(t) \frac{dt}{t^l} \tag{8.2.9}$$

holds for almost all $x > 0$. In particular, when $l = 1$,

$$f(x) = \frac{d}{dx}\int_0^\infty (xt)^{1/2} J_{1/2}(xt)\left(\mathfrak{F}_c f\right)(t)\frac{dt}{t}$$
$$= \left(\frac{2}{\pi}\right)^{1/2} \frac{d}{dx}\int_0^\infty \sin(xt)\left(\mathfrak{F}_c f\right)(t)\frac{dt}{t}. \tag{8.2.10}$$

(b) *Let $\gamma(r) \leqq \nu < 2$, or $r = 1$ and $1 \leqq \nu \leqq 2$. If $f \in \mathfrak{L}_{\nu,r}$, then the inversion relation*

$$f(x) = \left(\frac{1}{x}\frac{d}{dx}\right)^l x^l \int_0^\infty (xt)^{1/2} J_{l+1/2}(xt)\left(\mathfrak{F}_s f\right)(t)\frac{dt}{t^l} \tag{8.2.11}$$

holds for almost all $x > 0$. In particular, when $l = 1$,

$$f(x) = \frac{1}{x}\frac{d}{dx} x \int_0^\infty (xt)^{1/2} J_{3/2}(xt)\left(\mathfrak{F}_s f\right)(t)\frac{dt}{t}. \tag{8.2.12}$$

Remark 8.2. It follows from Theorems 8.5–8.8 that the range of the parameter η, for which the $\mathfrak{L}_{\nu,r}$-theory of the Fourier cosine and sine transforms (8.1.2) and (8.1.3) is valid, is wider for the former than for the latter. We also note that direct differentiation of the right sides in (8.2.10) and (8.2.12), by using (8.1.36) with $l = 1$ for the latter, yields the known inversion formulas

$$f(x) = \left(\frac{2}{\pi}\right)^{1/2} \int_0^\infty \cos(xt)\left(\mathfrak{F}_c f\right)(t)dt \tag{8.2.13}$$

and

$$f(x) = \left(\frac{2}{\pi}\right)^{1/2} \int_0^\infty \sin(xt)\left(\mathfrak{F}_s f\right)(t)dt \tag{8.2.14}$$

for the cosine and sine transforms (8.1.2) and (8.1.3), respectively.

8.3. Even and Odd Hilbert Transforms

Let us consider the even and odd Hilbert integral transforms $\boldsymbol{H}_+$ and $\boldsymbol{H}_-$ defined by

$$\boldsymbol{H}_+ = -\mathfrak{F}_s\mathfrak{F}_c \tag{8.3.1}$$

and

$$\boldsymbol{H}_- = \mathfrak{F}_c\mathfrak{F}_s, \tag{8.3.2}$$

where $\mathfrak{F}_c$ and $\mathfrak{F}_s$ are the Fourier cosine and Fourier sine transforms defined in (8.1.2) and (8.1.3), respectively. Since $\mathfrak{F}_c, \mathfrak{F}_s \in [\mathfrak{L}_{1/2,2}]$, $\boldsymbol{H}_+$ and $\boldsymbol{H}_-$ also belong to $[\mathfrak{L}_{1/2,2}]$. The fact that

$$\mathfrak{F}_c^2 = \mathfrak{F}_s^2 = I \tag{8.3.3}$$

on $\mathfrak{L}_{1/2,2} = L_2(\mathbb{R}_+)$ implies the relations

$$\boldsymbol{H}_+\boldsymbol{H}_- = \boldsymbol{H}_-\boldsymbol{H}_+ = -I \tag{8.3.4}$$

on $L_2(\mathbb{R}_+)$, where I is an identical operator.

Applying Theorem 8.1 twice, once with $\eta = -1/2$ and once with $\eta = 1/2$, it follows that if $f, g \in \mathfrak{L}_{1/2,2}$, then the relation of integration by parts (8.1.5) takes the forms

$$\int_0^\infty \left(\boldsymbol{H}_+ f\right)(t)g(t)dt = -\int_0^\infty f(t)\left(\boldsymbol{H}_- g\right)(t)dt \tag{8.3.5}$$

and

$$\int_0^\infty \left(\boldsymbol{H}_- f\right)(t)g(t)dt = -\int_0^\infty f(t)\left(\boldsymbol{H}_+ g\right)(t)dt. \tag{8.3.6}$$

Taking g to be the characteristic function

$$\chi_x(t) = \begin{cases} 1, & \text{if } 0 < t < x; \\ 0, & \text{if } t > x, \end{cases} \tag{8.3.7}$$

we obtain, by virtue of the simple equality

$$\int_0^\infty \frac{1}{x}\{\cos ax - \cos bx\}dx = \log\left(\frac{b}{a}\right) \quad (a, b > 0)$$

(see Prudnikov, Brychkov and Marichev [1, (2.5.29.16)]), the integral representations

$$\left(\boldsymbol{H}_+ f\right)(x) = -\frac{1}{\pi}\frac{d}{dx}\int_0^\infty \log\left|1 - \frac{x^2}{t^2}\right| f(t)dt \tag{8.3.8}$$

and

$$\left(\boldsymbol{H}_- f\right)(x) = -\frac{1}{\pi}\frac{d}{dx}\int_0^\infty \log\left|\frac{t-x}{t+x}\right| f(t)dt \tag{8.3.9}$$

for almost all $x > 0$. Comparing (8.3.8) and (8.3.9) with Theorem 90 in Titchmarsh [3], it is evident that $\boldsymbol{H}_-$ is the restriction on $\mathbb{R}_+$ of the Hilbert transform of even functions, while $\boldsymbol{H}_+$ is the restriction on $\mathbb{R}_+$ of the Hilbert transform of odd functions; hence the names, even and odd Hilbert transforms are given to them. It follow from (8.3.8) and (8.3.9) that for suitable functions f and almost all $x > 0$, $\boldsymbol{H}_+ f$ and $\boldsymbol{H}_- f$ can be represented in the forms

$$\left(\boldsymbol{H}_+ f\right)(x) = \frac{2x}{\pi}\ \text{p.v.}\int_0^\infty \frac{1}{t^2 - x^2} f(t)dt \tag{8.3.10}$$

and

$$\left(\boldsymbol{H}_- f\right)(x) = \frac{2}{\pi}\ \text{p.v.}\int_0^\infty \frac{t}{t^2 - x^2} f(t)dt, \tag{8.3.11}$$

where p.v. denotes the Cauchy principal value of the integral at $t = x$.

The action of the Mellin transform on $\boldsymbol{H}_+$ and $\boldsymbol{H}_-$ on $\mathfrak{L}_{1/2,2}$ is directly computed from (8.1.4). This yields that, if $f \in \mathfrak{L}_{1/2,2}$, then for $\text{Re}(s) = 1/2$ we have

$$\left(\mathfrak{M}\boldsymbol{H}_+ f\right)(s) = -\tan\left(\frac{\pi s}{2}\right)\left(\mathfrak{M}f\right)(s) \tag{8.3.12}$$

and

$$\left(\mathfrak{M}\boldsymbol{H}_- f\right)(s) = \cot\left(\frac{\pi s}{2}\right)\left(\mathfrak{M}f\right)(s). \tag{8.3.13}$$

According to (8.3.1) and (8.3.2) and Theorems 8.5 and 8.6, $\boldsymbol{H}_+$ and $\boldsymbol{H}_-$ can be extended to $\mathfrak{L}_{\nu,r}$ for $1 < r < \infty$. The properties of these operators on this space are given in the following result.

Theorem 8.9.

(a) *Let $1 < r < \infty$ and $-1 < \nu < 1$. The transform $\boldsymbol{H}_+$ belongs to $[\mathfrak{L}_{\nu,r}]$. When $\nu \neq 0$, $\boldsymbol{H}_+$ maps $\mathfrak{L}_{\nu,r}$ one-to-one onto itself. For $f \in \mathfrak{L}_{\nu,r}$ the representations* (8.3.8) *and* (8.3.10) *hold.*

If $1 < r \leqq 2$ and $f \in \mathfrak{L}_{\nu,r}$, the Mellin transform of $\boldsymbol{H}_+$ for $\mathrm{Re}(s) = \nu$ is given in (8.3.12).

(b) *Let $1 < r < \infty$ and $0 < \nu < 2$. The operator $\boldsymbol{H}_-$ belongs to $[\mathfrak{L}_{\nu,r}]$. When $\nu \neq 1$, $\boldsymbol{H}_-$ maps $\mathfrak{L}_{\nu,r}$ one-to-one onto itself. For $f \in \mathfrak{L}_{\nu,r}$ the representations* (8.3.9) *and* (8.3.11) *hold.*

If $1 < r \leqq 2$ and $f \in \mathfrak{L}_{\nu,r}$, the Mellin transform of $\boldsymbol{H}_-$ for $\mathrm{Re}(s) = \nu$ is given in (8.3.13).

(c) *Let $1 < r < \infty$, $f \in \mathfrak{L}_{\nu,r}$ and $g \in \mathfrak{L}_{1-\nu,r'}$. If $-1 < \nu < 1$, then the formula of integration by parts* (8.3.5) *holds, while when $0 < \nu < 2$,* (8.3.6) *holds.*

(d) *Let $1 < r < \infty$. On $\mathfrak{L}_{\nu,r}$ with $0 < \nu < 1$, the formula* (8.3.4) *holds. Further, the relation*

$$\boldsymbol{H}_+ = M_1 \boldsymbol{H}_- M_{-1} \tag{8.3.14}$$

holds for $-1 < \nu < 1$, while the relation

$$\boldsymbol{H}_- = M_{-1} \boldsymbol{H}_+ M_1 \tag{8.3.15}$$

holds for $0 < \nu < 2$, where the shift operator M_ζ is given in (3.3.11).

Proof. The function $m(s) = -\tan(\pi s/2)$ in (8.3.12) belongs to the class $\mathcal{A}$ (see Definition 3.2) with $\alpha(m) = -1$ and $\beta(m) = 1$. In fact, elementary arguments show that (a) $m(s)$ is analytic in $-1 < \mathrm{Re}(s) < 1$; (b) $m(s)$ is bounded in every closed substrip $\sigma_1 \leqq \mathrm{Re}(s) \leqq \sigma_2$, where $-1 < \sigma_1 \leqq \sigma_2 < 1$; (c) for $-1 < \sigma < 1$,

$$|m'(\sigma + it)| = \frac{\pi}{2}\left|\sec^2\left[\frac{\pi}{2}(\sigma + it)\right]\right| = O\left(\frac{1}{x}\right) \quad (|t| \to \infty).$$

Thus Theorem 3.1 yields that, since (8.3.12) holds on $\mathfrak{L}_{1/2,2}$, the operator $\boldsymbol{H}_+ \in [\mathfrak{L}_{\nu,r}]$ for $-1 < \nu < 1$, and if $f \in \mathfrak{L}_{\nu,r}$ with $1 < r < \infty$ and $-1 < \nu < 1$, then (8.3.12) holds for $\mathrm{Re}(s) = \nu$.

From

$$\frac{1}{m(s)} = -\cot\left(\frac{\pi s}{2}\right) = -\tan\left[\frac{\pi(1-s)}{2}\right] = m(1-s),$$

we have $1/m \in \mathcal{A}$ with $\alpha(1/m) = 0$ and $\beta(1/m) = 2$, which imply, by Theorem 3.1, that $\boldsymbol{H}_+$ maps $\mathfrak{L}_{\nu,r}$ one-to-one onto itself if $0 < \nu < 1$. Further, we have $1/m(s) = m(-1-s)$ and hence also $1/m \in \mathcal{A}$ with $\alpha(1/m) = -2$ and $\beta(1/m) = 0$. So, again by Theorem 3.1, $\boldsymbol{H}_+$ maps $\mathfrak{L}_{\nu,r}$ one-to-one onto itself if $-1 < \nu < 0$. The representation (8.3.10) follows from (8.3.8) by differentiating, while the representation (8.3.8) follow from the relation (8.3.5) by taking g as the characteristic function (8.3.7). Thus once (8.3.5) is proved for $f \in \mathfrak{L}_{\nu,r}$ and $g \in \mathfrak{L}_{\nu,r'}$, (a) is proved.

The proof of the statement (b) is exactly similar.

The relations of integration by parts (8.3.5) and (8.3.6) follow from the fact that these formulas hold for $f \in \mathfrak{L}_{1/2,2}$, $g \in \mathfrak{L}_{1/2,2}$ and that both sides of them represent bounded bilinear functionals on $\mathfrak{L}_{\nu,r} \times \mathfrak{L}_{1-\nu,r'}$.

The formula (8.3.4) holds since it is valid on $\mathfrak{L}_{1/2,2}$ and its both sides represent bounded operators on $\mathfrak{L}_{\nu,r}$. The equalities (8.3.14) and (8.3.15) follow on $\mathfrak{L}_{\nu,2}$ on taking the Mellin transform, and then on their respective $\mathfrak{L}_{\nu,r}$, since both sides of (8.3.14) and (8.3.15) represent bounded operators on those spaces. This completes the proof of Theorem 8.9.

According to (8.3.1) and (8.3.2), the inversion formulas for the even and odd Hilbert transforms $\boldsymbol{H}_+$ and $\boldsymbol{H}_-$ can be obtained from that for the Fourier cosine transform $\mathfrak{F}_c$ and the Fourier sine transform $\mathfrak{F}_s$ given in Theorem 8.8.

Theorem 8.10. *Let m and l be integers such that $m > 1/2$ and $l > 1/2$. Let $\mathfrak{F}_c^{(m)}$ and $\mathfrak{F}_s^{(l)}$ be given by*

$$\left(\mathfrak{F}_c^{(m)} f\right)(x) = x\left(\frac{1}{x}\frac{d}{dx}\right)^m x^{m-1} \int_0^\infty (xt)^{1/2} J_{m-1/2}(xt) \left(\mathfrak{F}_c f\right)(t) \frac{dt}{t^m}, \tag{8.3.16}$$

$$\left(\mathfrak{F}_s^{(l)} f\right)(x) = \left(\frac{1}{x}\frac{d}{dx}\right)^l x^l \int_0^\infty (xt)^{1/2} J_{l+1/2}(xt) \left(\mathfrak{F}_s f\right)(t) \frac{dt}{t^l}. \tag{8.3.17}$$

(a) *If $f \in L_2(\mathbb{R}_+)$, then the inversion relation*

$$f(x) = -\left(\mathfrak{F}_c^{(m)} \mathfrak{F}_s^{(l)} \boldsymbol{H}_+ f\right)(x) \tag{8.3.18}$$

holds for almost all $x > 0$.

(b) *If $f \in L_2(\mathbb{R}_+)$, then the inversion relation*

$$f(x) = \left(\mathfrak{F}_s^{(l)} \mathfrak{F}_c^{(m)} \boldsymbol{H}_- f\right)(x) \tag{8.3.19}$$

holds for almost all $x > 0$.

Proof. Let $f \in \mathfrak{L}_{\nu,r}$ with $1 < r < \infty$, $\nu \in \mathbb{R}$ and let $\gamma(r)$ be given by (3.3.9). According to (8.3.1)

$$\mathfrak{F}_s(\mathfrak{F}_s f) = -\mathbf{H}_+ f.$$

If $\gamma(r) \leqq \nu < 1$, then by Theorem 8.5(a) $\mathfrak{F}_s f \in \mathfrak{L}_{1-\nu,r}$. Applying Theorem 8.8(b) with ν being replaced by $1-\nu$ we have

$$\mathfrak{F}_s f = -\mathfrak{F}_s^{(l)} \mathbf{H}_+ f,$$

provided that $\gamma(r) \leqq 1-\nu < 2$ and l is an integer such that $l > 1-\nu$. Similarly, if $\gamma(r) \leqq \nu < 1$ and m is an integer such that $m > \nu$, then by Theorem 8.8(a) and the last relation we obtain

$$f = -\mathfrak{F}_c^{(m)} \mathfrak{F}_s^{(l)} \mathbf{H}_+ f,$$

and (8.3.18) is proved. From the above conditions this relation holds only when $\gamma(r) \leqq \nu \leqq 1-\gamma(r)$. In accordance with (3.3.9) such a fact is possible in the case $r = 2$ which yields $\nu = \gamma(2) = 1/2$. Hence the relation (8.3.18) holds for $f \in \mathfrak{L}_{1/2,2} \equiv L_2(\mathbf{R}_+)$. This completes the proof of (a). The assertion (b) is proved similarly.

Theorem 8.10 give the inversion formulas for the even and odd Hankel transforms defined by (8.3.1) and (8.3.2) in the space $L_2(\mathbb{R}_+)$. Another inversion formula for these transforms given in (8.3.10) and (8.3.11), was proved in the space $\mathfrak{L}_{\nu,r}$ by Heywood and Rooney [3, Theorem 4.1].

Theorem 8.11. *Let $1 < r < \infty$.*
(a) *If $-1 < \nu < 0$ and $f \in \mathfrak{L}_{\nu,r}$, then*

$$f(x) = \frac{2}{\pi} \text{ p.v.}\int_0^\infty \left[\frac{1}{t} - \frac{t}{t^2 - x^2}\right] \left(\boldsymbol{H}_+ f\right)(t)dt. \tag{8.3.20}$$

(b) *If $1 < \nu < 2$ and $f \in \mathfrak{L}_{\nu,r}$, then*

$$f(x) = \frac{2x}{\pi} \text{ p.v.}\int_0^\infty \left[\frac{x}{x^2 - t^2} - \frac{1}{x}\right] \left(\boldsymbol{H}_- f\right)(t)dt. \tag{8.3.21}$$

8.4. The Extended Hankel Transform

We consider the so-called extended Hankel integral transform defined by

$$\left(\mathbb{H}_{\eta,l} f\right)(x) = \int_0^\infty (xt)^{1/2} J_{\eta,l}(xt) f(t)dt \tag{8.4.1}$$

for $\text{Re}(\eta) > -1$, a non-negative integer l and $x \in \mathbb{R}_+$. The function $J_{\eta,k}(z)$ called a "cut" Bessel function,

$$J_{\eta,l}(z) = \sum_{k=l}^{\infty} \frac{(-1)^k (z/2)^{2k+\eta}}{\Gamma(\eta + k + 1)k!}. \tag{8.4.2}$$

The extended Hankel transform $\mathbb{H}_{\eta,l}$ is defined for $f \in C_0$ and for $\eta \in \mathbb{C}$ ($\eta \neq -1, -3, \cdots$) and for the least non-negative integer $l = l_\eta$ such that $\text{Re}(\eta) + 2l > -1$.

It is directly proved that if $f \in \mathfrak{L}_{1/2,2} = L_2(\mathbb{R}_+)$, then for $\text{Re}(s) = 1/2$ the Mellin transform of (8.4.1) is given by

$$\left(\mathfrak{M}\mathbb{H}_{\eta,l} f\right)(s) = \mathcal{G}_\eta(s)\left(\mathfrak{M} f\right)(1-s), \; \mathcal{G}_\eta(s) = 2^{s-1/2} \frac{\Gamma\left(\frac{1}{2}\left[\eta + s + \frac{1}{2}\right]\right)}{\Gamma\left(\frac{1}{2}\left[\eta - s + \frac{3}{2}\right]\right)}. \tag{8.4.3}$$

It is easy to see that if $1 < r \leqq 2$ and $\gamma(r) \leqq \nu < \text{Re}(\eta) + 2l + 3/2$, $\gamma(r)$ being given in (3.3.9), and if $f \in \mathfrak{L}_{\nu,r}$, then (8.4.3) is valid for $\text{Re}(s) = 1 - \nu$, because both sides of (8.4.3) represent the bounded linear transforms of $\mathfrak{L}_{\nu,r}$ into $L_{r'}(\mathbb{R}_+)$.

The boundedness and the representation for the extended Hankel transform (8.4.1) is given by the following result.

Theorem 8.12. *Let $1 \leqq r \leqq \infty$ and let $\gamma(r)$ be given by (3.3.9).*

(a) *If $1 < r < \infty$ and $\gamma(r) \leqq \nu < \mathrm{Re}(\eta) + 2l + 3/2$, then for all $s \geqq r$ such that $s' \geqq 1/\nu$ and $1/s + 1/s' = 1$, the operator $\mathbb{H}_{\eta,l}$ belongs to $[\mathfrak{L}_{\nu,r}, \mathfrak{L}_{1-\nu,s}]$. If $1 < r \leqq 2$ and $f \in \mathfrak{L}_{\nu,r}$, then the Mellin transform of (8.4.1) for (Re$(s) = 1 - \nu$ is given in (8.4.3).*

(b) *If $1 \leqq \nu \leqq \mathrm{Re}(\eta) + 2l + 3/2$, then $\mathbb{H}_{\eta,l} \in [\mathfrak{L}_{\nu,1}, \mathfrak{L}_{1-\nu,\infty}]$. If $1 < \nu < \mathrm{Re}(\eta) + 2l + 3/2$, then for all r, $1 \leqq r < \infty$, $\mathbb{H}_{\eta,l} \in [\mathfrak{L}_{\nu,1}, \mathfrak{L}_{1-\nu,r}]$.*

(c) *If $f \in \mathfrak{L}_{\nu,r}$ and $g \in \mathfrak{L}_{\nu,r'}$, where $1 < r < \infty$ and $\gamma(r) \leqq \nu < \mathrm{Re}(\eta) + 2l + 3/2$, then the following relation holds:*

$$\int_0^\infty f(x)\left(\mathbb{H}_{\eta,l}g\right)(x)dx = \int_0^\infty g(x)\left(\mathbb{H}_{\eta,l}f\right)(x)dx. \tag{8.4.4}$$

(d) *If $f \in \mathfrak{L}_{\nu,r}$, where $1 < r < \infty$, $\gamma(r) \leqq \nu < \mathrm{Re}(\eta)+2l+3/2$, then for almost all $x > 0$,*

$$\left(\mathbb{H}_{\eta,l}f\right)(x) = -x^{\eta-1/2}\frac{d}{dx}x^{1/2-\eta}\int_0^\infty (xt)^{1/2}J_{\eta-1,l+1}(xt)f(t)\frac{dt}{t}. \tag{8.4.5}$$

Proof. The asssertions (a)–(c) are proved similarly to those in Theorem 8.1(a)–(c). As for (d) it is proved similarly to that in Theorem 8.1(d), using the following statement similar to Lemma 8.1 in Section 8.1.

Lemma 8.3. *For $x > 0$ let*

$$g_{\eta,x}(t) = \begin{cases} t^{1/2-\eta}, & \text{if } 0 < t < x; \\ 0, & \text{if } t > x. \end{cases} \tag{8.4.6}$$

If either $1 \leqq r < \infty$ and $\nu > \mathrm{Re}(\eta) - 1/2$, or if $r = \infty$ and $\nu \geqq -\mathrm{Re}(\eta) - 1/2$, then the function $g_{\eta,x}(t) \in \mathfrak{L}_{\nu,r}$ and its extended Hankel transform is given by

$$\left(\mathbb{H}_{\eta,l}g_{\eta,x}\right)(t) = -x^{1-\eta}t^{-1/2}J_{\eta-1,l+1}(xt). \tag{8.4.7}$$

The inversion theory of the extended Hankel operator $\mathbb{H}_{\eta,l}$ on $\mathfrak{L}_{\nu,r}$ is different according as ν is less than, greater than, or equal to $-\mathrm{Re}(\eta) - 2l + 3/2$. The first two cases are given by the following statements.

Theorem 8.13 *Let $1 \leqq r < \infty$ and $\gamma(r) \leqq \nu < \mathrm{Re}(\eta)+2l+3/2$, and $\nu < -\mathrm{Re}(\eta)-2l+3/2$. If $f \in \mathfrak{L}_{\nu,r}$, then for almost all $x > 0$, the following inversion relation holds:*

$$f(x) = -x^{\eta+1/2}\left(\frac{1}{x}\frac{d}{dx}\right)^3 x^{5/2-\eta}\int_0^\infty (xt)^{1/2}J_{\eta-3,l+3}(xt)\left(\mathbb{H}_{\eta,l}f\right)(t)\frac{dt}{t^3}. \tag{8.4.8}$$

If, in addition, $\nu < 2$, then for almost all $x > 0$,

$$f(x) = x^{\eta+1/2}\left(\frac{1}{x}\frac{d}{dx}\right)^2 x^{3/2-\eta}\int_0^\infty (xt)^{1/2}J_{\eta-2,l+2}(xt)\left(\mathbb{H}_{\eta,l}f\right)(t)\frac{dt}{t^2}, \tag{8.4.9}$$

while, if in addition, $\nu < 1$, then for almost all $x > 0$

$$f(x) = -x^{\eta-1/2}\frac{d}{dx}x^{1/2-\eta}\int_0^\infty (xt)^{1/2}J_{\eta-1,l+1}(xt)\left(\mathbb{H}_{\eta,l}f\right)(t)\frac{dt}{t}. \tag{8.4.10}$$

Theorem 8.14. *Let $1 \leqq r < \infty$ and $\gamma(r) \leqq \nu < \mathrm{Re}(\eta) + 2l + 3/2$, and $\nu > -\mathrm{Re}(\eta) - 2l + 3/2$. If $f \in \mathfrak{L}_{\nu,r}$, then for almost all $x > 0$ the following inversion relation holds:*

$$f(x) = -x^{\eta+1/2}\left(\frac{1}{x}\frac{d}{dx}\right)^3 x^{5/2-\eta}\int_0^\infty (xt)^{1/2}J_{\eta-3,l+2}(xt)\left(\mathbb{H}_{\eta,l}f\right)(t)\frac{dt}{t^3}. \tag{8.4.11}$$

If, in addition, $\nu < 2$, then for almost all $x > 0$

$$f(x) = x^{\eta+1/2}\left(\frac{1}{x}\frac{d}{dx}\right)^2 x^{3/2-\eta}\int_0^\infty (xt)^{1/2}J_{\eta-2,l+1}(xt)\left(\mathbb{H}_{\eta,l}f\right)(t)\frac{dt}{t^2}, \tag{8.4.12}$$

while, if in addition, $\nu < 1$, then for almost all $x > 0$

$$f(x) = -x^{\eta-1/2}\frac{d}{dx}x^{1/2-\eta}\int_0^\infty (xt)^{1/2}J_{\eta-1,l}(xt)\left(\mathbb{H}_{\eta,l}f\right)(t)\frac{dt}{t}. \tag{8.4.13}$$

The proofs of Theorems 8.13 and 8.14 are based on the relation (8.4.4) in Theorem 8.12 and the auxiliary results for the three integrals I_1, I_2 and I_3 defined below for $x > 0$ and being taken in the principal value sense at infinity. These integrals are the inverse Mellin transforms for the function $\mathcal{G}_\eta(s)$ in (8.4.3).

Lemma 8.4. *There hold the following statements.*

(i) *If $\sigma < 3/2$, then*

$$I_1 = \frac{1}{2\pi i}\int_{\sigma-i\infty}^{\sigma+i\infty}\frac{\mathcal{G}_\eta(s)}{\dfrac{3}{2}-\eta-s}x^{-s}ds = -x^{-1/2}J_{\eta-1,l+1}(x) \tag{8.4.14}$$

for $-\mathrm{Re}(\eta) - 2l - 1/2 < \sigma < -\mathrm{Re}(\eta) - 2l + 3/2$, and

$$I_1 = \frac{1}{2\pi i}\int_{\sigma-i\infty}^{\sigma+i\infty}\frac{\mathcal{G}_\eta(s)}{\dfrac{3}{2}-\eta-s}x^{-s}ds = x^{-1/2}J_{\eta-1,l}(x) \tag{8.4.15}$$

for $-\mathrm{Re}(\eta) - 2l + 3/2 < \sigma < -\mathrm{Re}(\eta) - 2l + 7/2$.

(ii) *If $\sigma < 5/2$, then*

$$I_2 = \frac{1}{2\pi i}\int_{\sigma-i\infty}^{\sigma+i\infty}\frac{\mathcal{G}_\eta(s)}{\left(\dfrac{3}{2}-\eta-s\right)\left(\dfrac{7}{2}-\eta-s\right)}x^{-s}ds = x^{-3/2}J_{\eta-2,l+2}(x) \tag{8.4.16}$$

for $-\mathrm{Re}(\eta) - 2l - 1/2 < \sigma < -\mathrm{Re}(\eta) - 2l + 3/2$, and

$$I_2 = \frac{1}{2\pi i}\int_{\sigma-i\infty}^{\sigma+i\infty}\frac{\mathcal{G}_\eta(s)}{\left(\dfrac{3}{2}-\eta-s\right)\left(\dfrac{7}{2}-\eta-s\right)}x^{-s}ds = x^{-3/2}J_{\eta-2,l+1}(x) \tag{8.4.17}$$

for $-\mathrm{Re}(\eta) - 2l + 3/2 < \sigma < -\mathrm{Re}(\eta) - 2l + 7/2$.

(iii) *If $\sigma < 7/2$, then*

$$I_3 = \frac{1}{2\pi i}\int_{\sigma-i\infty}^{\sigma+i\infty} \frac{\mathcal{G}_\eta(s)}{\left(\frac{3}{2}-\eta-s\right)\left(\frac{7}{2}-\eta-s\right)\left(\frac{11}{2}-\eta-s\right)} x^{-s}ds$$

$$= -x^{-5/2}J_{\eta-3,l+3}(x) \tag{8.4.18}$$

for $-\mathrm{Re}(\eta) - 2l - 1/2 < \sigma < -\mathrm{Re}(\eta) - 2l + 3/2$, and

$$I_3 = \frac{1}{2\pi i}\int_{\sigma-i\infty}^{\sigma+i\infty} \frac{\mathcal{G}_\eta(s)}{\left(\frac{3}{2}-\eta-s\right)\left(\frac{7}{2}-\eta-s\right)\left(\frac{11}{2}-\eta-s\right)} x^{-s}ds$$

$$= -x^{-5/2}J_{\eta-3,l+2}(x) \tag{8.4.19}$$

for $-\mathrm{Re}(\eta) - 2l + 3/2 < \sigma < -\mathrm{Re}(\eta) - 2l + 7/2$.

For example, to prove (8.4.8) two functions are introduced:

$$h_{\eta,x}(t) = \begin{cases} \frac{1}{8}t^{1/2-\eta}(x^2-t^2)^2, & \text{if } 0 < t < x; \\ 0, & \text{if } t > x \end{cases} \tag{8.4.20}$$

and

$$\phi_{\eta,x}(t) = -x^{3-\eta}t^{-5/2}J_{\eta-3,l+3}(xt). \tag{8.4.21}$$

By using Lemma 8.4 it is proved that their Mellin transforms are given by

$$\left(\mathfrak{M}h_{\eta,x}\right)(s) = \frac{x^{s-\eta+9/2}}{\left(s-\eta+\frac{1}{2}\right)\left(s-\eta+\frac{5}{2}\right)\left(s-\eta+\frac{9}{2}\right)}$$

and

$$\left(\mathfrak{M}\phi_{\eta,x}\right)(s) = \frac{x^{11/2-\eta-s}}{\left(\frac{3}{2}-\eta-s\right)\left(\frac{7}{2}-\eta-s\right)\left(\frac{11}{2}-\eta-s\right)}.$$

From these formulas for $\mathrm{Re}(s) = 1-\nu$ the following relation follows

$$\left(\mathfrak{M}\mathbb{H}_{\eta,l}\phi_{\eta,x}\right)(s) = \mathcal{G}_\eta(s)\left(\mathfrak{M}\phi_{\eta,x}\right)(1-s) = \left(\mathfrak{M}h_{\eta,x}\right)(s),$$

so that

$$\mathbb{H}_{\eta,l}\phi_{\eta,x} = h_{\eta,x}. \tag{8.4.22}$$

Then in accordance with (8.4.4)

$$-x^{5/2-\eta}\int_0^\infty (xt)^{1/2}J_{\eta-3,l+3}(xt)\left(\mathbb{H}_{k,\eta}f\right)(t)\frac{dt}{t^3} = \int_0^\infty \phi_{\eta,x}(t)\left(\mathbb{H}_{\eta,l}f\right)(t)dt$$

$$= \int_0^\infty \left(\mathbb{H}_{\eta,l}\phi_{\eta,x}\right)(t)f(t)dt = \int_0^\infty h_{\eta,x}(t)f(t)dt$$

$$= \frac{1}{8}\int_0^x t^{1/2-\eta}(x^2-t^2)^2f(t)dt. \tag{8.4.23}$$

A direct calculation shows that

$$\frac{1}{8}\left(\frac{1}{x}\frac{d}{dx}\right)^3 \int_0^x t^{1/2-\eta}(x^2-t^2)^2 f(t)dt = x^{-1/2-\eta}f(x)$$

and so differentiating (8.4.23), (8.4.8) follows. The validity of the above are justified by the conditions of the theorem.

The third case when $\nu = -\mathrm{Re}(\eta) - 2l + 3/2$ is much harder, and the results are given in terms of the integral in the Cauchy sense defined by

$$\int_0^{\to\infty} f(t)dt = \lim_{R\to\infty}\int_0^R f(t)dt, \tag{8.4.24}$$

provided that $f(x)$ is integrable over $(0, R)$ for every $R > 0$ (see Titchmarsh [3, Section 1.7]).

Theorem 8.15. *Let* $1 < r < \infty$ *and* $\nu = -\mathrm{Re}(\eta) - 2l + 3/2$.

(a) *If* $0 < \mathrm{Re}(\eta) + 2l < 3/2 - \gamma(r)$ *and* $f \in \mathfrak{L}_{\nu,r}$, *then for almost all* $x > 0$, *the following inversion relation holds:*

$$f(x) = x^{\eta+1/2}\left(\frac{1}{x}\frac{d}{dx}\right)^2 x^{3/2-\eta}\int_0^{\to\infty}(xt)^{1/2}J_{\eta-2,l+2}(xt)\left(\mathbb{H}_{\eta,l}f\right)(t)\frac{dt}{t^2}. \tag{8.4.25}$$

(b) *If* $1/2 < \mathrm{Re}(\eta) + 2l \leqq 3/2 - \gamma(r)$ *and* $f \in \mathfrak{L}_{\nu,r}$, *then for almost all* $x > 0$, *the following inversion relation holds:*

$$f(x) = -x^{\eta-1/2}\frac{d}{dx}x^{1/2-\eta}\int_0^{\to\infty}(xt)^{1/2}J_{\eta-1,l+1}(xt)\left(\mathbb{H}_{\eta,l}f\right)(t)\frac{dt}{t}. \tag{8.4.26}$$

Proof. For real η the statements (a) and (b) were proved by Heywood and Rooney [6, Theorems 6.6 and 6.9]. They are directly extended to complex η by the same arguments which were used in Heywood and Rooney [6].

In conclusion we note that the extended Hankel transform (8.4.1) has constancy of the range similar to that for the Hankel transform in (8.1.20): if $1 < r < \infty$, $\gamma(r) \leqq \nu < \mathrm{Re}(\eta) + 2l + 3/2$ and $\mathrm{Re}(\xi) = |\mathrm{Re}(\eta) + 2l|$, then except when $\nu = -\mathrm{Re}(\eta) - 2l + 3/2$ and $\mathrm{Re}(\eta) < -1$,

$$\mathbb{H}_{\eta,l}(\mathfrak{L}_{\nu,r}) = \mathbb{H}_{\xi,l}(\mathfrak{L}_{\nu,r}). \tag{8.4.27}$$

The proof of this property is given by the same arguments that were used by Rooney [5, Theorem 2] to prove such a result for real η and ξ.

8.5. The Hankel Type Transform

We consider the integral transform defined by

$$\left(\mathbb{H}_{\eta;\sigma,\varpi;k,\lambda}f\right)(x) = x^\sigma\int_0^\infty J_\eta\left(\lambda(xt)^{1/k}\right)t^\varpi f(t)dt \quad (x > 0), \tag{8.5.1}$$

where $\mathrm{Re}(\eta) > -1$, $\sigma \in \mathbb{R}$, $\varpi \in \mathbb{R}$, $k > 0$, $\lambda > 0$ and $J_\eta(x)$ is the Bessel function of the first kind of order η. This transform is clearly defined for $f \in C_0$. When $\lambda = k$ and

$\sigma = \varpi = 1/k - 1/2$, the transform coincides with the generalized Hankel transform $\mathbb{H}_{k,\eta}$ given in (3.3.4), which, if further $k = 1$, is the Hankel transform $\mathbb{H}_\eta$ in (8.1.1). So, $\mathbb{H}_{\eta;\sigma,\varpi;k,\lambda}$ is named a Hankel type transform.

As in Section 8.1, we first give the results characterizing the boundedness and the representation of the Hankel type transform (8.5.1) in the space $\mathfrak{L}_{\nu,r}$.

Theorem 8.16. *Let $1 \leqq r \leqq \infty$ and let $\gamma(r)$ be given in* (3.3.9).

(a) *If $1 < r < \infty$ and $\gamma(r) \leqq k(\nu - \varpi - 1) + 3/2 < \mathrm{Re}(\eta) + 3/2$, then for all $s \geqq r$ such that $s' > [k(\nu - \varpi - 1) + 3/2]^{-1}$ and $1/s + 1/s' = 1$, the operator $\mathbb{H}_{\eta;\sigma,\varpi;k,\lambda}$ belongs to $[\mathfrak{L}_{\nu,r}, \mathfrak{L}_{1-\nu+\varpi-\sigma,s}]$ and is a one-to-one transform from $\mathfrak{L}_{\nu,r}$ onto $\mathfrak{L}_{1-\nu+\varpi-\sigma,s}$. If $1 < r \leqq 2$ and $f \in \mathfrak{L}_{\nu,r}$, then the Mellin transform of* (8.5.1) *for $\mathrm{Re}(s) = 1 - \nu + \varpi - \sigma$ is given by*

$$\left(\mathfrak{M}\mathbb{H}_{\eta;\sigma,\varpi;k,\lambda}f\right)(s)$$
$$= \frac{k}{2}\left(\frac{2}{\lambda}\right)^{k(\sigma+s)} \frac{\Gamma\left(\frac{1}{2}\{\eta + k\sigma + ks\}\right)}{\Gamma\left(\frac{1}{2}\{\eta - k\sigma - ks\} + 1\right)} \left(\mathfrak{M}f\right)(1 + \varpi - \sigma - s). \tag{8.5.2}$$

(b) *If $1 \leqq k(\nu - \varpi - 1) + 3/2 \leqq \mathrm{Re}(\eta) + 3/2$, then $\mathbb{H}_{\eta;\sigma,\varpi;k,\lambda} \in [\mathfrak{L}_{\nu,1}, \mathfrak{L}_{1-\nu+\varpi-\sigma,\infty}]$. If $1 < k(\nu-\varpi-1)+3/2 < \mathrm{Re}(\eta)+3/2$, then $\mathbb{H}_{\eta;\sigma,\varpi;k,\lambda} \in [\mathfrak{L}_{\nu,1}, \mathfrak{L}_{1-\nu+\varpi-\sigma,r}]$ for all r $(1 < r < \infty)$.*

(c) *If $f \in \mathfrak{L}_{\nu,r}$ and $g \in \mathfrak{L}_{\nu+\varpi-\sigma,s}$, where $1 < r < \infty$ and $1 < s < \infty$ are such that $1/r + 1/s \geqq 1$ and $\max[\gamma(r), \gamma(s)] \leqq k(\nu - \varpi - 1) + 3/2 < \mathrm{Re}(\eta) + 3/2$, then the relation*

$$\int_0^\infty f(x)\left(\mathbb{H}_{\eta;\sigma,\varpi;k,\lambda}g\right)(x)dx = \int_0^\infty \left(\mathbb{H}_{\eta;\varpi,\sigma;k,\lambda}f\right)(x)g(x)dx \tag{8.5.3}$$

holds.

(d) *If $f \in \mathfrak{L}_{\nu,r}$, where $1 < r < \infty$ and $\gamma(r) \leqq k(\nu - \varpi - 1) + 3/2 < \mathrm{Re}(\eta) + 3/2$, then the relation*

$$\left(\mathbb{H}_{\eta;\sigma,\varpi;k,\lambda}f\right)(x) = \frac{k}{\lambda}x^{\sigma+1-(\eta+2)/k}\frac{d}{dx}x^{(\eta+1)/k-\sigma}\left(\mathbb{H}_{\eta+1;\sigma,\varpi;k,\lambda}M_{-1/k}f\right)(x)$$
$$= \frac{k}{\lambda}x^{\sigma+1-(\eta+2)/k}\frac{d}{dx}x^{(\eta+1)/k}\int_0^\infty J_{\eta+1}\left(\lambda(xt)^{1/k}\right)t^\varpi f(t)\frac{dt}{t^{1/k}} \tag{8.5.4}$$

holds for almost all $x > 0$.

Proof. It is directly verified that the Hankel type transform (8.5.1) can be represented via the Hankel transform (8.1.1) as

$$\left(\mathbb{H}_{\eta;\sigma,\varpi;k,\lambda}f\right)(x) = k\lambda^{-k(\varpi+1)}\left(M_{\sigma-1/(2k)}N_{1/k}\mathbb{H}_\eta M_{k(\varpi+1)-3/2}W_\lambda N_k f\right)(x) \tag{8.5.5}$$

or

$$\left(\mathbb{H}_{\eta;\sigma,\varpi;k,\lambda}f\right)(x) = k\lambda^{-k(\varpi+1)}\left(N_{1/k}M_{k\sigma-1/2}\mathbb{H}_\eta M_{k(\varpi+1)-3/2}W_\lambda N_k f\right)(x), \tag{8.5.6}$$

where the operators M_ζ and W_δ are given in (3.3.11) and (3.3.12) and the operator N_a is defined by

$$\left(N_a f\right)(x) = f(x^a) \quad (a \in \mathbb{R},\ a \neq 0). \tag{8.5.7}$$

Here we used the clear operator equality

$$N_a M_\zeta = M_{a\zeta} N_a. \tag{8.5.8}$$

Then the assertion (a) follows from Theorem 8.1(a), Lemma 3.1(i),(ii) and from the following similar result for the operator (8.5.7):

Lemma 8.5 *For $\nu \in \mathbb{R}$ and $1 \leqq r < \infty$, N_a is an isometric isomorphism of $\mathfrak{L}_{\nu,r}$ onto $\mathfrak{L}_{a\nu,r}$. If $f \in \mathfrak{L}_{\nu,r}$ $(1 \leqq r \leqq 2)$, then for $\mathrm{Re}(s) = a\nu$*

$$\left(\mathfrak{M} N_a f\right)(s) = \frac{1}{|a|}\left(\mathfrak{M} f\right)\left(\frac{s}{a}\right). \tag{8.5.9}$$

Assertion (b) follows from (8.5.5) and Theorem 8.1(b) if we note that Lemmas 3.1(i) and 8.5 remain true for the space $\mathfrak{L}_{\nu,\infty}$. (c) is proved similarly to that in Theorem 8.1(c). The statement (d) is proved on the basis of (8.5.6), if we use Theorem 8.1(d), the equality (8.5.8) and the directly verified operator relation

$$N_a D = \frac{1}{a} M_{1-a} D N_a \quad \text{with} \quad D = \frac{d}{dx}. \tag{8.5.10}$$

The range of $\mathbb{H}_{\eta;\sigma,\varpi;k,\lambda}(\mathfrak{L}_{\nu,r})$ of the Hankel type transform (8.5.1) in $\mathfrak{L}_{\nu,r}$-space is characterized in terms of the operators M_ζ, N_a in (3.3.11), (8.5.7), the Erdélyi–Kober operator $I_{\alpha,\xi}$ in (8.1.21) and the Fourier cosine transform $\mathfrak{F}_c$ in (8.1.2).

Theorem 8.17. *Let $1 < r < \infty$ and let $\gamma(r)$ be given by (3.3.9). If $\gamma(r) \leqq k(\nu - \varpi - 1) + 3/2 < \mathrm{Re}(\eta) + 3/2$, then there holds*

$$\mathbb{H}_{\eta;\sigma,\varpi;k,\lambda}(\mathfrak{L}_{\nu,r}) = \left(N_{1/k} M_{k(\sigma+\nu-\varpi-1)+1-\gamma(r)} I_{k(\nu-\varpi-1)-\gamma(r)+3/2,-1/2}\mathfrak{F}_c\right)\left(\mathfrak{L}_{\gamma(r),r}\right). \tag{8.5.11}$$

In particular, if $1/2 \leqq k(\nu - \varpi - 1) + 3/2 < \mathrm{Re}(\eta) + 3/2$, then

$$\mathbb{H}_{\eta;\sigma,\varpi;k,\lambda}(\mathfrak{L}_{\nu,2}) = \left(N_{1/k} M_{k(\sigma+\nu-\varpi-1)+1/2} I_{k(\nu-\varpi-1)+1,-1/2}\mathfrak{F}_c\right)\left(L_2(\mathbb{R}_+)\right). \tag{8.5.12}$$

Proof. Using the representation (8.5.6), Lemma 3.1(i),(ii) and Lemma 8.5 and applying Lemma 8.2 with ν being replaced by $k(\nu-\varpi-1)+3/2$, we have for $f \in \mathfrak{L}_{\nu,r}$ the representation

$$\begin{aligned}\mathbb{H}_{\eta;\sigma,\varpi;k,\lambda} f = {} & k\lambda^{-k(\varpi+1)} 2^{\gamma(r)-k(\nu-\varpi-1)-3/2} N_{1/k} M_{k(\sigma+\nu-\varpi-1)+1-\gamma(r)} \\ & \cdot I_{k(\nu-\varpi-1)-\gamma(r)+3/2,[\eta-k(\nu-\varpi-1)+\gamma(r)-2]/2} \\ & \cdot \mathbb{H}_{\eta-k(\nu-\varpi-1)+\gamma(r)-3/2} M_{k\nu-\gamma(r)} W_\lambda N_k f. \end{aligned} \tag{8.5.13}$$

From here, taking the same arguments as in Theorem 8.2, we obtain (8.5.11) and (8.5.12).

Finally we present the inversion for the Hankel type transform (8.5.1) in the space $\mathfrak{L}_{\nu,r}$.

Theorem 8.18. *Let $1 < r < \infty$ and $\gamma(r) \leqq k(\nu - \varpi - 1) + 3/2 < \mathrm{Re}(\eta) + 3/2$, or $r = 1$ and $1 \leqq k(\nu - \varpi - 1) + 3/2 \leqq \mathrm{Re}(\eta) + 3/2$, where $\gamma(r)$ is given in (3.3.9). If we choose an integer $m > k(\nu - \varpi - 1) + 3/2$ and if $f \in \mathfrak{L}_{\nu,r}$, then the inversion relation*

$$f(x) = \left(\frac{\lambda}{k}\right)^{2-m} x^{(2-\eta)/k-\varpi-1} \left(x^{1-2/k}\frac{d}{dx}\right)^{m} x^{(\eta+m)/k}$$
$$\cdot \int_0^\infty J_{\eta+1}\left(\lambda(xt)^{1/k}\right) t^{2/k-\sigma-1}\left(\mathbb{H}_{\eta;\sigma,\varpi;k,\lambda}f\right)(t)\frac{dt}{t^{m/k}} \qquad (8.5.14)$$

holds for almost all $x > 0$. In particular, if $1 < r < \infty$ and $\gamma(r) \leqq k(\nu - \varpi - 1) + 3/2 < \min[1, \mathrm{Re}(\eta) + 3/2]$, then

$$f(x) = \frac{\lambda}{k} x^{-\varpi-\eta/k}\frac{d}{dx} x^{(\eta+1)/k}$$
$$\cdot \int_0^\infty J_{\eta+1}\left(\lambda(xt)^{1/k}\right) t^{2/k-\sigma-1}\left(\mathbb{H}_{\eta;\sigma,\varpi;k,\lambda}f\right)(t)\frac{dt}{t^{1/k}}. \qquad (8.5.15)$$

Proof. According to (3.3.11), (3.3.12), (8.5.7), Lemma 3.1(i),(ii) and Lemma 8.5 the operators inverse to M_ζ, W_δ and N_a have the forms

$$M_\zeta^{-1} = M_{-\zeta}, \quad W_\delta^{-1} = W_{1/\delta} \quad \text{and} \quad N_a^{-1} = N_{1/a}, \qquad (8.5.16)$$

respectively, and they are isometric isomorphisms in the corresponding spaces. It also follows from Theorem 8.4 that under the conditions of our theorem the operator $\mathbb{H}_\eta$ can be inverted in the space $\mathfrak{L}_{k(\nu-\varpi-1)+3/2,r}$.

Then from the relation (8.5.6) we obtain the inversion formula for (8.5.1):

$$f(x) = \left(\left(k\lambda^{-k(\varpi+1)}N_{1/k}M_{k\sigma-1/2}\mathbb{H}_\eta M_{k(\varpi+1)-3/2}W_\lambda N_k\right)^{-1}\mathbb{H}_{\eta;\sigma,\varpi;k,\lambda}f\right)(x)$$
$$= \frac{1}{k}\lambda^{k(\varpi+1)}\left(N_k^{-1}W_\lambda^{-1}M_{k(\varpi+1)-3/2}^{-1}\mathbb{H}_\eta^{-1}M_{k\sigma-1/2}^{-1}N_{1/k}^{-1}\mathbb{H}_{\eta;\sigma,\varpi;k,\lambda}f\right)(x)$$
$$= \frac{1}{k}\lambda^{k(\varpi+1)}\left(N_{1/k}W_{1/\lambda}M_{3/2-k(\varpi+1)}\mathbb{H}_\eta^{-1}M_{1/2-k\sigma}N_k\mathbb{H}_{\eta;\sigma,\varpi;k,\lambda}f\right)(x), \qquad (8.5.17)$$

where $\mathbb{H}_\eta^{-1}$ is the operator inverse to the Hankel transform (8.1.1). Using the relation (8.1.34), the equalities (8.5.8), (3.3.22) and the directly verified formulas

$$W_\delta(M_\zeta D)^m = \delta^{(1-\zeta)m}(M_\zeta D)^m W_\delta \quad (m = 1, 2, \cdots), \qquad (8.5.18)$$

$$N_a(M_\zeta D)^m = a^{-m}(M_{a(\zeta-1)+1}D)^m N_a \quad (m = 1, 2, \cdots), \qquad (8.5.19)$$

we arrive at (8.5.14). In particular, if $k(\nu - \varpi - 1) + 1/2 < 0$, we obtain (8.5.15).

Setting $\sigma = \varpi = 1/k - 1/2$ and $\lambda = k$, from Theorems 8.16–8.18 we have the corresponding statements for the generalized Hankel transform $\mathbb{H}_{k,\eta}$ given in (3.3.4).

Theorem 8.19. *Let $1 \leqq r \leqq \infty$ and let $\gamma(r)$ be given in (3.3.9).*

(a) *If $1 < r < \infty$ and $\gamma(r) \leqq k(\nu - 1/2) + 1/2 < \mathrm{Re}(\eta) + 3/2$, then for all $s \geqq r$ such that $s' > [k(\nu - 1/2) + 1/2]^{-1}$ and $1/s + 1/s' = 1$, the operator $\mathbb{H}_{k,\eta}$ belongs to $[\mathfrak{L}_{\nu,r}, \mathfrak{L}_{1-\nu,s}]$ and* *is a one-to-one transform from $\mathfrak{L}_{\nu,r}$ onto $\mathfrak{L}_{1-\nu,s}$. If $1 < r \leqq 2$ and $f \in \mathfrak{L}_{\nu,r}$, then the Mellin transform of $\mathbb{H}_{k,\eta}f$ for $\mathrm{Re}(s) = 1-\nu$ is given in (3.3.10) :*

$$\left(\mathfrak{M}\mathbb{H}_{k,\eta}f\right)(s) = \left(\frac{2}{k}\right)^{k(s-1/2)} \frac{\Gamma\left(\frac{1}{2}\left[\eta + k\left(s-\frac{1}{2}\right)+1\right]\right)}{\Gamma\left(\frac{1}{2}\left[\eta - k\left(s-\frac{1}{2}\right)+1\right]\right)} \left(\mathfrak{M}f\right)(1-s). \tag{8.5.20}$$

(b) *If $1 \leqq k(\nu - 1/2) + 1/2 \leqq \mathrm{Re}(\eta) + 3/2$, then $\mathbb{H}_{k,\eta} \in [\mathfrak{L}_{\nu,1}, \mathfrak{L}_{1-\nu,\infty}]$. If $1 < k(\nu-1/2) + 1/2 < \mathrm{Re}(\eta) + 3/2$, then $\mathbb{H}_{k,\eta} \in [\mathfrak{L}_{\nu,1}, \mathfrak{L}_{1-\nu,r}]$ for all r $(1 < r < \infty)$.*

(c) *If $f \in \mathfrak{L}_{\nu,r}$ and $g \in \mathfrak{L}_{\nu,s}$, where $1 < r < \infty$ and $1 < s < \infty$ such that $1/r + 1/s \geqq 1$ and $\max[\gamma(r), \gamma(s)] \leqq k(\nu - 1/2) + 1/2 < \mathrm{Re}(\eta) + 3/2$, then the relation*

$$\int_0^\infty f(x)\left(\mathbb{H}_{k,\eta}g\right)(x)dx = \int_0^\infty \left(\mathbb{H}_{k,\eta}f\right)(x)g(x)dx \tag{8.5.21}$$

holds.

(d) *If $f \in \mathfrak{L}_{\nu,r}$, where $1 < r < \infty$ and $\gamma(r) \leqq k(\nu - 1/2) + 1/2 < \mathrm{Re}(\eta) + 3/2$, then the relation*

$$\left(\mathbb{H}_{k,\eta}f\right)(x) = x^{1/2-(\eta+1)/k}\frac{d}{dx}x^{\eta/k+1/2}\left(\mathbb{H}_{k,\eta+1}M_{-1/k}f\right)(x)$$
$$= x^{1/2-(\eta+1)/k}\frac{d}{dx}x^{(\eta+1)/k}\int_0^\infty J_{\eta+1}\left(k(xt)^{1/k}\right)t^{-1/2}f(t)dt \tag{8.5.22}$$

holds for almost all $x > 0$.

Theorem 8.20. *Let $1 < r < \infty$ and $\gamma(r)$ be given by (3.3.9). If $\gamma(r) \leqq k(\nu-1)+1/2 < \mathrm{Re}(\eta) + 3/2$, then there holds*

$$\mathbb{H}_{k,\eta}(\mathfrak{L}_{\nu,r}) = \left(N_{1/k}M_{k(\nu-1)+1-\gamma(r)}I_{k(\nu-1/2)-\gamma(r)+1/2,-1/2}\mathfrak{F}_c\right)\left(\mathfrak{L}_{\gamma(r),r}\right). \tag{8.5.23}$$

In particular, if $1/2 \leqq k(\nu - 1) + 1/2 < \mathrm{Re}(\eta) + 3/2$, then

$$\mathbb{H}_{k,\eta}(\mathfrak{L}_{\nu,2}) = \left(N_{1/k}M_{k(\nu-1)+1/2}I_{k(\nu-1/2),-1/2}\mathfrak{F}_c\right)\left(L_2(\mathbb{R}_+)\right). \tag{8.5.24}$$

Theorem 8.21. *Let $1 < r < \infty$ and $\gamma(r) \leqq k(\nu - 1/2) + 1/2 < \mathrm{Re}(\eta) + 3/2$, or $r = 1$ and $1 \leqq k(\nu-1/2)+1/2 \leqq \mathrm{Re}(\eta)+3/2$, where $\gamma(r)$ is given in (3.3.9). If we choose an integer $m > k(\nu - 1/2) + 1/2$ and if $f \in \mathfrak{L}_{\nu,r}$, then the inversion relation*

$$f(x) = x^{(1-\eta)/k-1/2}\left(x^{1-2/k}\frac{d}{dx}\right)^m x^{(\eta+m-1)/k+1/2}$$
$$\cdot \int_0^\infty (xt)^{1/k-1/2}J_{\eta+1}\left(k(xt)^{1/k}\right)\left(\mathbb{H}_{k,\eta}f\right)(t)\frac{dt}{t^{m/k}} \tag{8.5.25}$$

holds for almost all $x > 0$. In particular, if $1 < r < \infty$ and $\gamma(r) \leqq k(\nu - 1/2) + 1/2 < \min[1, \mathrm{Re}(\eta) + 3/2]$, then

$$f(x) = x^{1/2-(\eta+1)/k}\frac{d}{dx}x^{\eta/k+1/2}$$
$$\cdot \int_0^\infty (xt)^{1/k-1/2}J_{\eta+1}\left(k(xt)^{1/k}\right)\left(\mathbb{H}_{k,\eta}f\right)(t)\frac{dt}{t^{1/k}}. \tag{8.5.26}$$

8.6. Hankel–Schwartz and Hankel–Clifford Transforms

Let us investigate the integral transforms

$$\left(\mathbf{h}_{\eta;1}f\right)(x) = x^{2\eta+1}\int_0^\infty \mathcal{J}_\eta(xt)f(t)dt \quad (x > 0); \tag{8.6.1}$$

$$\left(\mathbf{h}_{\eta;2}f\right)(x) = \int_0^\infty \mathcal{J}_\eta(xt)t^{2\eta+1}f(t)dt \quad (x > 0), \tag{8.6.2}$$

and

$$\left(\mathbf{b}_{\eta;1}f\right)(x) = x^{\eta}\int_0^\infty C_\eta(xt)f(t)dt \quad (x > 0); \tag{8.6.3}$$

$$\left(\mathbf{b}_{\eta;2}f\right)(x) = \int_0^\infty C_\eta(xt)t^{\eta}f(t)dt \quad (x > 0) \tag{8.6.4}$$

with the kernels

$$\mathcal{J}_\eta(z) = z^{-\eta}J_\eta(z), \qquad C_\eta(z) = z^{-\eta/2}J_\eta(2\sqrt{z}) \tag{8.6.5}$$

for $\eta \in \mathbb{R}$ $(\eta > -1)$.

The transforms (8.6.1), (8.6.2) and (8.6.3), (8.6.4) defined for $f \in C_0$ are called the Hankel–Schwartz and Hankel–Clifford transforms, respectively. They are special cases of the general Hankel transform $\mathbb{H}_{\eta;\sigma,\varpi;k,\lambda}$ given in (8.5.1):

$$\left(\mathbf{h}_{\eta;1}f\right)(x) \equiv x^{\eta+1}\int_0^\infty J_\eta(xt)t^{-\eta}f(t)dt = \left(\mathbb{H}_{\eta;\eta+1,-\eta;1,1}f\right)(x); \tag{8.6.6}$$

$$\left(\mathbf{h}_{\eta;2}f\right)(x) \equiv x^{-\eta}\int_0^\infty J_\eta(xt)t^{\eta+1}f(t)dt = \left(\mathbb{H}_{\eta;-\eta,\eta+1;1,1}f\right)(x), \tag{8.6.7}$$

and

$$\left(\mathbf{b}_{\eta;1}f\right)(x) \equiv x^{\eta/2}\int_0^\infty J_\eta(2(xt)^{1/2})t^{-\eta/2}f(t)dt = \left(\mathbb{H}_{\eta;\eta/2,-\eta/2;2,2}f\right)(x); \tag{8.6.8}$$

$$\left(\mathbf{b}_{\eta;2}f\right)(x) \equiv x^{-\eta/2}\int_0^\infty J_\eta(2(xt)^{1/2})t^{\eta/2}f(t)dt = \left(\mathbb{H}_{\eta;-\eta/2,\eta/2;2,2}f\right)(x) \tag{8.6.9}$$

for $\eta > -1$ and $x > 0$. These relations show that $\mathbf{h}_{\eta;1}$ and $\mathbf{h}_{\eta;2}$ as well as $\mathbf{b}_{\eta;1}$ and $\mathbf{b}_{\eta;2}$ are mutually conjugate operators.

Using the results in Theorems 8.16–8.18, we obtain the following statements giving the theory of the transforms $\mathbf{h}_{\eta;1}, \mathbf{h}_{\eta;2}, \mathbf{b}_{\eta;1}$ and $\mathbf{b}_{\eta;2}$ in the space $\mathfrak{L}_{\nu,r}$.

Theorem 8.22. *Let $1 \leqq r \leqq \infty$ and let $\gamma(r)$ be given in* (3.3.9).

(a) *If $1 < r < \infty$ and $\gamma(r)-\eta-1/2 \leqq \nu < 1$, then for all $s \geqq r$ such that $s' > [\nu+\eta+1/2]^{-1}$ and $1/s+1/s' = 1$, the operator $\mathbf{h}_{\eta;1}$ belongs to $[\mathfrak{L}_{\nu,r}, \mathfrak{L}_{-\nu-2\eta,s}]$ and is a one-to-one transform from $\mathfrak{L}_{\nu,r}$ onto $\mathfrak{L}_{-\nu-2\eta,s}$. If $1 < r \leqq 2$ and $f \in \mathfrak{L}_{\nu,r}$, then the Mellin transform of $\mathbf{h}_{\eta;1}f$ for $\mathrm{Re}(s) = -\nu - 2\eta$ is given by*

$$\left(\mathfrak{M}\mathbf{h}_{\eta;1}f\right)(s) = 2^{\eta+s}\frac{\Gamma\left(\eta+\dfrac{s+1}{2}\right)}{\Gamma\left(\dfrac{1-s}{2}\right)}\left(\mathfrak{M}f\right)(-2\eta-s). \tag{8.6.10}$$

(b) *If $1/2-\eta \leqq \nu \leqq 1$, then $\mathbf{h}_{\eta;1} \in [\mathfrak{L}_{\nu,1}, \mathfrak{L}_{-\nu-2\eta,\infty}]$. If $1/2-\eta < \nu < 1$, then $\mathbf{h}_{\eta;1} \in [\mathfrak{L}_{\nu,1}, \mathfrak{L}_{-\nu-2\eta,r}]$ for all r $(1<r<\infty)$.*

(c) *If $f \in \mathfrak{L}_{\nu,r}$ and $g \in \mathfrak{L}_{\nu-2\eta-1,s}$, where $1<r<\infty$ and $1<s<\infty$ are such that $1/r+1/s \geqq 1$ and $\max[\gamma(r),\gamma(s)]-\eta-1/2 \leqq \nu < 1$, then the relation*

$$\int_0^\infty f(x)\left(\mathbf{h}_{\eta;1}g\right)(x)dx = \int_0^\infty \left(\mathbf{h}_{\eta;2}f\right)(x)g(x)dx \tag{8.6.11}$$

holds, where $\mathbf{h}_{\eta;2}$ is the operator (8.6.2) conjugate to $\mathbf{h}_{\eta;1}$.

(d) *If $f \in \mathfrak{L}_{\nu,r}$, where $1<r<\infty$ and $\gamma(r)-\eta-1/2 \leqq \nu < 1$, then the relation*

$$\left(\mathbf{h}_{\eta;1}f\right)(x) = \frac{d}{dx}x^{\eta+1}\int_0^\infty J_{\eta+1}(xt)t^{-\eta}f(t)\frac{dt}{t} \tag{8.6.12}$$

holds for almost all $x>0$.

(e) *If $1<r<\infty$ and $\gamma(r)-\eta-1/2 \leqq \nu < 1$, then there holds the relation*

$$\mathbf{h}_{\eta;1}(\mathfrak{L}_{\nu,r}) = \left(M_{\nu+2\eta+1-\gamma(r)}I_{\nu+\eta-\gamma(r)+1/2,-1/2}\mathfrak{F}_c\right)\left(\mathfrak{L}_{\gamma(r),r}\right). \tag{8.6.13}$$

In particular,

$$\mathbf{h}_{\eta;1}(\mathfrak{L}_{\nu,2}) = \left(M_{\nu+2\eta+1/2}I_{\nu+\eta,-1/2}\mathfrak{F}_c\right)\left(L_2(\mathbb{R}_+)\right). \tag{8.6.14}$$

(f) *Let $1<r<\infty$ and $\gamma(r)-\eta-1/2 \leqq \nu < 1$, or $r=1$ and $-\eta+1/2 \leqq \nu \leqq 1$. If we choose an integer $m>\nu+\eta+1/2$ and if $f \in \mathfrak{L}_{\nu,r}$, then the inversion relation*

$$f(x) = x\left(\frac{1}{x}\frac{d}{dx}\right)^m x^{\eta+m}\int_0^\infty J_{\eta+1}(xt)t^{-(\eta+m)}\left(\mathbf{h}_{\eta;1}f\right)(t)dt \tag{8.6.15}$$

holds for almost all $x>0$. In particular, if $1<r<\infty$ and $\gamma(r)-\eta-1/2 \leqq \nu < \min[1,-\eta+1/2]$, then

$$f(x) = \frac{d}{dx}x^{\eta+1}\int_0^\infty J_{\eta+1}(xt)t^{-\eta-1}\left(\mathbf{h}_{\eta;1}f\right)(t)dt. \tag{8.6.16}$$

Theorem 8.23. *Let $1 \leqq r \leqq \infty$.*

(a) *If $1<r<\infty$ and $\gamma(r)+\eta+1/2 \leqq \nu < 2\eta+2$, then for all $s \geqq r$ such that $s' > [\nu-\eta-1/2]^{-1}$ and $1/s+1/s'=1$, the operator $\mathbf{h}_{\eta;2}$ belongs to $[\mathfrak{L}_{\nu,r}, \mathfrak{L}_{2\eta+2-\nu,s}]$ and is a one-to-one transform from $\mathfrak{L}_{\nu,r}$ onto $\mathfrak{L}_{2\eta+2-\nu,s}$. If $1<r\leqq 2$ and $f \in \mathfrak{L}_{\nu,r}$, then the Mellin transform of $\mathbf{h}_{\eta;2}f$ for $\mathrm{Re}(s)=2\eta+2-\nu$ is given by*

$$\left(\mathfrak{M}\mathbf{h}_{\eta;2}f\right)(s) = 2^{s-\eta-1}\frac{\Gamma\left(\dfrac{s}{2}\right)}{\Gamma\left(\eta+1-\dfrac{s}{2}\right)}\left(\mathfrak{M}f\right)(2\eta+2-s). \tag{8.6.17}$$

(b) *If $\eta+3/2 \leqq \nu \leqq 2\eta+2$, then $\mathbf{h}_{\eta;2} \in [\mathfrak{L}_{\nu,1}, \mathfrak{L}_{2\eta+2-\nu,\infty}]$. If $\eta+3/2 < \nu < 2\eta+2$, then $\mathbf{h}_{\eta;2} \in [\mathfrak{L}_{\nu,1}, \mathfrak{L}_{2\eta+2-\nu,r}]$ for all r $(1<r<\infty)$.*

(c) *If $f \in \mathfrak{L}_{\nu,r}$ and $g \in \mathfrak{L}_{\nu+2\eta+1,s}$, where $1<r<\infty$ and $1<s<\infty$ are such that $1/r+1/s \geqq 1$ and $\max[\gamma(r),\gamma(s)]+\eta+1/2 \leqq \nu < 2\eta+2$, then the relation*

$$\int_0^\infty f(x)\left(\mathbf{h}_{\eta;2}g\right)(x)dx = \int_0^\infty \left(\mathbf{h}_{\eta;1}f\right)(x)g(x)dx \tag{8.6.18}$$

holds, where $\mathbf{h}_{\eta;1}$ *is the operator* (8.6.1) *conjugate to* $\mathbf{h}_{\eta;2}$.

(d) *If* $f \in \mathfrak{L}_{\nu,r}$, *where* $1 < r < \infty$ *and* $\gamma(r) + \eta + 1/2 \leqq \nu < 2\eta + 2$, *then the relation*

$$\left(\mathbf{h}_{\eta;2} f\right)(x) = x^{-2\eta-1} \frac{d}{dx} x^{\eta+1} \int_0^\infty J_{\eta+1}(xt) t^\eta f(t) dt \tag{8.6.19}$$

holds for almost all $x > 0$.

(e) *If* $1 < r < \infty$ *and* $\gamma(r) + \eta + 1/2 \leqq \nu < 2\eta + 2$, *then there holds the relation*

$$\mathbf{h}_{\eta;2}(\mathfrak{L}_{\nu,r}) = \left(M_{\nu-2\eta-1-\gamma(r)} I_{\nu-\eta-\gamma(r)-1/2,-1/2} \mathfrak{F}_c\right) \left(\mathfrak{L}_{\gamma(r),r}\right). \tag{8.6.20}$$

In particular, if $\eta + 1 \leqq \nu < 2\eta + 2$, *then*

$$\mathbf{h}_{\eta;2}(\mathfrak{L}_{\nu,2}) = \left(M_{\nu-2\eta-3/2} I_{\nu-\eta-1,-1/2} \mathfrak{F}_c\right) \left(L_2(\mathbb{R}_+)\right). \tag{8.6.21}$$

(f) *Let* $1 < r < \infty$ *and* $\gamma(r) + \eta + 1/2 \leqq \nu < 2\eta + 2$, *or* $r = 1$ *and* $\eta + 3/2 \leqq \nu \leqq 2\eta + 2$. *If we choose an integer* $m > \nu - \eta - 1/2$ *and if* $f \in \mathfrak{L}_{\nu,r}$, *then the inversion relation*

$$f(x) = x^{-2\eta} \left(\frac{1}{x} \frac{d}{dx}\right)^m x^{\eta+m} \int_0^\infty J_{\eta+1}(xt) t^{\eta-m+1} \left(\mathbf{h}_{\eta;2} f\right)(t) dt \tag{8.6.22}$$

holds for almost all $x > 0$. *In particular, if* $1 < r < \infty$ *and* $\gamma(r) + \eta + 1/2 \leqq \nu < \min[\eta + 3/2, 2\eta + 2]$, *then*

$$f(x) = x^{-2\eta-1} \frac{d}{dx} x^{\eta+1} \int_0^\infty J_{\eta+1}(xt) t^\eta \left(\mathbf{h}_{\eta;2} f\right)(t) dt. \tag{8.6.23}$$

Theorem 8.24. *Let* $1 \leqq r \leqq \infty$.

(a) *If* $1 < r < \infty$ *and* $(\gamma(r) - \eta)/2 + 1/4 \leqq \nu < 1$, *then for all* $s \geqq r$ *such that* $s' > [2\nu + \eta - 1/2]^{-1}$ *and* $1/s + 1/s' = 1$, *the operator* $\mathbf{b}_{\eta;1}$ *belongs to* $[\mathfrak{L}_{\nu,r}, \mathfrak{L}_{1-\nu-\eta,s}]$ *and is a one-to-one transform from* $\mathfrak{L}_{\nu,r}$ *onto* $\mathfrak{L}_{1-\nu-\eta,s}$. *If* $1 < r \leqq 2$ *and* $f \in \mathfrak{L}_{\nu,r}$, *then the Mellin transform of* $\mathbf{b}_{\eta;1} f$ *for* $\mathrm{Re}(s) = 1 - \nu - \eta$ *is given by*

$$\left(\mathfrak{M}\mathbf{b}_{\eta;1} f\right)(s) = \frac{\Gamma(\eta + s)}{\Gamma(1 - s)} \left(\mathfrak{M} f\right)(1 - \eta - s). \tag{8.6.24}$$

(b) *If* $-\eta/2 + 3/4 \leqq \nu \leqq 1$, *then* $\mathbf{b}_{\eta;1} \in [\mathfrak{L}_{\nu,1}, \mathfrak{L}_{1-\nu-\eta,\infty}]$. *If* $-\eta/2 + 3/4 < \nu < 1$, *then* $\mathbf{b}_{\eta;1} \in [\mathfrak{L}_{\nu,1}, \mathfrak{L}_{1-\nu-\eta,r}]$ *for all* r $(1 < r < \infty)$.

(c) *If* $f \in \mathfrak{L}_{\nu,r}$ *and* $g \in \mathfrak{L}_{\nu-\eta,s}$, *where* $1 < r < \infty$ *and* $1 < s < \infty$ *are such that* $1/r + 1/s \geqq 1$ *and* $(\max[\gamma(r), \gamma(s)] - \eta)/2 + 1/4 \leqq \nu < 1$, *then the relation*

$$\int_0^\infty f(x) \left(\mathbf{b}_{\eta;1} g\right)(x) dx = \int_0^\infty \left(\mathbf{b}_{\eta;2} f\right)(x) g(x) dx \tag{8.6.25}$$

holds, where $\mathbf{b}_{\eta;2}$ *is the operator* (8.6.4) *conjugate to* $\mathbf{b}_{\eta;1}$.

(d) *If* $f \in \mathfrak{L}_{\nu,r}$, *where* $1 < r < \infty$ *and* $(\gamma(r) - \eta)/2 + 1/4 \leqq \nu < 1$, *then the relation*

$$\left(\mathbf{b}_{\eta;1} f\right)(x) = \frac{d}{dx} x^{(\eta+1)/2} \int_0^\infty J_{\eta+1}\left(2(xt)^{1/2}\right) t^{-(\eta+1)/2} f(t) dt \tag{8.6.26}$$

holds for almost all $x > 0$.

(e) *If* $1 < r < \infty$ *and* $(\gamma(r) - \eta)/2 + 1/4 \leqq \nu < 1$, *then there holds the relation*

$$\mathbf{b}_{\eta;1}(\mathfrak{L}_{\nu,r}) = \left(N_{1/2}M_{2(\nu+\eta)-1-\gamma(r)}I_{2\nu+\eta-\gamma(r)-1/2,-1/2}\mathfrak{F}_c\right)\left(\mathfrak{L}_{\gamma(r),r}\right). \tag{8.6.27}$$

In particular, if $(1-\eta)/2 \leqq \nu < 1$, *then*

$$\mathbf{b}_{\eta;1}(\mathfrak{L}_{\nu,2}) = \left(N_{1/2}M_{2(\nu+\eta)-3/2}I_{2\nu+\eta-1,-1/2}\mathfrak{F}_c\right)\left(L_2(\mathbb{R}_+)\right). \tag{8.6.28}$$

(f) *Let* $1 < r < \infty$ *and* $(\gamma(r) - \eta)/2 + 1/4 \leqq \nu < 1$, *or* $r = 1$ *and* $-\eta/2 + 3/4 \leqq \nu \leqq 1$. *If we choose an integer* $m > 2\nu + \eta - 1/2$ *and if* $f \in \mathfrak{L}_{\nu,r}$, *then the inversion relation*

$$f(x) = \left(\frac{d}{dx}\right)^m x^{(\eta+m)/2}\int_0^\infty J_{\eta+1}\left(2(xt)^{1/2}\right)t^{-(\eta+m)/2}\left(\mathbf{b}_{\eta;1}f\right)(t)dt \tag{8.6.29}$$

holds for almost all $x > 0$. *In particular, if* $1 < r < \infty$ *and* $(\gamma(r) - \eta)/2 + 1/4 \leqq \nu < \min[1, -\eta/2 + 3/4]$, *then*

$$f(x) = \frac{d}{dx}x^{(\eta+1)/2}\int_0^\infty J_{\eta+1}\left(2(xt)^{1/2}\right)t^{-(\eta+1)/2}\left(\mathbf{b}_{\eta;1}f\right)(t)dt. \tag{8.6.30}$$

Theorem 8.25. *Let* $1 \leqq r \leqq \infty$.

(a) *If* $1 < r < \infty$ *and* $(\gamma(r) + \eta)/2 + 1/4 \leqq \nu < \eta + 1$, *then for all* $s \geqq r$ *such that* $s' > [2\nu - \eta - 1/2]^{-1}$ *and* $1/s + 1/s' = 1$, *the operator* $\mathbf{b}_{\eta;2}$ *belongs to* $[\mathfrak{L}_{\nu,r}, \mathfrak{L}_{1-\nu+\eta,s}]$ *and is a one-to-one transform from* $\mathfrak{L}_{\nu,r}$ *onto* $\mathfrak{L}_{1-\nu+\eta,s}$. *If* $1 < r \leqq 2$ *and* $f \in \mathfrak{L}_{\nu,r}$, *then the Mellin transform of* $\mathbf{b}_{\eta;2}f$ *for* $\mathrm{Re}(s) = 1 - \nu + \eta$ *is given by*

$$\left(\mathfrak{M}\mathbf{b}_{\eta;2}f\right)(s) = \frac{\Gamma(s)}{\Gamma(1+\eta-s)}\left(\mathfrak{M}f\right)(1+\eta-s). \tag{8.6.31}$$

(b) *If* $\eta/2 + 3/4 \leqq \nu \leqq \eta + 1$, *then* $\mathbf{b}_{\eta;2} \in [\mathfrak{L}_{\nu,1}, \mathfrak{L}_{1-\nu+\eta,\infty}]$. *If* $\eta/2 + 3/4 < \nu < \eta + 1$, *then* $\mathbf{b}_{\eta;2} \in [\mathfrak{L}_{\nu,1}, \mathfrak{L}_{1-\nu+\eta,r}]$ *for all* r $(1 < r < \infty)$.

(c) *If* $f \in \mathfrak{L}_{\nu,r}$ *and* $g \in \mathfrak{L}_{\nu+\eta,s}$, *where* $1 < r < \infty$ *and* $1 < s < \infty$ *are such that* $1/r + 1/s \geqq 1$ *and* $(\max[\gamma(r), \gamma(s)] + \eta)/2 + 1/4 \leqq \nu < \eta + 1$, *then the relation*

$$\int_0^\infty f(x)\left(\mathbf{b}_{\eta;2}g\right)(x)dx = \int_0^\infty \left(\mathbf{b}_{\eta;1}f\right)(x)g(x)dx \tag{8.6.32}$$

holds, where $\mathbf{b}_{\eta;1}$ *is the operator* (8.6.3) *conjugate to* $\mathbf{b}_{\eta;2}$.

(d) *If* $f \in \mathfrak{L}_{\nu,r}$, *where* $1 < r < \infty$ *and* $(\gamma(r) + \eta)/2 + 1/4 \leqq \nu < \eta + 1$, *then the relation*

$$\left(\mathbf{b}_{\eta;2}f\right)(x) = x^{-\eta}\frac{d}{dx}x^{(\eta+1)/2}\int_0^\infty J_{\eta+1}\left(2(xt)^{1/2}\right)t^{(\eta-1)/2}f(t)dt \tag{8.6.33}$$

holds for almost all $x > 0$.

(e) *If* $1 < r < \infty$ *and* $(\gamma(r) + \eta)/2 + 1/4 \leqq \nu < \eta + 1$, *then there holds the relation*

$$\mathbf{b}_{\eta;2}(\mathfrak{L}_{\nu,r}) = \left(N_{1/2}M_{2(\nu-\eta)-1-\gamma(r)}I_{2\nu-\eta-\gamma(r)-1/2,-1/2}\mathfrak{F}_c\right)\left(\mathfrak{L}_{\gamma(r),r}\right). \tag{8.6.34}$$

In particular, if $(1+\eta)/2 \leqq \nu < \eta + 1$, *then*

$$\mathbf{b}_{\eta;2}(\mathfrak{L}_{\nu,2}) = \left(N_{1/2}M_{2(\nu-\eta)-3/2}I_{2\nu-\eta-1,-1/2}\mathfrak{F}_c\right)\left(L_2(\mathbb{R}_+)\right). \tag{8.6.35}$$

(f) *Let $1 < r < \infty$ and $(\gamma(r)+\eta)/2+1/4 \leqq \nu < \eta+1$, or $r=1$ and $\eta/2+3/4 \leqq \nu \leqq \eta+1$. If we choose an integer $m > 2\nu - \eta - 1/2$ and if $f \in \mathfrak{L}_{\nu,r}$, then the inversion relation*

$$f(x) = x^{-\eta}\left(\frac{d}{dx}\right)^m x^{(\eta+m)/2}\int_0^\infty J_{\eta+1}\left(2(xt)^{1/2}\right)t^{(\eta-m)/2}\left(\mathbf{b}_{\eta;2}f\right)(t)dt \qquad (8.6.36)$$

holds for almost all $x > 0$. In particular, if $1 < r < \infty$ and $(\gamma(r)+\eta)/2+1/4 \leqq \nu < \min[\eta/2+3/4, \eta+1]$, then

$$f(x) = x^{-\eta}\frac{d}{dx}x^{(\eta+1)/2}\int_0^\infty J_{\eta+1}\left(2(xt)^{1/2}\right)t^{(\eta-1)/2}\left(\mathbf{b}_{\eta;2}f\right)(t)dt. \qquad (8.6.37)$$

8.7. The Transform $\mathbb{Y}_\eta$

We consider the integral transform $\mathbb{Y}_\eta$ defined by

$$\left(\mathbb{Y}_\eta f\right)(x) = \int_0^\infty (xt)^{1/2}Y_\eta(xt)f(t)dt \quad (x>0), \qquad (8.7.1)$$

where $Y_\eta(z)$ is the Bessel function of the second kind (or Neumann function) given via the Bessel function of the first kind (2.6.2) by

$$Y_\eta(z) = \frac{1}{\sin(\eta\pi)}[J_\eta(z)\cos(\eta\pi) - J_{-\eta}(z)] \qquad (8.7.2)$$

(see Erdélyi, Magnus, Oberhettinger and Tricomi [2, 7.2(4)]). Since the Hankel transform (2.6.1) is defined for $\eta \in \mathbb{C}$ ($\mathrm{Re}(\eta) > -1$) the transform $\mathbb{Y}_\eta$ can be defined for $\eta \in \mathbb{C}$ ($|\mathrm{Re}(\eta)| < 1$).

First we give the representation of the transform $\mathbb{Y}_\eta$ in (8.7.1) in terms of the Hankel transform $\mathbb{H}_\eta$ in (8.1.1), the even Hilbert transform $\boldsymbol{H}_+$ in (8.3.1) and the elementary operator M_ξ in (3.3.11).

Theorem 8.26. *For $\eta \in \mathbb{C}$ with $|\mathrm{Re}(\eta)| < 1$ there holds the relation*

$$\mathbb{Y}_\eta = \mathbb{H}_\eta M_{\eta-1/2}\boldsymbol{H}_+M_{1/2-\eta} \qquad (8.7.3)$$

on C_0 being the collection of continuous functions with compact support on $\mathbb{R}_+$.

Proof. Since $M_{1/2-\eta}f \in C_0 \subseteq \mathfrak{L}_{\mathrm{Re}(\eta),2}$ and $|\mathrm{Re}(\eta)| < 1$, by Theorem 8.9(a) $\boldsymbol{H}_+M_{1/2-\eta}f \in \mathfrak{L}_{\mathrm{Re}(\eta),2}$ and thus Lemma 3.1(i) implies $M_{\eta-1/2}\boldsymbol{H}_+M_{1/2-\eta}f \in \mathfrak{L}_{1/2,2} = L_2(\mathbb{R}_+)$. Hence Theorem 8.1(d) and the use of the function $h_{\eta,x}$ in (8.1.13) imply that for almost all $x > 0$

$$\begin{aligned}&\left(\mathbb{H}_\eta M_{\eta-1/2}\boldsymbol{H}_+M_{1/2-\eta}f\right)(x)\\ &= x^{-(\eta+1/2)}\frac{d}{dx}x^{\eta+1/2}\int_0^\infty (xt)^{1/2}J_{\eta+1}(xt)t^{\eta-1/2}\left(\boldsymbol{H}_+M_{1/2-\eta}f\right)(t)\frac{dt}{t}\\ &= x^{-(\eta+1/2)}\frac{d}{dx}\int_0^\infty \left(M_{\eta-1/2}h_{\eta,x}\right)(t)\left(\boldsymbol{H}_+M_{1/2-\eta}f\right)(t)dt.\end{aligned} \qquad (8.7.4)$$

Since the intervals $(-\mathrm{Re}(\eta)-1/2, 1/2)$ and $(\mathrm{Re}(\eta)-1/2, \mathrm{Re}(\eta)+3/2)$ intersect, ν is taken from their intersection. From $-\mathrm{Re}(\eta)-1/2<\nu<1/2$, $h_{\eta,x}\in\mathfrak{L}_{\nu,r}$ for any r $(1<r<\infty)$ due to Lemma 8.1(b), we have

$$M_{\eta-1/2}h_{\eta,x}\in\mathfrak{L}_{\nu-\mathrm{Re}(\eta)+1/2,r}, \tag{8.7.5}$$

when $0<\nu-\mathrm{Re}(\eta)+1/2<2$ because of $\mathrm{Re}(\eta)-1/2<\nu<\mathrm{Re}(\eta)+3/2$.

Replacing $\nu-\mathrm{Re}(\eta)+1/2$ by $1-\nu_1$, we see that the new number ν_1 satisfies $-1<\nu_1<1$ and $M_{\eta-1/2}h_{\eta,x}\in\mathfrak{L}_{1-\nu_1,r}$, in particular, $M_{\eta-1/2}h_{\eta,x}\in\mathfrak{L}_{1-\nu_1,2}$. But since $f\in C_0$, we find $M_{1/2-\eta}f\in\mathfrak{L}_{\nu,2}$ for any ν. So

$$M_{1/2-\eta}f\in\mathfrak{L}_{\nu_1,2}, \qquad M_{\eta-1/2}h_{\eta,x}\in\mathfrak{L}_{1-\nu_1,2}$$

for such a ν_1 $(-1<\nu_1<1)$. Then we can apply Theorem 8.9(c) to (8.7.4) and in accordance with the formula of integration of parts (8.3.5), we have

$$\begin{aligned}&\left(\mathbb{H}_\eta M_{\eta-1/2}\boldsymbol{H}_+M_{1/2-\eta}f\right)(x)\\&\quad=-x^{-(\eta+1/2)}\frac{d}{dx}\int_0^\infty\left(\boldsymbol{H}_-M_{\eta-1/2}h_{\eta,x}\right)(t)t^{1/2-\eta}f(t)dt.\end{aligned} \tag{8.7.6}$$

To evaluate the inner integral in (8.7.6) we prove the following auxiliary result.

Lemma 8.6. *If $\eta\in\mathbb{C}$ with $|\mathrm{Re}(\eta)|<1$, then for almost all $x>0$,*

$$\left(\boldsymbol{H}_-M_{\eta-1/2}h_{\eta,x}\right)(t)=-x^{\eta+1}t^{\eta-1}\left[Y_{\eta+1}(xt)+\frac{\Gamma(\eta+1)}{\pi}\left(\frac{2}{xt}\right)^{\eta+1}\right]. \tag{8.7.7}$$

Proof. The proof is based on the technique of the Mellin transform (3.2.5) and residue theory. By (8.7.5) and Theorem 8.9(b), $(\boldsymbol{H}_-M_{\eta-1/2}h_{\eta,x})(t)$ exists almost everywhere on $\mathbb{R}_+$ and belongs to $\mathfrak{L}_{\nu-\mathrm{Re}(\eta)+1/2,r}$ for some η $(|\mathrm{Re}(\eta)|<1)$ and any r $(1<r<\infty)$. Then for $r=2$ we take the Mellin transform of the left-side of (8.7.7). Applying (8.3.13), (3.3.14), (8.1.14), (8.1.4) and the clear equality

$$\left(\mathfrak{M}g_{\eta,x}\right)(s)=\frac{x^{\eta+s+1/2}}{\eta+s+\dfrac{1}{2}}, \tag{8.7.8}$$

we have for $\mathrm{Re}(s)=\nu-\mathrm{Re}(\eta)+1/2$

$$\begin{aligned}\left(\mathfrak{M}\boldsymbol{H}_-M_{\eta-1/2}h_{\eta,x}\right)(s)&=\cot\left(\frac{\pi s}{2}\right)\left(\mathfrak{M}M_{\eta-1/2}h_{\eta,x}\right)(s)\\&=\cot\left(\frac{\pi s}{2}\right)\left(\mathfrak{M}h_{\eta,x}\right)\left(s+\eta-\frac{1}{2}\right)\\&=\cot\left(\frac{\pi s}{2}\right)\left(\mathfrak{M}\mathbb{H}_\eta g_{\eta,x}\right)\left(s+\eta-\frac{1}{2}\right)\\&=2^{s+\eta-1}\frac{\Gamma\left(\eta+\dfrac{s}{2}\right)}{\Gamma\left(1-\dfrac{s}{2}\right)}\cot\left(\frac{\pi s}{2}\right)\left(\mathfrak{M}g_{\eta,x}\right)\left(\frac{3}{2}-\eta-s\right)\end{aligned}$$

$$= 2^{\eta-2}x^2\left(\frac{x}{2}\right)^{-s}\frac{\Gamma\left(\eta+\frac{s}{2}\right)}{\Gamma\left(2-\frac{s}{2}\right)}\cot\left(\frac{\pi s}{2}\right).$$

Hence, by (3.2.6)

$$\left(\boldsymbol{H}_{-}M_{\eta-1/2}h_{\eta,x}\right)(t) = 2^{\eta-2}x^2 I(xt), \tag{8.7.9}$$

where

$$I(z) = \lim_{R\to\infty}\frac{1}{2\pi i}\int_{\nu_1-iR}^{\nu_1+iR}\left(\frac{z}{2}\right)^{-s}\frac{\Gamma\left(\eta+\frac{s}{2}\right)}{\Gamma\left(2-\frac{s}{2}\right)}\cot\left(\frac{\pi s}{2}\right)ds$$

and the limit is taken in the topology of $\mathfrak{L}_{\nu_1,2}$.

Closing the contour to the left and calculating the residues of the integrand at the simple poles $s=-2(k+\eta)$ $(k=0,1,2,\cdots)$ and $s=-2m$ $(m=0,1,2,\cdots)$ in view of the relations (1.3.3), (2.1.6) and (2.6.2), we have

$$\begin{aligned}
I(z) &= \sum_{k=0}^{\infty}\left(\frac{z}{2}\right)^{2(\eta+k)}\frac{2(-1)^k}{k!\Gamma(2+k+\eta)}\cot[-(\eta+k)\pi] \\
&\quad +\frac{2}{\pi}\sum_{m=-1}^{\infty}\left(\frac{z}{2}\right)^{2m}\frac{\Gamma(\eta-m)}{\Gamma(2+m)}-\frac{2}{\pi}\left(\frac{z}{2}\right)^{-2}\Gamma(\eta+1) \\
&= -2\left(\frac{z}{2}\right)^{\eta-1}\cot[(\eta+1)\pi]\sum_{k=0}^{\infty}\frac{(-1)^k}{k!\Gamma(\eta+k+2)}\left(\frac{z}{2}\right)^{2k+\eta+1} \\
&\quad +2\left(\frac{z}{2}\right)^{\eta-1}\frac{1}{\sin[(\eta+1)\pi]}\sum_{k=0}^{\infty}\frac{(-1)^k}{k!\Gamma(k-\eta)}\left(\frac{z}{2}\right)^{2k-\eta-1}-\frac{2}{\pi}\left(\frac{z}{2}\right)^{-2}\Gamma(\eta+1) \\
&= -2\left(\frac{z}{2}\right)^{\eta-1}\frac{1}{\sin[(\eta+1)\pi]}[\cos((\eta+1)\pi)J_{\eta+1}(z) \\
&\quad -J_{-(\eta+1)}(z)]-\frac{2}{\pi}\left(\frac{z}{2}\right)^{-2}\Gamma(\eta+1) \\
&= -2\left(\frac{z}{2}\right)^{\eta-1}Y_{\eta+1}(z)-\frac{2}{\pi}\left(\frac{z}{2}\right)^{-2}\Gamma(\eta+1)
\end{aligned}$$

according to (8.7.2). Substituting this result into (8.7.9) we obtain (8.7.7), and the lemma is proved.

We continue the proof of Theorem 8.26. Substituting (8.7.7) into (8.7.6), we have

$$\begin{aligned}
&\left(\mathbb{H}_\eta M_{\eta-1/2}\boldsymbol{H}_{+}M_{1/2-\eta}f\right)(x) \\
&= x^{-(\eta+1/2)}\frac{d}{dx}x^{\eta+1}\int_0^\infty t^{-1/2}\left[Y_{\eta+1}(xt)+\frac{\Gamma(\eta+1)}{\pi}\left(\frac{2}{xt}\right)^{\eta+1}\right]f(t)dt. \qquad (8.7.10)
\end{aligned}$$

Due to (8.7.2) and the formulas

$$\frac{d}{dz}[z^{\nu}J_{\nu}(z)] = z^{\nu}J_{\nu-1}(z), \quad \frac{d}{dz}[z^{-\nu}J_{\nu}(z)] = -z^{-\nu}J_{\nu+1}(z) \tag{8.7.11}$$

(see Erdélyi, Magnus, Oberhettinger and Tricomi [2, 7.2.51]), we have

$$\frac{d}{dz}[z^{\eta+1}Y_{\eta+1}(z)] = z^{\eta+1}Y_{\eta}(z). \tag{8.7.12}$$

Since $f \in C_0$, the differentiation may be taken under the integral sign of (8.7.10). Taking such a differentiation and using (8.7.12), we have for almost all $x > 0$,

$$\left(\mathbb{H}_{\eta}M_{\eta-1/2}\boldsymbol{H}_{+}M_{1/2-\eta}f\right)(x) = \int_0^{\infty}(xt)^{1/2}Y_{\eta}(xt)f(t)dt,$$

which completes the proof of Theorem 8.26.

Now we can present the results characterizing the boundedness, the range and the representation of the transform $\mathbb{Y}_{\eta}$ (8.7.1) in the space $\mathfrak{L}_{\nu,r}$.

Theorem 8.27. *Let $1 \leqq r \leqq \infty$ and let $\gamma(r)$ be given in (3.3.9).*

(a) *If $1 < r < \infty$ and $\gamma(r) \leqq \nu < 3/2 - |\mathrm{Re}(\eta)|$, then for all $s \geqq r$ such that $s' > 1/\nu$ and $1/s + 1/s' = 1$, the transform $\mathbb{Y}_{\eta}$ can be extended to $\mathfrak{L}_{\nu,r}$ as an element of $[\mathfrak{L}_{\nu,r}, \mathfrak{L}_{1-\nu,s}]$. If further $1 < r \leqq 2$ and $f \in \mathfrak{L}_{\nu,r}$, then the Mellin transform of (8.7.1) for $\mathrm{Re}(s) = 1 - \nu$ is given by*

$$\left(\mathfrak{M}\mathbb{Y}_{\eta}f\right)(s) = -2^{s-1/2}\frac{\Gamma\left(\frac{1}{2}\left[\eta+s+\frac{1}{2}\right]\right)}{\Gamma\left(\frac{1}{2}\left[\eta-s+\frac{3}{2}\right]\right)}\cot\left[\frac{\pi}{2}\left(s-\eta+\frac{1}{2}\right)\right](\mathfrak{M}f)(1-s). \tag{8.7.13}$$

(b) *If $1 < r < \infty$ and $\gamma(r) \leqq \nu < 3/2 - |\mathrm{Re}(\eta)|$, then except when $\nu = 1/2 - \mathrm{Re}(\eta)$, $\mathbb{Y}_{\eta}$ is a one-to-one transform from $\mathfrak{L}_{\nu,r}$ onto $\mathfrak{L}_{1-\nu,s}$, and*

$$\mathbb{Y}_{\eta}(\mathfrak{L}_{\nu,r}) = \mathbb{H}_{\eta}(\mathfrak{L}_{\nu,r}). \tag{8.7.14}$$

Further

$$\mathbb{Y}_{\eta} = \mathbb{H}_{\eta}M_{\eta-1/2}\boldsymbol{H}_{+}M_{1/2-\eta} \tag{8.7.15}$$

and

$$\mathbb{Y}_{\eta} = -M_{1/2-\eta}\boldsymbol{H}_{-}M_{\eta-1/2}\mathbb{H}_{\eta}. \tag{8.7.16}$$

(c) *For $\eta \neq 0$, if $1 \leqq \nu \leqq 3/2 - |\mathrm{Re}(\eta)|$, $\mathbb{Y}_{\eta} \in [\mathfrak{L}_{\nu,1}, \mathfrak{L}_{1-\nu,\infty}]$, and, if $1 < \nu < 3/2 - |\mathrm{Re}(\eta)|$, $\mathbb{Y}_{\eta} \in [\mathfrak{L}_{\nu,1}, \mathfrak{L}_{1-\nu,r}]$ for all r $(1 \leqq r < \infty)$. For $\eta = 0$, if $1 \leqq \nu < 3/2$, $\mathbb{Y}_0 \in [\mathfrak{L}_{\nu,1}, \mathfrak{L}_{1-\nu,\infty}]$, and, if $1 < \nu < 3/2$, $\mathbb{Y}_0 \in [\mathfrak{L}_{\nu,1}, \mathfrak{L}_{1-\nu,r}]$ for all r $(1 \leqq r < \infty)$.*

(d) *Let $1 < r < \infty$ and $1 < s < \infty$ such that $1/r + 1/s \geqq 1$ and $\max[\gamma(r), \gamma(s)] \leqq \nu < 3/2 - |\mathrm{Re}(\eta)|$, then for $f \in \mathfrak{L}_{\nu,r}$, $g \in \mathfrak{L}_{\nu,s}$ the following relation holds:*

$$\int_0^{\infty} f(x)\left(\mathbb{Y}_{\eta}g\right)(x)dx = \int_0^{\infty}\left(\mathbb{Y}_{\eta}f\right)(x)g(x)dx. \tag{8.7.17}$$

(e) *If $f \in \mathfrak{L}_{\nu,r}$, where $1 < r < \infty$ and $\gamma(r) \leqq \nu < 3/2 - |\text{Re}(\eta)|$, then for almost all $x > 0$ the following relation holds:*

$$\left(\mathbb{Y}_\eta f\right)(x) = x^{-(\eta+1/2)} \frac{d}{dx} x^{\eta+1/2}$$
$$\cdot \int_0^\infty (xt)^{1/2} \left[Y_{\eta+1}(xt) + \frac{\Gamma(\eta+1)}{\pi}\left(\frac{2}{xt}\right)^{\eta+1}\right] f(t)\frac{dt}{t}. \tag{8.7.18}$$

Proof. Since $\gamma(r) \geqq 1/2$, the assumption $\gamma(r) < 3/2 - |\text{Re}(\eta)|$ implies $|\text{Re}(\eta)| < 1$, and hence by Theorem 8.26 the relation (8.7.15) holds on C_0. It is directly verified by using Lemma 3.1(i), Theorem 8.9(a) and Theorem 8.1(a) that, since the assumption $\gamma(r) \leqq \nu < 3/2 - |\text{Re}(\eta)|$ implies $|\nu - 1/2 + \text{Re}(\eta)| < 1$, the transform on the right side of (8.7.15) is in $[\mathfrak{L}_{\nu,r}, \mathfrak{L}_{1-\nu,s}]$ for any $s \geqq r$ such that $s' > 1/\nu$. Thus we may extend $\mathbb{Y}_\eta$ to $\mathfrak{L}_{\nu,r}$ by defining it by (8.7.15) and then $\mathbb{Y}_\eta \in [\mathfrak{L}_{\nu,r}, \mathfrak{L}_{1-\nu,s}]$. The relation (8.7.13) is proved directly by using (8.7.15), (8.1.4), (3.3.14) and (8.3.12). Thus (a) is established.

Due to Lemma 3.1(i), $M_{\pm(\eta-1/2)}$ are isometric isomorphisms. By Theorem 8.9(a) $\boldsymbol{H}_+$ maps $\mathfrak{L}_{\nu+\text{Re}(\eta)-1/2,r}$ one-to-one onto itself except when $\nu = 1/2 - \text{Re}(\eta)$. The transform $\mathbb{H}_\eta$ is also one-to-one by Corollary 8.3.1. Then it follows from (8.7.15) that $\mathbb{Y}_\eta$ is one-to-one except when $\nu = 1/2 - \text{Re}(\eta)$, and hence (8.7.14) holds.

Further Theorem 8.1(a), Lemma 3.1(i) and Theorem 8.9 (b) yield under the conditions in (b) that the transform in the right side of (8.7.16) is in $[\mathfrak{L}_{\nu,r}, \mathfrak{L}_{1-\nu,s}]$ for some parameter ranges for $\mathbb{Y}_\eta$. Also, if $f \in \mathfrak{L}_{1/2,2}$, then (3.3.14), (8.3.13), (3.3.14) and (8.1.4) lead to the equality

$$-\left(\mathfrak{M} M_{1/2-\eta}\boldsymbol{H}_- M_{\eta-1/2}\mathbb{H}_\eta f\right)(s)$$
$$= -2^{s-1/2}\frac{\Gamma\left(\frac{1}{2}\left[\eta+s+\frac{1}{2}\right]\right)}{\Gamma\left(\frac{1}{2}\left[\eta-s+\frac{3}{2}\right]\right)} \cot\left(\frac{\pi}{2}\left[s-\eta+\frac{1}{2}\right]\right)\left(\mathfrak{M}f\right)(1-s).$$

Thus, by (8.7.13), the relation (8.7.16) holds on $\mathfrak{L}_{1/2,2}$, and hence on $\mathfrak{L}_{\nu,r}$, since both sides of (8.7.16) are in $[\mathfrak{L}_{\nu,r}, \mathfrak{L}_{1-\nu,s}]$. This completes the proof of assertion (b).

The results in (c) for $\eta \neq 0$ follow from Theorem 8.1(b), since by (8.7.2) $\mathbb{Y}_\eta = \cot(\eta\pi)\mathbb{H}_\eta - \csc(\eta\pi)\mathbb{H}_{-\eta}$. When $\eta = 0$, the results in (c) are proved by direct estimates similar to that in Theorem 8.1(b), if we take into account the asymptotic estimates of $Y_0(z)$ near zero and infinity:

$$Y_0(z) = O\left(\log(z)\right) \quad (z \to 0), \qquad Y_0(z) = O\left(z^{-1/2}\right) \quad (z \to \infty) \tag{8.7.19}$$

(see Erdélyi, Magnus, Oberhettinger and Tricomi [2, 7.2(33) and 7.13(4)]).

The assertion (d) is proved similarly to that in Theorem 8.1(c).

To prove (e) we note that, since $\nu \geqq \gamma(r) \geqq 1/2 > -\text{Re}(\eta) - 1/2$, $g_{\eta,x} \in \mathfrak{L}_{\nu,r'}$ by Lemma 8.1(a). Thus from (8.7.17) we have for all $x > 0$,

$$\int_0^x t^{\eta+1/2}\left(\mathbb{Y}_\eta f\right)(t)dt = \int_0^\infty g_{\eta,x}(t)\left(\mathbb{Y}_\eta f\right)(t)dt = \int_0^\infty \left(\mathbb{Y}_\eta g_{\eta,x}\right)(t)f(t)dt. \tag{8.7.20}$$

From (8.7.16), (8.1.14) and (8.7.7), we have

$$\left(\mathbb{Y}_\eta g_{\eta,x}\right)(t) = -\left(M_{1/2-\eta}\boldsymbol{H}_-M_{\eta-1/2}\mathbb{H}_\eta g_{\eta,x}\right)(t) = -\left(M_{1/2-\eta}\boldsymbol{H}_-M_{\eta-1/2}h_{\eta,x}\right)(t)$$

$$= x^{\eta+1/2}(xt)^{1/2}\left[Y_{\eta+1}(xt) + \frac{\Gamma(\eta+1)}{\pi}\left(\frac{2}{xt}\right)^{\eta+1}\right]\frac{1}{t},$$

and hence (8.7.20) takes the form

$$\int_0^x t^{\eta+1/2}\left(\mathbb{Y}_\eta f\right)(t)dt$$

$$= x^{\eta+1/2}\int_0^\infty (xt)^{1/2}\left[Y_{\eta+1}(xt) + \frac{\Gamma(\eta)+1}{\pi}\left(\frac{2}{xt}\right)^{\eta+1}\right]f(t)\frac{dt}{t}. \tag{8.7.21}$$

Then the result in (8.7.18) follows on differentiation. Thus Theorem 8.27 is proved.

Corollary 8.27.1. *If $1 < r < \infty$ and $\gamma(r) \leqq \nu < 3/2 - |\mathrm{Re}(\eta)|$, except when $\nu = 1/2 - \mathrm{Re}(\eta)$, then*

$$\mathbb{Y}_\eta(\mathfrak{L}_{\nu,r}) = \left(M_{\nu-\gamma(r)}I_{\nu-\gamma(r),-1/2}\mathfrak{F}_c\right)(\mathfrak{L}_{\nu,r}). \tag{8.7.22}$$

In particular, if $1/2 \leqq \nu < 3/2 - |\mathrm{Re}(\eta)|$, except when $\nu = 1/2 - \mathrm{Re}(\eta)$, then

$$\mathbb{Y}_\eta(\mathfrak{L}_{\nu,2}) = \left(M_{\nu-1/2}I_{\nu-1/2,-1/2}\mathfrak{F}_c\right)\left(L_2(\mathbb{R}_+)\right). \tag{8.7.23}$$

Corollary 8.27.2. *If $1 < r < \infty$ and $\gamma(r) \leqq \nu < 3/2 - |\mathrm{Re}(\eta)|$, except when $\nu = 1/2 - \mathrm{Re}(\eta)$, then the range $\mathbb{H}_\eta(\mathfrak{L}_{\nu,r})$ of the Hankel transform (8.1.1) is invariant under the operator $M_{1/2-\eta}\boldsymbol{H}_-M_{\eta-1/2}$.*

Corollaries 8.27.1 and 8.27.2 follow from Theorems 8.27 and 8.2, if we take into account (8.7.14), (8.1.27) and (8.7.14), (8.7.16), respectively.

Remark 8.3. The boundedness conditions in Theorem 8.27(a) can be extended for $\eta = -1/2$. In fact, if $\eta = -1/2$, (8.7.2) implies $Y_{-1/2} = J_{1/2}$ and hence $\mathbb{Y}_{-1/2} = \mathbb{H}_{1/2} = \mathfrak{F}_s$ by virtue of (8.1.3). Hence by Theorem 8.6(a), $\mathbb{Y}_{-1/2} = \mathfrak{F}_s$ is a one-to-one transform from $\mathfrak{L}_{\nu,r}$ onto $\mathfrak{L}_{1-\nu,s}$ provided that $\gamma(r) \leqq \nu < 2$.

Remark 8.4. The exceptional value $\nu = 1/2 - \mathrm{Re}(\eta)$, for which $\mathbb{Y}_\eta(\mathfrak{L}_{\nu,r}) \neq \mathbb{H}_\eta(\mathfrak{L}_{\nu,r})$ and the results in Corollaries 8.27.1 and 8.27.2 fail, is only possible if $-1/2 < \mathrm{Re}(\eta) \leqq 0$, since the condition $\gamma(r) \leqq 1/2 - \mathrm{Re}(\eta) < 3/2 - |\mathrm{Re}(\eta)|$ is equivalent to $-1/2 < \mathrm{Re}(\eta) \leqq 1/2 - \gamma(r)$ and $\gamma(r) \geqq 1/2$. Further, if $\mathrm{Re}(\eta) = 0$, then $r = 2$, and thus

$$\mathbb{Y}_{i\xi}(\mathfrak{L}_{\nu,r}) = \mathbb{H}_{i\xi}(\mathfrak{L}_{\nu,r}) \quad \left(\xi \in \mathbb{R};\ 1 < r < \infty,\ r \neq 2;\ \frac{1}{2} < \nu < \frac{3}{2}\right). \tag{8.7.24}$$

In particular,

$$\mathbb{Y}_0(\mathfrak{L}_{\nu,r}) = \mathbb{H}_0(\mathfrak{L}_{\nu,r}) \quad \left(1 < r < \infty,\ r \neq 2;\ \frac{1}{2} < \nu < \frac{3}{2}\right). \tag{8.7.25}$$

Remark 8.5. Since on $\mathfrak{L}_{1/2,2} = L_2(\mathbb{R}_+)$, $\mathbb{H}_\eta^2 = I$ (see Remark 8.1 in Section 8.1), then $\mathbb{H}_\eta\Big(L_2(\mathbb{R}_+)\Big) = L_2(\mathbb{R}_+)$ and in accordance with (8.7.14),

$$\mathbb{Y}_\eta\Big(L_2(\mathbb{R}_+)\Big) = \mathbb{H}_\eta\Big(L_2(\mathbb{R}_+)\Big) = L_2(\mathbb{R}_+) \quad (|\mathrm{Re}(\eta)| < 1). \tag{8.7.26}$$

Finally we present the inversion for the transform $\mathbb{Y}_\eta$ in (8.7.1) on $\mathfrak{L}_{\nu,r}$ in terms of the Struve function $\mathbf{H}_\eta(z)$ defined by

$$\mathbf{H}_\eta(z) = \sum_{k=0}^{\infty} \frac{(-1)^k}{\Gamma\left(k+\frac{3}{2}\right)\Gamma\left(k+\eta+\frac{3}{2}\right)} \left(\frac{z}{2}\right)^{2k+\eta+1} \tag{8.7.27}$$

(see Erdélyi, Magnus, Oberhettinger and Tricomi [2, 7.5(55)]).

Theorem 8.28. *Let $1 < r < \infty$ and $\gamma(r) \leqq \nu < \min[1/2 - \mathrm{Re}(\eta), 3/2 + \mathrm{Re}(\eta)]$, where $\gamma(r)$ is given in (3.3.9). For $f \in \mathfrak{L}_{\nu,r}$ and for almost all $x > 0$, the following inversion relation holds:*

$$f(x) = x^{-(\eta+1/2)} \frac{d}{dx} x^{\eta+1/2} \int_0^\infty (xt)^{1/2} \mathbf{H}_{\eta+1}(xt) \Big(\mathbb{Y}_\eta f\Big)(t) \frac{dt}{t}. \tag{8.7.28}$$

Proof. Since $\gamma(r) \geqq 1/2$, $1/2 < \min[1/2 - \mathrm{Re}(\eta), 3/2 + \mathrm{Re}(\eta)]$, and it follows that $-1 < \mathrm{Re}(\eta) < 0$, so that $\gamma(r) \leqq \nu < 3/2 + \mathrm{Re}(\eta) = 3/2 - |\mathrm{Re}(\eta)|$. Thus from Theorem 8.27(a), $\mathbb{Y}_\eta f$ exists and is in $\mathfrak{L}_{1-\nu,r}$. Hence, by Lemma 3.1(i), $M_{\eta+1/2}\mathbb{Y}_\eta f \in \mathfrak{L}_{1/2-\nu-\mathrm{Re}(\eta),r}$. Since $-1 < \mathrm{Re}(\eta) < 0$,

$$-\left[\mathrm{Re}(\eta) + \frac{1}{2}\right] < \frac{1}{2} \leqq \gamma(r) \leqq \nu < \min\left[\frac{1}{2} - \mathrm{Re}(\eta), \frac{3}{2} + \mathrm{Re}(\eta)\right] \leqq 1.$$

Hence, by Lemma 8.1(b), $h_{\eta,x} \in \mathfrak{L}_{\nu,r'}$, and thus $M_{-(\eta+1/2)}h_{\eta,x} \in \mathfrak{L}_{1/2+\nu+\mathrm{Re}(\eta),r'}$. We also note that $-1 < 1/2 + \nu + \mathrm{Re}(\eta) < 1$, since from $\mathrm{Re}(\eta) > -1$ and the assumption

$$-\mathrm{Re}(\eta) - \frac{3}{2} < -\frac{1}{2} < \nu < \frac{1}{2} - \mathrm{Re}(\eta).$$

Therefore we can apply Theorem 8.9(c) to obtain from (8.3.5) the relation

$$\begin{aligned} -\int_0^\infty &\left(M_{-(\eta+1/2)}h_{\eta,x}\right)(t) \left(\boldsymbol{H}_- M_{\eta+1/2}\mathbb{Y}_\eta f\right)(t)dt \\ &= \int_0^\infty \left(\boldsymbol{H}_+ M_{-(\eta+1/2)}h_{\eta,x}\right)(t) \left(M_{\eta+1/2}\mathbb{Y}_\eta f\right)(t)dt. \end{aligned} \tag{8.7.29}$$

Noting

$$\frac{1}{\pi}\int_{-\infty}^\infty J_\nu(a|x|)\frac{|x|^{-\nu}}{x-y}dx = -\mathrm{sign}\, y|y|^{-\nu}\mathbf{H}_\nu(a|y|) \quad \left(a > 0;\ \mathrm{Re}(\nu) > -\frac{3}{2}\right)$$

(see Erdélyi, Magnus, Oberhettinger and Tricomi [4, 15.3(15)]) and remembering that $\boldsymbol{H}_+$ is the restriction on $\mathbb{R}_+$ of the Hilbert transform of even functions, we find

$$\left(\boldsymbol{H}_+ M_{-(\eta+1/2)}h_{\eta,x}\right)(t) = -x^{\eta+1}t^{-(\eta+1)}\mathbf{H}_{\eta+1}(xt), \tag{8.7.30}$$

where $\mathbf{H}_{\eta+1}(z)$ is the Struve function (8.7.27). Also, by applying (8.7.16), (8.3.14) and (8.3.4), we have

$$\begin{aligned} \boldsymbol{H}_{-}M_{\eta+1/2}\mathbb{Y}_\eta f &= -\boldsymbol{H}_{-}M_{\eta+1/2}M_{1/2-\eta}\boldsymbol{H}_{-}M_{\eta-1/2}\mathbb{H}_\eta f \\ &= -\boldsymbol{H}_{-}M_1\boldsymbol{H}_{-}M_{-1}M_{\eta+1/2}\mathbb{H}_\eta f = -\boldsymbol{H}_{-}\boldsymbol{H}_{+}M_{\eta+1/2}\mathbb{H}_\eta f \\ &= M_{\eta+1/2}\mathbb{H}_\eta f. \end{aligned} \tag{8.7.31}$$

Substituting (8.7.30) and (8.7.31) into (8.7.29) and using (8.1.13), we obtain

$$\begin{aligned} &x^{\eta+1/2}\int_0^\infty (xt)^{1/2}J_{\eta+1}(xt)\left(\mathbb{H}_\eta f\right)(t)\frac{dt}{t} \\ &\quad = x^{\eta+1/2}\int_0^\infty (xt)^{1/2}\mathbf{H}_{\eta+1}(xt)\left(\mathbb{Y}_\eta f\right)(t)\frac{dt}{t}. \end{aligned} \tag{8.7.32}$$

Since, as noted, $\gamma(r) \leqq \nu < \mathrm{Re}(\eta)+3/2$ and $\nu < 1$, the result in (8.7.28) follows from Theorem 8.3. This completes the proof of Theorem 8.28.

In Theorem 8.28 the inversion formula for the transform $\mathbb{Y}_\eta$ was established for $\nu < 1/2 - \mathrm{Re}(\eta)$. In the case $\nu > 1/2 - \mathrm{Re}(\eta)$ there holds the following result.

Theorem 8.29. *Let $1 < r < \infty$ and $\gamma(r) \leqq \nu < 1$, where $\gamma(r)$ is given in (3.3.9), and let $1/2 < \nu + \mathrm{Re}(\eta) < 3/2$. If $f \in \mathfrak{L}_{\nu,r}$, then for almost all $x > 0$,*

$$f(x) = x^{-(\eta+1/2)}\frac{d}{dx}x^{\eta+1/2}\int_0^\infty (xt)^{1/2}\left[\mathbf{H}_{\eta+1}(xt) - A_\eta(xt)^\eta\right]\left(\mathbb{Y}_\eta f\right)(t)\frac{dt}{t}, \tag{8.7.33}$$

where

$$A_\eta = \frac{1}{2^\eta\pi^{1/2}\Gamma\left(\eta+\dfrac{3}{2}\right)}. \tag{8.7.34}$$

Proof. The proof is based on two preliminary lemmas for the auxiliary functions

$$h_\eta(t) = t^{-1/2}\left[\mathbf{H}_{\eta+1}(t) - A_\eta t^\eta\right] \tag{8.7.35}$$

and

$$r_{\eta,x}(t) = \begin{cases} t^{\eta+1/2}x^{-(\eta+3/2)}, & \text{if } 0 < t < x; \\ 0, & \text{if } t > x. \end{cases} \tag{8.7.36}$$

Lemma 8.7. *If $1/2 - \mathrm{Re}(\eta) < \nu < 1$, then $h_\eta \in \mathfrak{L}_{\nu,r}$ for $1 \leqq r < \infty$. If $1/2 \leqq \nu < 1$ and $1/2 - \nu < \mathrm{Re}(\eta) < 1$, then for $\mathrm{Re}(s) = \nu$,*

$$\left(\mathfrak{M}h_\eta\right)(s) = 2^{s-1/2}\frac{\Gamma\left(\dfrac{1}{2}\left[\eta+s+\dfrac{1}{2}\right]\right)}{\Gamma\left(\dfrac{1}{2}\left[\eta-s+\dfrac{3}{2}\right]\right)}\tan\left(\frac{\pi}{2}\left[s+\eta+\frac{1}{2}\right]\right)\frac{1}{\eta-s+\dfrac{3}{2}}. \tag{8.7.37}$$

Lemma 8.8. *If* $1/2 \leqq \nu < 1$ *and* $1/2 \leqq \nu + \mathrm{Re}(\eta) < 3/2$, *then for* $x > 0$,

$$\mathbb{Y}_\eta W_{1/x} h_\eta = r_{\eta,x}, \tag{8.7.38}$$

where the operator W_δ *is given in* (3.3.12).

For real η Lemmas 8.7 and 8.8 were proved by Heywood and Rooney [4, Lemmas 4.3 and 4.4]. Their proofs can be directly extended to complex η.

To prove Theorem 8.29, we first note that, from the assumption of the theorem, $\mathrm{Re}(\eta) < 1$ and $\nu < 3/2 - |\mathrm{Re}(\eta)|$. Thus by Lemma 8.7, $h_\eta \in \mathfrak{L}_{\nu,r'}$, and by Theorem 8.27(a) $\mathbb{Y}_\eta f$ exists. So we can apply Theorem 8.27(d) and Lemma 8.8, and by virtue of (8.7.35), (8.7.17), (8.7.38) and (8.7.36) we have

$$\begin{aligned} x^{\eta+1/2} &\int_0^\infty (xt)^{1/2}\Big[\mathbf{H}_{\eta+1}(xt) - A_\eta(xt)^\eta\Big]\Big(\mathbb{Y}_\eta f\Big)(t)\frac{dt}{t} \\ &= x^{\eta+3/2}\int_0^\infty (W_{1/x}h_\eta)(t)\Big(\mathbb{Y}_\eta f\Big)(t)dt = x^{\eta+3/2}\int_0^\infty \Big(\mathbb{Y}_\eta W_{1/x}h_\eta\Big)(t)f(t)dt \\ &= x^{\eta+3/2}\int_0^\infty r_{\eta,x}(t)f(t)dt = \int_0^x t^{\eta+1/2}f(t)dt, \end{aligned} \tag{8.7.39}$$

and the result in (8.7.33) follows on differentiation. So Theorem 8.29 is proved.

Corollary 8.29.1. *Under the hypotheses of Theorem 8.29, there holds the formula*

$$\int_0^x t^{\eta+1/2}f(t)dt = x^{\eta+1/2}\int_0^\infty (xt)^{1/2}\Big[\mathbf{H}_{\eta+1}(xt) - A_\eta(xt)^\eta\Big]\Big(\mathbb{Y}_\eta f\Big)(t)\frac{dt}{t} \tag{8.7.40}$$

for $x > 0$.

The relation (8.7.40) leads to another inversion result for the transform $\mathbb{Y}_\eta$.

Theorem 8.30. *Let* $f \in \mathfrak{L}_{\nu,r}$ *and let either of the following assumptions hold:*

(a) $1 < r < \infty$, $\gamma(r) \leqq \nu < 3/2 - |\mathrm{Re}(\eta)|$ *and* $\nu > 1/2 - \mathrm{Re}(\eta)$;

(b) $r = 1$, $\mathrm{Re}(\eta) \neq 0$ *and* $1 \leqq \nu \leqq 3/2 - |\mathrm{Re}(\eta)|$;

(c) $r = 1$, $\mathrm{Re}(\eta) = 0$ *and* $1 \leqq \nu < 3/2$.

Then for almost all $x > 0$,

$$\begin{aligned} f(x) = x^{-(\eta+1/2)}\left(\frac{1}{x}\frac{d}{dx}\right)^2 & x^{\eta+3/2} \\ \cdot \int_0^\infty (xt)^{1/2}&\Big[\mathbf{H}_{\eta+2}(xt) - A_{\eta+1}(xt)^{\eta+1}\Big]\Big(\mathbb{Y}_\eta f\Big)(t)\frac{dt}{t^2}. \end{aligned} \tag{8.7.41}$$

Proof. If $f \in C_0$ and η satisfies the hypotheses (a), (b) or (c), then by Corollary 8.29.1 the relation (8.7.40) holds for any $x > 0$. If we replace x by y, multiply both sides of the obtained result by y, integrate from 0 to x in y and interchange the order of integration, direct calculations show that for $f \in C_0$ the relation

$$\frac{1}{2}\int_0^x t^{\eta+1/2}(x^2-t^2)f(t)dt$$

$$= x^{\eta+3/2}\int_0^\infty (xt)^{1/2}\left[\mathbf{H}_{\eta+2}(xt) - A_{\eta+1}(xt)^{\eta+1}\right]\left(\mathbb{Y}_\eta f\right)(t)\frac{dt}{t^2} \tag{8.7.42}$$

holds, if we take into account the formula

$$\int z^\eta \mathbf{H}_{\eta-1}(z)dz = z^\eta \mathbf{H}_\eta(z) \tag{8.7.43}$$

for the Struve function (8.7.27) (see Erdélyi, Magnus, Oberhettiger and Tricomi [2, 7.5(48)]). It is easily verified that for each $x > 0$, under the hypotheses of the theorem both sides of (8.7.42) represent bounded linear functionals on $\mathfrak{L}_{\nu,r}$, and hence (8.7.42) is valid for $f \in \mathfrak{L}_{\nu,r}$. Then by differentiation of (8.7.42) twice, as in the proof of Theorem 8.29, we obtain the result in (8.7.41), and theorem is proved.

The inversion of the transform $\mathbb{Y}_\eta$ in (8.7.1) in the limiting case $\nu = 1/2 - \mathrm{Re}(\eta)$ is given in terms of the integral in the Cauchy sense defined in (8.4.24).

Theorem 8.31. *Let $1 < r < \infty$ and $-1/2 < \mathrm{Re}(\eta) \leqq 1/2 - \gamma(r)$, where $\gamma(r)$ is given in (3.3.9). If $f \in \mathfrak{L}_{1/2-\nu,r}$, then for almost all $x > 0$,*

$$f(x) = x^{-(\eta+1/2)}\frac{d}{dx}x^{\eta+1/2}\int_0^{\to\infty}(xt)^{1/2}\mathbf{H}_{\eta+1}(xt)\left(\mathbb{Y}_\eta f\right)(t)\frac{dt}{t}. \tag{8.7.44}$$

Proof. For real η the result in (8.7.44) was proved by Heywood and Rooney [4, Theorem 5.3]. It is directly verified that their proof can be extended to complex η. We only note that such a proof, which is more complicated than that in Theorem 8.29, is based on the representations of the form (8.7.28) and (8.7.33) for the characteristic function f_1 of $(1,\infty)$ and for $f_2 = f - f_1$, respectively, some auxiliary assertions and Theorem 8.27(d).

Using the notation, similar to (8.4.24),

$$\int_{\to 0}^\infty f(t)dt = \lim_{\delta\to 0}\int_\delta^\infty f(t)dt, \tag{8.7.45}$$

we have the clear corollary:

Corollary 8.31.1. *If $f \in \mathfrak{L}_{1/2-\nu,r}$, where $1 < r < \infty$ and $-1/2 < \mathrm{Re}(\eta) \leqq 1/2 - \gamma(r)$, then for almost all $x > 0$,*

$$f(x) = x^{-(\eta+1/2)}\frac{d}{dx}x^{\eta+1/2}\int_{\to 0}^\infty (xt)^{1/2}\left[\mathbf{H}_{\eta+1}(xt) - A_\eta(xt)^\eta\right]\left(\mathbb{Y}_\eta f\right)(t)\frac{dt}{t}. \tag{8.7.46}$$

Remark 8.6. The results in (8.7.44) and (8.7.46) are formally similar to those in (8.7.28) and (8.7.33). But the integrals in these relations have different convergences, that is, the former is convergent in the Cauchy sense (8.4.24) and (8.7.45), while the latter is convergent in the sense of $L_1(\mathbb{R}_+)$.

8.8. The Struve Transform

We consider the integral transform $\mathfrak{H}_\eta$ defined by

$$\left(\mathfrak{H}_\eta f\right)(x) = \int_0^\infty (xt)^{1/2}\mathbf{H}_\eta(xt)f(t)dt \quad (x > 0), \tag{8.8.1}$$

where $\mathbf{H}_\eta(z)$ is the Struve function (8.7.27).

First we show that the Struve transform $\mathfrak{H}_\eta$ is represented via the Hankel transform $\mathbb{H}_{\eta+1}$ in (8.1.1). Such a result is based on the following preliminary assertions.

Lemma 8.9. *Let $1 < r < \infty$ and*

$$m_\eta(s) = \frac{\Gamma\left(\frac{1}{2}\left[s+\eta+\frac{3}{2}\right]\right)\Gamma\left(\frac{1}{2}\left[s-\eta-\frac{1}{2}\right]\right)}{\Gamma\left(\frac{1}{2}\left[s+\eta+\frac{1}{2}\right]\right)\Gamma\left(\frac{1}{2}\left[s-\eta+\frac{1}{2}\right]\right)} \quad (\eta \in \mathbb{C},\ \mathrm{Re}(\eta) > -2). \tag{8.8.2}$$

(a) *If $\nu > \max[\mathrm{Re}(\eta)+1/2, -\mathrm{Re}(\eta)-3/2]$, then there is a transform $S_\eta \in [\mathfrak{L}_{\nu,r}]$ such that, for $f \in \mathfrak{L}_{\nu,r}$, $1 < r \leqq 2$ and $\mathrm{Re}(s) = \nu$, there holds*

$$\left(\mathfrak{M}S_\eta\right)(s) = m_\eta(s)\left(\mathfrak{M}f\right)(s). \tag{8.8.3}$$

If further $\nu > -\mathrm{Re}(\eta) - 1/2$, then the transform S_η is one-to-one on $\mathfrak{L}_{\nu,r}$ into itself.

(b) *If $\nu < \min[1/2 - \mathrm{Re}(\eta), \mathrm{Re}(\eta)+5/2]$, then there is a transform $T_\eta \in [\mathfrak{L}_{\nu,r}]$ such that, for $f \in \mathfrak{L}_{\nu,r}$, $1 < r \leqq 2$ and $\mathrm{Re}(s) = \nu$, there holds*

$$\left(\mathfrak{M}T_\eta f\right)(s) = m_\eta(1-s)\left(\mathfrak{M}f\right)(s). \tag{8.8.4}$$

If further $\nu < \mathrm{Re}(\eta) + 3/2$, then the transform T_η is one-to-one on $\mathfrak{L}_{\nu,r}$ into itself.

(c) *If $f \in \mathfrak{L}_{\nu,r}$ and $g \in \mathfrak{L}_{1-\nu,r'}$, where $\nu > \max[\mathrm{Re}(\eta)+1/2, -\mathrm{Re}(\eta)-3/2]$, then*

$$\int_0^\infty \left(S_\eta f\right)(x)g(x)dx = \int_0^\infty f(x)\left(T_\eta g\right)(x)dx. \tag{8.8.5}$$

(d) *If $\mathrm{Re}(\eta) > -2$, then for almost all $x > 0$,*

$$\left(T_\eta h_{\eta+1,x}\right)(t) = x^{\eta+3}t^{-(\eta+5/2)}\int_0^t y^{\eta+2}\mathbf{H}_\eta(xy)dy, \tag{8.8.6}$$

where $h_{\eta+1,x}(t)$ is given in (8.1.13).

Proof. The first assertion in (a) follows from Theorem 3.1, if we take $m(s) = m_\eta(s)$, $\alpha(m_\eta) = \max[\mathrm{Re}(\eta)+1/2, -\mathrm{Re}(\eta)-3/2]$ and $\beta(m_\eta) = \infty$. Using (1.2.4) and (3.4.19), we find $|m_\eta(\sigma+it)| \sim 1$ uniformly in $\alpha(m_\eta) < \sigma_1 \leqq \sigma \leqq \sigma_2 < \beta(m_\eta)$ and $m'_\eta(\sigma+it) = O(1/t)$ as $|t| \to \infty$. To prove the one-to-one property of S_η on $\mathfrak{L}_{\nu,r}$, we only note that $1/m_\eta(s)$ is holomorphic in the strip $\alpha_1(m_\eta) < \nu < \beta_1(m_\eta)$ with $\alpha_1(m_\eta) = \max[\mathrm{Re}(\eta)-1/2, -\mathrm{Re}(\eta)-1/2]$ and $\beta_1(m_\eta) = \infty$.

If we put $k_\eta(s) = m_\eta(1-s)$, then $m_\eta \in \mathcal{A}$ implies $k_\eta \in \mathcal{A}$ with $\alpha(k_\eta) = -\infty$ and $\beta(k_\eta) = 1 - \alpha(m_\eta)$, and (b) follows from (a).

For $r = 2$ the result in (8.8.5) is proved similarly to that in Theorem 3.5(c), and it may be extended to any $1 < r < \infty$, since under the hypotheses of (c) both sides of (8.8.5) are bounded linear functionals on $\mathfrak{L}_{\nu,r} \times \mathfrak{L}_{1-\nu,r'}$.

To prove (d) we note that, since $\text{Re}(\eta) > -2$, we can choose ν such that $-\text{Re}(\eta) - 3/2 < \nu < \min[1, -\text{Re}(\eta) + 1/2, \text{Re}(\eta) + 5/2]$. Thus, by Lemma 8.1 and the assertion in (b), we have $h_{\eta+1,x} \in \mathfrak{L}_{\nu,2}$, $h_{\eta+1,x} = \mathbb{H}_{\eta+1} g_{\eta+1,x}$ and $T_\eta h_{\eta+1,x} \in \mathfrak{L}_{\nu,2}$. Hence, in accordance with (8.8.4) and (8.1.4) and by using (8.8.2), (8.7.8) and (2.1.6), we have for $\text{Re}(s) = \nu$

$$\left(\mathfrak{M} T_\eta h_{\eta+1,x}\right)(s) = \left(\mathfrak{M} T_\eta \mathbb{H}_{\eta+1} g_{\eta+1,x}\right)(s)$$

$$= m_\eta(1-s) 2^{s-1/2} \frac{\Gamma\left(\frac{1}{2}\left[\eta + s + \frac{3}{2}\right]\right)}{\Gamma\left(\frac{1}{2}\left[\eta - s + \frac{5}{2}\right]\right)} (\mathfrak{M} g_{\eta+1,x})(1-s)$$

$$= 2^{s-1/2} \frac{\Gamma\left(\frac{1}{2}\left[\eta + s + \frac{3}{2}\right]\right) \Gamma\left(\frac{1}{2}\left[-\eta - s + \frac{1}{2}\right]\right)}{\left(\eta - s + \frac{5}{2}\right) \Gamma\left(\frac{1}{2}\left[\eta - s + \frac{3}{2}\right]\right) \Gamma\left(\frac{1}{2}\left[-\eta - s + \frac{3}{2}\right]\right)} x^{\eta - s + 5/2}$$

$$= 2^{s-1/2} \frac{\pi \sec\left(\frac{\pi}{2}\left[\eta + s + \frac{3}{2}\right]\right) x^{\eta - s + 5/2}}{\left(\eta - s + \frac{5}{2}\right) \Gamma\left(\frac{1}{2}\left[\eta - s + \frac{3}{2}\right]\right) \Gamma\left(\frac{1}{2}\left[-\eta - s + \frac{3}{2}\right]\right)}. \tag{8.8.7}$$

Hence by (3.2.6),

$$\left(T_\eta h_{\eta+1,x}\right)(t) = x^{\eta+5/2} I(xt),$$

where

$$I(z) = \lim_{R\to\infty} \frac{1}{2\pi i} \int_{\nu-iR}^{\nu+iR} 2^{s-1/2} \frac{\pi \sec\left(\frac{\pi}{2}\left[\eta + s + \frac{3}{2}\right]\right)}{\left(\eta - s + \frac{5}{2}\right) \Gamma\left(\frac{1}{2}\left[\eta - s + \frac{3}{2}\right]\right) \Gamma\left(\frac{1}{2}\left[-\eta - s + \frac{3}{2}\right]\right)} z^{-s} ds,$$

and the limit is taken in the topology of $\mathfrak{L}_{\nu,r}$. By closing the contour to the left, straightforward residue calculus (similar to that in the proof of Lemma 8.6) of the integrand at the simple poles $s = -\eta - 3/2 - 2k$ $(k = 0, 1, 2, \cdots)$, by taking into consideration the assumption $\nu < -\text{Re}(\eta) + 1/2$, yields the result in (8.8.6). Thus Lemma 8.9 is proved.

Now we study the representation of the Struve transform (8.8.1) via the Hankel transform (8.1.1)

Theorem 8.32. *For $\eta \in \mathbb{C}$ $(\text{Re}(\eta) > -2)$ and $f \in C_0$, there holds*

$$\mathfrak{H}_\eta f = \mathbb{H}_{\eta+1} S_\eta f, \tag{8.8.8}$$

where the transform S_η is given in Lemma 8.9.

Proof. For all $\nu > \max[\mathrm{Re}(\eta) + 1/2, -\mathrm{Re}(\eta) - 3/2]$ and $f \in C_0$, we find $S_\eta f \in \mathfrak{L}_{\nu,r}$ by Lemma 8.9(a). Since $\mathrm{Re}(\eta) > -2$, there exists ν such that

$$\frac{1}{2} \leqq \nu < \mathrm{Re}(\eta) + \frac{5}{2}, \qquad \nu > \max\left[\mathrm{Re}(\eta) + \frac{1}{2}, -\mathrm{Re}(\eta) - \frac{3}{2}\right],$$

and, according to Theorem 8.1(a), $\mathbb{H}_{\eta+1} S_\eta f$ can be defined. Lemma 8.1(b) assures $h_{\eta+1,x}(t) \in \mathfrak{L}_{1-\nu,2}$. Hence we can apply Theorem 8.1(d), and in accordance with (8.1.6) and (8.8.5) we have for almost all $x > 0$

$$\begin{aligned}\left(\mathbb{H}_{\eta+1} S_\eta f\right)(x) &= x^{-(\eta+3/2)} \frac{d}{dx} \int_0^\infty h_{\eta+1,x}(t) \left(S_\eta f\right)(t) dt \\ &= x^{-(\eta+3/2)} \frac{d}{dx} \int_0^\infty \left(T_\eta h_{\eta+1,x}\right)(t) f(t) dt.\end{aligned}$$

Then applying (8.8.6), we obtain for almost all $x > 0$

$$\begin{aligned}\left(\mathbb{H}_{\eta+1} S_\eta f\right)(x) &= x^{-(\eta+3/2)} \frac{d}{dx} x^{\eta+3} \int_0^\infty t^{-(\eta+5/2)} f(t) dt \int_0^t y^{\eta+2} \mathbf{H}_\eta(xy) dy \\ &= x^{-(\eta+3/2)} \frac{d}{dx} \int_0^\infty t^{-(\eta+5/2)} f(t) dt \int_0^{xt} s^{\eta+2} \mathbf{H}_\eta(s) ds \\ &= \int_0^t (xt)^{1/2} \mathbf{H}_\eta(xt) f(t) dt = \left(\mathfrak{H}_\eta f\right)(x),\end{aligned}$$

which completes the proof of the theorem.

Now, as in the previous section, we present the results characterizing the boundedness, range and representation of the Struve transform (8.8.1) in $\mathfrak{L}_{\nu,r}$-space.

Theorem 8.33. *Let $1 \leqq r \leqq \infty$ and let $\gamma(r)$ be given in (3.3.9).*

(a) *If $1 < r < \infty$, $\mathrm{Re}(\eta) + 1/2 < \nu < \mathrm{Re}(\eta) + 5/2$ and $\nu \geqq \gamma(r)$, then for all $s \geqq r$ such that $s' \geqq 1/\nu$ and $1/s + 1/s' = 1$, $\mathfrak{H}_\eta$ can be extended to $\mathfrak{L}_{\nu,r}$ as an element of $[\mathfrak{L}_{\nu,r}, \mathfrak{L}_{1-\nu,s}]$. If $1 < r \leqq 2$ and $f \in \mathfrak{L}_{\nu,r}$, then the Mellin transform of (8.8.1) for $\mathrm{Re}(s) = 1 - \nu$ is given by*

$$\left(\mathfrak{M}\mathfrak{H}_\eta f\right)(s) = \frac{2^{s-1/2}\pi \sec\left(\dfrac{\pi}{2}\left[s + \eta + \dfrac{1}{2}\right]\right)}{\Gamma\left(\dfrac{1}{2}\left[\eta - s + \dfrac{3}{2}\right]\right)\Gamma\left(\dfrac{1}{2}\left[-\eta - s + \dfrac{3}{2}\right]\right)} \left(\mathfrak{M}f\right)(1-s). \tag{8.8.9}$$

(b) *If $1 < r < \infty$, $|\mathrm{Re}(\eta) + 1/2| < \nu < \mathrm{Re}(\eta) + 5/2$ and $\nu \geqq \gamma(r)$, then $\mathfrak{H}_\eta$ is a one-to-one transform from $\mathfrak{L}_{\nu,r}$ onto $\mathfrak{L}_{1-\nu,s}$, and*

$$\mathfrak{H}_\eta(\mathfrak{L}_{\nu,r}) = \mathbb{H}_{\eta+1}(\mathfrak{L}_{\nu,r}). \tag{8.8.10}$$

Further, if $|\mathrm{Re}(\eta) + 1/2| < \nu < \mathrm{Re}(\eta) + 3/2$, then

$$\mathfrak{H}_\eta = \mathbb{H}_\eta M_{-\eta-1/2} \mathbf{H}_- M_{\eta+1/2} \tag{8.8.11}$$

and

$$\mathfrak{H}_\eta = -M_{\eta+1/2} \mathbf{H}_+ M_{-\eta-1/2} \mathbb{H}_\eta. \tag{8.8.12}$$

(c) *If* $\mathrm{Re}(\eta)+1/2 \leqq \nu \leqq \mathrm{Re}(\eta)+5/2$ *and* $\nu \geqq 1$, *then* $\mathfrak{H}_\eta \in [\mathfrak{L}_{\nu,1}, \mathfrak{L}_{1-\nu,\infty}]$, *and if* $\mathrm{Re}(\eta)+1/2 < \nu < \mathrm{Re}(\eta)+5/2$ *and* $\nu > 1$, *then for all* r $(1 \leqq r < \infty)$, $\mathfrak{H}_\eta \in [\mathfrak{L}_{\nu,1}, \mathfrak{L}_{1-\nu,r}]$.

(d) *If* $f \in \mathfrak{L}_{\nu,r}$ *and* $g \in \mathfrak{L}_{\nu,s}$, *where* $1 < r < \infty$ *and* $1 < s < \infty$ *such that* $1/r + 1/s \geqq 1$ *and* $\mathrm{Re}(\eta)+1/2 < \nu < \mathrm{Re}(\eta)+5/2$ *and* $\nu \geqq \max[\gamma(r), \gamma(s)]$, *then the following relation holds:*

$$\int_0^\infty f(x)\left(\mathfrak{H}_\eta g\right)(x)dx = \int_0^\infty \left(\mathfrak{H}_\eta f\right)(x)g(x)dx. \tag{8.8.13}$$

(e) *If* $f \in \mathfrak{L}_{\nu,r}$, *where* $1 < r < \infty$ *and* $\mathrm{Re}(\eta)+1/2 < \nu < \mathrm{Re}(\eta)+5/2$ *and* $\nu \geqq \gamma(r)$, *then for almost all* $x > 0$ *the following relations hold:*

$$\left(\mathfrak{H}_\eta f\right)(x) = x^{-\eta-1/2}\frac{d}{dx}x^{\eta+1/2}\int_0^\infty (xt)^{1/2}\mathbf{H}_{\eta+1}(xt)f(t)\frac{dt}{t}, \tag{8.8.14}$$

if $\mathrm{Re}(\eta) > -1$, *and*

$$\left(\mathfrak{H}_\eta f\right)(x) = -x^{\eta-1/2}\frac{d}{dx}x^{1/2-\eta}$$

$$\cdot \int_0^\infty (xt)^{1/2}\left[\mathbf{H}_{\eta-1}(xt) - \frac{2^{1-\eta}}{\pi^{1/2}\Gamma\left(\eta+\dfrac{1}{2}\right)}(xt)^\eta\right]f(t)\frac{dt}{t}, \tag{8.8.15}$$

if $-2 < \mathrm{Re}(\eta) < 1$.

Proof. Since $\mathrm{Re}(\eta) > -2$, Theorem 8.32 implies $\mathfrak{H}_\eta = \mathbb{H}_{\eta+1}S_\eta$ on C_0. Lemma 8.9(a) guarantees $S_\eta \in [\mathfrak{L}_{\nu,r}]$ under $\nu > \max[\mathrm{Re}(\eta)+1/2, -\mathrm{Re}(\eta)-3/2]$. The case $\mathrm{Re}(\eta)+1/2 \leqq -\mathrm{Re}(\eta) - 3/2$ occurs only if $-2 < \mathrm{Re}(\eta) \leqq -1$ and for η such that $-\mathrm{Re}(\eta)-3/2 < 1/2 \leqq \gamma(r) \leqq \nu$. Therefore for the value ν under consideration in (a), $S_\eta \in [\mathfrak{L}_{\nu,r}]$. Also, since $\gamma(r) \leqq \nu < \mathrm{Re}(\eta)+5/2$ by Theorem 8.1(a), $\mathbb{H}_{\eta+1} \in [\mathfrak{L}_{\nu,r}, \mathfrak{L}_{1-\nu,s}]$ for all $s \geqq r$ such that $s' \geqq 1/\nu$. Hence $\mathbb{H}_{\eta+1}S_\eta \in [\mathfrak{L}_{\nu,r}, \mathfrak{L}_{1-\nu,s}]$ for such s. Thus we can extend $\mathfrak{H}_\eta$ on C_0 to $\mathfrak{L}_{\nu,r}$ by defining it by (8.8.8) and then $\mathfrak{H}_\eta \in [\mathfrak{L}_{\nu,r}, \mathfrak{L}_{1-\nu,s}]$. The relation (8.8.9) is proved directly by using (8.1.4) and (8.8.3). Thus (a) is proved.

By Lemma 8.9(a), $S_\eta \in [\mathfrak{L}_{\nu,r}]$ is one-to-one for $\nu > |\mathrm{Re}(\eta)+1/2|$, and by Theorem 8.1(a), $\mathbb{H}_{\eta+1}$ is one-to-one from $\mathfrak{L}_{\nu,r}$ onto $\mathfrak{L}_{1-\nu,s}$ for $\gamma(r) \leqq \nu < \mathrm{Re}(\eta)+5/2$. Then $\mathfrak{H}_\eta = \mathbb{H}_{\eta+1}S_\eta$ is also one-to-one. Thus (8.8.10) holds.

Under the hypotheses of (b), Lemma 3.1(i), Theorems 8.9(b) and 8.1(a) yield that $\mathbb{H}_\eta M_{-\eta-1/2}\mathbf{H}_- M_{\eta+1/2}$ maps $\mathfrak{L}_{\nu,r}$ boundedly into $\mathfrak{L}_{1-\nu,s}$. Further, if $f \in \mathfrak{L}_{\nu,r}$ and $1/2 \leqq \nu < \mathrm{Re}(\nu)+3/2$, then by using (8.1.4), (3.3.14) and (8.3.13) for $\mathrm{Re}(s) = 1-\nu$ we obtain the relation

$$\left(\mathfrak{M}\mathbb{H}_\eta M_{-\eta-1/2}\mathbf{H}_- M_{\eta+1/2}f\right)(s)$$

$$= \frac{2^{s-1/2}\pi\sec\left(\dfrac{\pi}{2}\left[s+\eta+\dfrac{1}{2}\right]\right)}{\Gamma\left(\dfrac{1}{2}\left[\eta-s+\dfrac{3}{2}\right]\right)\Gamma\left(\dfrac{1}{2}\left[-\eta-s+\dfrac{3}{2}\right]\right)}\left(\mathfrak{M}f\right)(1-s),$$

which yields (8.8.11) according to (8.8.9). But both sides of (8.8.11) are in $[\mathfrak{L}_{\nu,r}, \mathfrak{L}_{1-\nu,s}]$ if $1 < r < \infty$, $\mathrm{Re}(\eta)+1/2 < \nu < \mathrm{Re}(\eta)+3/2$ and $\nu \geqq \gamma(r)$, and hence we have the relation (8.8.11). Similarly (8.8.12) is shown.

(c) is proved similarly to Theorem 8.1(b) by using the asymptotic estimates for the Struve function (8.7.27) at zero and infinity:

$$\mathbb{H}_\eta(z) = O(z^{\eta+1}) \quad (z \to 0), \qquad \mathbb{H}_\eta(z) = O(z^{-\min[1-\mathrm{Re}(\eta),1/2]}) \quad (z \to \infty) \tag{8.8.16}$$

(see Erdélyi, Magnus, Oberhettinger and Tricomi [2, 7.5(55) and 7.5(63)]).

The assertion (d) is also shown as in Theorem 8.1(c).

To prove (e) we note that, if $\mathrm{Re}(\eta) > -1$, $-\mathrm{Re}(\eta) - 1/2 < 1/2 \leqq \gamma(r) \leqq \nu$, and hence by Lemma 8.1(a), $g_{\eta,x} \in \mathfrak{L}_{\nu,r'}$. Then from (8.8.13), if $x > 0$, we obtain

$$\int_0^x t^{\eta+1/2}\left(\mathfrak{H}_\eta f\right)(t)dt = \int_0^\infty g_{\eta,x}(t)\left(\mathfrak{H}_\eta f\right)(t)dt = \int_0^\infty \left(\mathfrak{H}_\eta g_{\eta,x}\right)(t)f(t)dt. \tag{8.8.17}$$

We evaluate $(\mathfrak{H}_\eta g_{\eta,x})(t)$ as in Lemma 8.9(d) by using the technique of the Mellin transform and residue theory. Namely, since $g_{\eta,x} \in \mathfrak{L}_{\nu,2}$, (8.8.9) and (8.7.8) imply for $\mathrm{Re}(s) = 1 - \nu$

$$\left(\mathfrak{M}\mathfrak{H}_\eta g_{\eta,x}\right)(s) = \frac{\pi 2^{s-1/2} x^{\eta-s+3/2} \sec\left(\frac{\pi}{2}\left[s+\eta+\frac{1}{2}\right]\right)}{\Gamma\left(\frac{1}{2}\left[\eta - s + \frac{7}{2}\right]\right)\Gamma\left(\frac{1}{2}\left[-\eta - s + \frac{3}{2}\right]\right)}.$$

Applying (3.2.6) and closing the contour to the left and calculating the residues of the integrand at the poles $s = -\eta - 2m - 3/2$ $(m = 0, 1, 2, \cdots)$ by noting that $\mathrm{Re}(s) = 1 - \nu < -\mathrm{Re}(\eta) + 1/2$, we obtain

$$\left(\mathfrak{H}_\eta g_{\eta,x}\right)(t) = x^{\eta+3/2}\frac{1}{2\pi i}\lim_{R\to\infty}\int_{1-\nu-iR}^{1-\nu+iR} \frac{\pi 2^{s-1/2}\sec\left(\frac{\pi}{2}\left[s+\eta+\frac{1}{2}\right]\right)}{\Gamma\left(\frac{1}{2}\left[\eta - s + \frac{7}{2}\right]\right)\Gamma\left(\frac{1}{2}\left[-\eta - s + \frac{3}{2}\right]\right)}(xt)^{-s}ds$$

$$= x^{\eta+1/2}(xt)^{1/2}\mathbf{H}_{\eta+1}(xt)\frac{1}{t}.$$

Thus the result in (8.8.14) follows on differentiation.

If $\mathrm{Re}(\eta) < 1$, (8.8.15) is proved in a similar manner by using $g_{-\eta,x} \in \mathfrak{L}_{\nu,r'}$. Thus Theorem 8.33 is established.

In view of Theorems 8.2 and 8.33, we have

Corollary 8.33.1. *Let* $1 < r < \infty$, $|\mathrm{Re}(\eta) + 1/2| < \nu < \mathrm{Re}(\eta) + 3/2$ *and* $\nu \geqq \gamma(r)$. *Then*

$$\mathfrak{H}_\eta(\mathfrak{L}_{\nu,r}) = \left(M_{\eta+1/2}\mathbf{H}_+ M_{\nu-\eta-\gamma-1/2} I_{\nu-\gamma,-1/2}\mathfrak{F}_c\right)\left(\mathfrak{L}_{\gamma(r),r}\right). \tag{8.8.18}$$

In particular, if $|\mathrm{Re}(\eta) + 1/2| < \nu < \mathrm{Re}(\eta) + 3/2$ *and* $\nu \geqq 1/2$,

$$\mathfrak{H}_\eta(\mathfrak{L}_{\nu,2}) = \left(M_{\eta+1/2}\mathbf{H}_+ M_{\nu-\eta-1} I_{\nu-1/2,-1/2}\mathfrak{F}_c\right)\left(L_2(\mathbb{R}_+)\right). \tag{8.8.19}$$

Corollary 8.33.2. *If* $1 < r < \infty$, $|\mathrm{Re}(\eta) + 1/2| < \nu < \mathrm{Re}(\eta) + 3/2$ *and* $\nu \geqq \gamma(r)$, *then the range* $\mathbb{H}_\eta(\mathfrak{L}_{\nu,r})$ *of the Hankel transform (8.1.1) is invariant under the operator* $M_{\eta+1/2}\mathbf{H}_+ M_{-\eta-1/2}$.

Proof. From relations (8.1.20), (8.8.10), (8.8.12) and (8.1.28), we have

$$\mathbb{H}_\eta(\mathfrak{L}_{\nu,r}) = \mathbb{H}_{\eta+1}(\mathfrak{L}_{\nu,r}) = \mathfrak{H}_\eta(\mathfrak{L}_{\nu,r}) = (M_{\eta+1/2}\mathbf{H}_+M_{-\eta-1/2}\mathbb{H}_\eta)(\mathfrak{L}_{\nu,r})$$
$$= (M_{\eta+1/2}\mathbf{H}_+M_{-\eta-1/2})\Big(\mathbb{H}_\eta(\mathfrak{L}_{\nu,r})\Big).$$

Remark 8.7. The exceptional value $\nu = -\mathrm{Re}(\eta) - 1/2$, for which $\mathfrak{H}_\eta(\mathfrak{L}_{\nu,r}) \neq \mathbb{H}_{\eta+1}(\mathfrak{L}_{\nu,r})$, can only occur for $-3/2 < \mathrm{Re}(\eta) \leqq -1$ since $1/2 \leqq \gamma(r) \leqq \nu < \mathrm{Re}(\eta) + 5/2$. Further, if $\mathrm{Re}(\eta) = -1$, then $r = 2$ and thus

$$\mathfrak{H}_{-1+i\xi}(\mathfrak{L}_{\nu,r}) = \mathbb{H}_{i\xi}(\mathfrak{L}_{\nu,r}) \quad \left(\xi \in \mathbb{R};\ 1 < r < \infty,\ r \neq 2;\ -\frac{1}{2} < \nu < \frac{3}{2}\right). \tag{8.8.20}$$

In particular,

$$\mathfrak{H}_{-1}(\mathfrak{L}_{\nu,r}) = \mathbb{H}_0(\mathfrak{L}_{\nu,r}) \quad \left(1 < r < \infty,\ r \neq 2;\ -\frac{1}{2} < \nu < \frac{3}{2}\right). \tag{8.8.21}$$

Remark 8.8. Since on $\mathfrak{L}_{1/2,2} = L_2(\mathbb{R}_+)$, $\mathbf{H}_\eta^2 = I$ (see Remark 8.1 in Section 8.1), then $\mathbb{H}_{\eta+1}(L_2(\mathbb{R}_+)) = L_2(\mathbb{R}_+)$ and in accordance with (8.8.10),

$$\mathfrak{H}_\eta\Big(L_2(\mathbb{R}_+)\Big) = L_2(\mathbb{R}_+) \quad (-2 < \mathrm{Re}(\eta) < 0,\ \mathrm{Re}(\eta) \neq -1). \tag{8.8.22}$$

Theorems 8.28–8.31 show that the $\mathfrak{H}_\eta$-transform (8.8.1) is inverse to $\mathbb{Y}_\eta$-transform (8.7.1). Moreover these transforms are inverse to each other. First we prove such a result for the space $\mathfrak{L}_{1/2,2} = L_2(\mathbb{R}_+)$.

Theorem 8.34. *If $-1 < \mathrm{Re}(\eta) < 0$, then on $\mathfrak{L}_{1/2,2}$,*

$$\mathfrak{H}_\eta\mathbb{Y}_\eta = \mathbb{Y}_\eta\mathfrak{H}_\eta = I. \tag{8.8.23}$$

Proof. Since $-1 < \mathrm{Re}(\eta) < 0$, $1/2 < 3/2 + \mathrm{Re}(\eta) = 3/2 - |\mathrm{Re}(\eta)|$ and $\mathrm{Re}(\eta) + 1/2 < 1/2 < \mathrm{Re}(\eta) + 3/2$. Then we can apply Theorems 8.27(b) and 8.33(b). Using (8.8.12) and (8.7.15), the equality $\mathbb{H}_\eta^2 = I$ being held on $\mathfrak{L}_{1/2,2}$, and (8.3.15) and (8.3.4), we have

$$\mathfrak{H}_\eta\mathbb{Y}_\eta = -M_{\eta+1/2}\mathbf{H}_+M_{-\eta-1/2}\mathbb{H}_\eta\mathbb{H}_\eta M_{\eta-1/2}\mathbf{H}_+M_{1/2-\eta}$$
$$= -M_{\eta+1/2}\mathbf{H}_+M_{-1}\mathbf{H}_+M_{1/2-\eta} = M_{\eta+1/2}M_{-\eta-1/2} = I.$$

Thus the first relation in (8.8.23) is valid. The second one is shown similarly by using (8.8.11) and (8.7.16).

Theorem 8.28 in Section 8.7 is a considerable extension of the first result in Theorem 8.34. The next result gives such an extension of the second one.

Theorem 8.35. *Let* $1 < r < \infty$ *and* $|\mathrm{Re}(\eta) + 1/2| < \nu < \min[1, \mathrm{Re}(\eta) + 3/2]$ *and* $\nu \geqq \gamma(r)$. *For* $f \in \mathfrak{L}_{\nu,r}$ *and almost all* $x > 0$, *the inversion relation*

$$f(x) = x^{-\eta-1/2}\frac{d}{dx}x^{\eta+1/2} \cdot \int_0^\infty (xt)^{1/2}\left[Y_{\eta+1}(xt) + \frac{\Gamma(\eta+1)}{\pi}\left(\frac{2}{xt}\right)^{\eta+1}\right]\left(\mathfrak{H}_\eta f\right)(t)\frac{dt}{t} \qquad (8.8.24)$$

holds.

Proof. Since $\mathrm{Re}(\eta) + 1/2 < 1$ and $\gamma(r) \leqq \nu < \mathrm{Re}(\eta) + 3/2$, $-1 < \mathrm{Re}(\eta) < 1/2$, then under the hypotheses of the theorem

$$M_{1/2-\eta}\mathfrak{H}_\eta f \in \mathfrak{L}_{1/2-\nu+\mathrm{Re}(\eta),r}, \qquad M_{\eta-1/2}h_{\eta,x} \in \mathfrak{L}_{1/2+\nu-\mathrm{Re}(\eta),r'}$$

in accordance with Theorem 8.33(a), Lemmas 3.1(i), 8.1(b) and 3.1(i). The assumption implies $-1 < (1/2) - \nu + \mathrm{Re}(\eta) < 1$ and therefore we can apply Theorem 8.9(c) to obtain from (8.3.5) the relation

$$\int_0^\infty \left(M_{\eta-1/2}h_{\eta,x}\right)(t)\left(\mathbf{H}_+ M_{1/2-\eta}\mathfrak{H}_\eta f\right)(t)dt = -\int_0^\infty \left(\mathbf{H}_- M_{\eta-1/2}h_{\eta,x}\right)(t)\left(M_{1/2-\eta}\mathfrak{H}_\eta f\right)(t)dt. \qquad (8.8.25)$$

From (8.8.12), (8.3.14) and (8.3.4)

$$\mathbf{H}_+ M_{1/2-\eta}\mathfrak{H}_\eta f = -\mathbf{H}_+ M_{1/2-\eta}M_{\eta+1/2}\mathbf{H}_+ M_{-\eta-1/2}\mathbb{H}_\eta f = -M_1\mathbf{H}_-\mathbf{H}_+ M_{-\eta-1/2}\mathbb{H}_\eta f = M_{1/2-\eta}\mathbb{H}_\eta f. \qquad (8.8.26)$$

Substituting this and the relation (8.7.7) for $(\mathbf{H}_- M_{\eta-/2}h_{\eta,x})(t)$ in (8.8.25) and using (8.1.13), we obtain

$$x^{\eta+1/2}\int_0^\infty (xt)^{1/2}J_{\eta+1}(xt)\left(\mathbb{H}_\eta f\right)(t)\frac{dt}{t} = x^{\eta+1/2}\int_0^\infty (xt)^{1/2}\left[Y_{\eta+1}(xt) + \frac{\Gamma(\eta+1)}{\pi}\left(\frac{2}{xt}\right)^{\eta+1}\right]\left(\mathfrak{H}_\eta f\right)(t)\frac{dt}{t},$$

and the result in (8.8.24) follows from Theorem 8.3. This completes the proof of Theorem 8.35.

In Theorem 8.35 the inversion formula for the transform $\mathfrak{H}_\eta$ was proved for $|\mathrm{Re}(\eta)+1/2| < \nu < \min[1, \mathrm{Re}(\eta) + 3/2]$ and $\nu \geqq \gamma(r)$. The result below gives a formula for the inverse of $\mathfrak{H}_\eta$ in the nearly full range of boundedness of the transform.

Theorem 8.36. *Let* $1 \leqq r < \infty$, $m \in \mathbb{N}$ *and*

$$\left|\mathrm{Re}(\eta) + \frac{1}{2}\right| < \nu < \mathrm{Re}(\eta) + \frac{5}{2}, \qquad \gamma(r) \leqq \nu < m.$$

If $f \in \mathfrak{L}_{\nu,r}$, *then for almost all* $x > 0$,

$$f(x) = 2^{-m}x^{-\eta-5/2}\left(\frac{1}{x}\frac{d}{dx}\right)^m x^{\eta+2m+3/2} \cdot \int_0^\infty (xt)^{1/2}[\mathbf{\Phi}_{\eta,m}(xt) - m\mathbf{\Phi}_{\eta,m+1}(xt)]\left(\mathfrak{H}_\eta f\right)(t)dt, \qquad (8.8.27)$$

where

$$\Phi_{\eta,m}(x) = \left(\frac{2}{x}\right)^m \left[Y_{\eta+m}(x) + \frac{1}{\pi}\sum_{k=0}^{m-1}\left(\frac{x}{2}\right)^{2k-\eta-m}\frac{\Gamma(\eta+m-k)}{k!}\right]. \tag{8.8.28}$$

Proof. The proof of this theorem is based on two preliminary assertions for two auxiliary functions

$$\lambda_{\eta,m}(x) = x^{1/2}[\Phi_{\eta,m}(x) - m\Phi_{\eta,m+1}(x)] \tag{8.8.29}$$

and

$$q_{\eta,x}(t) = \frac{2}{\Gamma(m)}\begin{cases} t^{\eta+5/2}x^{-\eta-2m-3/2}(x^2-t^2)^{m-1}, & \text{if } 0 < t < x; \\ 0, & \text{if } t > x. \end{cases} \tag{8.8.30}$$

Lemma 8.10. *Let* $1/2 \leqq \nu < m$.
(a) *If* $|\mathrm{Re}(\eta) + 1/2| < \nu < \mathrm{Re}(\eta) + 7/2$, *then* $\lambda_{\eta,m} \in \mathfrak{L}_{\nu,r}$ *for* $1 \leqq r < \infty$.
(b) *If* $|\mathrm{Re}(\eta) + 1/2| < \nu < \mathrm{Re}(\eta) + 5/2$, *then for* $\mathrm{Re}(s) = \nu$,

$$\left(\mathfrak{M}\lambda_{\eta,m}\right)(s) = 2^{s-3/2}\left(\eta - s + \frac{3}{2}\right)\frac{\Gamma\left(\frac{1}{2}\left[\eta + s + \frac{1}{2}\right]\right)}{\Gamma\left(\frac{1}{2}\left[\frac{7}{2} + \eta - s + 2m\right]\right)}\cot\left(\frac{\pi}{2}\left[\eta - s + \frac{7}{2}\right]\right). \tag{8.8.31}$$

Lemma 8.11. *If* $|\mathrm{Re}(\eta) + 1/2| < \nu < \mathrm{Re}(\eta) + 5/2$ *and* $1/2 \leqq \nu < m$, *then for almost all* $x > 0$,

$$\mathfrak{H}_\eta W_{1/x}\lambda_{\eta,m} = q_{\eta,x}, \tag{8.8.32}$$

where the operator W_δ *is given in* (3.3.12).

For real η Lemmas 8.10 and 8.11 were proved by Heywood and Rooney [4, Lemmas 6.3 and 6.4]. Their proofs are directly extended to complex η.

Using Lemmas 8.10 and 8.11 and taking the same arguments as in Theorems 8.29 and 8.30 with direct calculations, we obtain (8.8.27).

Remark 8.9. For $m = 1$ the result in Theorem 8.36 is an improvement of that in Theorem 8.35, whereas $\nu < \mathrm{Re}(\eta) + 3/2$ is replaced by $\nu < \mathrm{Re}(\eta) + 5/2$. Also the hypothesis $\mathrm{Re}(\eta) > -1$ assures that $-[\mathrm{Re}(\eta) + 1/2] < 1/2 \leqq \nu$.

Remark 8.10. When $\nu < \mathrm{Re}(\eta)+3/2$, direct calculation show that the inversion relation (8.8.27) is simplified by using only $\Phi_{\eta,m}$:

$$f(x) = \frac{x^{-\eta-5/2}}{2^m}\left(\frac{1}{x}\frac{d}{dx}\right)^m x^{\eta+2m+3/2}\int_0^\infty (xt)^{1/2}\Phi_{\eta,m}(xt)\left(\mathfrak{H}_\eta f\right)(t)dt. \tag{8.8.33}$$

8.9. The Meijer $\mathfrak{K}_\eta$-Transform

We consider the integral transform, named the Meijer transform, for $\eta \in \mathbb{C}$:

$$\left(\mathfrak{K}_\eta f\right)(x) = \int_0^\infty (xt)^{1/2} K_\eta(xt) f(t)dt \qquad (x > 0), \tag{8.9.1}$$

where $K_\eta(z)$ is the modified Bessel function of the third kind, or Macdonald function defined by

$$K_\eta(z) = \frac{\pi}{2\sin(\eta\pi)}[I_{-\eta}(z) - I_\eta(z)] = \left(\frac{\pi}{2z}\right)^{1/2} W_{0,\eta}(2z), \tag{8.9.2}$$

in terms of the modified Bessel function of the first kind

$$I_\eta(z) = e^{-i\eta\pi/2} J_\eta\left(ze^{i\pi/2}\right) = \sum_{k=0}^\infty \frac{1}{k!\Gamma(k+\eta+1)}\left(\frac{z}{2}\right)^{2k+\eta} \tag{8.9.3}$$

or the Whittaker function $W_{0,\eta}(z)$ in (7.2.2), for which see the book by Erdélyi, Magnus, Oberhettinger and Tricomi [2, 7.2 (11)].

When $\eta = -1/2$, $K_{-1/2}(z) = \pi^{1/2}(2z)^{-1/2}e^{-z}$ and hence the Meijer transform $\mathfrak{K}_{-1/2}$ coincides with the Laplace transform (2.5.2) within a constant multiplier:

$$\left(\mathfrak{K}_{-1/2} f\right)(x) = \left(\frac{\pi}{2}\right)^{1/2} \int_0^\infty e^{-xt} f(t)dt = \left(\frac{\pi}{2}\right)^{1/2} \left(\mathbb{L}f\right)(x) \quad (x > 0). \tag{8.9.4}$$

The $\mathfrak{L}_{\nu,r}$-theory for the Meijer transform (8.9.1) is based on the Mellin transform of the kernel function $x^{1/2}K_\eta(x)$.

Lemma 8.12. *If $\eta \in \mathbb{C}$, then for $s \in \mathbb{C}$ with $\mathrm{Re}(s) > |\mathrm{Re}(\eta)| - 1/2$, there holds*

$$\begin{aligned}\left(\mathfrak{M}x^{1/2}K_\eta(x)\right)(s) &= 2^{\eta-1}\pi^{1/2}\frac{\Gamma\left(\frac{1}{4}+\frac{\eta}{2}+\frac{s}{2}\right)\Gamma\left(\frac{1}{2}-\eta+s\right)}{\Gamma\left(\frac{3}{4}-\frac{\eta}{2}+\frac{s}{2}\right)} \\ &= 2^{-\eta-1}\pi^{1/2}\frac{\Gamma\left(\frac{1}{4}-\frac{\eta}{2}+\frac{s}{2}\right)\Gamma\left(\frac{1}{2}+\eta+s\right)}{\Gamma\left(\frac{3}{4}+\frac{\eta}{2}+\frac{s}{2}\right)}.\end{aligned} \tag{8.9.5}$$

Proof. It is well known that

$$\left(\mathfrak{M}K_\eta(x)\right)(s) = 2^{s-2}\Gamma\left(\frac{s+\eta}{2}\right)\Gamma\left(\frac{s-\eta}{2}\right) \qquad (\mathrm{Re}(s) > |\mathrm{Re}(\eta)|) \tag{8.9.6}$$

holds, for which see, for example, Prudnikov, Bruchkov and Marichev [2, 2.16.2.2]. Multiplying the numerator and the denominator of the left-hand side of (8.9.6) by $\Gamma([s-\eta]/2+1/2)$ and using the Legendre duplication formula (see Erdélyi, Magnus, Oberhettinger and Tricomi [1, 1.2(15)]):

$$\Gamma(2z) = 2^{2z-1}\pi^{-1/2}\Gamma(z)\Gamma\left(z+\frac{1}{2}\right), \tag{8.9.7}$$

we have

$$\left(\mathfrak{M}K_\eta\right)(s) = 2^{\eta-1}\pi^{1/2}\frac{\Gamma\left(\dfrac{s+\eta}{2}\right)\Gamma(s-\eta)}{\Gamma\left(\dfrac{s-\eta+1}{2}\right)}. \tag{8.9.8}$$

Applying the property (3.3.14) with $\xi = 1/2$, we obtain

$$\left(\mathfrak{M}x^{1/2}K_\eta(x)\right)(s) = \left(\mathfrak{M}M_{1/2}K_\eta\right)(s) = \left(\mathfrak{M}K_\eta\right)\left(s+\frac{1}{2}\right)$$
$$= 2^{\eta-1}\pi^{1/2}\frac{\Gamma\left(\dfrac{1}{4}+\dfrac{\eta}{2}+\dfrac{s}{2}\right)\Gamma\left(\dfrac{1}{2}-\eta+s\right)}{\Gamma\left(\dfrac{3}{4}-\dfrac{\eta}{2}+\dfrac{s}{2}\right)},$$

which gives the first relation in (8.9.5). The second one is proved similarly if we multiply the numerator and the denominator in the right-hand side of (8.9.6) by $\Gamma[(s+\eta)/2+1/2]$ and apply (8.9.7) and (3.3.14).

Employing (8.9.5), we can apply the results in Chapters 3 and 4 to characterize the $\mathfrak{L}_{\nu,r}$-properties of the Meijer transform (8.9.1). Indeed, by (8.9.5) and (1.1.2), $\left(\mathfrak{M}x^{1/2}K_\eta\right)(s)$ is an $\mathcal{H}$-function of the forms

$$\left(\mathfrak{M}x^{1/2}K_\eta(x)\right)(s) = 2^{\eta-1}\pi^{1/2}\mathcal{H}_{1,2}^{2,0}\left[\begin{matrix}\left(\dfrac{3}{4}-\dfrac{\eta}{2},\dfrac{1}{2}\right)\\ \left(\dfrac{1}{4}+\dfrac{\eta}{2},\dfrac{1}{2}\right),\left(\dfrac{1}{2}-\eta,1\right)\end{matrix}\ \middle|\ s\right] \tag{8.9.9}$$

and

$$\left(\mathfrak{M}x^{1/2}K_\eta(x)\right)(s) = 2^{-\eta-1}\pi^{1/2}\mathcal{H}_{1,2}^{2,0}\left[\begin{matrix}\left(\dfrac{3}{4}+\dfrac{\eta}{2},\dfrac{1}{2}\right)\\ \left(\dfrac{1}{4}-\dfrac{\eta}{2},\dfrac{1}{2}\right),\left(\dfrac{1}{2}+\eta,1\right)\end{matrix}\ \middle|\ s\right]. \tag{8.9.10}$$

Hence in view of (3.1.5), the transform $\mathfrak{K}_\eta$ is a special $\boldsymbol{H}$-transform:

$$\left(\mathfrak{K}_\eta f\right)(x) = 2^{\eta-1}\pi^{1/2}\int_0^\infty H_{1,2}^{2,0}\left[xt\ \middle|\ \begin{matrix}\left(\dfrac{3}{4}-\dfrac{\eta}{2},\dfrac{1}{2}\right)\\ \left(\dfrac{1}{4}+\dfrac{\eta}{2},\dfrac{1}{2}\right),\left(\dfrac{1}{2}-\eta,1\right)\end{matrix}\right]f(t)dt \quad (x>0) \tag{8.9.11}$$

or

$$\left(\mathfrak{K}_\eta f\right)(x) = 2^{-\eta-1}\pi^{1/2}\int_0^\infty H_{1,2}^{2,0}\left[xt\ \middle|\ \begin{matrix}\left(\dfrac{3}{4}+\dfrac{\eta}{2},\dfrac{1}{2}\right)\\ \left(\dfrac{1}{4}-\dfrac{\eta}{2},\dfrac{1}{2}\right),\left(\dfrac{1}{2}+\eta,1\right)\end{matrix}\right]f(t)dt \quad (x>0). \tag{8.9.12}$$

By the notation in (3.4.1), (3.4.2), (1.1.7), (1.1.8), (1.1.10), (1.1.11) and (1.1.12), we have for the transform $\mathfrak{K}_\eta$ in (8.9.11) or (8.9.12) that

$$\alpha = |\text{Re}(\eta)| - \frac{1}{2}, \quad \beta = +\infty, \quad a^* = \Delta = a_1^* = 1, \quad a_2^* = 0, \quad \mu = -\frac{1}{2}. \tag{8.9.13}$$

We denote by $\mathcal{E}^1_{\mathcal{H}}$ and $\mathcal{E}^2_{\mathcal{H}}$ the exceptional sets of two functions $\mathcal{H}^{2,0}_{1,2}(s)$ in (8.9.9) and in (8.9.10) (see Definition 3.4). Let $\nu < 3/2 - |\text{Re}(\eta)|$. Then according to the property (1.3.3) of the gamma function, ν is not in the exceptional set $\mathcal{E}^1_{\mathcal{H}}$, if $\nu = 1 - \text{Re}(s)$ such that

$$s \neq \eta - 2k - \frac{3}{2} \quad (k \in \mathbb{N}_0) \tag{8.9.14}$$

and ν is not in the exceptional set $\mathcal{E}^2_{\mathcal{H}}$, if $\nu = 1 - \text{Re}(s)$ such that

$$s \neq -\eta - 2m - \frac{3}{2} \quad (m \in \mathbb{N}_0). \tag{8.9.15}$$

Applying Theorems 3.6 and 3.7, we obtain the following results for the Meijer transform (8.9.1) in the space $\mathfrak{L}_{\nu,2}$.

Theorem 8.37. *Let $\eta \in \mathbb{C}$ and $\nu < 3/2 - |\text{Re}(\eta)|$.*

(i) *There is a one-to-one transform $\mathfrak{K}_\eta \in [\mathfrak{L}_{\nu,2}, \mathfrak{L}_{1-\nu,2}]$ such that the relations*

$$\left(\mathfrak{M}\mathfrak{K}_\eta f\right)(s) = 2^{\eta-1}\pi^{1/2}\frac{\Gamma\left(\frac{1}{4}+\frac{\eta}{2}+\frac{s}{2}\right)\Gamma\left(\frac{1}{2}-\eta+s\right)}{\Gamma\left(\frac{3}{4}-\frac{\eta}{2}+\frac{s}{2}\right)}\left(\mathfrak{M}f\right)(1-s) \tag{8.9.16}$$

and

$$\left(\mathfrak{M}\mathfrak{K}_\eta f\right)(s) = 2^{-\eta-1}\pi^{1/2}\frac{\Gamma\left(\frac{1}{4}-\frac{\eta}{2}+\frac{s}{2}\right)\Gamma\left(\frac{1}{2}+\eta+s\right)}{\Gamma\left(\frac{3}{4}+\frac{\eta}{2}+\frac{s}{2}\right)}\left(\mathfrak{M}f\right)(1-s) \tag{8.9.17}$$

hold for $\text{Re}(s) = 1 - \nu$ and $f \in \mathfrak{L}_{\nu,2}$. If the condition (8.9.14) or (8.9.15) is valid, then the transform $\mathfrak{K}_\eta$ maps $\mathfrak{L}_{\nu,2}$ onto $\mathfrak{L}_{1-\nu,2}$.

(ii) *For $f, g \in \mathfrak{L}_{\nu,2}$ there holds the relation*

$$\int_0^\infty \left(\mathfrak{K}_\eta f\right)(x)g(x)dx = \int_0^\infty f(x)\left(\mathfrak{K}_\eta g\right)(x)dx. \tag{8.9.18}$$

(iii) *Let $f \in \mathfrak{L}_{\nu,2}$, $\lambda \in \mathbb{C}$ and $h \in \mathbb{R}_+$. If $\text{Re}(\lambda) > (1-\nu)h - 1$, then $\mathfrak{K}_\eta f$ is given by*

$$\left(\mathfrak{K}_\eta f\right)(x) = 2^{\eta-1}\pi^{1/2}hx^{1-(\lambda+1)/h}\frac{d}{dx}x^{(\lambda+1)/h}$$
$$\cdot\int_0^\infty H^{2,1}_{2,3}\left[xt\,\middle|\,\begin{matrix}(-\lambda,h),\left(\frac{3}{4}-\frac{\eta}{2},\frac{1}{2}\right)\\ \left(\frac{1}{4}+\frac{\eta}{2},\frac{1}{2}\right),\left(\frac{1}{2}-\eta,1\right),(-\lambda-1,h)\end{matrix}\right]f(t)dt$$
$$(x > 0) \tag{8.9.19}$$

or

$$\left(\mathfrak{K}_\eta f\right)(x) = 2^{-\eta-1}\pi^{1/2}hx^{1-(\lambda+1)/h}\frac{d}{dx}x^{(\lambda+1)/h}$$

$$\cdot \int_0^\infty H_{2,3}^{2,1}\left[xt \left| \begin{matrix} (-\lambda, h), \left(\frac{3}{4}+\frac{\eta}{2}, \frac{1}{2}\right) \\ \left(\frac{1}{4}-\frac{\eta}{2}, \frac{1}{2}\right), \left(\frac{1}{2}+\eta, 1\right), (-\lambda-1, h) \end{matrix} \right.\right] f(t)dt$$

$$(x > 0). \qquad (8.9.20)$$

If $\mathrm{Re}(\lambda) < (1-\nu)h - 1$, *then*

$$\left(\mathfrak{K}_\eta f\right)(x) = -2^{\eta-1}\pi^{1/2}hx^{1-(\lambda+1)/h}\frac{d}{dx}x^{(\lambda+1)/h}$$

$$\cdot \int_0^\infty H_{2,3}^{3,0}\left[xt \left| \begin{matrix} \left(\frac{3}{4}-\frac{\eta}{2}, \frac{1}{2}\right), (-\lambda, h) \\ (-\lambda-1, h), \left(\frac{1}{4}+\frac{\eta}{2}, \frac{1}{2}\right), \left(\frac{1}{2}-\eta, 1\right) \end{matrix} \right.\right] f(t)dt$$

$$(x > 0) \qquad (8.9.21)$$

or

$$\left(\mathfrak{K}_\eta f\right)(x) = -2^{-\eta-1}\pi^{1/2}hx^{1-(\lambda+1)/h}\frac{d}{dx}x^{(\lambda+1)/h}$$

$$\cdot \int_0^\infty H_{2,3}^{3,0}\left[xt \left| \begin{matrix} \left(\frac{3}{4}+\frac{\eta}{2}, \frac{1}{2}\right), (-\lambda, h) \\ (-\lambda-1, h), \left(\frac{1}{4}-\frac{\eta}{2}, \frac{1}{2}\right), \left(\frac{1}{2}+\eta, 1\right) \end{matrix} \right.\right] f(t)dt$$

$$(x > 0). \qquad (8.9.22)$$

(iv) *The transform* $\mathfrak{K}_\eta$ *is independent of* ν *in the sense that, if* ν_1 *and* ν_2 *are such that* $\nu_i < 3/2 - |\mathrm{Re}(\eta)|$ $(i = 1, 2)$ *and if the transforms* $\mathfrak{K}_{\eta,1}$ *and* $\mathfrak{K}_{\eta,2}$ *are given on* $\mathfrak{L}_{\nu_1,2}$ *and* $\mathfrak{L}_{\nu_2,2}$, *respectively, by* (8.9.16) *or* (8.9.17), *then* $\mathfrak{K}_{\eta,1}f = \mathfrak{K}_{\eta,2}f$ *for* $f \in \mathfrak{L}_{\nu_1,2}\bigcap\mathfrak{L}_{\nu_2,2}$.

(v) *For* $f \in \mathfrak{L}_{\nu,2}$, $\mathfrak{K}_\eta f$ *is given in* (8.9.1) *and* (8.9.11) *or* (8.9.12).

Now we present the results on the boundedeness, range and representation of the Meijer transform (8.9.1) in the space $\mathfrak{L}_{\nu,r}$, for which we need two lemmas.

Lemma 8.13. *The modified Bessel function of the third kind of complex order* η, $K_\eta(x)$, *has the following asymptotic estimates at zero:*

$$K_\eta(x) \sim 2^{-\eta-1}\Gamma(-\eta)x^\eta \quad (x \to +0;\ \mathrm{Re}(\eta) < 0); \qquad (8.9.23)$$

$$K_\eta(x) \sim 2^{\eta-1}\Gamma(\eta)x^{-\eta} \quad (x \to +0;\ \mathrm{Re}(\eta) > 0); \qquad (8.9.24)$$

$$K_\eta(x) \sim \frac{\pi}{2\sin(\eta\pi)}\left[\frac{(x/2)^{-\eta}}{\Gamma(1-\eta)} - \frac{(x/2)^\eta}{\Gamma(1+\eta)}\right] \quad (x \to +0;\ \mathrm{Re}(\eta) = 0,\ \eta \neq 0); \qquad (8.9.25)$$

$$K_0(x) \sim -\gamma - \log\left(\frac{x}{2}\right) \quad (x \to +0), \qquad (8.9.26)$$

and at infinity

$$K_\eta(x) \sim \sqrt{\frac{\pi}{2x}}\,e^{-x} \qquad (x \to +\infty), \qquad (8.9.27)$$

where $\gamma = -\psi(1)$ *is the Euler–Mascheroni constant* (see Erdélyi, Magnus, Oberhettiger and Tricomi [1, 1.1(4) and 1.7(7)]).

Proof. The estimates (8.9.23) and (8.9.24) follow from (8.9.2) if we take into account (7.2.3) and (7.2.4) for $\mathrm{Re}(\eta) > 0$ and the relation $\Psi(a, c; z) = z^{1-c}\Psi(a - c + 1, 2 - c; z)$ (see Erdélyi, Magnus, Oberhettinger and Tricomi [1, 6.5(6)]) and (7.2.3) and (7.2.4) when $\mathrm{Re}(\eta) < 0$. The estimate (8.9.25), being clear by (8.9.2) and (8.9.3), yields (8.9.26) in the limiting case when $\eta \to 0$. The asymptotic estimate (8.9.27) is the particular case of the relation given in Erdélyi, Magnus, Oberhettinger and Tricomi [2, 7.13(7)].

In the next lemma we treat two auxiliary functions $g_{\eta,x}(t)$ given in (8.1.12) and

$$h_{\eta,x}(t) = -t^{-\eta-3/2}\left[(xt)^{\eta+1}K_{\eta+1}(xt) - 2^\eta\Gamma(\eta+1)\right] \quad (\eta \in \mathbb{C},\ \mathrm{Re}(\eta) > 0). \tag{8.9.28}$$

We have

Lemma 8.14. *Let* $1 < r < \infty$ *and* $\mathrm{Re}(\eta) > 0$. *The following statements hold:*
(a) $g_{\eta,x}(t) \in \mathfrak{L}_{\nu,r}$ *if and only if* $\nu > -\mathrm{Re}(\eta) - 1/2$.
(b) $h_{\eta,x}(t) \in \mathfrak{L}_{\nu,r}$ *if and only if* $\nu > \mathrm{Re}(\eta) + 1/2$.
(c) *Further, there holds the relation*

$$\mathfrak{K}_\eta g_{\eta,x} = h_{\eta,x}. \tag{8.9.29}$$

Proof. The assertion (a) was proved in Lemma 8.1(a). The statement (b) is proved similarly to those in Lemma 8.1(b) on the basis of the asymptotic estimates (8.9.23), if we take into accout that by (8.9.2) and (8.9.3) the function $z^\eta K_\eta(z)$ is analytic for $z \in \mathbb{C}$. To prove (c), we use the property

$$\frac{d}{dz}\left[z^{\eta+1}K_{\eta+1}(z)\right] = -z^{\eta+1}K_\eta(z) \tag{8.9.30}$$

(see Erdélyi, Magnus, Obrerhettinger and Tricomi [2, 7.11(21)]) and the asymptotic estimate (8.9.23), and we have

$$\begin{aligned}
\left(\mathfrak{K}_\eta g_{\eta,x}\right)(t) &= \int_0^\infty (t\tau)^{1/2}K_\eta(t\tau)g_{\eta,x}(\tau)d\tau = \int_0^x (t\tau)^{1/2}K_\eta(t\tau)\tau^{\eta+1/2}d\tau \\
&= t^{-\eta-3/2}\int_0^{xt} u^{\eta+1}K_\eta(u)du = -t^{-\eta-3/2}\int_0^{xt}\frac{d}{du}[u^{\eta+1}K_{\eta+1}(u)]du \\
&= -t^{-\eta-3/2}\left[(xt)^{\eta+1}K_{\eta+1}(xt) - 2^\eta\Gamma(\eta+1)\right] = h_{\eta,x}(t).
\end{aligned}$$

Thus the relation (8.9.29) is proved, which completes the proof of Lemma 8.14.

Theorem 8.38. *Let* $\eta \in \mathbb{C}$, $1 \leqq r \leqq s \leqq \infty$ *and* $\nu < 3/2 - |\mathrm{Re}(\eta)|$.

(a) *The transform* $\mathfrak{K}_\eta$ *defined on* $\mathfrak{L}_{\nu,2}$ *can be extended to* $\mathfrak{L}_{\nu,r}$ *as an element of* $[\mathfrak{L}_{\nu,r}, \mathfrak{L}_{1-\nu,s}]$. *If either* (i) $1 < r \leqq 2$ *or* (ii) $1 < r < \infty$ *and the conditions in* (8.9.14) *or in* (8.9.15) *are satisfied, then* $\mathfrak{K}_\eta$ *is a one-to-one transform from* $\mathfrak{L}_{\nu,r}$ *onto* $\mathfrak{L}_{1-\nu,s}$.

(b) *If* $f \in \mathfrak{L}_{\nu,r}$ *and* $g \in \mathfrak{L}_{\nu,s'}$ *with* $1/s + 1/s' = 1$, *then the relation* (8.9.18) *holds.*

(c) *For* $\nu \leqq 3/2 - |\mathrm{Re}(\eta)|$ *except when* $\eta = 0$ *and* $\nu = 3/2$, $\mathfrak{K}_\eta$ *belongs to* $[\mathfrak{L}_{\nu,1}, \mathfrak{L}_{1-\nu,\infty}]$.

(d) *Let $1 < r < \infty$, $|\mathrm{Re}(\eta)| \geqq 1/2$ and the condition (8.9.14) or (8.9.15) be satisfied. Then*

$$\mathfrak{K}_\eta(\mathfrak{L}_{\nu,r}) = \mathbb{L}(\mathfrak{L}_{\nu,r}). \tag{8.9.31}$$

If neither (8.9.14) nor (8.9.15) is satisfied, then $\mathfrak{K}_\eta(\mathfrak{L}_{\nu,r})$ is a subset of the right-hand side of (8.9.31), where $\mathbb{L}$ is the Laplace transform (2.5.2).

Let $|\mathrm{Re}(\eta)| < 1/2$ and the condition (8.9.14) or (8.9.15) be satisfied. Then

$$\mathfrak{K}_\eta(\mathfrak{L}_{\nu,r}) = \mathfrak{K}_{-\eta}(\mathfrak{L}_{\nu,r}) = \left(I^{1/2-|\mathrm{Re}(\eta)|}_{-;1,1/2-|\mathrm{Re}(\eta)|}\mathbb{L}_{1,|\mathrm{Re}(\eta)|-1/2}\right)(\mathfrak{L}_{\nu,r}), \tag{8.9.32}$$

where $I^{1/2-|\mathrm{Re}(\eta)|}_{-;1,1/2-|\mathrm{Re}(\eta)|}$ and $\mathbb{L}_{1,|\mathrm{Re}(\eta)|-1/2}$ are the Erdélyi–Kober type operator (3.3.2) and the generalized Laplace transform (3.3.3), respectively. If neither (8.9.14) nor (8.9.15) is satisfied, then $\mathfrak{K}_\eta(\mathfrak{L}_{\nu,r})$ is a subset of the right-hand side of (8.9.32).

(e) *Let $\mathrm{Re}(\eta) > 0$, $1 < r < \infty$ and $-1/2 - \mathrm{Re}(\eta) < \nu < 3/2 - \mathrm{Re}(\eta)$. If $f \in \mathfrak{L}_{\nu,r}$, then the relation*

$$\left(\mathfrak{K}_\eta f\right)(x) = -x^{-\eta-1/2}\frac{d}{dx}\int_0^\infty t^{-\eta-1/2}\left[(xt)^{\eta+1}K_{\eta+1}(xt) - 2^\eta\Gamma(\eta+1)\right]f(t)\frac{dt}{t} \tag{8.9.33}$$

holds for almost all $x > 0$.

Proof. The assertion (a) follows from Theorems 4.5(a) and Theorem 4.7(a), and (b) is derived from Theorem 4.5(b). The assertion (c) is proved similarly to the proof of Theorem 8.1(b) on the basis of the asymptotic behavior of $K_\eta(x)$ established in Lemma 8.13. The assertion (d) follows from Theorem 4.7(b)–(c), if we use the relation

$$K_\eta(z) = K_{-\eta}(z) \tag{8.9.34}$$

(see Erdélyi, Magnus, Oberhettinger and Tricomi [2, 7.2(14)]), which is also clear by (8.9.6).

To prove (e) we employ Lemma 8.14 and Theorem 8.38(e). By Lemma 8.14(a), $g_{\eta,x} \in \mathfrak{L}_{\nu,r'}$, and we can appply Theorem 8.38(c)

$$\begin{aligned}\int_0^x t^{\eta+1/2}\left(\mathfrak{K}_\eta f\right)(t)dt &= \int_0^\infty g_{\eta,x}(t)\left(\mathfrak{K}_\eta f\right)(t)dt = \int_0^\infty \left(\mathfrak{K}_\eta g_{\eta,x}\right)(t)f(t)dt \\ &= \int_0^\infty h_{\eta,x}(t)f(t)dt \\ &= -t^{-\eta-3/2}\int_0^\infty \left[(xt)^{1/2}K_{\eta+1}(xt) - 2^\eta\Gamma(\eta+1)\right]f(t)dt,\end{aligned}$$

and the resullt in (8.9.33) follows on differentiation. Thus Theorem 8.38 is proved.

Remark 8.11. The result in (8.9.31) shows that in the case $|\mathrm{Re}(\eta)| \geqq 1/2$ the range $\mathfrak{K}_\eta(\mathfrak{L}_{\nu,r})$ of the Meijer transform (8.9.1) coincides with the range $\mathbb{L}(\mathfrak{L}_{\nu,r})$ of the Laplace transform (2.5.2).

8.10. Bessel Type Transforms

We consider the integral transforms

$$\left(\mathbf{K}_{\eta}^{\rho}f\right)(x)=\int_{0}^{\infty}Z_{\rho}^{\eta}(xt)f(t)dt\quad(x>0)\tag{8.10.1}$$

and

$$\left(\mathbf{L}_{\eta}^{(m)}f\right)(x)=\int_{0}^{\infty}\lambda_{\eta}^{(m)}(xt)f(t)dt\quad(x>0)\tag{8.10.2}$$

with the kernels

$$Z_{\rho}^{\eta}(z)=\int_{0}^{\infty}t^{\eta-1}\exp\left(-t^{\rho}-\frac{z}{t}\right)dt\quad(\rho\in\mathbb{R}_{+};\ \eta\in\mathbb{C})\tag{8.10.3}$$

and

$$\lambda_{\eta}^{(m)}(z)=\frac{(2\pi)^{(m-1)/2}\sqrt{m}}{\Gamma\left(\eta+1-\dfrac{1}{m}\right)}\left(\frac{z}{m}\right)^{\eta m}\int_{1}^{\infty}(t^{m}-1)^{\eta-1/m}e^{-zt}dt\tag{8.10.4}$$

$$\left(m\in\mathbb{N};\ \eta\in\mathbb{C},\ \mathrm{Re}(\eta)>\frac{1}{m}-1\right)$$

being analytic functions of $z\in\mathbb{C}$ for $\mathrm{Re}(z)>0$.

Particular cases of the functions $Z_{\rho}^{\eta}(z)$ and $\lambda_{\eta}^{(n)}(z)$ coincide with the Macdonald function $K_{\eta}(z)$ (see Section 7.2.2 in Erdélyi, Magnus, Oberhettinger and Tricomi [2] and the previous Section 8.9). Namely, when $\rho=1,\ z=t^{2}/4$ and $m=2$, then in accordance with the relation [2, 7.12(23)]

$$K_{\eta}(z)=\frac{1}{2}\int_{0}^{\infty}\exp\left[-\frac{z}{2}\left(t+\frac{1}{t}\right)\right]t^{-\eta-1}dt\quad(\mathrm{Re}(z)>0)\tag{8.10.5}$$

and the formula [2, 7.12(19)]

$$\Gamma\left(\frac{1}{2}-\eta\right)K_{\eta}(z)=\sqrt{\pi}\left(\frac{2}{z}\right)^{\eta}\int_{1}^{\infty}(t^{2}-1)^{-\eta-1/2}e^{-zt}dt\tag{8.10.6}$$

$$\left(\mathrm{Re}(z)>0,\ \mathrm{Re}(\eta)<\frac{1}{2}\right),$$

(8.10.3) and (8.10.4) take the forms

$$Z_{1}^{\eta}\left(\frac{z^{2}}{4}\right)=2\left(\frac{z}{2}\right)^{\eta}K_{-\eta}(z)\quad\left(\mathrm{Re}(z)>0,\ \mathrm{Re}(\eta)>-\frac{1}{2}\right)\tag{8.10.7}$$

and

$$\lambda_{\eta}^{(2)}(z)=2\left(\frac{2}{z}\right)^{\eta}K_{-\eta}(z)\quad\left(\mathrm{Re}(z)>0,\ \mathrm{Re}(\eta)>-\frac{1}{2}\right),\tag{8.10.8}$$

respectively. In these cases (8.10.1) and (8.10.2) are reduced to the Meijer type integral transform considered in the previous section. In particular, when $m=1$ and $\eta=1$, $\lambda_{1}^{(1)}(z)=e^{-z}$ and (8.10.2) is the Laplace transform (2.5.2):

$$\left(\mathbf{L}_{1}^{(1)}f\right)(x)=\int_{0}^{\infty}e^{-xt}f(t)dt\equiv\left(\mathbb{L}f\right)(x)\quad(x>0).\tag{8.10.9}$$

The $\mathfrak{L}_{\nu,r}$-theory for the Bessel type transforms (8.10.1) and (8.10.2), as well as such a theory for the Meijer transform (8.9.1), is based on the Mellin transforms of the functions (8.10.3) and (8.10.4).

Lemma 8.15. *Let* $\eta \in \mathbb{C}$.
(a) *If* $\rho \in \mathbb{R}_+$ *and* $\mathrm{Re}(s) > -\min[0, \mathrm{Re}(\eta)]$, *then*

$$\left(\mathfrak{M}Z_\rho^\eta\right)(s) = \frac{1}{\rho}\Gamma(s)\Gamma\left(\frac{\eta+s}{\rho}\right). \tag{8.10.10}$$

(b) *If* $\mathrm{Re}(\eta) > 1/m - 1$ *and* $\mathrm{Re}(s) > -\min[0, m\mathrm{Re}(\eta)]$, *then*

$$\left(\mathfrak{M}\lambda_\eta^{(m)}\right)(s) = \frac{(2\pi)^{(m-1)/2}}{m^{m\eta+1/2}} \frac{\Gamma(m\eta+s)\Gamma\left(\dfrac{s}{m}\right)}{\Gamma\left(\eta+1-\dfrac{1}{m}+\dfrac{s}{m}\right)}. \tag{8.10.11}$$

Proof. (8.10.10) and (8.10.11) are proved directly by applying (2.5.1) to (8.10.3) and (8.10.4) by virtue of properties of the gamma and beta functions (see Erdélyi, Magnus, Oberhettinger and Tricomi [1, 1.1(1), 1.5(1) and 1.5(5)]).

Now we can apply the results in Chapters 3 and 4 to characterize $\mathfrak{L}_{\nu,r}$-properties of the Bessel type transforms (8.10.1) and (8.10.2). According to (8.10.10) and (8.10.11) and (1.1.2), $\left(\mathfrak{M}Z_\rho^\eta\right)(s)$ and $\left(\mathfrak{M}\lambda_\eta^{(m)}\right)(s)$ are $\mathcal{H}$-functions of the form

$$\left(\mathfrak{M}Z_\rho^\eta\right)(s) = \frac{1}{\rho}\mathcal{H}_{0,2}^{2,0}\left[\begin{array}{c|c} \overline{\qquad\qquad} & \\ (0,1), \left(\dfrac{\eta}{\rho}, \dfrac{1}{\rho}\right) & s \end{array}\right] \tag{8.10.12}$$

and

$$\left(\mathfrak{M}\lambda_\eta^{(m)}\right)(s) = \frac{(2\pi)^{(m-1)/2}}{m^{m\eta+1/2}}\mathcal{H}_{1,2}^{2,0}\left[\begin{array}{c|c} \left(\eta+1-\dfrac{1}{m}, \dfrac{1}{m}\right) & \\ (m\eta,1), \left(0, \dfrac{1}{m}\right) & s \end{array}\right], \tag{8.10.13}$$

respectively. Hence in accordance with (3.1.5), the transform $\mathbf{K}_\eta^\rho$ in (8.10.1) and the transform $\mathbf{L}_\eta^{(m)}$ in (8.10.2) are special $\boldsymbol{H}$-transforms in (3.1.1):

$$\left(\mathbf{K}_\eta^\rho f\right)(x) = \frac{1}{\rho}\int_0^\infty H_{0,2}^{2,0}\left[xt \,\middle|\, \begin{array}{c} \overline{\qquad\qquad} \\ (0,1), \left(\dfrac{\eta}{\rho}, \dfrac{1}{\rho}\right) \end{array}\right] f(t)dt \tag{8.10.14}$$

and

$$\left(\mathbf{L}_\eta^{(m)} f\right)(x) = \frac{(2\pi)^{(m-1)/2}}{m^{m\eta+1/2}}\int_0^\infty H_{1,2}^{2,0}\left[xt \,\middle|\, \begin{array}{c} \left(\eta+1-\dfrac{1}{m}, \dfrac{1}{m}\right) \\ (m\eta,1), \left(0, \dfrac{1}{m}\right) \end{array}\right]. \tag{8.10.15}$$

According to (3.4.1), (3.4.2), (1.1.7), (1.1.8), (1.1.10), (1.1.11) and (1.1.12) we have

$$\alpha = -\min[0, \mathrm{Re}(\eta)],\ \beta = +\infty,\ a^* = \Delta = a_1^* = 1 + \frac{1}{\rho},\ a_2^* = 0,\ \mu = -1 + \frac{\eta}{\rho} \tag{8.10.16}$$

for the $\mathcal{H}$-function in (8.10.12) and

$$\alpha = -\min[0, m\mathrm{Re}(\eta)],\ \beta = +\infty,\ a^* = \Delta = a_1^* = 1,\ a_2^* = 0,$$
$$\mu = \eta(m-1) + \frac{1}{m} - \frac{3}{2} \tag{8.10.17}$$

for the $\mathcal{H}$-function in (8.10.13).

Let $\mathcal{E}_{\mathcal{H}}^1$ and $\mathcal{E}_{\mathcal{H}}^2$ be the exceptional sets of the $\mathcal{H}$-functions in (8.10.12) and (8.10.13) (see Definition 3.4 in Section 3.6). According to (8.10.10) and (8.10.12) the set $\mathcal{E}_{\mathcal{H}}^1$ is empty (because the gamma function $\Gamma(z)$ is not equal to zero), while by (8.10.11) and (8.10.13) ν is not in the exceptional set $\mathcal{E}_{\mathcal{H}}^2$, if

$$s \neq 1 - m(\eta + 1 + k) \quad (k = 0, -1, -2, \cdots;\ \mathrm{Re}(s) = 1 - \nu). \tag{8.10.18}$$

Applying Theorems 3.6 and 3.7 in Section 3.6, we obtain the following results for the Bessel type transforms (8.10.1) and (8.10.2) in the space $\mathfrak{L}_{\nu,2}$.

Theorem 8.39. *Let $\rho \in \mathbb{R}_+, \eta \in \mathbb{C}$ and ν be such that $\nu < 1 + \min[0, \mathrm{Re}(\eta)]$.*

(i) *There is a one-to-one transform $\mathbf{K}_\eta^\rho \in [\mathfrak{L}_{\nu,2}, \mathfrak{L}_{1-\nu,2}]$ such that the relation*

$$\left(\mathfrak{M}\mathbf{K}_\eta^\rho f\right)(s) = \frac{1}{\rho}\Gamma(s)\Gamma\left(\frac{\eta+s}{\rho}\right)\left(\mathfrak{M}f\right)(1-s) \tag{8.10.19}$$

holds for $\mathrm{Re}(s) = 1 - \nu$ and $f \in \mathfrak{L}_{\nu,2}$.

(ii) *If $f, g \in \mathfrak{L}_{\nu,2}$, then*

$$\int_0^\infty \left(\mathbf{K}_\eta^\rho f\right)(x) g(x)dx = \int_0^\infty f(x)\left(\mathbf{K}_\eta^\rho g\right)(x)dx. \tag{8.10.20}$$

(iii) *Let $f \in \mathfrak{L}_{\nu,2}$, $\lambda \in \mathbb{C}$ and $h \in \mathbb{R}_+$. If $\mathrm{Re}(\lambda) > (1-\nu)h - 1$, then $\mathbf{K}_\eta^\rho f$ is given by*

$$\left(\mathbf{K}_\eta^\rho f\right)(x) = \frac{h}{\rho} x^{1-(\lambda+1)/h} \frac{d}{dx} x^{(\lambda+1)/h}$$
$$\cdot \int_0^\infty H_{1,3}^{2,1}\left[xt \,\middle|\, \begin{matrix} (-\lambda, h) \\ (0,1), \left(\frac{\eta}{\rho}, \frac{1}{\rho}\right), (-\lambda-1, h) \end{matrix}\right] f(t)dt \quad (x > 0). \tag{8.10.21}$$

If $\mathrm{Re}(\lambda) < (1-\nu)h - 1$, then

$$\left(\mathbf{K}_\eta^\rho f\right)(x) = -\frac{h}{\rho} x^{1-(\lambda+1)/h} \frac{d}{dx} x^{(\lambda+1)/h}$$
$$\cdot \int_0^\infty H_{1,3}^{3,0}\left[xt \,\middle|\, \begin{matrix} (-\lambda, h) \\ (-\lambda-1, h), (0,1), \left(\frac{\eta}{\rho}, \frac{1}{\rho}\right) \end{matrix}\right] f(t)dt \quad (x > 0). \tag{8.10.22}$$

(iv) *$\mathbf{K}_\eta^\rho$ is independent of ν in the sense that if ν and $\widetilde{\nu}$ are such that $\max[\nu, \widetilde{\nu}] < 1 + \min[0, \mathrm{Re}(\eta)]$ and if the transforms $\mathbf{K}_\eta^\rho$ and $\widetilde{\mathbf{K}}_\eta^\rho$ are given in (8.10.19), then $\mathbf{K}_\eta^\rho f = \widetilde{\mathbf{K}}_\eta^\rho f$ for $f \in \mathfrak{L}_{\nu,2} \bigcap \mathfrak{L}_{\widetilde{\nu},2}$.*

(v) *$\mathbf{K}_\eta^\rho f$ is given in (8.10.1) and (8.10.14) for $f \in \mathfrak{L}_{\nu,2}$.*

Theorem 8.40. *Let $m \in \mathbb{N}$ and let $\eta \in \mathbb{C}$ be such that $\mathrm{Re}(\eta) > 1/m - 1$ and $\nu < 1 + \min[0, m\mathrm{Re}(\eta)]$.*

(i) *There is a one-to-one transform $\mathbf{L}_\eta^{(m)} \in [\mathfrak{L}_{\nu,2}, \mathfrak{L}_{1-\nu,2}]$ such that the relation*

$$\left(\mathfrak{M}\mathbf{L}_\eta^{(m)} f\right)(s) = \frac{(2\pi)^{(m-1)/2}}{m^{m\eta+1/2}} \frac{\Gamma(m\eta + s)\Gamma\left(\dfrac{s}{m}\right)}{\Gamma\left(\eta + 1 - \dfrac{1}{m} + \dfrac{s}{m}\right)} \left(\mathfrak{M}f\right)(1-s) \tag{8.10.23}$$

holds for $\mathrm{Re}(s) = 1 - \nu$ and $f \in \mathfrak{L}_{\nu,2}$.

(ii) *If $f, g \in \mathfrak{L}_{\nu,2}$, then*

$$\int_0^\infty \left(\mathbf{L}_\eta^{(m)} f\right)(x) g(x) dx = \int_0^\infty f(x) \left(\mathbf{L}_\eta^{(m)} g\right)(x) dx. \tag{8.10.24}$$

(iii) *Let $f \in \mathfrak{L}_{\nu,2}$, $\lambda \in \mathbb{C}$ and $h \in \mathbb{R}_+$. If $\mathrm{Re}(\lambda) > (1-\nu)h - 1$, then $\mathbf{L}_\eta^{(m)} f$ is given by*

$$\left(\mathbf{L}_\eta^{(m)} f\right)(x) = \frac{(2\pi)^{(m-1)/2}}{m^{n\eta+1/2}} h x^{1-(\lambda+1)/h} \frac{d}{dx} x^{(\lambda+1)/h}$$

$$\cdot \int_0^\infty H_{2,3}^{2,1}\left[xt \,\middle|\, \begin{matrix} (-\lambda, h), \left(\eta + 1 - \dfrac{1}{m}, \dfrac{1}{m}\right) \\ (m\eta, 1), \left(0, \dfrac{1}{m}\right), (-\lambda - 1, h) \end{matrix}\right] f(t) dt \quad (x > 0). \tag{8.10.25}$$

If $\mathrm{Re}(\lambda) < (1-\nu)h - 1$, then

$$\left(\mathbf{L}_\eta^{(m)} f\right)(x) = -\frac{(2\pi)^{(m-1)/2}}{m^{m\eta+1/2}} h x^{1-(\lambda+1)/h} \frac{d}{dx} x^{(\lambda+1)/h}$$

$$\cdot \int_0^\infty H_{2,3}^{3,0}\left[xt \,\middle|\, \begin{matrix} \left(\eta + 1 - \dfrac{1}{m}, \dfrac{1}{m}\right), (-\lambda, h) \\ (-\lambda - 1, h), (m\eta, 1), \left(0, \dfrac{1}{m}\right) \end{matrix}\right] f(t) dt \quad (x > 0). \tag{8.10.26}$$

(iv) *$\mathbf{L}_\eta^{(m)}$ is independent of ν in the sense that if ν and $\widetilde{\nu}$ are such that $\max[\nu, \widetilde{\nu}] < 1 + \min[0, m\mathrm{Re}(\eta)]$ and if the transforms $\mathbf{L}_\eta^{(m)}$ and $\widetilde{\mathbf{L}}_\eta^{(m)}$ are given in (8.10.23) on $\mathfrak{L}_{\nu,2}$ and $\mathfrak{L}_{\widetilde{\nu},2}$, respectively, then $\mathbf{L}_\eta^{(m)} f = \widetilde{\mathbf{L}}_\eta^{(m)} f$ for $f \in \mathfrak{L}_{\nu,2} \bigcap \mathfrak{L}_{\widetilde{\nu},2}$.*

(v) *$\mathbf{L}_\eta^{(m)} f$ is given in (8.10.2) and (8.10.15) for $f \in \mathfrak{L}_{\nu,2}$.*

Corollary 8.40.1 *Let $f \in \mathfrak{L}_{\nu,2}$, $\lambda \in \mathbb{C}$ and $h \in \mathbb{R}_+$. If $\mathrm{Re}(\lambda) > (1-\nu)h - 1$, then the Laplace transform $\mathbb{L}f$ is given by*

$$\left(\mathbb{L}f\right)(x) = h x^{1-(\lambda+1)/h} \frac{d}{dx} x^{(\lambda+1)/h}$$

$$\cdot \int_0^\infty H_{1,2}^{1,1}\left[xt \,\middle|\, \begin{matrix} (-\lambda, h) \\ (0, 1), (-\lambda - 1, h) \end{matrix}\right] f(t) dt \quad (x > 0), \tag{8.10.27}$$

while if $\mathrm{Re}(\lambda) < (1-\nu)h - 1$, then

$$\left(\mathbb{L}f\right)(x) = -h x^{1-(\lambda+1)/h} \frac{d}{dx} x^{(\lambda+1)/h}$$

$$\cdot \int_0^\infty H_{1,2}^{2,0}\left[xt \,\middle|\, \begin{matrix} (-\lambda, h) \\ (-\lambda - 1, h), (0, 1) \end{matrix}\right] f(t) dt \quad (x > 0). \tag{8.10.28}$$

Proof. The corollary follows from Theorem 8.40(c) if we take into account (8.10.9) and Property 2.2 of the H-function in Section 2.1.

Now we present the $\mathfrak{L}_{\nu,r}$-theory of the Bessel type transforms $\mathbf{K}_\eta^\rho$ and $\mathbf{L}_\eta^{(m)}$. First we characterize the boundedeness, range and representation of the Bessel type transform (8.10.1).

Theorem 8.41. *Let $\rho \in \mathbb{R}_+$, $\eta \in \mathbb{C}$ and $\nu < 1 + \min[0, \mathrm{Re}(\eta)]$.*

(a) *If $1 \leqq r \leqq s \leqq \infty$, then the transform $\mathbf{K}_\eta^\rho$ defined on $\mathfrak{L}_{\nu,2}$ can be extended to $\mathfrak{L}_{\nu,r}$ as an element of $[\mathfrak{L}_{\nu,r}, \mathfrak{L}_{1-\nu,s}]$. When $1 \leqq r \leqq 2$, then $\mathbf{K}_\eta^\rho$ is a one-to-one transform from $\mathfrak{L}_{\nu,r}$ onto $\mathfrak{L}_{1-\nu,s}$.*

(b) *If $1 \leqq r \leqq s \leqq \infty$, $f \in \mathfrak{L}_{\nu,r}$ and $g \in \mathfrak{L}_{\nu,s'}$ with $1/s + 1/s' = 1$, then the relation (8.10.20) holds.*

(c) *Let $1 < r < \infty$ and let $I^\alpha_{-;\sigma,\eta}$ and $\mathfrak{L}_{k,\alpha}$ be the Erdélyi–Kober type operator (3.1.1) and the generalized Laplace operator (3.3.3), respectively. Then*

$$\mathbf{K}_\eta^\rho(\mathfrak{L}_{\nu,r}) = \mathbb{L}_{(\rho+1)/\rho,(\rho-2\eta)/2(\rho+1)}(\mathfrak{L}_{\nu,r}), \tag{8.10.29}$$

if $\mathrm{Re}(\eta) \leqq -1/2$ or $\mathrm{Re}(\eta) \geqq \rho/2$;

$$\mathbf{K}_\eta^\rho(\mathfrak{L}_{\nu,r}) = \left(I^{1/2-\eta/\rho}_{-;\rho/(\rho+1),0}\mathbb{L}_{1+1/\rho,0}\right)(\mathfrak{L}_{\nu,r}), \tag{8.10.30}$$

if $0 \leqq \mathrm{Re}(\eta) < \rho/2$, and

$$\mathbf{K}_\eta^\rho(\mathfrak{L}_{\nu,r}) = \left(I^{1/2-\eta/\rho+(1+1/\rho)\mathrm{Re}(\eta)}_{-;\rho/(\rho+1),(1+1/\rho)\mathrm{Re}(\eta)}\mathbb{L}_{1+1/\rho,-\mathrm{Re}(\eta)}\right)(\mathfrak{L}_{\nu,r}), \tag{8.10.31}$$

if $-1/2 < \mathrm{Re}(\eta) < 0$.

(d) *If $f \in \mathfrak{L}_{\nu,r}$, where $1 < r < \infty$ and $0 < \nu < 1$, and $\mathrm{Re}(\eta) > 0$, then for almost all $x > 0$ the following relation holds:*

$$\left(\mathbf{K}_\eta^\rho f\right)(x) = -\frac{d}{dx}\int_0^\infty \left[Z_\rho^{\eta+1}(xt) - \frac{1}{\rho}\Gamma\left(\frac{\eta+1}{\rho}\right)\right] f(t)\frac{dt}{t}. \tag{8.10.32}$$

Proof. The assertion (a) follows from Theorems 4.5(a). The assertions (b) and (c) follow from Theorem 4.5(b) and Theorem 4.7(b)(c), respectively, with $\omega = \eta/\rho - (1 + 1/\rho)\min[0, \mathrm{Re}(\eta)] - 1/2$ in the latter.

To prove (d) we need preliminary results for two auxiliary functions of the forms

$$\chi_x(t) = \begin{cases} 1, & \text{if } 0 < t < x; \\ 0, & \text{if } t > x, \end{cases} \tag{8.10.33}$$

and

$$h_{\eta,x}(t) = \frac{1}{t}\left[Z_\rho^{\eta+1}(xt) - \frac{1}{\rho}\Gamma\left(\frac{\eta+1}{\rho}\right)\right]. \tag{8.10.34}$$

Lemma 8.16. *Let $1 < r < \infty$, $\rho \in \mathbb{R}_+$ and $\eta \in \mathbb{C}$. The following statements hold.*

(a) *$\chi_x(t) \in \mathfrak{L}_{\nu,r}$ if and only if $\nu > 0$.*

(b) *$h_{\eta,x}(t) \in \mathfrak{L}_{\nu,r}$ if and only if $\nu > 0$ for $\mathrm{Re}(\eta) > -1$, $\nu > 1$ for $\mathrm{Re}(\eta) = -1$ and $\mathrm{Im}(\eta) \neq -1$ or $\eta = -1$, and $\nu > -\mathrm{Re}(\eta)$ for $\mathrm{Re}(\eta) < -1$.*

(c) *Further, if $\mathrm{Re}(\eta) > 0$, then*

$$\left(\mathbf{K}_{\eta}^{\rho}\chi_x\right)(t) = -h_{\eta,x}(t). \tag{8.10.35}$$

Proof. The assertions (a) and (b) are proved similarly to those in Lemma 8.1(a)(b) on the basis of the asymptotic estimates of $Z_{\rho}^{\eta}(x)$ at zero and infinity (Krätzel [5]):

$$Z_{\rho}^{\eta}(x) \sim \begin{cases} \dfrac{1}{\rho}\Gamma\left(\dfrac{\eta}{\rho}\right), & \text{if } \mathrm{Re}(\eta) > 0; \\[2mm] \dfrac{1}{\rho}\Gamma\left(\dfrac{\eta}{\rho}\right) + \Gamma(-\eta)x^{\eta}, & \text{if } \mathrm{Re}(\eta) = 0,\ \mathrm{Im}(\eta) \neq 0; \\[2mm] -\log x, & \text{if } \eta = 0; \\[2mm] \Gamma(-\eta)x^{\eta}, & \text{if } \mathrm{Re}(\eta) < 0 \end{cases} \tag{8.10.36}$$

as $x \to +0$, and

$$Z_{\eta}^{\rho}(x) \sim \gamma x^{(2\eta-\rho)/(2\rho+2)} \exp\left[-\beta x^{\rho/(\rho+1)}\right], \tag{8.10.37}$$

$$\text{with} \quad \gamma = \left(\frac{2\pi}{\rho+1}\right)\rho^{-(2\eta+1)/(2\rho+2)}, \qquad \beta = \left(1 + \frac{1}{\rho}\right)^{1/(\rho+1)},$$

as $x \to +\infty$. Further we note the formula in Kilbas and Shlapakov [2, (22)]

$$\left(-\frac{d}{dz}\right)^m Z_{\rho}^{\eta}(z) = Z_{\rho}^{\eta-m}(z) \quad (m = 1, 2, \cdots). \tag{8.10.38}$$

Taking into account the first asymptotic estimate in (8.10.36) and the relation (8.10.34), we have

$$\begin{aligned} \left(\mathbf{K}_{\eta}^{\rho}\chi_x\right)(t) &= \int_0^{\infty} Z_{\rho}^{\eta}(t\tau)\chi_x(\tau)d\tau = \int_0^{x} Z_{\rho}^{(\eta)}(t\tau)d\tau \\ &= \frac{1}{t}\int_0^{xt} Z_{\rho}^{\eta}(u)du = \frac{1}{t}\int_0^{xt} \frac{d}{du}Z_{\rho}^{\eta+1}(u)du \\ &= -\frac{1}{t}\left[Z_{\rho}^{\eta+1}(xt) - \frac{1}{\rho}\Gamma\left(\frac{\eta+1}{\rho}\right)\right] = -h_{\eta,x}(t), \end{aligned}$$

and thus (8.10.35) is shown, which completes the proof of Lemma 8.16.

Returning to the proof of the assertion (d) of Theorem 8.41, we find from Lemma 8.16(a) that $\chi_x \in \mathfrak{L}_{\nu,r'}$ and we can appply Theorem 8.41(b) with $s = r$. Then (8.10.33), (8.10.20), (8.10.35), (8.10.38) (with $m = 1$) and (8.10.34) yield

$$\begin{aligned} \int_0^{x} \left(\mathbf{K}_{\eta}^{\rho}f\right)(t)dt &= \int_0^{\infty} \chi_x(t)\left(\mathbf{K}_{\eta}^{\rho}f\right)(t)dt = \int_0^{\infty} \left(\mathbf{K}_{\eta}^{\rho}\chi_x\right)(t)f(t)dt \\ &= -\int_0^{\infty} h_{\eta,x}(t)f(t)dt = -\int_0^{\infty} \frac{1}{t}\left[Z_{\rho}^{\eta+1}(xt) - \frac{1}{\rho}\Gamma\left(\frac{\eta+1}{\rho}\right)\right] f(t)dt, \end{aligned}$$

and the resullt in (8.10.32) follows on differentiation. Thus Theorem 8.41 is proved.

Now we present the results characterizing the boundedeness and the range of the Bessel type transform (8.10.2) in the space $\mathfrak{L}_{\nu,r}$.

Theorem 8.42. *Let* $m \in \mathbb{N}$, $\mathrm{Re}(\eta) > 1/m - 1$, *and* $\nu < 1 + \min[0, m\mathrm{Re}(\eta)]$.

(a) *If* $1 \leqq r \leqq s \leqq \infty$, *then the transform* $\mathbf{L}_\eta^{(m)}$ *defined on* $\mathfrak{L}_{\nu,2}$ *can be extended to* $\mathfrak{L}_{\nu,r}$ *as an element of* $[\mathfrak{L}_{\nu,r}, \mathfrak{L}_{1-\nu,s}]$. *If* $1 \leqq r \leqq 2$ *or if* $1 < r < \infty$ *and the condition* (8.10.18) *is satisfied, then* $\mathbf{L}_\eta^{(m)}$ *is a one-to-one transform from* $\mathfrak{L}_{\nu,r}$ *onto* $\mathfrak{L}_{1-\nu,s}$.

(b) *If* $1 \leqq r \leqq s \leqq \infty$, $f \in \mathfrak{L}_{\nu,r}$ *and* $g \in \mathfrak{L}_{\nu,s'}$ *with* $1/s + 1/s' = 1$, *then the relation* (8.10.24) *holds.*

(c) *Let* $1 < r < \infty$ *and let* $I^{\alpha}_{-;\sigma,\eta}$ *and* $\mathbb{L}_{k,\alpha}$ *be the Erdélyi–Kober type operator* (3.1.1) *and the generalized Laplace operator* (3.3.3), *respectively. If the condition* (8.10.18) *is satisfied, then*

$$\mathbf{L}_\eta^{(m)}(\mathfrak{L}_{\nu,r}) = \mathbb{L}_{1,(1-m)(\eta-1/m)}(\mathfrak{L}_{\nu,r}), \tag{8.10.39}$$

if $\mathrm{Re}(\eta) \geqq 1/m$;

$$\mathbf{L}_\eta^{(m)}(\mathfrak{L}_{\nu,r}) = \left(I^{(1-m)(\eta-1/m)}_{-;1,0}\mathbb{L}_{1,0}\right)(\mathfrak{L}_{\nu,r}), \tag{8.10.40}$$

if $0 \leqq \mathrm{Re}(\eta) < 1/m$; *and*

$$\mathbf{L}_\eta^{(m)}(\mathfrak{L}_{\nu,r}) = \left(I^{\eta+1-1/m}_{-;1,m\eta}\mathbb{L}_{1,-m\eta}\right)(\mathfrak{L}_{\nu,r}), \tag{8.10.41}$$

if $1/m - 1 < \mathrm{Re}(\eta) < 0$.

If the condition (8.10.18) *is not satisfied, then* $\mathbf{L}_\eta^{(m)}(\mathfrak{L}_{\nu,r})$ *is a subset of the right-hand sides of* (8.10.39), (8.10.40) *and* (8.10.41) *in the cases when* $\mathrm{Re}(\eta) \geqq 1/m$, $0 \leqq \mathrm{Re}(\eta) < 1/m$ *and* $1/m - 1 < \mathrm{Re}(\eta) < 0$, *respectively.*

(d) *If* $f \in \mathfrak{L}_{\nu,r}$, *where* $1 < r < \infty$ *and* $\nu > m - 1$, *and* $\mathrm{Re}(\eta) > 0$, *then for almost all* $x > 0$ *the following relation holds*:

$$\left(\mathbf{L}_\eta^{(m)} f\right)(x) = -\left(\frac{x}{m}\right)^{1-m} \frac{d}{dx}\int_0^\infty \left[\lambda_{\eta+1}^{(m)}(xt)f(t) - \prod_{k=0}^{m-2}\Gamma\left(\eta+1+\frac{k}{m}\right)\right]\frac{dt}{t^m}. \tag{8.10.42}$$

Proof. The assertions (a), (b) and (c) follow from Theorems 4.5 and Theorem 4.7 with $\omega = \eta(m-1) + 1/m - 1 - \min[0, m\mathrm{Re}(\eta)]$.

In order to show (d) we prepare a lemma for the auxiliary functions:

$$g_{m,x}(t) = \begin{cases} t^{m-1}, & \text{if } 0 < t < x; \\ 0, & \text{if } t > x \end{cases} \tag{8.10.43}$$

and

$$h_{\eta,x}(t) = m^{m-1}t^{-m}\left[\lambda_{\eta+1}^{(m)}(xt) - \prod_{k=0}^{m-2}\Gamma\left(\eta+1+\frac{k}{m}\right)\right]. \tag{8.10.44}$$

Lemma 8.17. *Let $1 < r < \infty$, $m = 2, 3, \cdots$ and $\mathrm{Re}(\eta) > 1/m - 1$. Then the following statements hold:*

(a) $g_{m,x}(t) \in \mathfrak{L}_{\nu,r}$ *if and only if* $\nu > 1 - m$.

(b) $h_{\eta,x}(t) \in \mathfrak{L}_{\nu,r}$ *if and only if* $\nu > m - 1$.

(c) *Further,*

$$\left(\mathbf{L}_{\eta}^{(m)} g_{m,x}\right)(t) = -h_{\eta,x}(t). \tag{8.10.45}$$

Proof. The assertions (a) and (b) are proved similarly to those in Lemma 8.1(a),(b) on the basis of the asymptotic estimates of $\lambda_{\eta}^{(m)}(x)$ at zero and infinity (see Krätzel [1, (8a) and (7)]):

$$\lambda_{\eta}^{(m)}(x) \sim A_{\eta} \quad \text{with} \quad A_{\eta} = \prod_{k=0}^{m-2} \Gamma\left(\eta + \frac{k}{m}\right) \quad (\mathrm{Re}(\eta) > 0) \tag{8.10.46}$$

as $x \to +0$, and

$$\lambda_{\eta}^{(m)}(x) \sim B x^{(m-1)\eta - 1 + 1/m} e^{-x}, \qquad B = (2\pi)^{(m-1)/2} m^{(1-m)\eta + 1/2 - 1/m} \tag{8.10.47}$$

as $x \to +\infty$ (see Krätzel [1, § 3]). To prove (c), by using the formula in Krätzel [1, (5)]

$$\frac{d}{dz}\lambda_{\eta}^{(m)}(z) = -\left(\frac{z}{m}\right)^{m-1} \lambda_{\eta-1}^{(m)}(z), \tag{8.10.48}$$

and by the asymptotic estimate (8.10.46) and the relation (8.10.44), we have

$$\begin{aligned}
\left(\mathbf{L}_{\eta}^{(m)} g_{m,x}\right)(t) &= \int_0^{\infty} \lambda_{\eta}^{(m)}(t\tau) g_{m,x}(\tau) d\tau = \int_0^{x} \lambda_{\eta}^{(m)}(t\tau) \tau^{m-1} d\tau \\
&= t^{-m} \int_0^{xt} \lambda_{\eta}^{(m)}(u) u^{m-1} du = -m^{m-1} t^{-m} \int_0^{xt} \frac{d}{du}\left[\lambda_{\eta+1}^{(m)}(u)\right] du \\
&= -m^{m-1} t^{-m} \left[\lambda_{\eta+1}^{(m)}(xt) - A_{\eta+1}\right] = -h_{\eta,x}(t),
\end{aligned}$$

and (8.10.45) is proved as well as Lemma 8.17.

Thus the assertion (d) of Theorem 8.42 follows. In fact, by Lemma 8.17(a), we have $g_{m,x} \in \mathfrak{L}_{\nu,r'}$, and we can appply Theorem 8.42(b) with $s = r$. Using (8.10.43), (8.10.24), (8.10.45), (8.10.48) and (8.10.44), we have

$$\begin{aligned}
\int_0^{x} t^{m-1} \left(\mathbf{L}_{\eta}^{(m)} f\right)(t) dt &= \int_0^{\infty} g_{m,x}(t) \left(\mathbf{L}_{\eta}^{(m)} f\right)(t) dt = \int_0^{\infty} \left(\mathbf{L}_{\eta}^{(m)} g_{m,x}\right)(t) f(t) dt \\
&= -\int_0^{\infty} h_{\eta,x}(t) f(t) dt \\
&= -m^{m-1} \int_0^{\infty} \left[\lambda_{\eta+1}^{(m)}(xt) - \prod_{k=0}^{m-2} \Gamma\left(\eta + 1 + \frac{k}{m}\right)\right] f(t) \frac{dt}{t^m},
\end{aligned}$$

and after differentiation we obtain (8.10.42). This completes the proof of Theorem 8.42.

8.11. The Modified Bessel Type Transform

We consider the integral transform

$$\left(\boldsymbol{L}_{\eta,\sigma}^{(\beta)}f\right)(x) = \int_0^\infty \lambda_{\eta,\sigma}^{(\beta)}(xt)f(t)dt \quad (x > 0) \tag{8.11.1}$$

with the kernel

$$\lambda_{\eta,\sigma}^{(\beta)}(z) = \frac{\beta}{\Gamma\left(\eta+1-\dfrac{1}{\beta}\right)}\int_1^\infty (t^\beta - 1)^{\eta-1/\beta}t^\sigma e^{-zt}dt \tag{8.11.2}$$

$$\left(\beta > 0;\ \sigma \in \mathbb{R};\ \mathrm{Re}(\eta) > \frac{1}{\beta} - 1\right)$$

being an analytic function of $z \in \mathbb{C}$ for $\mathrm{Re}(z) > 0$.

When $\sigma = 0$,

$$\lambda_{\eta,0}^{(\beta)}(z) = \frac{\beta}{\Gamma\left(\eta+1-\dfrac{1}{\beta}\right)}\int_1^\infty (t^\beta - 1)^{\eta-1/\beta}e^{-zt}dt \tag{8.11.3}$$

$$\left(\beta > 0;\ \mathrm{Re}(\eta) > \frac{1}{\beta} - 1\right)$$

and, hence, for $\beta = m \in \mathbb{N}$ the transform (8.11.1) is reduced to the transform (8.10.2):

$$\left(\boldsymbol{L}_{\eta,0}^{(m)}f\right)(x) = (2\pi)^{(1-m)/2}m^{m\eta+1/2}x^{-m\eta}\left(\boldsymbol{L}_\eta^{(m)}t^{\eta m}f\right)(x). \tag{8.11.4}$$

So, the transform $\boldsymbol{L}_{\eta,\sigma}^{(\beta)}$ is called the modified Bessel type transform. We note that, when $\sigma = 0$ and $\beta = 2$, by the relation (8.10.8), (8.11.1) is the Meijer type transform discussed in Section 8.9.

The $\mathfrak{L}_{\nu,r}$-theory of the modified Bessel type transform (8.11.1) is based on the Mellin transform of the function (8.11.2).

The following lemma can be obtained similarly to Lemma 8.15.

Lemma 8.18. *Let* $\beta > 0$, $\sigma \in \mathbb{R}$, *and* $\mathrm{Re}(s) > \max[0, \beta\mathrm{Re}(\eta) + \sigma]$. *Then*

$$\left(\mathfrak{M}\lambda_{\eta,\sigma}^{(\beta)}\right)(s) = \frac{\Gamma(s)\Gamma\left(-\eta - \dfrac{\sigma}{\beta} + \dfrac{s}{\beta}\right)}{\Gamma\left(1 - \dfrac{\sigma+1}{\beta} + \dfrac{s}{\beta}\right)}. \tag{8.11.5}$$

Now we characterize the $\mathfrak{L}_{\nu,r}$-properties of the modified Bessel type transform (8.11.1). The Mellin transform of $\lambda_{\eta,\sigma}^{(\beta)}(t)$ in (8.11.5) is written as

$$\left(\mathfrak{M}\lambda_{\eta,\sigma}^{(\beta)}\right)(s) = \mathcal{H}_{1,2}^{2,0}\left[\begin{array}{c|c} \left(1 - \dfrac{\sigma+1}{\beta}, \dfrac{1}{\beta}\right) & \\ (0,1), \left(-\eta - \dfrac{\sigma}{\beta}, \dfrac{1}{\beta}\right) & s \end{array}\right], \tag{8.11.6}$$

and hence the transform $\boldsymbol{L}_{\eta,\sigma}^{(\beta)}$ is a special $\boldsymbol{H}$-transform:

$$\left(\boldsymbol{L}_{\eta,\sigma}^{(\beta)} f\right)(x) = \int_0^\infty H_{1,2}^{2,0}\left[xt \left| \begin{matrix} \left(1-\frac{\sigma+1}{\beta}, \frac{1}{\beta}\right) \\ (0,1), \left(-\eta-\frac{\sigma}{\beta}, \frac{1}{\beta}\right) \end{matrix} \right.\right] f(t)dt \quad (x>0). \tag{8.11.7}$$

The parameters defined in (3.4.1), (3.4.2), (1.1.7), (1.1.8), (1.1.10), (1.1.11) and (1.1.12) are given with the suffix 0 in the forms

$$\begin{gathered} \alpha_0 = \max[0, \beta\mathrm{Re}(\eta)+\sigma], \quad \beta_0 = +\infty, \\ a_0^* = \Delta_0 = a_{1,0}^* = 1, \quad a_{2,0}^* = 0, \quad \mu_0 = \frac{1}{\beta} - \eta - \frac{3}{2} \end{gathered} \tag{8.11.8}$$

for the function $\mathcal{H}_{1,2}^{2,0}(s)$ in (8.11.6).

Let $\mathcal{E}_{\mathcal{H}}$ be the exceptional set of this function (cf. Definition 3.4). Then ν is not in the exceptional set $\mathcal{E}_{\mathcal{H}}$, if

$$\nu \neq k\beta - \sigma \quad (k \in \mathbb{N}). \tag{8.11.9}$$

Applying Theorems 3.6 and 3.7, we obtain the following results for the modified Bessel type transform (8.11.1) in the space $\mathfrak{L}_{\nu,2}$.

Theorem 8.43. *Let $\beta > 0$, $\sigma \in \mathbb{R}$ and $\mathrm{Re}(\eta) > 1/\beta - 1$ be such that $\nu < 1 - \max[0, \beta\mathrm{Re}(\eta)+\sigma]$.*

(i) *There is a one-to-one transform $\boldsymbol{L}_{\eta,\sigma}^{(\beta)} \in [\mathfrak{L}_{\nu,2}, \mathfrak{L}_{1-\nu,2}]$ such that the relation*

$$\left(\mathfrak{M}\boldsymbol{L}_{\eta,\sigma}^{(\beta)} f\right)(s) = \frac{\Gamma(s)\Gamma\left(-\eta-\frac{\sigma}{\beta}+\frac{s}{\beta}\right)}{\Gamma\left(1-\frac{\sigma+1}{\beta}+\frac{s}{\beta}\right)} \left(\mathfrak{M}f\right)(1-s) \tag{8.11.10}$$

holds for $\mathrm{Re}(s) = 1-\nu$ and $f \in \mathfrak{L}_{\nu,2}$.

(ii) *If $f, g \in \mathfrak{L}_{\nu,2}$, then*

$$\int_0^\infty \left(\boldsymbol{L}_{\eta,\sigma}^{(\beta)} f\right)(x) g(x)dx = \int_0^\infty f(x)\left(\boldsymbol{L}_{\eta,\sigma}^{(\beta)} g\right)(x)dx. \tag{8.11.11}$$

(iii) *Let $f \in \mathfrak{L}_{\nu,2}$, $\lambda \in \mathbb{C}$ and $h \in \mathbb{R}_+$. If $\mathrm{Re}(\lambda) > (1-\nu)h - 1$, then $\boldsymbol{L}_{\eta,\sigma}^{(\beta)} f$ is given by*

$$\left(\boldsymbol{L}_{\eta,\sigma}^{(\beta)} f\right)(x) = hx^{1-(\lambda+1)/h}\frac{d}{dx}x^{(\lambda+1)/h}$$

$$\cdot \int_0^\infty H_{2,3}^{2,1}\left[xt \left| \begin{matrix} (-\lambda,h), \left(1-\frac{\sigma+1}{\beta}, \frac{1}{\beta}\right) \\ (0,1), \left(-\eta-\frac{\sigma}{\beta}, \frac{1}{\beta}\right), (-\lambda-1,h) \end{matrix} \right.\right] f(t)dt \quad (x>0). \tag{8.11.12}$$

When $\mathrm{Re}(\lambda) < (1-\nu)h - 1$,

$$\left(\boldsymbol{L}_{\eta,\sigma}^{(\beta)} f\right)(x) = -hx^{1-(\lambda+1)/h}\frac{d}{dx}x^{(\lambda+1)/h}$$

$$\cdot \int_0^\infty H_{2,3}^{3,0}\left[xt \left| \begin{matrix} \left(1-\frac{\sigma+1}{\beta}, \frac{1}{\beta}\right), (-\lambda,h) \\ (-\lambda-1,h), (0,1), \left(-\eta-\frac{\sigma}{\beta}, \frac{1}{\beta}\right) \end{matrix} \right.\right] f(t)dt \quad (x>0). \tag{8.11.13}$$

(iv) *$\boldsymbol{L}_{\eta,\sigma}^{(\beta)}$ is independent of ν in the sense that, if ν and $\widetilde{\nu}$ are such that* $\max[\nu,\widetilde{\nu}] < 1 - \max[0, \beta\mathrm{Re}(\eta) + \sigma]$ *and if the transforms $\boldsymbol{L}_{\eta,\sigma}^{(\beta)}$ and $\widetilde{\boldsymbol{L}}_{\eta,\sigma}^{(\beta)}$ are given on $\mathfrak{L}_{\nu,2}$ and $\mathfrak{L}_{\widetilde{\nu},2}$, respectively in* (8.11.10), *then $\boldsymbol{L}_{\eta,\sigma}^{(\beta)}f = \widetilde{\boldsymbol{L}}_{\eta,\sigma}^{(\beta)}f$ for $f \in \mathfrak{L}_{\nu,2}\bigcap\mathfrak{L}_{\widetilde{\nu},2}$.*

(v) *$\boldsymbol{L}_{\eta,\sigma}^{(\beta)}f$ is given in* (8.11.1) *and* (8.11.7) *for $f \in \mathfrak{L}_{\nu,2}$.*

Now we proceed to characterize the mapping properties of the modified Bessel type transform (8.11.1) in the space $\mathfrak{L}_{\nu,r}$ owing to Theorems 4.5 and 4.7. First we prepare a lemma on the characteristic function $\chi_x(t)$ in (8.10.33) and

$$g_{\eta,x}(t) = \frac{1}{t}\left[\lambda_{\eta,\sigma-1}^{(\beta)}(xt) - \frac{\Gamma\left(-\eta - \dfrac{\sigma-1}{\beta}\right)}{\Gamma\left(1 - \dfrac{\sigma}{\beta}\right)}\right]. \tag{8.11.14}$$

Lemma 8.19. *Let $1 < r < \infty$ and let $\beta > 0$, $\sigma \in \mathbb{C}$ and $\eta \in \mathbb{C}$ be such that $1/\beta - 1 < \mathrm{Re}(\eta) < (1-\sigma)/\beta$. Then the following statements hold:*

(a) *$\chi_x(t) \in \mathfrak{L}_{\nu,r}$ if and only if $\nu > 0$.*

(b) *$g_{\eta,x}(t) \in \mathfrak{L}_{\nu,r}$ if and only if $\nu > 0$.*

(c) *Further,*

$$\mathbf{L}_{\eta,\sigma}^{(\beta)}\chi_x = -g_{\eta,x}. \tag{8.11.15}$$

Proof. The assertion (a) coincides with that in Lemma 8.15(a). (b) is proved similarly to that in Lemma 8.1(b) on the basis of the asymptotic estimates of $\lambda_{\eta,\sigma}^{(\beta)}(x)$ at zero and infinity (see Glaeske, Kilbas and Saigo [1]):

$$\lambda_{\eta,\sigma}^{(\beta)}(x) \sim A \quad \text{with} \quad A = \frac{\Gamma\left(-\eta - \dfrac{\sigma}{\beta}\right)}{\Gamma\left(1 - \dfrac{\sigma+1}{\beta}\right)} \quad \text{for} \quad \mathrm{Re}(\eta) < -\frac{\sigma}{\beta} \tag{8.11.16}$$

as $x \to +0$, and

$$\lambda_{\eta,\sigma}^{(\beta)}(z) \sim Be^{-x}x^{(1/\beta)-\eta-1} \quad \text{with} \quad B = \beta^{\eta+1-1/\beta} \quad \text{for} \quad \mathrm{Re}(\eta) > \frac{1}{\beta} - 1 \tag{8.11.17}$$

as $x \to +\infty$. As for (c), using the directly verified relation

$$\left(\frac{d}{dz}\right)^m \lambda_{\eta,\sigma}^{(\beta)}(z) = (-1)^m\lambda_{\eta,\sigma+m}^{(\beta)}(z) \quad (m = 1, 2, \cdots) \tag{8.11.18}$$

and the asymptotic estimate (8.11.16), we have

$$\begin{aligned}\left(\mathbf{L}_{\eta,\sigma}^{(\beta)}\chi_x\right)(t) &= \int_0^\infty \lambda_{\eta,\sigma}^{(\beta)}(t\tau)\chi_x(\tau)d\tau = \int_0^x \lambda_{\eta,\sigma}^{(\beta)}(t\tau)d\tau \\ &= \frac{1}{t}\int_0^{xt} \lambda_{\eta,\sigma}^{(\beta)}(u)du = -\frac{1}{t}\int_0^{xt} \frac{d}{du}\lambda_{\eta,\sigma-1}^{(\beta)}(u)du \\ &= -\frac{1}{t}\left[\lambda_{\eta,\sigma-1}^{(\beta)}(xt) - \frac{\Gamma\left(-\eta - \dfrac{\sigma-1}{\beta}\right)}{\Gamma\left(1 - \dfrac{\sigma}{\beta}\right)}\right].\end{aligned}$$

So the relation (8.11.15) is proved as well as Lemma 8.19.

Now we can state the $\mathfrak{L}_{\nu,r}$-theory of the transform $\boldsymbol{L}_{\eta,\sigma}^{(\beta)}$.

Theorem 8.44. *Let $\beta > 0$, $\sigma \in \mathbb{R}$ and $\eta \in \mathbb{C}$ be such that $\mathrm{Re}(\eta) > 1/\beta - 1$ and $\nu < 1 - \max[0, \beta\mathrm{Re}(\eta) + \sigma]$.*

(a) *If $1 \leqq r \leqq s \leqq \infty$, then the transform $\boldsymbol{L}_{\eta,\sigma}^{(\beta)}$ defined on $\mathfrak{L}_{\nu,2}$ can be extended to $\mathfrak{L}_{\nu,r}$ as an element of $[\mathfrak{L}_{\nu,r}, \mathfrak{L}_{1-\nu,s}]$. If $1 \leqq r \leqq 2$ or if $1 < r < \infty$ and the condition in (8.11.9) is satisfied, then $\boldsymbol{L}_{\eta,\sigma}^{(\beta)}$ is a one-to-one transform from $\mathfrak{L}_{\nu,r}$ onto $\mathfrak{L}_{1-\nu,s}$.*

(b) *If $1 \leqq r \leqq s \leqq \infty$, $f \in \mathfrak{L}_{\nu,r}$ and $g \in \mathfrak{L}_{\nu,s'}$ with $1/s + 1/s' = 1$, then the relation (8.11.11) holds.*

(c) *Let $1 < r < \infty$ and $\omega = \max[0, \beta\mathrm{Re}(\eta) + \sigma] - \eta - 1 + 1/\beta$. If the condition in (8.11.9) is satisfied, then*

$$\boldsymbol{L}_{\eta,\sigma}^{(\beta)}(\mathfrak{L}_{\nu,r}) = \mathbb{L}_{1,\alpha_0-\omega}(\mathfrak{L}_{\nu,r}), \tag{8.11.19}$$

when $\mathrm{Re}(\omega) \geqq 0$, and

$$\boldsymbol{L}_{\eta,\sigma}^{(\beta)}(\mathfrak{L}_{\nu,r}) = \left(I_{-;1,-\alpha_0}^{-\omega}\mathbb{L}_{1,\alpha_0}\right)(\mathfrak{L}_{\nu,r}), \tag{8.11.20}$$

when $\mathrm{Re}(\omega) < 0$, where α_0 is given in (8.11.8). If the condition (8.11.9) is not satisfied, then $\mathbf{L}_{\eta,\sigma}^{(\beta)}(\mathfrak{L}_{\nu,r})$ is a subset of the right-hand sides of (8.11.19) and (8.11.20) in the respective cases when $\mathrm{Re}(\omega) \geqq 0$ and $\mathrm{Re}(\omega) < 0$.

(d) *Let $1 < r < \infty$, $\nu > 0$ and $\mathrm{Re}(\eta) < (1-\sigma)/\beta$, then for $f \in \mathfrak{L}_{\nu,r}$ the relation*

$$\left(\boldsymbol{L}_{\eta,\sigma}^{(\beta)}f\right)(x) = -\frac{d}{dx}\int_0^\infty \left[\lambda_{\eta,\sigma-1}^{\beta}(xt) - \frac{\Gamma\left(-\eta - \dfrac{\sigma-1}{\beta}\right)}{\Gamma\left(1 - \dfrac{\sigma}{\beta}\right)}\right] f(t)\frac{dt}{t} \tag{8.11.21}$$

holds for almost all $x > 0$.

Proof. The assertion (a)–(c) can be obtained by appealing to Theorems 4.5 and 4.7.

For the assertion (d), since $\nu > 0$ and by Lemma 8.19(a), $\chi_x \in \mathfrak{L}_{\nu,r'}$ and we can apply Theorem 8.44(b) with $s = r$. Using (8.10.33), (8.11.11), (8.11.15) and (8.11.14), we have

$$\begin{aligned}\int_0^x \left(\boldsymbol{L}_{\eta,\sigma}^{(\beta)}f\right)(t)dt &= \int_0^\infty \chi_x(t)\left(\boldsymbol{L}_{\eta,\sigma}^{(\beta)}f\right)(t)dt \\ &= \int_0^\infty \left(\boldsymbol{L}_{\eta,\sigma}^{(\beta)}\chi_x\right)(t)f(t)dt = -\int_0^\infty g_{\eta,x}(t)f(t)dt \\ &= -\int_0^\infty \frac{1}{t}\left[\lambda_{\eta,\sigma-1}^{(\beta)}(xt) - \frac{\Gamma\left(-\eta - \dfrac{\sigma-1}{\beta}\right)}{\Gamma\left(1 - \dfrac{\sigma}{\beta}\right)}\right] f(t)dt,\end{aligned}$$

and the resullt in (8.11.21) follows on differentiation. Theorem 8.44 is proved.

8.12. The Generalized Hardy–Titchmarsh Transform

Let $a, b, c, \omega \in \mathbb{C}$ with $\mathrm{Re}(a) > 0$. We consider the integral transform

$$\left(\mathbb{J}_{a,b,c;\omega} f\right)(x) = \int_0^\infty J_{a,b,c;\omega}(xt) f(t)dt \quad (x > 0) \tag{8.12.1}$$

with the kernel

$$J_{a,b,c;\omega}(z) = \frac{\Gamma(a)}{\Gamma(b)\Gamma(c)} z^\omega \, {}_1F_2\left(a; b, c; -\frac{z^2}{4}\right) \tag{8.12.2}$$

containing the hypergeometric function ${}_1F_2(a; b, c; x)$. When $a = 1$, $b = 3/2$ and $c = \omega = \eta + 3/2$, then in accordance with (8.7.27) $J_{1,3/2,3/2;\eta+3/2}(z) = 2^{\eta+1}\sqrt{z}\mathbf{H}_\eta(z)$ and this transform coincides with the transform $\mathfrak{H}_\eta$ in (8.8.1) (apart from the constant multiplication factor $2^{\eta+1}$):

$$\left(\mathbb{J}_{1,3/2,\eta+3/2;\eta+3/2} f\right)(x) = 2^{\eta+1}\left(\mathfrak{H}_\eta f\right)(x). \tag{8.12.3}$$

If $a = 1$, $b = \sigma + 1$, $c = \sigma + \eta + 1$ and $\omega = 2\sigma + \eta + 1/2$, (8.12.2) takes the form

$$J_{1,\sigma+1,\sigma+\eta+1;2\sigma+\eta+1/2}(z) = 2^{\eta+2\sigma}\sqrt{z} J_{\eta,\sigma}(z), \tag{8.12.4}$$

where $J_{\eta,\sigma}(z)$ is the Lommel function

$$J_{\eta,\sigma}(z) = \sum_{k=0}^\infty \frac{(-1)^k}{\Gamma(1+\sigma+k)\Gamma(1+\sigma+\eta+k)} \left(\frac{z}{2}\right)^{\eta+2k+2\sigma}$$
$$= \frac{2^{2-2\sigma-\eta}}{\Gamma(\sigma)\Gamma(\sigma+\eta)} s_{2\sigma+\eta-1,\eta}(z) \tag{8.12.5}$$

with $\sigma, \eta \in \mathbb{C}$ ($\mathrm{Re}(\sigma) > -1$, $\mathrm{Re}(\eta + \sigma) > -1$), where for the function $s_{\sigma,\eta}(z)$ one may refer to Erdélyi, Magnus, Oberhettinger and Tricomi [2, 7.5(69)]. The transform with such a kernel function

$$\left(\mathbb{J}_{\eta,\sigma} f\right)(x) = \int_0^\infty (xt)^{1/2} J_{\eta,\sigma}(xt) f(t)dt \quad (x > 0) \tag{8.12.6}$$

is known as the modified Hardy transform. When $a = \sigma + \eta + 1/2$, $b = \sigma + \eta + 1$ and $c = \omega = \sigma + 2\eta + 1$, (8.12.1) is known as the transform given by Titchmarsh [2]. This transform as well as (8.12.3) and (8.12.4) are particular cases of the more general transform (8.12.1) with $\omega = b + c - a - 1/2$, indicated by Titchmarsh [3, Section 8.4, Example 3]. Therefore we call the transform (8.12.1) the generalized Hardy–Titchmarsh transform.

When $\sigma = l = 0, 1, 2, \cdots$, then the modified Hardy transform (8.12.6) coincides with the extended Hankel transform (8.4.1) (apart from the constant multiplication factor $(-1)^l$):

$$\left(\mathbb{J}_{\eta,l} f\right)(x) = (-1)^l \left(\mathbb{H}_\eta^l f\right)(x). \tag{8.12.7}$$

In particular, when $\sigma = 0$, the Lommel function $J_{\eta,0}(z)$ in (8.12.5) coincides with the Bessel function $J_\eta(z)$, and then the modified Hardy transform (8.12.6) reduces to the Hankel transform (8.1.1):

$$\left(\mathbb{J}_{\eta,0} f\right)(x) = \left(\mathbb{H}_\eta f\right)(x). \tag{8.12.8}$$

The $\mathfrak{L}_{\nu,r}$-theory of the generalized Hardy–Titchmarsh transform (8.12.1) and of the modified Hardy transform (8.12.6) are based on the Mellin transforms of the function $J_{a,b,c;\omega}(x)$ and $x^{1/2}J_{\eta,\sigma}(x)$ in (8.12.2) and (8.12.5), respectively.

Lemma 8.20. **(a)** *Let* $a,b,c,\omega \in \mathbb{C}$ $(a \neq 0,-1,\cdots)$ *and* $s \in \mathbb{C}$ *be such that*

$$0 < \mathrm{Re}(\omega+s) < 2\mathrm{Re}(a), \quad \mathrm{Re}(\omega+s) < \frac{1}{2} + \mathrm{Re}(b+c-a). \tag{8.12.9}$$

Then

$$\left(\mathfrak{M}J_{a,b,c;\omega}(x)\right)(s) = \pi^{1/2}\frac{\Gamma(\omega+s)\Gamma\left(a-\frac{\omega}{2}-\frac{s}{2}\right)}{\Gamma\left(\frac{1}{2}+\frac{\omega}{2}+\frac{s}{2}\right)\Gamma\left(b-\frac{\omega}{2}-\frac{s}{2}\right)\Gamma\left(c-\frac{\omega}{2}-\frac{s}{2}\right)}. \tag{8.12.10}$$

(b) *Let* $\sigma,\eta \in \mathbb{C}$ $(\mathrm{Re}(\sigma) > -1,\ \mathrm{Re}(\eta+\sigma) > -1)$ *and* $s \in \mathbb{C}$ *be such that*

$$-\frac{1}{2} - \mathrm{Re}(\eta+2\sigma) < \mathrm{Re}(s) < \frac{3}{2} - \mathrm{Re}(\eta+2\sigma), \quad \mathrm{Re}(s) < 1. \tag{8.12.11}$$

Then

$$\left(\mathfrak{M}[x^{1/2}J_{\eta,\sigma}(x)]\right)(s)$$

$$= 2^{-\eta-2\sigma}\pi^{1/2}\frac{\Gamma\left(\frac{1}{2}+\eta+2\sigma+s\right)\Gamma\left(\frac{3}{4}-\frac{\eta}{2}-\sigma-\frac{s}{2}\right)}{\Gamma\left(\frac{3}{4}+\frac{\eta}{2}+\sigma+\frac{s}{2}\right)\Gamma\left(\frac{3}{4}-\frac{\eta}{2}-\frac{s}{2}\right)\Gamma\left(\frac{3}{4}+\frac{\eta}{2}-\frac{s}{2}\right)}. \tag{8.12.12}$$

Proof. It is known (see, for example, Prudnikov, Brychkov and Marichev [3, 8.4.48.1]) that the Mellin transform of the hypergeometric function ${}_1F_2(a;b,c;-x)$ is given by

$$\left(\mathfrak{M}\ {}_1F_2(a;b,c;-x)\right)(s) = \frac{\Gamma(b)\Gamma(c)}{\Gamma(a)}\frac{\Gamma(s)\Gamma(a-c)}{\Gamma(b-s)\Gamma(c-s)}, \tag{8.12.13}$$

provided that

$$0 < \mathrm{Re}(s) < \mathrm{Re}(a), \quad \mathrm{Re}(s) < \frac{1}{4} + \frac{\mathrm{Re}(b+c-a)}{2} \quad (b,c \neq 0,-1,-2,\cdots). \tag{8.12.14}$$

Using the elementary operators M_ξ, W_δ and N_a, in (3.3.11), (3.3.12) and (8.5.7), we may represent the function $J_{a,b,c;\omega}(x)$ in (8.12.2) in the form

$$J_{a,b,c;\omega}(x) = \frac{\Gamma(a)}{\Gamma(b)\Gamma(c)}M_\omega W_2 N_2\left\{{}_1F_2(a;b,c;-x)\right\}. \tag{8.12.15}$$

Applying (3.3.14), (3.3.15) and (8.5.9), we have

$$\left(\mathfrak{M}J_{a,b,c;\omega}(x)\right)(s) = \frac{\Gamma(a)}{\Gamma(b)\Gamma(c)}\left(\mathfrak{M}\left[M_\omega W_2 N_2\left\{{}_1F_2(a;b,c;-x)\right\}\right]\right)(s)$$

$$= \frac{\Gamma(a)}{\Gamma(b)\Gamma(c)}\left(\mathfrak{M}\left[W_2 N_2\left\{{}_1F_2(a;b,c;-x)\right\}\right]\right)(s+\omega)$$

$$= \frac{\Gamma(a)}{\Gamma(b)\Gamma(c)}2^{s+\omega}\left(\mathfrak{M}\left[N_2\left\{{}_1F_2(a;b,c;-x)\right\}\right]\right)(s+\omega)$$

$$= \frac{\Gamma(a)}{\Gamma(b)\Gamma(c)}2^{s+\omega-1}\left(\mathfrak{M}\left[{}_1F_2(a;b,c;-x)\right]\right)\left(\frac{s+\omega}{2}\right).$$

Now we can apply (8.12.13) to obtain the relation

$$\left(\mathfrak{M}J_{a,b,c;\omega}(x)\right)(s) = 2^{s+\omega-1}\frac{\Gamma\left(\frac{\omega}{2}+\frac{s}{2}\right)\Gamma\left(a-\frac{\omega}{2}-\frac{s}{2}\right)}{\Gamma\left(b-\frac{\omega}{2}-\frac{s}{2}\right)\Gamma\left(c-\frac{\omega}{2}-\frac{s}{2}\right)}. \tag{8.12.16}$$

Multiplying the numerator and denominator in the right-hand side of (8.12.16) by $\Gamma(\omega/2+s/2+1/2)$ and using the Legendre duplication formula (8.9.7), we obtain (8.12.10). We note that the first two conditions in (8.12.4) with s being repalcing by $(\omega+s)/2$ yield conditions in (8.12.9) while the conditions $b, c \neq 0, -1, -2, \cdots$ can be omitted by analytic continuation of both sides of (8.12.16) in b and c.

Putting $a=1$, $b=\sigma+1$, $c=\sigma+\eta+1$ and $\omega=2\sigma+\eta+1/2$ in (8.12.10) and taking into account (8.12.5), we arrive at (8.12.12) under the conditions in (8.12.11), which completes the proof of Lemma 8.19.

Remark 8.12. When $\sigma=0$, (8.12.12) takes the form

$$\left(\mathfrak{M}[x^{1/2}J_\eta(x)]\right)(s) = 2^{-\eta}\pi^{1/2}\frac{\Gamma\left(\frac{1}{2}+\eta+s\right)}{\Gamma\left(\frac{3}{4}+\frac{\eta}{2}+\frac{s}{2}\right)\Gamma\left(\frac{3}{4}+\frac{\eta}{2}-\frac{s}{2}\right)}. \tag{8.12.17}$$

Applying the Legendre duplication formula (8.9.7) for $\Gamma(1/2+\eta+s)$, we have

$$\left(\mathfrak{M}[x^{1/2}J_\eta(x)]\right)(s) = 2^{s-1/2}\frac{\Gamma\left(\frac{1}{4}+\frac{\eta}{2}+\frac{s}{2}\right)}{\Gamma\left(\frac{3}{4}+\frac{\eta}{2}-\frac{s}{2}\right)}. \tag{8.12.18}$$

This formula is well known (see, for example, Prudnikov, Brychkov and Marichev [2, (2.12.2.2)]).

Now we can apply the results in Chapters 3 and 4 to characterize the $\mathfrak{L}_{\nu,r}$-properties of the generalized Hardy–Titchmarsh transform (8.12.1) and the modified Hardy transform (8.12.6). If $b \neq a$ and $c \neq a$, then according to (8.12.10) and (1.1.2), $\left(\mathfrak{M}J_{a,b,c;\omega}(x)\right)(s)$ is represented by the function $\mathcal{H}$ in the form

$$\left(\mathfrak{M}J_{a,b,c;\omega}(x)\right)(s) = \pi^{1/2}\mathcal{H}_{2,3}^{1,1}\left[\begin{array}{c|c} \left(1-a+\frac{\omega}{2},\frac{1}{2}\right), \left(\frac{1}{2}+\frac{\omega}{2},\frac{1}{2}\right) & \\ (\omega,1), \left(1-b+\frac{\omega}{2},\frac{1}{2}\right), \left(1-c+\frac{\omega}{2},\frac{1}{2}\right) & s \end{array}\right] \tag{8.12.19}$$

and hence, in accordance with (3.1.5), the transform $\mathbb{J}_{a,b,c;\omega}$ is a special $\boldsymbol{H}$-transform:

$$\left(\mathbb{J}_{a,b,c;\omega}f\right)(x)$$

$$= \pi^{1/2}\int_0^\infty H_{2,3}^{1,1}\left[xt \,\middle|\, \begin{array}{c} \left(1-a+\frac{\omega}{2},\frac{1}{2}\right), \left(\frac{1}{2}+\frac{\omega}{2},\frac{1}{2}\right) \\ (\omega,1), \left(1-b+\frac{\omega}{2},\frac{1}{2}\right), \left(1-c+\frac{\omega}{2},\frac{1}{2}\right) \end{array}\right] f(t)dt. \tag{8.12.20}$$

Similarly, by (8.12.12) and (1.1.2), $\left(\mathfrak{M}[x^{1/2}J_{\eta,\sigma}(x)]\right)(s)$ has the form

$$\left(\mathfrak{M}[x^{1/2}J_{\eta,\sigma}(x)]\right)(s)$$
$$= 2^{-\eta-2\sigma}\pi^{1/2}\mathcal{H}^{1,1}_{2,3}\left[\begin{array}{c|c} \left(\frac{1}{4}+\frac{\eta}{2}+\sigma,\frac{1}{2}\right),\left(\frac{3}{4}+\frac{\eta}{2}+\sigma,\frac{1}{2}\right) & \\ \left(\frac{1}{2}+\eta+2\sigma,1\right),\left(\frac{1}{4}+\frac{\eta}{2},\frac{1}{2}\right),\left(\frac{1}{4}-\frac{\eta}{2},\frac{1}{2}\right) & s \end{array}\right], \tag{8.12.21}$$

and the transform $\mathbb{J}_{\eta,\sigma}$ in (8.12.6) is an $\boldsymbol{H}$-transform of the form:

$$\left(\mathbb{J}_{\eta,\sigma}f\right)(x) = 2^{-\eta-2\sigma}\pi^{1/2}$$
$$\cdot\int_0^\infty H^{1,1}_{2,3}\left[xt\left|\begin{array}{c} \left(\frac{1}{4}+\frac{\eta}{2}+\sigma,\frac{1}{2}\right),\left(\frac{3}{4}+\frac{\eta}{2}+\sigma,\frac{1}{2}\right) \\ \left(\frac{1}{2}+\eta+2\sigma,1\right),\left(\frac{1}{4}+\frac{\eta}{2},\frac{1}{2}\right),\left(\frac{1}{4}-\frac{\eta}{2},\frac{1}{2}\right) \end{array}\right.\right]f(t)dt. \tag{8.12.22}$$

In particular, when $\sigma = 0$,

$$\left(\mathfrak{M}[x^{1/2}J_{\eta}(x)]\right)(s) = 2^{-\eta}\pi^{1/2}\mathcal{H}^{1,0}_{1,2}\left[\begin{array}{c|c} \left(\frac{3}{4}+\frac{\eta}{2},\frac{1}{2}\right) & \\ \left(\frac{1}{2}+\eta,1\right),\left(\frac{1}{4}-\frac{\eta}{2},\frac{1}{2}\right) & s \end{array}\right], \tag{8.12.23}$$

and hence we have the new representation for the Hankel transform

$$\left(\mathbb{H}_{\eta}f\right)(x) = 2^{-\eta}\pi^{1/2}\int_0^\infty H^{1,0}_{1,2}\left[xt\left|\begin{array}{c} \left(\frac{3}{4}+\frac{\eta}{2},\frac{1}{2}\right) \\ \left(\frac{1}{2}+\eta,1\right),\left(\frac{1}{4}-\frac{\eta}{2},\frac{1}{2}\right) \end{array}\right.\right]f(t)dt. \tag{8.12.24}$$

According to (3.4.1), (3.4.2), (1.1.7), (1.1.8), (1.1.9), (1.1.10), (1.1.11) and (1.1.12) we have

$$\alpha = -\mathrm{Re}(\omega),\quad \beta = 2\mathrm{Re}(a) - \mathrm{Re}(\omega),\quad a^* = 0,\quad \Delta = \delta = 1,$$
$$\mu = \omega + a - b - c,\quad a_1^* = \frac{1}{2},\quad a_2^* = -\frac{1}{2} \tag{8.12.25}$$

for the H-function in (8.12.19),

$$\alpha = -\mathrm{Re}(\eta+2\sigma) - \frac{1}{2},\quad \beta = -\mathrm{Re}(\eta+2\sigma) + \frac{3}{2},\quad a^* = 0,\quad \Delta = \delta = 1,$$
$$\mu = -\frac{1}{2},\quad a_1^* = \frac{1}{2},\quad a_2^* = -\frac{1}{2} \tag{8.12.26}$$

for the H-function in (8.12.21), and

$$\alpha = -\mathrm{Re}(\eta) - \frac{1}{2},\quad \beta = +\infty,\quad a^* = 0,\quad \Delta = \delta = 1,$$
$$\mu = -\frac{1}{2},\quad a_1^* = \frac{1}{2},\quad a_2^* = -\frac{1}{2} \tag{8.12.27}$$

for the H-function in (8.12.23).

Let $\mathcal{E}_{\mathcal{H}}^1$, $\mathcal{E}_{\mathcal{H}}^2$ and $\mathcal{E}_{\mathcal{H}}^3$ be the exceptional sets of the functions $\mathcal{H}$ in (8.12.19), (8.12.21) and (8.12.23), respectively (see Definition 3.4). Then ν is not in the sets $\mathcal{E}_{\mathcal{H}}^1$, $\mathcal{E}_{\mathcal{H}}^2$ and $\mathcal{E}_{\mathcal{H}}^3$, if $\nu = 1 - \mathrm{Re}(s)$ satisfies

$$s \neq 2b - \omega + 2k, \quad s \neq 2c - \omega + 2l \quad (k, l \in \mathbb{N}_0), \tag{8.12.28}$$

$$s \neq \frac{3}{2} - \eta + 2k, \quad s \neq \frac{3}{2} + \eta + 2l \quad (k, l \in \mathbb{N}_0) \tag{8.12.29}$$

and

$$s \neq \frac{3}{2} + \eta + 2k \quad (k \in \mathbb{N}_0), \tag{8.12.30}$$

respectively

Applying Theorems 3.6 and 3.7, we obtain the following results for the generalized Hardy–Titchmarsh transform $\mathbb{J}_{a,b,c;\omega}$ and the modified Hardy transform $\mathbb{J}_{\eta,\sigma}$ in the space $\mathfrak{L}_{\nu,2}$.

Theorem 8.45. *Let $a, b, c, \omega \in \mathbb{C}$, $(b \neq a,\ c \neq a)$ and $\nu \in \mathbb{R}$ be such that*

$$\mathrm{Re}(\omega) + 1 - 2\mathrm{Re}(a) < \nu < \mathrm{Re}(\omega) + 1, \qquad \nu \geqq 1 + \mathrm{Re}(\omega + a - b - c). \tag{8.12.31}$$

(i) *There is a one-to-one transform $\mathbb{J}_{a,b,c;\omega} \in [\mathfrak{L}_{\nu,2}, \mathfrak{L}_{1-\nu,2}]$ such that the relation*

$$\left(\mathfrak{M}\mathbb{J}_{a,b,c;\omega} f\right)(s)$$

$$= \pi^{1/2} \frac{\Gamma(\omega + s)\Gamma\left(a - \frac{\omega}{2} - \frac{s}{2}\right)}{\Gamma\left(\frac{1}{2} + \frac{\omega}{2} + \frac{s}{2}\right)\Gamma\left(b - \frac{\omega}{2} - \frac{s}{2}\right)\Gamma\left(c - \frac{\omega}{2} - \frac{s}{2}\right)} \left(\mathfrak{M}f\right)(1 - s) \tag{8.12.32}$$

holds for $\mathrm{Re}(s) = 1 - \nu$ and $f \in \mathfrak{L}_{\nu,2}$. If the conditions in (8.12.28) are fulfilled, then the transform $\mathbb{J}_{a,b,c;\omega}$ maps $\mathfrak{L}_{\nu,2}$ onto $\mathfrak{L}_{1-\nu,2}$.

(ii) *If $f, g \in \mathfrak{L}_{\nu,2}$, then*

$$\int_0^\infty \left(\mathbb{J}_{a,b,c;\omega} f\right)(x) g(x) dx = \int_0^\infty f(x) \left(\mathbb{J}_{a,b,c;\omega} g\right)(x) dx. \tag{8.12.33}$$

(iii) *Let $f \in \mathfrak{L}_{\nu,2}$, $\lambda \in \mathbb{C}$ and $h \in \mathbb{R}_+$. If $\mathrm{Re}(\lambda) > (1 - \nu)h - 1$, then $\mathbb{J}_{\eta,\sigma}^{\gamma} f$ is given by*

$$\left(\mathbb{J}_{a,b,c;\omega} f\right)(x) = \pi^{1/2} h x^{1-(\lambda+1)/h} \frac{d}{dx} x^{(\lambda+1)/h}$$

$$\cdot \int_0^\infty H_{3,4}^{1,2}\left[xt \,\middle|\, \begin{matrix} (-\lambda, h), \left(1 - a + \frac{\omega}{2}, \frac{1}{2}\right), \left(\frac{1}{2} + \frac{\omega}{2}, \frac{1}{2}\right) \\ (\omega, 1), \left(1 - b + \frac{\omega}{2}, \frac{1}{2}\right), \left(1 - c + \frac{\omega}{2}, \frac{1}{2}\right), (-\lambda - 1, h) \end{matrix}\right]$$

$$\cdot f(t) dt \tag{8.12.34}$$

for $x > 0$. When $\mathrm{Re}(\lambda) < (1 - \nu)h - 1$, for $x > 0$

$$\left(\mathbb{J}_{a,b,c;\omega} f\right)(x) = -\pi^{1/2} h x^{1-(\lambda+1)/h} \frac{d}{dx} x^{(\lambda+1)/h}$$

$$\cdot \int_0^\infty H_{3,4}^{2,1}\left[xt \left| \begin{matrix} \left(1-a+\frac{\omega}{2},\frac{1}{2}\right),\left(\frac{1}{2}+\frac{\omega}{2},\frac{1}{2}\right),(-\lambda,h) \\ (-\lambda-1,h),(\omega,1),\left(1-b+\frac{\omega}{2},\frac{1}{2}\right),\left(1-c+\frac{\omega}{2},\frac{1}{2}\right) \end{matrix} \right.\right] \cdot f(t)dt. \tag{8.12.35}$$

(iv) *$\mathbb{J}_{a,b,c;\omega}$ is independent of ν in the sense that if ν and $\widetilde{\nu}$ satisfy (8.12.31) and if the transforms $\mathbb{J}_{a,b,c;\omega}$ and $\widetilde{\mathbb{J}}_{a,b,c;\omega}$ are given in (8.12.32) on the respective spaces $\mathfrak{L}_{\nu,2}$ and $\mathfrak{L}_{\widetilde{\nu},2}$, then $\mathbb{J}_{a,b,c;\omega}f = \widetilde{\mathbb{J}}_{a,b,c;\omega}f$ for $f \in \mathfrak{L}_{\nu,2} \bigcap \mathfrak{L}_{\widetilde{\nu},2}$.*

(v) *If $f \in \mathfrak{L}_{\nu,2}$ and $\nu > 1+\mathrm{Re}(\omega+a-b-c)$, then $\mathbb{J}_{a,b,c;\omega}f$ is given in (8.12.1) and (8.12.20).*

Theorem 8.46. *Let $\sigma,\eta \in \mathbb{C}$ and $\nu \in \mathbb{R}$ be such that*

$$\begin{gathered} \mathrm{Re}(\sigma) > -1, \quad \mathrm{Re}(\eta+\sigma) > -1, \\ \mathrm{Re}(\eta+2\sigma) - \frac{1}{2} < \nu < \mathrm{Re}(\eta+2\sigma) + \frac{3}{2}, \quad \nu \geqq \frac{1}{2}. \end{gathered} \tag{8.12.36}$$

(i) *There is a one-to-one transform $\mathbb{J}_{\eta,\sigma} \in [\mathfrak{L}_{\nu,2}, \mathfrak{L}_{1-\nu,2}]$ such that the relation*

$$\left(\mathfrak{M}\{x^{1/2}\mathbb{J}_{\eta,\sigma}f\}\right)(s) = 2^{-\eta-2\sigma}\pi^{1/2} \frac{\Gamma\left(\frac{1}{2}+\eta+2\sigma+s\right)\Gamma\left(\frac{3}{4}-\frac{\eta}{2}-\sigma-\frac{s}{2}\right)}{\Gamma\left(\frac{3}{4}+\frac{\eta}{2}+\sigma+\frac{s}{2}\right)\Gamma\left(\frac{3}{4}-\frac{\eta}{2}-\frac{s}{2}\right)\Gamma\left(\frac{3}{4}+\frac{\eta}{2}-\frac{s}{2}\right)} \left(\mathfrak{M}f\right)(1-s) \tag{8.12.37}$$

holds for $\mathrm{Re}(s) = 1-\nu$ and $f \in \mathfrak{L}_{\nu,2}$. If the conditions in (8.12.29) are fulfilled, then the transform $\mathbb{J}_{\eta,\sigma}$ maps $\mathfrak{L}_{\nu,2}$ onto $\mathfrak{L}_{1-\nu,2}$.

(ii) *If $f, g \in \mathfrak{L}_{\nu,2}$, then*

$$\int_0^\infty \left(\mathbb{J}_{\eta,\sigma}f\right)(x)g(x)dx = \int_0^\infty f(x)\left(\mathbb{J}_{\eta,\sigma}g\right)(x)dx. \tag{8.12.38}$$

(iii) *Let $f \in \mathfrak{L}_{\nu,2}$, $\lambda \in \mathbb{C}$ and $h \in \mathbb{R}_+$. If $\mathrm{Re}(\lambda) > (1-\nu)h-1$, then $\mathbb{J}_{\eta,\sigma}f$ is given by*

$$\left(\mathbb{J}_{\eta,\sigma}f\right)(x) = 2^{-\eta-2\sigma}\pi^{1/2}hx^{1-(\lambda+1)/h}\frac{d}{dx}x^{(\lambda+1)/h} \cdot \int_0^\infty H_{3,4}^{1,2}\left[xt \left| \begin{matrix} (-\lambda,h),\left(\frac{1}{4}+\frac{\eta}{2}+\frac{\sigma}{2},\frac{1}{2}\right),\left(\frac{3}{4}+\frac{\eta}{2}+\sigma,\frac{1}{2}\right) \\ \left(\frac{1}{2}+\eta+\sigma,1\right),\left(\frac{1}{4}+\frac{\eta}{2},\frac{1}{2}\right),\left(\frac{1}{4}-\frac{\eta}{2},\frac{1}{2}\right),(-\lambda-1,h) \end{matrix} \right.\right] \cdot f(t)dt \quad (x>0). \tag{8.12.39}$$

When $\mathrm{Re}(\lambda) < (1-\nu)h-1$,

$$\left(\mathbb{J}_{\eta,\sigma}f\right)(x) = -2^{-\eta-2\sigma}\pi^{1/2}hx^{1-(\lambda+1)/h}\frac{d}{dx}x^{(\lambda+1)/h} \cdot \int_0^\infty H_{3,4}^{2,1}\left[xt \left| \begin{matrix} \left(\frac{1}{4}+\frac{\eta}{2}+\frac{\sigma}{2},\frac{1}{2}\right),\left(\frac{3}{4}+\frac{\eta}{2}+\sigma,\frac{1}{2}\right),(-\lambda,h) \\ (-\lambda-1,h),\left(\frac{1}{2}+\eta+\sigma,1\right),\left(\frac{1}{4}+\frac{\eta}{2},\frac{1}{2}\right),\left(\frac{1}{4}-\frac{\eta}{2},\frac{1}{2}\right) \end{matrix} \right.\right] \cdot f(t)dt \quad (x>0). \tag{8.12.40}$$

(iv) *$\mathbb{J}_{\eta,\sigma}$ is independent of ν in the sense that if ν and $\widetilde{\nu}$ satisfy* (8.12.36) *and if the transforms $\mathbb{J}_{\eta,\sigma}$ and $\widetilde{\mathbb{J}}_{\eta,\sigma}$ are given in* (8.12.37) *on the respective spaces $\mathfrak{L}_{\nu,2}$ and $\mathfrak{L}_{\widetilde{\nu},2}$, then $\mathbb{J}_{\eta,\sigma}f = \widetilde{\mathbb{J}}_{\eta,\sigma}f$ for $f \in \mathfrak{L}_{\nu,2} \bigcap \mathfrak{L}_{\widetilde{\nu},2}$.*

(v) *If $f \in \mathfrak{L}_{\nu,2}$ and $\nu > 1/2$, then $\mathbb{J}_{\eta,\sigma}f$ is given in* (8.12.6) *and* (8.12.22).

From Theorem 8.46 we obtain the corresponding statement for the Hankel transform $\mathbb{H}_\eta$ considered in Section 8.1.

Theorem 8.47. *Let $\eta \in \mathbb{C}$ and $\nu \in \mathbb{R}$ be such that $\mathrm{Re}(\eta) > -1$ and $1/2 \leqq \nu < \mathrm{Re}(\eta) + 3/2$.*

(i) *There is a one-to-one transform $\mathbb{H}_\eta \in [\mathfrak{L}_{\nu,2}, \mathfrak{L}_{1-\nu,2}]$ such that the relation*

$$\left(\mathfrak{M}\mathbb{H}_\eta f\right)(s) = 2^{-\eta}\pi^{1/2}\frac{\Gamma\left(\frac{1}{2}+\eta+s\right)}{\Gamma\left(\frac{3}{4}+\frac{\eta}{2}+\frac{s}{2}\right)\Gamma\left(\frac{3}{4}+\frac{\eta}{2}-\frac{s}{2}\right)}\left(\mathfrak{M}f\right)(1-s)$$

$$= 2^{s-1/2}\frac{\Gamma\left(\frac{1}{4}+\frac{\eta}{2}+\frac{s}{2}\right)}{\Gamma\left(\frac{3}{4}+\frac{\eta}{2}-\frac{s}{2}\right)}\left(\mathfrak{M}f\right)(1-s) \tag{8.12.41}$$

holds for $\mathrm{Re}(s) = 1-\nu$ and $f \in \mathfrak{L}_{\nu,2}$. If the conditions in (8.12.30) *are fulfilled, then the transform $\mathbb{H}_\eta$ maps $\mathfrak{L}_{\nu,2}$ onto $\mathfrak{L}_{1-\nu,2}$.*

(ii) *If $f, g \in \mathfrak{L}_{\nu,2}$, then*

$$\int_0^\infty \left(\mathbb{H}_\eta f\right)(x)g(x)dx = \int_0^\infty f(x)\left(\mathbb{H}_\eta g\right)(x)dx. \tag{8.12.42}$$

(iii) *Let $f \in \mathfrak{L}_{\nu,2}$, $\lambda \in \mathbb{C}$ and $h \in \mathbb{R}_+$. If $\mathrm{Re}(\lambda) > (1-\nu)h - 1$, then $\mathbb{H}_\eta f$ is given by*

$$\left(\mathbb{H}_\eta f\right)(x) = hx^{1-(\lambda+1)/h}\frac{d}{dx}x^{(\lambda+1)/h}$$

$$\cdot \int_0^\infty H_{2,3}^{1,1}\left[xt \,\middle|\, \begin{matrix} (-\lambda, h), \left(\frac{3}{4}+\frac{\eta}{2}, \frac{1}{2}\right) \\ \left(\eta+\frac{1}{2}, \frac{1}{2}\right), \left(\frac{1}{4}-\frac{\eta}{2}, \frac{1}{2}\right), (-\lambda-1, h) \end{matrix}\right] f(t)dt$$

$$(x > 0). \tag{8.12.43}$$

When $\mathrm{Re}(\lambda) < (1-\nu)h - 1$,

$$\left(\mathbb{H}_\eta f\right)(x) = -hx^{1-(\lambda+1)/h}\frac{d}{dx}x^{(\lambda+1)/h}$$

$$\cdot \int_0^\infty H_{2,3}^{2,0}\left[xt \,\middle|\, \begin{matrix} \left(\frac{3}{4}+\frac{\eta}{2}, \frac{1}{2}\right), (-\lambda, h) \\ (-\lambda-1, h), \left(\eta+\frac{1}{2}, \frac{1}{2}\right), \left(\frac{1}{4}-\frac{\eta}{2}, \frac{1}{2}\right) \end{matrix}\right] f(t)dt$$

$$(x > 0). \tag{8.12.44}$$

(iv) *$\mathbb{H}_\eta$ is independent of ν in the sense that if ν and $\widetilde{\nu}$ are such that $1/2 \leqq \nu < \mathrm{Re}(\eta)+3/2$ and $1/2 \leqq \widetilde{\nu} < \mathrm{Re}(\eta) + 3/2$, and if the transforms $\mathbb{H}_\eta$ and $\widetilde{\mathbb{H}}_\eta$ are given in* (8.12.41) *on the respective spaces $\mathfrak{L}_{\nu,2}$ and $\mathfrak{L}_{\widetilde{\nu},2}$, then $\mathbb{H}_\eta f = \widetilde{\mathbb{H}}_\eta f$ for $f \in \mathfrak{L}_{\nu,2} \bigcap \mathfrak{L}_{\widetilde{\nu},2}$.*

(v) *If $f \in \mathfrak{L}_{\nu,2}$ and $\nu > 1/2$, then $\mathbb{H}_\eta f$ is given in* (8.1.1) *and* (8.12.24).

Now we can apply Theorem 4.3 to obtain the boundedeness, range and representation of the generalized Hardy–Titchmarsh transform (8.12.1), the modified Hardy transform (8.12.6) and the Hankel transform (8.1.1) in the space $\mathfrak{L}_{\nu,r}$. By (8.12.20), the first one is characterized in the following:

Theorem 8.48. *Let $1 < r < \infty$ and let $a, b, c \in \mathbb{C}$ $(\mathrm{Re}(a) > 0, b \neq a, c \neq a)$, $\omega \in \mathbb{C}$ and $\nu \in \mathbb{R}$ be such that*

$$\mathrm{Re}(\omega) + 1 - 2\mathrm{Re}(a) < \nu < \mathrm{Re}(\omega) + 1, \quad \nu \geqq \frac{1}{2} + \mathrm{Re}(\omega + a - b - c) + \gamma(r), \tag{8.12.45}$$

where $\gamma(r)$ is given in (3.3.9).

(a) *The transform $\mathbb{J}_{a,b,c;\omega}$ defined on $\mathfrak{L}_{\nu,2}$ can be extended to $\mathfrak{L}_{\nu,r}$ as an element of $[\mathfrak{L}_{\nu,r}, \mathfrak{L}_{1-\nu,s}]$ for all s with $r \leqq s < \infty$ such that $s' \geqq [\nu - 1/2 + \mathrm{Re}(b + c - a - \omega)]^{-1}$ with $1/s + 1/s' = 1$.*

(b) *If $1 < r \leqq 2$, then the transform $\mathbb{J}_{a,b,c;\omega}$ is one-to-one on $\mathfrak{L}_{\nu,r}$ and there holds the equality* (8.12.32) *for $f \in \mathfrak{L}_{\nu,r}$ and $\mathrm{Re}(s) = 1 - \nu$.*

(c) *Let $f \in \mathfrak{L}_{\nu,r}$ and $g \in \mathfrak{L}_{\nu,s}$ with $1 < s < \infty$ and $1/r + 1/s \geqq 1$. If $\nu \geqq 1/2 + \mathrm{Re}(\omega + a - b - c) + \max[\gamma(r), \gamma(s)]$, then the relation* (8.12.33) *holds.*

(d) *Let $\eta = -\mathrm{Im}(\omega) - a + b + c - 1$. If the conditions in* (8.12.28) *are satisfied, then*

$$\mathbb{J}_{a,b,c;\omega}(\mathfrak{L}_{\nu,r}) = \left(M_{\omega+a-b-c+1/2}\mathbb{H}_\eta\right)(\mathfrak{L}_{\nu-\mathrm{Re}(\omega+a-b-c)-1/2,r}), \tag{8.12.46}$$

where $\mathbb{H}_\eta$ is the Hankel transform (8.12.1) *and M_ζ is the operator in* (3.3.11). *When any condition in* (8.12.28) *is not satisfied, $\mathbb{J}_{a,b,c;\omega}(\mathfrak{L}_{\nu,r})$ is a subset of the right-hand side of* (8.12.46).

(e) *If $f \in \mathfrak{L}_{\nu,r}$, $\lambda \in \mathbb{C}$ and $h \in \mathbb{R}_+$, then $\mathbb{J}_{a,b,c;\omega}f$ is given in* (8.12.34) *for $\mathrm{Re}(\lambda) > (1-\nu)h - 1$, while by* (8.12.35) *for $\mathrm{Re}(\lambda) < (1-\nu)h - 1$. If $\nu > 2 + \mathrm{Re}(\omega + a - b - c)$, then $\mathbb{J}_{a,b,c;\omega}f$ is given in* (8.12.1) *and* (8.12.20).

The replacement $a = 1, b = \sigma + 1, c = \sigma + \eta + 1, \omega = 2\sigma + \eta + 1/2$ in Theorem 8.48 yields $\mathfrak{L}_{\nu,r}$-theory of the modified Hardy transform (8.12.6).

Theorem 8.49. *Let $1 < r < \infty$ and let $\sigma, \eta \in \mathbb{C}$ and $\nu \in \mathbb{R}$ be such that*

$$\begin{gathered}\mathrm{Re}(\sigma) > -1, \quad \mathrm{Re}(\eta + \sigma) > -1, \\ \mathrm{Re}(\eta + 2\sigma) - \frac{1}{2} < \nu < \mathrm{Re}(\eta + 2\sigma) + \frac{3}{2}, \quad \nu \geqq \gamma(r),\end{gathered} \tag{8.12.47}$$

where $\gamma(r)$ is given in (3.3.9).

(a) *The transform $\mathbb{J}_{\eta,\sigma}$ defined on $\mathfrak{L}_{\nu,2}$ can be extended to $\mathfrak{L}_{\nu,r}$ as an element of $[\mathfrak{L}_{\nu,r}, \mathfrak{L}_{1-\nu,s}]$ for all s with $r \leqq s < \infty$ such that $s' \geqq 1/\nu$ with $1/s + 1/s' = 1$.*

(b) *If $1 < r \leqq 2$, then the transform $\mathbb{J}_{\eta,\sigma}$ is one-to-one on $\mathfrak{L}_{\nu,r}$ and there holds the equality* (8.12.37) *for $f \in \mathfrak{L}_{\nu,r}$ and $\mathrm{Re}(s) = 1 - \nu$.*

(c) *Let $f \in \mathfrak{L}_{\nu,r}$ and $g \in \mathfrak{L}_{\nu,s}$ with $1 < s < \infty$ and $1/r + 1/s \geqq 1$. If $\nu \geqq \max[\gamma(r), \gamma(s)]$, then the relation* (8.12.38) *holds.*

(d) *If the conditions in* (8.12.29) *are satisfied, then*

$$\mathbb{J}_{\eta,\sigma}(\mathfrak{L}_{\nu,r}) = \mathbb{H}_{\mathrm{Re}(\eta+2\sigma)}(\mathfrak{L}_{\nu,r}). \tag{8.12.48}$$

When any condition in (8.12.29) *is not satisfied,* $\mathbb{J}_{\eta,\sigma}(\mathfrak{L}_{\nu,r})$ *is a subset of the right-hand side of* (8.12.48).

(e) *If* $f \in \mathfrak{L}_{\nu,r}$, $\lambda \in \mathbb{C}$ *and* $h \in \mathbb{R}_+$, *then* $\mathbb{J}_{\eta,\sigma}f$ *is given in* (8.12.39) *for* $\mathrm{Re}(\lambda) > (1-\nu)h-1$, *while by* (8.12.40) *for* $\mathrm{Re}(\lambda) < (1-\nu)h-1$. *If* $\nu > 5/2$, *then* $\mathbb{J}_{\eta,\sigma}f$ *is given in* (8.12.6) *and* (8.12.22).

From Theorem 8.49 we obtain the corresponding statement for the Hankel transform $\mathbb{H}_\eta$.

Theorem 8.50. *Let* $1 < r < \infty$ *and let* $\eta \in \mathbb{C}$ *and* $\nu \in \mathbb{R}$ *be such that* $\mathrm{Re}(\eta) > -1$, $\mathrm{Re}(\eta) - 1/2 < \nu < \mathrm{Re}(\eta) + 3/2$ *and* $\nu \geqq \gamma(r)$.

(a) *The transform* $\mathbb{H}_\eta$ *defined on* $\mathfrak{L}_{\nu,2}$ *can be extended to* $\mathfrak{L}_{\nu,r}$ *as an element of* $[\mathfrak{L}_{\nu,r}, \mathfrak{L}_{1-\nu,s}]$ *for all* s *with* $r \leqq s < \infty$ *such that* $s' \geqq 1/\nu$ *with* $1/s + 1/s' = 1$.

(b) *If* $1 < r \leqq 2$, *then the transform* $\mathbb{H}_\eta$ *is one-to-one on* $\mathfrak{L}_{\nu,r}$ *and there holds the equality* (8.12.41) *for* $f \in \mathfrak{L}_{\nu,r}$ *and* $\mathrm{Re}(s) = 1-\nu$.

(c) *Let* $f \in \mathfrak{L}_{\nu,r}$ *and* $g \in \mathfrak{L}_{\nu,s}$ *with* $1 < s < \infty$ *and* $1/r + 1/s \geqq 1$. *If* $\nu \geqq \max[\gamma(r), \gamma(s)]$, *then the relation* (8.12.42) *holds.*

(d) *If the conditions in* (8.12.30) *are satisfied, then*

$$\mathbb{H}_\eta(\mathfrak{L}_{\nu,r}) = \mathbb{H}_{\mathrm{Re}(\eta)}(\mathfrak{L}_{\nu,r}). \tag{8.12.49}$$

When any condition in (8.12.30) *is not satisfied,* $\mathbb{H}_\eta(\mathfrak{L}_{\nu,r})$ *is a subset of the right-hand side of* (8.12.49).

(e) *If* $f \in \mathfrak{L}_{\nu,r}$, $\lambda \in \mathbb{C}$ *and* $h \in \mathbb{R}_+$, *then* $\mathbb{H}_\eta f$ *is given in* (8.12.43) *for* $\mathrm{Re}(\lambda) > (1-\nu)h-1$, *while by* (8.12.44) *for* $\mathrm{Re}(\lambda) < (1-\nu)h-1$. *If* $\nu > 5/2$, *then* $\mathbb{H}_\eta f$ *is given in* (8.1.1) *and* (8.12.24).

Remark 8.13. It follows from Theorems 8.49, that the range $\mathbb{J}_{\eta,\sigma}(\mathfrak{L}_{\nu,r})$ of the modified Hardy transform (8.12.6) coincides with the range $\mathbb{H}_{\mathrm{Re}(\eta+2\sigma)}(\mathfrak{L}_{\nu,r})$ of the Hankel transform, while these transforms have different representations: (8.12.6), (8.12.22) and (8.1.1), (8.12.24).

Since $a^* = 0$ for the transforms $\mathbb{J}_{a,b,c;\omega}$, $\mathbb{J}_{\eta,\sigma}$ and $\mathbb{H}_\eta$, we can apply the results from Sections 4.9 and 4.10 to obtain their inversions. The constants α and β are given in (8.12.25), (8.12.26) and (8.12.27), while the constants α_0 and β_0 in (4.9.6) and (4.9.7) take the forms

$$\alpha_0 = \mathrm{Re}(\omega) + 1 - 2\min[\mathrm{Re}(b), \mathrm{Re}(c)], \qquad \beta_0 = \mathrm{Re}(\omega) + 2 \tag{8.12.50}$$

for $\mathbb{J}_{a,b,c;\omega}$,

$$\alpha_0 = |\mathrm{Re}(\eta)| - \frac{1}{2}, \qquad \beta_0 = \mathrm{Re}(\eta + 2\sigma) + \frac{5}{2} \tag{8.12.51}$$

for $\mathbb{J}_{\eta,\sigma}$ and, in particular,

$$\alpha_0 = |\mathrm{Re}(\eta)| - \frac{1}{2}, \qquad \beta_0 = \mathrm{Re}(\eta) + \frac{5}{2} \tag{8.12.52}$$

for $\mathbb{H}_\eta$.

In the case of the transform $\mathbb{J}_{a,b,c;\omega}$ the inversion formulas (4.9.1) and (4.9.2) take the forms

$$f(x) = \pi^{-1/2} h x^{1-(\lambda+1)/h} \frac{d}{dx} x^{(\lambda+1)/h}$$

$$\cdot \int_0^\infty H_{3,4}^{2,2}\left[xt \left| \begin{matrix} (-\lambda, h), \left(-\frac{\omega}{2}, \frac{1}{2}\right), \left(a - \frac{\omega}{2} - \frac{1}{2}, \frac{1}{2}\right) \\ \left(b - \frac{\omega}{2} - \frac{1}{2}, \frac{1}{2}\right), \left(c - \frac{\omega}{2} - \frac{1}{2}, \frac{1}{2}\right), (-\omega, 1), (-\lambda - 1, h) \end{matrix} \right. \right]$$

$$\cdot \left(\mathbb{J}_{a,b,c;\omega} f\right)(t)dt \quad (x > 0); \tag{8.12.53}$$

$$f(x) = -\pi^{-1/2} h x^{1-(\lambda+1)/h} \frac{d}{dx} x^{(\lambda+1)/h}$$

$$\cdot \int_0^\infty H_{3,4}^{3,1}\left[xt \left| \begin{matrix} \left(-\frac{\omega}{2}, \frac{1}{2}\right), \left(a - \frac{\omega}{2} - \frac{1}{2}, \frac{1}{2}\right), (-\lambda, h) \\ (-\lambda - 1, h), \left(b - \frac{\omega}{2} - \frac{1}{2}, \frac{1}{2}\right), \left(c - \frac{\omega}{2} - \frac{1}{2}, \frac{1}{2}\right), (-\omega, 1) \end{matrix} \right. \right]$$

$$\cdot \left(\mathbb{J}_{a,b,c;\omega,\sigma} f\right)(t)dt \quad (x > 0). \tag{8.12.54}$$

For the modified Hardy transform $\mathbb{J}_{\eta,\sigma}$ the inversion relations are given by

$$f(x) = 2^{\eta+2\sigma} \pi^{-1/2} h x^{1-(\lambda+1)/h} \frac{d}{dx} x^{(\lambda+1)/h}$$

$$\cdot \int_0^\infty H_{3,4}^{2,2}\left[xt \left| \begin{matrix} (-\lambda, h), \left(-\frac{1}{4} - \frac{\eta}{2} - \sigma, \frac{1}{2}\right), \left(\frac{1}{4} - \frac{\eta}{2} - \sigma, \frac{1}{2}\right) \\ \left(\frac{1}{4} - \frac{\eta}{2}, \frac{1}{2}\right), \left(\frac{1}{4} + \frac{\eta}{2}, \frac{1}{2}\right), \left(-\frac{1}{2} - \eta - 2\sigma, 1\right), (-\lambda - 1, h) \end{matrix} \right. \right]$$

$$\cdot \left(\mathbb{J}_{\eta,\sigma} f\right)(t)dt \quad (x > 0); \tag{8.12.55}$$

$$f(x) = -2^{\eta+2\sigma} \pi^{-1/2} h x^{1-(\lambda+1)/h} \frac{d}{dx} x^{(\lambda+1)/h}$$

$$\cdot \int_0^\infty H_{3,4}^{3,1}\left[xt \left| \begin{matrix} \left(-\frac{1}{4} - \frac{\eta}{2} - \sigma, \frac{1}{2}\right), \left(\frac{1}{4} - \frac{\eta}{2} - \sigma, \frac{1}{2}\right), (-\lambda, h) \\ (-\lambda - 1, h), \left(\frac{1}{4} - \frac{\eta}{2}, \frac{1}{2}\right), \left(\frac{1}{4} + \frac{\eta}{2}, \frac{1}{2}\right), \left(-\frac{1}{2} - \eta - 2\sigma, 1\right) \end{matrix} \right. \right]$$

$$\cdot \left(\mathbb{J}_{\eta,\sigma} f\right)(t)dt \quad (x > 0). \tag{8.12.56}$$

In particular, from (8.12.55) and (8.12.56) we have the inversion formulas for the Hankel transform by putting $\sigma = 0$:

$$f(x) = 2^\eta \pi^{-1/2} h x^{1-(\lambda+1)/h} \frac{d}{dx} x^{(\lambda+1)/h}$$

$$\cdot \int_0^\infty H_{2,3}^{1,2}\left[xt \left| \begin{matrix} (-\lambda, h), \left(-\frac{\eta}{2} - \frac{1}{4}, \frac{1}{2}\right) \\ \left(\frac{\eta}{2} + \frac{1}{4}, \frac{1}{2}\right), \left(-\eta - \frac{1}{2}, 1\right), (-\lambda - 1, h) \end{matrix} \right. \right] \left(\mathbb{H}_\eta f\right)(t)dt \tag{8.12.57}$$

$$(x > 0);$$

$$f(x) = -2^{\eta}\pi^{-1/2}hx^{1-(\lambda+1)/h}\frac{d}{dx}x^{(\lambda+1)/h}$$

$$\cdot \int_0^\infty H_{2,3}^{2,1}\left[xt \left| \begin{matrix} \left(-\frac{\eta}{2}-\frac{1}{4},\frac{1}{2}\right), (-\lambda, h) \\ (-\lambda-1, h), \left(\frac{\eta}{2}+\frac{1}{4},\frac{1}{2}\right), \left(-\eta-\frac{1}{2},1\right) \end{matrix} \right.\right] \left(\mathbb{H}_\eta f\right)(t)dt \quad (8.12.58)$$

$$(x > 0).$$

Thus, we obtain the inversion results from Sections 4.9 and 4.10 for the transforms $\mathbb{J}_{a,b,c;\omega}$, $\mathbb{J}_{\eta,\sigma}$ and $\mathbb{H}_\eta$ in the spaces $\mathfrak{L}_{\nu,r}$ and $L_2(\mathbb{R}_+)$.

Theorem 8.51. *Let $a, b, c, \omega \in \mathbb{C}$ ($b \neq a, c \neq a$), $\nu \in \mathbb{R}$, and let $\lambda \in \mathbb{C}$ and $h > 0$.*
(a) *Let*

$$\nu = 1 + \mathrm{Re}(\omega + a - b - c), \ 0 < \mathrm{Re}(b + c - a) < 2\min[\mathrm{Re}(a), \mathrm{Re}(b), \mathrm{Re}(c)]. \quad (8.12.59)$$

If $f \in \mathfrak{L}_{\nu,2}$, then the relation (8.12.53) *holds for* $\mathrm{Re}(\lambda) > \nu h - 1$, *while* (8.12.54) *for* $\mathrm{Re}(\lambda) < \nu h - 1$.
(b) *Let $1 < r < \infty$ and*

$$\begin{gathered} -2\min[\mathrm{Re}(a), \mathrm{Re}(b), \mathrm{Re}(c)] < \nu - \mathrm{Re}(\omega) - 1 < \min\left[0, \mathrm{Re}(a - b - c) + \frac{1}{2}\right], \\ \nu \geqq \frac{1}{2} + \mathrm{Re}(\omega + a - b - c) + \gamma(r), \end{gathered} \quad (8.12.60)$$

where $\gamma(r)$ is given in (3.3.9). *If $f \in \mathfrak{L}_{\nu,r}$, then the relation* (8.12.53) *holds for* $\mathrm{Re}(\lambda) > \nu h - 1$, *while* (8.12.54) *for* $\mathrm{Re}(\lambda) < \nu h - 1$.

Proof. The results (a) and (b) follow from Theorems 4.11 and 4.13, respectively. Indeed, by (8.12.25) and (8.12.50) the conditions for ν in Theorems 4.11 and 4.13 for the transform $\mathbb{J}_{a,b,c;\omega}$ take the forms

$$\begin{gathered} 1 + \mathrm{Re}(\omega) - 2\mathrm{Re}(a) < \nu < 1 + \mathrm{Re}(\omega); \\ \mathrm{Re}(\omega) + 1 - 2\min[\mathrm{Re}(b), \mathrm{Re}(c)] < \nu < \mathrm{Re}(\omega) + 2; \\ \nu = 1 + \mathrm{Re}(\omega + a - b - c), \end{gathered} \quad (8.12.61)$$

and

$$\begin{gathered} 1 + \mathrm{Re}(\omega) - 2\mathrm{Re}(a) < \nu < 1 + \mathrm{Re}(\omega); \\ \mathrm{Re}(\omega) + 1 - 2\min[\mathrm{Re}(b), \mathrm{Re}(c)] < \nu \\ < \min\left[\mathrm{Re}(\omega) + 2, \mathrm{Re}(\omega + a - b - c) + \frac{3}{2}\right]; \\ \nu \geqq \frac{1}{2} + \mathrm{Re}(\omega + a - b - c) + \gamma(r). \end{gathered} \quad (8.12.62)$$

The existence of such a value ν in each case is guaranteed from the assumptions (8.12.59) and (8.12.60).

Similarly, in view of (8.12.26), the conditions of Theorems 4.11 and 4.13 for the transform $\mathbb{J}_{\eta,\sigma}$ can be given by

$$-\frac{1}{2}+\mathrm{Re}(\eta+2\sigma)<\nu<\frac{3}{2}+\mathrm{Re}(\eta+2\sigma),$$
$$|\mathrm{Re}(\eta)|-\frac{1}{2}<\nu<\mathrm{Re}(\eta+2\sigma)+\frac{5}{2},\qquad \nu=\frac{1}{2}, \tag{8.12.63}$$

and

$$-\frac{1}{2}+\mathrm{Re}(\eta+2\sigma)<\nu<\frac{3}{2}+\mathrm{Re}(\eta+2\sigma),$$
$$|\mathrm{Re}(\eta)|-\frac{1}{2}<\nu<\min\left[\mathrm{Re}(\eta+2\sigma)+\frac{5}{2},1\right],\qquad \nu\geqq\gamma(r), \tag{8.12.64}$$

which are certified by (8.12.65) and (8.12.66) below, respectively. Then by noting that $\mathfrak{L}_{1/2,2}=L_2(\mathbb{R}_+)$, we have

Theorem 8.52. *Let $\eta,\sigma\in\mathbb{C}$ be such that $\mathrm{Re}(\sigma)>-1$, $\mathrm{Re}(\eta+\sigma)>-1$, and let $\lambda\in\mathbb{C}$ and $h>0$.*

(a) *Let*

$$|\mathrm{Re}(\eta)|<1,\qquad 0<\mathrm{Re}(\eta+2\sigma)<1. \tag{8.12.65}$$

If $f\in L_2(\mathbb{R}_+)$, then the relation (8.12.55) holds for $\mathrm{Re}(\lambda)>h/2-1$, while (8.12.56) for $\mathrm{Re}(\lambda)<h/2-1$.

(b) *Let $1<r<\infty$ and*

$$\max\left[\mathrm{Re}(\eta+2\sigma),|\mathrm{Re}(\eta)|\right]-\frac{1}{2}<\nu<\min\left[1,\mathrm{Re}(\eta+2\sigma)+\frac{3}{2}\right],\qquad \nu\geqq\gamma(r), \tag{8.12.66}$$

where $\gamma(r)$ is given in (3.3.9). If $f\in\mathfrak{L}_{\nu,r}$, then the relation (8.12.55) holds for $\mathrm{Re}(\lambda)>\nu h-1$, while (8.12.56) for $\mathrm{Re}(\lambda)<\nu h-1$.

Taking $\sigma=0$ in Theorem 8.52 and using (8.12.8), we obtain the corresponding results for the Hankel transform $\mathbb{H}_\eta$ in the spaces $L_2(\mathbb{R}_+)$ and $\mathfrak{L}_{\nu,r}$.

Theorem 8.53. *Let $\eta\in\mathbb{C}$, $\mathrm{Re}(\eta)>-1$, and let $\lambda\in\mathbb{C}$ and $h>0$.*

(a) *Let $|\mathrm{Re}(\eta)|<1$. If $f\in L_2(\mathbb{R}_+)$, then the relation (8.12.57) holds for $\mathrm{Re}(\lambda)>h/2-1$, while (8.12.58) for $\mathrm{Re}(\lambda)<h/2-1$.*

(b) *Let $1<r<\infty$ and $\gamma(r)\leqq\nu<\max[\mathrm{Re}(\eta)+3/2,1]$, where $\gamma(r)$ is given in (3.3.9). If $f\in\mathfrak{L}_{\nu,r}$, then the relation (8.12.57) holds for $\mathrm{Re}(\lambda)>\nu h-1$, while (8.12.58) for $\mathrm{Re}(\lambda)<\nu h-1$.*

Remark 8.14. The relations (8.12.57) and (8.12.58) are new inversion formulas for the Hankel transform $\mathbb{H}_\eta$ compared with those in (8.1.33) and (8.1.34) proved in Section 8.1.

8.13. The Lommel–Maitland Transform

We consider the integral transform

$$\left(\mathbb{J}_{\eta,\sigma}^{\gamma} f\right)(x) = \int_0^\infty (xt)^{1/2} J_{\eta,\sigma}^{\gamma}(xt) f(t) dt \quad (x > 0) \tag{8.13.1}$$

with the generalized Bessel–Maitland function (2.9.26):

$$J_{\eta,\sigma}^{\gamma}(z) = \sum_{k=0}^{\infty} \frac{(-1)^k}{\Gamma(1+\sigma+k)\Gamma(1+\sigma+\eta+\gamma k)} \left(\frac{z}{2}\right)^{\eta+2k+2\sigma} \tag{8.13.2}$$

$$(0 < \gamma \leqq 1;\ \sigma, \eta \in \mathbb{C},\ \mathrm{Re}(\sigma) > -1,\ \mathrm{Re}(\eta+\sigma) > -1)$$

as the kernel. Since, when $\gamma = 1$, (8.13.2) is the Lommel function (8.12.6) and the transform (8.13.1) coincides with the modified Hardy transform (8.12.6) studied in the previous section, we shall treat here the Lommel–Maitland transform $\mathbb{J}_{\eta,\sigma}^{\gamma}$ only for $0 < \gamma < 1$.

When $\sigma = 0$,

$$J_{\eta}^{\gamma}(z) \equiv J_{\eta,0}^{\gamma}(z) = \sum_{k=0}^{\infty} \frac{(-1)^k}{k!\Gamma(1+\eta+\gamma k)} \left(\frac{z}{2}\right)^{\eta+2k} \tag{8.13.3}$$

for $\mathrm{Re}(\eta) > -1$, and (8.13.1) takes the form

$$\left(\mathbb{J}_{\eta}^{\gamma} f\right)(x) \equiv \left(\mathbb{J}_{\eta,0}^{\gamma} f\right)(x) = \int_0^\infty (xt)^{1/2} J_{\eta}^{\gamma}(xt) f(t) dt \quad (x > 0). \tag{8.13.4}$$

We also note that the Bessel–Maitland function $J^{\mu,\nu}(z)$ in (2.9.24) is expressed via (8.13.3) by

$$J^{\mu,\nu}(z) = \sum_{k=0}^{\infty} \frac{(-z)^k}{k!\Gamma(1+\nu+\mu k)} = z^{-\nu/2} J_{\nu}^{\mu}(2\sqrt{z}). \tag{8.13.5}$$

The following $\mathfrak{L}_{\nu,r}$-theory of the Lommel–Maitland transform (8.13.1) is based on the Mellin transform of the function (8.13.2).

Lemma 8.21. *Let* $0 < \gamma < 1$ *and* $\sigma, \eta \in \mathbb{C}$.

(a) *If* $\mathrm{Re}(\sigma) > -1$, $\mathrm{Re}(\eta+\sigma) > -1$ *and* $s \in \mathbb{C}$ *satisfies*

$$-\frac{1}{2} < \mathrm{Re}(s) + \mathrm{Re}(\eta+2\sigma) < \frac{3}{2}, \tag{8.13.6}$$

then

$$\left(\mathfrak{M}[x^{1/2} J_{\eta,\sigma}^{\gamma}(x)]\right)(s) = 2^{-\eta-2\sigma}\pi^{1/2}$$

$$\cdot \frac{\Gamma\left(\frac{1}{2}+\eta+2\sigma+s\right)\Gamma\left(\frac{3}{4}-\frac{\eta}{2}-\sigma-\frac{s}{2}\right)}{\Gamma\left(\frac{3}{4}+\frac{\eta}{2}+\sigma+\frac{s}{2}\right)\Gamma\left(\frac{3}{4}-\frac{\eta}{2}-\frac{s}{2}\right)\Gamma\left(1+\eta+\sigma-\gamma\left[\frac{1}{4}+\frac{\eta}{2}+\sigma\right]-\frac{\gamma s}{2}\right)}. \tag{8.13.7}$$

(b) *Let* $\mathrm{Re}(\eta) > -1$ *and* $s \in \mathbb{C}$ *satisfy* $\mathrm{Re}(s) > -\mathrm{Re}(\eta) - 1/2$. *Then*

$$\left(\mathfrak{M}[x^{1/2}J_\eta^\gamma(x)]\right)(s) = 2^{-\eta}\pi^{1/2}\frac{\Gamma\left(\frac{1}{2}+\eta+s\right)}{\Gamma\left(\frac{3}{4}+\frac{\eta}{2}+\frac{s}{2}\right)\Gamma\left(1+\eta-\gamma\left[\frac{1}{4}+\frac{\eta}{2}\right]-\frac{\gamma s}{2}\right)}. \tag{8.13.8}$$

Proof. It is known (see Betancor [1]) that

$$\left(\mathfrak{M}[x^{1/2}J_{\eta,\sigma}^\gamma(x)]\right)(s)$$
$$= 2^{s-1/2}\frac{\Gamma\left(\frac{1}{4}+\frac{\eta}{2}+\sigma+\frac{s}{2}\right)\Gamma\left(\frac{3}{4}-\frac{\eta}{2}-\sigma-\frac{s}{2}\right)}{\Gamma\left(\frac{3}{4}-\frac{\eta}{2}-\frac{s}{2}\right)\Gamma\left(1+\eta+\sigma-\gamma\left[\frac{1}{4}+\sigma+\frac{\eta}{2}\right]-\frac{\gamma s}{2}\right)}. \tag{8.13.9}$$

Then, (8.13.7) is deduced from (8.13.9) if we multiply both sides of (8.13.9) by $\Gamma(\eta/2+\sigma+3/4+s/2)$ and use the Legendre duplication formula (8.9.7). The equality (8.13.8) follows from (8.13.7) when $\sigma = 0$, which completes the proof of Lemma 8.20.

Now we can apply the results in Chapters 3 and 4 to characterize the $\mathfrak{L}_{\nu,r}$-properties of the Lommel–Maitland transform (8.13.1). In view of (8.13.7) and (1.1.2), $(\mathfrak{M}[x^{1/2}J_{\eta,\sigma}^\gamma(x)])(s)$ is a special case of the function $\mathcal{H}$ of the form

$$\left(\mathfrak{M}[x^{1/2}J_{\eta,\sigma}^\gamma(x)]\right)(s) = 2^{-\eta-2\sigma}\pi^{1/2}$$
$$\cdot\,\mathcal{H}_{2,3}^{1,1}\left[\left.\begin{matrix}\left(\frac{\eta}{2}+\sigma+\frac{1}{4},\frac{1}{2}\right),\left(\frac{\eta}{2}+\sigma+\frac{3}{4},\frac{1}{2}\right)\\ \left(\eta+2\sigma+\frac{1}{2},1\right),\left(\frac{\eta}{2}+\frac{1}{4},\frac{1}{2}\right),\left(\gamma\left[\frac{\eta}{2}+\sigma+\frac{1}{4}\right]-\eta-\sigma,\frac{\gamma}{2}\right)\end{matrix}\right|\, s\right] \tag{8.13.10}$$

and hence, in accordance with (3.1.5), the transform $\mathbb{J}_{\eta,\sigma}^\gamma f$ in (8.13.1) is regarded as a special $\boldsymbol{H}$-transform (3.1.1):

$$\left(\mathbb{J}_{\eta,\sigma}^\gamma f\right)(x) = 2^{-\eta-2\sigma}\pi^{1/2}$$
$$\cdot\int_0^\infty H_{2,3}^{1,1}\left[xt\left|\begin{matrix}\left(\frac{\eta}{2}+\sigma+\frac{1}{4},\frac{1}{2}\right),\left(\frac{\eta}{2}+\sigma+\frac{3}{4},\frac{1}{2}\right)\\ \left(\eta+2\sigma+\frac{1}{2},1\right),\left(\frac{\eta}{2}+\frac{1}{4},\frac{1}{2}\right),\left(\gamma\left[\frac{\eta}{2}+\sigma+\frac{1}{4}\right]-\sigma-\eta,\frac{\gamma}{2}\right)\end{matrix}\right.\right]$$
$$\cdot\, f(t)dt. \tag{8.13.11}$$

If $\sigma = 0$, then by (8.13.8), (1.1.2) and Property 2.2 of the H-function, we find

$$\left(\mathfrak{M}[x^{1/2}J_\eta^\gamma(x)]\right)(s) = 2^{-\eta}\pi^{1/2}\mathcal{H}_{1,2}^{1,0}\left[\left.\begin{matrix}\left(\frac{\eta}{2}+\frac{3}{4},\frac{1}{2}\right)\\ \left(\eta+\frac{1}{2},1\right),\left(\gamma\left[\frac{\eta}{2}+\frac{1}{4}\right]-\eta,\frac{\gamma}{2}\right)\end{matrix}\right|\, s\right] \tag{8.13.12}$$

and

$$\left(\mathbb{J}_\eta^\gamma f\right)(x)$$
$$= 2^{-\eta}\pi^{1/2}\int_0^\infty H_{1,2}^{1,0}\left[xt\left|\begin{matrix}\left(\frac{\eta}{2}+\frac{3}{4},\frac{1}{2}\right)\\ \left(\eta+\frac{1}{2},1\right),\left(\gamma\left[\frac{\eta}{2}+\frac{1}{4}\right]-\eta,\frac{\gamma}{2}\right)\end{matrix}\right.\right]f(t)dt. \tag{8.13.13}$$

For the Lommel–Maitland transform $\mathbb{J}^{\gamma}_{\eta,\sigma}$ the parameters can be evaluated in view of their definitions (3.4.1), (3.4.2), (1.1.7)–(1.1.12), and we have

$$\begin{gathered}\alpha = -\mathrm{Re}(\eta+2\sigma) - \frac{1}{2}, \quad \beta = \frac{3}{2} - \mathrm{Re}(\eta+2\sigma), \quad a^* = \frac{1-\gamma}{2}, \quad \Delta = \frac{1+\gamma}{2}, \\ \delta = 2\left(\frac{\gamma}{2}\right)^{\gamma/2}, \quad \mu = (\gamma-1)\left(\sigma + \frac{\eta}{2} + \frac{1}{4}\right) - \frac{1}{2}, \quad a_1^* = \frac{1}{2}, \quad a_2^* = -\frac{\gamma}{2}\end{gathered} \tag{8.13.14}$$

for the transform (8.13.11), and

$$\begin{gathered}\alpha = -\mathrm{Re}(\eta) - \frac{1}{2}, \quad \beta = +\infty, \quad a^* = \frac{1-\gamma}{2}, \quad \Delta = \frac{1+\gamma}{2}, \\ \delta = 2\left(\frac{\gamma}{2}\right)^{\gamma/2}, \quad \mu = (\gamma-1)\left(\frac{\eta}{2} + \frac{1}{4}\right) - \frac{1}{2}, \quad a_1^* = \frac{1}{2}, \quad a_2^* = -\frac{\gamma}{2}\end{gathered} \tag{8.13.15}$$

for the transform (8.13.13).

Let $\mathcal{E}^1_{\mathcal{H}}$ be the exceptional set of the H-function $H^{1,1}_{2,3}(z)$ in (8.13.11) (see Definition 3.4). According to (8.13.7), ν is not in the exceptional set $\mathcal{E}^1_{\mathcal{H}}$, if $\nu = 1 - \mathrm{Re}(s)$ with

$$\begin{gathered}s \neq -\eta - 2\sigma - \frac{3}{2} - 2k \quad (k \in \mathbb{N}_0), \quad s \neq \frac{3}{2} - \eta + 2m \quad (m \in \mathbb{N}_0), \\ s \neq -\left(\eta + 2\sigma + \frac{1}{2}\right) + \frac{2}{\gamma}(1 + \eta + \sigma + n) \quad (n \in \mathbb{N}_0).\end{gathered} \tag{8.13.16}$$

If $\sigma = 0$, then in accordance with (8.13.8) ν is not in the exceptional set $\mathcal{E}^2_{\mathcal{H}}$ of the H-function $H^{1,0}_{1,2}(z)$ in (8.13.13), if $\nu = 1 - \mathrm{Re}(s)$ with

$$s \neq -\eta - \frac{3}{2} - 2k \quad (k \in \mathbb{N}_0), \quad s \neq -\left(\eta + \frac{1}{2}\right) + \frac{2}{\gamma}(1 + \eta + n) \quad (n \in \mathbb{N}_0). \tag{8.13.17}$$

Applying Theorems 3.6 and 3.7, we obtain the following results for the Lommel–Maitland transforms (8.13.1) and (8.13.4) in the space $\mathfrak{L}_{\nu,2}$.

Theorem 8.54. *Let $\sigma \in \mathbb{C}$ $(\sigma \neq 0)$, $\eta \in \mathbb{C}$ and $\nu \in \mathbb{R}$ be such that the conditions $\mathrm{Re}(\sigma) > -1, \mathrm{Re}(\eta+\sigma) > -1$ and*

$$\mathrm{Re}(\eta+2\sigma) - \frac{1}{2} < \nu < \mathrm{Re}(\eta+2\sigma) + \frac{3}{2} \tag{8.13.18}$$

are satisfied and let $0 < \gamma < 1$.

(i) *There is a one-to-one transform $\mathbb{J}^{\gamma}_{\eta,\sigma} \in [\mathfrak{L}_{\nu,2}, \mathfrak{L}_{1-\nu,2}]$ such that the relation*

$$\begin{aligned}\left(\mathfrak{M}\mathbb{J}^{\gamma}_{\eta,\sigma} f\right)(s) &= 2^{-\eta-2\sigma}\pi^{1/2} \frac{\Gamma\left(\eta + 2\sigma + \frac{1}{2} + s\right)\Gamma\left(\frac{9}{4} - \frac{\eta}{2} - \sigma - \frac{s}{2}\right)}{\Gamma\left(\frac{3}{4} + \sigma + \frac{\eta}{2} + \frac{s}{2}\right)\Gamma\left(\frac{9}{4} - \frac{\eta}{2} - \frac{s}{2}\right)} \\ &\quad \cdot \frac{1}{\Gamma\left(1 + \eta + \sigma - \gamma\left[\frac{1}{4} + \sigma + \frac{\eta}{2}\right] - \frac{\gamma s}{2}\right)} \left(\mathfrak{M} f\right)(1-s)\end{aligned} \tag{8.13.19}$$

holds for $\mathrm{Re}(s) = 1 - \nu$ and $f \in \mathfrak{L}_{\nu,2}$. If the conditions in (8.13.16) are fulfilled, then the transform $\mathbb{J}^{\gamma}_{\eta,\sigma}$ maps $\mathfrak{L}_{\nu,2}$ onto $\mathfrak{L}_{1-\nu,2}$.

(ii) *If $f, g \in \mathfrak{L}_{\nu,2}$, then*

$$\int_0^\infty \left(\mathbb{J}_{\eta,\sigma}^\gamma f\right)(x) g(x) dx = \int_0^\infty f(x) \left(\mathbb{J}_{\eta,\sigma}^\gamma g\right)(x) dx. \tag{8.13.20}$$

(iii) *Let $f \in \mathfrak{L}_{\nu,2}$, $\lambda \in \mathbb{C}$ and $h \in \mathbb{R}_+$. If $\mathrm{Re}(\lambda) > (1-\nu)h - 1$, then $\mathbb{J}_{\eta,\sigma}^\gamma f$ is given by*

$$\left(\mathbb{J}_{\eta,\sigma}^\gamma f\right)(x) = 2^{-\eta-2\sigma}\pi^{1/2} h x^{1-(\lambda+1)/h} \frac{d}{dx} x^{(\lambda+1)/h}$$

$$\cdot \int_0^\infty H_{3,4}^{2,1}\left[xt \left| \begin{matrix} (-\lambda, h), \left(\frac{\eta}{2}+\sigma+\frac{1}{4}, \frac{1}{2}\right), \left(\frac{\eta}{2}+\sigma+\frac{3}{4}, \frac{1}{2}\right) \\ \left(\eta+2\sigma+\frac{1}{2}, 1\right), \left(\frac{\eta}{2}+\frac{1}{4}, \frac{1}{2}\right), \left(\gamma\left[\frac{\eta}{2}+\sigma+\frac{1}{4}\right]-\eta-\sigma, \frac{\gamma}{2}\right), (-\lambda-1, h) \end{matrix} \right.\right] f(t) dt \quad (x > 0). \tag{8.13.21}$$

When $\mathrm{Re}(\lambda) < (1-\nu)h - 1$,

$$\left(\mathbb{J}_{\eta,\sigma}^\gamma f\right)(x) = -2^{-\eta-2\sigma}\pi^{1/2} h x^{1-(\lambda+1)/h} \frac{d}{dx} x^{(\lambda+1)/h}$$

$$\cdot \int_0^\infty H_{3,4}^{1,2}\left[xt \left| \begin{matrix} \left(\frac{\eta}{2}+\sigma+\frac{1}{4}, \frac{1}{2}\right), \left(\frac{\eta}{2}+\sigma+\frac{3}{4}, \frac{1}{2}\right), (-\lambda, h) \\ (-\lambda-1, h), \left(\eta+2\sigma+\frac{1}{2}, 1\right), \left(\frac{\eta}{2}+\frac{1}{4}, \frac{1}{2}\right), \left(\gamma\left[\frac{\eta}{2}+\sigma+\frac{1}{4}\right]-\eta-\sigma, \frac{\gamma}{2}\right) \end{matrix} \right.\right] f(t) dt \quad (x > 0). \tag{8.13.22}$$

(iv) *$\mathbb{J}_{\eta,\sigma}^\gamma$ is independent of ν in the sense that, if ν and $\tilde{\nu}$ satisfy the assumption and if the transforms $\mathbb{J}_{\eta,\sigma}^\gamma$ and $\widetilde{\mathbb{J}}_{\eta,\sigma}^\gamma$ are given on $\mathfrak{L}_{\nu,2}$ and $\mathfrak{L}_{\tilde{\nu},2}$ respectively by (8.13.19), then $\mathbb{J}_{\eta,\sigma}^\gamma f = \widetilde{\mathbb{J}}_{\eta,\sigma}^\gamma f$ for $f \in \mathfrak{L}_{\nu,2} \bigcap \mathfrak{L}_{\tilde{\nu},2}$.*

(v) *For $f \in \mathfrak{L}_{\nu,2}$, $\mathbb{J}_{\eta,\sigma}^\gamma f$ is given in (8.13.1) and (8.13.11).*

Theorem 8.55. *Let $\eta \in \mathbb{C}$ and $\nu \in \mathbb{R}$ be such that $\mathrm{Re}(\eta) > -1$ and $\nu < \mathrm{Re}(\eta) + 3/2$ and let $0 < \gamma < 1$.*

(i) *There is a one-to-one transform $\mathbb{J}_\eta^\gamma \in [\mathfrak{L}_{\nu,2}, \mathfrak{L}_{1-\nu,2}]$ such that the relation*

$$\left(\mathfrak{M}\mathbb{J}_\eta^\gamma f\right)(s) = 2^{-\eta}\pi^{1/2} \frac{\Gamma\left(\frac{1}{2}+\eta+s\right)}{\Gamma\left(\frac{3}{4}+\frac{\eta}{2}+\frac{s}{2}\right)\Gamma\left(1+\eta-\gamma\left[\frac{1}{4}+\frac{\eta}{2}\right]-\frac{\gamma s}{2}\right)} \left(\mathfrak{M}f\right)(1-s) \tag{8.13.23}$$

holds for $\mathrm{Re}(s) = 1-\nu$ and $f \in \mathfrak{L}_{\nu,2}$. If the conditions in (8.13.17) are fulfilled, then the transform $\mathbb{J}_\eta^\gamma$ maps $\mathfrak{L}_{\nu,2}$ onto $\mathfrak{L}_{1-\nu,2}$.

(ii) *If $f \in \mathfrak{L}_{\nu,2}$ and $g \in \mathfrak{L}_{\nu,2}$, then*

$$\int_0^\infty \left(\mathbb{J}_\eta^\gamma f\right)(x) g(x) dx = \int_0^\infty f(x) \left(\mathbb{J}_\eta^\gamma g\right)(x) dx. \tag{8.13.24}$$

(iii) *Let $f \in \mathfrak{L}_{\nu,2}, \lambda \in \mathbb{C}$ and $h \in \mathbb{R}_+$. If* $\mathrm{Re}(\lambda) > (1-\nu)h - 1$, *then $\mathbb{J}_\eta^\gamma f$ is given by*

$$\left(\mathbb{J}_\eta^\gamma f\right)(x) = 2^{-\eta}\pi^{1/2} h x^{1-(\lambda+1)/h} \frac{d}{dx} x^{(\lambda+1)/h}$$

$$\cdot \int_0^\infty H_{2,3}^{1,1}\left[xt \,\middle|\, \begin{matrix} (-\lambda, h), \left(\frac{\eta}{2}+\frac{3}{4}, \frac{1}{2}\right) \\ \left(\eta+\frac{1}{2}, 1\right), \left(\gamma\left[\frac{\eta}{2}+\frac{1}{4}\right]-\eta, \frac{\gamma}{2}\right), (-\lambda-1, h) \end{matrix}\right] f(t)dt \quad (8.13.25)$$

$$(x > 0).$$

When $\mathrm{Re}(\lambda) < (1-\nu)h - 1$,

$$\left(\mathbb{J}_\eta^\gamma f\right)(x) = -2^{-\eta}\pi^{1/2} h x^{1-(\lambda+1)/h} \frac{d}{dx} x^{(\lambda+1)/h}$$

$$\cdot \int_0^\infty H_{2,3}^{2,0}\left[xt \,\middle|\, \begin{matrix} \left(\frac{\eta}{2}+\frac{3}{4}, \frac{1}{2}\right), (-\lambda, h) \\ (-\lambda-1, h), \left(\eta+\frac{1}{2}, 1\right), \left(\gamma\left[\frac{\eta}{2}+\frac{1}{4}\right]-\eta, \frac{\gamma}{2}\right) \end{matrix}\right] f(t)dt \quad (8.13.26)$$

$$(x > 0).$$

(iv) *$\mathbb{J}_\eta^\gamma$ is independent of ν in the sense that, if ν and $\widetilde{\nu}$ satisfy the assumption and if the transforms $\mathbb{J}_\eta^\gamma$ and $\widetilde{\mathbb{J}}_\eta^\gamma$ are given on $\mathfrak{L}_{\nu,2}$ and $\mathfrak{L}_{\widetilde{\nu},2}$ respectively by* (8.13.22), *then $\mathbb{J}_\eta^\gamma f = \widetilde{\mathbb{J}}_\eta^\gamma f$ for $f \in \mathfrak{L}_{\nu,2} \bigcap \mathfrak{L}_{\widetilde{\nu},2}$.*

(v) *For $f \in \mathfrak{L}_{\nu,2}$, $\mathbb{J}_\eta^\gamma f$ is given in* (8.13.4) *and* (8.13.13).

Now we characterize the boundedeness, range and representation of the Lommel–Maitland transform (8.13.1) and (8.13.4) in the space $\mathfrak{L}_{\nu,r}$, for which since $a^* > 0, a_1^* > 0, a_2^* < 0$, we may refer to Theorems 4.5 and 4.9.

Theorem 8.56. *Let $\sigma \in \mathbb{C}$ ($\sigma \neq 0$), $\eta \in \mathbb{C}$ and $\nu \in \mathbb{R}$ be such that the conditions $\mathrm{Re}(\sigma) > -1, \mathrm{Re}(\eta+\sigma) > -1$ and* (8.13.18) *are satisfied and let $0 < \gamma < 1$.*

(a) *If $1 \leqq r \leqq s \leqq \infty$, then the transform $\mathbb{J}_{\eta,\sigma}^\gamma$ defined on $\mathfrak{L}_{\nu,2}$ can be extended to $\mathfrak{L}_{\nu,r}$ as an element of $[\mathfrak{L}_{\nu,r}, \mathfrak{L}_{1-\nu,s}]$. If $1 < r \leqq 2$ or if $1 < r < \infty$ and the conditions in* (8.13.16) *are satisfied, then $\mathbb{J}_{\eta,\sigma}^\gamma$ is a one-to-one transform from $\mathfrak{L}_{\nu,r}$ onto $\mathfrak{L}_{1-\nu,s}$.*

(b) *If $1 \leqq r \leqq s \leqq \infty$, $f \in \mathfrak{L}_{\nu,r}$ and $g \in \mathfrak{L}_{\nu,s'}$ with $1/s + 1/s' = 1$, then the relation* (8.13.20) *holds.*

(c) *Let $1 < r < \infty$ and let $\omega, \xi, \zeta \in \mathbb{C}$ be chosen as*

$$\omega = (1-\gamma)\left(\frac{\xi+\eta}{2}+\sigma+\frac{1}{4}\right),$$

$$\mathrm{Re}(\xi) \geqq \frac{1}{1-\gamma}\left\{2\max\left[\frac{1}{r}, 1-\frac{1}{r}\right] + 2\gamma(1-\nu) - 1\right\} - \mathrm{Re}(2\sigma+\eta) - \frac{1}{2}, \quad (8.13.27)$$

$$\mathrm{Re}(\xi) > \nu - 1, \qquad \mathrm{Re}(\zeta) < 1 - \nu.$$

If the conditions in (8.13.16) *are satisfied, then*

$$\mathbb{J}_{\eta,\sigma}^\gamma(\mathfrak{L}_{\nu,r}) = \left(M_{1/2-\omega/\gamma}\mathbb{H}_{\gamma,\omega-\gamma\zeta-1}\mathbb{L}_{(\gamma-1)/2,\xi+1/2+\omega/\gamma}\right)(\mathfrak{L}_{3/2-\mathrm{Re}(\omega)/\gamma-\nu,r}), \quad (8.13.28)$$

where $\mathbb{L}_{k,\alpha}$ and $\mathbb{H}_{k,\eta}$ are the generalized Laplace and generalized Hankel transforms defined in (3.3.3) *and* (3.3.4). *If any condition in* (8.13.16) *is not satisfied, then $\mathbb{J}^{\gamma}_{\eta,\sigma}(\mathfrak{L}_{\nu,r})$ is a subset of the right-hand side of* (8.13.28).

Theorem 8.57. *Let $\eta \in \mathbb{C}$ and $\nu \in \mathbb{R}$ be such that* $\mathrm{Re}(\eta) > -1$ *and* $\nu < \mathrm{Re}(\eta) + 3/2$ *and let* $0 < \gamma < 1$.

(a) *If* $1 \leqq r \leqq s \leqq \infty$, *then the transform $\mathbb{J}^{\gamma}_{\eta}$ defined on $\mathfrak{L}_{\nu,2}$ can be extended to $\mathfrak{L}_{\nu,r}$ as an element of* $[\mathfrak{L}_{\nu,r}, \mathfrak{L}_{1-\nu,s}]$. *If* $1 < r \leqq 2$ *or if* $1 < r < \infty$ *and the conditions in* (8.13.17) *are satisfied, then $\mathbb{J}^{\gamma}_{\eta}$ is a one-to-one transform from $\mathfrak{L}_{\nu,r}$ onto $\mathfrak{L}_{1-\nu,s}$.*

(b) *If* $1 \leqq r \leqq s \leqq \infty$, $f \in \mathfrak{L}_{\nu,r}$ *and* $g \in \mathfrak{L}_{\nu,s'}$ *with* $1/s + 1/s' = 1$, *then the relation* (8.13.24) *holds.*

(c) *Let* $1 < r < \infty$. *Let* $\omega, \xi, \zeta \in \mathbb{C}$ *be chosen as in Theorem* 8.56. *If the conditions in* (8.13.17) *with* $\sigma = 0$ *are satisfied, then*

$$\mathbb{J}^{\gamma}_{\eta}(\mathfrak{L}_{\nu,r}) = \left(M_{1/2-\omega/\gamma}\mathbb{H}_{\gamma,\omega-\gamma\zeta-1}\mathbb{L}_{(\gamma-1)/2,\xi+1/2+\omega/\gamma}\right)\left(\mathfrak{L}_{3/2-\mathrm{Re}(\omega)/\gamma-\nu,r}\right). \tag{8.13.29}$$

If any condition in (8.13.17) *with* $\sigma = 0$ *is not satisfied, then $\mathbb{J}^{\gamma}_{\eta}(\mathfrak{L}_{\nu,r})$ is a subset of the right-hand side of* (8.13.29).

Proof. By virtue of (8.13.14) and (8.3.15), the assertions (a) of Theorems 8.56 and 8.57 follow from Theorems 4.5(a) and Theorem 4.9(a), while (b) and (c) are deduced from Theorem 4.5(b) and Theorem 4.9(b), respectively.

Remark 8.15. The boundedness of the Lommel–Maitland transform $\mathbb{J}^{\gamma}_{\eta,\sigma}$ for real η from $\mathfrak{L}_{\nu,r}$ into $\mathfrak{L}_{1-\nu,s}$ was given by Betancor [4, Theorem 2(a)] under more restrictive conditions than that of Theorem 8.56(a) with real η (see Section 8.14).

8.14. Bibliographical Remarks and Additional Information on Chapter 8

For Section 8.1. The classical results in the theory of the Hankel transform $\mathbb{H}_{\eta}$ in (8.1.1) are well known (see, for example, the books by Titchmarsh [3, Chapter VIII], Sneddon [1, Chapter II], Ditkin and Prudnikov [1, Part I, Chapter 3]) and Zayed [1, Section 27]. We only note that Titchmarsh [3, Theorem 135] first proved the inversion formula for the Hankel transform in the form

$$\frac{1}{2}[f(x+0) + f(x-0)] = \int_0^{\infty} (xt)^{1/2} J_{\nu}(xt)\left(\mathbb{H}_{\eta} f\right)(t)dt, \tag{8.14.1}$$

provided that $\eta > -1/2$ and $f(x)$ belongs to $L_1(\mathbb{R}_+)$, and is a bounded variation near the point x.

Kober [1] was probably the first to consider the mapping properties of the Hankel type operators in the space $L_r(\mathbb{R}_+)$ (see the remarks on Section 8.4 below).

The results presented in Theorems 8.1–8.4 and Lemmas 8.1–8.2 are extensions of those proved by Rooney [3], [4] and Heywood and Rooney [4] from real η to the complex case. In proving Theorems 8.1(b), 8.1(c,d), 8.2, 8.3 and 8.4 we followed the proofs of Heywood and Rooney [4, Theorem 2.1], Rooney [4, Theorems 2.1, 2.2], Rooney [3, Theorem 2], Rooney [4, Theorem 2.3] and Heywood and Rooney [4, Theorem 3.1].

From other results we note those by Rooney [8] who proved another inversion formula, in addition to (8.1.33) and (8.1.34), for the Hankel transform (8.1.1) in the form

$$f(x) = \lim_{k\to\infty}\left(\mathbb{H}_{k,\eta}\mathbb{H}_{\eta} f\right)(x), \tag{8.14.2}$$

where

$$\left(\mathbb{H}_{k,\eta}g\right)(x) = \frac{2k^{k+1}}{k!}x^{-(2k+\eta+3/2)}\int_0^\infty t^{2k+\eta+3/2}\exp\left(-k\frac{t^2}{x^2}\right)g(t)dt \tag{8.14.3}$$

and which is valid for $f \in \mathfrak{L}_{\nu,r}$, provided that either $1 \leqq r < \infty$, $\gamma(r) \leqq \nu \leqq \eta + 3/2$ or $r = \infty$, $1 < \nu < \eta + 3/2$ (Rooney [8, Theorem 4.1]). In terms of the operator $\mathbb{H}_{k,\eta}$ he also gave necessary and sufficient conditions for f to be in the space of functions

$$\mathbb{H}_\eta(\mathfrak{L}_{\nu,r}) = \{f : f = \mathbb{H}_\eta g, \ g \in \mathfrak{L}_{\nu,r}\} \tag{8.14.4}$$

characterized by different statements in the cases when $\eta > -1$ and $\eta < -1$ ($\eta \neq -3, -5, \cdots$) (Rooney [8, Theorem 5.1, 5.5, 6.1 and 6.3]). We also mention the paper by Heywood and Rooney [2] where conditions were given on pairs of non-negative functions $U(x)$ and $V(x)$ which are sufficient for the validity of the so-called two-weighted estimate

$$\left[\int_0^\infty |U(x)\left(\mathbb{H}_\eta f\right)(x)|^s dx\right]^{1/s} \leqq \left[\int_0^\infty |V(x)f(x)|^r dx\right]^{1/r} \tag{8.14.5}$$

with $1 < r \leqq s < \infty$ for the Hankel transform $\mathbb{H}_\eta$. Heywood and Rooney [8] showed that the Hankel transform (8.1.1) in an appropriate space $\mathfrak{L}_{\nu,r}$ satisfies ordinary and integral Lipschitz conditions.

Tricomi [1] proved the connection between the Hankel transform (8.1.1) and the Laplace transform (2.5.2) in the form

$$\left(\mathbb{L}t^{\eta/2}\mathbb{H}_\eta f(t)\right)(x) = x^{-\eta-1}\left(\mathbb{L}t^{\eta/2}f(t)\right)\left(\frac{1}{x}\right),$$

and in [2] he established some Abelian and Tauberian theorems for $\mathbb{H}_\eta$.

Relations between the Hankel transform $\mathbb{H}_\eta$, the Laplace transform $\mathbb{L}$ and the Varma transform $\boldsymbol{V}_{k,m}$ given in (8.1.1), (2.5.2) and (7.2.15) of $t^\mu f(t)$ were obtained by Bhonsle [2], [3] and by R.K. Saxena [4], respectively. Soni [1] derived various expansions for the Hankel transform (8.1.1) of any order $\eta > 0$ on $L_2(\mathbb{R}_+)$ by applying the fractional integration operators to the known expressions for $\mathbb{H}_0$. Relations between the fractional integration operators (2.7.1), (2.7.2) and the Hankel transform (8.1.1) were obtained by Bora and R.K. Saxena [1], and Martić [1], [2] gave an alternative proof of these results.

The Hankel transform $\mathbb{H}_\eta$ in various spaces of tested and generalized functions was studied by Lions [1], Zemanian [1]–[3], [5], [6], Fenyö [1], [2], Koh and Zemanian [1], Koh [1]–[4], Lee [1], [2], Braaksma and Schuitman [1], McBride [3], Pathak [10], Pathak and A.B. Pandey [1], Pathak and Sahoo [1], Koh and Li [1], [2], Pathak and Upadhyay [1], Malgonde and Chaudhary [1], Betancor, Linares and Méndez [2], [3], Betancor and Marrero [1], Betancor and Rodriguez-Mesa [3], [4], Betancor and Jerez Diaz [3], Vu Kim Tuan [3], Kanjin [1], et al. We also mention the papers by Bhonsle and Chaudhary [1] who investigated the Laplace–Hankel transform with the kernel $e^{-xt}(xt)^{1/2}J_\eta(xt)$ in a certain space of generalized functions, and by Betancor, Linares and Méndez [1], where Paley–Wiener theorems for the Hankel transform $\mathbb{H}_\eta f$ were proved. See in this connection the books by Zemanian [6], Brychkov and Prudnikov [1, Section 6.2] and Zayed [1, Section 21.7.1].

For Section 8.2. The well-known classical results in the theory of the Fourier cosine and sine transforms $\mathfrak{F}_c$ and $\mathfrak{F}_s$, defined by (8.1.2) and (8.1.3), on the spaces $L_1(\mathbb{R}_+)$ and $L_2(\mathbb{R}_+)$ were first presented by Titchmarsh [3]. The results of this section given in Theorems 8.5–8.8 and concerning the properties of $\mathfrak{F}_c$ and $\mathfrak{F}_s$ on the space $\mathfrak{L}_{\nu,r}$ were not mentioned earlier.

For Section 8.3. The even and odd Hilbert transforms (8.3.1) and (8.3.2) were first studied by Hardy and Littlewood [1] and Babenko [1], who proved that $\mathbf{H}_+ \in [\mathfrak{L}_{\nu,r}]$ for $-r < \nu < r$ and $\mathbf{H}_- \in [\mathfrak{L}_{\nu,r}]$ for $0 < \nu < r$, respectively. Rooney [1, Corollary 8.1.2] extended the latter result to $0 < \nu < 2r$ and showed that $\mathbf{H}_+$ and $\mathbf{H}_-$ are one-to-one from $\mathfrak{L}_{\nu,r}$ onto $\mathfrak{L}_{\nu,r}$ if $1 < r < 2$ or $0 < \nu < r$. The statements of Theorem 8.9 except (8.3.6) were also given by Rooney [4, Theorem 3.1]. The results in Theorem 8.11 were obtained by Heywood and Rooney [3, Theorem 4.1]. Using a technique of the Mellin transform for some auxiliary operators, they obtained inversion relations (8.3.20) and (8.3.21) and the characterization of the ranges $\boldsymbol{H}_+(\mathfrak{L}_{\nu,r})$ and $\boldsymbol{H}_-(\mathfrak{L}_{\nu,r})$.

Prooving Theorem 8.9, we followed Rooney [4, Theorem 3.1]. The inversion formulas (8.3.18) and (8.3.19) for the even and odd Hilbert transforms (8.3.1) and (8.3.2) are not mentioned earlier.

For Section 8.4. As noted above in the comments on Section 8.1, the extended Hankel transform $\mathbb{H}_{\eta,l}$ in (8.4.1) with real η was first considered in the space $L_r(\mathbb{R}_+)$ by Kober [1]. He proved sufficient conditions for the transform $\mathbb{H}_{\eta,l}$ to be bounded from $L_r(\mathbb{R}_+)$ into $L_{r'}(\mathbb{R}_+)$, provided that $1 < r \leqq 2$ $(1/r + 1/r' = 1)$ and $f(x)$ satisfy some additional conditions. Busbridge [1] obtained simpler conditions for such a result. Erdélyi and Kober [1] investigated the modified extended Hankel transform, obtained from (8.4.1) by replacing the kernel $(xt)^{1/2}J_{\eta,l}(xt)$ by $J_{\eta,l}[2(xt)^{1/2}]$. They gave conditions for the boundedeness of this operator from $L_r(\mathbb{R}_+)$ for $1 \leqq r \leqq 2$ and $-1/r' < l + \text{Re}(\eta)/2 < 1/r$, proved a Parseval theorem and discussed connections between two such operators with different η and l. Erdélyi [2] investigated connections between the extended Hankel transforms of different order.

The results presented in Section 8.4 are extensions of those proved by Rooney [5] and Heywood and Rooney [6] from real η to the complex case. In the proofs of Theorem 8.12(a) and Theorems 8.12(b), 8.13, 8.14 and 8.15 we follow that by Rooney [5, Theorem 1] and Heywood and Rooney [6, Theorem 2.1], [6, Lemma 3.3], [6, Theorem 4.1], [6, Theorem 5.1], [6, Theorem 5.2] and [6, Theorems 6.6 and 6.9], respectively.

We also mention Ahuja [1], [2] who investigated the extended Hankel transform $\mathbb{H}_{\eta,l}$ in McBride spaces of tested and generalized functions $\mathsf{F}_{p,\mu}$ and $\mathsf{F}'_{p,\mu}$ (see (5.7.8) and McBride [2]), and Vu Kim Tuan [3] who studied $\mathbf{H}_{\eta,l}$ in a space of rapidly decreasing functions and in the spaces L_p and infinitely differentiable functions with bounded support.

For Section 8.5. The results presented in this section were proved by Kilbas and Trujillo [1]. They generalize those considered by many authors in the particular cases of the Hankel type transform $\mathbb{H}_{\eta;\sigma,\varpi;k,\lambda}$ in (8.5.1).

Information concerning the investigation of the Hankel–Schwartz transforms $\mathbf{h}_{\eta;1} = \mathbb{H}_{\eta;\eta+1,-\eta;1,1}$ and $\mathbf{h}_{\eta;2} = \mathbb{H}_{\eta;-\eta,\eta+1;1,1}$ in (8.6.1) and (8.6.2) as well of the Hankel–Clifford transforms $\mathbf{b}_{\eta;1} = \mathbb{H}_{\eta;\eta/2,-\eta/2;2,2}$ and $\mathbf{b}_{\eta;2} = \mathbb{H}_{\eta;-\eta/2,\eta/2;2,2}$ in (8.6.3) and (8.6.4), are given below in the comments on Section 8.6.

As for other types of such transforms, we first note the investigations by Sneddon who studied the transform

$$\mathcal{H}_\nu[f(at);x] \equiv \int_0^\infty tJ_\eta(xt)f(at)dt = a^{-2}\left(\mathbb{H}_{\eta;0,1;1,1/a}f\right)(x) \tag{8.14.6}$$

with $a > 0$ and $\eta > -1/2$, provided that $x^{1/2}f(x)$ is a piecewise continuously and absolutely integrable function on $\mathbb{R}_+$. In particular, he proved the inversion formula of the form (8.13.1):

$$\frac{1}{2}[f(x+0) + f(x-0)] = \int_0^\infty tJ_\nu(xt)\mathcal{H}_\nu[f(t);x]dt. \tag{8.14.7}$$

His results were presented in his book Sneddon [4, Chapter 5], where one can find many references. Using the results obtained, Sneddon gave in terms of the integral transform (8.14.6) the exact solutions of axisymmetric Dirichlet problems for a half-space, for a thick plate and for the biharmonic equation, and applied these representations to problems in symmetrical vibrations of a large membrane and of a thin elastic plate, and in the motion of a viscous fluid under a surface load (Sneddon [4, Chapter 5.10]). He also applied his results to the solution of dual integral equations and mixed boundary value problems [2], [4, Chapter 5.11] and of generalized axisymmetric potentials (Sneddon [4, Chapter 5.12]).

We note that the inversion formula (8.14.7) was first obtained by Cooke [1] for special functions f and $\eta \geqq -1/2$, and Bradley [1] extended this result to $\eta > -1$.

Rooney [1, Sections 8 and 9] and Kilbas and Trujillo [2] considered the Hankel type transform defined for $f \in C_0$ by

$$\left(\mathcal{H}_{\rho,\eta}f\right)(x) = x^{1-\rho}\int_0^\infty J_\eta(xt)t^\rho f(t)dt = \left(\mathbb{H}_{\eta;1-\rho,\rho;1,1}f\right)(x) \quad (\eta > -1). \tag{8.14.8}$$

Rooney showed that $\mathcal{H}_{\rho,\eta}$ can be extended on $\mathfrak{L}_{\rho,2}$ as a unitary transform of the space into itself, i.e.

$\mathcal{H}_{\rho,\eta}^{-1} = \mathcal{H}_{\rho,\eta}$, and that, if $f \in \mathfrak{L}_{2\rho,2}$, its Mellin transform is given by

$$\left(\mathfrak{M}\mathcal{H}_{\rho,\eta}f\right)(s) = 2^{s-\rho}\frac{\Gamma\left(\dfrac{\eta-\rho+1+s}{2}\right)}{\Gamma\left(\dfrac{\eta+\rho+1-s}{2}\right)}\left(\mathfrak{M}f\right)(2\rho-s) \quad (\mathrm{Re}(s)=\rho). \tag{8.14.9}$$

Rooney also proved that $\mathcal{H}_{\rho,\eta}$ belongs to $[\mathfrak{L}_{2\rho/r,r}, \mathfrak{L}_{2\rho/r',r'}]$, provided that $1 \leqq r \leqq 2$ and $1/2 \leqq \rho \leqq \eta+1$, and characterized the constancy of its range as

$$\mathcal{H}_{\rho,\eta+\gamma}(\mathfrak{L}_{2\rho/r,r}) = \mathcal{H}_{\rho,\eta}(\mathfrak{L}_{2\rho/r,r}), \tag{8.14.10}$$

provided that $1 < r < 2$ and $1/2 \leqq \rho \leqq \min[\eta+1, \eta+\gamma+1]$. The former results were applied to study the product of two transforms (8.14.8) defined by $\mathbb{H}_{\rho,\eta,\gamma} = \mathcal{H}_{\rho,\eta+\gamma}\mathcal{H}_{\rho,\eta}$. The representations

$$\mathbb{H}_{\rho,\eta,\gamma} = \left(I^{\gamma/2}_{-;2,(\eta-\rho+1)/2}\right)^{-1} I^{\gamma/2}_{0+;2,(\eta+\rho-1)/2} \quad (\gamma>0,\ \eta>-1) \tag{8.14.11}$$

and

$$\mathbb{H}_{\rho,\eta,\gamma} = \left(I^{-\gamma/2}_{0+;2,(\eta+\gamma+\rho+1)/2}\right)^{-1} I^{-\gamma/2}_{-;2,(\eta+\gamma-\rho+1)/2} \quad (\gamma<0,\ \eta+\gamma>-1) \tag{8.14.12}$$

were obtained in terms of the Erdélyi–Kober type fractional integration operators (3.3.1) and (3.3.2) and inverses to them, and on the basis of these relations the results on the boundedness and the one-to-one nature of $H_{\rho,\eta,\gamma}$ in $\mathfrak{L}_{\nu,r}$ were proved.

Kilbas and Trujillo [2] generalized the boundedness results by Rooney and gave the representation for $\mathcal{H}_{\rho,\eta}$.

Theorem 8.58. (Kilbas and Trujillo [2, Theorem 3.1]) *Let $1 \leqq r \leqq \infty$ and let $\gamma(r)$ be given by* (3.3.9).

(a) *If $1 < r < \infty$ and $\gamma(r) \leqq \nu - \mathrm{Re}(\rho) + 1/2 < \mathrm{Re}(\eta) + 3/2$, then for all $s \geqq r$ such that $s' > [\nu - \mathrm{Re}(\rho) + 1/2]^{-1}$ and $1/s + 1/s' = 1$, the operator $\mathcal{H}_{\rho,\eta}$ in (8.14.8) belongs to $[\mathfrak{L}_{\nu,r}, \mathfrak{L}_{2\mathrm{Re}(\rho)-\nu,s}]$ and is a one-to-one transform from $\mathfrak{L}_{\nu,r}$ onto $\mathfrak{L}_{2\mathrm{Re}(\rho)-\nu,s}$. If $1 < r \leqq 2$ and $f \in \mathfrak{L}_{\nu,r}$, then the Mellin transform of $\mathcal{H}_{\rho,\eta}f$ for $\mathrm{Re}(s) = -\nu + 2\mathrm{Re}(\rho)$ is given by (8.14.9).*

(b) *If $1 \leqq \nu - \mathrm{Re}(\rho) + 1/2 \leqq \mathrm{Re}(\eta) + 3/2$, then $\mathcal{H}_{\rho,\eta} \in [\mathfrak{L}_{\nu,1}, \mathfrak{L}_{2\mathrm{Re}(\rho)-\nu,\infty}]$. If $1 < \nu - \mathrm{Re}(\rho) + 1/2 < \mathrm{Re}(\eta) + 3/2$, then for all r $(1 < r < \infty)$ $\mathcal{H}_{\rho,\eta} \in [\mathfrak{L}_{\nu,1}, \mathfrak{L}_{2\mathrm{Re}(\rho)-\nu,r}]$.*

(c) *If $f \in \mathfrak{L}_{\nu,r}$ and $g \in \mathfrak{L}_{2\mathrm{Re}(\rho)+\nu-1,s}$ for $1 < r < \infty, 1 < s < \infty$ such that $1/r + 1/s \geqq 1$ and $\max[\gamma(r), \gamma(s)] \leqq \nu - \mathrm{Re}(\rho) + 1/2 < \mathrm{Re}(\eta) + 3/2$, then the relation*

$$\int_0^\infty f(x)\left(\mathcal{H}_{\rho,\eta}g\right)(x)dx = \int_0^\infty g(x)\left(\mathcal{H}_{1-\rho,\eta}f\right)(x)dx \tag{8.14.13}$$

holds.

(d) *If $f \in \mathfrak{L}_{\nu,r}$ for $1 < r < \infty, \gamma(r) \leqq \nu - \mathrm{Re}(\rho) + 1/2 < \mathrm{Re}(\eta) + 3/2$, then for almost all $x > 0$ the relation*

$$\left(\mathcal{H}_{\rho,\eta}f\right)(x) = x^{-\rho-\eta}\frac{d}{dx}x^{\eta+1}\int_0^\infty J_{\eta+1}(xt)t^\rho f(t)\frac{dt}{t} \tag{8.14.14}$$

holds.

They also gave, in addition to (8.14.14), another representation for $\mathcal{H}_{\rho,\eta}f$ and characterized the range $\mathcal{H}_{\rho,\eta}(\mathfrak{L}_{\nu,r})$ of the Hankel type transform (8.14.8) in $\mathfrak{L}_{\nu,r}$-space in terms of the elementary operators (3.3.11) and (3.3.12), the Erdélyi–Kober operator (8.1.21) and the Fourier cosine transform (8.1.2) by

$$\mathcal{H}_{\rho,\eta}(\mathfrak{L}_{\nu,r}) = \left(M_{\nu-2\rho+1-\gamma}I_{\nu-\rho-\gamma+1/2,-1/2}\mathfrak{F}_c\right)(\mathfrak{L}_{\gamma,r}), \tag{8.14.15}$$

provided that $1 < r < \infty$ and $\gamma \equiv \gamma(r) \leqq \nu - \mathrm{Re}(\rho) + 1/2 < \mathrm{Re}(\eta) + 3/2$. The inversion theorem was also proved.

Theorem 8.59. (Kilbas and Trujillo [2, Theorem 3.3]) *Let $1 < r < \infty$ and $\gamma(r) \leqq \nu - \mathrm{Re}(\rho) + 1/2 < \mathrm{Re}(\eta) + 3/2$ or $r = 1$ and $1 \leqq \nu - \mathrm{Re}(\rho) + 1/2 \leqq \mathrm{Re}(\eta) + 3/2$, where $\gamma(r)$ is given by (3.3.9). If we choose the integer $m > \nu - \mathrm{Re}(\rho) + 1/2$ and if $f \in \mathfrak{L}_{\nu,r}$, then for almost all $x > 0$ the inversion relation*

$$f(x) = x^{1-\eta-\rho}\left(\frac{1}{x}\frac{d}{dx}\right)^m x^{\eta+m}\int_0^\infty J_{\eta+1}(xt)t^\rho\left(\mathcal{H}_{\rho,\eta}f\right)(t)\frac{dt}{t^m} \tag{8.14.16}$$

holds. In particular, if $1 < r < \infty$ and $\gamma(r) \leqq \nu - \mathrm{Re}(\rho) + 1/2 < \min[1, \mathrm{Re}(\eta) + 3/2]$, then

$$f(x) = x^{-\eta-\rho}\frac{d}{dx}x^{\eta+1}\int_0^\infty J_{\eta+1}(xt)t^\rho\left(\mathcal{H}_{\rho,\eta}f\right)(t)\frac{dt}{t}. \tag{8.14.17}$$

Kilbas and Marichev studied the generalized Hankel transform

$$\left(S_{\eta,\alpha,\sigma}f\right)(x) = \sigma^\alpha x^{-\alpha\sigma/2}\int_0^\infty t^{-\alpha\sigma/2+\sigma-1}J_{2\eta+\alpha}\left(\frac{2}{\sigma}(xt)^{\sigma/2}\right)f(t)dt$$

$$\equiv \sigma^\alpha\left(\mathbb{H}_{2\eta+\alpha;-\alpha\sigma/2,-\alpha\sigma/2+\sigma-1;2/\sigma,2/\sigma}f\right)(x) \tag{8.14.18}$$

$$\left(\mathrm{Re}(2\eta+\alpha) \geqq -\frac{1}{2},\ \sigma > 0\right)$$

and established its inversion formula and connections with the Erdélyi–Kober type operators (3.3.1) and (3.3.2), and Marichev applied the results to solve dual and triple integral equations with the Bessel function $J_\mu(x)$ in the kernels (see Samko, Kilbas and Marichev [1, Sections 18.1 and 38.1-38.2]). These investigations generalize those by Sneddon [2], who first suggested such a method by using the operators (8.14.18) with $\sigma = 2$: $S_{\eta,\alpha}f = S_{\eta,\alpha,2}f$ to solve dual integral equations. We also mention that Kalla and R.K. Saxena [2] obtained the relations between the operator $S_{\eta,\alpha}$ and the generalized fractional integration operators containing the Gauss hypergeometric function in the kernels.

K. Soni and R.P. Soni [1] obtained the Tauberian theorems for the generalized Hankel transform

$$\int_0^\infty (xt)^{-\eta}J_\eta(xt)dF(t) \tag{8.14.19}$$

with $\eta \geqq -1/2$ and a probability measure F on $[0, +\infty)$.

Betancor [2], [3] studied the Hankel type transform defined by

$$\left(\mathbf{F}_{\alpha_0,\alpha_1,\alpha_2}f\right)(x) = x^{\alpha_0-\alpha_2}\int_0^\infty (xt)^{(1-\alpha_1)/2-\alpha_2}J_\eta\left(\frac{2}{2+k}(xt)^{(2+k)/2}\right)f(t)dt$$

$$= \left(\mathbb{H}_{\eta;\alpha_0-2\alpha_2+(1-\alpha_1)/2,(1-\alpha_1)/2-\alpha_2;2/(2+k),2/(2+k)}f\right)(x) \tag{8.14.20}$$

with real α_0, α_1, α_2 and k and $\eta = (\alpha_1-1)/(2+k) \geqq -1/2$. He proved the inversion formula for such a transform provided that f is of bounded variation in a neighborhood of the point $x_0 > 0$ and belongs to $\mathfrak{L}_{\nu,1}$ with $\nu = (1-k-\alpha_1)/2-\alpha_2$, and Parseval's type relations for such a transform. Betancor in [10] and [11] investigated the transform (8.14.20) in some spaces of generalized functions and gave applications to solve a Cauchy problem involving the Bessel type operator $B_{\alpha_0,\alpha_1,\alpha_2} = x^{\alpha_0}Dx^{\alpha_1}Dx^{\alpha_2}$ $(D = d/dx)$ in [10] and to solve several differential equations involving such a Bessel type operator in [11].

Marichev and Vu Kim Tuan [1], [2] studied the isomorphism and factorization properties of the operator

$$\int_0^\infty J_\eta\left(2\sqrt{\frac{x}{t}}\right)f(t)\frac{dt}{t}$$

in special spaces of functions (see Samko, Kilbas and Marichev [1, Theorem 36.11]). We also note that Okikiolu [1], [2] studied the boundedness of the operator of the form (8.5.1) $\mathbb{H}_{\eta;\sigma-\eta+1/2,1/2-\eta;1,1}$ from

$L_r(\mathbb{R}_+)$, and Duran [1] proved the isomorphism of some subspaces of the space of infinitely differentiable functions $C^\infty(\mathbb{R}_+)$ with respect to the Hankel type operator $\mathbb{H}_{\eta;\sigma,\omega;1,1}$.

For Section 8.6. The results presented in this section were proved by Kilbas and Trujillo [3].

Concerning other results on the Hankel–Schwartz transforms $\mathbf{h}_{\eta;1}$ and $\mathbf{h}_{\eta;2}$ in (8.6.1) and (8.6.2) and the Hankel–Clifford transforms $\mathbf{b}_{\eta;1}$ and $\mathbf{b}_{\eta;2}$ in (8.6.3) and (8.6.4), we first indicate Heywood and Rooney [1, Lemma 3] who proved that the Hankel–Clifford operator $\mathbf{b}_{\eta;1}$ ($\eta > -1$) defined on $\mathbf{C}_0$ can be extended to a bounded operator in $[\mathfrak{L}_{\nu,r}, \mathfrak{L}_{1-\nu-\eta,s}]$, provided that $1 < r \leqq s < \infty$, $\max[1/r, 1-1/s] \leqq 2\nu + \eta - 1/2$ and $\nu < 1$. They defined the operator

$$\left(\boldsymbol{R}_{k,\mu,\alpha}f\right)(x) = x^{-\alpha}\left(\mathbf{b}_{\mu+\alpha;1}T_k\mathbf{b}_{\mu;1}f\right)(x) \quad (x>0;\ k>0;\ \alpha>0;\ \mu\in\mathbb{R}) \tag{8.14.21}$$

being the composition of two Hankel–Clifford operators $\mathbf{b}_{\mu+\alpha;1}$ and $\mathbf{b}_{\mu;1}$ with the translation operator T_k such that $(T_kf)(x) = f(x-k^2/4)$ for $x > k^2/4$ and $f(x) = 0$ for $0 \leqq x \leqq k^2/4$, and they showed it is bounded in $\mathfrak{L}_{\nu,r}$, provided that $1 < r < \infty$, $\nu < 1$ and

$$\max\left[\gamma(r) - 2\nu + \frac{1}{2}, \frac{1}{r'} - \nu\right] \leqq \mu \leqq \frac{3}{2} + \alpha - 2\nu - \gamma(r),$$

where $\gamma(r)$ is given by (3.3.9). Heywood and Rooney applied the latter statement to investigate the integral operators

$$\left(\mathcal{J}_k(\eta,\alpha)f\right)(x) = 2^\alpha k^{1-\alpha}x^{-2(\eta+\alpha)}\int_0^x t^{2\eta+1}(x^2-t^2)^{(\alpha-1)/2}J_{\alpha-1}\left(k(x^2-t^2)^{1/2}\right)f(t)dt \tag{8.14.22}$$

and

$$\left(\mathcal{R}_k(\eta,\alpha)f\right)(x) = 2^\alpha k^{1-\alpha}x^{2\eta}\int_x^\infty t^{1-2(\eta+\alpha)}(t^2-x^2)^{(\alpha-1)/2}J_{\alpha-1}\left(k(t^2-x^2)^{1/2}\right)f(t)dt \tag{8.14.23}$$

in $\mathfrak{L}_{\nu,r}$ with $k > 0$ and $\alpha > 0$ and real η, containing the Bessel function of the first kind (2.6.2) in the kernels. They proved the following result.

Theorem 8.60. (Heywood and Rooney [1, Theorem 4]) *Let $0 < \alpha < 1/2$. Then*

(i) *If $2/(1+\alpha) \leqq r \leqq 2$, then $\mathcal{J}_k(\eta,\alpha) \in [\mathfrak{L}_{\nu,r}]$ for $\nu \leqq 1 + 2(\alpha+\eta)$, and $\mathcal{R}_k(\eta,\alpha) \in [\mathfrak{L}_{\nu,r}]$ for $\nu \leqq 4p^{-1} - 2(\alpha+\eta) - 1$;*

(ii) *If $2 \leqq r \leqq 2(1-\alpha)$, then $\mathcal{J}_k(\eta,\alpha) \in [\mathfrak{L}_{\nu,r}]$ for $\nu \leqq 4p^{-1} + 2(\alpha+\eta) - 1$, and $\mathcal{R}_k(\eta,\alpha) \in [\mathfrak{L}_{\nu,r}]$ for $\nu \geqq 1 - 2(\alpha+\eta)$.*

Soni [3] proved the criterion of invertibility of a simpler modified operator of the form (8.14.22)

$$k^{(1-\alpha)/2}\int_0^x (x^2-t^2)^{(\alpha-1)/2}J_{\alpha-1}\left(k(x^2-t^2)^{1/2}\right)f(t)dt \quad (x>0) \tag{8.14.24}$$

in the space $L_2(\mathbb{R}_+)$. This operator was considered by many authors (see Samko, Kilbas and Marichev [1, (37.41) and Section 39.1, To Section 37.2]). In Sections 37 and 39 of this book one may find the results and bibliographical remarks concerning other types of convolution and non-convolution operators generalizing the fractional integral operators (2.7.1) and (2.7.2) and containing Bessel functions $J_\eta(z)$ and $I_\eta(z)$ in the kernels. We only indicate the paper by Soni [2], who considered the simplest non-convolution operators

$$\int_0^x J_0\left(kt^{1/2}(x^2-t^2)^{1/2}\right)f(t)dt, \quad \int_x^\infty J_0\left(kx^{1/2}(t^2-x^2)^{1/2}\right)f(t)dt \quad (x>0), \tag{8.14.25}$$

proved the boundedness of these operators from $L_p(\mathbb{R}_+)$ to $L_{p'}(\mathbb{R}_+)$ when $1 \leqq p \leqq 2$ and $1/p'+1/p = 1$, and obtained the inversion formulas for the first operator in the cases $2/3 < p \leqq 2$ and $p = 1$.

It should be noted that the operators (8.14.22) and (8.14.23), being generalizations of the Erdélyi–Kober type operators (3.3.1) and (3.3.2), were introduced by Lowndes [1] and given his name, who defined the generalized Hankel transform

$$\left(S\left(\begin{smallmatrix}a,b,y\\ \eta,\alpha,\sigma\end{smallmatrix}\right)f\right)(x)$$

$$= 2^{\alpha}x^{2\sigma-\alpha}(x^2-a^2)^{-\sigma}\int_y^{\infty} t^{1-\alpha-2\sigma}(t^2-y^2)^{\sigma}J_{2\eta+\alpha}\left((x^2-a^2)^{1/2}(t^2-y^2)^{1/2}\right)f(t)dt \quad (8.14.26)$$

and in [5] a new unsymmetrical generalized operator of fractional integration (see the relation (37.68) in Samko, Kilbas and Marichev [1]). Then he gave applications of the operators to solving dual and triple integral equations in [1] and boundary value problems involving Helmholtz type partial differential equations in [2]–[5]. In this connection see Samko, Kilbas and Marichev [1; Sections 37.2, 37.3, 38.1 (Examples 38.2 and 38.5), 39.1, 40.2 (Lemmas 40.1–40.3) and 43.2, note 40.1].

We also indicate some other results. R.K. Saxena and Sethi [1] investigated relations between the Lowndes operators (8.14.22), (8.14.23) and the generalized fractional integration operators involving the Gaussian hypergeometric function, and Ahuja [3], [4] studied Lowndes operators in McBride spaces $F_{p,\mu}$ and $F'_{p,\mu}$ (see McBride [2] and (5.7.8)). Malgonde and Raj.K. Saxena [1] extended the operators of the form (8.14.22) and (8.14.23), in which x^2 and t^2 are replaced by x^m and t^m with $m > 0$, to spaces of generalized functions.

Gasper and Trebels [1] studied in $\mathfrak{L}_{\nu,r}$ the operator $\mathcal{H}_\eta^{(\alpha)} = D_-^{\alpha}\mathbf{h}_{\eta;2}$ of the composition of the Riemann–Liouville fractional derivative (2.7.4) and the Hankel–Schwartz operator (8.6.2). They proved the estimate

$$||\mathcal{H}_\eta^{(\alpha)}f||_{\eta+1/2+1/r',r'} \leqq c\,||f||_{\eta+\alpha+1/2+1/r,r} \quad \left(\frac{1}{r}+\frac{1}{r'}=1\right), \quad (8.14.27)$$

provided that $1 < r \leqq 2$, $0 < \alpha < \eta + 3/2$ and f belongs to the space $S_0(k)$ $(k = [2\eta]+4)$ of functions continuous on $[0,\infty)$, rapidly decreasing and infinitely differentiable away from the origin, and such that $\mathcal{H}_\eta^{(\alpha)}$ has compact support in $[0,\infty)$ for all $\eta \in (-1/2, k]$.

We note that the Hankel–Schwartz transforms $\mathbf{h}_{\eta;1}$ and $\mathbf{h}_{\eta;2}$ and the Hankel–Clifford transforms $\mathbf{b}_{\eta;1}$ and $\mathbf{b}_{\eta;2}$ were studied in different spaces of tested and generalized functions. The Hankel–Schwartz transform $\mathbf{h}_{\eta;2}$, introduced by Schwartz [1], and its conjugate transform $\mathbf{h}_{\eta;1}$ were investigated by Lee [3], Dube and Pandey [1], Schuitman [1], Altenburg [1], Chaudhary [1], van Eijndhoven and de Graaf [1], Méndez [1], [2], Méndez and Sánchez Quintana [1], [2], Sánchez Quintana and Méndez [1], Betancor [1], [9], Betancor and Negrin [1], van Eijndhoven and van Berkel [1], Linares and Méndez [1] and others (see the papers above and the book by Brychkov and Prudnikov [1, Section 6.4]). The Hankel–Clifford transforms $\mathbf{b}_{\eta;1}$ and $\mathbf{b}_{\eta;2}$ were considered by Betancor [5]–[7], Méndez and Socas Robayna [1]. Duran [2] proved the isomorphism of some spaces of functions with respect to the modified Hankel–Clifford transform

$$\left(\mathbf{H}_\eta f\right)(x) = \frac{1}{2}x^{-\eta/2}\int_0^{\infty} J_\eta\left((xt)^{1/2}\right)t^{\eta/2}f(t)dt$$

$$= \frac{1}{2}\left(\mathbb{H}_{\eta;-\eta/2,\eta/2;2,1}f\right)(x) \quad (\eta > -1). \quad (8.14.28)$$

Mahato and Mahato [1] proved some Abelian theorems for the distributional Hankel–Clifford transform (8.6.8). We also mention Betancor and Rodriguez-Mesa [1] who proved necessary and sufficient conditions for a measurable function $f(x)$ on $\mathbb{R}_+$ to satisfy the relation

$$f(x) = \lim_{T\to\infty}\int_0^T t^{2\eta+1}(xt)^{-\eta}J_\eta(xt)\left(\mathbf{b}_{\eta,2}f\right)(t)dt \quad \left(-\frac{1}{2} < \eta < \frac{1}{2}\right) \quad (8.14.29)$$

for almost all $x \in \mathbb{R}_+$, and used this formula to obtain a new inversion formula for the Hankel–Clifford transform $\mathbf{b}_{\eta,2}$ in (8.6.9).

For Sections 8.7 and 8.8. The transform $\mathbb{Y}_\eta$ in (8.7.1) and the Struve transform $\mathfrak{H}_\eta$ in (8.8.1) were first considered by Titchmarsh [1], [3, §§8.1 and 8.5] as a pair of reciprocal transforms:

$$g(x) = \int_0^\infty k(xt)f(t)dt, \quad f(x) = \int_0^\infty h(xt)g(t)dt \tag{8.14.30}$$

with

$$k(x) = x^{1/2}Y_\eta(x), \quad h(x) = x^{1/2}H_\eta(x). \tag{8.14.31}$$

These transforms are given as a pair of transforms in Erdélyi, Magnus, Oberhettinger and Tricomi [4, Chapter IX and X], where it is also indicated that these transformations are the inverses of each other, provided that $-1/2 < \eta < 1/2$.

It should be noted that Hardy [1] presented the similar reciprocal relations

$$g(x) = \int_0^\infty Y_\eta(xt)tf(t)dt, \quad f(x) = \int_0^\infty H_\eta(xt)tf(t)dt, \tag{8.14.32}$$

as a particular case of more general formulas given below in (8.14.52) and (8.14.53) with $\sigma = 1/2$.

For real η the results presented in Sections 8.7 and 8.8 were proved by Rooney [4] and Heywood and Rooney [4]. Theorems 8.26–8.36 and Lemmas 8.6–8.11 are extensions of those proved by Rooney [4] and Heywood and Rooney [4] from real η to the complex case. In our proofs of Theorems 8.26, 8.27(a,b), 8.27(c), 8.27(d), 8.27(e), 8.28, 8.29, 8.30 and 8.31 we follow Rooney [4, Theorem 4.1], Rooney [4, Theorem 4.2], Heywood and Rooney [4, Theorem 2.2], Rooney [4, Theorem 4.3], Rooney [4, Theorem 4.4], Rooney [4, Theorem 6.2], Heywood and Rooney [4, Theorem 4.1], Heywood and Rooney [4, Theorem 4.2] and Heywood and Rooney [4, Theorem 5.3], respectively. Similarly we follow Rooney [4, Theorem 5.1], Rooney [4, Theorem 5.2], Heywood and Rooney [4, Theorem 2.3], Rooney [4, Theorem 5.3], Rooney [4, Theorem 5.4], Rooney [4, Theorem 6.1], Rooney [4, Theorem 6.3] and Heywood and Rooney [4, Theorem 6.1] while proving Theorems 8.32, Theorem 8.33(a,b), Theorem 8.33(c), 8.33(d), 8.33(e), 8.34, 8.35 and 8.36, respectively.

We note that the inversion formula for the Struve transform $\mathfrak{H}_\eta$ was obtained in Theorem 8.36 for $\nu \in \mathbb{C}$ such that $\nu > -\mathrm{Re}(\eta) - 1/2$. Heywood and Rooney [7] proved the inversion formulas for the Struve transform with real η for $\nu \leqq -\eta - 1/2$:

Theorem 8.61. (Heywood and Rooney [7, Theorems 2.1, 2.3 and 3.2]) *Let $1 < r < \infty$, $\nu \geqq \gamma(r)$, where $\gamma(r)$ is given by (3.3.9), and $f \in \mathfrak{L}_{\nu,r}$. Then for almost all $x > 0$*

$$f(x) = x^{-(\eta+1/2)}\frac{d}{dx}x^{\eta+1/2}\int_0^\infty (xt)^{1/2}Y_{\eta+1}(xt)\left(\mathfrak{H}_\eta f\right)(t)\frac{dt}{t} \tag{8.14.33}$$

for $\max[-(\eta+1/2), \eta+3/2] < \nu < 1$,

$$f(x) = x^{-(\eta+1/2)}\frac{d}{dx}x^{\eta+1/2}$$

$$\cdot\int_0^\infty (xt)^{1/2}\left[Y_{\eta+1}(xt) - \frac{\cot(\pi\eta)}{\Gamma(\eta+2)}\left(\frac{xt}{2}\right)^{\eta+1}\right]\left(\mathfrak{H}_\eta f\right)(t)\frac{dt}{t} \tag{8.14.34}$$

for $\nu < \min[1, -(\eta+1/2), \eta+5/2]$, and

$$f(x) = x^{-(\eta+1/2)}\frac{d}{dx}x^{\eta+1/2}\int_{\to 0}^\infty (xt)^{1/2}Y_{\eta+1}(xt)(\mathfrak{H}_\eta f)(t)\frac{dt}{t} \tag{8.14.35}$$

for $\nu = -(\eta+1/2)$ and $\eta+3/2 < \nu < 1$, where the integral in (8.14.35) is given in (8.7.46).

The $\mathfrak{L}_{\nu,r}$-theory of the transforms $\mathbb{Y}_\eta$ and $\mathfrak{H}_\eta$ can be also constructed by using their representations as special cases of the $\boldsymbol{H}$-transform (3.1.1). Such an approach was developed by Kilbas and Gromak [1]. In this way new representations and new inversion relations for $\mathbb{Y}_\eta$- and $\mathfrak{H}_\eta$-transforms were established. Rooney [8, Corollaries 5.3 and 5.4 and Theorem 6.2] gave the necessary and sufficient conditions for f to be in the spaces of functions

$$\mathbb{Y}_\eta(\mathfrak{L}_{\nu,r}) = \{f : f = \mathbb{Y}_\eta g, \ g \in \mathfrak{L}_{\nu,r}\} \quad (\eta \in \mathbb{R}) \tag{8.14.36}$$

and

$$\mathfrak{H}_\eta(\mathfrak{L}_{\nu,r}) = \{f : f = \mathfrak{H}_\eta g, \ g \in \mathfrak{L}_{\nu,r}\} \quad (\eta \in \mathbb{R}) \tag{8.14.37}$$

in terms of the operator $\mathbb{H}_{k,\eta}$ defined by (8.14.3). Rooney [7, Theorem 2] and Heywood and Rooney [7, Theorem 3.4] characterized the ranges $\mathbb{Y}_\eta(\mathfrak{L}_{\nu,r})$ and $\mathfrak{H}_\eta(\mathfrak{L}_{\nu,r})$ in the exceptional cases when $\nu = -\eta + 1/2$ and $\nu = -\eta - 1/2$, respectively. We also indicate the paper by Vu Kim Tuan [2] in which the range of the transform $\mathbb{Y}_\eta$ was studied in some spaces of functions.

Okikiolu [1], [2] studied the modified $\mathbb{Y}_{\eta;\rho}$ and $\mathfrak{H}_{\eta;\rho}$ transforms of the form

$$\left(\mathbb{Y}_{\eta;\rho} f\right)(x) = x^\rho \int_0^\infty (xt)^{1/2-\eta} Y_{\eta-1/2}(xt) f(t) dt \tag{8.14.38}$$

and

$$\left(\mathfrak{H}_{\eta;\rho} f\right)(x) = x^\rho \int_0^\infty (xt)^{1/2-\eta} \mathbf{H}_{\eta-1/2}(xt) f(t) dt \tag{8.14.39}$$

in the space $L_p(\mathbb{R}_+)$. He showed that the above operators can be represented as the compositions of the Erdélyi–Kober type operators (3.3.1) and (3.3.2) with the cosine- and sine-transforms (8.1.2) and (8.1.3).

Love [4] studied the modified Struve transform defined for $\eta \in \mathbb{C}$ ($\mathrm{Re}(\eta) > -1/2$) and $x > 0$ by

$$\left(\mathcal{S}_\eta f\right)(x) = \int_0^\infty (x\sqrt{t})^\eta \mathbf{H}_\eta(x\sqrt{t}) f(t) dt \tag{8.14.40}$$

in the space $\mathfrak{L}_{\eta-1,1}$ and established the inversion relation of the form

$$\left(I_-^{\eta+1/2} f\right)(x) = \frac{2^\eta}{\sqrt{\pi}} \int_0^\infty \sin(t\sqrt{x}) t^{-2\eta} \left(\mathcal{S}_\eta f\right)(t) dt, \tag{8.14.41}$$

where $I_-^{\eta+1/2} f$ is the fractional integral (2.7.2). McKellar, Box and Love [1] proved the inversion theorems for the modified Struve transform

$$\left(\frac{\pi}{2}\right)^{1/2} \int_0^\infty (xt)^{1/2-n} \mathbf{H}_{n+1/2}(xt) f(t) dt \quad (n \in \mathbb{N}_0). \tag{8.14.42}$$

Isomorphism and factorization properties of the operators

$$\int_0^\infty Y_\eta\left(2\sqrt{\frac{x}{t}}\right) f(t) \frac{dt}{t}, \quad \int_0^\infty \mathbf{H}_\eta\left(2\sqrt{\frac{x}{t}}\right) f(t) \frac{dt}{t} \tag{8.14.43}$$

in special spaces of functions were investigated by Marichev and Vu Kim Tuan [1], [2] (see Samko, Kilbas and Marichev [1, Theorems 36.12 and 36.13]).

For Section 8.9. The transform $\mathfrak{K}_\eta$ in (8.9.1) was introduced by Meijer [1] and therefore given his name. Meijer [1], [2] considered the reciprocal formulas

$$g(x) = \left(\frac{2}{\pi}\right)^{1/2} \int_0^\infty (xt)^{1/2} K_\eta(xt) f(t) dt, \tag{8.14.44}$$

$$f(x) = \frac{1}{i} \left(\frac{2}{\pi}\right)^{1/2} \int_{\gamma-i\infty}^{\gamma+i\infty} (xt)^{1/2} I_\eta(xt) g(t) dt, \tag{8.14.45}$$

where $I_\eta(z)$ is the modified Bessel function of the first kind (8.9.3). Meijer [1], [2] proved that if $f(t)$ is a bounded variation function and the integral $\int_0^\infty e^{-\gamma t}|g(t)|dt$ is convergent, then (8.14.44) implies (8.14.45) provided that $-1/2 \leqq \eta \leqq 1/2$. Such a result is also true for $\eta \in \mathbb{C}$, $\mathrm{Re}(\eta) \geqq -1/2$ (see Zayed [1, Section 23.5]).

When $\eta = -1/2$, the right side of (8.14.44) coincides with the Laplace transform (2.5.2) (see (8.9.4)), while the right side of (8.14.45) coincides with the complex inversion formula for the Laplace transform (see, for example, Titchmarsh [3, Sections 1.4 and 11.7]). Boas [1] gave an inversion formula for the transform

$$g(x) = \left(\frac{2}{\pi}\right)^{1/2} \int_0^\infty (xt)^{1/2} K_\eta(xt) d\alpha(t), \tag{8.14.46}$$

which generalizes the Post–Widder inversion operator for the Laplace transform (see Widder [1]).

Using the Erdélyi–Kober fractional integration operators (3.3.1) and (3.3.2), Fox [3] reduced the modified Meijer transform

$$\left(\mathfrak{K}_{\eta,\eta} f\right)(x) = \int_0^\infty (xt)^\eta K_\eta(xt) f(t) dt \quad (x > 0) \tag{8.14.47}$$

to the form of a Laplace transform and obtained its inversion formula. In [4] he gave such a solution as an illustration of a method, developed by him and based on direct and inverse Laplace transforms, to find a formal solution of the first equation in (8.14.30) with the kernel $k(x)$ such that its Mellin transform (2.5.1) has the form

$$(\mathfrak{M}k)(s) = \prod_{i=1}^n \Gamma(a_i + \alpha_i s) \left[\prod_{j=1}^m \Gamma(b_j + \beta_j s)\right]^{-1}. \tag{8.14.48}$$

González de Galindo and Kalla [1] used this method to find the inversion for the modified Meijer transform

$$\int_0^\infty e^{xt} K_\eta(xt) f(t) dt. \tag{8.14.49}$$

It should be noted that earlier Conolly [1] on the basis of the direct and inverse Laplace transforms suggested a formal process to obtain the inversion formula for the modified Meijer transform

$$\left(\mathfrak{K}_{\eta;\alpha,\omega} f\right)(x) = x^\omega \int_0^\infty (xt)^{\alpha-1/2} K_{\eta-1/2}(xt) f(t) dt \tag{8.14.50}$$

with $\omega = 0$ and $\alpha = 1/2$. Such a modified Meijer transform in the space $L_p(\mathbb{R}_+)$ was studied by Saksena [1] for $p = 2$, $\omega = 0$ and $\alpha = \eta > 1/2$, Okikiolu [1], [2] and Manandhar [1] for $p \geqq 1$ and $\eta \in \mathbb{R}$. They applied the Erdélyi–Kober type fractional integrals (3.3.1) and (3.3.2) and the Mellin transform to find the inversion relations for such a modified Meijer transform. Okikiolu [1], [2] showed that the operator $\mathfrak{K}_{\eta;\alpha,\omega}$ can be represented as a composition of the generalized Laplace transform and Erdélyi–Kober type operator (3.3.2), and that $K_{\eta;\alpha,\omega}$ is bounded from $L_p(\mathbb{R}_+)$ into $L_q(\mathbb{R}_+)$, where $1/q = 1 - \omega - 1/p > 0$.

K.C. Sharma [1] investigated the compositions of the modified Meijer transforms $K_{\eta+1/2,1,1}$ in (8.14.50) with different indices η, and in [2] he gave four inversion formulas for such a transform in terms of direct and inverse Mellin transforms and by means of the inversion formulas for Mellin and $\mathbb{Y}$-transforms, Mellin and Hankel transforms and a formal process suggested by Conolly [1]. Nasim [1] obtained the inversion formula for the transform $K_{\eta+1/2;\alpha+1/2,0}$ with $|\eta| \leqq 1/2$ and $\alpha > 0$ in terms of the Hankel transform (8.1.1) by using an infinite-order differential operator.

Bora and R.K. Saxena [1] showed that the Meijer K_η-transform (8.9.1) can be represented as the Laplace transform (2.5.2) of the functions $I^\eta_{0+} f(x^2)$ and $x^{2\eta-2} I^\eta_- f(x^{-2})$, where I^η_{0+} and I^η_- are the fractional integration operators given in (2.7.1) and (2.7.2). Martić [1], [2] gave an alternative proof of these results. Connections of the Meijer transform (8.9.1) with the Laplace transform (2.5.2) and the generalized Varma transform (7.3.1) were studied by Bhonsle [1] and Bhise [2], with the Varma transform (7.2.15) by Bhise [2] and Gupta [1], and with the Hankel transform (8.1.1) by Bhise [2].

Zemanian [3], [4], Fenyö [1], [2], Koh and Zemanian [1], Koh [1], Barrios and Betancor [4], [5], Betancor and Barrios [1], Malgonde [2], Mahato and Agrawal [1], and Betancor and Rodriguez-Mesa [2] studied the Meijer transform K_μ in some spaces of tested and generalized functions (see in this connection the books by Zemanian [5], Brychkov and Prudnikov [1, Section 6.7]) and Zayed [1, Section 23.7]. Conlan and Koh [1], Koh, Deeba and Ali [1], [2], Deeba and Koh [1] investigated the Meijer type transform

$$\left(\mathbf{K}_\eta f\right)(x) = \int_0^\infty (xt)^{\eta/2} K_\eta\left((xt)^{1/2}\right) f(t)dt \tag{8.14.51}$$

in some spaces of generalized functions and gave applications to construct an operational calculus for the Bessel operator $B_\eta = t^{-\eta} D t^{1+\eta} D$ $(D = d/dt)$, and to solve a boundary value problem for the one-dimensional wave equation.

The results presented in Section 8.9 were obtained by Kilbas, Saigo and Trujillo [1].

For Section 8.10. The Bessel type transforms $\mathbf{K}_\eta^\rho$ and $\mathbf{L}_\eta^{(m)}$ in (8.10.1) and (8.10.2) were introduced by Krätzel in [5] and [1], respectively. For the transform $\mathbf{K}_\eta^\rho$ an inversion and convolution theorem were given by Krätzel [5], and an operational calculus was constructed by Rodriguez, Trujillo and Rivero [1]. Compositions of $\mathbf{K}_\eta^\rho$ with the Riemann–Liouville fractional integration (2.7.1), (2.7.2) and fractional differentiation operators (2.7.3), (2.7.4) in certain weighted spaces of locally integrable and finite differentiable functions were proved by Kilbas and Shlapakov [1], [3] with applications to the solution of ordinary linear differential equations of the second kind. Such compositions in McBride spaces of tested and generalized functions $\mathsf{F}_{p,\mu}$ and $\mathsf{F}'_{p,\mu}$ (McBride [2]) were given by Kilbas, Bonilla, Rodriguez, Trujillo and Rivero [1].

The properties of $\mathbf{L}_\eta^{(m)}$ such as an inversion and convolution theorem, operational rules, differentiation relations, and connections with differential operators were investigated by Krätzel [1]–[4]. Barrios and Betancor [1] obtained two real inversion formulas and discuss Abelian and Tauberian theorems for the transform $\mathbf{L}_\eta^{(m)}$. Rao and Debnath [1], Barrios and Betancor [2], [3], and Malgonde [2] investigated this transform in some spaces of generalized functions.

The results presented in Section 8.10 were proved by Kilbas and Glaeske [1] and Glaeske and Kilbas [1].

For Section 8.11. The modified Bessel type transform $\mathbf{L}_{\eta,\sigma}^{(\beta)}$ in (8.11.1) was introduced by Kilbas, Saigo and Glaeske [1] and Glaeske, Kilbas and Saigo [1] and they proved its Mellin transform (8.11.11), the asymptotic relations (8.11.20) and (8.11.21) and the composition formulas of $\mathbf{L}_{\eta,\sigma}^{(\beta)}$ with the left- and right-sided Liouville fractional integrals and derivatives on McBride spaces $\mathsf{F}_{p,\mu}$ and $\mathsf{F}'_{p,\mu}$.

The results presented in Section 8.11 were proved by Bonilla, Kilbas, Rivero, Rodriguez and Trujillo [1].

For Section 8.12. The transform of the form (8.12.6) with $(xt)^{1/2} J_{\eta,\sigma}(xt)$ being replaced by $J_{\eta,\sigma}(xt)$ is known as the Hardy transform. Hardy [1] established a pair of reciprocal relations (8.14.30) of the form

$$g(x) = \int_0^\infty C_{\eta,\sigma}(xt) t f(t) dt, \quad C_{\eta,\sigma}(z) = \cos(\sigma\pi) J_\eta(z) + \sin(\sigma\pi) Y_\eta(z) \tag{8.14.52}$$

and

$$f(x) = \int_0^\infty J_{\eta,\sigma}(xt) t g(t) dt, \tag{8.14.53}$$

where $J_{\eta,\sigma}(z)$ is given by (8.12.4). Because of this both transforms in (8.14.52) and (8.14.53) are called Hardy transforms.

Cooke [1] obtained the following inversion formula for the transforms (8.14.52) and (8.14.53).

Theorem 8.62. (Cooke [1]) *Let $\sigma > -1$, $\sigma + \eta > -1$ and let $f(t)$ be a function of bounded variation in the neighborhood of the point $t = x$.*

(i) *If $|\eta + 2\sigma| < 3/2$, $t^{\rho} f(t) \in L(0, \delta)$ and $\sqrt{t} f(t) \in L(\delta, \infty)$, where $\delta > 0$ and $\rho = \min[1 - |\eta|, (1 - \eta - \sigma)/2]$, then*

$$\frac{1}{2}[f(x+0) + f(x-0)] = \int_0^\infty J_{\eta,\sigma}(xt) t dt \int_0^\infty C_{\eta,\sigma}(t\tau) \tau f(\tau) d\tau. \tag{8.14.54}$$

(ii) *If $\eta + 2\sigma < 3/2$, $|\eta| \leqq 3/2$, $t^{\rho} f(t) \in L(0, \delta)$ and $\sqrt{t} f(t) \in L(\delta, \infty)$, where $\delta > 0$ and $\rho = \min[1 + \eta + 2\sigma, 1/2]$, then*

$$\frac{1}{2}[f(x+0) + f(x-0)] = \int_0^\infty C_{\eta,\sigma}(xt) t dt \int_0^\infty J_{\eta,\sigma}(t\tau) \tau f(\tau) d\tau. \tag{8.14.55}$$

Srivastava [3] proved certain theorems for the Hardy transform (8.14.53). Connections of this transform with the Varma transform (7.2.15) were investigated by Srivastava [4], with the generalized Varma transform (7.3.1) by Srivastava [3], and with the integral operator $(Kf)(x) = x \int_0^\infty k(xt) f(t) dt$ of the general form by Kalla [9].

Moiseev, Prudnikov and Skurnik [1] considered the particular case of the transform (8.12.1) with $a = 1$, $b = \sigma + 1$, $c = \sigma + 3/2$ and $\omega = 2\sigma + 1$ in the form

$$F(x) = \int_0^\infty \frac{(xt)^{2\sigma+1}}{\Gamma(2\sigma+2)} {}_1F_2\left(1; \sigma+1, \sigma+\frac{3}{2}; -\frac{(xt)^2}{4}\right) f(t) dt \tag{8.14.56}$$

in the space $L_r(\mathbb{R}_+)$. They proved [1, Theorem 1] that this transform is one-to-one in $L_r(\mathbb{R}_+)$ for $a \in [-1/2, 1/2)$, and established its inversion formula

$$f(x) = \frac{2}{\pi} \int_0^\infty \sin(xt - \sigma\pi) F(t) dt, \tag{8.14.57}$$

provided that $a \in [-1/2, 1/4)$.

Pathak and J.N. Pandey in [1], [3] extended the Hardy transform (8.14.52) to special spaces of generalized functions and, in particular, obtained the inversion formula for such a distributional Hardy transform (see some of their results in Zayed [1, Section 22.6]). In [2] Pathak and J.N. Pandey proved four Abelian theorems for the Hardy transforms (8.14.52) and (8.14.53).

The generalized Hardy–Titchmarsh transform (8.12.1) with $\omega = b + c - a - 1/2$

$$\left(\mathbf{J}_{a,b,c} f\right)(x) = \int_0^\infty J_{a,b,c}(xt) f(t) dt, \tag{8.14.58}$$

where

$$\begin{aligned} J_{a,b,c}(x) &= \sum_{k=0}^\infty \frac{(-1)^k \Gamma(a+k)}{\Gamma(b+k)\Gamma(c+k)k!} \left(\frac{x}{2}\right)^{2k+b+c-a-1/2} \\ &= \frac{\Gamma(a)}{\Gamma(b)\Gamma(c)} \left(\frac{x}{2}\right)^{b+c-a-1/2} {}_1F_2\left(a; b, c; -\frac{x^2}{4}\right) \end{aligned} \tag{8.14.59}$$

was studied by Titchmarsh [3, Section 8.4, Example 3]. Considering (8.14.58) as the first reciprocal transform in (8.14.30) with the kernel $k(x) = J_{a,b,c}(x)$, Titchmarsh proved the inversion formula in the form of the second reciprocal transform in (8.14.30) with

$$\begin{aligned} h(x) = {} & \frac{\sin(a-b)\pi}{\sin(c-b)\pi} \left(\frac{x}{2}\right)^{a+b-c-1/2} {}_1F_2\left(1-a+b; 1+b-c, b; -\frac{x^2}{4}\right) \\ & + \frac{\sin(a-c)\pi}{\sin(b-c)\pi} \left(\frac{x}{2}\right)^{a-b+c-1/2} {}_1F_2\left(1-a+c; 1-b+c, c; -\frac{x^2}{4}\right). \end{aligned} \tag{8.14.60}$$

Titchmarsh noted that these reciprocal transforms can also be obtained from the results by Fox [1]. The particular case $a = \eta + 3/2$, $b = \eta + 1$ and $c = 2\eta + 1$ with

$$k(x) = \frac{\sqrt{\pi}}{2}\frac{d}{dx}\left[xJ_\eta^2\left(\frac{x}{2}\right)\right], \quad h(x) = -\sqrt{\pi}J_\eta\left(\frac{x}{2}\right)Y_\eta\left(\frac{x}{2}\right) \tag{8.14.61}$$

was investigated earlier by Bateman [1], [2]. The case $a = \eta+\sigma+1/2$, $b = \eta+\sigma+1$, $c = 2\eta+\sigma+1$ yields a more general transform considered by Titchmarsh [2]; the inversion relation for such a transform was proved by Cook [1]. Bhise and Dighe [1] proved Tauberian theorems for the transform of the form (8.14.58), while Dighe and Bhise [1] established Abelian theorems for such a transform.

H.M. Srivastava [1] introduced the transform more general than (8.14.58):

$$\left(\mathbf{J}_\eta^{\lambda,\mu}f\right)(x) = \int_0^\infty t\psi_{\eta,\lambda,\mu}\left(\frac{x^2t^2}{4}\right)f(t)dt \tag{8.14.62}$$

with the kernel

$$\psi_{\eta,\lambda,\mu}(z) = \sqrt{\pi}\sum_{k=0}^\infty \frac{(-1)^k(\eta+k+1)_k}{\Gamma(\lambda+k+1)\Gamma(\mu+k+1)k!}\left(\frac{x}{4}\right)^{k+\eta/2} \tag{8.14.63}$$

and investigated the relations of this transform with the Laplace transform (2.5.2) and the Meijer transform (8.9.1) in [1] and [2], respectively. Srivastava and Vyas [1] obtained a relationship between the $\mathbf{J}_\eta^{\lambda,\mu}$-transform and the generalized Whittaker transform (7.3.1).

The results presented in Section 8.12 were obtained in Kilbas, Saigo and Borovco [2], [3]. Some of these assertions in the particular case $\omega = b+c-a-1/2$ for real η were given by Kilbas and Borovco [1].

For Section 8.13. The Lommel–Maitland transform $\mathbb{J}_{\eta,\sigma}^\gamma$ in (8.13.1) was first considered by Pathak [1]–[3], who obtained some elementary properties of this transform, proved the inversion formula and indicated the relation of this transform with the Laplace transform and applied the results obtained to evaluate a number of infinite integrals involving Meijer's G-function.

The Lommel–Maitland transform $\mathbb{J}_{\eta,\sigma}^\gamma$ with real η in the spaces $\mathfrak{L}_{\nu,r}$ with $1 < r < \infty$, $\nu \in \mathbb{R}$, $\max[1/r, 1/r'] \leqq \nu < 1$ and $r \leqq s < 1/(1-\nu)$ was considered by Betancor [4]. He proved the boundedness of $\mathbb{J}_{\eta,\sigma}^\gamma$ from $\mathfrak{L}_{\nu,r}$ into $\mathfrak{L}_{1-\nu,s}$ provided that $0 < \gamma < 1$ and the conditions

$$\sigma > -1, \quad \eta+\sigma > -1, \quad 1 + \max\left[1, \eta + 2\sigma - \frac{9}{2}\right] < \nu < \eta + 2\sigma + \frac{3}{2}. \tag{8.14.64}$$

These conditions are harder than that of Theorem 8.56(a) for real η. Betancor also gave conditions for characterizing the range of the Lommel–Maitland transform (8.13.1) in terms of the Fourier cosine transform (8.1.2) by $\mathbb{J}_{\eta,\sigma}^\gamma(\mathfrak{L}_{\nu,r}) \subseteq \mathfrak{F}_c(\mathfrak{L}_{\nu,r})$ and $\mathbb{J}_{\eta,\sigma}^\gamma(\mathfrak{L}_{\nu,r}) = \mathfrak{F}_c(\mathfrak{L}_{\nu,r})$, and for the imbedding $\mathbb{J}_{\eta_1,\sigma_1}^{\gamma_1}(\mathfrak{L}_{\nu,r}) \subseteq \mathbb{J}_{\eta_2,\sigma_2}^{\gamma_2}(\mathfrak{L}_{\nu,r})$ with different γ_i, η_i and σ_i $(i = 1, 2)$.

The results presented in Section 8.13 were obtained by the authors together with Borovco in Kilbas, Saigo and Borovco [1].

The particular case of the Lommel–Maitland transform $\mathbb{J}_\eta^\gamma$ in (8.13.4) in the form

$$\left(\mathbb{J}^{\gamma,\eta}f\right)(x) = \int_0^\infty (xt)^{\eta+1/2}J^{\gamma,\eta}\left(\frac{x^2t^2}{4}\right)f(t)dt \quad (x > 0), \tag{8.14.65}$$

where $J^{\gamma,\eta}(z)$ is the Bessel–Maitland function (8.13.5), was considered by Agarwal [1]–[3], who gave the Parseval relation and some other properties of this transform proving three inversion formulas: in a form similar to (8.13.4), in terms of a double integral and as a differential operator of infinite order. Betancor [12] obtained another inversion formula and proved Abelian theorems for the transform (8.14.65).

We mention several results concerning integral transforms with the Bessel–Maitland function (8.13.5) in the kernel. Marichev [1] studied the transform

$$\left(\mathbf{J}^{\gamma,\eta}f\right)(x) = \int_0^\infty (xt)^{1/2}J^{\gamma,\eta}(xt)f(t)dt \quad (x > 0) \tag{8.14.66}$$

in $L_2(\mathbb{R}_+)$. Kumar [1]–[4] studied properties of the modified transform

$$\left(\mathbf{J}_{\sigma}^{\gamma,\eta} f\right)(x) = \int_0^\infty (xt)^\sigma J^{\gamma,\eta}(xt) f(t)dt \quad (x > 0) \tag{8.14.67}$$

including the existence, recurrence relations, and connections with Laplace and Hankel transforms and gave applications to evaluate some integrals. R. Gupta and Jain [1], [2] investigated such a modified transform in some spaces of generalized functions.

Betancor [8] proved the boundedness and the range in $\mathfrak{L}_{\nu,r}$ of the so-called Watson–Wright transform defined by

$$\left(\mathbf{H}_{\eta,\eta',\lambda}^{\gamma,\gamma',\sigma} f\right)(x) = \int_0^\infty (xt)^\sigma w_{\eta,\eta',\lambda}^{\gamma,\gamma'}\left(x^2t^2\right) f(t)dt \quad (x > 0) \tag{8.14.68}$$

with $\eta > -1$, $\eta' > -1$, and $0 < \gamma < 1$ or $\gamma = 1$, $\eta' - \lambda > -1/2$, and $0 < \gamma' < 1$ or $\gamma' = 1$, $\eta + \lambda > -1/2$. Such a transform, introduced by Olkha and Rathie [1], [2], contains the function

$$w_{\eta,\eta',\lambda}^{\gamma,\gamma'}(x) = x^{1/2} \int_0^\infty t^{\lambda-1} J^{\gamma,\eta}(xt) J^{\gamma',\eta'}\left(\frac{1}{t}\right) dt \tag{8.14.69}$$

in the kernel, which for $\gamma = \gamma' = 1$ and $\lambda = 0$ coincides with the function

$$w_{\eta,\eta'}(x) = x^{1/2} \int_0^\infty J_\eta(xt) J_{\eta'}\left(\frac{1}{t}\right) \frac{dt}{t} \quad \left(\eta > -\frac{1}{2},\ \eta' > -\frac{1}{2}\right) \tag{8.14.70}$$

defined by Watson [1]. Bhatnagar [1], [2] studied the functions of this type as the kernels $k(x)$ and $h(x)$ of a pair of reciprocal transforms (8.14.30). Using the Kober operators (7.12.5) and (7.12.6), K.J. Srivastava [2], [3] investigated the integral transform

$$\left(\boldsymbol{W}_{\eta,\eta'} f\right)(x) = \int_0^\infty w_{\eta,\eta'}(xt) f(t)dt \tag{8.14.71}$$

with such a function kernel in the space $L_2(\mathbb{R}_+)$.

BIBLIOGRAPHY

Agarwal R.P. [1] Sur une généralisation de la transformation de Hankel (French), *Ann. Soc. Sci. Bruxelles. Sér. I* **64**(1950), 164–168.

[2] Some properties of generalised Hankel transform, *Bull. Calcutta Math. Soc.* **43**(1951), 153–167.

[3] Some inversion formulae for the generalised Hankel transform, *Bull. Calcutta Math. Soc.* **45**(1953), 69–73.

Agrawal B.M. [1] On generalized Meijer's H functions satisfying the Truesdell F-equations, *Proc. Nat. Acad. Sci. India Sect. A* **38**(1968), 259–264.

Ahuja G. [1] On extended Hankel transformation for generalized functions, *Ranchi Univ. Math. J.* **14**(1983), 113–120 (1984).

[2] A study of the distributional extended Hankel transformation, *Indian J. Math.* **28**(1986), 155–162.

[3] A study of Lowndes' operators for generalized function. I, *J. Maulana Azad College Tech.* **19**(1986), 115–124.

[4] A study of Lowndes' operators for generalized function, *J. Maulana Azad College Tech.* **21**(1988), 9–15.

Altenburg G. [1] Bessel-Transformationen in Räumen von Grundfunktionen über dem Intervall $\Omega = (0, \infty)$ und deren Dualräumen, *Math. Nachr.* **108**(1982), 197–218.

Anandani P. [1] Some expansion formulae for H-function. III, *Proc. Nat. Acad. Sci. India Sect. A* **39**(1969), 23–34.

[2] On some recurrence formulae for the H-function, *Ann. Polon. Math.* **21**(1969), 113–117.

[3] On finite summation, recurrence relations and identities of H-functions, *Ann. Polon. Math.* **21**(1969), 125–137.

[4] Some infinite series of H-function. I, *Math. Student* **37**(1969), 117–123.

[5] Some infinite series of H-functions. II, *Vijnana Parishad Anusandhan Patrika* **13**(1970), 57–66.

[6] On the derivative of the H-function, *Rev. Roumaine Math. Pures Appl.* **15**(1970), 189–191.

[7] Some finite series of H-functions, *Gaṇita* **23**(1972), 11–17.

[8] Expansion theorem for the H-function involving associated Legendre functions, *Bull. Soc. Math. Phys. Macédoine* **24**(1973), 39–43.

Arya S.C. [1] Abelian theorem for a generalised Stieltjes transform, *Boll. Un. Mat. Ital.* (3) **13**(1958), 497–504.

[2] Inversion theorem for a generalized Stieltjes transform, *Riv. Mat. Univ. Parma* **9**(1958), 139–148.

[3] On two generalized Laplace transforms, *Boll. Un. Mat. Ital.* (3) **14**(1959), 307–317.

[4] Abelian theorem for a generalization of Laplace transformation, *Collect. Math.* **11**(1959), 3–12.

[5] A complex inversion formula for a generalized Stieltjes transform, *Agra Univ. J. Res. Sci.* **9**(1960), 233–242.

Babenko K.I. [1] On conjugate functions (Russian), *Doklady Akad. Nauk SSSR* **62**(1948), 157–160.

Bajpai S.D. [1] On some results involving Fox's H-function and Jacobi polynomials, *Proc. Cambridge Philos. Soc.* **65**(1969), 697–701.

[2] Fourier series of generalized hypergeometric functions, *Proc. Cambridge Philos. Soc.* **65**(1969), 703–707.

[3] An expansion formula for Fox's H-function involving Bessel functions, *Labdev J. Sci. Tech. Part A* **7**(1969), 18–20.

[4] Transformation of an infinite series of Fox's H-function, *Portugal Math.* **29**(1970), 141–144.

Banerjee D.P. [1] Generalised Meijer transforms, *J. London Math. Soc.* **36**(1961), 433–435.

Barrios J.A. and Betancor J.J. [1] On a generalization of Laplace transform due to E. Krätzel, *J. Inst. Math. Comput. Sci. Math. Ser.* **3**(1990), 273–291.

[2] The Krätzel integral transformation of distributions, *Math. Nachr.* **154**(1991), 11–26.

[3] A Krätzel's integral transformation of distributions, *Collect. Math.* **42**(1991), 11–32.

[4] On K_μ transformation of distributions, *Appl. Anal.* **42**(1991), 175–197.

[5] A Parseval equation and the K_μ and I_μ transforms of distributions, *Math. Japon.* **36**(1991), 1063–1083.

Bateman H. [1] The inversion of a definite integral, *Proc. London Math. Soc.* (2) **4**(1906), 461–498.

[2] Report on the history and present state of the theory of integral equations, *Reports of the British Association* (1910), 354–424.

Betancor J.J. [1] The Hankel–Schwartz transform for function of compact support, *Rend. Mat. Appl.* (7) **7**(1987), 399–409 (1988).

[2] A mixed Parseval's equation and a generalized Hankel transformation of distributions, *Canad. J. Math.* **41**(1989), 274–284.

[3] A generalized Hankel transformation, *Comment. Math. Univ. Carolin.* **30**(1989), 1–15.

[4] On the boundedness and the range of the Lommel–Maitland transformation, *Bull. Soc. Roy. Sci. Liège* **58**(1989), 3–11.

[5] On Hankel–Clifford transforms of ultradistributions, *Pure Appl. Math. Sci.* **29**(1989), 21–43.

[6] Two complex variants of a Hankel type transformation of generalized functions, *Portugal. Math.* **46**(1989), 229–243.

[7] The Hankel–Clifford transformation on certain spaces of ultradistributions, *Indian J. Pure Appl. Math.* **20**(1989), 583–603.

[8] On the Watson–Wright transformation on certain weighted L_p-spaces, *Rev. Acad. Canaria Cienc.* **1**(1990), 27–38.

[9] On the Hankel–Schwartz integral transform (Spanish), *Rev. Colombiana Mat.* **24**(1990), 15–23.

[10] A generalized Hankel transformation of distributions and a Cauchy problem involving the Bessel type operator $B_{\alpha_0,\alpha_1,\alpha_2} = x^{\alpha_0}Dx^{\alpha_1}Dx^{\alpha_2}$, *Jñānābha* **20**(1990), 111–136.

[11] A generalized Hankel transformation of certain spaces of distributions, *Rev. Roumaine Math. Pures Appl.* **37**(1992), 445–463.

[12] A new inversion formula and Abelian theorems for a generalized Hankel transform, *Bol. Soc. Mat. Mexicana* (2) **35**(1990), 71–78.

Betancor J.J. and Barrios J.A. [1] On K_μ and I_μ transforms of generalized functions, *J. Math. Anal. Appl.* **165**(1992), 12–35.

Betancor J.J. and Jerez Diaz C. [1] Boundedness and range of H-transformation on certain weighted L_p-spaces, *Serdica* **20**(1994), 269–297.

[2] Weighted norm inequalities for the H-transformation, *Internat. J. Math. Math. Sci.* **20**(1997), 647–656.

[3] New inversion formulas for the Hankel transformations, *Rend. Circ. Mat. Palermo* (2) **46**(1997), 287–308.

Betancor J.J., Linares M. and Méndez J.M.R. [1] Paley–Wiener theorems for the Hankel transformation, *Rend. Circ. Mat. Palermo* (2) **44**(1995), 293–300.

[2] The Hankel transform of integrable Boehmians, *Appl. Anal.* **58**(1995), 367–382.

[3] The Hankel transform of tempered Boehmians, *Boll. Un. Mat. Ital. B*(7) **10**(1996), 325–340.

Betancor J.J. and Marrero I. [1] New spaces of type H_μ and the Hankel transformation, *Integral Transform. Spec. Funct.* **3**(1995), 175–200.

Betancor J.J. and Negrin E.R. [1] The generalized Bessel transformations on the spaces $L'_{p,\nu}$ of distributions, *J. Korean Math. Soc.* **27**(1990), 129–135.

Betancor J.J. and Rodriguez-Mesa L. [1] Pointwise convergence and a new inversion theorem for Hankel transforms, *Publ. Math. Debrecen* **50**(1997), 235–247.

[2] The K_μ-transformation on McBride's spaces of generalized functions, *Math. Nachr.* **185**(1997), 21–31.

[3] Hankel transformation of Colombeau type tempered generalized functions, *J. Math. Anal. Appl.* **217**(1998), 293–320.

[4] A Paley–Wiener theorem for the Hankel transform of Colombeau type generalized functions, *J. Math. Anal. Appl.* **230**(1999), 11–29.

Bhatnagar K.P. [1] On certain theorems on self-reciprocal functions, *Acad. Roy. Belgique. Bull. Cl. Sci.* (5) **39**(1953), 42–69.

[2] On self-reciprocal functions, *Gaṇita* **4**(1953), 19–37.

[3] Two theorems on self-reciprocal functions and a new transform, *Bull. Calcutta Math. Soc.* **45**(1953), 109–112.

[4] On self-reciprocal functions and a new transform, *Bull. Calcutta Math. Soc.* **46**(1954), 179–199.

Bhise V.M. [1] Inversion formulae for a generalised Laplace integral, *Vikram-Quart. Res. J. Vikram Univ.* **3**(1959), 57–63.

[2] Some relations involving Meijer's K-transform, *Vikram-Quart. Res. J. Vikram Univ.* **4**(1960), 68–77.

[3] Certain rules and recurrence relations for Meijer–Laplace transform, *Proc. Nat. Acad. Sci. India Sect. A* **32**(1962), 389–404.

[4] Some finite and infinite series of Meijer–Laplace transform, *Math. Ann.* **154**(1964), 267–272.

[5] Operators of fractional integration and a generalised Hankel transform, *Collect. Math.* **16**(1964), 201–209.

[6] Certain properties of Meijer–Laplace transform, *Compositio Math.* **18**(1967), 1–6.

Bhise V.M. and Dighe M. [1] Tauberian theorems for a Fourier type integral operator. I, *J. Indian Acad. Math.* **1**(1979), 25–32.

[2] On composition of integral operators with Fourier-type kernels, *Indian J. Pure Appl. Math.* **11**(1980), 1183–1187.

Bhonsle B.R. [1] On some results involving generalised Laplace's transforms, *Bull. Calcutta Math. Soc.* **48**(1956), 55–63.

[2] A relation between Laplace and Hankel transforms, *Proc. Glasgow Math. Assoc.* **5**(1962), 114–115.

[3] A relation between Laplace and Hankel transforms, *Math. Japon* **10**(1965), 85–89.

Bhonsle B.R. and Chaudhary M.S. [1] The complex Laplace–Hankel transformation of generalized functions, *Indian J. Pure Appl. Math.* **7**(1976), 205–211.

Boas R.P., Jr. [1] Inversion of generalized Laplace integral, *Proc. Nat. Acad. Sci. U.S.A.* **28**(1942), 21–24.

Bochner S. [1] Some properties of modular relations, *Ann. of Math.* (2) **53**(1951), 332–363.

Boersma J. [1] On a function which is a special case of Meijer's G-function, *Compositio Math.* **15**(1961), 34–63.

Bonilla B., Kilbas A.A., Rivero M., Rodriguez L. and Trujillo J.J. [1] Modified Bessel-type integral transform in $\mathcal{L}_{\nu,r}$-space, *Rev. Acad. Canaria Cienc.* **10**(1998), 45–55.

[2] Modified Bessel-type function and solution of differential and integral equations, *Indian J. Pure Appl. Math.* **31**(2000), 93–109.

Bora S.L. [1] Some theorems on generalized Meijer transform, *Kyungpook Math. J.* **12**(1972), 55–60.

Bora S.L. and Kalla S.L. [1] Some recurrence relations for the H-function (Hindi), *Vijnana Parishad Anusandhan Patrika* **14**(1971), 9–12.

Bora S.L. and Saxena R.K. [1] On fractional integration, *Publ. Inst. Math.* (Beograd) (N.S.) **11(25)**(1971), 19–22.

Braaksma B.L.J. [1] Asymptotic expansions and analytic continuations for a class of Barnes-integrals, *Compositio Math.* **15**(1964), 239–341.

Braaksma B.L.J. and Schuitman A. [1] Some classes of Watson transforms and related integral equations for generalized functions, *SIAM J. Math. Anal.* **7**(1976), 771–798.

Bradley F.W. [1] An extension of the Fourier–Bessel integral, *Proc. London Math. Soc.* (2) **41**(1936), 209–214.

Brychkov Yu.A., Glaeske H.-J. and Marichev O.I. [1] Factorization of integral transformations of convolution type (Russian), *Mathematical Analysis*, Vol. 21, 3–41, Itogi Nauki i Tekhniki, Akad. Nauk SSSR, Vsesoyuz. Inst. Nauchn. i Tekhn. Inform., Moscow, 1983.

[2] The product structure of a class of integral transformations (German), *Z. Anal. Anwendungen* **5**(1986), 119–123.

Brychkov Yu.A., Glaeske H.-J., Prudnikov A.P. and Vu Kim Tuan [1] *Multidimensional Integral Transformations*, Gordon and Breach Science Publ., Philadelphia, 1992.

Brychkov Yu.A. and Prudnikov A.P. [1] *Integral Transforms of Generalized Functions*, Translated and revised from the second Russian edition, Gordon and Breach Science Publ., New York, 1989.

Busbridge W. [1] A theory of general transformations for functions of class $L_p(0,\infty)$, *Quart. J. Math. Oxford Ser.* **9**(1938), 148–160.

Buschman R.G. [1] An inversion integral, *Proc. Amer. Math. Soc.* **13**(1962), 675–677.

[2] Contiguous relations and related formulas for the H-function of Fox, *Jñānābha Sect. A* **2**(1972), 39–47.

[3] Partial derivatives of the H-function with respect to parameters expressed as finite sums and as integrals, *Univ. Nac. Tucumán Rev. Ser. A* **24**(1974), 149–155.

Buschman R.G. and Srivastava H.M. [1] Inversion formulas for the integral transformation with the H-function as kernel, *Indian J. Pure Appl. Math.* **6**(1975), 583–589.

Carmichael R.D. and Pathak R.S. [1] Abelian theorems for Whittaker transforms, *Internat. J. Math. Math. Sci.* **10**(1987), 417–431.

[2] Asymptotic behaviour of the H-transform in the complex domain, *Math. Proc. Cambridge Philos. Soc.* **102**(1987), 533–552.

[3] Asymptotic analysis of the H-function transform, *Glas. Mat. Ser. III* **25(45)**(1990), 103–127.

Chaudhary M.S. [1] Hankel type transform of distribution, *Ranchi Univ. Math. J.* **12**(1981), 9–16.

Chaurasia V.B.L. [1] On some integrals involving Kampé de Fériet function and the H-function (Hindi), *Vijnana Parishad Anusandhan Patrika*, **19**(1976), 163–167.

Conlan J. and Koh E.L. [1] On the Meijer transformation, *Internat. J. Math. Math. Sci.* **1**(1978), 145–159.

Conolly B.W. [1] On integral transforms, *Proc. Edinburgh Math. Soc. (2)* **10**(1956), 125–128.

Cooke R.G. [1] The inversion formulae of Hardy and Titchmarsh, *Proc. London Math. Soc.* (2) **24**(1925), 381–420.

Dange S. and Chaudhary M.S. [1] Distributional Abelian theorems for the generalized Stieltjes transform, *J. Math. Anal. Appl.* **128**(1987), 125–137.

de Amin L.H. and Kalla S.L. [1] Relation between the Hardy transform and the transform whose kernel is the H-function (Spanish), *Univ. Nac. Tucumán Rev. Ser. A* **23**(1973), 45–50.

Debnath L. [1] *Integral Transforms and Their Applications*, CRC Press, Boca Raton, Florida, 2000.

Deeba E.Y. and Koh E.L. [1] A characterization of the generalized Meijer transform, *Internat. J. Math. Math. Sci.* **15**(1992), 823–827.

Dighe M. [1] Composition of fractional integral operator and an operator with Fourier type kernel, *Bull. Univ. Brasov* C **20**(1978/79), 3–8.

Dighe M. and Bhise V.M. [1] On composition of fractional integral operators as an integral operator with Fourier type kernel, *Math. Notae* **27**(1979/80), 23–30.

[2] Some abelian theorems for a Fourier type integral operator (Hindi), *Vijnana Parishad Anusandhan Patrika* **26**(1983), 11–20.

Ditkin V.A. and Prudnikov A.P. [1] *Integral Transforms and Operational Calculus* (Russian), Moscow, Gosudarstv. Izdat. Fiz.-Mat. Lit., 1961; English translation published by Pergamon Press, Oxford, 1965; Second Russian edition, Nauka, Moscow, 1974.

Dixon A.L. and Ferrar W.L. [1] A class of discontinuous integrals, *Quart. J. Math.* **7**(1936), 81–96.

Doetsch G. [1] *Theorie und Anwendung der Laplace-Transformation* (German), Springer, Berlin, 1937.

[2] *Handbuch der Laplace-Transformation.* Band I, *Theorie der Laplace-Transformation* (German), Birkhäuser, Basel, 1950.

Dube L.S. and Pandey J.N. [1] On the Hankel transform of distributions, *Tōhoku Math. J.* (2) **27**(1975), 337–354.

Duran A.J. [1] On Hankel transform, *Proc. Amer. Math. Soc.* **110**(1990), 417–424.

[2] Gel'fand–Shilov spaces for the Hankel transform, *Indag. Math.* (N.S.) **3**(1992), 137–151.

van Eijndhoven S.J.L. and van Berkel C.A.M. [1] Hankel transformations and spaces of type S, *Indag. Math.* (N.S.) **2**(1991), 29–38.

van Eijndhoven S.J.L. and de Graaf J. [1] Analyticity spaces of selfadjoint operators subjected to perturbations with applications to Hankel invariant distribution spaces, *SIAM J. Math. Anal.* **17**(1986), 485–494.

Erdélyi A. [1] A class of hypergeometric transforms, *J. London Math. Soc.* **15**(1940), 209–212.

[2] On the connection between Hankel transforms of different order, *J. London Math. Soc.* **16**(1941), 113–117.

[3] On some functional transformations, *Univ. e Politecnico Torino. Rend. Sem. Mat.* **10**(1951), 217–234.

[4] Stieltjes transforms of generalised functions, *Proc. Royal Soc. Edinburgh Sect. A* **77**(1977), 231–249.

Erdélyi A. and Kober H. [1] Some remarks on Hankel transforms, *Quart. J. Math., Oxford Ser.* **11**(1940), 212–221.

Erdélyi A, Magnus W., Oberhettinger F. and Tricomi F.G. [1] *Higher Transcendental Functions, Vol. 1*, McGraw-Hill, New York, 1953.

[2] *Higher Transcendental Functions, Vol. 2*, McGraw-Hill, New York, 1953.

[3] *Higher Transcendental Functions, Vol. 3*, McGraw-Hill, New York, 1955.

[4] *Tables of Integral Transforms, Vol. 2*, McGraw-Hill, New York, 1954.

Fenyö I. [1] Hankel-Transformation verallgemeinerter Funktionen (German), *Mathematica (Cluj)* **8** (**31**)(1966), 235–242.

[2] On the generalized Hankel transformation, *Studia Sci. Math. Hungar.* **23**(1988), 7–14.

Fox C. [1] A generalization of the Fourier–Bessel transform, *Proc. London Math. Soc.* (2) **29**(1929), 401–452.

[2] The G and H functions as symmetrical Fourier kernels, *Trans. Amer. Math. Soc.* **98**(1961), 395–429.

[3] An inversion formula for the kernel $K_\nu(x)$, *Proc. Cambridge Philos. Soc.* **61**(1965), 457–467.

[4] Solving integral equations by L and L^{-1} operators, *Proc. Amer. Math. Soc.* **29**(1971), 299–306.

[5] Applications of Laplace transforms and their inverses, *Proc. Amer. Math. Soc.* **35**(1972), 193–200.

Galué L., Kalla S.L. and Srivastava H.M. [1] Further results on an H-function generalized fractional calculus, *J. Fract. Calc.* **4**(1993), 89–102.

Gasper G. and Trebels W. [1] Necessary conditions for Hankel multipliers, *Indiana Univ. Math. J.* **31**(1982), 403–414.

Glaeske H.-J. and Kilbas A.A. [1] Bessel-type integral transforms on $\mathcal{L}_{\nu,r}$-spaces, *Results Math.* **34**(1998), 320–329.

Glaeske H.-J., Kilbas A.A. and Saigo M. [1] A modified Bessel-type integral transform and its compositions with fractional calculus operators on spaces $\mathsf{F}_{p,\mu}$ and $\mathsf{F}'_{p,\mu}$, *J. Comput. Appl. Math.* **118**(2000), 151–168.

Glaeske H.-J., Kilbas A.A., Saigo M. and Shlapakov S.A. [1] $\mathcal{L}_{\nu,r}$-theory of integral transformations with the H-function in the kernel (Russian), *Dokl. Akad. Nauk Belarusi* **41**(1997), *no.2*, 10–15.

[2] Integral transforms with H-function kernels on $\mathcal{L}_{\nu,r}$-spaces, *Appl. Anal.*, **79**(2001), 443–474.

Glaeske H.-J. and Saigo M. [1] Products of Laplace transform and fractional integrals on spaces of generalized functions, *Math. Japon.* **37**(1992), 373–382.

[2] Stieltjes transform and fractional integrals on spaces of generalized functions, *Math. Japon.* **39**(1994), 127–135.

Golas P.C. [1] Certain convergence theorems for an integral transform (Hindi. English summary), *Vijnana Parishad Anusandhan Patrika* **10**(1967), 189–196.

[2] On a generalised Stieltjes transform, *Bull. Calcutta Math. Soc.* **59**(1967), 73–80.

[3] Inversion and representation theorems for a generalised Stieltjes transform, *Ranchi Univ. Math. J.* **1**(1970), 27–32.

González de Galindo S.E. and Kalla S.L. [1] The inversion formulae for some Bessel and hypergeometric transforms, *Rev. Colombiana Mat.* **10**(1976), 83–92.

Goyal A.N. and Goyal G.K. [1] On the derivatives of the H-function, *Proc. Nat. Acad. Sci. India Sect. A* **37**(1967), 56–59.

Goyal G.K. [1] A finite integral involving H-function, *Proc. Nat. Acad. Sci. India Sect. A* **39**(1969), 201–203.

Goyal R.P. [1] Convergence of the generalized Whittaker transform, *Riv. Mat. Univ. Parma (2)* **6**(1965), 83–93.

[2] Convergence of Meijer–Laplace transform, *Gaṇita* **17**(1966), 57–67.

Goyal S.P. [1] On transformations of infinite series of Fox's H-function, *Indian J. Pure Appl. Math.* **2**(1971), 684–691.

[2] On chains for Meijer's G-function transform, *Univ. Nac. Tucumán Rev. Ser. A* **22**(1972), 17–30.

Goyal S.P. and Jain R.M. [1] Fractional integral operator and the generalized hypergeometric functions, *Indian J. Pure Appl. Math.* **18**(1987), 251–259.

Goyal S.P., Jain R.M. and Gaur N. [1] Fractional integral operators involving a product of generalized hypergeometric functions and a general class of polynomials, *Indian J. Pure Appl. Math.* **22**(1991), 403–411.

[2] Fractional integral operators involving a product of generalized hypergeometric functions and a general class of polynomials. II, *Indian J. Pure Appl. Math.* **23**(1992), 121–128.

Grin'ko A.P. and Kilbas A.A. [1] Generalized fractional integrals in weighted Hölder spaces. (Russian) *Dokl. Akad. Nauk BSSR* **34**(1990), 493–496.

[2] On compositions of generalized fractional integrals, *J. Math. Res. Exposition* **11**(1991), 165–171.

Gupta K.C. [1] A theorem concerning Meijer and Varma transforms, *Proc. Nat. Acad. Sci. India Sect. A* **34**(1964), 163–168.

[2] On the H-function, *Ann. Soc. Sci. Bruxelles Sér. I* **79**(1965), 97–106.

[3] Some theorems on integral transforms, *Riv. Mat. Univ. Parma* (4) **2**(1976), 1–14.

Gupta K.C. and Jain U.C. [1] The H-function II, *Proc. Nat. Acad. Sci. India Sect. A* **36**(1966), 594–609.

[2] On the derivative of the H-function, *Proc. Nat. Acad. Sci. India Sect. A* **38**(1968), 189–192.

[3] The H-function IV (Hindi), *Vijnana Parishad Anusandhan Patrika* **12**(1969), 25–30.

Gupta K.C. and Mittal P.K. [1] The H-function transform, *J. Austral. Math. Soc.* **11**(1970), 142–148.

[2] The H-function transform. II , *J. Austral. Math. Soc.* **12**(1971), 444–450.

[3] On a chain for the G-function transform, *Univ. Nac. Tucumán Rev. Ser. A* **22**(1972), 101–107.

Gupta K.C. and Mittal S.S. [1] On Gauss's hypergeometric transform, *Proc. Nat. Acad. Sci. India Sect. A* **37**(1967), 49–55.

[2] On Gauss's hypergeometric function transform. II (Hindi), *Vijnana Parishad Anusandhan Patrika* **10**(1967), 69–79.

[3] On Gauss's hypergeometric transform. III (Hindi), *Vijnana Parishad Anusandhan Patrika* **12**(1969), 133–137.

Gupta K.C. and Srivastava A. [1] On finite expansions for the H-function, *Indian J. Pure Appl. Math.* **3**(1972), 322–328.

Gupta R. and Jain U.C. [1] A representation of a generalised Hankel transformable distribution, *Ranchi Univ. Math. J.* **14**(1983), 31–38.

[2] A generalized Hankel transformation of generalized functions, *Bull. Calcutta Math. Soc.* **77**(1985), 191–198.

Habibullah G.M. [1] Some integral equations involving confluent hypergeometric functions, *Yokohama Math. J.* **19**(1971), 35–43.

[2] A note on a pair of integral operators involving Whittaker functions, *Glasgow Math. J.* **18**(1977), 99–100.

[3] Some integral operators with hypergeometric functions as kernels, *Bull. Math. Soc. Sci. Math. R. S. Roumanie* (N.S.) **21**(**69**)(1977), 293–300.

Halmos P.R. [1] *A Hilbert Space Problem Book*, Van Nostrand, Princeton, 1967.

Hardy G.H. [1] Some formulae in the theory of Bessel functions, *Proc. London Math. Soc. (2)* **23**(1924), 61–63.

Hardy G.H. and Littlewood J.E. [1] Some theorems concerning Fourier series and Fourier power series, *Duke Math. J.* **2**(1936), 354–382.

Heywood P. and Rooney P.G. [1] On the boundedness of Lowndes' operators, *J. London Math. Soc.* (2) **10**(1975), 241–248.

[2] A weighted norm inequality for the Hankel transformation, *Proc. Roy. Soc. Edinburgh Sect. A* **99**(1984), 45–50.

[3] On the inversion of the even and odd Hilbert transformations, *Proc. Roy. Soc. Edinburgh Sect. A* **109**(1988), 201–211.

[4] On the Hankel and some related transformations, *Canad. J. Math.* **40**(1988), 989–1009.

[5] On the Gegenbauer transformation, *Proc. Roy. Soc. Edinburgh Sect. A* **115**(1990), 151–166.

[6] On the inversion of the extended Hankel transformation, *J. Math. Anal. Appl.* **160**(1991), 284–302.

[7] On the Struve transformation, *SIAM J. Math. Anal.* **25**(1994), 450–461.

[8] Lipschitz conditions satisfied by Hankel transforms, *Proc. Roy. Soc. Edinburgh. Sect. A* **125**(1995), 847–858.

Higgins T.P. [1] An inversion integral of a Gegenbauer transform, *J. Soc. Indust. Appl. Math.* **11**(1963), 886–893.

[2] A hypergeometric function transform, *J. Soc. Indust. Appl. Math.* **12**(1964), 601–612.

Jain R.N. [1] General series involving H-functions, *Proc. Cambridge Philos. Soc.* **65**(1969), 461–465.

Jain U.C. [1] On generalized Laplace transforms I, *Proc. Nat. Acad. Sci. India Sect. A* **36**(1966), 661–674.

[2] Certain recurrence relations for the H-function, *Proc. Nat. Inst. Sci. India Part A* **33**(1967), 19–24.

Joshi J.M.C. [1] Real inversion theorems for a generalised Laplace transform, *Collect. Math.* **14**(1962), 217–225.

[2] On a generalized Stieltjes transform, *Pacific J. Math.* **14**(1964), 969–975.

[3] Inversion and representation theorems for a generalized Laplace transform, *Pacific J. Math.* **14**(1964), 977–985.

[4] Abelian theorem for a generalization of Laplace transform, *Collect. Math.* **17**(1965), 95–99.

[5] On Joshi's generalized Stieltjes transform, *Gaṇita* **28**(1977), 15–24.

[6] On the generalized Laplace transform of Lorentz spaces, *Jñānābha* **12**(1982), 113–119.

[7] S.M.Joshi generalized Laplace transform in the space $M(\delta)$, *Bull. Inst. Math. Acad. Sinica* **17**(1989), 235–242.

Joshi N. and Joshi J.M.C. [1] A real inversion theorem for H-transform, *Gaṇita* **33**(1982), 67–73.

Joshi V.G. and Saxena Raj.K. [1] Abelian theorems for distributional H-transform, *Math. Ann.* **256**(1981), 311–321.

[2] Structure theorems for H-transformable generalized functions, *Indian J. Pure Appl. Math.* **13**(1982), 25–29.

[3] Complex inversion and uniqueness theorem for the generalized H-transform, *Indian J. Pure Appl. Math.* **14**(1983), 322–329.

Kalla S.L. [1] Some theorems of fractional integration, *Proc. Nat. Acad. Sci. India Sect. A* **36**(1966), 1007–1012.

[2] A study of Gauss hypergeometric transform, *Proc. Nat. Acad. Sci. India Sect. A* **36**(1966), 675–686.

[3] Some theorems on fractional integration. II, *Proc. Nat. Acad. Sci. India Sect. A* **39**(1969), 49–56.

[4] Integral operators involving Fox's H-function. *Acta Mexicana Ci. Tecn.* **3**(1969), 117–122.

[5] Integral operators involving Fox's H-function. II, *Notas Ci. Ser. M Mat.* **7**(1969), 72–79.

[6] Fractional integration operators involving generalized hypergeometric functions. II, *Acta Mexicana Ci. Tecn.* **3**(1969), 1–5.

[7] Fractional integration operators involving generalized hypergeometric functions, *Univ. Nac. Tucumán Rev. Ser. A* **20**(1970), 93–100.

[8] On the solution of an integral equation involving a kernel of Mellin–Barnes type integral, *Kyungpook Math. J.* **12**(1972), 93–101.

[9] On a relation involving Hardy's transform, *An. Şti. Univ. "Al. I. Cuza" Iasi Sect. I a Mat.* (N.S.) **18**(1972), 113–117.

[10] Operators of fractional integration, *Analytic functions, Kozubnik* 1979 (Proc. Seventh Conf., Kozubnik, 1979), 258–280, Lecture Notes in Math. 798, Springer, Berlin, 1980.

Kalla S.L. and Kiryakova V.S. [1] An H-function generalized fractional calculus based upon compositions of Erdélyi–Kober operators in L_p, *Math. Japon.* **35**(1990), 1151–1171.

Kalla S.L. and Munot P.C. [1] Some theorems on Laplace and Varma transforms, *Rev. Ci. Mat. Univ. Lourenço Marques Sér. A* **2**(1971), 27–37.

Kalla S.L. and Saxena R.K. [1] Integral operators involving hypergeometric functions, *Math. Z.* **108**(1969), 231–234.

[2] Relations between Hankel and hypergeometric function operators, *Univ. Nac. Tucumán Rev. Ser. A* **21**(1971), 231–234.

[3] Integral operators involving hypergeometric functions. II, , *Univ. Nac. Tucumán Rev. Ser. A* **24**(1974), 31–36.

Kanjin Y. [1] On Hardy-type inequalities and Hankel transforms, *Monatsh. Math.* **127**(1999), 311–319.

Kapoor V.K. [1] On a generalized Stieltjes transform, *Proc. Cambridge Philos. Soc.* **64**(1968), 407–412.

[2] Some theorems on integral transforms, *Gaṇita* **22**(1971), 17–26.

[3] Some theorems on a generalized Stieltjes transform, *J. Sci. Res. Banaras Hindu Univ.* **22**(1971/72), 41–49.

Kapoor V.K. and Masood S. [1] On a generalized $L - H$ transform, *Proc. Cambridge Philos. Soc.* **64**(1968), 399–406.

Kesarwani R.N. (Narain Roop) [1] On a generalization of Hankel transform and self-reciprocal functions, *Univ. e Politec. Torino. Rend. Sem. Mat.* **16**(1956/1957), 269–300.

[2] Certain properties of generalized Laplace transform involving Meijer's G-function, *Math. Z.* **68**(1957), 272–281.

[3] Some properties of generalized Laplace transform. I, *Riv. Mat. Univ. Parma* **8**(1957), 283–306.

[4] Some properties of generalized Laplace transform. II, *Riv. Mat. Univ. Parma* **10**(1959), 167–170.

[5] Some properties of generalized Laplace transform. III, *Univ. e Politec. Torino. Rend. Sem. Mat.* **17**(1957/1958), 85–93.

[6] Some properties of generalized Laplace transform. IV, *Univ. e Politec. Torino. Rend. Sem. Mat.* **18**(1958/1959), 35–41.

[7] Some properties of generalized Laplace transform. V, *Vikram-Quart. Res. J. Vikram Univ.* **3**(1959), 33–39.

[8] The G-functions as unsymmetrical Fourier kernels. I, *Proc. Amer. Math. Soc.* **13**(1962), 950–959.

[9] The G-functions as unsymmetrical Fourier kernels. II, *Proc. Amer. Math. Soc.* **14**(1963), 18–28.

[10] The G-functions as unsymmetrical Fourier kernels. III, *Proc. Amer. Math. Soc.* **14**(1963), 271–277.

[11] A pair of unsymmetrical Fourier kernels, *Trans. Amer. Math. Soc.* **115**(1965), 356–369.

[12] On an integral transform involving G-functions, *SIAM J. Appl. Math.* **20**(1971), 93–98.

Kilbas A.A., Bonilla B., Rodriguez J., Trujillo J.J. and Rivero M. [1] Compositions of Bessel-type integral transform with fractional calculus operators on spaces $F_{p,\mu}$ and $F'_{p,\mu}$. *Fractional Calculus and Applied Analysis* **1**(1998), 135–150.

Kilbas A.A. and Borovco A.N. [1] Hardy–Titchmarsh and Hankel-type transforms in $\mathcal{L}_{\nu,r}$-spaces, *Integral Transform. Spec. Funct.* **10**(2000), 239–266.

Kilbas A.A. and Glaeske H.-J. [1] Bessel type integral transforms in weighted L_p-spaces. (Russian. English summary) *Dokl. Nats. Akad. Nauk Belarusi* **42**(1998), *no.4*, 33–39.

Kilbas A.A. and Gromak E.V. [1] $\mathcal{Y}_\nu$ and $\mathcal{H}_\nu$ transforms in $\mathcal{L}_{\nu,r}$-spaces, *Integral Transform. Spec. Funct.*, **13**(2002), 259–275.

Kilbas A.A., Repin O.A. and Saigo M. [1] Generalized fractional integral transforms with Gauss function kernels as G-transforms, *Integral Transform. Spec. Funct.*, **13**(2002), 285–307.

Kilbas A.A. and Saigo M. [1] On asymptotics of Fox's H-function at zero and infinity, *Transforms Methods and Special Functions, Proc. Intern. Workshop, 12-17 August 1994*, 99–122, Science Culture Techn. Publ., Singapore, 1995.

[2] Generalized fractional integrals with the Fox H-function in the spaces $\mathsf{F}_{\mu,p}$ and $\mathsf{F}'_{\mu,p}$ (Russian), *Dokl. Akad. Nauk Belarusi* **40**(1996), *no.4*, 9–14.

[3] On generalized fractional integration operators with Fox's H-function on spaces $\mathsf{F}_{\mu,p}$ and $\mathsf{F}'_{\mu,p}$. Proc. Conf. "Different Aspects of Differentiability, II" (Warsaw, 1995) *Integral Transform. Spec. Funct.* **4**(1996), 103–114.

[4] Fractional integrals and derivatives of the H-function (Russian), *Dokl. Akad. Nauk Belarusi* **41**(1997), *no.4*, 34–39.

[5] Fractional calculus of the H-function, *Fukuoka Univ. Sci. Rep.* **28**(1998), 41–51.

[6] On the H-function, *J. Appl. Math. Stochast. Anal.* **12**(1999), 191–204.

[7] Modified H-transforms in $\mathcal{L}_{\nu,r}$-spaces, *Demonstratio Mathematica* **33**(2000), 603–625.

Kilbas A.A., Saigo M. and Borovco A.N. [1] On the Lommel–Maitland transform in $\mathcal{L}_{\nu,r}$-space, *Fractional Calc. Appl. Anal.* **2**(1999), 429–444.

[2] On the generalized Hardy–Titchmarsh transform in $\mathcal{L}_{\nu,r}$-space, *Fukuoka Univ. Sci. Rep.* **30**(2000), 67–85.

[3] The generalized Hardy–Titchmarsh transform in the space of summable functions (Russian), *Dokl. Akad. Nauk* **372**(2000), 451–454.

Kilbas A.A., Saigo M. and Glaeske H.-J. [1] A modified transform of Bessel type and its compositions with operators of fractional integration and differentiation (Russian), *Dokl. Nats. Akad. Nauk Belarusi* **43**(1999), no.4, 26–30.

Kilbas A.A., Saigo M. and Shlapakov S.A. [1] Integral transforms with Fox's H-function in spaces of summable functions, *Integral Transform. Spec. Funct.* **1**(1993), 87–103.

[2] Integral transforms with Fox's H-function in $\mathcal{L}_{\nu,r}$-spaces, *Fukuoka Univ. Sci. Rep.* **23**(1993), 9–31.

[3] Integral transforms with Fox's H-function in $\mathcal{L}_{\nu,r}$-spaces, II. *Fukuoka Univ. Sci. Rep.* **24**(1994), 13–38.

Kilbas A.A., Saigo M. and Trujillo J.J. [1] On the Meijer transform in $\mathcal{L}_{\nu,r}$-space, *Integral Transform. Spec. Funct.* **10**(2000), 267–282.

Kilbas A.A., Saigo M. and Zhuk V.A. [1] On the composition of operators of generalized fractional integration with a differential operator in axisymmetric potential theory (Russian), *Differential'nye Uravnenija* **27**(1991), 1640–1642.

Kilbas A.A. and Shlapakov S.A. [1] On a Bessel type integral transformation and its compositions with integral and differential operators (Russian), *Dokl. Akad. Nauk Belarusi* **37**(1993), *no.4*, 10–14.

[2] On an integral transform with the Fox H-function (Russian), *Dokl. Akad. Nauk Belarusi* **38**(1994), *no.1*, 12–15.

[3] On the composition of a Bessel-type integral operator with operators of fractional integro-differentiation and the solution of differential equations (Russian), *Differentsial'nye Uravneniya* **30**(1994), 256–268, 364–365; *translation in Differential Equations* **30**(1994), 235–246.

Kilbas A.A. and Trujillo J.J. [1] On the Hankel-type integral transform in $\mathcal{L}_{\nu,r}$-spaces, *Fract. Calc. Appl. Anal.* **2**(1999), 343–353.

[2] Generalized Hankel transforms on $\mathcal{L}_{\nu,r}$-spaces, *Integral Transform. Spec. Funct.* **9**(2000), 271–286.

[3] Hankel–Schwartz and Hankel–Clifford transforms on $\mathcal{L}_{\nu,r}$-spaces, Preprint, University of La Laguna, 2000, 21 p.

Kiryakova V.S. [1] On operators of fractional integration involving Meijer's G-function, *C.R. Acad. Bulgare Sci.* **39**(1986), 25–28.

[2] A generalized fractional calculus and integral transforms, *Generalized Functions, Convergence Structures, and Their Applications* (Dubrovnik, 1987), 205–217, Plenum, New York, 1988.

[3] Fractional integration operators involving Fox's $H^{m,0}_{m,m}$-function, *C.R. Acad. Bulgare Sci.* **41**(1988), 11–14.

[4] Generalized $H^{m,0}_{m,m}$-function fractional integration operators in some classes of analytic functions, *Mat. Vesnik* **40**(1988), 259–266.

[5] *Generalized Fractional Calculus and Applications*, Pitman Res. Notes Math. 301, Longman Scientific & Technical, Harlow; copublished with John Wiley & Sons, New York, 1994.

Kiryakova V.S., Raina R.K. and Saigo M. [1] Representation of generalized fractional integrals in terms of Laplace transforms on spaces L_p, *Math. Nachr.* **176**(1995), 149–158.

Kober H. [1] Hankelsche Transformationen, *Quart. J. Math. Ser. 2* **8**(1937), 186–199.

[2] On fractional integrals and derivatives, *Quart. J. Math. Oxford Ser.* **11**(1940), 193–211.

Koh E.L. [1] On the generalized Hankel and K transformations, *Canad. Math. Bull.* **12**(1969), 733–740.

[2] A representation of Hankel transformable generalized functions, *SIAM J. Math. Anal.* **1**(1970), 33–36.

[3] The Hankel transformation of negative order for distributions of rapid growth, *SIAM J. Math. Anal.* **1**(1970), 322–327.

[4] The Hankel transformation of negative order for distributions of rapid growth, Ordinary and partial differential equations (Dundee, 1976), 291–300, *Lecture Notes in Math.* 564, 1970.

Koh E.L., Deeba E.Y. and Ali M.A. [1] The Meijer transformation of generalized functions, *Internat J. Math. Math. Sci.* **10**(1987), 267–286.

[2] On the generalized Meijer transformation, *Generalized Functions, Convergence Structures, and Their Applications* (Dubrovnik, 1987), 219–225, Plenum, New York, 1988.

Koh E.L. and Li C.K. [1] The Hankel transformation on M'_μ and its representation, *Proc. Amer. Math. Soc.* **122**(1994), 1085–1094.

[2] On the inverse of the Hankel transform, *Integral Transform. Spec. Funct.* **2**(1994), 279–282.

Koh E.L. and Zemanian A.H. [1] The complex Hankel and I transformations of generalized functions, *SIAM J. Appl. Anal.* **16**(1968), 945–957.

Krätzel E. [1] Eine Verallgemeinerung der Laplace- und Meijer-Transformation (German), *Wiss. Z. Friedrich–Schiller–Univ. Jena/ Thüringen* **14**(1965), 369–381.

[2] Die Faltung der L-Transformation (German), *Wiss. Z. Friedrich–Schiller–Univ. Jena/ Thüringen* **14**(1965), 383–390.

[3] Bemerkungen zur Meijer-Transformation und Anwendungen (German), *Math. Nachr.* **30**(1965), 327–334.

[4] Differentiationssätze der L-Transformation und Differentialgleichungen nach dem Operator $(d/dt)t^{1/n-\nu}(t^{1-1/n}d/dt)^{n-1}t^{\nu+1-2/n}$ (German), *Math. Nachr.* **35**(1967), 105–114.

[5] Integral transformations of Bessel-type, *Generalized Functions and Operational Calculus* (Proc. Conf. Varna 1975), 148–155, Bulgar. Acad. Sci., Sofia, 1979.

Kumar R. [1] Some theorems connected with generalised Hankel-transform, *Riv. Mat. Univ. Parma* **7**(1956), 321–332.

[2] On generalised Hankel-transform. I, II, *Bull. Calcutta Math. Soc.* **49**(1957), 105–118.

[3] Certain convergence theorems connected with a generalized Hankel-transform, *J. Indian Math. Soc.* (N.S.) **23**(1959), 125–132 (1961).

[4] On generalised Hankel-transform. III, *Bull. Calcutta Math. Soc.* **53**(1961), 7–13.

Kumbhat R.K. [1] An inversion formula for an integral transform, *Indian J. Pure Appl. Math.* **7**(1976), 368–375.

Laddha R.K. [1] General fractional integral formulas involving the generalized polynomial sets and Fox's H-function, *Jñānābha* **28**(1998), 89–95.

Lawrynowicz J. [1] Remarks on the preceding paper of P. Anandani, *Ann. Polon. Math.* **21**(1969), 120–123.

Lee W.Y. [1] On spaces of type $\mathcal{H}_\mu$ and their Hankel transformations, *SIAM J. Math. Anal.* **5**(1974), 336–348.

[2] On the Cauchy problem of the differential operator S_μ, *Proc. Amer. Math. Soc.* **51**(1975), 149–154.

[3] On Schwartz's Hankel transformation of certain spaces of distributions, *SIAM J. Math. Anal.* **6**(1975), 427–432.

Linares M. and Méndez J.M.R. [1] Hankel complementary integral transformations of arbitrary order, *Internat. J. Math. Math. Sci.* **15**(1992), 323–332.

Lions J.L. [1] Opérateurs de transmutation singuliers et équations d'Euler–Poisson–Darboux généralisées (French), *Rend. Sem. Math. Fis., Milano* **28**(1959), 124–137.

Love E.R. [1] Some integral equations involving hypergeometric functions, *Proc. Edinburgh Math. Soc.* (2) **15**(1967), 169–198.

[2] Two more hypergeometric integral equations, *Proc. Cambridge Philos. Soc.* **63**(1967), 1055–1076.

[3] A hypergeometric integral equation, *Fractional Calculus and Its Applications*, (Proc. Internat. Conf., New Haven 1974), 272–288, Lecture Notes in Math. 457, Springer, Berlin, 1975.

[4] Inversion of the Struve transform, *Fractional Calculus* (Glasgow, 1984), 75–86, Res. Notes in Math. 138, Pitman, Boston, 1985.

Love E.R., Prabhakar T.R. and Kashyap N.K. [1] A confluent hypergeometric integral equation, *Glasgow Math. J.* **23**(1982), 31–40.

Lowndes J.S. [1] A generalisation of the Erdélyi–Kober operators, *Proc. Edinburgh Math. Soc.* (2) **17** (1970/71), 139–148.

[2] An application of some fractional integrals, *Glasgow Math. J.* **20**(1979), 35–41.

[3] On some generalisations of the Riemann–Liouville and Weyl fractional integrals and their applications, *Glasgow Math. J.* **22**(1981), 173–180.

[4] On some fractional integrals and their applications, *Proc. Edinburgh Math. Soc.* (2) **28**(1985), 97–105.

[5] On two new operators of fractional integration, *Fractional Calculus* (Glasgow, 1984), 87–98, Res. Notes in Math. 138, Pitman, Boston, 1985.

Luchko Yu.F. and Kiryakova V.S. [1] Hankel type integral transforms connected with the hyper-Bessel differential operators, *Algebraic Analysis and Related Topics* (Warsaw 1999), 155–165, Banach Center Publ., 53, Polish Acad. Sci., Warsaw, 2000.

Luke Y.L. [1] *The Special Functions and their Approximations, Vol. I*, Math. Sci. Engng. Vol. 53, Academic Press, New York, 1969.

Mahato A.K. and Agrawal N.K. [1] Generalised convolution for K-transformation, *Jñānābha* **26** (1996), 67–73.

Mahato A.K. and Saksena K.M. [1] Some results for a generalised integral transform, *Rend. Mat. Appl.* (7) **11**(1991), 761–775.

[2] A generalized Laplace transform of generalized functions, *Anal. Math.* **18**(1992), 139–151.

[3] An integral transform of generalized functions. I, *Pure Appl. Math. Sci.* **37**(1993), 67–75.

Mahato R.M. and Mahato A.K. [1] Some Abelian theorems for distributional Hankel–Clifford transformation, *Jñānābha* **26**(1996), 57–60.

Mainra V.P. [1] A new generalization of the Laplace transform, *Bull. Calcutta Math. Soc.* **53**(1961), 23–31.

Malgonde S.P. [1] On a generalized Stieltjes transformation of Banach-space-valued distributions, *Rev. Acad. Canaria Cienc.* **6**(1994), 9–18.

[2] On the distributional generalized Meijer transformation, *Rev. Acad. Canaria Cienc.* **7**(1995), 31–42.

Malgonde S.P. and Chaudhary M.S. [1] A representation of Hankel type transformable generalized functions, *Proc. Nat. Acad. Sci. India Sect. A* **58**(1988), 387–391.

Malgonde S.P. and Saxena Raj.K. [1] A representation of H-transformable generalized functions, *Ranchi Univ. Math. J.* **12**(1981), 1–8 (1982).

[2] An inversion formula for the distributional H-transformation, *Math. Ann.* **258**(1981/82), 409–417.

[3] A representation of generalized Meijer–Laplace transformable generalized functions, *J. Indian Inst. Sci.* **64**(1983), 291–297.

[4] Some abelian theorems for the distributional H-transformation, *Indian J. Pure Appl. Math.* **15**(1984), 365–370.

[5] Generalized Meijer–Laplace transformation of generalized functions, *Proc. Nat. Acad. Sci. India Sect. A* **55**(1985), 235–244.

[6] Some abelian theorems for the distributional generalized Meijer–Laplace transformation, *Proc. Nat. Acad. Sci. India Sect. A* **56**(1986), 265–271.

[7] A generalization of Erdélyi–Kober operators of generalized functions, *J. Indian Acad. Math.* **12**(1990), 129–141.

Manandhar R.P. [1] Fractional integrals and Meijer Bessel transform, *Ranchi Univ. Math. J.* **3**(1972), 36–43.

Marichev O.I. [1] *Handbook of Integral Transforms of Higher Transcendental Functions. Theory and Algorithmic Tables*, Ellis Horwood, Chichester; John Wiley & Sons, New York, 1983.

Marichev O.I. and Vu Kim Tuan [1] Composition structure of some integral transformations of convolution type (Russian), *Reports of the Extended Sessions of a Seminar of the I.N. Vekua* Institute of Applied Mathematics, Vol. I, no.1 (Russian) (Tbilisi, 1985), 139–142, Tbilis. Gos. Univ., Tbilisi, 1985.

[2] The factorization of G-transform in two spaces of functions, *Complex Analysis and Applications* '85 (Varna, 1985), 418–433, Bulgar. Acad. Sci., Sofia, 1986.

Martić B. [1] A note on fractional integration, *Publ. Inst. Math.* (Beograd) (N.S.) **16 (30)**(1973), 111–113.

[2] The connection between the Riemann–Liouville fractional integral, the Meijer and the Hankel transform (Serbo-Croatian), *Akad. Nauka i Umjet. Bosne i Hercegov. Rad. Knj. {45} Odjelj. Prirod. Mat. Nauka Knj* **12**(1973), 145–148.

Masood S. and Kapoor V.K. [1] On a generalization of Hankel transform and self-reciprocal functions, *Indian J. Pure Appl. Math.* **3**(1972), 32–40.

Mathai A.M. [1] A few remarks on the exact distributions of certain multivariate statistics-II, *Multivariate Statistical Inference* (Proc. Res. Sem. Dalhousie Univ., Halifax, N.S., 1972), 169–181. North–Holland, Amsterdam; American Elsevier, New York, 1973.

Mathai A.M. and Saxena R.K. [1] *Generalized Hypergeometric Functions with Applications in Statistics and Physical Sciences*, Lecture Notes in Math. 348, Springer–Verlag, Berlin, 1973.

[2] *The H-Function with Applications in Statistics and other Disciplines*, Halsted Press [John Wiley and Sons], New York, 1978.

Mathur S.L. [1] Certain recurrence relations for the H-function, *Math. Education* **4**(1970), A132–A136.

McBride A.C. [1] Solution of hypergeometric integral equations involving generalised functions, *Proc. Edinburgh Math. Soc.* (2) **19**(1974/75), 265–285.

[2] *Fractional Calculus and Integral Transforms of Generalized Functions*, Research Notes in Math. 31, Pitman, Boston, 1979.

[3] The Hankel transform of some classes of generalized functions and connections with fractional integration, *Proc. Roy. Soc. Edinburgh Sect. A* **81**(1978), 95–117.

[4] Connections between fractional calculus and some Mellin multiplier transforms, *Univalent Functions, Fractional Calculus, and Their Applications* (Koriyama, 1988), 121–138, Ellis Horwood Ser. Math. Appl., Horwood, Chichester, 1989.

[5] The range and invertibility of a class of Mellin multiplier transforms, *Direct and Inverse Boundary Value Problems* (Oberwolfach, 1989), 169–186, Methoden Verfahren Math. Phys. 37 Lang, Frankfurt/Main, 1991.

McBride A.C. and Spratt W.J. [1] On the range and invertibility of a class of Mellin multiplier transforms. I, *J. Math. Anal. Appl.* **156**(1991), 568–587.

[2] On the range and invertibility of a class of Mellin multiplier transforms. III, *Canad. J. Math.* **43**(1991), 1323–1338.

McKellar B.H.J., Box M.A. and Love E.R. [1] Inversion of the Struve transform of half integer order, *J. Austral. Math. Soc. Ser. B* **25**(1983), 161–174.

Mehra A.N. [1] Some properties of Meijer transform, *Gaṇita* **20**(1969), 101–111.

[2] On a generalized Stieltles transform, *Univ. Nac. Tucumán Rev. Ser. A* **20**(1970), 25–32.

[3] On Meijer transform, *Riv. Mat. Univ. Parma* (2) **11**(1970), 183–189.

Mehra K.N. and Saxena R.K. [1] On a generalized Stieltles transform (Hindi), *Vijnana Parishad Anusandhan Patrika* **10**(1967), 121–126.

Meijer C.S. [1] Ueber eine Erweiterung der Laplace-Transformation. I (German), *Nederl. Akad. Wetensch., Proc.* **43**(1940), 599–608 = *Indag. Math.* **2**(1940), 229–238.

[2] Ueber eine Erweiterung der Laplace-Transformation. II (German), *Nederl. Akad. Wetensch., Proc.* **43**(1940), 702–711 = *Indag. Math.* **2**(1940), 269–278.

[3] Eine neue Erweiterung der Laplace-Transformation. I (German), *Nederl. Akad. Wetensch., Proc.* **44**(1941), 727–737 = *Indag. Math.* **3**(1941), 338–348.

[4] Eine neue Erweiterung der Laplace-Transformation. II (German), *Nederl. Akad. Wetensch., Proc.* **44**(1941), 831–839.

[5] Multiplikationstheoreme für die Funktion $G_{p,q}^{m,n}(z)$ (German), *Nederl. Akad. Wetensch., Proc.* **44**(1941), 1062–1070 = *Indag. Math.* **3**(1941), 486–490.

Mellin H. [1] Abris einer einhaitlichen Theorie der Gamma und der hypergeometrischen Funktionen, *Math. Ann.* **68**(1910), 305–337.

Méndez J.M.R. [1] On the Bessel transforms, *Jñānābha* **17**(1987), 79–88.

[2] A mixed Parseval equation and the generalized Hankel transformations, *Proc. Amer. Math. Soc.* **102**(1988), 619–624.

Méndez J.M.R. and Sánchez Quintana A.M. [1] On the Bessel transformation of rapidly increasing generalized functions, *Ranchi Univ. Math. J.* **19**(1988), 13–24.

[2] On the Schwartz's Hankel transformation of distributions, *Analysis* **13**(1993), 1–18.

Méndez J.M.R. and Socas Robayna M.-M. [1] A pair of generalized Hankel–Clifford transformations and their applications, *J. Math. Anal. Appl.* **154**(1991), 543–557.

Misra O.P. [1] Abelian and Tauberian theorems for the generalized Whittaker–Meijer transform, *Gaṇita* **19**(1968), 93–102.

[2] Some abelian theorems for the distributional Meijer–Laplace transformation, *Indian J. Pure Appl. Math.* **3**(1972), 241–247.

[3] Generalised convolution for G-transformations, *Bull. Calcutta Math. Soc.* **64**(1972), 137–142.

[4] Distributional G-transformation, *Bull. Calcutta Math. Soc.* **73**(1981), 247–255.

Mittal P.K. [1] On the H-function transform. III, *Univ. Nac. Tucumán Rev. Ser. A* **20**(1970), 7–16.

[2] Certain properties of Meijer's G-function transform involving the H-function (Hindi), *Vijnana Parishad Anusandhan Patrika* **14**(1971), 29–38.

Moharir S.K. and Saxena Raj.K. [1] Abelian theorems for the generalized Whittaker transform, *J. Shivaji Univ.* (*Sci.*) **18**(1978), 1–6.

Moiseev E.I., Prudnikov A.P. and Skurnik U. [1] On the integral expansion of a function in degenerate hypergeometric functions, *Differential Equations* **34**(1998), 765–774; translation from *Differentsial'nye Uravneniya* **34**(1998), 762–771.

Nair V.C. [1] Differentiation formulae for the H-function. I, *Math. Student* **40A**(1972), 74–78.

[2] Differentiation formulae for the H-function. II, *J. Indian Math. Soc.* (N.S.) **37**(1973), 329–334.

Nasim C. [1] On K-transform, *Internat. J. Math. Math. Sci.* **4**(1981), 493–501.

[2] An integral equation involving Fox's H-function, *Indian J. Pure Appl. Math.* **13**(1982), 1149–1162.

[3] Integral operators involving Whittaker functions, *Glasgow Math. J.* **24**(1983), 139–148.

Nguyen Thanh Hai and Yakubovich S.B. [1] *The Double Mellin–Barnes Type Integrals and Their Applications to Convolution Theory*, World Scientific Publ., River Edge, 1992.

Okikiolu G.O. [1] On integral operators with kernels involving Bessel functions, *Proc. Cambridge Philos. Soc.* **62**(1966), 477–484.

[2] On integral operators with kernels involving Bessel functions (Corrections and addendum), *Proc. Cambridge Philos. Soc.* **67**(1970), 583–586.

Oliver M.L. and Kalla S.L. [1] On the derivative of Fox's H-function (Spanish), *Acta Mexicana Ci. Tecn.* **5**(1971), 3–5.

Olkha G.S. [1] Some finite expansions for the H-function, *Indian J. Pure Appl. Math.* **1**(1970), 425–429.

Olkha G.S. and Rathie P.N. [1] On some new generalized Bessel functions and integral transforms. II, *Univ. Nac. Tucumán Rev. Ser. A* **19**(1969), 45–53.

[2] On a generalized Bessel function and an integral transform, *Math. Nachr.* **51**(1971), 231–240.

Parashar B.P. [1] Domain and range of fractional integration operators, *Math. Japan.* **12**(1967), 141–145.

[2] Some theorems on a generalised Laplace transform and results involving H-function of Fox, *Riv. Mat. Univ. Parma* (2) **8**(1967), 375–384.

Pathak R.S. [1] Some properties of a generalization of Lommel and Maitland transforms, *Proc. Nat. Acad. Sci. India. Sect. A* **36**(1966), 557–565.

[2] Two theorems on a generalization of Lommel and Maitland transforms, *Proc. Nat. Acad. Sci. India. Sect. A* **36**(1966), 809–816.

[3] An inversion formula for a generalization of Lommel and Maitland transforms, *J. Sci. Res. Banaras Hindu Univ.* **17**(1966/1967), 65–69.

[4] A general differential equation satisfied by special functions, *Progr. Math. (Allahabad)* **6**(1972), 46–50.

[5] Some theorems on Whittaker transforms, *Indian J. Pure Appl. Math.* **4**(1973), 308–317.

[6] An inversion of the Varma transform by means of integral operators, *Mathematika (Cluj)* **15(38)**(1973), 229–240.

[7] Transformée de Varma des fonctions généralisées, *Bull. Sci. Math. (2)* **99**(1975), 3–16.

[8] On the Meijer transform of generalized functions, *Pacif. J. Math.* **80**(1979), 523–536.

[9] Abelian theorems for the G-transformation, *J. Indian Math. Soc. (N.S.)* **45**(1981), 243–249.

[10] On Hankel transformable spaces and a Cauchy problem, *Canad. J. Math.* **37**(1985), 84–106.

[11] *Integral Transforms of Generalized Functions and Their Applications*, Gordon and Breach Science Publ., Amsterdam, 1997.

Pathak R.S. and Pandey J.N. [1] A distributional Hardy transformation, *Proc. Cambridge Philos. Soc.* **76**(1974), 247–262.

[2] Abelian theorems for Hardy transformations, *Canad. Math. Bull.* **20**(1977), 331–335.

[3] A distributional Hardy transformation, *Internat. J. Math. Math. Sci.* **2**(1979), 693–701.

[4] The G-transform of generalized functions, *Rocky Montain J. Math.* **9**(1979), 307–325.

Pathak R.S. and Pandey A.B. [1] On Hankel transforms of ultradistributions, *Applicable Anal.* **20**(1985), 245–268.

Pathak R.S. and Rai R.B. [1] Abelian theorems for Meijer transform (Hindi), *Vijnana Parishad Anusandhan Patrika* **26**(1983), 273–279.

Pathak R.S. and Sahoo H.K. [1] A generalization of H_μ-spaces and Hankel transforms, *Anal. Math.* **12**(1986), 129–142.

Pathak R.S. and Upadhyay S.K. [1] U_μ^p-spaces and Hankel transform, *Integral Transform. Spec. Funct.* **3**(1995), 285–300.

Pincherle S. [1] Sulle funzioni geometriche generalizzate. Nota I, *Atti della Reale Accademia dei Lincei, Rendiconti della classe di Scienza Fisiche, Mathematiche e Naturali (Roma)* (4) **4**(1888), 694–700.

Prabhakar T.R. [1] Two singular integral equations involving confluent hypergeometric functions, *Proc. Cambridge Philos. Soc.* **66**(1969), 71–89.

[2] Some integral equations with Kummer's functions in the kernels, *Canad. Math. Bull.* **14**(1971), 391–404.

[3] A class of integral equations with Gauss functions in the kernels, *Math. Nachr.* **52**(1972), 71–83.

[4] Hypergeometric integral equations of a general kind and fractional integration, *SIAM J. Math. Anal.* **3**(1972), 422–425.

[5] A general class of operators involving $\varphi_1(a,b;c;z,w)$ and related integral equations, *J. Indian Math. Soc.* (N.S.) **41**(1977), 163–179.

Prabhakar T.R. and Kashyap N.K. [1] A new class of hypergeometric integral equations, *Indian J. Pure Appl. Math.* **11**(1980), 92–97.

Prudnikov A.P., Brychkov Yu.A. and Marichev O.I. [1] *Integrals and Series. Vol.1, Elementary Functions*, Gordon and Breach Science Publ., New York, 1986.

[2] *Integrals and Series. Vol.2, Special Functions*, Gordon and Breach Science Publ., New York, 1986.

[3] *Integrals and Series. Vol.3, More Special Functions*, Gordon and Breach Science Publ., New York, 1990.

Raina R.K. [1] Some recurrence relations for the H-function, *Math. Education* **10**(1976), A45–A49.

[2] The H-function transform and the moments of probability distribution functions of an arbitrary order, *Simon Stevin* **60**(1986), 97–103.

Raina R.K. and Koul C.L. [1] Fractional derivatives of the H-functions, *Jñānābha* **7**(1977), 97–105.

[2] On Weyl fractional calculus, *Proc. Amer. Math. Soc.* **73**(1979), 188–192.

[3] On Weyl fractional calculus and H-function transform, *Kyungpook Math. J.* **21**(1981), 275–279.

Raina R.K. and Saigo M. [1] A note on fractional calculus operators involving Fox's H-function on space $\mathsf{F}_{p,\mu}$, *Recent advances in fractional calculus*, 219–229, *Global Res. Notes Ser. Math., Global, Sauk Rapids*, 1993.

[2] On inter-connection properties associated with H-transform and certain fractional integrals on spaces of generalized functions, *J. Fract. Calc.* **12**(1997), 83–94.

Rao G.L.N. [1] The generalized Laplace transform of generalized functions, *Ranchi Univ. Math. J.* **5**(1974), 76–88.

[2] Abelian theorems for a distributional generalized Stieltjes transform, *Rev. Real Acad. Ci. Exact. Fis. Natur. Madrid* **70**(1976), 97–108.

[3] A complex inversion theorem for a distributional ${}_1F_1$-transform, *Bull. Calcutta Math. Soc.* **68**(1976), 267–273.

[4] A note on a representation of ${}_1F_1$-transformable distributions, *J. Math. Phys. Sci.* **11**(1977), 183–190; *Errata* **12**(1978), 405.

[5] Another real inversion theorem for a distributional ${}_1F_1$-transform, *Riv. Mat. Univ. Parma* (4) **4**(1978), 63–72.

[6] The distributional ${}_1F_1$-transform, *Collect. Math.* **29**(1978), 119–131.

[7] Complex inversion and representation theory for certain distributional generalized integral transformations, *J. Indian Math. Soc.* (N.S.) **43**(1979), 161–174.

[8] Some abelian theorems for the distributional ${}_1F_1$-transform, *Rev. Real Acad. Cienc Exact. Fis. Natur. Madrid* **74**(1980), 901–912.

[9] A unification of generalization of the Laplace transform and generalized functions, *Acta Math. Hungar.* **41**(1983), 119–126.

Rao G.L.N. and Debnath L. [1] A generalized Meijer transformation, *Internat. J. Math. Math. Sci.* **8**(1985), 359–365.

Rodriguez J., Trujillo J.J. and Rivero M. [1] Operational fractional calculus of Kratzel integral transformation, *Differential Equations* (Xanthi, 1987), 613–620, Lecture Notes in Pure and Appl. Math. **118**, Dekker, New York, 1989.

Rooney P.G. [1] On the ranges of certin fractional integrals, *Canad. J. Math.* **24**(1972), 1198–1216.

[2] A technique for studying the boundedness and extendability of certain types of operators, *Canad. J. Math.* **25**(1973), 1090–1102.

[3] On the range of the Hankel transformation, *Bull. London Math. Soc.* **11**(1979), 45–48.

[4] On the $\mathcal{Y}_\nu$ and $\mathcal{H}_\nu$ transformations, *Canad. J. Math.* **32**(1980), 1021–1044.

[5] On the boundedness and range of the extended Hankel transformation, *Canad. Math. Bull.* **23**(1980), 321–325

[6] On integral transformations with G-function kernels, *Proc. Royal Soc. Edinburgh Sect. A* **93**(1982/83), 265–297.

[7] On the range of an integral transformation, *Canad. Math. Bull.* **37**(1994), 545–548.

[8] On the representation of functions by the Hankel and some related transformations, *Proc. Royal Soc. Edinburgh Sect. A* **125**(1995), 449–463.

Saigo M. [1] A remark on integral operators involving the Gauss hypergeometric functions, *Math. Rep. Kyushu Uinv.* **11**(1977/78), 135–143.

Saigo M. and Glaeske H.-J. [1] Fractional calculus operators involving the Gauss function in spaces $\mathsf{F}_{p,\mu}$ and $\mathsf{F}'_{p,\mu}$, *Math. Nachr.* **147**(1990), 285–306.

Saigo M., Goyal S.P. and Saxena S. [1] A theorem relating a generalized Weyl fractional integral, Laplace and Varma transforms with applications, *J. Fract. Calc.* **13**(1998), 43–56.

Saigo M. and Kilbas A.A. [1] Generalized fractional integrals and derivatives in Hölder spaces, *Transform Methods Special Functions* (Proc. Intern. Workshop), Sofia, 12-17 August, 1994, 282–293, Science Culture Techn. Publ., Singapore, 1995.

[2] Compositions of generalized fractional calculus operators with axisymmetric differential operator of potential theory on spaces $\mathsf{F}_{\mu,p}$ and $\mathsf{F}'_{\mu,p}$, *Boundary Value Problems, Special Functions and Fractional Calculus* (Russian), Proc. Internat. Conference (Minsk, 1996), 335–350, Belorus. Gos. Univ., Minsk, 1996.

[3] Compositions of generalized fractional integration operators with the Fox H-function and a differential operator in axisymmetric potential theory (Russian), *Dokl. Akad. Nauk Belarusi* **40**(1996), *no.6*, 12–17.

[4] Generalized fractional calculus of the H-function, *Fukuoka Univ. Sci. Rep.* **29**(1999), 31–45.

[5] Modified G-transforms and generalized Stieltjes transform in $\mathcal{L}_{\nu,r}$-spaces, *Fukuoka Univ. Sci. Rep.* **30**(2000), 181–200.

Saigo M. and Maeda N. [1] More generalization of fractional calculus, *Transform Methods & Special Functions*, 386–400, Varna '96, Proc. 2nd Intern. Workshop, Bulgar. Acad. Sci., Sofia, 1998.

Saigo M. and Raina R.K. [1] On the fractional calculus operator involving Gauss's series and its application to certain statistical distributions, *Rev. Técn. Fac. Ingr. Univ. Zulia* **14**(1991), 53–62.

Saigo M., Raina R.K. and Kilbas A.A. [1] On generalized fractional calculus operators and their compositions with the axisymmetric differential operator of the potential theory on spaces $\mathsf{F}_{p,\mu}$ and $\mathsf{F}'_{p,\mu}$, *Fukuoka Univ. Sci. Rep.* **23**(1993), 133–154.

Saigo M., Saxena R.K. and Ram J. [1] On the fractional calculus operator associated with H-function, *Gaṇita Sandesh* **6**(1992), 36–47.

[2] Certain properties of operators of fractional integration associated with Mellin and Laplace transformations, *Current Topics in Analytic Function Theory*, 291–304, World Sci. Publ., River Edge, 1992.

Saksena K.M. [1] Inversion formulae for a generalized Laplace integral, *Proc. Nat. Inst. Sci. India* **19**(1953), 173–181.

[2] On a generalized Laplace integral, *Math. Z.* **68**(1957), 267–271.

[3] Inversion and representation theorems for a generalization of Laplace transformation, *Nieuw Arch. Wisk.* (3) **6**(1958), 1–9.

[4] Inversion and representation theorems for a generalized Laplace integral, *Pacific. J. Math.* **8**(1958), 597–607.

[5] An inversion theory for the Laplace integral, *Nieuw Arch. Wisk.* (3) **15**(1967), 218–224.

[6] Extension of certain integral transforms to generalized functions, *Generalized Functions and Their Applications in Mathematical Physics* (Moscow, 1980), 462–475, Akad. Nauk SSSR, Vychisl. Tsentr, Moscow, 1981.

Samko S.G., Kilbas A.A. and Marichev O.I. [1] *Fractional Integrals and Derivatives. Theory and Applications* (Edited and with a foreword by S.M. Nikol'skii), Gordon and Breach Science Publ., Yverdon, 1993.

Sánchez Quintana A.M. and Méndez J.M.R. [1] The Schwartz's Hankel transformations of disributions of rapid growth and a mixed Parseval equation, *Bull. Soc. Roy. Sci. Liege* **59**(1990), 183–204.

Saxena Raj. K. [1] Relation between Whittaker transform and modified $\chi_{v,k,m}$-transform, *Math. Ann.* **154**(1964), 301–306.

Saxena Raj.K., Koranne V.D. and Malgonde S.P. [1] On a distributional generalized Stieltjes transformation, *J. Indian Acad. Math.* **7**(1985), 105–110.

Saxena R.K. [1] Some theorems on generalized Laplace transform. I, *Proc. Nat. Inst. Sci. India Part A* **26**(1960), 400–413.

[2] Some theorems on generalized Laplace transform. II, *Riv. Mat. Univ. Parma* (2) **2**(1961), 287–299.

[3] A theorem on generalised Laplace transform, *Bull. Calcutta Math. Soc.* **53**(1961), 155–160.

[4] A relation between generalized Laplace and Hankel transforms, *Math. Z.* **81**(1963), 414–415.

[5] Certain properties of Varma transform involving Whittaker functions, *Collect. Math.* **16**(1964), 193–200.

[6] Some theorems on generalized Laplace transform. III, *Riv. Mat. Univ. Parma* (2) **6**(1965), 135–146.

[7] An inversion formula for the Varma transform, *Proc. Cambridge Philos. Soc.* **62**(1966), 467–471.

[8] An inversion formula for a kernel involving a Mellin–Barnes type integral, *Proc. Amer. Math. Soc.* **17**(1966), 771–779.

[9] On fractional integration operators, *Math. Z.* **96**(1967), 288–291.

[10] Abelian theorems for distributional H-transform, *Acta Mexicana Ci. Tecn.* **7**(1973), 66–76.

Saxena R.K. and Gupta K.C. [1] Certain properties of generalized Stieltjes transform involving Meijer's G-function, *Proc. Nat. Inst. Sci. India Part A* **30**(1964), 707–714.

Saxena R.K. and Gupta N. [1] On the asymptotic expansion of generalized Stieltjes transform, *Math. Student* **64**(1995), 51–56.

Saxena R.K. and Kumbhat R.K. [1] A generalization of Kober operators (Hindi), *Vijnana Parishad Anusandhan Patrika* **16**(1973), 31–36.

[2] Integral operators involving H-function, *Indian J. Pure Appl. Math.* **5**(1974), 1–6.

[3] Some properties of generalized Kober operators (Hindi), *Vijnana Parishad Anusandhan Patrika* **18**(1975), 139–150.

Saxena R.K. and Kushwaha R.S. [1] An integral transform associated with a kernel of Fox, *Math. Student* **40**(1972), 201–206 (1974).

Saxena R.K. and Mathur S.N. [1] A finite series for the H-functions, *Univ. Nac Tacumán Rev. Ser. A* **21**(1971), 49–52.

Saxena R.K. and Saigo M. [1] Fractional integral formula for the H-function. II, *J. Fract. Calc.* **13**(1998), 37–41.

[2] Generalized fractional calculus of the H-function associated with the Appell function F_3, *J. Fract. Calc.* **19**(2001), 89–104.

Saxena R.K. and Sethi P.L. [1] Relations between generalised Hankel and modified hypergeometric function operators, *Proc. Indian Acad. Sci. Sect. A* **78**(1973), 267–273.

Saxena R.K. and Singh Y. [1] Integral operators involving generalized H-function, *Indian J. Math.* **35**(1993), 177–188.

Saxena V.P. [1] Inversion formulae to certain integral equations involving H-function, *Portugal. Math.* **29**(1970), 31–42.

Schuitman A. [1] On a certain test function space for Schwartz's Hankel transform, *Delft Progress Rep. Ser. F* **2**(1976/77), 193–206.

Schwartz A.L. [1] An inversion theorem for Hankel transforms, *Proc. Amer. Math. Soc.* **22**(1969), 713–717.

Shah M. [1] On some results involving H-functions and associated Legendre functions, *Proc. Nat. Acad. Sci. India Sect. A* **39**(1969), 503–507.

[2] On some results on H-functions associated with orthogonal polynomials, *Math. Scand.* **30**(1972), 331–336.

[3] On expansion theorem for the H-function, *J. Reine Angew. Math.* **285**(1976), 1–6.

[4] Some extensions of the multiplication theorems, *Gac. Mat.* (Madrid) (1) **28**(1976), 64–69.

Sharma C.K. [1] A generalized Laplace transform (Hindi), *Vijnana Parishad Anusandhan Patrika* **20**(1977), 227–235.

Sharma K.C. [1] Theorems on Meijer's Bessel-function transform, *Proc. Nat. Inst. Sci. India Part A* **30**(1964), 360–366.

[2] A few inversion formulae for Meijer transform, *Proc. Nat. Inst. Sci. India Part A* **30**(1964), 736–742.

[3] On an integral transform, *Math. Z.* **89**(1965), 94–97.

Sharma O.P. [1] On the Hankel transformation of H-functions, *J. Math. Sci.* **3**(1968), 17–26.

[2] On generalised Hankel and K-transforms, *Math. Student* **37**(1969), 109–116.

Shlapakov S.A. [1] An integral transformation with the Fox H-function in the space of summable functions (Russian), *Dokl. Akad. Nayk Belarusi* **38**(1994), *no.2*, 14–18.

Shlapakov S.A., Saigo M. and Kilbas A.A. [1] On inversion of H-transform in $\mathcal{L}_{\nu,r}$-space, *Internat. J. Math. Math. Sci.* **21**(1998), 713–722.

Singh R. [1] An inversion formula for Fox-H-transform, *Proc. Nat. Acad. Sci. India Sect A* **40**(1970), 57–64.

Skibinski P. [1] Some expansion theorems for the H-function, *Ann. Polon. Math.* **23**(1970/71), 125–138.

Sneddon I.N. [1] *Fourier Transforms*, McGraw-Hill, New York, 1951.

[2] *Fractional Integrals and Derivatives and Dual Integral Equations*, North Carolina State College, Appl. Math. Res. Group Raleigh, RSR-6, 1962.

[3] *Mixed Boundary Value Problems in Potential Theory*, North–Holland Publishing Co., Amsterdam; Interscience Publ. John Wiley & Sons, New York, 1966.

[4] *The Use of Integral Transforms*, THM Edition, McGraw-Hill, New Delhi, 1974.

[5] *Application of Integral Transforms in the Theory of Elasticity* (Edited), International Centre for Mechanical Sciences, Courses and Lectures, No. 220, Springer–Verlag, Vienna, 1975.

[6] Recent applications of integral transforms in the linear theory of elasticity, *Trends in Applications of Pure Mathematics in Mechanics, Vol.II* (Second Sympos., Kozubnik, 1977), 307–325; Monographs Stud. Math., 5, Pitman, Boston, 1979.

Soni K. [1] Fractional integrals and Hankel transforms, *Duke Math. J.* **35**(1968), 313–319.

[2] A Sonine transform, *Duke Math. J.* **37**(1970), 431–438.

[3] An integral equation with Bessel function kernel, *Duke Math. J.* **38**(1971), 175–180.

Soni K. and Soni R.P. [1] A Tauberian theorem related to the modified Hankel transform, *Bull. Austral. Math. Soc.* **11**(1974), 167–180.

Soni S.L. [1] Fourier series of H-function involving orthogonal polynomial, *Math. Education* **4**(1970), A80–A84.

Srivastava A. and Gupta K.C. [1] On certain recurrence relations, *Math. Nachr.* **46**(1970), 13–23.

Srivastava H.M. [1] On a relation between Laplace and Hankel transforms, *Matematiche (Catania)*, **21**(1966), 199–202.

[2] A relation between Meijer and generalized Hankel transforms, *Math. Japon.*, **11**(1966), 11–13.

[3] Some theorems on Hardy transform, *Nederl. Akad. Wetensch. Proc. Ser. A 71=Indag. Math.* **30**(1968), 316–320.

[4] A relation between generalized Laplace and Hardy transforms, *Nieuw Arch. Wisk.* (3), **16**(1968), 79–81.

[5] Fractional integration and inversion formulae associated with the generalized Whittaker transform, *Pacific J. Math.* **26**(1968), 375–377.

[6] Certain properties of generalized Whittaker transform, *Mathematica (Cluj)*, **10 (33)**(1968), 385–390.

[7] A class of integral equations involving the H function as kernel, *Nederl. Akad. Wetensch. Proc. Ser. A 75=Indag. Math.* **34**(1972), 212–220.

Srivastava H.M. and Buschman R.G. [1] Composition of fractional integral operators involving Fox's H-function, *Acta Mexicana Ci. Tecn.* **7**(1973), 21–28.

[2] Some convolution integral equations, *Nederl. Akad. Wetensch. Proc. Ser. A 77=Indag. Math.* **36**(1974), 211–216.

[3] Mellin convolutions and H-function transformations, *Rocky Mountain J. Math.* **6**(1976), 331–343.

[4] *Theory and Applications of Convolution Integral Equations*, Math. Appl. 79, Kluwer Academic Publ., Dordrecht, 1992.

Srivastava H.M., Goyal S.P. and Jain R.M. [1] A theorem relating a certain generalized Weyl fractional integral with the Laplace transform and a class of Whittaker transforms, *J. Math. Anal. Appl.* **153**(1990), 407–419.

Srivastava H.M., Gupta K.C. and Goyal S.P. [1] *The H-Functions of One and Two Variables with Applications*, South Asian Publ., New Delhi, 1982.

Srivastava H.M. and Karlsson P.W. [1] *Multiple Gaussian Hypergeometric Series*, Halsted Press (Ellis Horwood Limited, Chichester), John Wiley and Sons, New York, 1985.

Srivastava H.M. and Saigo M. [1] Multiplication of fractional calculus operators and boundary value problems involving the Euler–Darboux equation, *J. Math. Anal. Appl.* **128**(1987), 325–369.

Srivastava H.M., Saigo M. and Raina R.K. [1] Some existence and connection theorems associated with the Laplace transform and a certain class of integral operators, *J. Math. Anal. Appl.* **172**(1993), 1–10.

Srivastava H.M. and Vyas O.D. [1] A theorem relating generalized Hankel and Whittaker transforms, *Nederl. Akad. Wetensch. Proc. Ser. A 72=Indag. Math.* **31**(1969), 140–144.

Srivastava K.J. [1] Fractional integration and Meijer transform, *Math. Z.* **67**(1957), 404–412.

[2] Fractional integration and the $\widetilde{w}_{\mu,\nu}$-transform, *Univ. e Politec. Torino. Rend. Sem. Mat.* **17**(1957/1958), 201–208.

[3] Self-reciprocal function and $\widetilde{w}_{\mu,\nu}$-transform, *Bull. Calcutta Math. Soc.* **51**(1959), 57–65.

Srivastava K.N. [1] On some integral transforms, *Math. Japon.* **6**(1961/62), 65–72.

Srivastava M.M. [1] Infinite series of H-functions, *Istanbul Üniv. Fen. Fak. Mecm. Ser. A* **34**(1969), 79–81.

Stein E.M. [1] Interpolation of linear spaces, *Trans. Amer. Math. Soc.* **87**(1956), 482–492.

Swaroop R. [1] On a generalization of the Laplace and Stieltjes transformations, *Ann. Soc. Sci. Bruxelles Sér. I* **78**(1964), 105–112.

Taxak R.L. [1] Some series for the Fox's H-function, *Defence Sci. J.* **23**(1973), 33–36.

Titchmarsh E.C. [1] A pair of inversion formulae, *Proc. London Math. Soc. (2)* **22**(1923), 34–35.

[2] An inversion formula involving Bessel functions, *Proc. London Math. Soc. (2)* **24**(1925), 6–7.

[3] *Introduction to the Theory of Fourier Transforms*, Chelsea Publishing Company, New York, 1986; the first edition in Oxford Univ. Press, Oxford, 1937.

Tiwari A.K. [1] On the Laplace transform of generalized functions, *Math. Student* **46**(1978), 208–217.

[2] Some theorems on a distributional generalized Stieltjes transform, *J. Indian Math. Soc.* (N.S.) **43**(1979), 241–251.

[3] A distributional generalized Stieltjes transformation, *Indian J. Pure Appl. Math.* **11**(1980), 1045–1054.

[4] On a distributional generalized Laplace transform, *Indian J. Pure Appl. Math.* **11**(1980), 1609–1616.

[5] On a distributional generalized Laplace transform, *Math. Student* **49**(1981), 201–206 (1985).

[6] On the generalized Stieltjes transform of distributions, *Indian J. Pure Appl. Math.* **17**(1986), 1396–1404.

Tiwari A.K. and Ko A. [1] Certain properties of distributional generalized Whittaker transforms, *Indian J. Pure Appl. Math.* **13**(1982), 348–361.

Tiwari A.K. and Koranne P.S. [1] Abelian theorems for a distributional generalized Stieltjes transform, *Indian J. Pure Appl. Math.* **16**(1985), 383–394.

Tricomi F.G. [1] Sulla transformazione e il teorema di reciprocità, di Hankel (Italian), *Atti Accad. Nat. Lincei, Rend.* VI **22**(1935), 564–571.

[2] Un teorema abeliano per la transformazione di Hankel e aleune nuove applicazioni di una formula sulle funzioni di Bessel (Italian), *Atti Accad. Nat. Lincei, Rend.* VI **22**(1935), 572–576.

Varma R.S. [1] A generalisation of Laplace's transform, *Current Sci.* **16**(1947), 17–18.

[2] An inversion formula for the generalised Laplace transform, *Proc. Edinburgh Math. Soc.* (2) **8**(1949), 126–127.

[3] On a generalization of Laplace integral, *Proc. Nat. Acad. Sci. India Sect. A* **20**(1951), 209–216.

Verma C.B.L. [1] On H-function of Fox, *Proc. Nat. Acad. Sci. India Sect. A* **36**(1966), 637–642.

Verma R.U. [1] On a generalised transform of one variable, *Gaṇita* **19**(1968), 73–80.

[2] On certain properties of generalised Hankel transform. I, *Repûb. Venezuela Bol. Acad. Ci. Fis. Mat. Natur.* **32**(1972), 77–81.

[3] On symmetrical Fourier kernel. I, *Gaz. Mat.* (*Lisboa*) **34–35**(1973/74), 13–16.

[4] On symmetrical Fourier kernel. II, *Bull. Math. Soc. Sci. Math. R.S. Roumanie* (N.S.) **17 (65)**(1973), 101–111 (1974).

[5] Application of Laplace transforms in the solution of integral equations, *Repûb. Venezuela Bol. Acad. Ci. Fís. Mat. Natur.* **34**(1974), 97–106.

[6] Application of gama functions in solving certain integral equations, *Acta Mexicana Ci. Tecn.* **8**(1974), 56–60.

[7] A formal solution of an integral equation by L and L^{-1} operators, *Ghana J. Sci.* **15**(1975), 225–237.

[8] Application of L and L^{-1} operators in solving integral equations, *An. Univ. Timişoara Ser. Sti. Mat.* **13**(1975), 155–161 (1977).

[9] Application of L-operator in the solution of certain integral equation, *Indian J. Pure Appl. Math.* **7**(1976), 104–109.

[10] Integral equations involving the G-function as kernel, *An. Univ. Bucuresti Mat.* **27**(1978), 107–110.

Virchenko N.O. and Haidey V.O. [1] On generalized m-Bessel functions, *Integral Transform. Spec. Funct.* **8**(1999), 275–286.

Vu Kim Tuan [1] On the factorization of integral transformations of convolution type in the space L_2^{Φ} (Russian), *Akad. Nauk Armyan. SSR Dokl.* **83**(1986), 7–10.

[2] On the range of the $\mathcal{Y}$-transform, *Bull. Austral. Math. Soc.* **54**(1996), 329–345.

[3] On the range of the Hankel and extended Hankel transforms, *J. Math. Anal. Appl.* **209**(1997), 460–478.

Vu Kim Tuan, Marichev O.I. and Yakubovich S.B. [1] Compositional structure of integral transformations (Russian), *Dokl. Akad. Nauk SSSR* **286**(1986), 786–790.

Watson G.N. [1] Some selfreciprocal functions, *Quart. J. Math. Oxford Series* **2**(1931), 298–309.

Widder D.V. [1] *The Laplace Transform*, Princeton Math, Series, 6, Princeton Univ. Press, Princeton, 1941.

Wimp J. [1] Two integral transform pairs involving hypergeometric functions, *Proc. Glasgow Math. Assoc.* **7**(1965), 42–44.

Wong C.F. and Kesarwani R.N. [1] On an integral transform involving Meijer's G-functions, *Kyungpook Math. J.* **13**(1973), 281–286.

Yakubovich S.B. and Luchko Yu.F. [1] *The Hypergeometric Approach to Integral Transforms and Convolutions*, Math. Appl. 287, Kluwer Academic Publ., Dordrecht, 1994.

Zayed A.I. [1] *Handbook of Function and Generalized Function Transformations*, CRC Press, Boca Raton, 1996.

Zemanian A.H. [1] A distributional Hankel transformation, *SIAM J. Appl. Math.* **14**(1966), 561–576.

[2] The Hankel transformation of certain distributions of rapid growth, *SIAM J. Appl. Math.* **14**(1966), 678–690.

[3] Some Abelian theorems for distributional Hankel and K-transformations, *SIAM J. Appl. Math.* **14**(1966), 1255–1265.

[4] A distributional K transformation, *SIAM J. Appl. Math.* **14**(1966), 1350–1365; **15**(1966), 765.

[5] Hankel transforms of arbitrary order, *Duke Math. J.* **34**(1967), 761–769.

[6] *Generalized Integral Transformations*, Pure Appl. Math. 18, Interscience Publ. [John Wiley & Sons], New York, 1968.

SUBJECT INDEX

AUTHOR INDEX

SYMBOL INDEX

Printed and bound by CPI Group (UK) Ltd, Croydon, CR0 4YY
23/10/2024
01778249-0020